Molecular Food Microbiology

Food Microbiology Series

Series Editor:
Dongyou Liu

Biology of Foodborne Parasites
edited by Lihua Xiao, Una Ryan, and Yaoyu Feng

Molecular Biology of Food and Water Borne Mycotoxigenic and Mycotic Fungi
edited by R. Russell M. Paterson, and Nelson Lima

Foodborne Viral Pathogens
edited by Peter A. White, Natalie E. Netzler, and Grant S. Hansman

Food Spoilage Microorganisms: Ecology and Control
edited by Yanbo Wang

Laboratory Models for Foodborne Infections
edited by Dongyou Liu

Foodborne Viral Pathogens
edited by Peter A. White, Natalie E. Netzler, and Grant S. Hansman

Food Spoilage Microorganisms: Ecology and Control
edited by Yanbo Wang, Wangang Zhang, and Linglin Fu

Handbook of Foodborne Diseases
edited by Dongyou Liu

Molecular Food Microbiology
edited by Dongyou Liu

For more information about this series, please visit: https://www.crcpress.com/Food-Microbiology/book-series/CRCFOOMIC

Molecular Food Microbiology

Edited by
Dongyou Liu

CRC Press is an imprint of the
Taylor & Francis Group, an **informa** business

First edition published 2021
by CRC Press
6000 Broken Sound Parkway NW, Suite 300, Boca Raton, FL 33487-2742

and by CRC Press
2 Park Square, Milton Park, Abingdon, Oxon, OX14 4RN

CRC Press is an imprint of Taylor & Francis Group, LLC

Library of Congress Cataloging-in-Publication Data
Names: Liu, Dongyou, editor.
Title: Molecular food microbiology / edited by Dongyou Liu.
Description: First edition. | Boca Raton : Taylor & Francis, 2021. |
Series: Food microbiology series | Includes bibliographical references and index.
Identifiers: LCCN 2020048389 (print) | LCCN 2020048390 (ebook) |
ISBN 9780815359500 (hardback) | ISBN 9781351120388 (ebook)
Subjects: LCSH: Food—Microbiology. | Nanotechnology.
Classification: LCC QR115 .M62 2021 (print) | LCC QR115 (ebook) | DDC 579/.16—dc23
LC record available at https://lccn.loc.gov/2020048389
LC ebook record available at https://lccn.loc.gov/2020048390

ISBN: 978-0-8153-5950-0 (hbk)
ISBN: 978-1-351-12038-8 (ebk)

Typeset in Times
by codeMantra

Contents

Series Preface

Microorganisms (including viruses, bacteria, molds, yeasts, protozoa, and helminthes) represent abundant and diverse forms of life that occupy various ecological niches of Earth. Those utilizing food and food products for growth and maintenance are important to human society due not only to their positive and negative impacts on food supply but also to their potential pathogenicity to human and animal hosts.

Foodborne microorganisms are known to play a critical role in the fermentation and modification of foods, leading to a variety of nutritious food products (e.g., bread, beverage, yogurt, cheese, etc.) that have contributed to the sustainment of human civilization since time immemorial. However, foodborne microorganisms may be responsible for food spoilage, which, albeit a necessary step in maintaining ecological balance, reduces the quality and quantity of foods for human and animal consumption. Additionally, some foodborne microorganisms are pathogenic to humans and animals, which, besides creating havoc on human health and animal welfare, decrease the availability of meat and other animal-related products.

Food microbiology is a continuously evolving field of biological sciences that addresses issues arising from the interactions between food-/water-borne microorganisms and foods. Topics of relevance to food microbiology include, but are not limited to, adoption of innovative fermentation and other techniques to improve food production; optimization of effective preservation procedures to reduce food spoilage; development of rapid, sensitive, and specific methods to identify and monitor foodborne microbes and toxins, helping alleviate food safety concerns among consumers; use of omic approaches to unravel the pathogenicity of foodborne microbes and toxins; selection of non-pathogenic foodborne microbes as probiotics to inhibit and eliminate pathogenic viruses, bacteria, fungi, and parasites; and design and implementation of novel control and prevention strategies against foodborne diseases in human and animal populations.

The **Food Microbiology** series aims to present a state-of-the-art coverage on topics central to the understanding of the interactions between food-/water-borne microorganisms and foods. The series consists of individual volumes, each of which focuses on a particular aspect/group of foodborne microbes and toxins, in relation to their biology, ecology, epidemiology, immunology, clinical features, pathogenesis, diagnosis, antibiotic resistance, stress responses, treatment and prevention, etc. The volume editors/authors are professionals with expertise in respective fields of food microbiology, and the chapter contributors are scientists directly involved in foodborne microbe and toxin research.

Extending the contents of classical textbooks on food microbiology, this series serves as an indispensable tool for food microbiology researchers, industry food microbiologists, and food regulation authorities, wishing to keep abreast with latest developments in food microbiology. The series also offers a reliable reference for undergraduate and graduate students in their pursuit to becoming competent and consummate future food microbiologists. Moreover, the series provides a trustworthy source of information to the general public interested in food safety and other related issues.

Preface

Food microbiology is a specialized field of biological science that deals with issues arising from the interactions between microorganisms and foods during processing, storage, or after consumption, with the ultimate goals to reduce microbial contamination in food products and to eliminate foodborne disease outbreaks worldwide. Thorough understanding of the molecular basis of foodborne microbial stress responses, adhesion and colonization, epidemiology and evolution, invasion, pathogenesis, immune evasion, drug resistance, and other aspects is crucial to achieving these goals.

Exploiting the power and versatility of molecular technologies, which have quietly revolutionized many fields of biological science, including food microbiology, after the elucidation of DNA double helix in 1953 and the establishment of DNA cloning protocol in 1973, molecular food microbiology extends and greatly improves on phenotype-based food microbiology. This has not only facilitated the development of speedy, sensitive, and specific diagnostic assays for foodborne infections and intoxication but also contributed to the design and implementation of innovative treatment and prevention measures against foodborne diseases.

Forming part of the Food Microbiology series, *Molecular Food Microbiology* focuses on molecular principles and techniques that are applicable to food microbiology research and development. While the introductory chapter presents a brief overview on DNA, RNA, and protein techniques currently available and highlights the utility of these molecular approaches in helping solve the lingering problems that food microbiology is facing now and in the future, the remaining chapters dwell on relevant molecular techniques and their practical values in the characterization, detection and control of individual foodborne viruses, bacteria, fungi, or parasites.

With chapters written by scientists in the forefront of molecular food microbiology research, *Molecular Food Microbiology* constitutes an informative textbook for undergraduates and postgraduates majoring in food, medical, and veterinary microbiology; offers an indispensable guide for food, medical, and veterinary scientists engaged in molecular food microbiology research and development; and provides an insightful update for scholars and educators in their attempts to keep abreast with the latest developments in molecular food microbiology.

In a rapidly advancing field like molecular food microbiology, in which the amounts of research findings on a single organism may overwhelm a scholar from a slightly different field, it is impossible to compile a state-of-the-art volume such as this on one's own. I am fortunate and honored to have a devoted group of international experts who are willing to share their in-depth knowledge and technical insights on molecular food microbiology. Further, the professionalism and dedication of senior editor Stephen Zollo have smoothened the process of getting this volume in print. Finally, the understanding and support from my family – Liling Ma, Brenda, and Cathy – have been crucial to help maintain my sanity during the undertaking of this work.

Editor

Dongyou Liu, Ph.D., studied veterinary science at Hunan Agricultural University, China and conducted postgraduate research on the generation and application of monoclonal antibodies for improved immunodiagnosis of human hydatidosis at the University of Melbourne, Australia. In the past three decades, he has worked at several research and clinical laboratories in Australia and the United States, with focuses on molecular characterization of microbial pathogens and detection of human genetic disorders and tumors/cancers. He is the first author of more than 50 original research and review articles in peer-reviewed international journals, the contributor of 294 book chapters, and the editor of "Handbook of *Listeria monocytogenes*" (2008), "Handbook of Nucleic Acid Purification" (2009), "Molecular Detection of Foodborne Pathogens" (2009), "Molecular Detection of Human Viral Pathogens" (2010), "Molecular Detection of Human Bacterial Pathogens" (2011), "Molecular Detection of Human Fungal Pathogens" (2011), "Molecular Detection of Human Parasitic Pathogens" (2012), "Manual of Security Sensitive Microbes and Toxins" (2014), "Molecular Detection of Animal Viral Pathogens" (2016), "Laboratory Models for Foodborne Infections" (2017), "Handbook of Foodborne Diseases" (2018), and "Handbook of Tumor Syndromes" (2020), all of which are published by CRC Press. He is also a co-editor for "Molecular Medical Microbiology 2nd edition" (2014), which is released by Elsevier. Further, he is the author of recent CRC books: "Pocket Guides to Biomedical Sciences: Tumors and Cancers – Central and Peripheral Nervous Systems" (2017), "Pocket Guides to Biomedical Sciences: Tumors and Cancers – Head, Neck, Heart, Lung and Gut" (2017), "Pocket Guides to Biomedical Sciences: Tumors and Cancers – Skin, Soft Tissue, Bone and Urogenitals" (2017), and "Pocket Guides to Biomedical Sciences: Tumors and Cancers – Endocrine glands, Blood, Marrow and Lymph" (2017).

Contributors

A. Ajayi
Department of Microbiology
University of Lagos
Lagos, Nigeria

Alberto Alía
Faculty of Veterinary Science, Food Hygiene and Safety
Meat and Meat Products Research Institute, University of Extremadura
Cáceres, Spain

Haoran An
School of Medicine
Tsinghua University
Beijing, China

María J. Andrade
Faculty of Veterinary Science, Food Hygiene and Safety
Meat and Meat Products Research Institute, University of Extremadura
Cáceres, Spain

Yasuhisa Ano
Kirin Central Research Institute
Kirin Holdings Co. Ltd.
Kanagawa, Japan

Miguel A. Asensio
Faculty of Veterinary Sciences, Food Hygiene and Safety
Institute of Meat Products, University of Extremadura
Cáceres, Spain

Andréa Rodrigues Ávila
Instituto Carlos Chagas (ICC)
Fundação Oswaldo Cruz
Curitiba, Paraná, Brazil

Abdul Mannan Baig
Department of Biological and Biomedical Sciences
Aga Khan University
Karachi, Pakistan

Elena Bermúdez
Faculty of Veterinary Sciences, Food Hygiene and Safety
Institute of Meat Products, University of Extremadura
Cáceres, Spain

S. Cavallero
Department of Public Health and Infectious Diseases
Section of Parasitology, Sapienza University of Rome
Rome, Italy

Juan J. Córdoba
Faculty of Veterinary Science, Food Hygiene and Safety
Meat and Meat Products Research Institute, University of Extremadura
Cáceres, Spain

Daniel Coronado-Velázquez
Department of Infectomics and Molecular Pathogenesis
Center for Research and Advanced Studies of the National Polytechnic Institute (Cinvestav-IPN)
Mexico City, Mexico

Renato Augusto DaMatta
Laboratório de Biologia Celular e Tecidual
Centro de Biociências e Biotecnologia, Universidade Estadual do Norte Fluminense Darcy Ribeiro
Rio de Janeiro, Brazil

S. D'Amelio
Department of Public Health and Infectious Diseases
Section of Parasitology, Sapienza University of Rome
Rome, Italy

Thiago Torres de Aguiar
Laboratório de Biologia Celular e Tecidual
Centro de Biociências e Biotecnologia, Universidade Estadual do Norte Fluminense Darcy Ribeiro
Rio de Janeiro, Brazil

Mireya de la Garza
Department of Cell Biology
Center for Research and Advanced Studies of the National Polytechnic Institute (Cinvestav-IPN)
Mexico City, Mexico

Josué Delgado
Heart Unit, Virgen de la Victoria University Clinic Hospital
Institute of Biomedical Research in Malaga, CIBERCV, University of Málaga
Málaga, Spain

Yaoyu Feng
Center for Emerging and Zoonotic Diseases, College of Veterinary Medicine
South China Agricultural University
Guangdong, China

Xi He
College of Animal Science and Technology
Hunan Agricultural University
Hunan, China

Alissa Hendricks
Department of Biomedical Sciences and Pathobiology
Virginia Polytechnic Institute and State University
Blacksburg, Virginia
and
Graduate Program in Translational Biology, Medicine, and Health
Virginia Polytechnic Institute and State University
Blacksburg, Virginia

Wanyi Huang
Center for Emerging and Zoonotic Diseases, College of Veterinary Medicine
South China Agricultural University
Guangdong, China

Julie Jean
Département des sciences des aliments
Université Laval
Québec, Canada

Pattara Khamrin
Faculty of Medicine, Department of Microbiology
Chiang Mai University
Chiang Mai, Thailand

Kattareeya Kumthip
Faculty of Medicine, Department of Microbiology
Chiang Mai University
Chiang Mai, Thailand

Fen Li
College of Veterinary Medicine
Hunan Agricultural University
Hunan, China

Jun Li
State Key Laboratory of Pathogenesis, Prevention and Treatment of High Incidence Diseases in Central Asia
WHO Collaborating Centre for Prevention and Care Management of Echinococcosis
Xinjiang, China
and
Clinical Medical Research Institute
The First Affiliated Hospital, Xinjiang Medical University
Xinjiang, China

Dongyou Liu
Royal College of Pathologists of Australasia Quality Assurance Programs
Sydney, NSW, Australia

Guohua Liu
College of Veterinary Medicine
Hunan Agricultural University
Hunan, China

Wei Liu
College of Veterinary Medicine
Hunan Agricultural University
Hunan, China

Niwat Maneekarn
Faculty of Medicine, Department of Microbiology
Chiang Mai University
Chiang Mai, Thailand

Moisés Martínez-Castillo
Department of Infectomics and Molecular Pathogenesis
Center for Research and Advanced Studies of the National Polytechnic Institute (Cinvestav-IPN)
Mexico City, Mexico

Elena Mendoza-Barberá
Department of Genetics, Microbiology and Statistics
University of Barcelona
Barcelona, Spain

Susana Merino
Department of Genetics, Microbiology and Statistics
University of Barcelona
Barcelona, Spain

Devendra T. Mourya
Maximum Containment Laboratory Microbial Containment Complex
National Institute of Virology
Maharashtra, India

Félix Núñez
Faculty of Veterinary Sciences, Food Hygiene and Safety
Institute of Meat Products, University of Extremadura
Cáceres, Spain

Takashi Onodera
Research Center for Food Safety, Graduate School of Agricultural and Life Sciences
University of Tokyo
Tokyo, Japan

Judith Pacheco-Yépez
Sección de Estudios de Posgrado e Investigación
Escuela Superior de Medicina, IPN
Ciudad de México, México

Brunella Posteraro
Istituto di Patologia Speciale Medica e Semeiotica Medica
Fondazione Policlinico Universitario A. Gemelli IRCCS
Università Cattolica del Sacro Cuore
Rome, Italy

Patrizia Posteraro
Laboratorio di Analisi Cliniche e Microbiologiche
GVM Ospedale San Carlo di Nancy
Rome, Italy

Si Qin
College of Food Science and Technology
Hunan Agricultural University
Hunan, China

Ashwin Ramesh
Department of Biomedical Sciences and Pathobiology
Virginia Polytechnic Institute and State University
Blacksburg, Virginia

Alicia Rodríguez
Faculty of Veterinary Science, Food Hygiene and Safety
Meat and Meat Products Research Institute, University of Extremadura
Cáceres, Spain

Mar Rodríguez
Faculty of Veterinary Science, Food Hygiene and Safety
Meat and Meat Products Research Institute, University of Extremadura
Cáceres, Spain

Akikazu Sakudo
School of Veterinary Medicine
Okayama University of Science
Imabari, Ehime, Japan

Idrissa Samandoulgou
Département des sciences des aliments
Université Laval
Québec, Canada

Maurizio Sanguinetti
Istituto di Microbiologia
Fondazione Policlinico Universitario A. Gemelli IRCCS
Università Cattolica del Sacro Cuore
Rome, Italy

A. Seriki
Department of Microbiology
University of Lagos
Lagos, Nigeria

Jesús Serrano-Luna
Department of Cell Biology
Center for Research and Advanced Studies of the National Polytechnic Institute (Cinvestav-IPN)
Mexico City, Mexico

Mineko Shibayama
Department of Infectomics and Molecular Pathogenesis
Center for Research and Advanced Studies of the National Polytechnic Institute (Cinvestav-IPN)
Mexico City, Mexico

Ioannis Sitaras
Exotic and Emerging Avian Viral Diseases Unit, Southeast Poultry Research Laboratory
United States National Poultry Research Center, Agricultural Research Service, United States Department of Agriculture
Athens, Georgia

S. I. Smith
Molecular Biology & Biotechnology Department
Nigerian Institute of Medical Research (NIMR)
Lagos, Nigeria

Xun Suo
National Animal Protozoa Laboratory, College of Veterinary Medicine
China Agricultural University
Beijing, China

Erica Spackman
Exotic and Emerging Avian Viral Diseases Unit, Southeast Poultry Research Laboratory
United States National Poultry Research Center, Agricultural Research Service, United States Department of Agriculture
Athens, Georgia

Xinming Tang
Institute of Animal Science
Chinese Academy of Agricultural Sciences
Beijing, China

Juan M. Tomás
Department of Genetics, Microbiology and Statistics
University of Barcelona
Barcelona, Spain

Padinjaremattathil Thankappan Ullas
Maximum Containment Laboratory, Microbial Containment Complex
National Institute of Virology
Maharashtra, India

Ming Wang
College of Veterinary Medicine
China Agricultural University
Beijing, China

Song Weining
State Key Laboratory of Crop Stress Biology in Arid Areas
College of Agronomy, Northwest A&F University
Shaanxi, China

Pragya D. Yadav
Maximum Containment Laboratory, Microbial Containment Complex
National Institute of Virology
Maharashtra, India

Lijuan Yuan
Department of Biomedical Sciences and Pathobiology
Virginia Polytechnic Institute and State University
Blacksburg, Virginia

Lihua Xiao
Center for Emerging and Zoonotic Diseases, College of Veterinary Medicine
South China Agricultural University
Guangdong, China

Wenbao Zhang
State Key Laboratory of Pathogenesis, Prevention and Treatment of High Incidence Diseases in Central Asia
WHO Collaborating Centre for Prevention and Care Management of Echinococcosis
Xinjiang, China
and
Clinical Medical Research Institute
The First Affiliated Hospital, Xinjiang Medical University
Xinjiang, China

1 Molecular Food Microbiology

An Overview

Dongyou Liu
Royal College of Pathologists of Australasia Quality Assurance Programs

Contents

1.1 INTRODUCTION

Earth is occupied by a diversity of biological organisms, ranging from viruses, bacteria, fungi, parasites, animals to plants. Among an estimated 1×10^{31} viral taxa (mostly phages), about 100 million occur in 1.74 million recognized species of vertebrates, invertebrates, plants, lichens, mushrooms, and brown algae (including ~320,000 viruses in mammals alone). Up to 1 billion bacterial taxa exist, of which ~4 million inhabit soil, ~2 million live in water (river, lake and sea), and ~4 million distribute in the atmosphere. The number of fungal taxa approaches 12 million, with 1,500 yeasts (out of estimated 150,000) and 120,000 filamentous fungi characterized so far. Although 6 million parasitic taxa are likely present, only 1,400 taxa are determined. Further, 1.4 million animal taxa (out of estimated 10-30 million in total) are known, including 1.3 million invertebrates and 66,000 vertebrates, whereas 307,000 plant taxa are identified.

Although a vast majority of microorganisms are free living and maintain symbiotic relationships with their hosts, some have the ability to take advantages of host temporary or long-term weaknesses for their own gains, and induce both innate and adaptive immune responses from as well as pathological changes in their hosts [1–3]. To date, 1,407 microorganisms are implicated in human infections, including 208 viruses or prions, 538 bacteria (representing about 10% of 5,000 recognized taxa), 317 fungi (i.e. ~200 yeasts, ~100 filamentous fungi, and 16 microsporidia) and 344 parasites (i.e. 287 helminthes and 57 protozoa) (Table 1.1) [4]. Over 200 human diseases are attributable to pathogenic microorganisms (e.g., hepatitis A virus, norovirus, *Bacillus cereus*, *Campylobacter*, *Clostridium perfringens*, *Escherichia coli*, *Listeria monocytogenes*, *Salmonella*, *Shigella*, *Staphylococcus aureus*, *Yersinia enterocolitica*, *Cryptosporidium parvum*, *Giardia lamblia*) or their toxins (e.g., *S. aureus* enterotoxins, *B. cereus* emetic toxin, *C. botulinum* neurotoxins) that dwell in food or water and enter the hosts mainly via oral ingestion and occasionally direct/wound exposure (Table 1.1) [5].

Indeed, foodborne infections were responsible for 33 million disability-adjusted life year (DALY) worldwide and 128,000 hospital stays and 3,000 deaths in the USA in 2010 alone. Furthermore, a 2015 survey in the USA recorded a total of 20,107 foodborne infections involving *Salmonella* spp. (7,728), *Campylobacter* spp. (6,309), *Shigella* spp. (2,688), *E. coli* non-O157 (796), *E. coli* O157 (463), *Vibrio* spp. (192), *Yersinia* spp. (139), *Listeria* spp (116), and parasites (1,676), highlighting the disease-causing capacity of food- or water-borne microbial pathogens [6].

As a branch of biological science, food microbiology examines the biology, epidemiology, pathology, immunology, molecular biology, genetics, control and prevention of food- and water-borne pathogens and toxins, with the goal to unravel their pathogenic mechanisms, to develop methods for their detection and tracking, and to implement appropriate measures for their control and prevention. For many years,

TABLE 1.1 Characteristics of pathogenic or toxigenic microbes implicated in foodborne infection or intoxication

CATEGORY	*CHARACTERISTICS*
Pathogenic viruses	**RNA viruses** may be single or double stranded (4–33 kb); relatively unstable with high error rate during transcription and high rate of recombination/reassortment during co-infection
	DNA viruses may be single stranded (3–6 kb) or double stranded (5–375 kb); relatively stable with low error rate during transcription and low rate of recombination/reassortment during co-infection
	Prions (or proteinaceous and infectious virions) are proteins with the ability to change the normal shape of host protein into the prion shape, which coverts even more host proteins into prions
Pathogenic bacteria	**Gram-positive bacteria** possess a cell wall composed of a thick layer (or several layers) of peptidoglycan attached to an inner cell membrane via lipoproteins and lipoteichoic acids, with a notable absence of an outer membrane
	Gram-negative bacteria have a cell wall consisting of a thin layer of peptidoglycan sandwiched between an inner cell membrane and an outer membrane, which contains lipopolysaccharides (LPS, made up of lipid-A, core polysaccharide, and O-antigen) in its outer leaflet and phospholipids in the inner leaflet, with a notable absence of teichoic acids and lipoids
Pathogenic fungi	**Yeasts** are single-celled organisms that reproduce by budding or binary fission; of ~700 known species, 200 are implicated in superficial, cutaneous, subcutaneous, and systemic infections
	Filamentous fungi (~100,000 species identified so far) generate tubular, elongated, and thread-like (filamentous) cellular structures (so-called hyphae), which contain multiple nuclei and extend at tips; filamentous fungi often cause superficial, cutaneous, and subcutaneous infections and produce mycotoxins that lead to food poisoning
	Microsporidia are relatives of zygomycetes (possession of chitin and trehalose; sequence similarity in α- and β-tubulin as well as Hsp70 genes), display features reminiscent of both prokaryotes (small genome, 16S and 23S RNA) and eukaryotes (nucleus, mitotic spindle-separated chromosome, cytoskeleton, polyadenylation on mRNA), and produce highly resistant oval or pyriform spores; of ~1,200 species identified, 16 are associated with human diseases
Pathogenic parasites	**Protozoa** are small (~50 μm), unicellular eukaryotes (~50,000 species identified); human pathogenic protozoa belong mainly to the phyla Sarcomastigophora (amoebae and flagellates, generally reproducing by asexual binary fission) and Apicomplexa (sporozoa, reproducing by both asexual sporogony/schizogony and sexual gamogony)
	Cestodes (tapeworms) have a head (scolex) with sucking organs, a segmented body, but lack alimentary canal; each segment is hermaphrodite
	Trematodes (flatworms or flukes) have a nonsegmented, usually leaf-like body, with two suckers but no distinct head; contain an alimentary canal (but no anus) and are hermaphrodite; however, schistosomes are thread like and form separate sexes
	Nematodes (round worms) appear round in cross section; have body cavities, a straight alimentary canal, and an anus; form separate sexes
Toxigenic bacteria	Endotoxin [i.e., lipopolysaccharide (LPS) or lipooligosaccharide (LOS), also called intracellular toxin] constitutes part of the outer membrane in Gram-negative cell wall and is mostly released upon bacterial disintegration
	Exotoxin (also called extracellular toxin) is a soluble, diffusible protein produced by virulent Gram-positive and occasionally Gram-negative bacterial strains

(*Continued*)

TABLE 1.1 (*Continued*) Characteristics of pathogenic or toxigenic microbes implicated in foodborne infection or intoxication

CATEGORY	*CHARACTERISTICS*
Toxigenic fungi	**Aflatoxins** (*Aspergillus flavus*, *A. parasiticus*, *A. fumigatus*, *A. bombycis*, *A. ochraceoroseus*, *A. nomius*, *A. pseudotamari*, and *Penicillum islandicum*), **fumonisins** [*Fusarium verticillioides* (formerly *F. moniliforme*=*Gibberella fujikuroi*), *F. proliferatum*, *F. nygamai*, and *Alternaria alternata* f. sp. *lycopersici*], **ochratoxins** (*Aspergillus alliaceus*, *A. auricomus*, *A. carbonarius*, *A. glaucus*, *A. melleus*, *A. niger*, *A. ochraceus*, and *P. verrucosum*), **patulin** [*P. griseofulvum* (formerly *P. patulum* or *P. urticae*)], **trichothecenes** (*Fusarium*, *Myrothecium*, *Phomopsis*, *Stachybotrys*, *Trichoderma*, and *Trichothecium*), **zearalenone** [*F. graminearum* (teleomorph *Gibberella zeae*), *F. culmorum*, *F. equiseti*, *F. crookwellense*, and *F. moniliformae*], **ergot alkaloids** (*Claviceps purpurea*), **3-nitropropionic acid** (*Aspergillus*), **mushroom toxins** [α-amanitin from death cap (*Amanita phalloides*); gyromitrin from *Helvella* mushrooms; muscarine from *Inocybe* and *Clitocybe* mushrooms]
Toxigenic marine bacteria, dinoflagellates, algae, and coral	**Domoic acid** (marine diatoms *Pseudo-nitzschia* and *Nitzschia*; red alga *Chondria armata*), **okadaic acid** (dinoflagellates *Prorocentrum lima* and *Dinophysis*), **pectenotoxin** (*Dinophysis*), **azaspiracid** (dinoflagellates *Azadinium spinosum*, *Azadinium dexteroporum*, *Amphidoma languida*, and *Protoperidinium crassipes*), **yessotoxin** (dinoflagellates *Protoceratium reticulatum*, *Lingulodinium polyedrum*, and *Gonyaulax spinifera*), **brevetoxin** (dinoflagellates *Karenia brevis*, *Chatonella marina*, *C. antiqua*, and *C. cf. verruculosa*), **ciguatoxin** (dinoflagellate *Gambierdiscus toxicus*), **palytoxin** (coral *Palythoa* and dinoflagellate *Ostreopsis*), **saxitoxin** (dinoflagellates *Alexandrium*, *Gymnodinium catenatum*, *Pyrodinium bahamense*; cyanobacteria *Anabaena*, *Aphanizomenon*, *Cylindrospermopsis*, *Lyngbya*, *Planktothrix*, *Oscillatoria*), **cyclic imine** (dinoflagellates *Karenia selliformis*, *Alexandrium ostenfeldii*, *A. peruvianum*, and *Vulcanodinium rugosum*), **scombrotoxin** (a histamine produced by spoilage bacteria *Vibrio*, *Pseudomonas*, *Photobacterium*, *Morganella*, *Raoultella*, *Hafnia*, etc.), **tetrodotoxin** (marine bacteria *Vibrio, Aeromonas, Pseudomonas, Shewanella, Alteromonas, Caulobacter, Roseobacter, Alcaligenes*)

phenotypical procedures (e.g., microscopy, *in vitro* isolation, *in vivo* bioassay) have played a fundamental role in the identification and characterization of foodborne pathogens and the development of diagnostic techniques for and intervention strategies against foodborne infections and intoxication. The initial description of nucleic acid in 1868, resolution of deoxyribonucleic acid (DNA) double helix in 1953, and deciphering of the genetic code in 1961 led to the establishment of the central dogma that transcription of DNA into messenger ribonucleic acid (mRNA) and subsequent translation into protein underscore the biological processes. The development of the DNA cloning procedure in 1973 has further empowered our efforts towards the elucidation of biological processes from a molecular perspective that was inconceivable before.

Through the adoption and incorporation of molecular principles and techniques in the analysis of foodborne microorganisms and toxins, molecular food microbiology extends and significantly improves phenotype-based food microbiology. This not only facilitates speedy, sensitive, and specific detection and typing of foodborne microorganisms and toxins but also offers unprecedented opportunity to develop innovative treatments for foodborne diseases. In the following sections, a concise overview of DNA, RNA, and protein techniques applicable to food microbiology is presented. Students and novice scientists in molecular food microbiology will find it valuable as it expands their technical knowhows and simplifies their selection and application of relevant molecular techniques for their specific needs.

1.2 DNA TECHNIQUES

1.2.1 DNA Structure and Function

Deoxyribonucleic acid (DNA) is made up of two polymerized strands of deoxyribonucleotides (forming the so-called double helix) that comprise deoxyribose (sugar), phosphate groups (the phosphate group attached to the 5′ carbon of deoxyribose in one deoxyribonucleotide is linked via a phosphodiester bond to the hydroxyl group of the 3′ carbon of deoxyribose in another deoxyribonucleotide), and nitrogenous bases [e.g., adenine (A) and guanine (G), which are double-ringed purines with a six-carbon ring fused to a five-carbon ring; cytosine (C) and thymine (T), which are single-ringed pyrimidines with only a six-carbon ring structure] involved in complementary base pairing between a purine and a pyrimidine (A–T via two hydrogen bonds or C–G via two hydrogen bonds and one N-glycosidic bond) in the deoxyribonucleotide strands [7].

As DNA double strands are antiparallel, the 3′ end of one strand containing a free hydroxyl group faces the 5′ end of the other strand containing a free phosphate group. Moreover, with approximately ten bases per turn in the DNA double helix (the bases located inside the helix and the sugar–phosphate backbone on the outer surface), major grooves (where the sugar–phosphate backbone is far apart) and minor grooves (where the sugar–phosphate backbone is close together) are produced by asymmetrical spacing of the sugar–phosphate backbones. These grooves facilitate protein binding to DNA for DNA replication or transcription of DNA into RNA.

DNA denaturation involves exposure of double-stranded DNA (dsDNA) to heat or certain chemical (strong alkaline), leading to breakup of the hydrogen bonds between complementary bases and separation into two single-stranded DNA (ssDNA). Conversely, cooling or removal of the chemical denaturant allows hydrogen bonds to reform and transforms ssDNA into dsDNA. Given that a C–G base pair has two hydrogen bonds and one N-glycosidic bond in comparison to an A–T base pair with two hydrogen bonds, DNA with a high GC content is more resistant to denaturation than DNA with a lower GC content.

Functionally, DNA carries and retains genetic/hereditary information in a cell and has the ability to use its base sequence for self-synthesis (with one strand serving as the template for the making of its copy) and also for synthesis of RNA and proteins (encoded by segments of DNA called genes). DNA may be degraded enzymatically for use as a source of nucleotides and nucleosides (consisting of deoxyribose and phosphate groups) in cellular metabolism. It is noteworthy that while genetic information usually flows from DNA to RNA to protein, some exceptions exist. For instance, retroviral RNA genome requires reverse transcription into DNA before RNA transcription and protein translation take place. Further, retrotransposons (mobile elements of genome) are transcribed into DNA to enable their movement from one site of the genome to other site [7].

Apart from some RNA viruses whose genetic information is contained in RNA, most microorganisms harbor DNA genome (which is often organized into chromosomes, with eukaryotic chromosomes housed in the membrane-bound nucleus and prokaryotic chromosomes located in the nucleoid, an area of the cytoplasm). The genotype of a cell refers to the full collection of genes within its genome, which remain largely constant throughout the cell cycle; the phenotype of a cell refers to the set of genes that are expressed at a given point in time (which may change in response to environmental signals) and that determine its activities and observable characteristics. Genes that are expressed all time are known as constitutive genes (e.g., housekeeping genes).

1.2.2 DNA Purification

As DNA is covered/protected by proteins (e.g., nucleoprotein in virus, DNA-binding proteins such as histones in eukaryotes) and embedded inside the nucleus (mitochondrion) of a cell (or host cell), it usually requires purification/extraction prior to molecular analysis. Typically, DNA purification involves cell lysis, digestion of RNA and cellular proteins, removal of degraded proteins and cell debris, and precipitation and concentration of DNA (Table 1.2). However, for optimal yield and purity, special considerations are given as to the type of starting material (e.g., virus, bacterium, fungus, parasite, or infected host cell) and output requirement (genomic, mitochondrial, or plasmid DNA) [8].

TABLE 1.2 Special considerations on DNA purification from viruses, bacteria, fungi, and parasites

STEP	*VIRUSES*	*BACTERIA*	*FUNGI*	*PARASITES*
Cell lysis	DNA virus-infected cells are lysed with three cycles of freeze (−20°C) and thaw (37°C) or mild detergent (which lyses cytoplasmic members but leaves nuclei intact), or glycerol shock procedure; destructive processes (e.g., sonication, excessive pipetting, and high-speed vortexing) with nonpackaged virus DNA are avoided if intact viral DNA of >50 kb is desired; centrifugation of lysed mixture at 10,000×*g* leaves nuclei (containing nuclear replicating viral DNA and host cell DNA) and cell debris in the pellet, and cytoplasmic extract in the supernatant; further centrifugation of cytoplasmic extract (the supernatant) at 100,000×*g* leaves cytoplasmic replicating virions in the pellet; centrifugation of lysed mixture at 1,500×*g* leaves nuclear	Gram-positive bacteria with multiple layers of peptidoglycan in the cell wall are generally susceptible to lysozyme; Gram-negative bacteria with only a thin layer of peptidoglycan in the cell wall are somewhat resistant to lysozyme and require ethylenediaminetetraacetic acid (EDTA) or other reducing agents [e.g., β-mercaptoethanol or dithiothreitol (DTT)] for optimal lysis; alternative lytic enzymes for lysozyme-resistant bacteria include mutanlysin, lysostaphin, achromopeptidase, labiase, and proteinase K; glass beads, sonication, and French press may be utilized to mechanically disrupt the bacterial cell wall (note, bead beating and sonication may break DNA down to <20 kb fragments)	Yeast cell wall (containing an inner layer of glucan and chitin and an outer layer of mannoproteins and polysaccharides) is lysed with lyticase, chitinase, zymolase, solication, glass beads, or mortar/pestle grinding; filamentous fungal cell wall is often broken up with mortar/pestle grinding, bead beating (using a vortex mixer) or lytic enzyme	Parasite is lysed by physical (grinder, blender, mortar, or bead beating) and/or chemical (alkaline treatment and neutralization) means

(*Continued*)

TABLE 1.2 (*Continued*) Special considerations on DNA purification from viruses, bacteria, fungi, and parasites

STEP	*VIRUSES*	*BACTERIA*	*FUNGI*	*PARASITES*
	replicating DNA virions in the supernatant and cell debris in the pellet; DNase 1 treatment of the concentrated DNA virions (whose capsid protects viral DNA) removes contaminating host mitochondrial DNA and genomic DNA			
Digestion of RNA and cellular proteins	Proteinase K treatment of the cytoplasmic extract (containing cytoplasmic replicating DNA viruses) in the presence of sodium dodecyl sulfate (SDS) eliminates host cell and viral capsid proteins	Treatment of bacterial lysate with RNase A (free of DNase) eliminates unwanted RNA; treatment with proteinase K in the presence of detergent [e.g., SDS, Triton X 100, or cetyltrimethyl-ammonium bromide (CTAB)] destroys cellular proteins	Treatment of yeast/ filamentous fungus lysate with RNase A (free of DNase) eliminates unwanted RNA; treatment with proteinase K in the presence of detergent (e.g., SDS, Triton X 100, or CTAB) destroys cellular proteins	Treatment of parasite lysate with RNase A (free of DNase) eliminates unwanted RNA; treatment with proteinase K in the presence of detergent (e.g., SDS, Triton X 100, or CTAB) destroys cellular proteins
Removal of degraded proteins and cell debris	Phenol/chloroform extraction leaves viral DNA in the upper phase and host cell and viral capsid proteins in the lower phase; alternative protocols free of hazardous chemicals are available and increasingly adopted	Phenol/chloroform extraction leaves bacterial DNA in the upper phase and degraded proteins and cell debris in the lower phase; alternatively, silica oxide (diatomaceous earth or silica-coated beads or glass beads) may be used to separate ds DNA from degraded proteins and cell debris	Phenol/chloroform extraction leaves yeast DNA in the upper phase and degraded proteins and cell debris in the lower phase	Phenol/ chloroform extraction leaves parasite DNA in the upper phase and degraded proteins and cell debris in the lower phase
Precipitation and concentration of DNA	Purification and concentration of viral DNA in the supernatant by ethanol precipitation, filter column, or magnetic beads	Precipitation and concentration of bacterial DNA by either two volumes of ethanol or 0.6 volumes of isopropanol in the presence of a monovalent cation (salt)	Precipitation and concentration of yeast/fungal DNA by either two volumes of ethanol or 0.6 volumes of isopropanol in the presence of a monovalent cation (salt)	Precipitation and concentration of parasite DNA by either two volumes of ethanol or 0.6 volumes of isopropanol in the presence of a monovalent cation (salt)

1.2.3 DNA Characterization

After purification, the quantity and quality (purity and intactness) of DNA are determined by UV spectroscopy (e.g., NanoDrop 2000™ on the basis of ratios at A_{260}/A_{280} and A_{260}/A_{230}, with A_{260}/A_{280} ratios of 1.8–2.0 indicating high DNA purity and those below indicating RNA/protein contamination, and with A_{260}/A_{230} ratios of 1.8–2.2 suggesting high DNA purity and those below suggesting phenol, salt, protein, or polysaccharide contamination), PCR inhibition assay (for the presence of inhibitors in purified DNA), and gel electrophoresis (for intactness and size distribution of extracted DNA fragments, including Southern blot). Further assessment of DNA quantity may be conducted with fluorometry (e.g., Qubit 2.0® and Bioanalyzer 2100©), quantitative PCR (qPCR), and serial analysis of gene expression (SAGE) [9]. It is noteworthy that NanoDrop 2000™ is nondiscriminatory between nucleic acid species, has assay ranges of 2–27,500 ng μl^{-1} for dsDNA; 1.3–18,150 ng μl^{-1} for ssDNA; 1.6–22,000 ng µl–1 for RNA; and 100–8,000 µg ml^{-1} for protein (Bradford), and tends to overestimate concentration; Qubit 2.0® fluorometer uses dyes to bind DNA, RNA, and protein, specifically, with an assay range of 200 pg-1000 ng, and sample starting concentration range of 10 pg – 1000 ng μl^{-1} for dsDNA; an assay range of 1-200 ng, and sample starting concentration range of, 50 pg–200 ng μl^{-1} for ssDNA; an assay range of 5-1000 ng, and sample starting concentration range of 250 pg–100 ng μl^{-1} for RNA; an assay range of 250 ng–5 µg, and sample starting concentration range of 12.5 µg–5 mg ml^{-1} for protein; and Bioanalyzer 2100© also uses dyes to specifically bind DNA, RNA, and protein, with detection limits of 100 pg μl^{-1} DNA, 5 ng μl^{-1} total RNA, 50 pg μl^{-1} purified miRNA, and 300 pg–3000 ng μl^{-1} protein.

1.2.4 DNA Cloning

First described in 1973, DNA (molecular) cloning is the process of inserting a DNA fragment (or gene of interest) into a plasmid (which is a small circle of DNA independent of bacterial chromosome and has the ability to replicate autonomously) for generating multiple, identical copies or expressing large quantities of protein *in vitro* [10]. This usually involves four steps: (i) restriction enzyme digestion of DNA insert and plasmid (which contains multiple cloning site, the origin of replication, and antibiotic/other selection markers), (ii) ligation of DNA insert and linearized plasmid with DNA ligase, (iii) introduction (or transformation) of recombinant plasmid (harboring DNA insert) into suitable bacterial strain (e.g., *E. coli*), (iv) selection of bacteria containing recombinant plasmid using antibiotics and other markers, and subsequent growth of bacterial strain for production of protein and biopharmaceuticals or for gene therapy [11].

Variations from the original DNA cloning protocol include blunt end ligation, TA cloning (i.e., ligation between vector with T overhang and PCR fragment with A overhang), and ligation independent cloning. In addition to plasmids, viruses (e.g., lambda phage, baculovirus, and adenovirus) are also employed as alternative vectors for protein expression in bacteria, insect and mammalian cells.

1.2.5 DNA Amplification

DNA amplification is an *in vitro* (literally in glass or plastic vessel) procedure that enables generation of multiple copies of a target from a single molecule of DNA for specific detection and analysis. Compared to DNA cloning involving multiples steps, *in vitro* DNA amplification is relatively simple and efficient.

Developed in 1985 and widely adopted in both research and clinical laboratories, polymerase chain reaction (PCR) utilizes a pair of short oligonucleotides as primers, thermostable, high-fidelity DNA polymerases (e.g., Taq DNA polymerase), and a thermocycler to produce billions of copies from a single DNA template in a matter of hours. Subsequent incorporation of specific DNA dyes facilitates simultaneous detection and quantification (qPCR) [12].

Another useful DNA amplification technique (i.e., loop-mediated isothermal amplification or LAMP) makes use of a strand-displacing DNA polymerase (e.g., Bst DNA polymerase) and two or three sets of primers for sequence-specific isothermal amplification at a constant temperature (even room temperature) and thus eliminates the thermocycling process altogether [13].

1.2.6 DNA Sequencing

DNA sequencing is a process that aims to determine the order of nucleotides (adenine, guanine, cytosine, and thymine) in DNA. The availability of DNA sequences is critical for advancing our knowledge in basic and applied fields, including medical diagnosis, biotechnology, forensic biology, virology, and biological systematics. Procedurally, DNA sequencing has evolved from early, relatively inefficient Maxam–Gilbert sequencing and Sanger sequencing to high-throughput sequencing [14].

The Maxam–Gilbert sequencing (also known as chemical sequencing), described by Allan Maxam and Walter Gilbert in 1977, relies on the use of radioactive substances to label at one 5′ end of ds DNA and chemical treatment to generate breaks (on average one modification per DNA molecule) in one or two of the four nucleotide bases in each of four reactions (G, A+G, C, C+T). Subsequent electrophoresis in denaturing acrylamide gel and autoradiography on X-ray film yield a series of dark bands each corresponding to a radiolabeled DNA fragment.

The Sanger sequencing (also known as chain termination) was developed by Frederick Sanger in 1977 from a location-specific primer extension strategy established by Ray Wu in 1970. Involving DNA polymerase extension and specific nucleotide labeling, this technique requires fewer toxic chemicals and lower amounts of radioactivity than the Maxam and Gilbert sequencing. Due to its ease of performance and reliability, this technique underlies the principles of the first-generation DNA sequencers that have been extensively utilized from the 1980s to the mid-2000s.

High-throughput sequencing (including second- or next-generation "short-read" sequencing and third-generation "long-read" sequencing) is a highly scalable technique that allows the entire genome to be sequenced at once. Conceptualized in the mid- to late 1990s, next-generation sequencing (NGS) involves fragmenting the genome into small pieces, randomly sampling for a fragment, and massively paralleled sequencing using one of several technical approaches [e.g., Illumina's Hi-Seq genome sequencer that combines base-by-base sequencing with removable 3′ blockers on DNA arrays (blots and single DNA molecules) and random surface-PCR arraying; massively parallel signature sequencing (MPSS) incorporating a parallelized, adapter/ligation-mediated, bead-based sequencing technology; pyrosequencing (454); and SOLiD sequencing based on emulsion PCR]. Third-generation sequencing reads the nucleotide sequences at the single molecule level, instead of current breaking long strands of DNA into small segments and then inferring nucleotide sequences by amplification and synthesis. High-throughput sequencing has found its way into exome sequencing, genome sequencing, transcriptome profiling (RNA-seq), DNA–protein interaction (ChIP sequencing), and epigenome characterization.

1.2.7 Mutagenesis and Genome Editing

Mutagenesis refers to a change in the nucleic acid sequence of an organism following exposure, either naturally or artificially, to physical (e.g., high temperature, X-ray), chemical (e.g., mustard gas, polycyclic aromatic hydrocarbons in soot and coal tar), and biological (e.g., cytochrome P450, antibiotic, glutathione *S*-transferase and microsomal epoxide hydrolase) agents.

Mechanistically, mutagenesis may evolve from spontaneous hydrolysis (inducing either depurination and preferential incorporation of adenine in apurinic site or deamination from cytidine to uridine and from 5-methylcytosine to thymine), base modification (e.g., methylation by *S*-adenosylmethionine and glycosylation by reducing sugars), replication error (e.g., cytochrome P450 catalyzes reactive oxygen species that cause errors in replication), cross-linking (induced by alkylating agents), dimerization (e.g., UV radiation promotes formation of pyrimidine dimers), intercalation between bases (e.g., ethidium bromide and proflavine may stretch the DNA backbone and produce forward slippage or deletion mutation and reverse slippage or insertion mutation in DNA), backbone damage (e.g., ionizing radiation induces highly reactive free radicals that break the bonds in DNA and causes chromosomal translocation and deletion; mustard gas may also cause breakages in DNA), inaccurate DNA repair (e.g., cadmium, chromium, and nickel may induce DNA repair errors), to insertion (addition of one or more base pairs in DNA mediated by transposons or other means).

More specifically, insertional mutagenesis consists of random mutation induced by transposons (so-called transposon mutagenesis, transposition mutagenesis, or signature-tagged mutagenesis) and site-specific (or directed) mutation created by PCR and other approaches (e.g., Delitto perfetto, CRISPR-Cas9). Notably, CRISPR-Cas9 involves the delivery of Cas9 nuclease complexed with a single-guide RNA (sgRNA) into a cell, so that the genome of the cell can be cut at a desired location, facilitating removal of existing genes and addition of new ones *in vivo*. As it requires no transposon insertion site, leaves no marker, and demonstrates high efficiency and relative simplicity, CRISPR-Cas9 has become the preferred method for genome editing in a variety of organisms since 2013 [15,16].

1.2.8 DNA Modification

Inside the cell, DNA is packaged (wrapped) by histones (DNA-binding proteins), which attach to scaffolding proteins to form so-called chromatin. Environmental factors that alter methyl groups on certain cytosine nucleotides (e.g., 5′-CpG-3′) of DNA (DNA methylation) may modify the packaging of DNA by histones (histone modification). Methylation is associated with gene function control and related to human diseases (e.g., cancer). Further, epigenetic changes may also influence gene expression without altering the sequence of nucleotides.

The 5-methylcytosine sites in genes at the sequence level are often determined by DNA methylation sequencing after selective conversion of cytosine to uracil with a bisulfite-mediated technique that has minimal effect on 5-methylcytosine. DNA methylation sequencing and chromatin immunoprecipitation followed by sequencing (ChIP-Seq) permit genomic localization of epigenetic markers and evaluation of gene activity and expression as well as chromatin state [17].

1.3 RNA TECHNIQUES

1.3.1 RNA Structure and Function

Ribonucleic acid (RNA) is usually a single chain (strand) of ribonucleotides [consisting of a ribose (pentose sugar), a phosphate group, and a nitrogenous base (A, U, G, or C)] linked by phosphodiester bonds. In comparison with DNA that utilizes thymine (T) to form a complementary base pair with adenine (A), RNA relies on pyrimidine uracil (U) instead. Further, DNA is notably long and double stranded, whereas RNA is short and single stranded. Nonetheless, despite its single strandedness, RNA often demonstrates intramolecular base pairing between complementary sequences, leading to predictable three-dimensional structure essential for its function. The key roles of RNA relate to its synthesis (translation) of protein and its regulation of cellular processes [e.g., RNA molecules called ribozymes are capable of cleaving RNA for viral RNA replication and catalyzing removal of introns from messenger RNA (mRNA) in the nucleus of eukaryotes].

Although ~66% of human genomic DNA are transcribed, only ~2% of the transcriptional products are protein-coding mRNA, and ~98% are noncoding RNA (ncRNA) involved in either housekeeping or

TABLE 1.3 Key features of RNA types/species

RNA TYPE/SPECIES	*KEY FEATURES*
Messenger RNA (mRNA)	As an intermediary between DNA and its protein product, mRNA is transcribed from DNA (antisense strand to produce so-called RNA transcript, which is identical to nontemplate sense strand, apart from replacement of T with U) and carries message (instruction) from DNA to translate into a protein needed at a particular point in time, through interaction with ribosomes and other cellular machinery. Being relatively unstable and short lived in the cell, mRNA ensures that protein is only made when needed
Ribosomal RNA (rRNA)	rRNA is transcribed from DNA as long RNA molecule, which is cut (in the nucleolus region of the nucleus of eukaryotes and the cytoplasm of prokaryotes) into smaller species (e.g., 5S, 16/18S, 23/26/28S) for assembly in ribosome. rRNA ensures proper alignment of mRNA, tRNA, and the ribosomes and catalyzes the formation of the peptide bonds between two aligned amino acids during protein synthesis through its peptidyl transferase activity
Transfer RNA (tRNA)	tRNA is also transcribed from DNA as long RNA molecule, which is cut into smaller species of 70–90 nucleotides long; through base pairing with mRNA, tRNA transports correct amino acid to the site of protein synthesis in the ribosome
Antisense RNA	Antisense RNA is polynucleotide with base sequences complementary to mRNA and has the capacity of binding to complementary mRNA and repressing its translation
Small-interfering RNA (siRNA)	siRNA (short-interfering RNA or silencing RNA) is exogenously derived, double-stranded, noncoding RNA of 20–25 bp in length and operates within the RNA interference (RNAi) pathway by degrading mRNA (with complementary nucleotide sequence) after transcription and preventing its translation
MicroRNA (miRNA)	miRNA is endogenously expressed, small noncoding RNA that regulates gene expression via induction of a steric hindrance obstruction (following miRNA attachment on target transcript) instead of target transcript degradation; through its involvement in apoptosis, immune response, cell differentiation, and cell metabolism, miRNA contributes to the pathogenesis of metabolic disorder and cancer
Piwi-interacting RNA (piRNA)	piRNA (as well as related subspecies repeat associated small-interfering RNA or rasiRNA) is small noncoding RNA expressed in animal cells and forms RNA–protein complexes through interaction with piwi-subfamily Argonaute proteins, which are involved in the epigenetic and post-transcriptional silencing of transposable elements and other spurious or repeat-derived transcripts, and also in the regulation of other genetic elements in germline cells
Small nuclear RNA (snRNA)	Found within the splicing speckles and Cajal bodies of the cell nucleus in eukaryotic cells, snRNA (about 150 nt) is involved in the processing of pre-messenger RNA (hnRNA) in the nucleus, the regulation of transcription factors or RNA polymerase II, and the maintenance of the telomeres
Small nucleolar RNA (snoRNA)	Derived mostly from introns, snoRNA takes part in the modification (e.g., 2′-*O*-methylation and pseudouridylation) of rRNA, snRNA, and possibly other RNA species; a small subset of snoRNA localized to Cajal bodies (conserved subnuclear organelles in the nucleoplasm) is termed scaRNA with the capacity to mediate both 2′-*O*-methylation and pseudouridylation
Long noncoding RNA (lincRNA)	lincRNA noncoding RNA with poorly defined functions

regulatory functions. Notable housekeeping ncRNA consist of ribosomal RNA (rRNA), transfer RNA (tRNA), and small nuclear RNA (snRNA), while regulatory ncRNA include microRNA (miRNA) and long noncoding RNA (lncRNA) (Table 1.3). Of these, mRNA, rRNA, and tRNA are known to participate in protein synthesis, while other RNA species appear to have specific or unspecified functions (Table 1.3) [18,19].

Most RNA species exist in the single-stranded state and form complex secondary and tertiary structures related to their functions. Specifically, the RNA secondary structure results from complementarity between various regions of one strand, producing both helical (stem) and nonhelical structures (e.g., bulges, inner loops, hairpin loops, multibranched loops/junctions, pseudoknots). The RNA tertiary structure is attributed to noncovalent forces (i.e., hydrogen bonds and stacking in addition to influences from proteins, water molecules, and counter-ions). Nonetheless, some RNA species may appear double stranded as exemplified by double-stranded RNA viruses (e.g., rotaviruses) and the genomes of mitochondria and chloroplasts.

1.3.2 RNA Purification

RNA is unstable and highly susceptible to ubiquitous RNAse activity. Therefore, successful purification of RNA from cells and tissues is dependent on (i) the rapid disruption/lysis of cell and subcellular compartments (for total RNA, mRNA, and rRNA), (ii) the total inhibition of nuclease activity (including removal of RNase from reagents, glassware, consumable plasticware, and equipment; inhibition of nuclease activity in cell lysate), (iii) the effective deproteinization (e.g., proteinase K digestion, repeated extraction with phenol and chloroform, solubilization in guanidinium buffers, silica column separation, salting-out), (iv) the efficient RNA concentration (e.g., ethanol precipitation in the presence of sodium acetate, silica column), and (v) the proper storage of purified RNA [8,20]. While it is important to inhibit and/or eliminate RNAse activity [through the use of strong denaturants such as guanidine salts, sodium dodecylsulfate (SDS), or phenol-based compounds] during RNA purification, care should also be taken to maintain RNA integrity (with the help of RNA stabilizers) prior to and after extraction.

1.3.3 RNA Characterization

Similar to DNA characterization (see Section 1.2.3), the quantity and quality (purity and intactness) of RNA may be assessed by UV spectroscopy, RT-PCR inhibition assay, and gel electrophoresis (including northern blot).

1.3.4 RNA Cloning

For RNA cloning, polyadenylated RNA (poly(A)+ RNA), which comprises mRNA rather than total RNA, may be cloned directly or converted into cDNA through the action of reverse transcriptase and DNA polymerase and then cloned into an appropriate plasmid or lambda vector [21].

For small RNA (e.g., miRNA and siRNA) cloning, gel-purified small RNA is ligated directly to a nonphosphorylated 5′-adapter oligonucleotide using T4 RNA ligase. After removal of excessive 5′-adapter from the ligation products on a 15% denaturing polyacrylamide gel, the small RNA is ligated to a 5′-phosphorylated 3′-adapter oligonucleotide with a blocked 3′-hydroxyl terminus. Upon removal of excessive 3′-adapter, the small RNA is amplified by reverse transcription (RT)-PCR. The PCR products are gel purified, digested with EcoRI and NcoI restriction enzymes, and concatamerized using T4 DNA ligase. The concatamers are ligated into an EcoRI-NcoI digested cloning vector, and the recombinant plasmids are transformed into TOP10 cells. Individual colonies are screened for the size of concatamer inserts by PCR, and selected PCR fragments are purified and sequenced. The small RNA sequences are extracted from the sequence using software tools (e.g., Staden Package) [18,22].

1.3.5 RNA Amplification

For RNA amplification, RNA is typically converted into cDNA with reverse transcriptases and amplified with DNA polymerase in a procedure called RT-PCR. Use of Tth DNA polymerase, which demonstrates both reverse transcriptase and DNA polymerase activities, further streamlines RNA amplification.

For single-cell RNA amplification, a limited number of PCR cycles and T7-based *in vitro* transcription (IVT) may be utilized to detect all regions of transcripts, including the recognition of mRNA with short or completely absent poly(A) tails, identify noncoding RNA, discover the full array of splice isoforms from any given gene product, and eliminate 3′ bias that is associated with other protocols [23].

1.3.6 RNA Sequencing

RNA-sequencing (RNA-seq) assesses the sequences as well as quantity of RNA (including mRNA, rRNA, tRNA, and miRNA) in biological sample with the use of the Sanger sequencing or more recently NGS. This typically involves the conversion of RNA into cDNA fragments (cDNA library). After addition of adapters (which contain functional elements such as amplification element and primary sequencing site) to each end of the fragments, the cDNA library is sequenced by NGS (software packages such as STAR; quality control tools such as Picard or Qualimap; alignment tools such as further Sailfish, RSEM, and BitSeq; alternatively spliced gene quantification tools such as MISO). Not limiting to genomic sequences, having low background signal and being more quantifiable, RNA-seq is clearly superior to microarray for detection of novel transcripts, SNPs, or other alterations and analysis of continuously changing cellular transcriptome (including alternative gene spliced transcripts, post-transcriptional modifications, gene fusion, mutations/SNPs, changes in gene expression over time, and differences in gene expression in different groups or treatments).

Single-cell RNA sequencing (scRNA-Seq) reveals the expression profiles of individual cells and helps uncover the existence of rare cell types within a cell population. Procedurally, scRNA-Seq involves isolation of single cell (by encapsulating individual cells in droplets in a microfluidic device) and RNA, reverse transcription, amplification, library generation, and sequencing [23,24].

1.3.7 RNA Interference

RNA interference (RNAi) refers to the degradation of complementary mRNA by endogenously expressed miRNA or exogenously derived small-interfering RNA (siRNA), leading to suppression of a target gene expression. Being 21–28 nucleotides long RNA duplexes derived from cleavage of double-stranded cellular RNA transcripts with hairpin structures (endogenous siRNA) or from viruses (exogenous siRNA), one strand of siRNA is incorporated in RNA-induced silencing complex (RISC, which is a multiprotein complex) that guides its enzymatic cleavage and degradation of complementary mRNA. Furthermore, siRNAs may function as primers for the synthesis of additional siRNA by RNA-dependent RNA polymerase, leading to an enhanced effect. Other small RNAs [e.g., piwi-interacting RNA (piRNA) and its subspecies repeat-associated small-interfering RNA (rasiRNA)] are incapable of inducing RNAi alone and need to form the core of RISC with Argonaute protein for the interfering effect to occur [25].

1.3.8 RNA Modification

While mRNA maturation typically involves 5′-capping, splicing, and polyadenylation, post-transcriptional modifications of epitranscriptomics (or RNA-epigenetics) tend to be variable, including pseudouridine (Ψ) [i.e., site-specific isomerization of uridine (U) to Ψ (5-ribosyluracil) irreversibly catalyzed via Ψ synthases, found in rRNA, tRNA, and mRNA)], *N*6-methyladenosine (m6A, a reversible internal modification in eukaryotic mRNA recognizable by YTH domain proteins or readers, which regulate mRNA processing and metabolism), 5-methyl cytidine (m5C) and *N*1-methyladenosine (m1A) [methylation of adenosine by methyltransferases (METTL3, METTL14, and WTAP) or writers], removal of methyl group [by demethylases (FTO and ALKBH5) or erasers], and RNA editing [which creates an irreversible insertion, deletion, and base substitution in all RNA species, such as adenosine (A) to inosine (I)]. To date, >140 diverse and distinct modifications of RNA (e.g., m6A, m6Am, m5C, m1A, A-to-I editing, Ψ, and NAD cap) are identified, and a majority of epitranscriptomic modifications occur in rRNA, tRNA, and snRNA. Disruptions of RNA modification mechanisms appear to be linked to disease [26,27].

RNA modifications are detected by transcriptome-wide and NGS-based RNA modification mapping techniques involving methyl-RNA immunoprecipitation and chemical-based methods. The most commonly used NGS-based RNA techniques include methyl-RNA immunoprecipitation and UV cross-linking (m6A-seq, MeRIP-seq, m6A-LAICIC-seq for m6A/m6Am modification), methyl-RNA immunoprecipitation and the inherent ability of m1A to stall reverse transcription (m1A-ID-seq for m1A modification), chemical conversion of modified nucleotides (bisulfite sequencing or BS-seq for m5C modification), cyanoethylation of RNA combined with reverse transcription [inosine chemical erasing (ICE)-seq for A-to-I editing modification], chemical modification to terminate reverse transcription in the pseudouridylated site (pseudo-seq, Ψ-seq for Ψ modification), and chemoenzymatic capture (NAD captureSeq for NAD modification).

1.4 PROTEIN TECHNIQUES

1.4.1 Protein Structure and Function

Proteins are composed of 20 amino acids (each defined within mRNA by a triplet of nucleotides called codon), which interlink by peptide bonds to form polypeptide chains. The three-dimensional structure of a protein is determined by its amino acid sequence (so-called primary protein structure). Local interactions between parts of a polypeptide chain may alter the three-dimensional shape and fold into a regular substructure (e.g., α-helix, β-pleated sheet, turns, and loops) (so-called secondary protein structure). Water-soluble protein may fold into a compact three-dimensional structure with a nonpolar core (so-called tertiary protein structure) after secondary interactions. Further, proteins with multiple polypeptide chains may display specific orientation and arrangement of subunits (so-called quaternary protein structure). Several functional groups (e.g., alcohols, carboxamines, carboxylic acids, thioesters, thiols, and other basic groups) may attach to protein and affect the folding and function of protein. Proteins may also interact with each other or other macromolecules to create complex assemblies with additional functions (e.g., DNA replication, cell signal transmission).

Directly related to their three-dimensional shapes, proteins demonstrate many different functions in the body, ranging from enzyme catalyst, transporter, storage, mechanical support (cytoskeleton or connective tissues), immune protection, movement facilitator (hinges, springs, or levers), transmission of nerve impulses, to controlling of cell growth and differentiation [28].

1.4.2 Protein Purification

Purifying one or a few proteins from a complex mixture (e.g., cells, tissues, or whole organisms) is essential for characterizing the function, structure, post-translational modification, and interactions of the protein(s) of interest. Typically, protein purification requires breakup of cells (in case that proteins of interest such as integral membrane proteins are not secreted into solution), removal of other substances, further separation of proteins, and final concentration. These steps should be ideally carried out at low temperature (4°C) to reduce protein denaturation that takes place after proteins are released from cells [29].

For disruption of the cells containing the protein, several procedures may be considered: (i) repeated freezing and thawing, (ii) sonication, (iii) homogenization by high pressure (e.g., French press), (iv) homogenization by grinding (e.g., bead mill), and (v) permeabilization by detergents [e.g., triton X-100, CHAPS or sodium dodecyl sulfate (SDS); the latter is somewhat destructive to cell membrane proteins] and/or enzymes (e.g., lysozyme). It is important to proceed cell disruption quickly and keep the cell lysate cool to slow down/prevent protein digestion by endogenous proteases. Further, one or more protease inhibitors may be included in the lysis buffer prior to cell disruption. Addition of DNAse may also help reduce the viscosity of cell lysate due to high DNA content.

For removal of other substances, centrifugation helps separate proteins and other soluble compounds (which remain in supernatant) from cell debris (in pellet). Use of sucrose gradient (e.g., sucrose, glycerol, or Percoll, which is a silica-based density gradient media) during centrifugation offers another option for separation of proteins from other substances.

For further separation of proteins, various procedures that exploit differences in protein size, physico-chemical properties, binding affinity, and biological activity can be utilized. These include ammonium sulfate [$(NH_4)_2SO_4$] precipitation and subsequent dialysis to remove ammonium sulfate (this technique is extremely helpful in reducing the overall preparation volume), chromatography (e.g., size exclusion, ion exchange, affinity based on lectin, antibody, His-tag/Strep-tag), high-performance liquid chromatography (HPLC)/reversed-phase chromatography/hydrophobic interaction chromatography (based on polarity/hydrophobicity), sodium dodecyl sulfate-polyacrylamide gel electrophoresis (SDS-PAGE, based on size and solubility), and immunoprecipitation.

Finally, purified proteins may be concentrated by lyophilization (commonly performed after HPLC), ultrafiltration (using selective permeable membranes), precipitation (e.g., isoelectric focus; miscible solvents such as ethanol or methanol; polymers such as dextrans and polyethylene glycols), and flocculation (polyelectrolytes such as alginate, carboxymethylcellulose, polyacrylic acid, tannic acid, and polyphosphates; polyvalent metallic ions such as Ca^{2+}, Mg^{2+}, Mn^{2+}, or Fe^{2+}) [8].

1.4.3 Protein Characterization

The quality (integrity) and quantity of purified proteins are determined using several techniques to ensure their suitability for subsequent studies on their function, structure, post-translational modification, and interactions.

The quantity of purified proteins is often evaluated by the Bradford total protein assay and spectrophotometry (at A280 nm, note that residual imidazole may absorb at 280 nm, and cause inaccurate reading of protein concentration) [30].

The size (molecular weight) and integrity of purified proteins are assessed by SDS-PAGE, immunoprecipitation, and western blot (if specific antibodies are available) using Coomassie blue dye or silver stain. Moreover, the molecular weight and purity of purified proteins may be ascertained by HPLC, ESI-MS, and MALDI-TOF-MS [30].

1.4.4 Protein Sequencing

Protein sequencing is a technique for analyzing the amino acid composition of a protein (or peptide), which helps uncover the protein identity and its post-translational modifications. A number of protocols are available, including N-terminal sequencing (by reacting the peptide with a reagent such as 1-fluoro-2,4-dinitrobenzene, dansyl chloride, or phenylisothiocyanate that selectively labels the terminal amino acid, hydrolyzing the protein, and determining the amino acid by thin-layer chromatography or high-pressure liquid chromatography and comparison with standards; this offers a more accurate approach for N-terminus analysis than the Edman degradation), C-terminal sequencing (by adding carboxypeptidases to a solution of the protein, taking samples at regular intervals, and determining the terminal amino acid by analyzing a plot of amino acid concentrations against time; this helps verify the primary structures of proteins predicted from DNA sequences and detect any post-translational processing of gene products from known codon sequences), Edman degradation (by breaking any disulfide bridges in the protein with

2-mercaptoethanol and preventing the bonds from re-forming with iodoacetic acid, separating, and purifying the individual chains of the protein complex if there are more than one, determining the amino acid composition and terminal amino acids of each chain, breaking each chain into fragments under 50 amino acids long with trypsin, pepsin, or cyanogen bromide, separating and purifying the fragments, determining the sequence of each fragment, repeating with a different pattern of cleavage, and constructing the sequence of the overall protein; this allows discovery of the ordered amino acid composition of a protein up to approximately 50 amino acids long), blocked N-terminal sequencing with MALDI-ISD, peptide mapping, peptide mass fingerprinting, de novo sequencing, amino acid composition analysis, extinction co-efficiency, and differential scanning calorimetry (DSC).

1.4.5 Polyclonal and Monoclonal Antibodies

Antibody (Ab, also known as immunoglobulin or Ig) is a Y-shaped glycoprotein generated mainly by plasma cells (which are differentiated from native B lymphocytes) as part of host humoral immune response against invading microbial pathogens. Made up of two large heavy chains and two small light chains, the Y-shaped antibody contains binding sites on the tips (variable domains in the antigen-binding fragment or Fab region consisting of light chain and part of heavy chain) with specificity for an epitope on an antigen of microbe, with the goal to inhibit microbial invasion and survival and/or activate macrophages to destroy invading microbe. The base (crystallizable fragment or Fc region consisting of heavy chain only) of the Y-shaped antibody determines its isotype (A, D, E, G, or M).

Secreted by different B-cell lineages, polyclonal antibodies (pAb) represent a collection of immunoglobulins that specifically recognize different epitopes of an antigen. Laboratory production of polyclonal antibodies involves (i) antigen preparation, (ii) adjuvant selection (e.g., Freund's, alum, Ribi adjuvant system, and Titermax), (iii) animal selection (e.g., mouse, rabbit, or goat), (iv) injection, and (v) blood serum extraction.

By contrast, monoclonal antibodies (mAb) come from a single B-cell lineage and recognize a single epitope of an antigen. In comparison with pAb that binds to multiple epitopes, mAb demonstrates monovalent affinity and binds to the same epitope of an antigen. Laboratory production of mAb, which was first described by Georges Köhler and César Milstein in 1975, involves (i) immunization of mice with antigen, (ii) fusion of mouse spleen cells (B cells) with myeloma cells, (iii) selection of hybridomas secreting antibodies with desired specificity, and (iv) bulk production of mAb in culture or mice (ascites fluid). Recent advances in mAb production include phage display, single B-cell culture, single-cell amplification from various B-cell populations, and single plasma cell interrogation [31–33].

1.4.6 Protein Synthesis and Expression

Proteins may be synthesized *in vitro* through a cell-free system or expressed in prokaryotic and eukaryotic cell-based systems.

For *in vitro* protein synthesis, cell extracts containing RNA polymerase, ribosomes, tRNA, and ribonucleotides are utilized. However, due to its low expression levels and high cost, this system has limited practical value.

For protein expression in cell-based systems, the gene encoding the protein of interest is cloned into a plasmid or other vectors (e.g., bacteriophage lambda, baculovirus, retrovirus, adenovirus, and artificial chromosome), transformed into prokaryotic or prokaryotic host [bacteria (e.g., *Escherichia coli* strain BL21, *Pseudomonas fluorescens*), yeast (e.g., *Saccharomyces cerevisiae, Pichia pastoris*), fungi (*Aspergillus, Trichoderma, Myceliophthora thermophila* C1), insect cells (e.g., Sf9, Sf21 from *Spodoptera frugiperda*, Hi-5 from *Trichoplusia ni*, and Schneider 2 and Schneider 3 cells from *Drosophila melanogaster*), and mammalian cells (Chinese hamster ovary cell, mouse myeloma lymphoblastoid NS0 cell, HeLa, human embryonic kidney HEK 293 cell, human embryonic retinal Crucell's Per.C6 cell, human amniocyte glycotope, and CEVEC cells)], and subsequently induced to produce the protein [34].

1.4.7 Protein Modification

Inside the cells, DNA is transcribed into mRNA, which undergoes post-transcriptional modifications (including the addition of a 5′ cap and a 3′ poly(A) tail to the 5′ and 3′ ends of the pre-mRNA, respectively, as well as the removal of introns via RNA splicing) before translation into protein.

After translation, protein (polypeptide) goes through post-translational modifications (>200 types known to date) to become biologically active. These include (i) cleavage (hydrolysis of peptide bonds by proteases leading to a shortened protein with altered function), (ii) addition of chemical groups (through methylation, acetylation, and phosphorylation), (iii) addition of complex molecules (through glycosylation, including N-linked glycosylation and O-linked glycosylation), and (iv) formation of intramolecular bonds (e.g., disulfide bond/bridge between two cysteine amino acids in the oxidizing environment of the endoplasmic reticulum).

Techniques for examining protein modifications range from glycosylation analysis (N-glycan profiling, O-glycan profiling, N-glycan site occupation analysis, O-glycan site occupation analysis, glycopeptide analysis, sialic acid analysis), phosphorylation analysis, deamidation and oxidation analysis, disulfide bridges and free sulfhydryl groups, N-acetylation analysis, methylation analysis, ubiquitination analysis, sumoylation analysis, lipidation analysis, S-nitrosylation analysis, N-myristoylation analysis, S-palmitoylation analysis to S-prenylation analysis [35].

1.4.8 Protein Interaction

Protein interactions with other proteins, DNA, RNA, and other molecules occur constantly within and between cells and underpin the proper running of various biological processes. In the past, studies on biological processes had been limited to single molecules and their individual interactions due to the lack of suitable technologies. Recent advances in molecular biology have enabled simultaneous analyses of multiple molecules and their interactions, uncovered valuable insights on the complex cellular mechanisms at both health and disease states, and empowered the discovery of putative protein targets for the therapeutic purpose [36].

Protein–protein interactions (PPI) are physical and biochemical events between two or more protein molecules, which are modulated by the electrostatic forces, hydrogen bonding, and hydrophobic effect of individual proteins. Current approaches for analyzing PPI are grouped into three broad categories: *in vitro* [tandem affinity purification-mass spectroscopy (TAP-MS), affinity chromatography, co-immunoprecipitation, protein microarray, protein-fragment complementation, phage display, X-ray crystallography, NMR spectroscopy], *in vivo* [Yeast 2 hybrid (Y2H), synthetic lethality], and *in silico* [ortholog-based sequence approach, domain-pair-based sequence approach, structure-based approach, gene neighborhood, gene fusion, *in silico* 2 hybrid (I2H), phylogenetic tree, phylogenetic profile, gene expression]. Use of these techniques facilitates *in silico* mapping of PPI revealed by *in vitro* or *in vivo* studies and also experimental confirmation of computationally identified protein interaction networks [37].

Protein–DNA interactions (PDI) arise from electrostatic interactions (salt bridges), dipolar interactions (hydrogen bonding, H-bonds), entropic effects (hydrophobic interactions), and dispersion forces (base stacking) between related proteins and DNA and play an indispensible part in gene activation and other biological processes. As DNA-binding proteins (e.g., transcription factors, polymerases, nucleases, histones) contain DNA-binding domains (e.g., zinc finger, helix-turn-helix, leucine zipper) with specific or general affinity for single- or double-stranded DNA, current approaches for PDI analysis center on binding site characterization and gene activation. Binding site characterization examines the binding preferences of a given protein for various DNA sequences *in vitro*, while gene activation study further investigates the unique sequences (e.g., transcription factors) bound by proteins in the cellular context. Techniques utilized frequently for PDI studies include electrophoretic mobility shift assay (EMSA, which assesses the degree of affinity or specificity of the interaction between protein and known DNA probes),

DNase footprinting assay (which identifies the specific site of binding of a protein to DNA), chromatin immunoprecipitation (ChIP, which identifies the sequence of DNA fragments that bind to a known transcription factor), ChIP-Seq (a combination of ChIP and high-throughput sequencing), ChIP-Seq (a combination of ChIP and microarrays), yeast one-hybrid system (Y1H, which identifies protein that binds to a particular DNA fragment), bacterial one-hybrid system (B1H, which identifies protein that binds to a particular DNA fragment), and X-ray crystallography (which gives a detailed atomic view of protein–DNA interactions) [38,39].

Protein–RNA interactions (PRI) result from electrostatic interactions, hydrogen bonding, hydrophobic interactions, and base stacking between related proteins and RNA and are necessary for transportation of mRNA into the cytoplasm of eukaryotic cells and for the formation of the translation machinery. Common techniques for PRI analysis are RNA electrophoretic mobility shift assay (which detects protein–RNA interactions through changes in migration speed during gel electrophoresis), RNA pull-down assay (which selectively extracts a protein–RNA complex from a sample), oligonucleotide-targeted RNase H protection assay (which detects RNA and RNA fragments in cell extracts and helps map protein–RNA interactions), and fluorescent in situ hybridization co-localization (which detects the position and abundance of a RNA and protein in a cell or tissue sample). It should be noted that both RNA and protein have to be correctly folded to allow proper binding, and special care is taken to avoid introduction of RNases into the assay during PRI analysis [40].

1.5 UTILITY OF MOLECULAR TECHNIQUES IN FOOD MICROBIOLOGY

1.5.1 Phylogenetic Analysis

Phylogenetic analysis infers the evolutionary history and relationships among or within groups of organisms through evaluation of observed heritable traits (e.g., morphology pre-1950s and DNA sequences from the 1950s) and construction of a phylogenetic tree/diagram (or phylogeny), leading to the taxonomic determination or classification (e.g., species or more inclusive taxa) of organisms concerned [41,42].

Prior to the 1950s, phylogenetic inferences were derived from phylogenetic trees constructed by distance matrix methods on the basis of overall similarity in morphology and other phenotypic traits as well as DNA (e.g., G+C content and DNA–DNA hybridization, but not DNA sequences). Due to their ambiguous nature and the absence of explicit criteria for evaluating alternative hypotheses, distance matrix methods have become largely obsolete now.

Currently, phylogenetic analysis is built on the comparison of DNA sequences from ribosomal, mitochondrial, and other house-keeping genes and whole-genome sequences. Subsequent construction of a phylogenetic diagram/tree (of which a rooted tree indicates the hypothetical common ancestor of the tree, whereas an unrooted tree makes no assumption about the ancestral lineage and displays no origin or root of the taxa in question) using parsimony, maximum likelihood, and Bayesian inference software helps ascertain the phylogenetic relationship and taxonomic status of organisms under investigation.

1.5.2 Species Identification, Strain Typing, and Population Genetics

Species identification forms an integral part of microbiological investigation that underpins laboratory diagnosis of infectious diseases. While morphological, biological, and biochemical examinations have contributed significantly to our understanding of microorganisms in the past, procedures targeting specific proteins and nucleic acids have further streamlined species-specific identification of microbial pathogens in both research and clinical contexts [43,44].

Strain typing offers an additional tool for investigation of microbial diversity and tracking of microbial types and subtypes implicated in epidemiological outbreaks. In this regard, a variety of procedures targeting phenotypic [e.g., biotyping, serotyping, bacteriocin typing, antibiogram typing, multilocus enzyme electrophoresis (MLEE) typing] and genotypic [e.g., phage typing, pulsed-field gel electrophoresis (PFGE), restriction fragment length polymorphism, ribotyping, random amplification of polymorphic DNA (RAPD), amplified fragment length polymorphism (AFLP), enterobacterial repetitive intergenic consensus sequence-PCR (ERIC-PCR), variable number of tandem repeats (VNTR), multilocus sequence typing (MLST), and multilocus variable tandem repeat analysis (MLVA)] features are available.

Population genetics examines the frequencies of alleles and genes (genotypes) that change within and between populations of organisms over time. By focusing on certain polymorphic and reproducible DNA or protein markers together with the use of relevant models, population genetics helps predict the occurrence of specific alleles in populations. Specifically, DNA markers with high mutation rates (e.g., microsatellites and minisatellites) reveal insights into recent divergence, whereas those with relatively low mutation rates (e.g., mitochondrial and nuclear loci) provide inference about distant evolutionary history. With data generated by genome sequencing technology and computer-based operation, evolutionary trees (or molecular clocks) can be constructed to pinpoint the time when populations diverge and form new species.

1.5.3 Antibiotic Resistance

The discovery of penicillin in the 1930s heralded a new era in our fight against microbial infections and diseases, which have inflicted the human society longer than any historic records could possibly reveal. With the help of antibiotics and other antimicrobial compounds, many persistent infections that were once beyond the capability of most talented physicians are cured with considerable ease, and many bed-ridden patients with previously untreatable diseases recover and enjoy their new life like never before. However, microbes are not innocent standbys in the face of seemingly all-conquering antibiotics. They have evolved and acquired a number of traits that help sabotage the efficiency of and confer resistance to antimicrobial compounds. Overuse, misuse, and inappropriate use of antimicrobials have hastened this evolution and buildup of antibiotic resistance.

Application of molecular techniques has uncovered the key mechanisms underlying the development of antibiotic resistance, including (i) enzymatic modification or inactivation of drug (acquisition of genes encoding enzymes that chemically modify or inactivate antimicrobial compound), (ii) antimicrobial target modification/mimicry (bacteria may undergo spontaneous mutations in the genes encoding antibacterial drug targets, making them no longer accessible to antimicrobial compound), (iii) prevention of drug penetration or accumulation (Gram-negative pathogens often alter their outer membrane lipid composition, porin channel selectivity, and/or porin channel concentrations that inhibit the accumulation of antimicrobial compound), and (iv) efflux pump expulsion of drug (many Gram-positive and Gram-negative pathogens produce efflux pumps that expel the antimicrobial compound out of the cell) [45]. This has facilitated the development of molecular tests for rapid, specific, and sensitive detection of antimicrobial resistance in pathogenic microbes, especially multidrug-resistant microbes [MDR, or so-called superbugs, including methicillin-resistant *Staphylococcus aureus* (MRSA), vancomycin-resistant enterococci (VRE), vancomycin-resistant *S. aureus* (VRSA), and vancomycin-intermediate *S. aureus* (VISA), extended-spectrum β-lactamase–producing Gram-negative bacteria, carbapenem-resistant Gram-negative bacteria, and multidrug-resistant *Mycobacterium tuberculosis*].

1.5.4 Biosensors

Biosensors are analytical devices that combine bioreceptors (e.g., antibodies, enzymes, DNA, cells, biomimetics, and phages for the recognition of analytes of interest), which are attached to the surface of the sensor (e.g., metal, polymer, or glass) or entrapped in three-dimensional lattices (e.g., hydrogel, xerogel), with biotransducers (e.g., optical, electrochemical, mass-based, and nano-based for signal amplification and display) for enhanced detection/identification of biological elements (e.g., proteins, antibodies, enzymes, toxins, nucleic acids) or living organisms (e.g., cells, organelles). The key features of bioreceptors and biotransducers commonly used in biosensors are summarized in Table 1.4 [46].

TABLE 1.4 Key features of bioreceptors and transducers commonly used in biosensors

		FEATURES	*COMMENTS*
Bioreceptors	Antibody	During its specific interaction with antigen, antibody containing a fluorescent molecule, enzyme, or radioisotope generates a signal that can be picked up by transducer	The binding capacity of antibody is dependent on assay conditions (e.g., pH and temperature); the antibody–antigen interaction is liable to disruption by chaotropic reagents, organic solvents, or ultrasonic radiation
	Enzyme	With specific binding capability and catalytic activity, enzyme helps convert the analyte into a sensor-detectable product, detect enzyme inhibition or activation by the analyte, or monitor modification of enzyme property due to interaction with the analyte	Enzyme is capable of catalyzing a large number of reactions, detecting a group of analytes (substrates, products, inhibitors, and modulators of the catalytic activity) and fitting into different transduction methods
	DNA	Through complementary base pairing, DNA immobilized on the sensor hybridizes with target DNA/RNA for subsequent optical detection; aptamers (specific nucleic acid-based antibody mimics) labeled with a fluorophore/metal nanoparticles recognizes target (e.g., cells and viruses) via specific noncovalent interactions and allows easy optical detection	DNA and aptamer are genosensors providing consistent detection independent of external influence
	Cell	Cells immobilized on surface/acetylcellulose membrane or entrapped on quartz microfiber are useful for the detection of global parameter (e.g., stress conditions, toxicity, and organic derivatives), monitoring of drug treatment effect of drugs	Cells are sensitive to surrounding environment, responsive to all kinds of stimulants, easily immobilized on surface, and controllable to no more than 5 min
	Biomimetic	Recombinant binding fragments (Fab, Fv, or scFv) or domains (VH, VHH) of antibodies and recombinant small protein scaffolds with favorable biophysical properties are examples of biomimetics that facilitate specific binding to different target proteins	Biomimetics (usually <100 amino acid residues) are smaller than antibodies, highly stable, free of disulfide bonds, and easily expressed in high yield
	Phage	Phage containing fluorescent or other label invades target bacteria and facilitates specific detection	Phages display strict host specificity

(*Continued*)

TABLE 1.4 (*Continued*) Key features of bioreceptors and transducers commonly used in biosensors

		FEATURES	*COMMENTS*
Biotransducers	Optical	Optical fiber, Raman infrared spectroscopy, surface plasmon resonance	High sensitivity, label-free detection, but high cost
	Electrochemical	Electrochemical (amperometric, potentiometric, impedimetric, conductometric) biosensor detects electrons produced or consumed by a redox enzyme-catalyzed reaction	Label-free detection, low cost, but low specificity
	Mass based	Piezoelectric (quartz crystal microbalance and surface acoustic wave), magnetoelastic	Low cost, easy operation, label-free detect, but low specificity and sensitivity
	Nanomaterial based	Modification of sensing electrode by a nanomaterial (e.g., gold) achieves a quick electron transfer due to the stimulation of different biomarkers	User-friendly measurement and measurement, moderate cost, but toxic nanomaterial

1.5.5 Phage Therapy

Phages (or bacteriophages) are viruses (24–400 nm in size) that specifically infect, replicate in, and cause the death of bacterial species or strains. Structurally, a bacteriophage comprises a single- or double-stranded DNA or RNA core encapsulated by an icosahedral protein or lipoprotein capsid, which is connected with a tail (hollow tube) for interaction with bacterial surface receptors and passage of nucleic acid to the bacterial host.

As a natural predator of bacteria, bacteriophages (numbered between 10^{30} and 10^{31} on Earth) are 10 times more common than their bacterial hosts. Abundant in the marine environment, bacteriophages mostly belong to the families *Siphoviridae* (icosahedral capsid with a filamentous noncontractile tail), *Myoviridae* (icosahedral symmetric head with a neck and a helical contractile tail), and *Podoviridae* (icosahedral symmetric head with a very short noncontractile tail) in the order *Caudovirales*. For example, a T4-like phage KVP40 of the family *Myoviridae* consists of a dsDNA genome of 244,835 bp, an icosahedral capsid, and a contractile tail with baseplate and tail fibers, and is infective to several *Vibrio* species.

Bacteriophages undergo two types of life cycles: lytic and lysogenic. During the lytic cycle (lasting from 20 min to 2 h), a bacteriophage attaches to a bacterial cell, injects the phage nucleic acid, reproduces intracellularly using bacterial synthesis machinery, and lyses the host cell wall to release daughter phage particles, causing direct damage to the host. During the lysogenic cycle, a bacteriophage replicates along with the host for several generations without destruction to the host cell until the lytic cycle is induced.

Given its high specificity and lack of adverse effects, bacteriophages have been employed following their initial description in 1896 as a therapeutic agent against infectious diseases in various countries prior to the discovery of broad-spectrum antibiotics. With the emergence of multidrug-resistant bacteria in recent years, there is a renewed interest in the utilization of bacteriophages for the control of bacterial pathogens [47]. This is evidenced by the approval of ListShield™ (a phage that targets *Listeria monocytogenes*) and Listex P100 by the USFDA for the treatment of food products.

1.5.6 Probiotics

Probiotics (from the Latin preposition "pro" meaning "for," and the Greek word "biotic" meaning "life") are living, nonpathogenic microorganisms (many of which are foodborne bacteria and yeasts) with the ability to enhance both innate and adaptive immunity, sustain homeostasis related to inflammatory and anti-inflammatory responses, and activate opioid and endocannabinoid receptors, thus providing pain relief [48].

Among most common probiotics are *Lactobacillus* spp. (found in yogurt and other fermented foods, which help to digest lactose and sugar in milk and fight diarrhea), *Bifidobacterium* spp. (present in some dairy products, which help to ease the symptoms of irritable bowel syndrome and other conditions), and *Saccharomyces boulardii* (which helps to treat diarrhea and other digestive problems). Other probiotics include *Pediococcus pentosaceus* and *Leuconostoc* spp.

With their beneficial effects being first recognized in 1908, probiotics have shown to help alleviate irritable bowel syndrome, inflammatory bowel disease (IBD), infectious diarrhea (caused by viruses, bacteria, or parasites), antibiotic-associated diarrhea (AAD), and skin conditions (e.g., eczema); promote oral, gut, urinary, and vaginal health; and prevent allergies and colds.

However, there is evidence that *Saccharomyces cerevisiae* and *S. boulardii* may be occasionally implicated in fungemia; *Lactobacillus* spp., *S. boulardii, S. cerevisiae, Bacillus subtilis*, and *Bifidobacterium breve* are potential causes of bacteremia and sepsis; *Lactobacillus* and *Streptococcus* are possibly involved in endocarditis, especially in individuals with compromised immune functions [49]. This necessitates further investigation on the molecular interactions between probiotics and host, and identification of novel probiotics with improved efficiency and reduced side effects.

1.5.7 Omics

Built upon the recent availability of efficient analytical and data mining methods, the "omics" approaches (including genomics, transcriptomics, proteomics, and metabolomics) provide a holistic, in-depth, high-throughput evaluation of genes/transcripts/proteins/metabolites under specific conditions and time points (Table 1.5) [50,51]. Overcoming the classical genetic/biochemical studies based on single or few target molecules, the "omics" approaches give a comprehensive perspective of cell biology and improve the diagnosis, treatment, and control of foodborne diseases [52].

1.5.8 Bioinformatics

Bioinformatics is an interdisciplinary field of science that integrates biology, computer science, information technology, mathematics, and statistics, and develops software and tools for *in silico* analysis and interpretation of large and complex biological data, including the structures and functions of genes, genomes, RNA, proteins, and metabolites. In the early days, research groups interested in conducting bioinformatic analysis had to make do with home-designed software that caused difficulty in data sharing with other groups. This is no longer the case with a variety of highly sophisticated bioinformatic software being freely accessible online [53,54].

For example, a number of valuable software packages are available on the Galaxy site (usegalaxy.org), ranging from **get data**, **collection operation**, **general text tools** (text manipulation, filler and sort, join, subtract and group, datamatch), **genomic file manipulation** (FASTA/FASTQ, FASTQ quality control, SAM/BAM, BED, VCF/BCF, nanopore, convert formats, left-over), **common genomics tools** (operate on genomic intervals, fetch sequence/alignments), **genomic analysis** (assembly, annotation, mapping, variant calling, ChIP-seq, RNA-seq, multiple alignments, phenotype association, evolution, regional variation, STR-FM: microsatellite analysis, chromosome confirmation, virology), **metagenomics** (metagenomics analysis, Mothur), **genomic toolkits** (Picard, deep tools, EMBOSSs, GALK, NCBI BLAST+, HyPhy, RseQC, MiModD, Du Novo, seqtk), **miscellaneous tools** (transposon insertion sequencing, IWTomis,

TABLE 1.5 Principles and technological platforms of omics approaches

OMICS	*PRINCIPLES*	*TECHNOLOGICAL PLATFORM*
Genomics	**Genomics** examines the structure, function, evolution, mapping, and editing of genome; specifically, **structural genomics** utilizes genomic sequencing and modeling to investigate the three-dimensional structure of every protein encoded by a genome rather than one particular protein; **functional genomics** employs microarrays and bioinformatics to conduct genome-wide instead of gene-by-gene analysis of the functions of genes, RNA transcripts, and protein products, particularly the patterns of gene expression under various conditions; **epigenomics** relies on genomic high-throughput assays to analyze the complete set of reversible epigenetic modifications (e.g., methylation) on DNA or histones that affect gene expression without alteration in DNA sequence; **metagenomics** uses 16S rRNA gene sequencing and massively parallel pyrosequencing to examine microbial diversity on genetic material recovered directly from environmental specimens	Next-generation sequencing (NGS), 16S rRNA gene sequencing
Transcriptomics	**Transcriptomics** studies transcriptome (the sum of all RNA transcripts) in an organism to find out whether a cellular process is active or dormant. Besides large-scale identification of transcriptional start sites, transcriptomics helps uncover alternative promoter usage and novel splicing alterations and enables disease profiling (e.g., disease-associated single-nucleotide polymorphisms or SNP, allele-specific expression, and gene fusions); evolving from low-throughput techniques [e.g., conversion of mRNA transcripts to cDNA, northern blot for displaying the size and amounts of small RNA, and serial analysis of gene expression (SAGE, involving Sanger sequencing of concatenated random transcript fragments and subsequent quantification by matching the fragments to known genes)], which capture only a tiny subsection of a transcriptome, transcriptomics is increasingly conducted with high-throughput techniques (e.g., RNA-seq, cDNA microarray, and quantitative PCR), which cover entire transcripts	RNA-seq (also known as massive parallel cDNA sequencing), cDNA microarray, nanoString
Proteomics	**Proteomics** detects, quantifies, and characterizes proteins, their expression, and post-translational modifications in a large scale during a disease state relative to a healthy state using two-dimensional gel electrophoresis and mass spectrophotometry (which involves protein extraction and purification, direct analysis or enzymatic digestion, optional protein/peptide separation based on liquid chromatography, mass-to-charge and intensity detection of protein/peptide and induced fragments, and protein identification and quantification by de novo or database-driven data analysis)	Mass spectrometry, protein array
Metabolomics	**Metabolomics** determines in a systematic, nonbiased manner the types and levels of low-molecular-weight small molecules (metabolites) secreted into body fluids or tissues by host and microbial cells in response to environmental, nutritional, and immunological stimuli, using high-throughput mass spectrometry nuclear or magnetic resonance spectroscopy followed by alignment against libraries of known biochemicals	Mass spectrometry, nuclear magnetic resonance (NMR)

pRESTO, PlantTribes, motif, single cell), **statistics and visualization** (statistics, graph/display data), and **deprecated tools** (genome diversity, cloud map, RNA analysis, get data).

Unlike the Galaxy (usegalaxy.org) site with a genomic focus, the SIB Bioinformatics Resource Portal (https://www.expasy.org/) include software for **proteomics** (protein sequences and identification, proteomics experiment, function analysis, sequence sites, features and motifs, protein modifications, protein structure, protein interactions, similarity search/alignment), **genomics** (sequence alignment, similarity search, characterization/annotation), **structure analysis** (SWISS-MODEL Workspace, SwissDock, CAMEO, Click2Drug, COILS, MARCOIL, Missense3D, OpenStructure, QMEAN, rBAN, Swiss-PdbViewer, SwissParam, SwissTargetPrediction), **systems biology** (arrayMap, Biochemical Pathways, CRUNCH, efmtool, Genome History, Genonets, iPtgxDBs, ISMARA, MetaNetX, Phylogibbs, Progenetix, STR similarity search tool, The Systems Biology Research Tool), **evolutionary biology** (Bgee, OMA, ALF, BayeScan, BIONJ, BLASTO, BUSCO, CEGA, Covid-19 Scenarios, CT-CBN, DendroUPGMA, Evolutionary Trace Server, fastsimcoal, Genonets, MLtree, Newick Utilities, Nextstrain, PHYLIP, Phylo.io, Phylogeny.fr, Phylogeny programs, RAxML, REALPHY, SuperTree, SwissTree, The PhylOgenetic Web Repeater), **population genetics** (Arlequin, BayeScan, BEAST2, Covid-19 Scenarios, CT-CBN, fastsimcoal, mOTUs, TriFLe, V-pipe), **transcriptomics** (Bgee, ESTscan, ExpressionView, ISA, ISMARA, mOTUs, Ping pong algorithm, QuasR, SIBsim4, The Miner Suite, Translate, tromer), **biophysics**, **imaging**, **IT infrastructure**, **medicinal chemistry**, and **glycomics**.

For metabolomics analysis, a number of software packages (BioSpider, COLMAR, FiD, HORA, MeltDB, MetaboloAnalyst, MetaboMiner, MolFind, MS-based Structure Elucidation Software, MVAPACK, OpenMS, Peak Alignment Software, PolySearch, SetupX, Seven Golden Rules Software, XCMS2) are accessible via http://metabolomicssociety.org/resources/metabolomics-software.

Further, https://www.genomics-online lists free tools and software for CRISPR/Cas9, flow cytometry, gel electrophoresis, gene expression microarray, gene suppression, genomics, high-throughput sequencing, imaging software, mass spectrophotometry, mass cytometry, metagenomics, microarray, PCR, proteomics, sequence alignment, transcriptomics, data management, and diverse tools.

1.6 FUTURE PERSPECTIVES

Breakthroughs in the second half of the 20th century, symbolized by the elucidation of DNA structure in 1953 and publication of the DNA cloning procedure in 1973, have added a spring into the wings of molecular biology, which has since quietly revolutionized many fields of biological science, including but not limited to cell biology, biochemistry, immunology, genetics, microbiology, medicine, veterinary science, animal husbandry, aquaculture, and agriculture. Food microbiology is no exception. A quick Pubmed survey of reports published during the past 10 years using the terms of "rotavirus" and "rotavirus molecular," "*Salmonella*" and "*Salmonella* molecular," "*Escherichia coli*" and "*Escherichia coli* molecular," "*Candida*" and "*Candida* molecular," and "*Giardia*" and "*Giardia* molecular" indicated that publications with a molecular aspect account for 20%–40% of total outputs on these foodborne pathogens. As this trend continues, there is a distinct possibility that molecular food microbiology may one day catch up with and surpass food microbiology in the number of published reports and in the role of delivering innovational outcomes, positively impacting on the human society, animal welfare, and environmental conservation on an even grander scale.

The overview of DNA, RNA, and protein techniques in the above sections is not meant to be all inclusive, and it rather represents, as the cliché says, the tip of the iceberg. Over the past few decades, the breadth and depth of molecular technologies have expanded in such a way that even a single technique (e.g., bioinformatics) will easily take someone's lifetime to acquire and use aptly, not alone to further improve and innovate. Similarly, the application of molecular technologies has led to so many important discoveries on the fundamental mechanisms of the biological processes and empowered our decision-making on the diagnosis, treatment, and prevention of foodborne infections. If one feels that the rapid taking-off and continuing advances in molecular studies since the establishment of the DNA cloning protocol in 1973 have meant that most low hanging fruits in the field of molecular food microbiology have been picked, just imagines what people 500 or 1,000 years from now will do when they have all the molecular findings accumulated till then in front of their eyes? In fact, witnessing the birth and gradual sophistication of powerful molecular technologies, our generations are fortunate, much like standing on the top of a newly found gold mine, and we only need to dig a little deep for some exciting rewards. Indeed, if one has the chance of choosing a good time to work for potential discoveries in molecular food microbiology, the odds will decisively favor the present over 500–1000 years later. Now is the time to sharpen our molecular skills for the next discovery in molecular food microbiology, for which the golden age has just begun.

REFERENCES

1. Cox FE. History of human parasitology. *Clin Microbiol Rev.* 2002;15(4):595–612. [correction: *Clin Microbiol Rev.* 2003;16(1):174.]
2. Köhler JR, Casadevall A, Perfect J. The spectrum of fungi that infects humans. *Cold Spring Harb Perspect Med.* 2014;5(1):a019273.
3. One health: Fungal pathogens of humans, animals, and plants: Report on an American Academy of Microbiology Colloquium held in Washington, DC, on October 18, 2017. Washington (DC): American Society for Microbiology; 2019. Available from: https://www.ncbi.nlm.nih.gov/books/NBK549988/ doi: 10.1128/AAMCol.18Oct.2017.
4. Woolhouse ME, Gowtage-Sequeria S. Host range and emerging and reemerging pathogens. *Emerg Infect Dis.* 2005;11(12):1842–7.
5. Kirk MD, Pires SM, Black RE, et al. World Health Organization estimates of the global and regional disease burden of 22 foodborne bacterial, protozoal, and viral diseases, 2010: A data synthesis. *PLoS Med.* 2015;12(12):e1001921.
6. Bintsis T. Foodborne pathogens. *AIMS Microbiol.* 2017;3(3):529–63.
7. Travers A, Muskhelishvili G. DNA structure and function. *FEBS J.* 2015;282(12):2279–95.
8. Tan SC, Yiap BC. DNA, RNA, and protein extraction: the past and the present. *J Biomed Biotechnol.* 2009;2009:574398. [correction: *J Biomed Biotechnol.* 2013;2013:628968.]
9. Olson ND, Morrow JB. DNA extract characterization process for microbial detection methods development and validation. *BMC Res Notes.* 2012;5:668.
10. Cohen SN, Chang ACY, Boyer HW, Helling RB. Construction of biologically functional bacterial plasmids in vitro. *Proc Natl Acad Sci USA.* 1973;70(11):3240–4.
11. Lodish H, Berk A, Zipursky SL, et al. *Molecular Cell Biology.* 4th edition. New York: W. H. Freeman; 2000. Section 7.1, DNA cloning with plasmid vectors. Available from: https://www.ncbi.nlm.nih.gov/books/NBK21498/.
12. Ghannam MG, Varacallo M. *Biochemistry, Polymerase Chain Reaction (PCR).* Treasure Island (FL): StatPearls Publishing; 2020.
13. Wong YP, Othman S, Lau YL, Radu S, Chee HY. Loop-mediated isothermal amplification (LAMP): A versatile technique for detection of micro-organisms. *J Appl Microbiol.* 2018;124(3):626–43.
14. Heather JM, Chain B. The sequence of sequencers: The history of sequencing DNA. *Genomics.* 2016;107(1):1–8.
15. Gilbert LA, Larson MH, Morsut L, Liu Z, Brar GA. CRISPR-mediated modular RNA-guided regulation of transcription in eukaryotes. *Cell.* 2013;154(2):442–51.
16. Liu C, Zhang L, Liu H, Cheng K. Delivery strategies of the CRISPR-Cas9 gene-editing system for therapeutic applications. *J Control Release.* 2017;266:17–26.
17. Gouil Q, Keniry A. Latest techniques to study DNA methylation. *Essays Biochem.* 2019;63(6):639–48.
18. Ro S, Yan W. Small RNA cloning. *Methods Mol Biol.* 2010;629:273–85.
19. Lu TX, Rothenberg ME. MicroRNA. *J Allergy Clin Immunol.* 2018;141(4):1202–7.
20. Wong RKY, MacMahon M, Woodside JV, Simpson DA. A comparison of RNA extraction and sequencing protocols for detection of small RNAs in plasma. *BMC Genomics.* 2019;20(1):446.

21. Zhang Z, Lee JE, Riemondy K, et al. High-efficiency RNA cloning enables accurate quantification of miRNA expression by deep sequencing. *Genome Biol.* 2013;14:R109.
22. Devor EJ, Huang L, Abdukarimov A, Abdurakhmonov IY. Methodologies for in vitro cloning of small RNAs and application for plant genome(s). *Int J Plant Genomics.* 2009;2009:915061.
23. Suslov O, Silver DJ, Siebzehnrubl FA, et al. Application of an RNA amplification method for reliable single-cell transcriptome analysis. *Biotechniques.* 2015;59(3):137–48.
24. Choi JR, Yong KW, Choi JY, Cowie AC. Single-cell RNA sequencing and its combination with protein and DNA analyses. *Cells.* 2020;9(5):1130.
25. Bajan S, Hutvagner G. RNA-based therapeutics: From antisense oligonucleotides to miRNAs. *Cells.* 2020;9(1):137.
26. Jonkhout N, Tran J, Smith MA, et al. The RNA modification landscape in human disease. RNA. 2017;23(12):1754–69.
27. Meier UT. RNA modification in Cajal bodies. *RNA Biol.* 2017;14(6):693–700.
28. Kuhlman B, Bradley P. Advances in protein structure prediction and design. *Nat Rev Mol Cell Biol.* 2019;20(11):681–97.
29. Berg JM, Tymoczko JL, Stryer L. *Biochemistry.* 5th edition. New York: W. H. Freeman; 2002. Section 4.1, The purification of proteins is an essential first step in understanding their function. Available from: https://www.ncbi.nlm.nih.gov/books/NBK22410/.
30. Lodish H, Berk A, Zipursky SL, et al. *Molecular Cell Biology.* 4th edition. New York: W. H. Freeman; 2000. Section 3.5, Purifying, detecting, and characterizing proteins. Available from: https://www.ncbi.nlm.nih.gov/books/NBK21589/Journal.
31. Frenzel A, Schirrmann T, Hust M. Phage display-derived human antibodies in clinical development and therapy. *MAbs.* 2016;8(7):1177–94.
32. Parola C, Neumeier D, Reddy ST. Integrating high-throughput screening and sequencing for monoclonal antibody discovery and engineering. *Immunology.* 2018;153(1):31–41.
33. Strohl WR. Current progress in innovative engineered antibodies. *Protein Cell.* 2018;9(1):86–120.
34. Jia B, Jeon CO. High-throughput recombinant protein expression in *Escherichia coli*: Current status and future perspectives. *Open Biol.* 2016;6(8):160196.
35. Hawkins CL, Davies MJ. Detection, identification, and quantification of oxidative protein modifications. *J Biol Chem.* 2019;294(51):19683–708.
36. Cho H, Wu M, Bilgin B, Walton SP, Chan C. Latest developments in experimental and computational approaches to characterize protein-lipid interactions. *Proteomics.* 2012;12(22):3273–85.
37. Rao VS, Srinivas K, Sujini GN, Kumar GN. Protein-protein interaction detection: Methods and analysis. *Int J Proteomics.* 2014;2014:147648.
38. Furey TS. ChIP-seq and beyond: New and improved methodologies to detect and characterize protein-DNA interactions. *Nat Rev Genet.* 2012;13(12):840–52.
39. Mahony S, Pugh BF. Protein-DNA binding in high-resolution. *Crit Rev Biochem Mol Biol.* 2015;50(4):269–83.
40. Wheeler EC, Van Nostrand EL, Yeo GW. Advances and challenges in the detection of transcriptome-wide protein-RNA interactions. *Wiley Interdiscip Rev RNA.* 2018;9(1):e1436.
41. Rahaman MH, Islam T, Colwell RR, Alam M. Molecular tools in understanding the evolution of *Vibrio cholerae. Front Microbiol.* 2015;6: 1040.
42. Ruggiero MA, Gordon DP, Orrell TM, et al. A higher level classification of all living organisms. *PLoS One.* 2015;10(4):e0119248.
43. Alahi MEE, Mukhopadhyay SC. Detection methodologies for pathogen and toxins: A review. *Sensors (Basel).* 2017;17(8):1885.
44. Forbes JD, Knox NC, Ronholm J, Pagotto F, Reimer A. Metagenomics: The next culture-independent game changer. *Front Microbiol.* 2017;8:1069.
45. Ogawara H. Comparison of antibiotic resistance mechanisms in antibiotic-producing and pathogenic bacteria. *Molecules.* 2019;24(19):3430.
46. Asal M, Özen Ö, Şahinler M, Polatoğlu İ. Recent developments in enzyme, DNA and immuno-based biosensors. *Sensors (Basel).* 2018;18(6):1924.
47. Divya Ganeshan S, Hosseinidoust Z. Phage therapy with a focus on the human microbiota. *Antibiotics (Basel).* 2019;8(3):131.
48. Jäger R, Mohr AE, Carpenter KC, et al. International society of sports nutrition position stand: Probiotics. *J Int Soc Sports Nutr.* 2019;16(1):62.
49. Shahrokhi M, Nagalli S. *Probiotics.* Treasure Island (FL): StatPearls Publishing; 2020.
50. Wang Z, Gerstein M, Snyder M. RNA-seq: A revolutionary tool for transcriptomics. *Nat Rev Genet.* 2009;10(1):57–63.

51. Preidis GA, Hotez PJ. The newest "omics" – Metagenomics and metabolomics – Enter the battle against the neglected tropical diseases. *PLoS Negl Trop Dis.* 2015;9(2):e0003382.
52. Marzano V, Mancinelli L, Bracaglia G, et al. "Omic" investigations of protozoa and worms for a deeper understanding of the human gut "parasitome". *PLoS Negl Trop Dis.* 2017; 11(11):e0005916.
53. Melbourne Bioinformatics. De novo genome assembly for illumina data. Available from: https://www.melbournebioinformatics.org.au/tutorials/tutorials/assembly/assembly-protocol/.
54. Sutton TDS, Clooney AG, et al. Choice of assembly software has a critical impact on virome characterisation. *Microbiome.* 2019;7:12.

SUMMARY

Food microbiology is a constantly evolving field of biological science that addresses issues arising from the interactions between microorganisms and foods during processing and storage or after consumption, with the ultimate goals to reduce microbial contamination in food products and to eliminate foodborne disease outbreaks worldwide. Exploiting the power and versatility of molecular technologies, which have quietly revolutionized many fields of biological science, including food microbiology, after the elucidation of the DNA double helix in 1953 and the establishment of the DNA cloning protocol in 1973, molecular food microbiology extends and significantly improves phenotypically based food microbiology. This has not only facilitated the development of speedy, sensitive, and specific diagnostic assays for foodborne infections and intoxication but also contributed to the design and implementation of innovative treatment and prevention measures against foodborne diseases. By first presenting a brief overview of DNA, RNA, and protein techniques currently available, this chapter highlights the utility of molecular approaches in helping to solve the lingering problems that food microbiology is facing now and in the future.

SECTION I

Molecular Analysis and Manipulation of Foodborne Viruses

Molecular Mechanisms of Norovirus Adhesion to Agri-Food Surfaces and Fresh Foods

2

Idrissa Samandoulgou and Julie Jean
Université Laval

Contents

2.1 INTRODUCTION

Noroviruses (NoVs) were the first viruses identified in association with human acute infectious nonbacterial gastroenteritis,[1,2] which typically causes diarrhea, nausea, vomiting, abdominal soreness, and prevalent discomfort.[3] The disease onset generally follows an incubation period of 24–48 h, and the infection may persist for 12–60 h, often longer in persons weakened by a pre-existing condition.[3,4] Because of its characteristic peak occurrence during winter months, norovirus (NoV) gastroenteritis was formerly called "winter vomiting disease."[3] Although generally less severe than other diarrheic diseases, NoV gastroenteritis can lead to severe dehydration in children aged under 5 years and in elders aged above 65 years.[5–7] In the United States, it is estimated that NoVs have caused as much as 58% of foodborne illnesses (approximately 5.5 million cases) in some years and annual economic losses evaluated at $3.7 billion.[2] In Canada, they were responsible for a yearly estimation of 65% of all foodborne illnesses between 2000 and 2010.[8] Outbreaks have been documented worldwide, including in the UK, Japan, the Netherlands, France, New Zealand, and Australia.[9–12]

Ingestion of 3.3–7 PCR units (1,320–2,800 genome equivalents) of NoV in the infectious state is sufficient to cause illness within 12–60 h.[13,14] The principal mode of transmission is from person to person, via fecal matter or vomitus, whereas the second mode is less direct, through fecal contamination of foods and drinking water or contact with contaminated inert surfaces such as doorknobs, water fountains, and kitchen devices.[3] Outbreaks occur generally in closed groups that are in contact with the same surfaces and supplies of water and food.[3] The most likely sites are hospitals, schools, restaurants, nursing homes, cruise ships, hotels, holidays, and military camps.[3,5,6,15,16] The foods most often involved are vegetables, salads, berries, fresh herbs, some bakery products, sandwiches, ice cream and similar products, and fresh and processed meats, while others such as fish, poultry, soups, cakes, lobster, shrimp, and dough have been reported at least once.[14,16–18]

The original strain was identified in 1972 by Kapikian et al. who obtained stool filtrates from Norwalk, Ohio (USA), following the infamous 1968 outbreak of nonbacterial gastroenteritis.[1] Initially named the "Norwalk virus," this strain was later classified in the genus *Norovirus* created in 2002.[19] Together with *Sapovirus*, *Lagovirus*, *Vesivirus*, *Nebovirus*, and the recently suggested genera *Recovirus* and *Valovirus*, this genus belongs to the *Caliciviridae* family.[3] NoVs are icosahedral, nonenveloped, and single-stranded (+) RNA type, with a capsid diameter ranging from 28 to 35 nm.[3] The genome consists of about 7,700 nucleotides organized into three open-reading frames: ORF1 encoding a set of six post-translational proteins involved in the replication process, namely N-terminal, NTPase, 3A-like protein, VPg protein, viral protease, and the polymerase;[5,20,21] ORF2 encoding a major capsid protein (VP1); and ORF3 encoding a minor structural capsid protein.[20] Known NoVs are divided currently into six genogroups (GI to GVI) plus one unconfirmed genogroup (GVII).[22] Genogroups I and II, which contain, respectively, nine genotypes (including the original "Norwalk virus" strain GI.1) and 22 genotypes,[23] are well known to infect humans. Exceptions are GII genotypes 11, 18, and 19, which infect swine asymptomatically.[3] GIV contains two strains, one infecting humans and one infecting animals.[24] Members of GIII, GV, and GIV/GVI/GVII infect cattle, mice, and canines/felines, respectively.[13,22]

NoVs evolve under selective pressure from immune systems.[25] Their antigenic diversity combined with the short-term nature of the elicited immunity maintains host susceptibility to both new and old strains.[3,5,14,18] GI strains are often responsible for infections traced to seafood consumption,[26] while GII.4 genotype is involved notoriously in person-to-person transmission.[27] Most of the outbreaks reported around the world during the current decade have been attributed to strains of GII.4, notably in the United States,[28] Europe,[9,12,25] Japan,[10] the Middle East, and Africa.[29,30]

In addition to being highly infectious,[13,17] NoVs defy food-processing extremes such as acidity, heat (60°C), and free chlorine (0.5–1 mg l^{-1}) for up to 30 min.[17] Moreover, NoV remains infectious longer on inert surfaces[31] and increases the disease burden through postsymptomatic food handlers that still shed the virus.[32]

NoVs are detected using the same methodology as for other enteric viruses, which distinguishes environmental and food samples from clinical samples.[18] In general, these methods comprise virion concentration (precipitation, centrifugation, filtration, etc.) followed by the use of cell-culture techniques, electron or fluorescence microscopy, immunological tests (ELISA), or molecular methods.[3,13,17,18,33–35]

It is clear that except for direct "person-to-person" transmission via vomitus aerosols, the spread of NoV infection usually requires attachment of viral particles to foods or other surfaces and persistence thereon, which in turn supposes the ability of the virus to adhere to such surfaces.[36] According to the updated X-DLVO theory of particle interactions in aqueous media,[37] these kinds of interactions are governed by three forces, namely the ubiquitous attractive hydrophobic effect, the apolar van der Waals forces, and the double-layer electrostatic interactions. The particles adhere one to another when the resulting interfacial free energy of interaction is less than zero ($\Delta G^{IF} < 0$). The intrinsic properties of the interacting entities (particles, surfaces, surrounding fluid), such as their surface tension, hydrophobic or hydrophilic character, ionic charge, secondary and tertiary structures, and size, are all relevant[38–53] as well as are extrinsic factors or environmental modifiers of adhesion such as pH, temperature, and ionic strength.[41,54–58] Detailed understanding of NoV adhesion to surfaces therefore requires a multidisciplinary approach. The present chapter addresses the physical chemistry and the molecular mechanisms of this phenomenon, specifically the adhesion theory and the contribution of well-known techniques including measurements of contact angle (interfacial free energy of interaction, surface free energy, hydrophobicity, or hydrophilicity) and zeta potential (surface electrical charge), zeta sizing, transmission electron microscopy (morphology), and circular dichroism (protein secondary and tertiary structures). NoVs GI.1 and GII.4 and feline calicivirus (FCV) are used as models to study the adhesion on vegetables and inert surfaces (lettuce, polyethylene, and stainless steel). The influences of external parameters such as pH, temperature, and ionic strength are also discussed. Since our center of interest is the intrinsic nature of particle adhesion, the effects of vomitus and environmental factors (fecal matter, soil, clay, biological and chemical flocculants in sewage)[59] that may modify viral adhesion are not examined here. It should be noted that the discussions are mainly based on virus-like particles (VLPs) studies. Despite lacking the infectious genome, NoV-like particles behave quite similarly to the native particles,[60] particularly in terms of the very important surface charge property.[61]

2.2 THERMODYNAMICS OF ADHESION

Unlike autonomous unicellular organisms, which often possess physiological means such as extracellular polymers or external organelles (e.g., fimbriae) to ensure their adhesion to surfaces, viruses behave as "inert" nanoparticles that interact passively with their environment and remain devoid of physiological activities as long as they are outside of host cells. They have been compared to amphoteric colloidal particles that adsorb to biological and nonbiological surfaces depending on the ionization of exposed amino acid side groups.[36] In the case of NoV and FCV, moisture is a major factor promoting their transfer from inert surfaces such as stainless steel to fresh foods such as lettuce.[62]

According to the classical DLVO theory (named after Derjaguin and Landau; Verway and Overbeek),[63,64] colloidal particles dispersed in a liquid medium spontaneously acquire an electrostatic double layer that introduces repulsive forces in opposition to the innate Lifshitz–van der Waals (LW) attractive forces between the particles, thereby stabilizing the colloidal system.[65,66] Agglomeration of particles occurs when the repulsive forces dissipate and the attractive forces finally predominate.[66] Such interactions have long been known to underlie the adhesion of biological particles to surfaces.[42,59] The distances, limits, and reverse effects of these interacting forces in the context of bacterial adhesion have been discussed for decades.[67] Figure 2.1 (adapted from the original work of Busscher and Weerkamp[67]) illustrates the same concept applied to viruses. At distances longer than 50 nm, van der Waals attractions are predominant, while at distances of 10–20 nm, electrostatic repulsion makes any adhesion reversible.

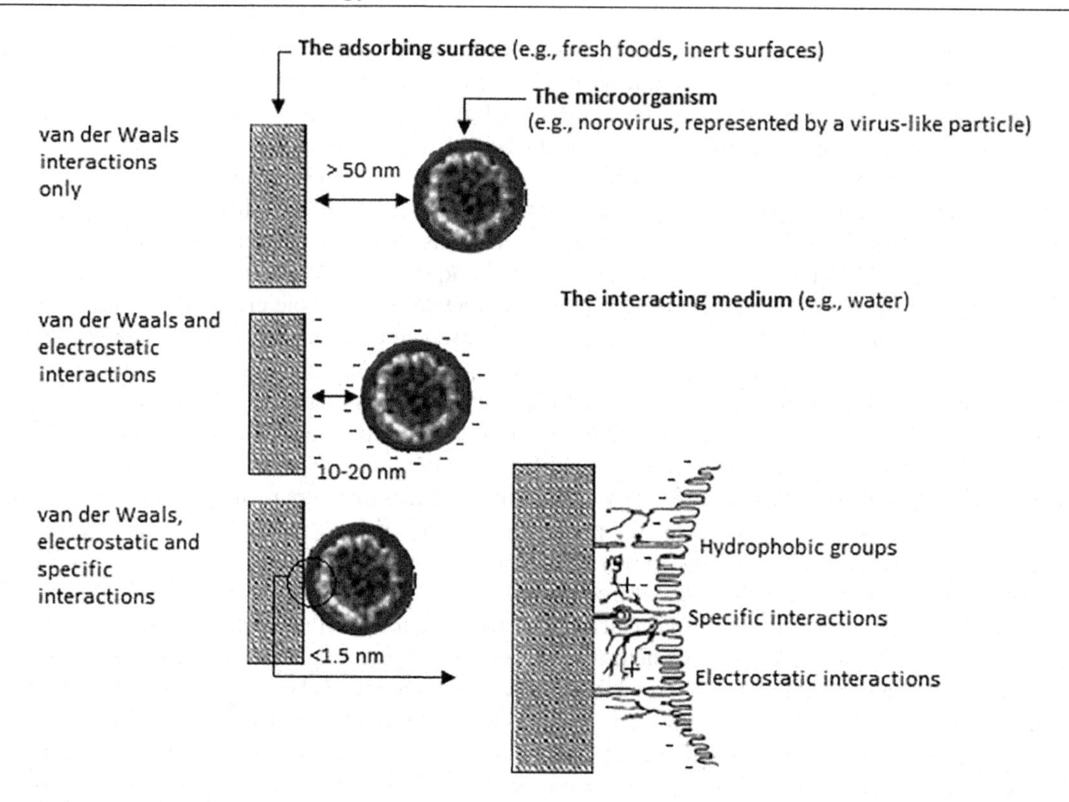

FIGURE 2.1 Schematic illustration of the steps and interactions involved in the adhesion process (adapted from Busscher and Weerkamp.[67])

Although irreversibility becomes possible at distances shorter than 1.5 nm, as short-range specific interactions (direct contact) arise due to rearrangement and dehydration of outward-oriented hydrophobic groups (strain-specific adhesion elements) or through specific short-range polar interactions (i.e., hydrogen bonding) that override electrostatic repulsion.

The finding that hydrogen bonding (electron acceptor–electron donor interactions, also called Lewis acid–base forces) prevails in polar solvents (notably water) led to the "extended-DLVO" theory for aqueous media. According to this theory, adhesion occurs only when the resultant force of the aforementioned three interactions or interfacial free energy of interaction (ΔG^{IF}) is below zero. Lewis acid–base interactions are the principal force underlying particle adhesion phenomena in aqueous media (in which long-range van der Waals attractive forces and the electrostatic interactions are limited) and account for 90% of all non-covalent forces, and 100% of all nonelectrostatic, noncovalent, and polar forces.[37,68,69] They can be attractive or repulsive, depending on the degree of hydrophobicity or hydrophilicity of the interacting entities.[37,68] Hydrophobic attraction is the largest component of the Lewis acid–base forces that contribute to adhesion and is prevalent in water, being generated by the omnipresent hydrogen-bonding free energy of cohesion (about –102 mJ m^{-2} at 20°C) of water molecules enveloping either apolar or polar molecules or moieties.[37] Lewis acid–base repulsions usually occur only between hydrophilic components that have smaller electron-acceptor and greater electron-donor characters than those of water.[70] The most important portion of the interfacial free energy of interaction between a colloidal particle and a surface or substratum via an interacting medium such as water (respectively, superscripts 1, 2, and 3) is obtained using the contact angle method (Figure 2.2) and the formula $\Delta G^{IF}_{132} = \Delta G^{LW}_{132} + \Delta G^{AB}_{132}$ (Equation 2.1).[68,71] The key parameters in this approach are the nonpolar components (Lifshitz–Van der Waals forces), polar electron-acceptor and electron-donor Lewis acid–base parameters (respectively, γ^{LW}, γ^{+}, and γ^{-}) of the different interacting entities (e.g., Virus-like Particles (VLPs), food or inert surfaces, and water). For any solid material, three different measuring liquids with known parameters and total surface free energy (γL) are used for contact

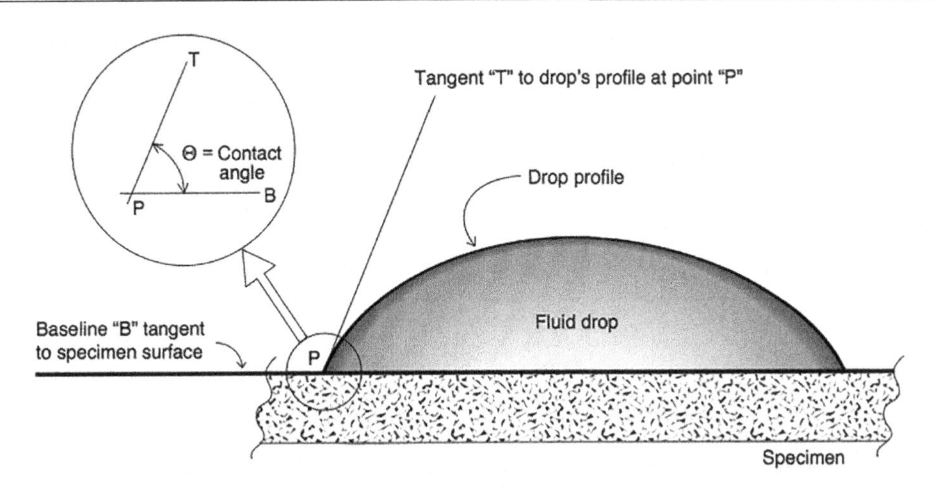

FIGURE 2.2 Schematic representation of contact angle measurement (Woodward, online; permission from First Ten Angstroms, Inc. by Charli Martucci.)

angle measurements, and the respective angles values are reported in Equation 2.2 to obtain and resolve a system of three equations with three variables corresponding to the key parameters to be determined: $(1+\cos\theta)\gamma_L = 2\left(\sqrt{\gamma_S^{LW} * \gamma_L^{LW}} + \sqrt{\gamma_S^{-} * \gamma_L^{+}} + \sqrt{\gamma_S^{+} * \gamma_L^{-}}\right)$ (2.2). In this equation (The complete Young–Good–Girifalco–Fowkes equation),[71] S refers to solid and L refers to liquid., Technically, contact angle measurements are performed using an apparatus called 'goniometer'. The liquid is deposited as a droplet (Figure 2.2) on the mounted test surface (e.g., lettuce, strawberry, stainless steel, polyethylene, dried layer of virus-like particles, etc.) and the contact angle is measured at the desired time (s) and registered. It is essential that one of the three liquids is non-polar, and the two others polar, one of the latter being necessarily water. For instance, diiodomethane (non-polar), and water and formamide (polar). The key parameters for these three liquids, i.e., their γ^{TOT}, γ^{LW}, γ^{+}, and γ^{-} are, respectively, 50.8, 0, 0.01, and 0 mJ m^{-2}; 72.8, 51.0, 25.5, and 25.5 mJ m^{-2}; and 58.8, 19.0, 2.28, and 39.6 mJ m^{-2}.[37] Once γ^{LW}, γ^{+}, and γ^{-} are obtained, the total interfacial free energy of interaction (ΔG^{IF}) between the two entities is determined using Equation 1 further elaborated as Equations 2.1 and 2.2[71,72]:

$$\Delta G_{132}^{IF} = \gamma_{12}^{LW} - \gamma_{13}^{LW} - \gamma_{23}^{LW} + 2\left[\sqrt{\gamma_3^{+}}\left(\sqrt{\gamma_1^{-}} + \sqrt{\gamma_2^{-}} - \sqrt{\gamma_3^{-}}\right) + \sqrt{\gamma_3^{-}}\left(\sqrt{\gamma_1^{+}} + \sqrt{\gamma_2^{+}} - \sqrt{\gamma_3^{+}}\right) - \sqrt{\gamma_1^{+} * \gamma_2^{-}} - \sqrt{\gamma_1^{-} * \gamma_2^{+}}\right] \quad (2.1)$$

$$\Delta G_{132}^{IF} = \left(\sqrt{\gamma_1^{LW}} - \sqrt{\gamma_2^{LW}}\right)^2 - \left(\sqrt{\gamma_1^{LW}} - \sqrt{\gamma_3^{LW}}\right)^2 - \left(\sqrt{\gamma_2^{LW}} - \sqrt{\gamma_3^{LW}}\right)^2 + 2\left[\sqrt{\gamma_3^{+}}\left(\sqrt{\gamma_1^{-}} + \sqrt{\gamma_2^{-}} - \sqrt{\gamma_3^{-}}\right) + \sqrt{\gamma_3^{-}}\left(\sqrt{\gamma_1^{+}} + \sqrt{\gamma_2^{+}} - \sqrt{\gamma_3^{+}}\right) - \sqrt{\gamma_1^{+} * \gamma_2^{-}} - \sqrt{\gamma_1^{-} * \gamma_2^{+}}\right. \quad (2.2)$$

where subscripts 1, 2, and 3 refer, respectively, to the particle, the substratum, and the medium.

According to a more recent concept, adhesion at distances of about 1 nm is an intermolecular cohesion phenomenon, primarily the result of van der Waals dispersion forces.[73] Contaminants and water reduce adhesion as their dielectric constants decrease the van der Waals forces. Generally, investigations are made using atomic force microscopy.[73] However, this concept is beyond the scope of the present chapter.

Using the contact angle method, Samandoulgou et al. (unpublished) found that the three VLPs (GI.1, GII.4, and FCV) showed interfacial free energies of interaction that favor ($\Delta G^{IF} < 0$) their adhesion to both fresh foods and inert surfaces, especially lettuce and polyethylene. For GI.1, GII.4, and FCV, these energies were, respectively, −10, −23, and −20 mJ m^{-2} on polyethylene and −2.3, −18, and

TABLE 2.1 Interfacial free energies of adhesion of virus-like particles (VLPs) of norovirus (NoV) genotypes GI.1 and GII.4 and feline calicivirus (FCV) on different types of surfaces (SS: stainless steel; PE: polyethylene; LT: Lettuce; SW: strawberry) (data from Samandoulgou et al. unpublished)

	ΔG^{IF} (mJ m^{-2})		
SURFACES	*GI VLP*	*GII4 VLP*	*FCV VLP*
SS	9	−7	−14
PE	−10	−23	−20
LT	−2.3	−18	−20
SW	10	−5	−12

−20 mJ m^{-2} on lettuce (Table 2.1), suggesting that spontaneous adhesion is favored thermodynamically, especially for GII.4 and FCV VLPs. This was confirmed for GII.4 VLPs in adhesion tests although no obvious effect of the magnitude of the interfacial free energy of interaction was observed, as had been noted previously with bacteria.[74]

2.3 CHARACTERIZATION OF NOROVIRUS STRUCTURE (INTRINSIC PROPERTIES)

2.3.1 Hydrophobic and Hydrophilic Character

The role of hydrophobicity has been investigated extensively in previous studies of adhesion of bacteria,[67] proteins,[38,75] phages, and viruses.[35,76] In surface chemistry theory, the surface of any material, whether biological or inert, can be characterized as hydrophobic or hydrophilic based on their Helmholtz surface free energy. This excess energy available on a surface due to breakage (e.g., by cutting, polishing, etc.) of the intermolecular interactions (e.g., van der Waals Keesom, Debye, and London, hydrogen bonding, attractive ionic bonding, etc.) that existed between neighboring molecules within the material.[77] Hydrophobic or nonpolar character is conferred when the resulting surface free energy is below 100 mJ m^{-2} (low-energy surfaces), while a resulting surface free energy anywhere from 200 to 5,000 mJ m^{-2} (high-energy surfaces) indicates hydrophilic or polar character.[37,78] Organic compounds and polymers (fluorocarbons, alkanes, polystyrene, polypropylene, polyethylene, some biopolymers, cells, etc.) are classified as hydrophobic materials, while solid dextran T150, polyethylene oxide or polyethylene glycol 6000, minerals, graphite, metals, glass, some biopolymers, and cells are classified as hydrophilic materials.[37,46,78,79]

The role of the Lewis acid–base component of the interfacial free energy of interaction (ΔG^{IF}) is largely tributary to the degree of hydrophobicity of the interacting entities.[37] Various techniques are used to characterize particles: the microbial adhesion to hydrocarbons using hexadecane or MATH test,[80] the cell partitioning test in a polyethylene glycol two-phase system,[81] and the well-known contact angle method, which allows determination of the surface free energy (γ^{TOT}),[82] and degree of hydrophobicity $\Delta G\ iwi\ IF$.[83] The latter corresponds to the interfacial free energy of cohesion between two particles of material *i* in water, the magnitude of this value being proportional to hydrophobic (if negative) or hydrophilic (if positive) character.[83] Besides the measured angle, the overall shape assumed by the contact-angle-measuring liquid itself (e.g., water) can be used to define the hydrophobic $(90° < \theta < 180°)$, semi-hydrophilic $(0° < \theta < 90°)$, or purely hydrophilic $(\theta = 0)$ character of test materials.[84–86] VLPs representing NoVs GI.1 and GII.4 and

FCV, polyethylene, stainless steel, and two fresh foods (lettuce and strawberry) have surface free energies of 52, 46, 44, 49, 48, 35, and 48 mJ m^{-2}, respectively (Samandoulgou et al. unpublished), and hence hydrophobic character based on the criteria proposed by Lavielle ($\gamma^{TOT} < 100$ mJ m^{-2}).[78] Degrees of hydrophobicity were consistent with these values (Samandoulgou et al., unpublished). According to these same measurements, FCV (−12 mJ m^{-2}) was far more hydrophobic than GII.4 (−2.27 mJ m^{-2}) and GI.1 (25 mJ m^{-2}), and the surfaces in order of most to least hydrophobic were polyethylene > lettuce > stainless steel > strawberry, with, respectively, −36.21, −32.11, −9.97, and −5.47 mJ m^{-2}. However, as already mentioned for the magnitudes of the interfacial free energies of interaction, these investigators did not find any correlation between adhesion of GII.4 and surface free energy or hydrophobicity of the adsorbing material. A similarity of trends in hydrophobicity and interfacial free energy of adhesion ΔG^{IF} was noted. Other investigators have reported similar surface free energies for hepatitis A virus and agri-food surfaces (including stainless steel, copper, polyethylene, and polyvinyl chloride), but no effect on adhesion was noted.[35] Despite the contradictory nature of GI.1, that is, having low surface energy but being hydrophilic, it should be kept in mind that adhesion can occur between hydrophilic and hydrophobic components,[37] and this could explain why these VLPs showed a favorable ΔG^{IF} (below 0) on lettuce and polyethylene. Furthermore, adhesion is much greater on low-energy materials than on high-energy materials if the surface free energy of the suspending medium is greater than that of the "microorganism,"[87] which was likely the case in the study by Samandoulgou et al. (unpublished). For water, γ^{TOT} is 72.8 mJ m^{-2}, which is considerably larger than that of any of the materials tested in that study. Differences between GI.1 and GII.4 could also be due to alteration of the VLPs during storage (in deionized water for GI.1 versus PBS for GII.4 and FCV) according to Samandoulgou et al. (unpublished). Other not less important factors are presumed to be involved in or to modify the adhesion process (see the following sections).

2.3.2 Electrical Charges and Isoelectric Point

According to the X-DLVO theory, the contribution of electrostatic forces to interactions between very hydrophobic components is negligible in aqueous media at physiological ionic strength, far behind that of Lewis acid–base interactions. However, these forces are the principal cause of interactions between highly charged entities.[37] Surface charges due primarily to ionization of exposed amino-acid side groups are reportedly significant factors in the adhesion of viruses,[36] for example, of echovirus 11 to lettuce.[51] In general, the adhesion of charged proteins is subject to the influence of many factors, including their structural stability and adsorbent nature (Table 2.2).

TABLE 2.2 Possible modes of adsorption based on protein particle and inert surface characteristics (adapted from Norde et al.[88])

			ADSORBENT (FRESH FOODS AND AGRISURFACES)			
			HYDROPHOBIC		*HYDROPHILIC*	
STRUCTURAL STABILITY		*ELECTRICAL CHARGE*	−	+	−	−
Stability	Stable	+	Yes	Yes	No	Yes
		−	Yes	Yes	Yes	No
	Unstable	+	Yes	Yes	Yes	Yes
		−	Yes	Yes	Yes	Yes
Viral protein (i.e., VLPs)			Adsorption by hydrophobic dehydration		Adsorption dominated by electrostatic interactions or structural rearrangement	

"Yes" indicates a possibility of adhesion, while "No" indicates no possibility.

The surface charge and isoelectric point (the pH at which the net charge is zero) of many colloidal particles such as VLPs can be determined using various techniques. These techniques combine electrophoretic mobility or related measurements with various illuminations or imaging technologies, some of which have been used for VLPs of NoVs GI.1 and GII.4, FCV, and bacteriophage MS2.[58,89–91]

Based on more recent studies, VLPs of NoVs GI.1, GII.4, and FCV have their isoelectric points at pH 4.25, 4.15, and 3.9, respectively, and bear a positive charge below these pHs and a negative charge above them, approximately in proportion with the differential. These electrical charges decreased with increasing ionic strength.[89] Although no adhesion may happen between NoV VLPs to agri-food materials such as polystyrene (isoelectric at pH 3.5–5.0[92]), polypropylene (isoelectric at pH 3.8,[93]), and stainless steel (isoelectric at pH 2.4–3.0 according to Lefèvre et al.[94], at pH 3.0–4.8 according to Boulanger-Petermann et al.[95]) at extreme pH because of similarity of surface charges, it might occur spontaneously at the isoelectric points on account of hydrophobic (Lewis acid–base) and van der Waals forces.[89]

The potential contribution of electrostatic forces to the adhesion of NoV to lettuce and strawberries cannot be analyzed until reliable data on surface isoelectric points of these foods become available.[89] Nevertheless, these forces are believed to allow Echovirus 11 (isoelectric at pH 6) to adhere to lettuce and could do likewise for NoV.[51] This would not occur at acidic pH including the isoelectric points, in which cases van der Waals forces and hydrophobic interactions would dominate, especially the latter because the outer surfaces of NoVs GI.1 and GII.4 bear, respectively, 94% and 85% of their total nonpolar amino acids, and the surface wax layer of lettuce is rich in hydrophobic components such as fatty acids, hydrocarbons, and esters.[89,96] VLP surface charges appear to be decreased at increased salt concentrations;[89] consequently, increasing the ionic strength might favor NoV adhesion by van der Waals and hydrophobic interactions.[55,56] Other early work has also shown the dominance of hydrophobic interactions over electrostatic forces in the adhesion of proteins at their isoelectric pH.[38,54,75] GII.4 VLP concrete adhesion test has shown that electrostatic interactions do not likely contribute to NoV adhesion to fresh foods and agri-food surfaces (Samandoulgou et al., unpublished).

2.3.3 Size

The adhesion capacity of proteinaceous particles to other materials appears to depend on the abundance of particle surface points of attachment.[37,97] For example, it has been shown that large phages (e.g., PRD1 and PM2) adhere better than small phages (ΦX174, MS-2, Qβ) and that adhesion strength is proportional to the total number of surface electrical charges.[47]

Using a measurement called dynamic light scattering (DLS), the size of particles in a suspension can be estimated from its known correlation with Brownian movement (Malvern Instruments Ltd). This technique has been used to measure the size of VLPs of NoV GI.1 and GII.4 and FCV and to determine the conditions under which they aggregate.[58,89] Data can be combined or compared with transmission electron microscopy (TEM) data for insightful analysis on particle shape and size. Using both of these techniques, the hydrodynamic diameters (hϕ) of these particles appear to be in the 20–45 nm range at pH far from the isoelectric point (Figure 2.3a) and NaCl concentrations in the 0–500 mM range (Figure 2.3b). Particle aggregation occurs at or near the isoelectric point (about 4) and low or high temperatures or at high temperatures and high salt concentrations.[89] In the case of GII.4, adhesion is extensive under these conditions and is thought to involve both van der Waals and hydrophobic forces (Samandoulgou et al, unpublished). In a study with GI.1, aggregation was observed at pH 4 and 8 at low salt concentration, but adhesion did not increase accordingly, perhaps because of the solution complexity under the test conditions.[58] In this same study, GII.4 VLPs aggregated at pH 5.2 and high or low salt concentrations and in deionized water, but adhesion was not tested under any of these conditions.[58]

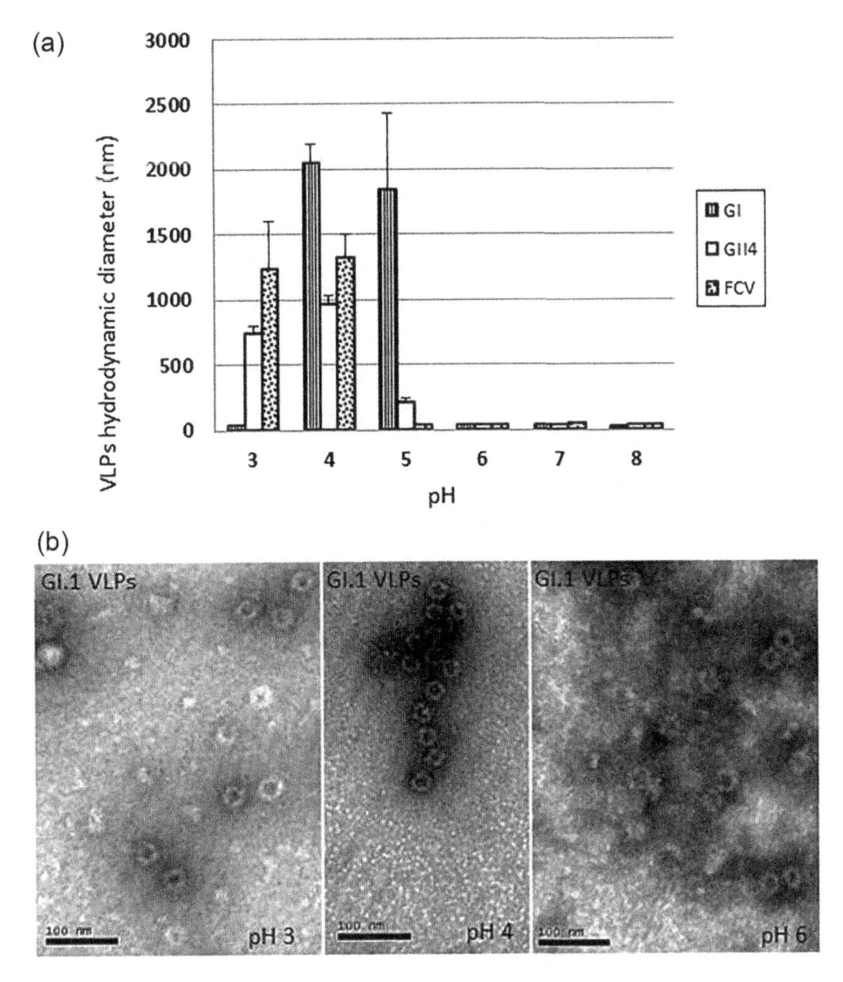

FIGURE 2.3 Size of VLP proxies of human NoVs and feline calicivirus (FCV) under different conditions of pH (a) and ionic strength (M NaCl), (b) (top) hydrodynamic diameters based on dynamic light scattering measurement and (bottom) visualization by transmission electron microscopy.[89] $G = 120{,}000\times$; bar = 100 nm.

2.3.4 Secondary Structure

Macromolecule secondary structures (α-helices and β-strands) are likely involved in the adhesion of protein particles to surfaces since they determine whether or not hydrophobic side chains can get involved, for example, following dehydration and the attendant structural rearrangements.[38,40] This has been illustrated with bovine and human serum albumin, chymotrypsin, cutinase, lysozyme, lactalbumin, γ-globulin, etc.[41,44,46,48–50] The most exposed part of human NoV (GI.1) VLPs, namely the P1 subdomain, is located in the protruding "P" domain and comprises three short-stretched β-strands, six β-strands, one α-helix, and four additional β-strands organized in a Greek-key antiparallel sheet.[98] To characterize macromolecule secondary and tertiary structures, an optical technique called circular dichroism is used, which measures the shift of plane-polarized light from a circular to an elliptical pattern upon absorption by chiral chromophores in solution. Far UV (<240 nm) is used for this purpose, which corresponds to the peptide-bond absorption region. Software allows the transformation of the readings into percentages of α-helices, β-strands, turns, and unordered sequences.[99] Macromolecule structure and stability can thus be measured under different conditions of pH, temperature, and ionic strength. This has been achieved for major NoV capsid proteins again using VLPs.[100–102] It has been shown that NoV GI.1 and GII,4, and FCV retain their configuration at 4°C at neutral pH while heating (≥65°C) affected all strains by shifting or altering their ordered structures in various ways. However, investigation of secondary structure alone does not provide unequivocal evidence of exposure of internal hydrophobic groups. Proof of such exposure requires analysis of tertiary structure stability.[102]

2.3.5 Tertiary Structure

Tertiary structures of proteinaceous compounds arise from interactions within the polypeptide chain, including noncovalent bonding between hydrophobic groups, such that any adhesion involving such groups would lead to structural unfolding or alteration.[40,44] Losses of tertiary structure in conjunction with adhesion are more likely to occur on low energy (hydrophobic) surfaces than on high energy (hydrophilic) surfaces and are very unlikely in aqueous media.[37] The stability of protein tertiary structures can be evaluated by measuring intrinsic UV fluorescence. For example, human plasma fibronectin unfolding and bovine serum albumin folding upon adhesion, respectively, to hydrophobic and hydrophilic quartz were revealed using this technique.[43] Intrinsic fluorescence due to tryptophan is the most useful for analyzing tertiary structures under different physicochemical conditions. A tryptophan residue absorbs at 280 nm and transmits in the 305–450 nm range, with a peak at 330 nm when it is buried in nonpolar surroundings inside the protein molecule or at 350 nm when it is exposed to an aqueous medium.[103] Investigation of the unfolding, folding, and stability of NoV GI.1, GII.4, and FCV VLPs has shown that at 10°C–22°C, neither the pH (3–8) nor the ionic strength (0–500 mM) affects tertiary structure noticeably. The unfolding starts at pasteurization temperatures, especially at 75°C, at which the secondary structure undergoes rearrangement.[100,102] This suggested that adhesion of NoV to hydrophobic surfaces would be favored more by heating than by any specific conditions of pH and/or ionic strength alone. It has since been observed that heating to 65°C does indeed increase the adhesion of GII.4 VLPs significantly, both to polyethylene (by about 33%) and to stainless steel (by about 21%) at neutral pH, thus confirming the endothermic character of this adhesion underlain by structural rearrangements (Samandoulgou et al., unpublished).

2.3.6 Morphology

There is considerable evidence that spherical particles are capable of adhering to surfaces in spite of unfavorable strong electrostatic repulsive forces, due in large part to protuberances such as those borne by the influenza virus.[37] Human NoV capsids are icosahedral particles about 38 nm in diameter, made of 180 VP1 monomers plus a few VP2 monomers, the intrinsic structures of which create domains called "shell" (S) and "protruding" (P). The P domain is further divided into P1 and P2 subdomains, the latter being the most peripheral[104] and the most direct contact for the surrounding medium. Such advanced characterizations were made possible mostly by electron cryo-microscopy and X-ray crystallography.[104] However, transmission electron microscopy has also provided useful basic information such as shape, size, and presence of protruding microstructures (Figure 2.4), aggregation patterns, etc. Bearing over 80% of the nonpolar amino acid residues in the principal capsid protein, at least in the cases of GI.1 and GII.4, the prominent P2 subdomain appears to be involved in the attachment of NoV to host-cell receptors.[5,98,105] Therefore, it could be a major element of contact with hydrophobic surfaces (e.g., polyethylene, polypropylene, fresh foods with waxy surfaces, etc.). However, this hypothesis remains to be tested in multidisciplinary investigations using a variety of specific techniques such as atomic force microscopy, electron microscopy, and binding assays.

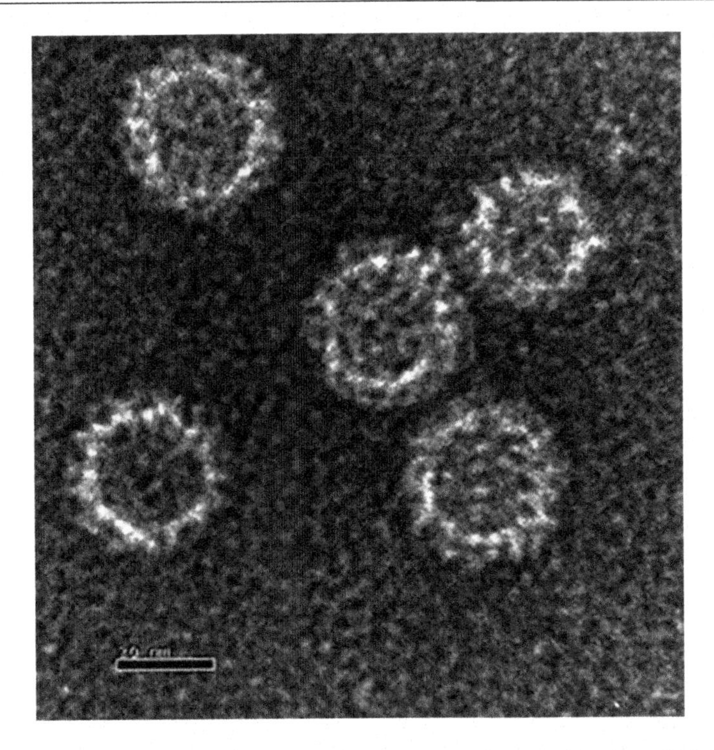

FIGURE 2.4 Electron micrograph of GII.4 VLPs in 0.025 M NaCl, magnification: 600,000 X. Bar=20 nm; protruding microstructures are somewhat visible on all sides of the particles.

2.3.7 Presence of Ligands

The adhesion of viruses to cells tends to be highly specific, due to the presence of matched ligands and receptor sites. Human NoV binds specifically to glycans A, B, H, and Lewis, which are histo-blood-antigen-group system (HBAGS) receptors on the host cells.[106,107] The P2 subdomain of the viral capsid bears the ligand,[106] and depending on the genogroup, the glycan will stand straight up (GI) or lie flat (GII) on the capsid surface.[107] Such specific binding is not likely involved in adhesion of virions to agri-food surfaces but has been shown very clearly to occur with seafood such as oysters and fresh foods such as lettuce, which reportedly express the receptor or a very similar structure for human NoV GI.1 and GII.4, respectively.[108,109]

2.3.8 Factors Influencing the Adhesion Phenomenon: Extrinsic Parameters or the Physicochemical Conditions of the Surrounding Medium

2.3.8.1 pH

The impact of the pH of the surrounding medium on the adhesion of biological particles to surfaces has been reported widely in the literature over the decades.[54,57–59,74] In the case of viruses, low pH is often reported to promote adhesion, while high pH does the opposite.[59] Adhesion of NoV GI.1 to silica at or below the isoelectric point is inconsistent but decreases clearly above.[58] The pH appears to have little impact on the adhesion of murine norovirus (MNV-1, a common surrogate used in research on NoV) to stainless steel.[57] In more recent studies, GII.4 VLPs adhered very well (94%–97% of capsids proteins) to stainless steel, polyethylene, and lettuce at pH 4 (near the isoelectric point) but poorly at pH 7 or 9 (Samandoulgou et al. unpublished). This finding corroborates other investigations of protein adhesion[54,75,110] and is attributable to aggregation at the acidic pH, possibly multiplying the points of attachment to the adsorbing surface as discussed above.[47,89,97,111] Poor adhesion at pH 7 and 9 has been reported also for GI.1 VLPs on silica.[58] It appears safe to say that pH well above the isoelectric point does not favor adhesion,[59] and this might be the result of repulsive negative charges between particles, as has been reported for proteins in suspension.[110] Since electrical charges are neutralized or reduced at or near the isoelectric point, adhesion at pH 4 is due most likely to nonelectrostatic interactions, for example, hydrophobic and van der Waals forces.

2.3.8.2 Ionic Strength

In general, the effects of ionic strength on the adhesion of proteins or viruses have been reported in relation to particle type and concentration, surface type, anions and cations nature, and pH[41,54,58] (Samandoulgou et al. unpublished). It thus has been observed that increasing the ionic strength (NaCl solutions) at pH 8 increases the adhesion of GI.1 and GII.4 NoV VLPs to silica.[58] Increasing the NaCl concentration increases the adhesion of GII.4 VLPs to stainless steel and lettuce but not to polyethylene (Samandoulgou et al. unpublished). Since the interfacial free energy of interaction is due mostly to the Liftshitz van der Waals component on stainless steel and to the Lewis acid–base component on lettuce, it is proposed that these were enhanced, respectively, by electronic double layer compression[55] and by the anti-chaotropic effect of NaCl.[56] Furthermore, it has been found that negative charges on GII.4 VLPs decrease with increasing ionic strength,[89] which clearly would reduce electrostatic repulsion. The negligible impact of ionic strength on adhesion to polyethylene compared to lettuce, which has similar attractive Lewis acid–base portions, was attributed to the presence of stomata on the leafy vegetable (Samandoulgou et al. unpublished).

2.3.8.3 Temperature

Based on studies of BSA, fibrinogen, and γ-globulin on polymer membranes,[54] two different and related types of adhesions are noted. Type 1 is exothermic, decreasing with temperature, hydrophilic, and easily reversible, while type 2 is endothermic, increasing with temperature and hydrophobic. In an assay of GII.4 VLP adhesion to stainless steel and polyethylene at refrigeration or near pasteurization temperatures (65°C) in acidic or neutral media at low or moderate ionic strength, it was found that the warmer temperature (65°C) increased adhesion on either surface (neutral pH only), while refrigeration did not have significant effects. Neither refrigeration, nor the temperature of 65°C had affected significantly GII.4 adhesion at zero ionic strength (deionized water). As for the moderate ionic strength, only the pasteurization temperature had a significant (increasing) effect, and only on polyethylene. The increases were attributed to heat-induced secondary and tertiary structural changes, thus supporting an endothermic mechanism of GII.4 VLPs adhesion (Samandoulgou et al. unpublished).

2.4 FUTURE PERSPECTIVES

Based on this overview of the adhesion of human noroviruses to surfaces, it is apparent that much of our current understanding has come from the use of virus-like particles. It is clear that this phenomenon must be studied using a multidisciplinary approach and that almost all of the intrinsic and extrinsic parameters mentioned here, as well as the general thermodynamic principles governing particle interactions, are involved to some extent. However, the phenomenon itself remains complex, and the procedures on which many deductions have been based are not only few and indirect but have shed so far only a little light on some aspects. These achievements are nevertheless very interesting, but more in-depth studies are needed for further clarification and complementary information, for example, nanoparticle binding combined with electron microscopy, crystallography, and cutting-edge techniques such as atomic force microscopy. Investigating protein particle adhesion phenomena using this approach is certainly worthwhile and would be a very good first step towards achieving the goal of defining practical conditions that reduce the transmission of viral gastroenteritis.

REFERENCES

1. Kapikian AZ, Wyatt RG, Dolin R, Thornhill TS, Kalica AR, Chanock RM. Visualization by immune electron microscopy of a 27-nm particle associated with acute infectious nonbacterial gastroenteritis. *J Virol* 1972;10(5):1075–81.
2. Scallan E, Hoekstra RM, Angulo FJ, et al. Foodborne illness acquired in the United States – Major pathogens. *Emerg Infect Dis* 2011;17(1):7–15.
3. Greening GE, Cannon JL. Human and animal viruses in food (Including taxonomy of enteric viruses) [Internet]. In: Goyal MS, Cannon JL, editors. *Viruses in Foods*, Second ed. Cham: Springer Nature; 2016. pp. 5–59.
4. Kaplan JE, Feldman R, Campbell DS, Lookabaugh C, Gary GW. The frequency of a Norwalk-like pattern of illness in outbreaks of acute gastroenteritis. *Am J Public Health* 1982;72(12):2–5.
5. Glass RI, Parashar UD, Estes MK, Mountain S. Norovirus gastroenteritis. *N Engl J Med* 2010; 362(6):557.
6. Fankhauser RL, Monroe SS, Noel JS, et al. Epidemiologic and molecular trends of "Norwalk-like viruses" associated with outbreaks of gastroenteritis in the United States. *J Infect Dis* 2002;186(1):1–7.
7. Patel MM, Widdowson M, Glass RI, Akazawa K, Vinjé J, Parashar UD. Systematic literature review of role of noroviruses in sporadic gastroenteritis. *Emerg Infect Dis* 2008;14(8):1224–31.
8. Kohn MA, Farley TA, Ando T, et al. An outbreak of Norwalk virus gastroenteritis associated with eating raw oysters Implications for maintaining safe oyster beds. *J Am Med Assoc* 1995;273(6):466–71.
9. Public Health England. PHE National Norovirus and Rotavirus Report. 2018. https://assets.publishing.service.gov.uk/government/uploads/system/uploads/attachment_data/file/733162/Norovirus_update_2018_weeks_27_to_30.pdf
10. National Institute of Infectious Diseases (Japan). Infectious Agents Surveillance Report [Internet]. 2017. Available from: https://www.niid.go.jp/niid/images/iasr/rapid/noro/160920/norosyue_170913.gif
11. Havelaar A, Haagasma J, Magen MJ, et al. Disease burden of foodborne pathogens in the Netherlands. *Int J Food Microbiol* 2012;156:231–8.
12. van Beek J, Botteldoorn N, Eden JS, et al. Indications for worldwide increased norovirus activity associated with emergence of a new variant of genotype II. 4, late 2012. *Eurosurveillance* 2013;18(1):20345.
13. Atmar RL. Noroviruses : State of the art. *Food Environ Virol* 2010;2:117–26.
14. Appleton H. Control of food-borne viruses. *Br Med Bull* 2000;56(1):172–83.
15. Norkin LC. Virology: Molecular biology and pathogenesis. In: Norkin LC, editor. Printed in Canada. *Miscellaneous RNA Viruses*. Herdon, VA: ASM Press; 2010. pp. 346–61.
16. Lopman BA, Reacher MH, van Duijnhoven Y, Hanon F, Brown D, Koopmans M. Viral gastroenteritis outbreaks in Europe, 1995–2000. *Emerg Infect Dis* 2003;9(1):90–6.
17. Koopmans M, Duizer E. Foodborne viruses: An emerging problem. *Int J Food Microbiol* 2004;90:23–41.
18. Carter MJ. Enterically infecting viruses: Pathogenicity, transmission and significance for food and waterborne infection. *J Appl Microbiol* 2005;98:1354–80.
19. Mayo MA. Virus taxonomy – Houston. *Arch Virol* 2002;147:1071–6.
20. Jiang X, Wang M, Kening W, Estes MK. Sequence and genome organisationof Norwalk virus. *Virology* 1993;195:51–61.
21. Green KY, Ando T, Balayan M, et al. Taxonomy of the calieiviruses. *J Infect Dis* 2000;181:322–30.
22. Vinjé J. Advances in laboratory methods for detection and typing of norovirus. *J Clin Microbiol* 2015;53(2):373–81.
23. Kroneman A, Vega E, Vennema H, et al. Proposal for a unified norovirus nomenclature and genotyping. 2013;158(10):2059–68.
24. Martella V, Campolo M, Lorusso E, et al. Norovirus in captive lion cub (Panthera leo). *Emerg Infect Dis* 2007;13(7):1071–3.
25. Siebenga JJ, Vennema H, Zheng DP, et al. Norovirus illness is a global problem: Emergence and spread of norovirus GII.4 variants, 2001–2007. *J Infect Dis* 2009;200:802–12.
26. Le Guyader FS, Loisy F, Atmar RL, et al. Norwalk virus–specific binding to oyster digestive tissues. *Emerg Infect Dis* 2006;12(6):931–6.
27. Siebenga JJ, Vennema H, Renckens B, et al. Epochal evolution of GGII.4 norovirus capsid proteins from 1995 to 2006. *J Virol* 2007;81(18):9932–41.
28. Vega E, Barclay L, Gregoricus N, Williams K, Lee D, Vinjé J. Novel surveillance network for norovirus gastroenteritis. *Emerg Infect Dis* 2011;17(8):1389–95.

29. Kreidieh K, Charide R, Dbaibo G, Melhem NM. The epidemiology of Norovirus in the Middle East and North Africa (MENA) region: A systematic review. *J Virol* 2017;14(1):220.
30. Mans J, Armah GE, Steele AD, Taylor MB. Norovirus epidemiology in Africa: A review. *PLoS One* 2016;11(4):e46280.
31. Lamhoujeb S, Fliss I, Ngazoa SE, Jean J. Molecular study of the persistence of infectious human norovirus on food-contact surfaces. *Food Environ Virol* 2009;1:51–6.
32. Patterson T, Hutchings P, Palmer S. Outbreak of SRSV gastroenteritis at an international conference traced to food handled by a post-symptomatic caterer. *Epidemiol Infect* 1993;111:157–62.
33. Atmar RL, Estes MK. Diagnosis of noncultivatable gastroenteritis viruses, the human caliciviruses. *Clin Microbiol Rev* 2001;14(1):15–37.
34. Jones KM, Watanabe M, Zhu S, et al. Enteric bacteria promote human and mouse norovirus infection of B cells. *Res Rep* 2014;346(6210):755–9.
35. Kukavica-Ibrulj I, Darveau A, Jean J, Fliss I. Hepatitis A virus attachment to agri-food surfaces using immunological, virological and thermodynamic assays. *J Appl Bacteriol* 2004;97:923–34.
36. Bitton G. Adsorption of viruses to surfaces: Technological and ecological implications. In: Bitton G, Marshall K, editors. *Adsorption of Microorganisms to Surfaces*. New York, Brisbane, Toronto, ON: A Wiley-Interscience, John Wiley & Sons; 1980. pp. 331–74.
37. van Oss CJ. Long-range and short-range mechanisms of hydrophobic attraction and hydrophilic repulsion in specific and aspecific interactions. *J Mol Biol* 2003;16(2002):177–90.
38. Norde W, Lyklema J. The adsorption of human plasma albumin and bovine pancreas ribonuclease at negatively charged polystyrene surfaces. *J Colloid Interface Sci* 1978;66(2):257–65.
39. Norde W, Lyklema J. Thermodynamics of protein adsorption theory with special reference to the adsorption of human plasma albumin and bovine pancreas ribonuclease at polystyrene surfaces. *J Colloid Interface Sci* 1979;71(2):350–66.
40. Norde W, Lyklema J. Protein adsorption and bacterial adhesion to solid surfaces: A colloid-chemical approach. *Colloids Surf* 1989;38:1–13.
41. Soderquist ME, Walton AG. Structural changes in proteins adsorbed on polymer surfaces. *J Colloid Interface Sci* 1980;75(2):386–97.
42. Murray JP, Parks GA. Poliovirus adsorption on oxide surfaces. *Adv Chem Ser* 1980; 189: 97–133.
43. Andrade JD, Hlady VL, Van Wagenen RA. Effects of plasma protein adsorption on protein conformation and activity. *Pure Appl Chem* 1984;56(10):1345–50.
44. Norde W. Adsorption of proteins from solution at the solid-liquid interface. *Adv Colloid Interface Sci* 1986;25:267–340.
45. Norde W, MacRitchie F, Nowicka G, Lyklema J. Protein adsorption at solid-liquid interfaces: Reversibility and conformation aspects ~ 1. 1. Reversibility: Comparison with adsorption of macromolecules. *J Colloid Interface Sci* 1986;112(2):447–56.
46. Norde W, Zoungrana T. Surface-induced changes in the structure and activity of enzymes physically immobilized at solid/liquid interfaces. *Biotechnol Appl Biochem* 1998;28:133–43.
47. Dowd SE, Pillai SD, Wang S. Delineating the specific influence of virus isoelectric point and size on virus adsorption and transport through sandy soils. *Appl Environ Microbiol* 1998;64(2):405–10.
48. Norde W. Protein adsorption at solid surfaces: A thermodynamic approach. *Pure Appl Chem* 1994;66(3):491–6.
49. Haynes CA, Norde W. Globular proteins at solid/liquid interfaces. *Colloids Surf B Biointerfaces* 1994;2:517–66.
50. Roach P, Farrar D, Perry CC. Interpretation of protein adsorption: Surface-induced conformational changes. J Am Chem Socoiety 2005;127(1):8168–73.
51. Vega E, Smith J, Garland JAY, Matos A. Variability of virus attachment patterns to butterhead lettuce. *J Food Prot* 2005;68(10):2112–7.
52. Tsai B. Penetration of nonenveloped viruses into the cytoplasm. *Annu Rev Cell Dev Biol* 2007;23:23–43.
53. Ossiboff RJ, Zhou Y, Lightfoot PJ, Prasad BVV, Parker JSL. Conformational changes in the capsid of a calicivirus upon interaction with its functional receptor. *J Virol* 2010;84(11):5550–64.
54. Dillman WJJ, Miller IF. On the adsorption of serum proteins on polymer membrane surfaces. *J Colloid Interface Sci* 1973;44(2):221–214.
55. Bitton G, Pancorbo O, Gifford GE. Factors affecting the adsorption of poliovirus to magnetite in water and wastewater. *Water Res* 1976;10:973–80.
56. Farrah S, Bitton G, Hoffmann EM, et al. Survival of enteroviruses and coliform bacteria in a sludge. *Appl Environ Microbiol* 1981;41(2):459–65.
57. Girard M, Ngazoa S, Mattison K, Jean J. Attachment of noroviruses to stainless steel and their inactivation, using household disinfectants. *J Food Prot* 2010;73(2):400–4.

58. da Silva AK, Kavanagh O V, Estes MK, Elimelech M. Adsorption and aggregation properties of norovirus GI and GII virus-like particles demonstrate differing responses to solution chemistry. *Environ Sci Technol* 2011;45:520–6.
59. Gerba CP. Applied and theoretical aspects of virus adsorption to surfaces. *Adv Appl Microbiol* 1984;30:133–68.
60. Jiang XI, Wang MIN, Graham DY, Estes MK. Expression, self-assembly, and antigenicity of the Norwalk virus capsid protein. *J Virol* 1992;66(11):6527–32.
61. Schaldach CM, Bourcier WL, Shaw HF, Viani BE, Wilson WD. The influence of ionic strength on the interaction of viruses with charged surfaces under environmental conditions. *J Colloid Interface Sci* 2006;294:1–10.
62. Souza DHD, Sair A, Williams K, et al. Persistence of caliciviruses on environmental surfaces and their transfer to food. *Int J Food Microbiol* 2006;108:84–91.
63. Derjaguin B, Landau L. Theory of stability of strongly charged lyophobic sols and of the adhesion of strongly charged particles in solutions of electrolytes. *Acta Phys Chem URSS*, 1941;14:633.
64. Verway EJW, Overbeek JTG. *Theory of Stability of Lyophobic Colloids.* New York, Amsterdam, London, Brussels: Elsevier Publishing Company Inc.; 1948.
65. Hunter RJ. The double layer in colloidal system [Internet]. In: Bockris JO, Conway BE, Yeager E, editors. *Comprehensive Treatise of Electrochemistry.* New York: Plenum Press; 1980. pp. 397–434.
66. Zeta-Meter Inc. Zeta-Potential: A complete course in 5 minutes. Tech Note [Internet] 1997;1–8 http://zetarod.com/wp-content/uploads/Zeta-Potential-A-Complete-Course-in-5-Minutes.pdf.
67. Busscher HJ, Weerkamp AH. Specific and non-specific interactions in bacterial adhesion to solid substrata. *FEMS Microbiol Rev* 1987;46:165–73.
68. van Oss CJ, Good RJ, Chaudhury MK. The role of van der Waals forces and hydrogen bonds in "hydrophobic interactions" between biopolymers and low energy surfaces. *J Colloid Interface Sci* 1986;111(2):378–90.
69. Grasso D, Subramaniam K, Butkus M, Strevett K, Bergendahl J. A review of non-DLVO interactions in environmental colloidal systems. *Rev Environ Sci Bio/Technol* 2002;1:17–38.
70. van Oss CJ. *Interfacial Forces in Aqueous Media* [Internet]. New York, Basel: Marcel Dekker, Inc; 1994.
71. van Oss CJ, Chaudhury MK, Good RJ. Monopolar surfaces. *Adv Colloid Interface Sci* 1987;28:35–64.
72. van Oss CJ, Chaudhury MK, Good RJ. Interfacial Lifshitz-van der Waals and polar interactions in macroscopic systems. *Chem Rev* 1988;88(6):927–41.
73. Kendall K, Kendall M, Rehfeldt F. *Adhesion of Cells, Viruses and Nanoparticles.* New York: Springer; 2011.
74. Mafu AA, Plumety C, Deschênes L, Goulet J. Adhesion of pathogenic bacteria to food contact surfaces: Influence of pH of culture. *Int J Microbiol* 2011;2011:10.
75. MacRitchie F. The adsorption of proteins at the solid/liquid interface. *J Colloid Interface Sci* 1972;38(2):484–8.
76. Chattopadhyay S, Puls RW. Adsorption of bacteriophages on clay minerals. *Environ Sci Technol* 1999; 33(20):3609–14.
77. Briant J, Tenebre L. Généralités sur les phénomènes de surface [Internet]. In: Briant J, editor. *Phénomènes d'interface, Agents de surfaces, Principes et modes d'actions.* Paris Cedex 15: Institut Français du Pétrole; 1989. pp. 1–38.
78. Lavielle L. Caractérisation des surfaces des matériaux. In: Leveau J-Y, Bouix M, editors. *Nettoyage, désinfection et hygiène dans les bio-industries, Première partie, Les matériaux et leurs propriétés de surfaces. Technique & Documentation, 1999.* Paris: Lavoisier; 2005. pp. 3–24.
79. Fowkes FM. Attractive forces at interfaces. *Ind Eng Chem* 1964;56:40–52.
80. Georgieva R, Danguleva A, Stefanova Todorova N, Karapetkov N, Rumyan N, Karaivanova E. Cell-surface hydrophobicity and adhesion ability to human epithelial cell line of industrially important lactic acid bacteria and bifidobacteria. In: Paper presented at the *International Scientific Conference on Probiotics and Prebiotics*, Budapest (p. 96). IPC2016. 2016. p. 1–2.
81. van Loosdrecht MCM, Lyklema J, Norde W, Schraa G, Zehnder AJB. The role of bacterial cell wall hydrophobicity in adhesion. *Appl Environ Microbiol* 1987;53(8):1893–7.
82. Good R. Contact angles and the surface free energy of solids. In: Good RJ, Stromberg RR, editors. *Surface and Colloid Science.* New York: Plenum Press; 1979. pp. 1–2.
83. van Oss CJ, Giese RF. The hydrophilicity and hydrophobicity of clay minerals. *Clays Clay Miner* 1995;43(4):474–7.
84. Adamson AW, Gast AP. *Physical Chemistry of Surfaces Sixth Edition* [Internet]. 6th ed. New York, Chichester, Weinheim, Brisbane, Singapore, Toronto, ON: Wiley-Interscience, John Wiley & Sons, Inc; 1997.
85. Bear J. *Hydraulic of Groundwater* [Internet]. Dover, New York: Dover Publications Inc.; 2007.
86. Childress A, Brant JA. Characterization of the hydrophobicity of polymeric reverse ssmosis and nanofiltration membranes: Implications to membrane fouling. 2000; Desalination and Water Purification Research and Development Program Report N° 57. U.S. Department of Interior Bureau of Reclamation. Denver, Colorado. .

87. Absolom DR, Lamberti F V, Policova Z, Zingg W, van Oss CJ, Neumann AW. Surface thermodynamics of bacterial adhesion. *Appl Environ Microbiol* 1983;46(1):90–7.
88. Norde W, Tan W, Koopal L. K-6 protein adsorption at solid surfaces and protein complexation with humic acids. In: *5th International Symposium ISMOM 2008*, Pucon, Chile. 2008. p. 11.
89. Samandoulgou I, Fliss I, Jean J. Zeta potential and aggregation of virus-like particle of human norovirus and feline calicivirus under different physicochemical conditions. *Food Environ Virol* 2015;7:249–60.
90. Redman JA, Grant SB, Olson TM, Hardy ME, Estes MK. Filtration of recombinant Norwalk virus particles and bacteriophage MS2 in quartz sand: Importance of electrostatic interactions. *Environ Sci Technol* 1997;31(12):3378–83.
91. Goodridge L, Goodridge C, Wu J, Griffiths M, Pawliszyn J. Isoelectric point determination of Norovirus virus-like particles by capillary isoelectric focusing with whole column imaging detection. *Anal Chem* 2004;76(1):48–52.
92. Bolt PS, Goodwin JW, Ottewill RH. Studies on the preparation and characterization of monodisperse polystyrene latices. VI. Preparation of zwitterionic latices. *Langmuir* 2005;21(12):9911–6.
93. Stawski D, Bellmann C. Electrokinetic properties of polypropylene textile fabrics containing deposited layers of polyelectrolytes. *Colloids Surf A: Physicochem Eng Asp* 2009;345:191–4.
94. Lefèvre G, Milonjic S, Fédoroff M, Finne J, Jaubertie A, Ljiljana C. Determination of isoelectric points of metals and metallic alloys by adhesion of latex particles. *J Colloid Interface Sci* 2009;337:449–55.
95. Boulangé-Petermann L, Doren A, Baroux B, Bellon-Fontaine M-N. Zeta potential measurements on passive métals. *J Colloid Interface Sci* 1995;171:179–86.
96. Baker E. Chemistry and morphology of plant epicuticular waxes. In: Cutler DF, Alvin KL, Price CE, editors. *The Plant Cuticle*. London: Academic Press; 1982. pp. 139–65.
97. Stuart MAC. Adsorbed polymers in colloidal systems – From static to dynamics. *Polym J* 1991;23(5): 669–82.
98. Prasad BV V, Prasad BVV, Hardy ME, et al. X-ray crystallographic structure of the Norwalk virus capsid. *Science* 1999;286:286–90.
99. Kelly SM, Jess TJ, Price NC. How to study proteins by circular dichroism. *Biochim Biophys Acta* 2005;1751:119–39.
100. Ausar SF, Foubert TR, Hudson MH, Vedvick TS, Middaugh CR. Conformational stability and disassembly of Norwalk virus-like particles. *J Biol Chem* 2006;281(28):19478–88.
101. Kissmann J, Ausar SF, Fouber TR, et al. Physical stabilization of Norwalk virus-like particles. *J Pharm Sci* 2008;97:4208–18.
102. Samandoulgou I, Hammami R, Morales Rayas R, Fliss I, Jean J. Stability of secondary and tertiary structures of virus-like particles representing noroviruses: Effects of pH, ionic strength, and temperature and implications for adhesion to surfaces. *Appl Environ Microbiol* 2015;81:7680–6.
103. Creighton T. *The Biophysique Chemistry of Nucleic Acids and Proteins*. Eastbourne: HP Pelvetian Press; 2010.
104. Prasad BVV, Rothnagel R, Jiang XI, Estes MK. Three-dimensional structure of baculovirus-expressed Norwalk virus capsids. *J Virol* 1994;68(8):5117–25.
105. Tan M, Huang P, Meller J, Zhong W, Farkas T, Jiang X. Mutations within the P2 domain of Norovirus capsid affect binding to human histo-blood group antigens: Evidence for a binding pocket. *J Virol* 2003;77(23):12562–71.
106. Marionneau S, Ruvoën N, Moullac-Vaidye B, et al. Norwalk virus binds to histo-blood group antigens present on gastroduodenal epithelial cells of secretor individuals. *Gastroenterology* 2002;122:1967–77.
107. Ruvoën-Clouet N, Le Pendu J. Sensibilité génétique aux infections à norovirus genetic susceptibility to norovirus infection. *Pathol Biol* 2013;61:28–35.
108. Tian P, Bates AH, Jensen HM, Mandrell RE. Norovirus binds to blood group A-like antigens in oyster gastrointestinal cells. *Lett Appl Microbiol* 2006;43:645–51.
109. Gao X, Esseili MA, Lu Z, Saif LJ, Wang, Q. Recognition of histo-blood group antigen-like carbohydrates in lettuce by human GII.4 norovirus. *Appl Environ Microbiol* 2016; 82: 2966-2974.
110. Needham J. Adsoption of bovine serum albumin to polyethylene tubing reversibility and pH-dependance. Doctoral dissertation, University of British Columbia; 1988.
111. van Oss CJ. Acid-base interfacial interactions in aqueous media. *Colloids Surf A Physiochem Eng Asp* 1993;78:1–49.

SUMMARY

This chapter provides a brief overview of how a phenomenon with annoying and costly implications has been studied so far and what knowledge has been gained for possible application in the agri-food field. In order to facilitate comprehension of this subject, the general theory of adhesion phenomena is first presented, followed by specific aspects including intrinsic and extrinsic parameters that influence the phenomena as they relate to noroviruses (NoVs). *Stricto sensu*, based on studies with virus-like particles, NoV adhesion in the food sector (to fresh foods and to inert surfaces) appears not to be promoted by the magnitude of the interfacial free energy of interaction but the proximity of the medium pH to the isoelectric point of the particles and elevated temperatures. On most surfaces, adhesion appears to be an endothermic process at pH far from the isoelectric point, and ionic strength appears to matter only on stainless steel and lettuce. Hydrophobic interactions and van der Waals forces appear to be the principal factors involved in adhesion, in particular when the isoelectric points of the particles and the surface are similar. Any strategy for reducing the transmission of human NoVs within the food production and distribution sectors and, subsequently, to the population could consider the use of safe chaotropic molecules for contact surface cleaning. Such molecules could also be used as additives for some fresh products that are normally rinsed before consumption.

Molecular Mechanisms of Prion Invasion and Analytical Methods for Detection

3

Akikazu Sakudo
Okayama University of Science

Yasuhisa Ano
Kirin Holdings Co. Ltd.

Takashi Onodera
The University of Tokyo

Contents

3.1 INTRODUCTION

Prions (proteinaceous infectious particles) are unique pathogens that lack nucleic acids and are thought to contain only protein [1]. The major component of a prion is an abnormal isoform of prion protein (PrP^{Sc}), which is generated by a conformational change from a cellular isoform of prion protein (PrP^{C}) (Figure 3.1). However, the precise mechanism by which PrP^{C} is converted to PrP^{Sc} remains unclear. PrP^{Sc} may be derived from PrP^{C}

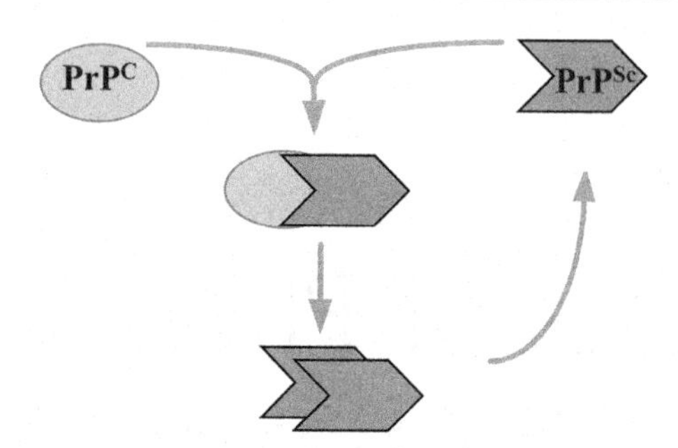

FIGURE 3.1 PrP^{C} and PrP^{Sc}. PrP^{Sc} is the main component of prions. After prion infection, PrP^{Sc} interacts with PrP^{C} leading to a conformational change of PrP^{C} into PrP^{Sc}. This conversion results in the accumulation of PrP^{Sc} and depletion of PrP^{C} in the neuron, resulting in neuronal cell loss. PrP^{C}: cellular isoform of prion protein; PrP^{Sc}: abnormal isoform of prion protein.

in a posttranslational process that seems to involve direct molecular interaction between PrP^{C} and PrP^{Sc} [2]. PrP^{C} is a glycosylphosphatidylinositol (GPI)-anchored glycoprotein that is predominantly found in the central nervous system (CNS), especially in neurons and glia [3–5]. PrP^{C} is also expressed in several tissues including heart, muscle, lymphoid tissues, kidney, gastrointestinal tract, skin, testis, placenta, uterus, and endothelium [6,7]. The crucial characteristic of prion diseases is the spongiform change accompanied by neuronal loss and gliosis along with PrP^{Sc} deposition and PrP^{C} reduction [8]. Both PrP^{Sc} accumulation and PrP^{C} deficiency in the CNS are thought to be the major pathological events associated with prion diseases [8].

Although the amino acid sequences of PrP^{C} and PrP^{Sc} are identical, they have markedly different biochemical characteristics (Table 3.1). PrP^{C} is a highly soluble protein with a short half-life (3–6 h) and is susceptible to proteolytic enzymes such as proteinase K (PK) [1,9,10]. By contrast, PrP^{Sc} is insoluble and readily forms aggregates. Moreover, PrP^{Sc} has a much longer half-life (>24 h) than PrP^{C}. Of note, PrP^{Sc} is partially resistant to PK digestion, which results in almost complete hydrolysis of the N-terminal region with the C-terminal portion of the protein remaining intact. These features are thought to be related to the difference in conformation between PrP^{C} and PrP^{Sc}, which is supported by analysis of the secondary structure based on the results of infrared spectroscopy [9] and nuclear magnetic resonance spectroscopy (NMR), revealing a high proportion of α-helix (~40% of the protein) and relatively little β-sheet (3% of the protein) in PrP^{C} [10]. By contrast, infrared spectroscopy [9] and circular dichroism (CD) spectroscopy [11] analyses demonstrate that PrP^{Sc} has a much higher β-sheet content of ~40% and a slightly reduced α-helix content of 30% compared to PrP^{C}.

In recent years, many analytical methods have been developed for prion detection. A representative example of such a technique is protein misfolding cyclic amplification (PMCA), which mimics the autocatalytic replication of PrP^{Sc} [12]. PMCA not only enables sensitive detection of prions but also provides useful information concerning the nature of these infectious agents. Studies using PMCA suggest that in addition to PrP^{Sc}, other cofactors or minor components in the prion agents are required for its

TABLE 3.1 Differences in the characteristics of PrP^{C} and PrP^{Sc}

	PrP^{C}	PrP^{SC}
Resistance against PK	Low	High
Percentage of secondary structure	α-Helix 42%, β-sheet 3%	α-Helix 30%, β-sheet 43%
Half-life	3–6 h	>24 h
Solubility	High	Low
Infectivity	No	Yes

PrP^{C}: Cellular isoform of prion protein.
PrP^{Sc}: Abnormal isoform of prion protein.
PK: Proteinase K.

proliferation. Specifically, other factors such as metal cations [13], RNA [14,15], DNA [16], lipids [17], plasminogen [18], and phosphatidylethanolamine [19] can enhance the amplification. Thus, PMCA studies indicate that in addition to PrP^{Sc}, these factors either are required or contribute to prion propagation. Nonetheless, this conclusion is contradicted by other studies using recombinant prion protein (PrP). Amyloid fibrils derived from recombinant PrP have been shown to be infectious without the addition of cofactors. Specifically, recombinant PrP aggregates were found to cause prion disease after inoculation in transgenic mice or wild-type hamsters [20,21]. These findings support the notion that PrP is the sole factor of a prion agent although infectivity appears to be very low in the absence of supplementary cofactors [20,21]. Size fractionation analysis suggests that most infectious prion particles have a diameter of 17–27 nm (300–600 kDa) [22], corresponding to 14–28 PrP molecules. Thus, in conclusion, these investigations show the main component of prion agent is composed of PrP although other components may be required for full infectivity.

Another unique characteristic of prions is the concept of a "species barrier," which means that the transmission of prions between species is less efficient than transmission within the same species [23]. The mechanisms involved in the species barrier remain to be fully understood. Nonetheless, the species barrier arises, at least in part, from differences between amino acid sequences of donor PrP^{Sc} and host (acceptor) PrP^{C}. However, the magnitude of the barrier seems to be determined by the compatibility of the conformation of PrP between PrP^{Sc} donor and PrP^{C} acceptor.

Intriguingly, even when the species of the prion agent and host are the same, different prion strains cause different biological reactions in the host [24]. Furthermore, these characters are transmitted to animals after infection. For example, strain-specific properties can be maintained after passage in animals. These characteristics include the accumulated brain sites of PrP^{Sc}, which depend on the type of prion strain. Therefore, prion strains can be identified by PrP^{Sc} accumulation patterns in brain lesions.

This chapter focuses on molecular principles and techniques used to investigate prions and the mechanism by which these agents act as foodborne pathogens. The risk of contracting a prion disease from ingesting contaminated food products is now a worldwide concern. As well as discussing fundamental aspects of prion diseases, this chapter also reviews recent advances in the development of molecular techniques for the detection of prion agents.

3.2 ZOONOTIC PRION DISEASES

Prion diseases, which are also known as transmissible spongiform encephalopathies (TSEs), are a family of fatal neurological zoonotic diseases [1]. These diseases include bovine spongiform encephalopathy (BSE) in cattle, chronic wasting disease (CWD) in cervids, feline spongiform encephalopathy (FSE) in cats, exotic ungulate encephalopathy (EUE) in zoo animals (e.g., kudu, nyala, gemsbok, eland and oryx), scrapie in goats and sheep, as well as transmissible mink encephalopathy (TME) in mink (Table 3.2). Prion agents are transmissible via ingestion, horizontal transmission, and vertical transmission, depending on each animal prion disease [1].

BSE, CWD, FSE, TME, and EUE are all thought to occur after the consumption of prion-infected materials such as meat and bone meal (MBM). Among these prion diseases, BSE is a foodborne disease that affects human with the emergence of variant Creutzfeldt–Jakob disease (vCJD) [25,26]. There are no reports that other animal prion agents besides BSE transmit to humans.

Nonetheless, following the BSE epidemic and demonstration of its zoonotic potential, general concerns were raised about the potential risk posed by other animal prions. A recent study has shown that scrapie agent can be transmitted via cerebral inoculation to macaque monkeys after a prolonged incubation time [27]. Moreover, ovine scrapie prion can be transmitted to transgenic mice expressing human PrP [28]. Recently, a prion disease in dromedary camels (*Camelus dromedarius*) in Algeria has been identified as camel spongiform encephalopathy (CSE) [29]. The implications of a broad spread of prion diseases

TABLE 3.2 Animal prion diseases

DISEASES	*AFFECTED ANIMALS*	*PROBABLE TRANSMISSION ROUTE*
Bovine spongiform encephalopathy (BSE)	Cattle	Ingestion of prion-contaminated food such as meat bone meal
Camel spongiform encephalopathy (CSE)	Camel	Unknown, but possibly due to ingestion of prion-contaminated food
Chronic wasting disease (CWD)	Elk Mule deer Moose	Acquired, horizontal, and vertical transmission as well as transmission by ingestion
Exotic ungulate encephalopathy (EUE)	Arabian oryx Eland Kudu Gemsbok Nyala	Ingestion of prion-contaminated food
Feline spongiform encephalopathy (FSE)	Albino tiger Cat Cheetah Puma	Ingestion of prion-contaminated food
Scrapie	Goat Sheep Mouflon	Acquired, horizontal, and vertical transmission as well as transmission by ingestion
Transmissible mink encephalopathy (TME)	Mink	Acquired and transmission by ingestion

suggest that active surveillance and control methods should be implemented for various animals on a worldwide basis. Furthermore, additional studies are needed to establish the type of animal prion agents that can be transmitted to humans.

3.3 CONVENTIONAL BIOCHEMICAL METHODS FOR PRION ANALYSIS

At present, definitive diagnosis of prion diseases can only be confirmed after death because this requires biochemical analysis of the brain [30]. These biochemical methods for prion diseases are mostly based on the characteristics of PrP^{Sc}, which is resistant to PK. Specifically, PK completely degrades PrP^{C} but only partially digests PrP^{Sc} because the latter forms protease-resistant aggregates. The portion of PrP remaining after PK treatment is named PrPres (PK-resistant PrP). Various biochemical techniques, including Western blotting and enzyme-linked immunosorbent assay (ELISA), can be used to detect PrPres after PK digestion. Thus, these techniques are capable of distinguishing PrP^{Sc} from PrP^{C} following treatment with PK.

Western blotting is a qualitative method in which proteins separated according to molecular weight by sodium dodecyl sulfate (SDS)-polyacrylamide gel electrophoresis are electroblotted onto a membrane for immunological detection (Figures 3.2 and 3.3). PrP can subsequently be detected on the membrane after probing with an anti-PrP antibody. If PK treatment is not performed (PK(−)), the same banding pattern is observed for prion infection and noninfection. However, because the N-terminal region of PrP is specifically degraded upon treatment with PK (PK(+)), only the C-terminal region is detected if PrPres is

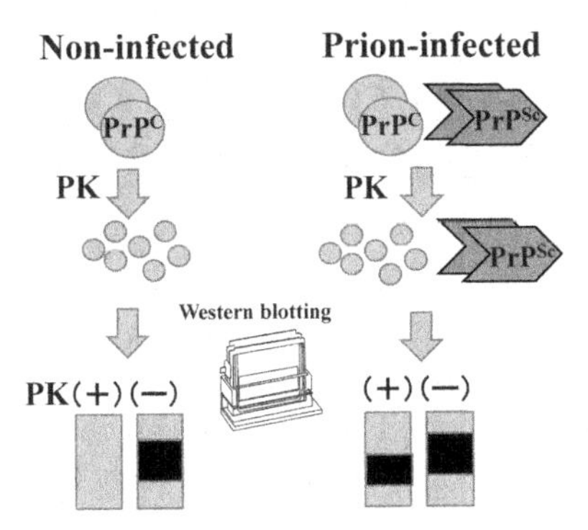

FIGURE 3.2 Schematic representation of Western blotting for the detection of PrP^{C} and PrP^{Sc}. Prion infection can be analyzed by Western blotting using an anti-prion protein (PrP) antibody after treatment of the sample with proteinase K (PK). Prion-infected brains contain both PrP^{Sc} and PrP^{C}, whereas noninfected brains contain only PrP^{C}. PrP^{C} is completely digested by PK, while PrP^{Sc} is only partially digested, resulting in the generation of a PK-resistant PrP fragment (PrPres). Specifically, after PK treatment (PK(+)), the C-terminal region of PrP^{Sc} is retained, whereas the N-terminal region is fully digested. Untreated control (PK(−)) is also shown.

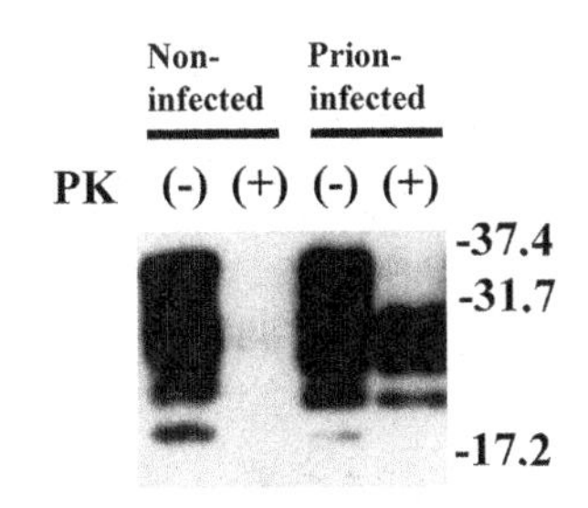

FIGURE 3.3 Representative data of a Western blotting experiment for the detection of PrP^{C} and PrP^{Sc}. Representative data using homogenates of mouse brains infected with prion (prion-infected) or uninfected (noninfected) after treatment with PK (+). Untreated control (PK(−)) is also included. The samples were subjected to Western blotting with the anti-PrP antibody to detect PK-resistant PrP (PrPres), which included PrP^{Sc}. Because the N-terminal region of PrP^{C} is fully digested, the PrPres band is shifted to a lower molecular weight compared to untreated PrP. The molecular weight markers (kDa) are shown on the right. (Modified from supplemental Figure 2B of Sakudo and Onodera [31] with permission from Elsevier.)

present. Thus, as the N-terminal region of PrP^{Sc} is digested with PK (+), PrPres displays a shift to a lower molecular weight compared to untreated PrP (−) in the Western blot. Interestingly, banding patterns vary depending on the specific type of prion strain and animal species. Therefore, these banding patterns provide useful qualitative information on the properties of the prion infection in a host.

ELISA is a quantitative analytical method for prions. The most commonly used test for prions involves performing a sandwich ELISA, for which various commercial kits are available. In sandwich ELISA, the capture antibody is adsorbed to the bottom of a microtiter plate. PrP bound to the capture antibody is subsequently detected using a different anti-PrP antibody. Therefore, the sandwich ELISA technique requires two antibodies, which recognize different epitopes of PrP.

The insolubility of PrP^{Sc} may also be used to distinguish between PrP^{Sc} and PrP^{C}. Immunohistochemistry (IHC) can be performed after treatment with formic acid, which efficiently removes the epitope of PrP^{C} but not that of PrP^{Sc}. Paraffin-embedded and formalin-fixed tissue sections are often used that are heated and pretreated with formic acid to eliminate PrP^{C} immunoreactivity and reveal PrP^{Sc} deposits. Histological analysis can also identify characteristics of prion diseases including astrocytosis, neuronal loss, spongiosis, and decomposition of PrP^{Sc}. In IHC, prion strains can often be discriminated from the lesion profiles, which result from strain-specific patterns of vacuolation and PrP^{Sc} accumulation in the brain.

3.4 RECENT DEVELOPMENT OF ANALYTICAL METHODS FOR PRIONS

In addition to conventional biochemical procedures, several novel methods have been developed to detect prions, which have recently undergone modification to enhance sensitivity. The archetypal method, known as PMCA, involves performing *in vitro* amplification of PrP^{Sc} [12,32]. In PMCA, sequential rounds of sonication and incubation enable amplification of PrP^{Sc} *in vitro* (Figures 3.4 and 3.5). Normal brain homogenate is used as a source of PrP^{C} in every cycle. The destruction of PrP^{Sc} aggregates by sonication causes

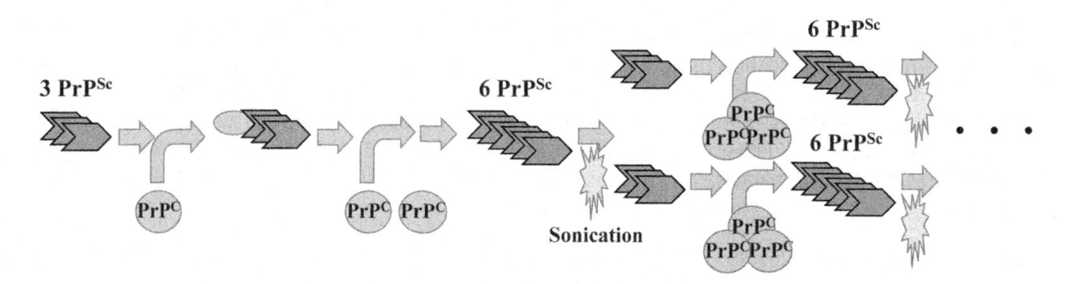

FIGURE 3.4 Schematic representation of protein misfolding cyclic amplification (PMCA) for the amplification of PrP^{Sc}. Aggregated PrP^{Sc} (3 PrP^{Sc}) binds to three molecules of PrP^{C}, resulting in the production of larger PrP^{Sc} aggregates (6 PrP^{Sc}). The aggregates (6 PrP^{Sc}) are broken down into two smaller aggregates (3 PrP^{Sc}) by sonication. Each aggregate then binds to more PrP^{C}, resulting in the stochastic generation of PrP^{Sc} *de novo*.

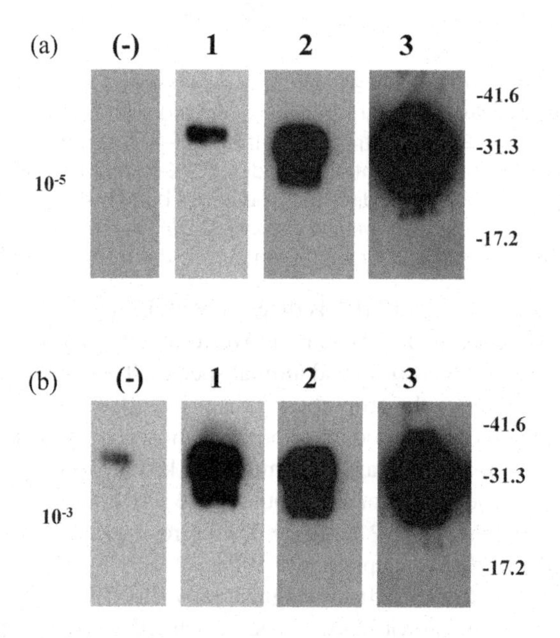

FIGURE 3.5 Representative data from PMCA PrP^{Sc} in brain homogenates of hamsters infected with 263K prion were diluted to an undetectable concentration at 10^{-5} (a) or detectable concentration at 10^{-3} (b) as shown by (–). PrP^{Sc} present in the samples was subsequently amplified with 1 cycle (1), 2 cycles (2), and 3 cycles (3) of PMCA. After amplification, the samples were subjected to PK treatment and Western blotting with anti-PrP antibody to detect PrPres. The molecular weight markers (kDa) are shown on the right.

fragmentation of PrP^{Sc} seed, which can drive the conversion of PrP^{C} to yield more PrP^{Sc}. Importantly, the generated PrP^{Sc} particles amplified by PMCA retain infectivity as well as physicochemical and structural properties similar to those of PrP^{Sc} derived from infected brains [33]. Unfortunately, PMCA does not yield quantitative data, and the assays take several days to complete.

However, recent developments of PMCA offer the prospect of a more highly sensitive detection and quantification procedure. These novel methods include recombinant PrP-PMCA (rPMCA), standard quaking-induced conversion (S-QuIC), amyloid seeding assay (ASA), real-time QuIC (RT-QuIC), and enhanced QuIC (eQuIC) [34]. In particular, RT-QuIC-based assays represent highly sensitive PrP^{Sc} detection methods (i.e., 1 fg levels) that achieve quantitative analysis [35,36]. Indeed, RT-QuIC shows promise in detecting prions present in body fluids including cerebrospinal fluid (CSF) in human [37] and deer [38].

RT-QuIC is based on the phenomena that shaking causes the conversion of recombinant PrP to β-sheet rich oligomers and fibrils [39]. Crucially, this conversion is enhanced by the presence of PrP^{Sc}. In RT-QuIC, a small amount (about 2–15 μl) of test sample is added to a reaction mixture containing recombinant PrP substrate and a detergent or chaotropic agent as well as the amyloid sensitive dye thioflavin T (ThT), whose fluorescent signal can be used to quantify amyloid formation. ThT fluorescence is monitored in real time using a fluorescence microplate reader under temperature-controlled conditions with several cycles of vigorous shaking. Because the reaction is performed in a microtiter plate with small sample volumes, large numbers of samples can be applied to this assay. A further advantage of this technique is that the concentration of prions can be estimated from the index of ThT fluorescence. Moreover, because CSF [40,41] and nasal brushing [42] can be used to diagnose prion diseases with high sensitivity and specificity, this method has the potential to enable preclinical and antemortem diagnosis of human and animal prion diseases using a universal procedure [43]. Therefore, RT-QuIC may be the most appropriate method for surveillance of prion diseases including animal prion diseases, such as CWD [44,45] and BSE [46,47], as well as human prion diseases [48–51].

3.5 INVASION ROUTE OF PRIONS

The mechanisms by which prions spread and reach the CNS after oral/intragastric exposure remain to be determined [52]. There are two main pathways of adsorption of PrP^{Sc} from the intestine and transport to the enteric nervous system (ENS) (Figure 3.6).

In the M cell-dependent pathway, M cells, which are specialized follicle-associated epithelium (FAE) that covers the luminal surface of Peyer's patch and plays an important role in the uptake of macromolecules from the enteric lumen via phagocytic activity, contribute to the uptake of PrP^{Sc}. By contrast, in the M cell-independent pathway, enterocytes of FAE contribute to the uptake of PrP^{Sc} from the intestinal lumen. Exosomes [53], 37 kDa-laminin receptor precursor (LRP) [54], and ferritin [55] as well as intracellular transfer of PrP^{Sc}-containing organelles through tunneling nanotubes [56] may be involved in enterocytic incorporation of PrP^{Sc} in the intestine.

After uptake or adsorption of PrP^{Sc} by the M cell-dependent or -independent pathways, PrP^{Sc} accumulates in gut-associated lymphoid tissue (GALT), particularly in the Payer's patches of the intestine. Dendritic cells (DCs) and macrophages are associated with the transportation of PrP^{Sc} to the lymphoid follicles of Peyer's patches. Follicular dendritic cells (FDCs) in the Peyer's patches are essential for proliferation of PrP^{Sc} to establish host infection [57]. Subsequently, PrP^{Sc} reaches the nerve fiber of mucosal plexus in the ENS endings in and around the follicles and are retrograde transported to the ENS, which acts as the gateway for PrP^{Sc} neuroinvasion. In enteric nerves, PrP^{Sc} replicates and spreads to the CNS including spinal cord and brain. Further studies to fully address the mechanism of prion invasion are needed to develop new strategies to prevent foodborne prion infection.

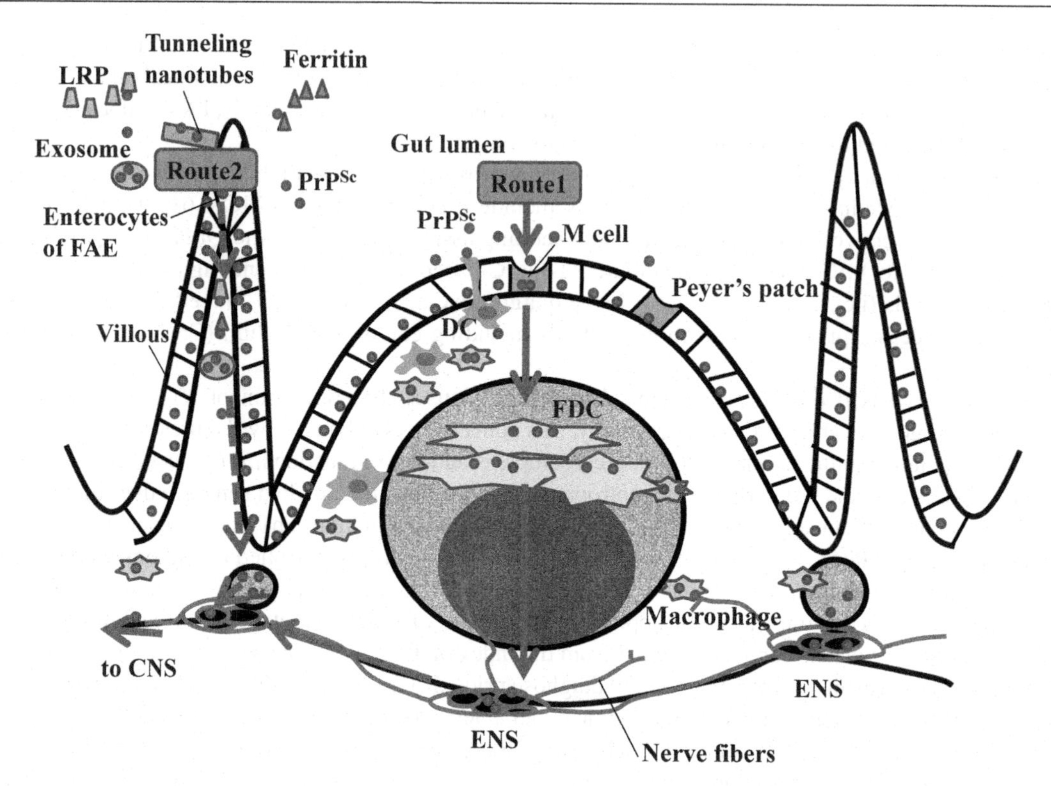

FIGURE 3.6 Proposed model of prion invasion from the intestine to the enteric nervous system (ENS). After oral administration, there are two main routes by which PrP^{Sc} can enter the intestinal epithelium. One route is mediated by transcytosis of PrP^{Sc} by M cells on Peyer's patches, while the other route is mediated by enterocytes of follicle-associated epithelium (FAE). Capture of PrP^{Sc} by ferritin and 37-kDa laminin receptor precursor (LRP) may assist in the enterocytic uptake of PrP^{Sc} and transport to dendritic cells (DCs) and macrophages in Peyer's patches. Exosome-mediated release of PrP^{Sc} and intracellular transfer of PrP^{Sc}-containing organelles through tunneling nanotubes may also contribute to the transportation of PrP^{Sc} from the gut lumen to the DCs and macrophages. Peyer's patches contain follicular dendritic cells (FDCs), which replicate PrP^{Sc}. Because some of the enteritic nerve endings are located in and around follicles, PrP^{Sc} spreads to the ENS and ultimately to the central nervous system (CNS). (Modified from Figure 1 of Ano et al. [68] with permission from the Food Safety Commission of Japan.)

3.6 FUTURE PERSPECTIVES

Recently, the "one health" concept has attracted considerable attention among national and international specialists. "One health" is the idea that human health and animal health are interdependent and bound to the health of an animal–human ecosystem interface [58]. It should be noted that zoonosis and animal diseases influencing food safety are closely related to animal and public health. Thus, a multidisciplinary approach should be adopted for the diagnosis and control of prion diseases at a local, national, and global level. In particular, we would like to emphasize that prion diseases should be dealt with using a "one health" concept, in combination with the following future perspectives.

First, the consumption of BSE agent-contaminated food is closely linked to vCJD. So far, more than 200 cases of vCJD and more than 190,000 cases of BSE have been recorded worldwide [59–61]. Although the prevalence of BSE and BSE-related vCJD has been reduced, the risk assessment from OIE (World Organisation for Animal Health) is based exclusively on classical BSE and does not include information on atypical BSE. However, recent studies have shown that atypical L-type BSE prion infects transgenic mice expressing human PrP more efficiently than typical BSE [62], while atypical H-type BSE prions do not infect these transgenic mice [63]. Moreover, another report has shown that a novel BSE prion could have emerged from atypical H-type BSE [64]. Therefore, information on atypical BSE should be incorporated into the risk assessment. However, because only limited information is currently available on atypical BSE, further studies are required to complete the risk assessment.

Second, there is the possibility of an outbreak of interspecies or intraspecies transmission of prion diseases at some time in the future. In particular, the high incidence and prevalence of CWD in North America should be monitored because it remains unclear whether CWD prion is transmissible to humans. Moreover, recent findings involving CSE suggest a broad spread of prion disease beyond and among species [29]. Thus, to prevent this disease from spreading, international active surveillance systems and improved diagnostic methods are needed. There are many potential sources of prion infection, including contaminated foods (especially beef and beef-related products), chemicals, and blood products. Because the quantity of prion agent is likely to be very low, a further increase in the sensitivity of the assays is required for detection. Recent developments in diagnostic methods, such as PMCA and RT-QuIC, have the potential to facilitate sensitive detection of prions in these samples. Prion-contaminated materials should be subjected to incineration as soon as possible to eliminate the risk of infection.

Significant progress has been made in the characterization of prion agents and prion invasion mechanisms, as well as in the development of analytical methods for prion diseases. However, the molecular basis for the neurodegenerative processes observed during the course of prion diseases is still poorly understood. Moreover, the mechanisms of prion conversion, as well as the maintenance of prion strains and species barriers, remain unknown. A better understanding of the basic biochemistry of these issues will help in the development of useful strategies for overcoming prion diseases.

ACKNOWLEDGMENTS

This work was supported by grant-in-aids for Scientific Research from the Ministry of Education, Science, Culture and Technology of Japan (20K03919, 16K04997, 25450447, 24110717, 23780299, 22110514, 20780219, 17780228). As the contents in this chapter partially overlap with those in previous books [65–67], permission was granted from CRC Press (Taylor & Francis Group) for their use here. Figure 3.6 was modified with permission from the Food Safety Commission of Japan [68].

REFERENCES

1. Prusiner, S.B. (1998). Prions. *Proc Natl Acad Sci USA. 95*, 13363–83.
2. Horiuchi, M., and Caughey, B. (1999). Specific binding of normal prion protein to the scrapie form via a localized domain initiates its conversion to the protease-resistant state. *EMBO J. 18*, 3193–203.
3. Kretzschmar, H.A., Prusiner, S.B., Stowring, L.E., and DeArmond, S.J. (1986). Scrapie prion proteins are synthesized in neurons. *Am J Pathol. 122*, 1–5.
4. Moser, M., Colello, R.J., Pott, U., and Oesch, B. (1995). Developmental expression of the prion protein gene in glial cells. *Neuron. 14*, 509–17.
5. Ford, M.J., Burton, L.J., Morris, R.J., and Hall, S.M. (2002). Selective expression of prion protein in peripheral tissues of the adult mouse. *Neuroscience. 113*, 177–92.
6. Bendheim, P.E., Brown, H.R., Rudelli, R.D., et al. (1992). Nearly ubiquitous tissue distribution of the scrapie agent precursor protein. *Neurology. 42*, 149–56.
7. Aguzzi, A., and Heikenwalder, M. (2006). Pathogenesis of prion diseases: current status and future outlook. *Nat Rev Microbiol. 4*, 765–75.

8. Sakudo, A., and Ikuta, K. (2009). Prion protein functions and dysfunction in prion diseases. *Curr Med Chem. 16*, 380–9.
9. Pan, K.M., Baldwin, M., Nguyen, J., et al. (1993). Conversion of alpha-helices into beta-sheets features in the formation of the scrapie prion proteins. *Proc Natl Acad Sci USA. 90*, 10962–6.
10. Riek, R., Hornemann, S., Wider, G., Billeter, M., Glockshuber, R., and Wuthrich, K. (1996). NMR structure of the mouse prion protein domain PrP(121-231). *Nature. 382*, 180–2.
11. Safar, J., Roller, P.P., Gajdusek, D.C., and Gibbs, C.J., Jr. (1993). Thermal stability and conformational transitions of scrapie amyloid (prion) protein correlate with infectivity. *Protein Sci. 2*, 2206–2216.
12. Soto, C., Saborio, G.P., and Anderes, L. (2002). Cyclic amplification of protein misfolding: application to prion-related disorders and beyond. *Trends Neurosci. 25*, 390–4.
13. Kim, N.H., Choi, J.K., Jeong, B.H., et al. (2005). Effect of transition metals (Mn, Cu, Fe) and deoxycholic acid (DA) on the conversion of PrP^C to PrPres. *FASEB J. 19*, 783–5.
14. Deleault, N.R., Kascsak, R., Geoghegan, J.C., and Supattapone, S. (2010). Species-dependent differences in cofactor utilization for formation of the protease-resistant prion protein in vitro. *Biochemistry. 49*, 3928–34.
15. Katorcha, E., Gonzalez-Montalban, N., Makarava, N., Kovacs, G.G., and Baskakov, I.V. (2018). Prion replication environment defines the fate of prion strain adaptation. *PLoS Pathog. 14*, e1007093.
16. Imamura, M., Kato, N., Okada, H., et al. (2013). Insect cell-derived cofactors become fully functional after proteinase K and heat treatment for high-fidelity amplification of glycosylphosphatidylinositol-anchored recombinant scrapie and BSE prion proteins. *PLoS One. 8*, e82538.
17. Srivastava, S., and Baskakov, I.V. (2015). Contrasting effects of two lipid cofactors of prion replication on the conformation of the prion protein. *PLoS One. 10*, e0130283.
18. Mays, C.E., and Ryou, C. (2010). Plasminogen stimulates propagation of protease-resistant prion protein in vitro. *FASEB J. 24*, 5102–12.
19. Deleault, N.R., Piro, J.R., Walsh, D.J., et al. (2012). Isolation of phosphatidylethanolamine as a solitary cofactor for prion formation in the absence of nucleic acids. *Proc Natl Acad Sci USA. 109*, 8546–51.
20. Legname, G., Baskakov, I.V., Nguyen, H.O., et al. (2004). Synthetic mammalian prions. *Science. 305*, 673–6.
21. Makarava, N., Kovacs, G.G., Bocharova, O., et al. (2010). Recombinant prion protein induces a new transmissible prion disease in wild-type animals. *Acta Neuropathol. 119*, 177–87.
22. Silveira, J.R., Raymond, G.J., Hughson, A.G., et al. (2005). The most infectious prion protein particles. *Nature. 437*, 257–61.
23. Hagiwara, K., Hara, H., and Hanada, K. (2013). Species-barrier phenomenon in prion transmissibility from a viewpoint of protein science. *J Biochem. 153*, 139–45.
24. Igel-Egalon, A., Beringue, V., Rezaei, H., and Sibille, P. (2018). Prion strains and transmission barrier phenomena. *Pathogens. 7*, E55.
25. Brown, P., Will, R.G., Bradley, R., Asher, D.M., and Detwiler, L. (2001). Bovine spongiform encephalopathy and variant Creutzfeldt-Jakob disease: background, evolution, and current concerns. *Emerg Infect Dis. 7*, 6–16.
26. Will, R.G., Ironside, J.W., Zeidler, M., et al. (1996). A new variant of Creutzfeldt-Jakob disease in the UK. *Lancet. 347*, 921–5.
27. Comoy, E.E., Mikol, J., Luccantoni-Freire, S., et al. (2015). Transmission of scrapie prions to primate after an extended silent incubation period. *Sci Rep. 5*, 11573.
28. Cassard, H., Torres, J.M., Lacroux, C., et al. (2014). Evidence for zoonotic potential of ovine scrapie prions. *Nat Commun. 5*, 5821.
29. Babelhadj, B., Di Bari, M.A., Pirisinu, L., et al. (2018). Prion disease in dromedary camels, Algeria. *Emerg Infect Dis. 24*, 1029–36.
30. Sakudo, A., Nakamura, I., Ikuta, K., and Onodera, T. (2007). Recent developments in prion disease research: diagnostic tools and in vitro cell culture models. *J Vet Med Sci. 69*, 329–37.
31. Sakudo, A., and Onodera, T. (2011). Tissue- and cell type-specific modification of prion protein (PrP)-like protein Doppel, which affects PrP endoproteolysis. *Biochem Biophys Res Commun. 404*, 523–7.
32. Soto, C. (2004). Diagnosing prion diseases: needs, challenges and hopes. *Nat Rev Microbiol. 2*, 809–19.
33. Castilla, J., Saa, P., Hetz, C., and Soto, C. (2005). In vitro generation of infectious scrapie prions. *Cell. 121*, 195–206.
34. Orru, C.D., Wilham, J.M., Vascellari, S., Hughson, A.G., and Caughey, B. (2012). New generation QuIC assays for prion seeding activity. *Prion. 6*, 147–52.
35. Atarashi, R., Satoh, K., Sano, K., et al. (2011). Ultrasensitive human prion detection in cerebrospinal fluid by real-time quaking-induced conversion. *Nat Med. 17*, 175–8.
36. Wilham, J.M., Orru, C.D., Bessen, R.A., et al. (2010). Rapid end-point quantitation of prion seeding activity with sensitivity comparable to bioassays. *PLoS Pathog. 6*, e1001217.
37. Green, A.J.E., and Zanusso, G. (2018). Prion protein amplification techniques. *Handb Clin Neurol. 153*, 357–70.

38. Haley, N.J., Van de Motter, A., Carver, S., et al. (2013). Prion-seeding activity in cerebrospinal fluid of deer with chronic wasting disease. *PLoS One. 8*, e81488.
39. Ladner-Keay, C.L., Griffith, B.J., and Wishart, D.S. (2014). Shaking alone induces de novo conversion of recombinant prion proteins to beta-sheet rich oligomers and fibrils. *PLoS One. 9*, e98753.
40. McGuire, L.I., Peden, A.H., Orru, C.D., et al. (2012). Real time quaking-induced conversion analysis of cerebrospinal fluid in sporadic Creutzfeldt-Jakob disease. *Ann Neurol. 72*, 278–85.
41. Orru, C.D., Groveman, B.R., Hughson, A.G., Zanusso, G., Coulthart, M.B., and Caughey, B. (2015). Rapid and sensitive RT-QuIC detection of human Creutzfeldt-Jakob disease using cerebrospinal fluid. *MBio. 6*, e02451–14.
42. Orru, C.D., Bongianni, M., Tonoli, G., et al. (2014). A test for Creutzfeldt-Jakob disease using nasal brushings. *N Engl J Med. 371*, 519–29.
43. Orru, C.D., Groveman, B.R., Raymond, L.D., et al. (2015). Bank vole prion protein as an apparently universal substrate for RT-QuIC-based detection and discrimination of prion strains. *PLoS Pathog. 11*, e1004983.
44. Haley, N.J., Henderson, D.M., Wycoff, S., et al. (2018). Chronic wasting disease management in ranched elk using rectal biopsy testing. *Prion. 12*, 93–108.
45. Cheng, Y.C., Hannaoui, S., John, T.R., Dudas, S., Czub, S., and Gilch, S. (2016). Early and non-invasive detection of chronic wasting disease prions in elk feces by real-time quaking induced conversion. *PLoS One. 11*, e0166187.
46. Levavasseur, E., Biacabe, A.G., Comoy, E., et al. (2017). Detection and partial discrimination of atypical and classical bovine spongiform encephalopathies in cattle and primates using real-time quaking-induced conversion assay. *PLoS One. 12*, e0172428.
47. Orru, C.D., Favole, A., Corona, C., et al. (2015). Detection and discrimination of classical and atypical L-type bovine spongiform encephalopathy by real-time quaking-induced conversion. *J Clin Microbiol. 53*, 1115–20.
48. Cramm, M., Schmitz, M., Karch, A., et al. (2016). Stability and reproducibility underscore utility of RT-QuIC for diagnosis of Creutzfeldt-Jakob disease. *Mol Neurobiol. 53*, 1896–904.
49. Park, J.H., Choi, Y.G., Lee, Y.J., et al. (2016). Real-time quaking-induced conversion analysis for the diagnosis of sporadic Creutzfeldt-Jakob disease in Korea. *J Clin Neurol. 12*, 101–6.
50. Hermann, P., Laux, M., Glatzel, M., et al. (2018). Validation and utilization of amended diagnostic criteria in Creutzfeldt-Jakob disease surveillance. *Neurology. 91*, e331–8.
51. Abu-Rumeileh, S., Redaelli, V., Baiardi, S., et al. (2018). Sporadic fatal insomnia in Europe: phenotypic features and diagnostic challenges. *Ann Neurol. 84*, 347–60 (doi: 10.1002/ana.25300).
52. Mabbott, N.A. (2017). How do PrP(Sc) prions spread between host species, and within hosts? *Pathogens. 6*, 60.
53. Fevrier, B., Vilette, D., Archer, F., et al. (2004). Cells release prions in association with exosomes. *Proc Natl Acad Sci USA. 101*, 9683–8.
54. Morel, E., Andrieu, T., Casagrande, F., et al. (2005). Bovine prion is endocytosed by human enterocytes via the 37 kDa/67 kDa laminin receptor. *Am J Pathol. 167*, 1033–42.
55. Mishra, R.S., Basu, S., Gu, Y., et al. (2004). Protease-resistant human prion protein and ferritin are cotransported across Caco-2 epithelial cells: implications for species barrier in prion uptake from the intestine. *J Neurosci. 24*, 11280–90.
56. Zhu, S., Victoria, G.S., Marzo, L., Ghosh, R., and Zurzolo, C. (2015). Prion aggregates transfer through tunneling nanotubes in endocytic vesicles. *Prion. 9*, 125–35.
57. Mabbott, N.A. (2017). Immunology of prion protein and prions. *Prog Mol Biol Transl Sci. 150*, 203–40.
58. OIE (2018). One Health (http://www.oie.int/en/for-the-media/onehealth/).
59. OIE (2018). Number of reported cases of bovine spongiform encephalopathy (BSE) in farmed cattle worldwide (excluding the United Kingdom) (http://www.oie.int/?id=505).
60. OIE (2018). Number of cases of bovine spongiform encephalopathy (BSE) reported in the United Kingdom (http://www.oie.int/index.php?id=504).
61. Creutzfeldt-Jakob Disease International Surveillance Network, formerly EuroCJD, CJD Surveillance Data 1993–2013, 2014 (http://www.eurocjd.ed.ac.uk/surveillance%20data%201.html).
62. Beringue, V., Herzog, L., Reine, F., et al. (2008). Transmission of atypical bovine prions to mice transgenic for human prion protein. *Emerg Infect Dis. 14*, 1898–901.
63. Comoy, E.E., Casalone, C., Lescoutra-Etchegaray, N., et al. (2008). Atypical BSE (BASE) transmitted from asymptomatic aging cattle to a primate. *PLoS One. 3*, e3017.
64. Masujin, K., Okada, H., Miyazawa, K., et al. (2016). Emergence of a novel bovine spongiform encephalopathy (BSE) prion from an atypical H-type BSE. *Sci Rep. 6*, 22753.
65. Sakudo, A., Onodera, T. Chapter 99. Bovine spongiform encephalopathy (BSE). In "*Molecular Detection of Animal Viral Pathogens* (Liu D, ed.)", 901–12, Boca Raton, FL: Taylor & Francis CRC Press, 2016.

66. Sakudo, A., Onodera, T. Chapter 7, Prions. In "*Laboratory Models for Foodborne Infections* (Liu D, ed.)", 117–27, Boca Raton, FL: Taylor & Francis CRC Press, 2017.
67. Sakudo, A., Onodera, T. Chapter 12, Prions. In "*Foodborne Viral Pathogens* (White PA, Netzler NE, eds.)", 237–56, Boca Raton, FL: Taylor & Francis CRC Press, 2017.
68. Ano Y, Sakudo A, Uraki R, Kono J, Yukawa M, Onodera T. (2013). Intestinal transmission of prion proteins and role of exosomes in enterocytes. *Food Safety 1*(1), 2013005.

SUMMARY

Prions are foodborne pathogens that cause neurodegenerative diseases in humans and animals, which are often referred to as transmissible spongiform encephalopathies (TSEs) or simply prion diseases. The main component of a prion is an abnormal isoform of prion protein (PrP^{Sc}), which enters the host through the gastrointestinal tract and subsequently migrates to the central nervous system (CNS). The mechanisms underlying the neuroinvasion of prions are not fully understood. However, PrP^{Sc} is taken up by M cells or enterocytes of follicle-associated epithelium (FAE) in the intestine. Exosomes, ferritin, and laminin receptor precursor as well as tunneling nanotubes may also be crucial in PrP^{Sc} transportation through the intestinal epithelium. After transportation of PrP^{Sc} by dendritic cells (DCs) and macrophages, follicular dendritic cells (FDCs) play an important role in PrP^{Sc} replication, which then spreads to the CNS. Various analytical methods have been developed to detect prion invasion, including conventional biochemical methods, such as Western blotting, enzyme-linked immunosorbent assay (ELISA), and immunohistochemistry (IHC), as well as more recently developed methods, such as protein misfolding cyclic amplification (PMCA) and real-time quaking-induced conversion (RT-QuIC). In this chapter, fundamental knowledge of prions as causative agents of foodborne zoonotic diseases is discussed, together with the mechanisms of prion invasion and recent developments of analytical methods for their detection.

Molecular Pathogenesis of Nipah Virus Infection

4

Devendra T. Mourya, Padinjaremattathil Thankappan Ullas, and Pragya D. Yadav
National Institute of Virology

Contents

4.1 INTRODUCTION

Virus transgressions across species boundaries pose an ongoing threat to human, animal, and plant health worldwide. The past decades have witnessed the emergence of several zoonotic viral pathogens associated with high human and animal morbidity and/or mortality, including West Nile, severe acute respiratory syndrome (SARS), influenza A (H1N1)pdm09, avian influenza A (H5N1), influenza A (H7N9), Middle East respiratory syndrome coronavirus, zika, Hendra, and Nipah viruses. Among these, Nipah virus (NiV) represents a unique category of paramyxoviruses that can cause a severe acute encephalitic illness in humans. Outbreaks of human NiV infections have occurred in five Asian countries so far, viz., Malaysia, Singapore, Bangladesh, the Philippines, and India. However, due to the widespread distribution and vast flight range of pteropid fruit bats, the primary reservoir of NiV in nature, the risk may be global. Recent findings of serological reactivity to NiV among many species of fruit bats suggest a wider circulation of NiV. Approximately 2 billion people worldwide are at risk of exposure to NiV. Of this, about 250 million susceptible individuals reside in the outbreak-prone areas of Bangladesh and India alone. Many cases of NiV infection may go undiagnosed in the low-resource countries, and the lack of systematic surveillance makes it difficult to estimate the exact burden of NiV disease.[1] Outbreaks associated with the virus can cause a considerable economic impact as far as human health is concerned, as evident from the requirement for intensive care treatment, high death rates, and debilitating sequelae in survivors. The risk for aerosol transmission and human-to-human spread, high rates of death and disabilities, absence of efficacious vaccination, and treatment strategies make NiV a priority pathogen.[1–3] The WHO has recently listed NiV as one of the nine priority pathogens capable of causing a future public health emergency and needing urgent vaccine research and development.[4] Studies on this virus throw light on novel pathways of spillover of viral pathogens from their natural reservoirs and how sociocultural practices often impact the occurrence of novel diseases. Despite considerable advances in the understanding of NiV epidemiology over the past two decades, many aspects of its molecular pathogenesis remain clouded, on which this chapter aims to address.

4.2 CLASSIFICATION AND MORPHOGENETICS OF NIV

NiV and the closely related Hendra virus (HeV) constitute the pathogenic members of the genus Henipavirus within the family *Paramyxoviridae* (Order *Mononegavirales*). As an enveloped pleomorphic virus, NiV measures 40–1,900 nm in size and possesses a nonsegmented, single-stranded, negative-sense RNA genome of about 18.2 kb in length. The genome carries six open reading frames (ORFs), coding for the nucleocapsid (N), phosphoprotein (P), matrix protein (M), fusion protein (F), glycoprotein (G), and large or polymerase protein (P). A noncoding leader (55 nt long) and trailer (33 nt long) sequence mark the 3′ and 5′ ends of the genome, respectively. Intergenic sequences consisting of three nucleotides are present at each gene junction.

The viral genome is tightly encapsidated by the nucleoprotein (yielding a herringbone-like appearance under the electron microscope) and the L and P proteins associated with it to complete the skeletal structure. The M protein plays important roles in maintaining virus structure and virus assembly and is localized between the viral envelope and the nucleocapsid core. The ORF for P yields three additional proteins, viz., V and W (produced by co-transcriptional mRNA editing) and C (produced by alternate sites of initiation of translation). The virus has a lipid envelope bearing multimeric spikes of surface glycoproteins F and G, and the M protein lies beneath it.[5]

4.3 GENETIC DIVERSITY OF NIV

NiV contains a longer genome than other paramyxoviruses. Two distinct genetic lineages of human pathogenic NiV exist, viz., NiV-Malaysia (NiV-MY) and NiV-Bangladesh (NiV-BD). The full genome of NiV-MY measures 18,246 nt, while the NiV-BD genome has a length of 18,252 nt.[6] The genome of NiV-BD strains bears six additional nucleotides and shares an amino acid homology of over 92% with the Malaysian strain. There is phylogenetic evidence of at least two separate introductions of NiV into Malaysia, while the sequence heterogeneity among the Bangladesh strains suggests the likelihood of multiple introductions there.[7] The two lineages exhibit subtle pathogenic differences in experimental animal models.[8–10] The Indian strain shows greater similarity to the Bangladesh strains than NiV-MY. The strains from Cambodia had close resemblance with the Malaysian strain. These observations may reflect the selection pressures induced by the different host/geographic systems in which the strains replicate.

4.4 HOST RANGE

NiV exhibits one of the broadest host ranges reported among paramyxoviruses. While fruit-eating bats of the *Megachiroptera* family (mainly, species *Pteropus*, *Eidolon* and *Rousettus*) have been identified as the natural reservoirs of NiV, reports also indicated the detection of NiV-specific antibodies in Microchiropteran bats, indicating their susceptibility.[11] Evidence for NiV infection has been reported in *Pteropus hypomelanus*, *P. vampyrus*, *Cynopterus brachyotis*, *Eonycteris spelaea*, and *Scotophilus kuhlii* (an insectivorous bat) in Malaysia[12] and *P. giganteus* in Bangladesh.[13] Studies have reported the isolation of NiV in *P. lylei* from Cambodia[14] and the detection of viral RNA in urine and saliva of *P. lylei* and *Hipposideros larvatus* bats in Thailand.[15] Yadav et al.[16] identified RNA sequences similar to earlier outbreak strains of NiV and NiV-IgG in *P. giganteus* bats from Myanaguri, West Bengal, India. Serological reactivity to the Nipah-like virus was observed in fruit bats from India, Indonesia, and Timor Leste.[17,18]

Natural infections with NiV have also been described in multiple species of domesticated animals, including pigs, sheep, goats, cats, dogs, and horses.[19]

4.5 DISCOVERY OF NIV AND MAJOR NIV OUTBREAKS

NiV, maintained in nature principally in pteropid and multiple other species of fruit bats, probably evolved within them over millions of years.[20] Time-scaled phylogenetic analyses suggest the probable time of its entry into the southeastern Asiatic regions as 1947. Two independent introductions of the virus may have occurred in 1985 and 1995, prior to its spread along the different flight paths of the pteropid bats.[21,22]

NiV was identified during investigations of an outbreak of acute encephalitic illness among the handlers of pigs showing severe respiratory illness in the Sungai Nipah village in Peninsular Malaysia in 1998. The outbreak, which caused more than 265 human infections and 105 deaths, subsequently spread to Singapore, probably through the transport of infected pigs, where it caused another outbreak involving 11 deaths in 1999.[23] The outbreak in Malaysia ended with the culling of over 1 million pigs and infections with NiV have not been reported from the country since then.

Bangladesh reported its first Nipah outbreak in 2001 in Meherpur, with 13 cases and 9 deaths.[13,24] The disease has been occurring in epidemic or sporadic forms almost annually in the country since then.[13,25–28] Subsequently, outbreaks were reported from Naogaon (2003), Goalando and Faridpur (2004), Tangail (2005), and Manikgonj and Rajbari districts (2008). The reported risk factors for the acquisition of NiV infection in these clusters included climbing trees, contact with sick animals, and contact with ill patients.[2] Consumption of raw date palm sap was the only risk factor most strongly associated with the development of illness in the outbreaks in Tangail and Manikgonj and Rajbari districts.

Siliguri district in the West Bengal state of India subsequently reported 66 cases of Nipah infection in 2001. The case fatality ratio in this outbreak was 74%, with about 75% cases occurring through patient, medical staff, and family contacts in four local hospitals.[29] Five fatal cases of human Nipah infections occurred in a subsequent outbreak in India in 2007, and four of these cases were attributable to contacts with the index patient.[30]

An outbreak associated with serious illness and high mortality in horses and humans, possibly caused by NiV, was also reported from the southern Philippines in 2014. A total of 17 cases was noted during this outbreak, including 11 acute encephalitis, 5 influenza-like illness, and 1 meningitis.[31]

4.6 CELLULAR PATHOGENESIS OF NIV

NiV, like morbilliviruses, employs protein receptors to gain entry into susceptible cells. Ephrin-B2 and ephrin-B3, members of the ephrin family of receptor tyrosine kinases, have been identified as the primary receptors for NiV in susceptible cells.[32–34] Ephrin-B2 is among the most ubiquitously expressed and highly conserved molecules in mammalian cells,[35] based on which NiV has developed a very successful evolutionary strategy across a diverse set of host species. Ephrin-B2 shows abundant expression in vascular endothelial, smooth muscle cells in arteries, brain, neuroepithelial cells, placenta, lungs, and prostate gland. Ephrin-B3 is mostly expressed in specific regions of the central nervous system (CNS), prostate, and heart.[36–38] These receptors, through interactions with their "ephrin" ligands, can induce bidirectional signaling in the cells.[39,40] They play key roles in chemotaxis, cell and neuronal migration, formation of junctions between veins and arteries, and immune activation.[41,42] The abolition of surface expression of ephrin-B2 by intramembrane proteases like rhomboid protease may render some cells refractory to NiV infection.

Several continuous and primary cell lines expressing ephrin-B2 are susceptible to NiV infection in vitro. These include HEK293T, Vero, HeLa-CCL2, and CHO expressing ephrin-B2 and B3, porcine brain microvascular endothelial cells (PBMEC), human brain endothelial cells (HBMEC), and human umbilical vein endothelial cells (HUVEC) (Figure 4.1). Endothelial cells that lack ephrin-B2 expression are resistant to NiV infection.[43]

The envelope glycoprotein (G) of NiV is the primary determinant of viral attachment to the cell surface. The NiV G, arranged as tetramers on the virion surface, possesses an N-terminal transmembrane domain, a stalk domain (consisting of an alpha-helical portion and a distal proline-rich portion), and a C-terminal receptor-binding domain, which has a six-bladed beta-propeller fold. The fusion (F) proteins possess N-terminal extracytoplasmic domain as well as a C-terminal transmembrane and cytoplasmic domain. The fusion of viral and target cell membrane occurs through an intricate interaction of G and F. The F, produced as F0, an inactive precursor, undergoes proteolytic activation to yield an active heterodimer consisting of disulfide-linked F1 and F2 subunits. The proteolytic activation of F differs from that of other paramyxoviruses in that it is mediated by the lysosomal endopeptidase, Cathepsin-L, and only after endocytosis of the nascent F0. Such processing by a ubiquitous intracellular protease may facilitate the systemic spread of the infection and may also represent a step that can be targeted for the development of antiviral strategies. Two alpha-helical regions named HR1 and HR2 are present in the extracellular domain of F1 subunit. Receptor binding causes conformational rearrangements in G tetramers, which

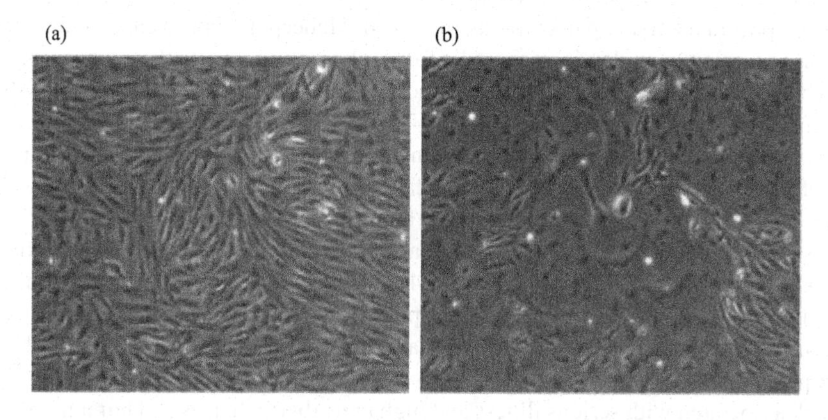

FIGURE 4.1 NiV infection of primary human umbilical vein endothelial cells (HUVECs). (a) Mock-infected HUVEC. (b) HUVEC infected with NiV for 24 h showing extensive syncytium formation. (Photo credit: Mathieu C, Guillaume V, Sabine A, et al. *PLoS One* 2012;7(2):e32157.)

expose a set of key residues in the stalk domain, due to which the F transitions from a pre-fusion state to a post-fusion state. The resultant apposition of the host and viral membranes promotes the fusion of the two. In the process, F undergoes a conformational change that causes the insertion of the hydrophobic N-terminus of F1 into the cell membrane. Upon another conformational change involving the folding and zipping up of HR1 and HR2 regions around each other, a six-helix bundle (6HB) is formed. The distinct conformations adopted by F during the membrane fusion process also offer potential targets for developing antivirals against NiV. Several stages in the fusion process can be targeted for NiV inhibition, viz., abrogating the proteolytic processing of the inactive F precursor, blocking F in the pre- or post-fusion step, inhibition of formation of the six-helix bundle, which is critical for membrane fusion.[5,44]

The interaction between NiV G and the ephrin-B2 has the highest affinity reported ever for a virus envelope–receptor interaction[45] and is dependent on a G–H loop (constituted by amino acid residues 120–125) on the latter.[34,46,47] Three residues on G, viz., E533, W504, and E505, are critical for receptor binding.[45,48] A salt bridge formed by E533 with the K60 residue on ephrin-B2 positions the G–H loop of the ephrin-B2 in the hydrophobic canyon of NiV G, which facilitates the interaction of the loop with the amino acids lining the surface of the canyon. The E533 residue also mediates a charge-based interaction with R57 of ephrin-B3, the alternate cellular receptor of NiV.[49]

In addition to a low pH-stimulated membrane fusion pathway, the macropinocytotic pathway may also aid in cellular entry of NiV. The NiV G-mediated activation of ephrin-B2 involves phosphorylation of specific tyrosine residues. The phosphorylation of the Y304 residue is essential to the reverse signaling process. The phosphorylated Y304 subsequently recruits Ras, which activates multiple downstream pathways including vesicle trafficking (Arf6), actin remodeling and membrane ruffling (Rab5/RNtre and PI3K/PKC/Rab34), and regulation of the switch between micropinocytosis and filopodia formation (Rac1/Cdc42).[46]

Peripheral blood lymphocytes and monocytes lack ephrin-B2 expression and are generally resistant to productive infection by NiV. However, recent work showed that these cells could capture NiV virions and pass them on to susceptible cells.[50] Glycoprotein-mediated binding of NiV to the cell surface may underlie such trans-infection of lymphocytes, which can lead to dissemination of infection to other cells or to a new host. Human dendritic cells (DCs) may also support a low-level infection with NiV, probably through a CD169-dependent mechanism.[51] Mathieu et al.[52] showed that heparan sulfate can mediate trans-infection with NiV in leucocytes that do not express CD169. Receptor-independent mechanisms may also lead to cellular binding of the virus. Inhibition of C-type lectin can effectively abolish such trans-infections.[50] These may offer potential therapeutic strategies against NiV.

4.7 NIV PATHOLOGY

A pronounced tropism for endothelial cells is a hallmark of NiV infection across species, while the neuronal, epithelial, and lymphoid affinity varies. Endothelia that express functional ephrin-B2, including those of the blood vessels, CNS, and the lower respiratory tract, are consistent targets for NiV infection. Organ-specific variation in endothelial susceptibility to NiV may reflect the relative absence of cellular receptors or variations in their post-translational modifications (e.g., glycosylation), which may affect virus binding and membrane fusion. NiV G can facilitate endothelial cell infection by interacting with LSECtin, the endothelial cell lectin.[53]

Endothelial cell infection is an important determinant of clinical features in the NiV-infected host. The infected endothelial cells may mediate contiguous spread of the virus to parenchymal cells of the organs, as well as facilitate virus dissemination by recruiting inflammatory and immune cells. Syncytia and multinucleated giant cells are typically seen in the vascular endothelial in humans, hamsters, cats, and guinea pigs infected with NiV. Notably, despite a prominent immunostaining for viral antigen, endothelial syncytia are absent in pigs infected naturally or experimentally with NiV (Figure 4.2). Endothelial

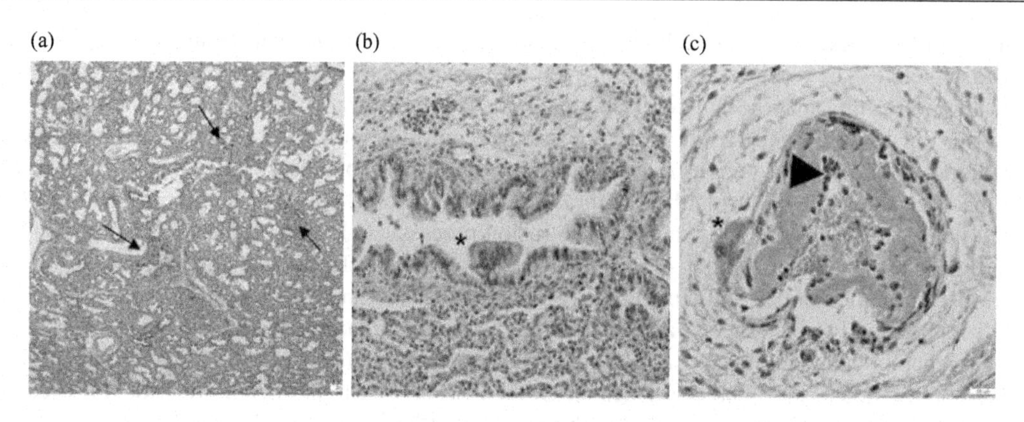

FIGURE 4.2 Histopathological changes during NiV infection in human lung xenografts. (a) Human lung with focal areas of necrosis and syncytia formation (arrow) on day 3 post infection (H&E, 10× magnification). (b) Bronchi with syncytia formation (*) on day 3 post infection (H&E, 20× magnification). (c) Pulmonary vasculature with syncytial formation (*) on day 3 post infection (H&E, 40× magnification), fibrinoid necrosis of the intima, and influx of granulocytes (arrowhead). (Photo credit: Valbuena G, Halliday H, Borisevich V, Goez Y, Rockx B. *PLoS Pathog* 2014;10(4):e1004063.)

apoptosis, necrosis, fibrinoid necrosis, and inflammatory cell infiltration may ensue NiV infection of the endothelial cells. Perivascular cuffing is occasionally seen, with infiltration of neutrophils, monocytes, or occasionally lymphocytes. Organs with abundant expression of ephrin-B2, such as the lungs, kidneys, and the heart, show prominent vascular inflammatory changes. Spleen and other highly vascular organs may show focal perivascular necrosis.[54]

The oronasal and respiratory epithelia appear to be the primary sites of NiV replication during natural infections in cats and pigs, with subsequent spread of the virus to the vascular and lymphoreticular systems. The smooth muscle cells in the tunica media of the blood vessels are permissive to NiV infection, with subsequent spread of infection occurring to organs including the brain, lungs, and the spleen by hematogenous route. The viremia is thought to be of low level, and its exact duration is unclear. NiV also specifically targets the lymphoreticular system, and lymphoid depletion and necrosis have been reported in studies of infected humans, swines, and cats.

Hematogenous dissemination of NiV often results in infection of the kidneys and urinary tract. Infected bats often show intermittent shedding of NiV in their urine.[55] In contrast, urinary shedding of NiV is not usually observed during natural infections in pigs. During the NiV outbreak in Malaysia, NiV was isolated from 3 of 20 patients with a confirmed diagnosis of NiV infection.[56] Histopathological studies in experimentally infected guinea pigs showed lymphohistiocytic vasculitis, necrosis, and ulceration of transitional epithelium and occurrence of syncytial cells, in the urinary bladder and mild vasculitis and syncytial cells in kidneys. Fogarty et al.[57] showed that NiV has a half-life of about 17.8 h in the urine (pH about 7) of *Pteropus vampyrus* bats at a temperature of 22°C, which dropped to 1.8 h at 37°C. In urine samples from the same species, the virus half-life decreased to less than 0.5 h at both these temperatures. The patterns of urinary shedding of NiV in naturally infected hosts and its roles in transmission of infection warrant detailed investigations.

A significant association with consumption of the sap of date palm tree (*Phoenix sylvestris*), a local delicacy, was observed in several recent outbreaks of NiV disease in Bangladesh and India.[25,58–61] It is known that fruit bats frequently raid the clay pots used for the collection of the date palm sap in the affected regions and contaminate the contents with their saliva, urine, or feces. The near-neutral pH of date palm sap may maintain the viability of NiV, and transmission can occur if the contaminated product is consumed raw, within a few hours of collection. In addition to this, practices such as the repeated use of the same collection pot without cleaning and the pooling of sap from multiple sources increase the risk for contamination of sap with virus-laden bat secretions.[62]

Considerable ambiguity exists about the ability of NiV to infect cells of the gastrointestinal epithelia, which lack expression of ephrin-B2 and -B3. The types of cells in the gastrointestinal tract, which are susceptible to NiV and the mechanisms of viral entry and dissemination from them, remain unclear. It is probable that the virus might enter the tissue-resident macrophages and/or dendritic cells through microscopic lesions that commonly occur in the mouth and throat. Such opportunities for entry into macrophages may also be provided by inflamed tonsils, diseased gingival tissue, etc. The role of comorbidities, such as acid reflux disease, gastric or duodenal ulceration, malnutrition, etc., in modulating the risk of NiV infection of gastrointestinal epithelia following its ingestion is also unknown.

The virus can also cross the placental barrier to infect the developing fetus in animals like bats, guinea pigs, and cats.[5] This may contribute to the maintenance of NiV in its natural reservoir species.

NiV also shows considerable lymphoid tropism and may cause disruption in the architecture of lymphoid tissue and lymphoid necrosis. These appear to indicate that NiV infection may cause immunosuppression in the infected host.[5]

Species-specific differences are observed in the organ targeted by NiV in natural infections. Brain is invariably involved in NiV infections in several species and immunohistochemical staining reveals the viral antigen distribution in vascular endothelial cells, surrounding smooth muscles of tunica media, neurons, glial cells, ependymal cells, and the epithelial cells of the choroid plexus. Significant pathology is also seen in lungs in natural NiV infections of cats, pigs, and humans. Borisevich et al.[63] in a recent *in vitro* study reported the susceptibility of human primary olfactory epithelial cells to infection by NiV-Malaysia and NiV-Bangladesh strains. Both the strains replicated efficiently in the olfactory epithelial cells and induced progressive syncytiation and cell death. Notably, immunohistochemical staining revealed the presence of viral antigen only in olfactory sensory neurons in the olfactory epithelium. These facts indicate that human olfactory epithelium may serve as a point of entry of NiV into the CNS. NiV can enter the CNS through disruptions in the blood–brain barrier (BBB), infected microvascular endothelial cells, or a combination of both. Breaches in the BBB resulting from extensive inflammatory changes may facilitate the direct entry of the virus into the CNS and infection of cells expressing ephrin-B2/B3. In the alternate pathway, infected brain microvascular endothelial cells can release virions through the basolateral surface, which can subsequently infect other cells, even with an intact BBB. Weise et al.[64] have reported the presence of basolateral-targeting signals in the cytoplasmic domains of G and F proteins of NiV.

Primary epithelial cell monolayers derived from bronchi and small airways also exhibit high susceptibility to NiV infection. Notably, NiV-induced inflammation occurred predominantly in the small airways but not the bronchi.[65] Bronchiolar epithelial cells are infected early in human NiV infections, and virus shedding occurs in nasopharyngeal and tracheal secretions.[56] NiV can replicate to high titers in cells of the lower respiratory tract. There is a greater chance of aerosol transmission of the infection from patients with a symptomatic respiratory infection. Histopathological changes consistent with necrotizing alveolitis, hemorrhage, pulmonary edema, and aspiration pneumonia are often evident. The other findings include multinucleated giant cells in the alveolar septum and alveolar spaces, as well as intra-alveolar inflammatory cells. Infected airway epithelia may release inflammatory cytokines including IL-6, IL-8, IL-1α, MCP-1, G-CSF, and CXCL-10, and the recruitment of immune cells by these may produce acute respiratory distress syndrome (ARDS). NiV infection is also known to induce the IFN-β and IP-10.[66–68] With progressive infection, the viral replication spreads from respiratory epithelia to the endothelia of lungs and causes prominent pathology in small vessels and capillaries. Subsequent viremic spread can lead to multiorgan failure.

NiV-Malaysia and NiV-Bangladesh seem to be associated with subtle differences in virulence properties despite a nucleotide sequence similarity of about 91.8%. In comparison to NiV-M, the NiV-B shows a shorter incubation period as well as a narrower range for it. Respiratory symptoms predominated in most cases with NiV-B infections, while only a few of those with NiV-M infections experienced them. Myoclonus was reported in only a few of the Bangladeshi and Indian patients infected with NiV-B, while a significant number developed it during infection with NiV-M. Some of the fatal cases during the NiV outbreak caused by NiV-M in the Philippines presented with acute encephalitis syndrome.[8]

4.8 CLINICAL FEATURES OF HUMAN NIV INFECTION

The incubation period in most human cases is about 2 weeks or less (ranging from 4 days to 2 months).[6,69] After a febrile prodrome with a headache, dizziness, and vomiting, there was a rapid progression to severe encephalitis, with altered sensorium, signs of brainstem dysfunction (e.g., abnormal doll's eye reflex and papillary responses) vasomotor disturbances, seizures, and myoclonus), cerebellar signs, etc. The spectrum of presentations included aseptic meningitis, diffuse encephalitis, and focal brainstem involvement. Respiratory disturbances developed in 14%–20% of cases in the Malaysian outbreak and two cases presented with respiratory manifestations alone in the Singapore outbreak. Respiratory involvement was prominently observed during the outbreaks of NiV infection in Bangladesh and India, with a few cases of acute respiratory distress syndrome.

Magnetic resonance imaging often reveals multiple hyperintense lesions in the cortex, pons, putamen, and cerebral and cerebellar peduncles in NiV infections in humans. This is thought to correspond to the diffuse expression of ephrin-B2 and often correlates with the neurological symptoms seen. In contrast, the expression of ephrin-B3 is restricted to the brainstem and spinal cord. Brainstem dysfunction carries a bad prognosis in patients with Nipah encephalitis and is a frequent cause of death.[69]

A relapse of the infection and late-onset encephalitis occurred in many patients who survived an initial infection with NiV, months to years later.[6] Psychiatric disturbances, including depression, personality changes, and persistent neurocognitive deficits, were reported in a significant proportion of the cases from Malaysia and Singapore.[70–72]

4.9 ANTIVIRAL RESEARCH ON NIV

A clinically effective antiviral against NiV infections in humans is still lacking. Several approaches have been employed in developing an efficacious therapy against NiV, with varying success. Inhibition of viral entry has been explored as an antiviral strategy against NiV, since the NiV G protein exhibits a high affinity for its cognate receptor, ephrin-B2/B3. In *in vitro* experiments, soluble forms of NiV G protein showed potent inhibition of virus entry.[33,73] This may work as a form of passive immunotherapy, which blocks acute viral replication and generates protective antibody responses. Fusing the G moiety to Fc region of human immunoglobulin G (IgG) molecules can enhance the stability, half-life, and bioavailability of the formulation.

Studies have also described the utility of fusion inhibitors derived from the HR2 region of NiV F protein in inhibiting NiV replication *in vitro*.[74,75] Several modifications of these molecules with better solubility and half-lives, such as shorter, capped, or polyethylene glycol-attached forms, have also been reported.[75] NiV-mediated membrane fusion can also be inhibited by peptides derived from the HR2 domain of human parainfluenza virus-3 (HPIV-3)[76,77] although the mechanisms of action remain unknown and the in vivo efficacy remains to be evaluated. Combination therapies employing inhibitors of membrane fusion and viral entry may be more promising.[5]

There is research currently focused on the development of inhibitors of NiV entry and fusion. HR2-derived peptides, when fused with the Fc region of human IgG1, are capable of inhibiting NiV fusion as much as the free peptides[78] and can enhance their plasma half-life.

Considerable research is also conducted on the use of nonpeptidic small molecule inhibitors targeting specific regions or conformations of the fusion protein.[79,80] Molecules that dock into specific regions on the N-terminus of HR1 and impact the formation of the critical 6HB structure (which drives membrane fusion) may also be a promising approach.[80,81] Neidermieier et al.[82] reported the utility of several quinolone-based small molecule inhibitors of henipavirus fusion.

4.10 IMMUNOPROPHYLAXIS OF NIV INFECTIONS

Immunoprophylaxis of NiV using convalescent sera, monoclonal antibodies, and humanized monoclonal antibodies has shown promising results in experimental studies.

Immune sera from hamsters vaccinated against NiV-F, NiV-G, or both protected hamsters from NiV when administered intraperitoneally 1 h before or 24 h after a lethal challenge.[83] The protected animals, however, developed antibodies against other NiV proteins, indicating the occurrence of viral replication. Sterilizing immunity could also be achieved in the animals using murine monoclonal antibodies (MAb) when administered at the same time points after a lethal challenge with NiV. The protective effect, however, was concentration dependent. MAb specific to G and F proteins protected 50% of the animals when administered up to 24 or 96 h post-challenge, respectively, and delayed death from 11 to 21 days. Anti-F (but not anti-G) MAb yielded 100% protection when administered 1 h after a lethal challenge and neutralized HeV as well.[48]

Human monoclonal antibodies with cross-neutralizing abilities against henipaviruses have also been identified. Two human MAbs, m102 and m106, were reported to have cross-neutralizing effect against henipavirus-mediated cell fusion. An affinity-matured, IgG1 antibody, m102.4 Ab was reported, with inhibitory concentrations in the range of 0.04–0.6 μg ml^{-1}. This antibody, which recognizes conformational epitopes and inhibits henipavirus binding with ephrin-B2 receptors, protected ferrets when administered at a concentration of 50 mg, 10 h after an oronasal challenge with NiV.[84]

The high cost associated with the production of monoclonal antibody-based therapy may restrict its use in low-resource settings. However, this approach can be of use in emergency post-exposure prophylaxis of laboratory workers, healthcare staff, family contacts, and other susceptible individuals especially during outbreaks.[5] Avian-derived immunoglobulin Y (IgY) may be a cost-effective alternative to the monoclonal antibodies and presents several advantages, including the feasibility for large scale and continuous production, identification of novel epitope specificities, and absence of cross-reactivity with mammalian epitopes. Technology for the generation of recombinant IgY using mammalian monoclonal antibodies has also been reported, further enhancing the utility of IgY. Several studies have shown the protective efficacy of IgY in post-exposure prophylaxis against pathogens.[85] A recent study has shown the protective efficacy of goose-derived IgY against dengue virus type 2 in AG129 mice, at 24 h after a lethal challenge.[86] The utility of similar approaches needs to be explored in post-exposure prophylaxis against NiV in animals and/or humans.

4.11 PERSPECTIVES

NiV is a highly pathogenic paramyxovirus that presents a serious risk to human, veterinary, and agricultural health worldwide. A broad host range, the potential for aerosol and human-to-human transmission, and the absence of efficacious prophylaxis and treatment call for urgent attention from the global biomedical community. Despite advances over the past 20 years since its discovery, many aspects of its disease biology remain unclear. The development of several animal models of infection and advances in reverse genetics, high-throughput sequencing, in-silico modeling, and viral pseudotyping tools have yielded fresh insights into the pathogenesis of NiV in the recent years. Several promising approaches for the development of efficacious vaccines and treatment of Nipah infections have been identified and await clinical evaluation. The intimate association of sociocultural factors with the transmission of NiV in the endemic countries emphasizes the importance of creating public awareness about the disease and its prevention.

REFERENCES

1. Satterfield BA, Dawes BE, Milligan GN. Status of vaccine research and development of vaccines for Nipah virus. *Vaccine* 2016;34:2971–5.
2. Luby SP, Hossain MJ, Gurley ES, et al. Recurrent zoonotic transmission of Nipah virus into humans, Bangladesh, 2001–2007. *Emerg Infect Dis* 2009;15(8):1229–35.
3. Hayman DT, Wang LF, Barr J, et al. Antibodies to henipavirus or henipa-like viruses in domestic pigs in Ghana, West Africa. *PLoS One* 2011;6(9):e25256.
4. WHO. Annual review of diseases prioritized under the Research and Development Blueprint. Meeting Report. 2017 [Available at http://www.who.int/blueprint/what/research-development/2017-Prioritization-Long-Report.pdf?ua=1].
5. Vigant F, Lee B. Hendra and Nipah infection: pathology, models and potential therapies. *Infect Disord Drug Targets* 2011;11(3):315–36.
6. Ang BSP, Lim TCC, Wang L. Nipah virus infection. *J Clin Microbiol* 2018;56(6):e01875-17.
7. Harcourt BJ, Lowe L, Tamin A, et al. Genetic characterization of Nipah virus, Bangladesh. *Emerg Infect Dis* 2005;11(2004):1594–7.
8. Mire CE, Satterfield BA, Geisbert JB, et al. Pathogenic differences between Nipah virus Bangladesh and Malaysia strains in primates: implications for antibody therapy. *Sci Rep* 2016;6:30916.
9. Clayton BA, Middleton D, Bergfeld J, et al. Transmission routes for Nipah virus from Malaysia and Bangladesh. *Emerg Infect Dis* 2012;19(12):1983–93.
10. Clayton BA, Middleton D, Arkinstall R, Frazer L, Wang L-F, Marsh GA. The nature of exposure drives transmission of Nipah viruses from Malaysia and Bangladesh in ferrets. *PLoS Negl Trop Dis* 2016;10(6):e0004775.
11. Li Y, Wang J, Hickey AC, et al. Antibodies to Nipah or Nipah-like viruses in bats, China. *Emerg Infect Dis* 2008;14(12):1974–6.
12. Chua KB, Koh CL, Hooi PS, et al. Isolation of Nipah virus from Malaysian Island flying-foxes. *Microbes Infect* 2002; 4:145–51
13. Hsu, VP, Hossain, MJ, Parashar UD et al. Nipah virus encephalitis reemergence, 7. Bangladesh. *Emerg Infect Dis* 2004;10:2082–7.
14. Reynes JM, Counor D, Ong S, et al. Nipah virus in Lyle's flying foxes, Cambodia. *Emerg Infect Dis* 2005;11:1042–7.
15. Wacharapluesadee S, Lumlertdacha B, Boongird K, et al. Bat Nipah virus, Thailand. *Emerg Infect Dis* 2005;11:1949–51.
16. Yadav PD, Raut CG, Shete AM, et al. Detection of Nipah virus RNA in fruit bat (*Pteropus giganteus*) from India. *Am J Trop Med Hyg* 2012;87(3):576–8.
17. Heymann DL. *Henipavirus: Hendra and Nipah Viral Diseases. Control of Communicable 10. Diseases Manual.* 2008. 19th Edition. Washington, DC: American Public Health Association: 275–8.
18. Sendow I, Field HE, Adjid A, et al. Screening for Nipah virus infection in West Kalimantan province, Indonesia. *Zoonoses Public Health.* 2010;57(7–8):499–503.
19. Glennon EE, Restif O, Sbarboro SR, et al. Domesticated animals as hosts of henipaviruses and filoviruses: a systematic review. *Vet J* 2018;233:25–34.
20. Field HE, Mackenzie JS, Daszak P. Henipaviruses: emerging paramyxoviruses associated with fruit bats. *Curr Top Microbiol Immunol* 2007;315:133–59.
21. Lo Presti A, Cella E, Giovanetti M, et al. Origin and evolution of Nipah virus. *J Med Virol* 2016;88:380–8.
22. Angeletti S, Lo Presti A, Cella E, Ciccozzi M. Molecular epidemiology and phylogeny of Nipah virus infection: a mini review. *Asian Pacific J Trop Med* 2016; 9(7):630–4.
23. Chua KB, Bellini WJ, Rota PA, et al. Nipah virus: a recently emergent deadly paramyxovirus. *Science* 2000;288:1432–5.
24. ICDDR B. Outbreaks of encephalitis due to Nipah/Hendra-like viruses, Western Bangladesh. *Health Sci Bull.* 2003;1(5):1–6.
25. Luby SP, Rahman M, Hossain MJ, et al. Foodborne transmission of Nipah virus, Bangladesh. *Emerg Infect Dis* 2006;12:1888–94.
26. International Centre for Diarrheal Diseases Research Bangladesh. Person-to-person transmission of Nipah infection in Bangladesh, 2007. *Health Sci Bull.* 2007;5:1–6.

27. Gurley ES, Montgomery JM, Hossain MJ, et al. Person-to-person transmission of Nipah virus in a Bangladeshi community. *Emerg Infect Dis* 2007;13:1031–7.
28. Montgomery JM, Hossain MJ, Gurley E, et al. Risk factors for Nipah virus encephalitis in Bangladesh. *Emerg Infect Dis* 2008;14:1526–32.
29. Chadha MS, Comer JA, Lowe L, et al. Nipah virus-associated encephalitis outbreak, Siliguri, India. *Emerg Infect Dis* 2006;12:235–40.
30. Arankalle VA, Bandyopadhyay BT, Ramdasi AY, et al. Genomic characterization of Nipah Virus, West Bengal, India. *Emerg Infect Dis* 2011;17(5):907–9.
31. Ching PK, de los Reyes VC, Sucaldito MN, et al. Outbreak of henipavirus infection, Philippines, 2014. *Emerg Infect Dis* 2015;21:328.
32. Bonaparte MI, Dimitrov AS, Bossart KN, et al. Ephrin-B2 ligand is a functional receptor for Hendra virus and Nipah virus. *Proc Natl Acad Sci USA* 2005;102(30):10652–7.
33. Negrete OA, Levroney EL, Aguilar HC, et al. EphrinB2 is the entry receptor for Nipah virus, an emergent deadly paramyxovirus. *Nature* 2005;7049:401–5.
34. Negrete OA, Wolf MC, Aguilar HC, et al. Two key residues in ephrinB3 are critical for its use as an alternate receptor for Nipah virus. *PLoS Pathog* 2006;2:e7.
35. Bossart KN, Tachedjian M, McEachern JA, et al. Functional studies of host-specific ephrin-B ligands as henipavirus receptors. *Virology* 2008;2:357–71.
36. Benson MD, Romero MI, Lush ME, et al. Ephrin-b3 is a myelin-based inhibitor of neurite outgrowth. *Proc Natl Acad Sci USA* 2005;102:10694–9.
37. Liebl DJ, Morris CJ, Henkemeyer M, Parada LF. mRNA expression of ephrins and eph receptor tyrosine kinases in the neonatal and adult mouse central nervous system. *J Neurosci Res* 2003;71:7–22.
38. Su Z, Xu P, Ni F. Single phosphorylation of tyr304 in the cytoplasmic tail of ephrin b2 confers high affinity and bifunctional binding to both the sh2 domain of grb4 and the pdz domain of the pdzrgs3 protein. *Eur J Biochem* 2004;271:1725–36.
39. Cowan CA, Henkemeyer M. The sh2/sh3 adaptor grb4 transduces b-ephrin reverse signals. *Nature* 2001;413:174–9.
40. Zhao C, Irie N, Takada Y, et al. Bidirectional ephrinb2-ephb4 signaling controls bone homeostasis. *Cell Metab* 2006;4:111–21.
41. Meyer S, Hafner C, Guba M, et al. Ephrin-b2 overexpression enhances integrinmediated ecm-attachment and migration of b16 melanoma cells. *Int J Oncol* 2005;27:1197–206.
42. Wang HU, Chen ZF, Anderson DJ. Molecular distinction and angiogenic interaction between embryonic arteries and veins revealed by ephrin-b2 and its receptor eph-b4. *Cell* 1998;93:741–53.
43. Erbar S, Diederich S, Maisner A. Selective receptor expression restricts Nipah virus infection of endothelial cells. *Virol J* 2008;5:142.
44. Wong JJW, Young TA, Zhang J, et al. Monomeric ephrinB2 binding induces allosteric changes in Nipah virus G that precede its full activation. *Nat Commun* 2017;8:781.
45. Negrete OA, Chu D, Aguilar HC, Lee B. Single amino acid changes in the Nipah and Hendra virus attachment glycoproteins distinguish ephrinb2 from ephrinb3 usage. *J Virol* 2007;81:10804–14.
46. Pernet O, Pohl C, Ainouze M, Kweder H, Buckland R. Nipah virus entry can occur by macropinocytosis. *Virology* 2009;395:298–311.
47. Yuan J, Marsh G, Khetawat D, et al. Mutations in the g-h loop region of ephrinb2 can enhance Nipah virus binding and infection. *J Gen Virol* 2011;92:2142–52.
48. Guillaume V, Aslan H, Ainouze M, et al. Evidence of a potential receptor binding site on the Nipah virus g protein (niv-g): identification of globular head residues with a role in fusion promotion and their localization on a niv-g structural model. *J Virol* 2006;80:7546–54.
49. Pernet O, Pohl C, Ainouze M, Kweder H, Buckland R. Nipah virus entry can occur by macropinocytosis. *Virology* 2009;395:298–311.
50. Mathieu C, Pohl C, Szecsi J, et al. Nipah virus uses leukocytes for efficient dissemination within a host. *J Virol* 2011;85:7863–71.
51. Akiyama H, Miller C, Patel HV, et al. Virus particle release from glycosphingolipid-enriched microdomains is essential for dendritic cell-mediated capture and transfer of HIV-1 and henipavirus. *J Virol* 2014;88(16):8813–25.
52. Mathieu C, Dhondt KP, Chalons M, et al. Heparan sulfate-dependent enhancement of henipavirus infection. *mBio* 2015;6(2):e02427-14.
53. Bowden TA, Crispin M, Harvey DJ, et al. Crystal structure and carbohydrate analysis of Nipah virus attachment glycoprotein: a template for antiviral and vaccine design. *J Virol* 2008;82(23):11628–36.

54. Wong KT, Shieh WJ, Kumar S, et al. Nipah virus infection: pathology and pathogenesis of an emerging paramyxoviral zoonosis. *Am J Pathol* 2002;6:2153–67.
55. Wacharapluesadee S, Boongird K, Wanghongsa S, et al. A longitudinal study of the prevalence of Nipah virus in *Pteropus lylei* Bats in Thailand: evidence for seasonal preference in disease transmission. *Vector Borne Zoonotic Dis* 2010;10(2):183–90.
56. Chua KB, Lam SK, Goh KJ, et al. The presence of Nipah virus in respiratory secretions and urine of patients during an outbreak of Nipah virus encephalitis in Malaysia. *J Infect* 2001;42(1):40–3.
57. Fogarty R, Halpin K, Hyatt AD, Daszak P, Mungall BA. Henipavirus susceptibility to environmental variables. *Virus Res* 2008;132(1–2):140–4.
58. Rahman MA, Hossain MJ, Sultana S, et al. Date palm sap linked to Nipah virus outbreak in Bangladesh, 2008. *Vector Borne Zoonotic Dis* 2012;12(1):65–72.
59. Chakraborty A. Nipah outbreak in Lalmonirhat district, 2011. *Health Sci Bull* 2011;9:13–19.
60. Sazzad HM, Hossain MJ, Gurley E, et al. Nipah virus infection outbreak with nosocomial and corpse-to-human transmission, Bangladesh. *Emerg Infect Dis* 2013;19:210–7.
61. Islam MS, Sazzad HM, Satter SM, et al. Traditional liquor made from date palm sap, Bangladesh, 2011–2014. *Emerg Infect Dis* 2016;22(4):664–70.
62. Luby SP, Nahar N, Gurley ES. Reducing the risk of foodborne transmission of Nipah virus. In: M Jay-Russel, MP Doyle, editors. *Food Safety Risks from Wildlife. Food Microbiology and Food Safety.* Cham: Springer International Publishing. 2016;151–67.
63. Borisevich V, Ozdener MH, Malik B, Rockx B. Hendra and Nipah virus infection in cultured human olfactory epithelial cells. *mSphere* 2017;2(3):e00252-17.
64. Weise C, Erbar S, Lamp B, Vogt C, Diederich S, Maisner A. Tyrosine residues in the cytoplasmic domains affect sorting and fusion activity of the Nipah virus glycoproteins in polarized epithelial cells. *J Virol* 2010;84(15):7634–41.
65. Escaffre O, Borisevich V, Carmical R, et al. Henipavirus pathogenesis in human respiratory epithelial cells. *J Virol* 2013;87(6):3284–94.
66. Lo MK, Miller D, Aljofan M, et al. Characterization of the antiviral and inflammatory responses against Nipah virus in endothelial cells and neurons. *Virol* 2010;404(1):78–88.
67. Rockx B, Brining D, Kramer J, et al. Clinical outcome of henipavirus infection in hamsters is determined by the route and dose of infection. *J Virol* 2011;85(15):7658–71.
68. Mathieu C, Guillaume V, Sabine A, et al. Lethal Nipah virus infection induces rapid overexpression of CXCL10. *PLoS One* 2012;7:e32157.
69. Goh KJ, Tan CT, Chew NK, et al. Clinical features of Nipah virus encephalitis among pig farmers in Malaysia. *N Engl J Med* 2000;342:1229.
70. Abdulla S, Chang LY, Kartini R, Jin Goh K, Tan CT. Late-onset Nipah virus encephalitis 11 years after the initial outbreak: a case report. *Neurol Asia* 2012;17(1):71–4.
71. Ng BY, Lim CC, Yeoh A, Lee WL. Neuropsychiatric sequelae of Nipah virus encephalitis. *J Neuropsychiatry Clin Neurosci* 2004;16(4):500–4.
72. Sejvar JJ, Hossain J, Saha SK, et al. Long-term neurological and functional outcome in Nipah virus infection. *Ann Neurol* 2007;62:235–42.
73. Bossart KN, Crameri G, Dimitrov AS, et al. Receptor binding, fusion inhibition, and induction of cross-reactive neutralizing antibodies by a soluble G glycoprotein of Hendra virus. *J Virol* 2005;11:6690–702.
74. Bossart KN, Wang LF, Eaton BT, Broder CC. Functional expression and membrane fusion tropism of the envelope glycoproteins of Hendra virus. *Virology* 2001;1:121–35.
75. Bossart KN, Wang LF, Flora MN, et al. Membrane fusion tropism and heterotypic functional activities of the Nipah virus and Hendra virus envelope glycoproteins. *J Virol* 2002;22:11186–98.
76. Porotto M, Doctor L, Carta P, et al. Inhibition of Hendra virus fusion. *J Virol* 2006;19:9837–49.
77. Porotto M, Carta P, Deng Y, et al. Molecular determinants of antiviral potency of paramyxovirus entry inhibitors. *J Virol* 2007;19:10567–74.
78. Aguilar HC, Matreyek KA, Filone CM, et al. N-glycans on Nipah virus fusion protein protect against neutralization but reduce membrane fusion and viral entry. *J Virol* 2006;10:4878–89.
79. Jiang S, Zhao Q, Debnath AK. Peptide and non-peptide HIV fusion inhibitors. *Curr Pharm Des* 2002;8:563–80.
80. Cianci C, Meanwell N, Krystal M. Antiviral activity and molecular mechanism of an orally active respiratory syncytial virus fusion inhibitor. *J Antimicrob Chemother* 2005;3:289–92.
81. Weissenhorn W, Dessen A, Calder LJ, Harrison SC, Skehel JJ, Wiley DC. Structural basis for membrane fusion by enveloped viruses. *Mol Membr Biol* 1999;1:3–9.
82. Neidermeier S, Singethan K, Rohrer SG, et al. A small-molecule inhibitor of Nipah virus envelope protein-mediated membrane fusion. *J Med Chem* 2009;14:4257–65.

83. Guillaume V, Contamin H, Loth P, et al. Nipah virus: vaccination and passive protection studies in a hamster model. *J Virol* 2004;2:834–40.
84. Bossart KN, Zhu Z, Middleton D, et al. A neutralizing human monoclonal antibody protects against lethal disease in a new ferret model of acute Nipah virus infection. *PLoS Pathog* 2009;5(10): e1000642.
85. Haese N, Brocato RL, Henderson T, et al. Antiviral biologic produced in DNA vaccine/goose platform protects hamsters against Hantavirus Pulmonary Syndrome when administered post-exposure. *PLoS Negl Trop Dis* 2015;9(6):e0003803.
86. Fink AL, Williams KL, Harris E, et al. Dengue virus specific IgY provides protection following lethal dengue virus challenge and is neutralizing in the absence of inducing antibody dependent enhancement. *PLoS Negl Trop Dis* 2017;11(7):e0005721.

SUMMARY

Nipah virus (NiV) is a nonsegmented, single-stranded, negative-sense RNA virus that affects fruit bats (natural reservoir) and a broad range of other hosts (e.g., pigs, sheep, goats, cats, dogs, and horses) and has the potential for aerosol and human-to-human transmission. Despite research efforts over the past 20 years since its discovery, many aspects of NiV biology remain unclear. The development of several animal models of infection and advances in reverse genetics, high-throughput sequencing, in-silico modeling, and viral pseudotyping tools have yielded valuable insights into the pathogenesis of NiV in the recent years. In the absence of efficacious prophylaxis and treatment, several promising vaccine candidates for NiV infections have been identified and await clinical evaluation. The intimate association of sociocultural factors with the transmission of NiV in the endemic countries emphasizes the importance of creating public awareness about the disease and its prevention.

Molecular Identification of Human Bocavirus

5

Pattara Khamrin, Kattareeya Kumthip, and Niwat Maneekarn
Chiang Mai University

Contents

5.1 INTRODUCTION

Human bocavirus (HBoV) was first described in 2005 by Allander et al. [1] from nasopharyngeal aspirations of patients with respiratory tract infection using novel molecular virus screening approaches, which involved DNase treatment of the samples, random amplification and cloning, high-throughput sequencing, and bioinformation sequence analysis. Since then, this virus has been shown to cause both respiratory tract infection and gastrointestinal diseases in young children worldwide. Further development and application of specific molecular methods such as PCR and quantitative PCR assays have allowed for the detection and quantification of HBoV genome in various clinical samples (e.g., upper and lower respiratory secretions, blood, stool, urine, saliva, and tonsillar lymphocytes) [2–8], as well as environmental/food samples (e.g., urban sewage, river water, and bivalve shellfish) [9–11]. Following a brief overview on the classification, genome organization, route of transmission, clinical feature, pathogenesis, and epidemiology of HBoV, this chapter focuses on the molecular approaches that have proven valuable for the detection and identification of HBoV in clinical specimens.

5.1.1 Classification, Morphology, and Genome Organization

According to the current classification of the International Committee on Taxonomy of Viruses (ICTV), human bocavirus (HBoV) belongs to the single-stranded DNA virus genus *Bocaparvovirus*, which represents one of the eight genera (i.e., *Protoparvovirus*, *Amdoparvovirus*, *Aveparvovirus*, *Bocaparvovirus*, *Dependoparvovirus*, *Erythroparvovirus*, *Copiparvovirus*, and *Tetraparvovirus*) within the subfamily *Parvovirinae*, family *Parvoviridae* [12].

Interestingly, within the subfamily *Parvovirinae*, the genus *Erythroparvovirus* contains a well-known human-infecting parvovirus commonly referred to as B19 virus, which is transmitted by respiratory droplets and blood and causes erythema infectiosum (also called fifth disease or slapped cheek syndrome) in children, arthralgias and arthritis in adults, chronic anemia in patients with AIDS, aplastic crisis (or reticulocytopenia) in patients with hemolytic syndromes (e.g., sickle cell anemia, hereditary spherocytosis), and hydrops fetalis in fetuses; the genus *Tetraparvovirus* includes human parvoviruses 4 (PARV4) and 5 (PARV5), whose pathogenic roles remain unclear; the genus *Dependoparvovirus* consists of another human parvovirus, that is, apathogenic adeno-associated virus; and the genus *Protoparvovirus* comprises bufavirus (BuV) and tusavirus (TuV), which are detected only in patients with diarrhea [13].

Based on full genome nucleotide and amino acid sequences of HBoV prototype strains (ST1 and ST2), the viruses are most similar to the other known bocaviruses "bovine parvovirus" (BPV) and "canine minute virus or minute virus of canine" (MVC). The genomic organization of both ST1 and ST2 also closely resembles to that of the BPV and MVC. The name "bocavirus" was derived from a combination of bovine parvovirus and canine minute virus [1]. Being a small non-enveloped virus with an icosahedral symmetry of approximately 25 nm in diameter, HBoV contains a linear single-stranded DNA (ssDNA) genome of about 5.3 kb in length, which is flanked by two terminal hairpin structures [14] (Figure 5.1). The HBoV genome harbors three open-reading frames (ORFs). The ORF1 and ORF2 encode nonstructural protein 1 (NS1) and nuclear phosphoprotein 1 (NP1), respectively. The ORF3 encodes structural viral capsid proteins 1 and 2 (VP1 and VP2), which is generated as a result of alternative splicing of precursor genome. NS1 plays a role in DNA replication and DNA packaging [15]. NP1 can induce cell cycle arrest and apoptosis in HeLa cells [16]. VP1 and VP2 capsid proteins play important roles in antigenicity of the virus, bind to host cell receptors, and transport the genome into the nucleus [17]. Currently, HBoV consists of four genotypes, HBoV1, HBoV2, HBoV3, and HBoV4, of which HBoV2 is further distinguished into two subtypes 2A and 2B (Figure 5.2).

5.1.2 Transmission, Clinical Feature, and Pathogenesis

The mode and route of transmissions of HBoV are unknown. It is most likely transmitted by the respiratory route, as it causes respiratory illness in humans. Moreover, the transmission may occur through several other routes, such as the fecal–oral route, person-to-person direct contact with the infectious sputum,

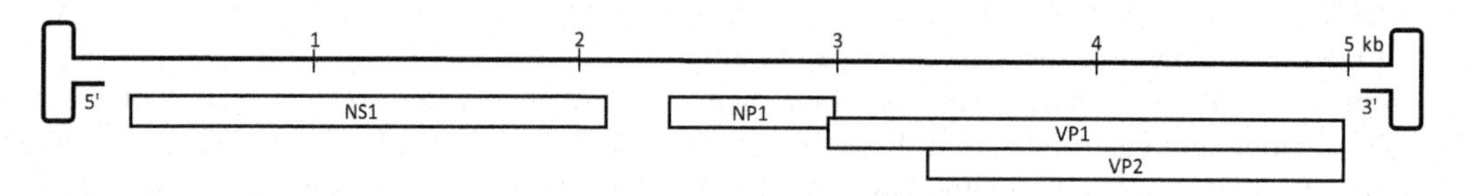

FIGURE 5.1 Genomic organization of human bocavirus. NS1 and NP1 represent nonstructural proteins 1 and nuclear phosphoprotein 1, respectively. VP1 and VP2 represent structural viral capsid proteins 1 and 2, respectively.

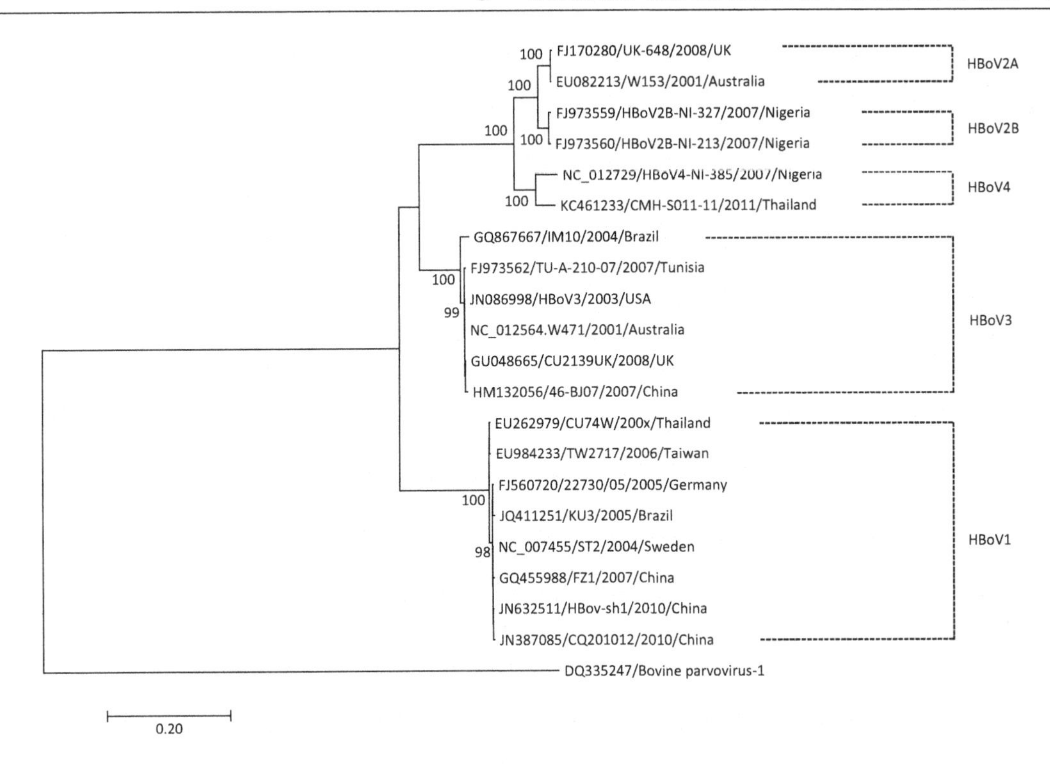

FIGURE 5.2 Phylogenetic relationships of HBoV1–4 genotypes based on nucleotide sequences of the complete genome. The scale bar indicates the number of nucleotide substitutions per site, and bootstrap values (>80) are indicated for the corresponding nodes.

saliva, and urine or by ingestion of HBoV contaminated food and water. There are several reports confirming that water is an important source for HBoV contamination and may be the route of HBoV transmission to humans [10,11]. Furthermore, bivalve shellfish has also been reported as a possible important source of HBoV infection [9]. Similar to other parvoviruses, HBoV may enter the host through inhalation or ingestion, and then the virus can escape from the entry tracts into the circulatory system and spread to other organs such as heart tissue, CSF, and plasma/serum [18,19]. Also, the studies from China, Italy, and Saudi Arabia reported the evidence of HBoV viremia by detecting viral DNAs in the plasma/serum of healthy blood donors [2,20,21].

The roles of HBoV as a causative agent of diseases, as well as biology, life cycle, and pathogenesis have not been clearly established, due to the lack of appropriate cell culture system and animal models for the cultivation of the virus. Moreover, coinfections of HBoV with other well-defined viral pathogens have frequently been reported. Several studies have identified HBoV in respiratory secretion together with influenza virus, respiratory syncytial virus, rhinovirus, parainfluenza virus, enterovirus, and human metapneumovirus [22,23]. Particularly, a high rate of coinfection with respiratory syncytial virus has been frequently reported [24]. Similar to the studies of respiratory illness, HBoV is commonly found to be coinfected with other enteric viral pathogens, such as group A rotavirus, norovirus GII, and adenovirus [3,25–27]. In contrast, several molecular epidemiological studies demonstrated that HBoV has also been detected in asymptomatic controls [28,29]. These findings raise concern about the etiologic roles of HBoV in human diseases. The question whether or not HBoV might be the cofactor for other viral agents or may require the presence of other helper viruses to establish diseases in human remain to be investigated.

HBoV has been identified in patients from both respiratory tract illness and gastroenteritis disease with a wide range of clinical presentations. Clinical findings in patients with respiratory illness include cough, wheezing, fever, asthma, bronchiolitis, and pneumonia [30]. Moreover, the most common gastrointestinal symptoms are diarrhea, vomiting, and nausea [31]. Most recently, HBoV infections have been documented in association with myocarditis, meningitis/encephalitis, and post-liver transplantation [18,19,32]. Most of these studies have been conducted in infants and young children, and the highest infection rates are found in children under 2 years of age [27,33,34]. To gain a full spectrum of clinical symptoms, HBoV infection in other groups of people and immunocompromised hosts is needed for further clarification.

5.1.3 Epidemiology, Seroprevalence, and Genetic Recombination

Since the discovery of the first HBoV in 2005, extensive epidemiological studies on HBoV infections have been carried out. HBoV has been reported from several countries worldwide including the countries in Europe [40–35] Asia [26,27,41–45], Australia [46–48], Africa [49], and America [29–50, 53]. HBoV infections mainly involve the respiratory and gastrointestinal tracts illness in children and adults. HBoV1 has been reported to involve mostly respiratory tract infection [30,39,54,55], while HBoV2, HBoV3, and HBoV4 are generally detected in stool samples and possibly associate with gastroenteritis [3,26,56–58]. HBoV infection occurs all year round with a peak in winter and spring months [40,42,50,59–62]. Seroprevalence of HBoV has been reported with variable rates, ranging from 25% to 100% depending on the study population [63–71]. In adults, the seroprevalence of IgG has been reported at 78.7%, 76%, and more than 90% in China [71], Germany [70], and Japan and Italy [63,64], respectively. In children, the seropositive rate of HBoV is age related, 25%–50% in children between 4 and 11 months, and up to 100% in children older than 2 years [63,64,67]. However, the seroprevalence is generally high (90.5%–91.8%) in infants at the age of 0–3 months and assume to be associated with maternal antibodies [63,67].

The global prevalence of HBoV infection showed variable positive rates ranged from 1.0% to 56.8% and 1.3% to 63% in respiratory secretions and stool samples, respectively [23]. The high prevalence of HBoV infection has been reported from Egypt (56.8%) [72] and Nicaragua (33.3%) [73] in respiratory-infected cases, while those of gastrointestinal-infected cases were reported from Bangladesh (63%) [74], Tunisia (33%), and Nigeria (29.2%) [75]. Recently, the presence of HBoV in food and environmental samples including shellfish, river water, wastewater, and sewages has also been reported with high prevalence between 37.5% and 79.1% [9–11,76–80]. Moreover, some patients were referred to hospital with acute gastroenteritis due to the consumption of drinking water contaminated with HBoV [81], suggesting a potential role of environmental waters in HBoV transmission.

Recombination is a well-known phenomenon for parvovirus evolution [82]. Several studies of full-genome sequences revealed that HBoV genomes are highly mosaic and consistent with the occurrence of frequent recombination events. Increasing numbers of inter- and intragenotype recombinant HBoV strains are detected. Recently, various types of HBoV intergenotype recombinations have been documented, such as between HBoV1 and HBoV2, HBoV2 and HBoV4, and HBoV3 and HBoV4 [25,83,84]. In fact, HBoV has been shown to be highly mosaic, and the potential recombination breakpoints are mostly located at the NP1/VP1 junction and also near the upstream of VP2 region [83,84]. Moreover, intragenotype recombination within HBoV1 and HBoV2 (2A/2B) genotypes has also been reported [85,86]. Further studies on recombination of other HBoV strains may help a better understanding of their genetic diversity and virus evolution.

5.2 MOLECULAR IDENTIFICATION

5.2.1 General Considerations

To date, no cell culture systems for the *in vitro* study of HBoV replication have been established. The HBoV DNA genome is frequently detected in respiratory secretion [1,29,35–38,42,43,46,47,49,51,52,87–89], serum [6,90], stool [3,26,54,55,57,58,62,91–98], and urine [62] samples. Currently, diagnosis of HBoV infection is based mainly on the detection of viral genome present in respiratory secretion, serum and stool samples using conventional PCR [1,46,48,50,51], quantitative real-time PCR [6,42,49,99], and viral metagenomics analysis [76,100]. PCR amplifications of NP1, NS1, or VP1/2 genes are used for the detection of viral genome fragments in clinical specimens, of which NP1 and NS1 are more common targets for PCR-based detection. Moreover, the Luminex RVP assay and RespiFinder assay have been developed and commercially available for the detection of different HBoV genotypes in respiratory secretion [101,102]. For PCR-based detection of HBoV, a number of different sets of primer pairs have been described. The molecular detection methods and oligonucleotide primers used for the detection of HBoV are listed in Table 5.1. The genotype of HBoV can be determined by nucleotide sequence analysis of the VP1 and/or VP2 regions. Furthermore, the restriction enzyme fragment length polymorphism (RFLP) method for the identification of HBoV genotype has also been developed by amplification of a 309-bp fragment of VP1/VP2 region and subsequent digestion of the amplicon with the *BstAPI* endonuclease. The DNA fragment of genotype I HBoV isolate is cut into two fragments of 150 and 159 bp, whereas the fragment of genotype II isolate remains uncut [103].

5.2.2 Sample Preparation

Specimens used for the identification of HBoV can be nasopharyngeal aspirate (NAP), serum, and stool samples. Viral nucleic acids were extracted from 400 μl of the NAP samples using the MagAttract virus mini M48 kit (Qiagen), while 100 μl of serum samples was processed using the QIAamp blood mini kit (Qiagen) [6]. Stool samples were prepared as 10% suspension in phosphate-buffered saline, and HBoV DNA genome was extracted with the QIAamp viral RNA mini kit (Qiagen) [105] according to the manufacturer's instructions. The extracted DNA genome can either be used immediately or kept at −70°C.

5.2.3 Conventional PCR

A single-round PCR detection of the NP1 region of HBoV has been described by Allander et al. [1]. Briefly, 5 μl of extracted DNA genome of HBoV was used as a template for the PCR amplification. The 50 μl reaction mix consisted of 1×GeneAmp PCR buffer II (100mM Tris-HCl, pH 8.3/500mM KCl), 2.5mM $MgCl_2$, each of dNTP at 0.2mM, 0.2μM each of the primers 188F and 542R (Table 5.1), and 2.5 units of AmpliTaq Gold DNA polymerase (Applied Biosystems). The following PCR amplification cycling conditions: 1 cycle at 94°C for 10min and 35 cycles at 94°C for 1min, 54°C for 1min, and 72°C for 2min were used. The PCR product of a 354-bp fragment was visualized by agarose gel electrophoresis.

5.2.4 Nested PCR

Kapoor et al. [75] had established a pan-bocavirus PCR primers targeting the VP1–VP2 region of HBoV genome. In the first-round PCR, primers AK-VP-F1 and AK-VP-R1 were used, whereas in the second-round PCR, primers AK-VP-F2 and AK-VP-R2 were used (Table 5.1). PCR reaction mix contained 2.5 units of Taq DNA polymerase (NEB) in 1×Thermopol reaction buffer with 2.0mM of $MgCl_2$, 0.2μM each of the primers, 0.2mM of dNTPs, and 2.5 μl of extracted DNA genome of HBoV or 1 μl of the first-round PCR product as a template for first-round or second-round PCR, respectively, in a 50 μl total volume. First-round PCR conditions were 10 cycles of 95°C for 35s, 58°C for 1min, and 72°C for 1min, with a decrease of 0.5°C in annealing temperature each cycle; 30 cycles of 95°C for 30s, 54°C for 45s, and 72°C for 45s; and a final extension at 72°C for 10min. Similar conditions were used for the second-round PCR, excepted that the initial annealing temperatures were 60°C and 58°C in the first and second sets of PCR cycles, respectively. Amplicons of the first- and second-round PCR were 609 and 576 bp, respectively.

TABLE 5.1 Molecular assays and primer sets used for the detection of human bocavirus

DETECTION METHOD	*TARGET REGION*	*PRIMER NAME (DIRECTION)*	*SEQUENCE (5′–3′)*	*AMPLICON SIZE (BP)*	*REFERENCES*
Conventional PCR	NS1	OS1 (+)	CCCAAGAAACGTCGTCTAAC	400	Simon et al. [87]
		OS2 (–)	GTGTTGACTGAATACAGTGT		
	NP1	188F (+)	GAGCTCTGTAAGTACTATTAC	354	Allander et al. [1]
		542R (–)	CTCTGTGTTGACTGAATACAG		
	NP1	HBoV 01.2 (+)	TATGGCCAAGGCAATCGTCCAAG	291	Arden et al. [46]
		HBoV 02.2 (–)	GCTCTCTCCTCCCAGTGACAT		
	NP1-VP1	HBoV-c1 (+)	CTTYGAAGAYCTCAGACC	690	Zhao et al. [86]
		HBoV-c2 (–)	TKGAKCCAATAATKCCAC		
	VP1	(+)	GATAACTGACGAGGAAATG	2193	Qu et al. [104]
		(–)	GAGACGGTAACACCACTA		
	VP1	VPF2 (+)	TTCAGAATGGTCACCTCTACA	648	Chieochansin et al. [89]
		VPR2 (–)	CTGTGCTTCCGTTTTGTCTTA		
	VP2	(+)	CAGTGGTACCAGACACCAGAAG	403	Catalano-Pons et al. [90]
		(–)	GCCAGTTCTTTGTTGCGTATCT		
	VP2	(+)	AGACAAAACGGAAGCACAGC	200	Catalano-Pons et al. [90]
		(–)	TCAAAGCCAGATCCAAATCC		
	VP2	BoV3885s (+)	ACAATGACCTCACAGCTGGCGT	422	Neske et al. [105]
		BoV4456a (+)	TCCAAATCCTGCAGCACCTGTG		
	VP2	BoV4287s (–)	CAGCCAGCACAGGCAGAATT	505	Neske et al. [105]
		BoV4939a (–)	TGCAGTATGTCTTCTTTCTGGACG		
	VP2	BocaGenoFW (+)	CCAAAAAAGACACTTTTACTTTGCTAACTCA	1564	Canducci et al. [106]
		BocaGenoRW (–)	TGGACGCCAGTTCTTTGTTGCGTATCTTTC		
	VP1-VP2	(+)	GGACCACAGTCATCAGAC	820	Kesebir et al. [51]
		(–)	CCACTACCATCGGGCTG		
	VP1-VP2	VP1/VP2F (+)	GCAAACCCATCACTCTCAATGC	404	Bastien et al. [50]
		VP1/VP2R (–)	GCTCTCTCCTCCCAGTAGACAT		
	VP1-VP2	VP/VP2–1017F (+)	GTGACCACCAAGTACTTAGAACTGG	842	Bastien et al. [107]
		VP/VP2–1020R (–)	GCTCTCTCCTCCCAGTGACAT		

(*Continued*)

TABLE 5.1 (*Continued*) Molecular assays and primer sets used for the detection of human bocavirus

DETECTION METHOD	*TARGET REGION*	*PRIMER NAME (DIRECTION)*	*SEQUENCE (5′–3′)*	*AMPLICON SIZE (BP)*	*REFERENCES*
Semi-nested PCR	NP1	NP-1 s1 (+)	Outer: TAACTGCTCCAGCAAGTCCTCCA	370	Smuts et al. [49]
		NP-1 as1 (–)	Outer: GGAAGCTCTGTGTTGACTGAAT		
		NP-1 s2 (+)	Inner: CTCACCTGCGAGCTCTGTAAGTA	368	
	VP1-VP2	VP s1 (+)	Outer: GCACTTCTGTATCAGATGCCTT	904	Smuts et al. [49]
		VP as1 (–)	Outer: CGTGGTATGTAGGCCGTGTAG		
		VP s2 (+)	Inner: CTTAGAACTGGTGAGAGCACTG	850	
Nested PCR	NS1	(+)	Outer: TATGGGTGTGTTAATCATTTGAAYA	216	Manning et al. [108]
		(–)	Outer: GTAGATATCGTGRTTRGTKGATAT		
		(+)	Inner: AACAAAGGATTTGTWTTYAATGAYTG	104	
		(–)	Inner: CCCAAGATACACTTTGCWKGTTCCACCC		
	NS1	Adel-OF (+)	Outer: AGGTAAAACAAATATTGCAAAGGCCATAGTC	732	Arthur et al. [109]
		Adel-OR (–)	Outer: TGGGAGTTCTCTCCGTCCGTATC		
		Adel-IF (+)	Inner: AGGGTTTGTCTTTAACGATTGCAGACAAC	518	
		Adel-IR (–)	Inner: TATACACAGAGTCGTCAGCACTATGAG		
	NP1	(+)	Outer: CCAGCAAGTCCTCCAAACTCACCTGC	399	Manning et al. [108]
		(–)	Outer: GGAGCTTCAGGATTGGAAGCTCTGTG		
		188F (+)	Inner: GAGCTCTGTAAGTACTATTAC	354	
		542R (–)	Inner: CTCTGTGTTGACTGAATACAG		
	NP1	Boca1N (+)	Outer: GAAGACACCGAGCCTGAGAC	328	Villa et al. [110]
		Boca2N (–)	Outer: GCTGATTGGGTTGTTCCTGAT		
		Boca3N (+)	Inner: AAACGTCGTCTAACTGCTCCA	259	
		Boca4N (–)	Inner: ATATGAGCCCGAGCCTCTCT		
	NP1	nestF1 (+)	Outer: AAGTACTATTACTTTCTTTAACACTTGGCA	377	Tozer et al. [99]
		nestR1 (–)	Outer: CCCACACCACCCTGGAGC		
		nestF2 (+)	Inner: GCACAGCCACGTGACGAAG	269	

(Continued)

TABLE 5.1 (*Continued*) Molecular assays and primer sets used for the detection of human bocavirus

DETECTION METHOD	*TARGET REGION*	*PRIMER NAME (DIRECTION)*	*SEQUENCE (5′–3′)*	*AMPLICON SIZE (BP)*	*REFERENCES*
		nestR2 (–)	Inner: TTTTCCCCGATGTACTCTCCC		
	VP1-VP2	AK-VP-F1 (+)	Outer: CGCCGTGGCTCCTGCTCT	609	Kapoor et al. [75]
		AK-VP-R1 (–)	Outer: TGTTCGCCATCACAAAAGATGTG		
		AK-VP-F2 (+)	Inner: GGCTCCTGCTCTAGGAAATAAAGAG	576	
		AK-VP-R2 (–)	Inner: CCTGCTGTTAGGTCGTTGTTGTATGT		
Nested PCR for HBoV2/3/4	VP1-VP2	2028 (+)	Outer: GAAATGCTTTCTGCTGYTGAAA	543	La Rosa et al. [94]
		2029 (–)	Outer: GTGGATATACCCACAYCAGAA		
		2030 (+)	Inner: GGTGGGTGCTTCCTGGTTA	382	
		2031 (–)	Inner: TCTTGRATTTCATTTTCAGACAT		
Nested PCR for HBoV1	VP1-VP2	2028 (+)	Outer: GAAATGCTTTCGCTGYTGAAAG	543	Iaconelli et al. [11]
		2045 (–)	Outer: GTGGAAATCCCCACACCAGAT		
		2046 (+)	Inner: GGTGGGTGCTGCCTGGATA	382	
		2047 (–)	Inner: TCTTGAATGTCAGTGTCAGACAT		
Real-time PCR	NS1	(+)	TGCAGACAACGCYTAGTTGTTT	88	Lu et al. [88]
		(–)	CTGTCCCGCCCAAGATACA		
		Probe	FAM-CCAGGATTGGGTGGAACCTGCAAA-BHQ		
	NS1	(+)	AGCTTTTGTTGATTCAAGGCTATAATC	84	Lin et al. [43]
		(–)	TGTTTCCCGAATTGTTTGTTCA		
		Probe	FAM-TCTAGCCGTTGGTCACGCCCTGTG-TMR		
	NS1	(+)	TAATGACTGCAGACAACGCCTAG	94	Qu et al. [104]
		(–)	TGTCCCGCCCAAGATACACT		
		Probe	FAM-TTCCACCCAATCCTGGT-MGB		
	NS1	(+)	CACTGGCAGACAACTCATCACA	78	Regamey et al. [38]
		(–)	GATATGAGCCCGAGCCTCTCT		
		Probe	AGCAGGAGCCGCAGCCCGA		

(*Continued*)

TABLE 5.1 (*Continued*) Molecular assays and primer sets used for the detection of human bocavirus

DETECTION METHOD	*TARGET REGION*	*PRIMER NAME (DIRECTION)*	*SEQUENCE (5′–3′)*	*AMPLICON SIZE (BP)*	*REFERENCES*
Real-time PCR for HBoV1	NS1	HBoV1F (+)	CTGCTGCACTTCCTGATTCAAT	69	Ligozzi et al. [111]
		HBoV1R (–)	GGAGCTTCTTCCAGAGATGTTC		
		Probe	FAM-ACTGCATCCGGTCTC-TMR		
Real-time PCR for HBoV2/3/4	NS1	HBoV2/3/4F (+)	ATGCACTTCCGCATYTCGTC	69	Ligozzi et al. [111]
		HBoV2/3/4R (–)	GGAGCTCTTYCCAGAGATGTTC		
		Probe	FAM-ACTGCATCCGGTCTC-TMR		
	NP1	BocaRT1 (+)	CGAAGATGAGCTCAGGGAAT	163	Foulongne et al. [35]
		BocaRT2 (–)	GCTGATTGGGTGTTCCTGAT		
		Probe	FAM-CACAGGAGCAGGAGCCGCAG-TMR		
	NP1	(+)	AGAGGCTCGGGCTCATATCA	81	Lu et al. [88]
		(–)	CACTTGGTCTTGAGGTCTTCGAA		
		Probe	FAM-AGGAACACCCAATCARCCACCTATCGTCT-BHQ		
	NP1	Boca-forward (+)	GGAAGAGACACTGGCAGACAA	100	Allander et al. [6]
		Boca-reverse (–)	GGGTGTTCCTGATGATATGAGC		
		Probe	FAM-CTGCGGCTCCTGCTCCTGTGAT-TMR		
	NP1	188F (+)	GAGCTCTGTAAGTACTATTAC	354	Kleines et al. [112]
		542R (–)	CTCTGTGTTGACTGAATACAG		
		Probe	GGAAGAGACATGGCAGACAAC		
	NP1	BoV2391s (+)	GCACAGCCACGTGACGAA	76	Neske et al. [105]
		BoV2466a (–)	TGGACTCCCTTTTCTTTTGTAGGA		
		BoV2411s (probe)	FAM-TGAGCTCAGGGAATATGAAAGACAAGCATCG-TMR		
	NP1	(+)	CCACGTGACGAAGATGAGCTC	196	Christensen et al. [113]
		(–)	TAGGTGGCTGATTGGGTGTTC		
		Probe	FAM-CCGAGCCTCTCTCCCCACTGTGTCG-TMR		

(*Continued*)

TABLE 5.1 (*Continued*) Molecular assays and primer sets used for the detection of human bocavirus

DETECTION METHOD	*TARGET REGION*	*PRIMER NAME (DIRECTION)*	*SEQUENCE (5′–3′)*	*AMPLICON SIZE (BP)*	*REFERENCES*
	NP1	HBoV-UP (+)	AGGAGCAGGAGCCGCAGCC	97	Schenk et al. [114]
		HBoV-DP (–)	CAGTGCAAGACGATAGGTGGC		
		HBoV-P (probe)	FAM-ATGAGCCCGAGCCTCT CTCCCCACTGTGTC-TMR		
	NP1	(+)	GGAAGAGACACTGGCAGACAA	157	von Linstow et al. [40]
		(–)	GTCTTCATCACTTGGTCTGAGGTCT		
		Probe	FAM-CTCATATCATCAGGAACAC-MGB		
	NP1	STBoNP-1f (+)	AGCATCGCTCCTACAAAAGAAAAG	201	Tozer et al. [99]
		STBoNP-1r (–)	TCTTCATCACTTGGTCTGAGGTCT		
		STBoNP-1pr (probe)	FAM-AGGCTCGGGCTCATATCATCAGGAACA-BHQ1		
	NP1	HBoV 2693f (+)	GTCAACACAGAGCTTCCAATCC	89	Kaida et al. [115]
		HBoV 2781r (–)	TGAATTAGTACCATCTCTAGCAATGC		
		HBoV 2724-TP (probe)	FAM-AGTGCC AGTAGAACCCACACCACCCT-BHQ1		
	VP1	STBoVP-1f (+)	GGCAGAATTCAGCCATACTCAAA	122	Tozer et al. [99]
		STBoVP-1r (–)	TCTGGGTTAGTGCAAACCATGA		
		STBoVP-1pr (probe)	JOE-AGAGTAGGACCACAGTCATCAGACACTGCTCC-BHQ1		
Multiplex-real time PCR	5′UTR-NS1	HBoV1F (+)	CCTATATAAGCTGCTGCACTTCCTG	107	Kantola et al. [116]
		HBoV1R (–)	AAGCCATAGTAGACTCACCACAAG		
		HBoV234F (+)	GCACTTCCGCATYTCGTCAG		
		HBoV3R (–)	GTGGATTGAAAGCCATAATTTGA	102	
		HBoV24R (–)	AGCAGAAAAGGCCATAGTGTCA	101	
		Probe	FAM-CCAGAGATGTTCACTCGCCG-BHQ1		

(+): Forward primer; (–): Reverse primer; FAM: 6-carboxyfluorescein; TMR: 6-carboxytetramethylrhodamine; MGB: minor groove binder; BHQ: black hole quencher; JOE: carboxy 4′5 dichloro-2′, 7′-dimethoxyfluorecein.

5.2.5 Real-Time PCR

Neske et al. [105] had developed the quantitative real-time PCR for the detection of HBoV DNA genome. Primers and probe for the real-time PCR targeted the NP1 gene. The real-time PCR was carried out in a final volume of 20 μl, which consisted of 5 μl of extracted DNA, primers and probe at a final concentration of 200 nM each, and 1×Quantitect probe master mix (Qiagen). Amplification was performed on an ABI7500 real-time PCR system (Applied Biosystems). After a preheating step for 15 min at 95°C, the cycling conditions were 50 cycles of 30 s at 95°C and 60 s at 60°C.

5.3 FURTHER PERSPECTIVES

Human bocavirus is a parvovirus that was discovered only about a decade ago. Based on the current data, several aspects of HBoV remain to be fully elucidated. Most of these problems stem from a lack of an *in vitro* culture system and animal models for the study of this virus. Further research to develop novel permissive cell lines and animal models for the cultivation of HBoV is needed. Since HBoV is often detected in the clinical specimens in the presence of other well-defined pathogens both in symptomatic and asymptomatic cases, the pathogenic roles of HBoV infections remain unresolved and awaiting for clarification. Most of the studies reported HBoV infections of respiratory and gastrointestinal tracts in infants and young children and HBoV infections in adults, elderly people, and immunocompromised hosts are limited and needed to be extensively explored. The information of HBoV infections in these groups of people will provide a better understanding of epidemiology and probably pathogenic role of HBoV infection. From the clinical standpoint, the detections of HBoV in different clinical specimens including the respiratory secretions, stool, urine, saliva, whole blood, blood leukocytes, plasma, serum, etc., remain an important diagnostic goal of HBoV infection. All of these aspects remain further investigation.

REFERENCES

1. Allander T, Tammi MT, Eriksson M, et al. (2005) Cloning of a human parvovirus by molecular screening of respiratory tract samples. *Proc Natl Acad Sci USA* 102: 12891–6.
2. Abdel-Moneim AS, Mahfouz ME, Zytouni DM (2018) Detection of human bocavirus in Saudi healthy blood donors. *PLoS One 13*: e0193594.
3. Khamrin P, Thongprachum A, Shimizu H, et al. (2012) Detection of human bocavirus 1 and 2 from children with acute gastroenteritis in Japan. *J Med Virol* 84: 901–5.
4. Salvo M, Lizasoain A, Castells M, et al. (2018) Human bocavirus: detection, quantification and molecular characterization in sewage and surface waters in Uruguay. *Food Environ Virol* 10: 193–200.
5. Wang K, Wang W, Yan H, et al. (2010) Correlation between bocavirus infection and humoral response, and co-infection with other respiratory viruses in children with acute respiratory infection. *J Clin Virol* 47: 148–55.
6. Allander T, Jartti T, Gupta S, et al. (2007) Human bocavirus and acute wheezing in children. *Clin Infect Dis* 44: 904–10.
7. Martin ET, Taylor J, Kuypers J, et al. (2009) Detection of bocavirus in saliva of children with and without respiratory illness. *J Clin Microbiol* 47: 4131–2.
8. Lu X, Gooding LR, Erdman DD (2008) Human bocavirus in tonsillar lymphocytes. *Emerg Infect Dis* 14: 1332–4.
9. La Rosa G, Purpari G, Guercio A, et al. (2018) Detection of human bocavirus species 2 and 3 in bivalve shellfish in Italy. *Appl Environ Microbiol* 84: pii:e02754-17.
10. Hamza IA, Jurzik L, Wilhelm M, Uberla K (2009) Detection and quantification of human bocavirus in river water. *J Gen Virol* 90: 2634–7.

11. Iaconelli M, Divizia M, Della Libera S, Di Bonito P, La Rosa G (2016) Frequent detection and genetic diversity of human bocavirus in urban sewage samples. *Food Environ Virol* 8: 289–295.
12. Cotmore SF, Agbandje-McKenna M, Chiorini JA, et al. (2014) The family *Parvoviridae. Arch Virol* 159: 1239–47.
13. Phan TG, Vo NP, Bonkoungou IJ, et al. (2012) Acute diarrhea in West African children: diverse enteric viruses and a novel parvovirus genus. *J Virol* 86: 11024–30.
14. Schildgen O, Qiu J, Soderlund-Venermo M (2012) Genomic features of the human bocaviruses. *Future Virol* 7: 31–9.
15. Tewary SK, Zhao H, Shen W, Qiu J, Tang L (2013) Structure of the NS1 protein N-terminal origin recognition/nickase domain from the emerging human bocavirus. *J Virol* 87: 11487–93.
16. Sun B, Cai Y, Li Y, et al. (2013) The nonstructural protein NP1 of human bocavirus 1 induces cell cycle arrest and apoptosis in Hela cells. *Virology* 440: 75–83.
17. Gurda BL, Parent KN, Bladek H, et al. (2010) Human bocavirus capsid structure: insights into the structural repertoire of the *Parvoviridae. J Virol* 84: 5880–9.
18. Yu JM, Chen QQ, Hao YX, et al. (2013) Identification of human bocaviruses in the cerebrospinal fluid of children hospitalized with encephalitis in China. *J Clin Virol* 57: 374–7.
19. Brebion A, Vanlieferinghen P, Dechelotte P, et al. (2014) Fatal subacute myocarditis associated with human bocavirus 2 in a 13-month-old child. *J Clin Microbiol* 52: 1006–8.
20. Bonvicini F, Manaresi E, Gentilomi GA, et al. (2011) Evidence of human bocavirus viremia in healthy blood donors. *Diagn Microbiol Infect Dis* 71: 460–2.
21. Li H, He M, Zeng P, et al. (2015) The genomic and seroprevalence of human bocavirus in healthy Chinese plasma donors and plasma derivatives. *Transfusion* 55: 154–63.
22. Chow BD, Esper FP (2009) The human bocaviruses: a review and discussion of their role in infection. *Clin Lab Med* 29: 695–713.
23. Guido M, Tumolo MR, Verri T, et al. (2016) Human bocavirus: current knowledge and future challenges. *World J Gastroenterol* 22: 8684–97.
24. Ghietto LM, Majul D, Ferreyra Soaje P, et al. (2015) Comorbidity and high viral load linked to clinical presentation of respiratory human bocavirus infection. *Arch Virol* 160: 117–27.
25. Cashman O, O'Shea H (2012) Detection of human bocaviruses 1, 2 and 3 in Irish children presenting with gastroenteritis. *Arch Virol* 157: 1767–73.
26. Khamrin P, Malasao R, Chaimongkol N, et al. (2012) Circulating of human bocavirus 1, 2, 3, and 4 in pediatric patients with acute gastroenteritis in Thailand. *Infect Genet Evol* 12: 565–9.
27. Lasure N, Gopalkrishna V (2017) Molecular epidemiology and clinical severity of Human Bocavirus (HBoV) 1–4 in children with acute gastroenteritis from Pune, Western India. *J Med Virol* 89: 17–23.
28. Kim S (2014) Prevalence of human bocavirus 1 among people without gastroenteritis symptoms in South Korea between 2008 and 2010. *Arch Virol* 159: 2741–4.
29. Sousa TT, Almeida TNV, Fiaccadori FS, et al. (2017) Identification of Human bocavirus type 4 in a child asymptomatic for respiratory tract infection and acute gastroenteritis – Goiania, Goias, Brazil. *Braz J Infect Dis* 21: 472–6.
30. Broccolo F, Falcone V, Esposito S, Toniolo A (2015) Human bocaviruses: possible etiologic role in respiratory infection. *J Clin Virol* 72: 75–81.
31. De R, Liu L, Qian Y, et al. (2017) Risk of acute gastroenteritis associated with human bocavirus infection in children: a systematic review and meta-analysis. *PLoS One* 12: e0184833.
32. Tan MY, Tan LN, Aw MM, Quak SH, Karthik SV (2017) Bocavirus infection following paediatric liver transplantation. *Pediatr Transplant* 21 (1).
33. Silva PE, Figueiredo CA, Luchs A, et al. (2018) Human bocavirus in hospitalized children under 5 years with acute respiratory infection, Sao Paulo, Brazil, 2010. *Arch Virol* 163: 1325–30.
34. Wang Y, Li Y, Liu J, et al. (2016) Genetic characterization of human bocavirus among children with severe acute respiratory infection in China. *J Infect* 73: 155–63.
35. Foulongne V, Olejnik Y, Perez V, et al. (2006) Human bocavirus in French children. *Emerg Infect Dis* 12: 1251–3.
36. Maggi F, Andreoli E, Pifferi M, et al. (2007) Human bocavirus in Italian patients with respiratory diseases. *J Clin Virol* 38: 321–5.
37. Monteny M, Niesters HG, Moll HA, Berger MY (2007) Human bocavirus in febrile children, The Netherlands. *Emerg Infect Dis* 13: 180–2.
38. Regamey N, Frey U, Deffernez C, Latzin P, Kaiser L (2007) Isolation of human bocavirus from Swiss infants with respiratory infections. *Pediatr Infect Dis J* 26: 177–9.

39. Vicente D, Cilla G, Montes M, Perez-Yarza EG, Perez-Trallero E (2007) Human bocavirus, a respiratory and enteric virus. *Emerg Infect Dis* 13: 636–7.
40. von Linstow ML, Hogh M, Hogh B (2008) Clinical and epidemiologic characteristics of human bocavirus in Danish infants: results from a prospective birth cohort study. *Pediatr Infect Dis J* 27: 897–902.
41. Alam MM, Khurshid A, Shaukat S, et al. (2015) Human bocavirus in Pakistani children with gastroenteritis. *J Med Virol* 87: 656–63.
42. Choi EH, Lee HJ, Kim SJ, et al. (2006) The association of newly identified respiratory viruses with lower respiratory tract infections in Korean children, 2000–2005. *Clin Infect Dis* 43: 585–92.
43. Lin F, Zeng A, Yang N, et al. (2007) Quantification of human bocavirus in lower respiratory tract infections in China. *Infect Agent Cancer* 2: 3.
44. Ma X, Endo R, Ishiguro N, et al. (2006) Detection of human bocavirus in Japanese children with lower respiratory tract infections. *J Clin Microbiol* 44: 1132–4.
45. Margaret IP, Nelson EA, Cheuk ES, et al. (2008) Pediatric hospitalization of acute respiratory tract infections with human bocavirus in Hong Kong. *J Clin Virol* 42: 72–4.
46. Arden KE, McErlean P, Nissen MD, Sloots TP, Mackay IM (2006) Frequent detection of human rhinoviruses, paramyxoviruses, coronaviruses, and bocavirus during acute respiratory tract infections. *J Med Virol* 78: 1232–40.
47. Redshaw N, Wood C, Rich F, Grimwood K, Kirman JR (2007) Human bocavirus in infants, New Zealand. *Emerg Infect Dis* 13: 1797–9.
48. Sloots TP, McErlean P, Speicher DJ, et al. (2006) Evidence of human coronavirus HKU1 and human bocavirus in Australian children. *J Clin Virol* 35: 99–102.
49. Smuts H, Hardie D (2006) Human bocavirus in hospitalized children, South Africa. *Emerg Infect Dis* 12: 1457–8.
50. Bastien N, Brandt K, Dust K, Ward D, Li Y (2006) Human bocavirus infection, Canada. *Emerg Infect Dis* 12: 848–50.
51. Kesebir D, Vazquez M, Weibel C, et al. (2006) Human bocavirus infection in young children in the United States: molecular epidemiological profile and clinical characteristics of a newly emerging respiratory virus. *J Infect Dis* 194: 1276–82.
52. Arnold JC, Singh KK, Spector SA, Sawyer MH (2006) Human bocavirus: prevalence and clinical spectrum at a children's hospital. *Clin Infect Dis* 43: 283–8.
53. Gagliardi TB, Iwamoto MA, Paula FE, et al. (2009) Human bocavirus respiratory infections in children. *Epidemiol Infect* 137: 1032–6.
54. Lau SK, Yip CC, Que TL, et al. (2007) Clinical and molecular epidemiology of human bocavirus in respiratory and fecal samples from children in Hong Kong. *J Infect Dis* 196: 986–93.
55. Lee JI, Chung JY, Han TH, Song MO, Hwang ES (2007) Detection of human bocavirus in children hospitalized because of acute gastroenteritis. *J Infect Dis* 196: 994–7.
56. Santos N, Peret TC, Humphrey CD, et al. (2010) Human bocavirus species 2 and 3 in Brazil. *J Clin Virol* 48: 127–30.
57. Paloniemi M, Lappalainen S, Salminen M, et al. (2014) Human bocaviruses are commonly found in stools of hospitalized children without causal association to acute gastroenteritis. *Eur J Pediatr* 173: 1051–7.
58. Han TH, Kim CH, Park SH, et al. (2009) Detection of human bocavirus-2 in children with acute gastroenteritis in South Korea. *Arch Virol* 154: 1923–7.
59. Chow BD, Huang YT, Esper FP (2008) Evidence of human bocavirus circulating in children and adults, Cleveland, Ohio. *J Clin Virol* 43: 302–6.
60. Brieu N, Guyon G, Rodiere M, Segondy M, Foulongne V (2008) Human bocavirus infection in children with respiratory tract disease. *Pediatr Infect Dis J* 27: 969–73.
61. Hengst M, Hausler M, Honnef D, et al. (2008) Human bocavirus-infection (HBoV): an important cause of severe viral obstructive bronchitis in children. *Klin Padiatr* 220: 296–301.
62. Pozo F, Garcia-Garcia ML, Calvo C, et al. (2007) High incidence of human bocavirus infection in children in Spain. *J Clin Virol* 40: 224–8.
63. Endo R, Ishiguro N, Kikuta H, et al. (2007) Seroepidemiology of human bocavirus in Hokkaido prefecture, Japan. *J Clin Microbiol* 45: 3218–23.
64. Guido M, Zizza A, Bredl S, et al. (2012) Seroepidemiology of human bocavirus in Apulia, Italy. *Clin Microbiol Infect* 18: E74–6.
65. Hao Y, Gao J, Zhang X, et al. (2015) Seroepidemiology of human bocaviruses 1 and 2 in China. *PLoS One* 10: e0122751.
66. Hustedt JW, Christie C, Hustedt MM, Esposito D, Vazquez M (2012) Seroepidemiology of human bocavirus infection in Jamaica. *PLoS One* 7: e38206.

67. Kahn JS, Kesebir D, Cotmore SF, et al. (2008) Seroepidemiology of human bocavirus defined using recombinant virus-like particles. *J Infect Dis* 198: 41–50.
68. Kantola K, Hedman L, Arthur J, et al. (2011) Seroepidemiology of human bocaviruses 1–4. *J Infect Dis* 204: 1403–12.
69. Meriluoto M, Hedman L, Tanner L, et al. (2012) Association of human bocavirus 1 infection with respiratory disease in childhood follow-up study, Finland. *Emerg Infect Dis* 18: 264–71.
70. Neske F, Prifert C, Scheiner B, et al. (2010) High prevalence of antibodies against polyomavirus WU, polyomavirus KI, and human bocavirus in German blood donors. *BMC Infect Dis* 10: 215.
71. Zhao LQ, Qian Y, Zhu RN, et al. (2009) Human bocavirus infections are common in Beijing population indicated by sero-antibody prevalence analysis. *Chin Med J (Engl)* 122: 1289–92.
72. Abdel-Moneim AS, Kamel MM, Hamed DH, et al. (2016) A novel primer set for improved direct gene sequencing of human bocavirus genotype-1 from clinical samples. *J Virol Methods* 228: 108–13.
73. Salmon-Mulanovich G, Sovero M, Laguna-Torres VA, et al. (2011) Frequency of human bocavirus (HBoV) infection among children with febrile respiratory symptoms in Argentina, Nicaragua and Peru. *Influenza Other Respir Viruses* 5: 1–5.
74. Mitui MT, Bozdayi G, Ahmed S, et al. (2014) Detection and molecular characterization of diarrhea causing viruses in single and mixed infections in children: a comparative study between Bangladesh and Turkey. *J Med Virol* 86: 1159–68.
75. Kapoor A, Simmonds P, Slikas E, et al. (2010) Human bocaviruses are highly diverse, dispersed, recombination prone, and prevalent in enteric infections. *J Infect Dis* 201: 1633–43.
76. Bibby K, Peccia J (2013) Identification of viral pathogen diversity in sewage sludge by metagenome analysis. *Environ Sci Technol* 47: 1945–51.
77. Blinkova O, Rosario K, Li L, et al. (2009) Frequent detection of highly diverse variants of cardiovirus, cosavirus, bocavirus, and circovirus in sewage samples collected in the United States. *J Clin Microbiol* 47: 3507–13.
78. Hamza H, Leifels M, Wilhelm M, Hamza IA (2017) Relative abundance of human bocaviruses in urban sewage in Greater Cairo, Egypt. *Food Environ Virol* 9: 304–13.
79. La Rosa G, Sanseverino I, Della Libera S, et al. (2017) The impact of anthropogenic pressure on the virological quality of water from the Tiber River, Italy. *Lett Appl Microbiol* 65: 298–305.
80. Myrmel M, Lange H, Rimstad E (2015) A 1-year quantitative survey of noro-, adeno-, human boca-, and hepatitis E viruses in raw and secondarily treated sewage from two plants in Norway. *Food Environ Virol* 7: 213–23.
81. Rasanen S, Lappalainen S, Kaikkonen S, et al. (2010) Mixed viral infections causing acute gastroenteritis in children in a waterborne outbreak. *Epidemiol Infect* 138: 1227–34.
82. Lau SK, Woo PC, Yip CC, et al. (2011) Co-existence of multiple strains of two novel porcine bocaviruses in the same pig, a previously undescribed phenomenon in members of the family *Parvoviridae*, and evidence for inter- and intra-host genetic diversity and recombination. *J Gen Virol* 92: 2047–59.
83. Khamrin P, Okitsu S, Ushijima H, Maneekarn N (2013) Complete genome sequence analysis of novel human bocavirus reveals genetic recombination between human bocavirus 2 and human bocavirus 4. *Infect Genet Evol* 17: 132–6.
84. Tyumentsev AI, Tikunova NV, Tikunov AY, Babkin IV (2014) Recombination in the evolution of human bocavirus. *Infect Genet Evol* 28: 11–4.
85. Abdel-Moneim AS, Kamel MM, Hassan NM (2017) Evolutionary and genetic analysis of human bocavirus genotype-1 strains reveals an evidence of intragenomic recombination. *J Med Microbiol* 66: 245–54.
86. Zhao M, Zhu R, Qian Y, et al. (2014) Prevalence analysis of different human bocavirus genotypes in pediatric patients revealed intra-genotype recombination. *Infect Genet Evol* 27: 382–8.
87. Simon A, Groneck P, Kupfer B, et al. (2007) Detection of bocavirus DNA in nasopharyngeal aspirates of a child with bronchiolitis. *J Infect* 54: e125–7.
88. Lu X, Chittaganpitch M, Olsen SJ, et al. (2006) Real-time PCR assays for detection of bocavirus in human specimens. *J Clin Microbiol* 44: 3231–5.
89. Chieochansin T, Samransamruajkit R, Chutinimitkul S, et al. (2008) Human bocavirus (HBoV) in Thailand: clinical manifestations in a hospitalized pediatric patient and molecular virus characterization. *J Infect* 56: 137–42.
90. Catalano-Pons C, Giraud C, Rozenberg F, et al. (2007) Detection of human bocavirus in children with Kawasaki disease. *Clin Microbiol Infect* 13: 1220–2.
91. Tymentsev A, Tikunov A, Zhirakovskaia E, et al. (2016) Human bocavirus in hospitalized children with acute gastroenteritis in Russia from 2010 to 2012. *Infect Genet Evol* 37: 143–9.

92. Pham NT, Trinh QD, Chan-It W, et al. (2011) Human bocavirus infection in children with acute gastroenteritis in Japan and Thailand. *J Med Virol* 83: 286–90.
93. Campos GS, Silva Sampaio ML, Menezes AD, et al. (2016) Human bocavirus in acute gastroenteritis in children in Brazil. *J Med Virol* 88: 166–70.
94. La Rosa G, Della Libera S, Iaconelli M, et al. (2016) Human bocavirus in children with acute gastroenteritis in Albania. *Biomed Res Int* 88: 906–10.
95. Lee EJ, Kim HS (2016) Human bocavirus in Korean children with gastroenteritis and respiratory tract infections. *Biomed Res Int* 2016: 7507895.
96. Proenca-Modena JL, Martinez M, Amarilla AA, et al. (2013) Viral load of human bocavirus-1 in stools from children with viral diarrhoea in Paraguay. *Epidemiol Infect* 141: 2576–80.
97. Wang X, Zhang X, Tian H, et al. (2012) Complete genomes of three human bocavirus strains from children with gastroenteritis and respiratory tract illnesses in Jiangsu, China. *J Virol* 86: 13826–7.
98. Zhou T, Chen Y, Chen J, et al. (2017) Prevalence and clinical profile of human bocavirus in children with acute gastroenteritis in Chengdu, West China, 2012–2013. *J Med Virol* 89: 1743–8.
99. Tozer SJ, Lambert SB, Whiley DM, et al. (2009) Detection of human bocavirus in respiratory, fecal, and blood samples by real-time PCR. *J Med Virol* 81: 488–93.
100. Prachayangprecha S, Schapendonk CM, Koopmans MP, et al. (2014) Exploring the potential of next-generation sequencing in detection of respiratory viruses. *J Clin Microbiol* 52: 3722–30.
101. Babady NE, Mead P, Stiles J, et al. (2012) Comparison of the Luminex xTAG RVP fast assay and the Idaho Technology FilmArray RP assay for detection of respiratory viruses in pediatric patients at a cancer hospital. *J Clin Microbiol* 50: 2282–8.
102. Balada-Llasat JM, LaRue H, Kelly C, Rigali L, Pancholi P (2011) Evaluation of commercial ResPlex II v2.0, MultiCode-PLx, and xTAG respiratory viral panels for the diagnosis of respiratory viral infections in adults. *J Clin Virol* 50: 42–5.
103. Ditt V, Viazov S, Tillmann R, Schildgen V, Schildgen O (2008) Genotyping of human bocavirus using a restriction length polymorphism. *Virus Genes* 36: 67–9.
104. Qu XW, Duan ZJ, Qi ZY, et al. (2007) Human bocavirus infection, People's Republic of China. *Emerg Infect Dis* 13: 165–8.
105. Neske F, Blessing K, Tollmann F, et al. (2007) Real-time PCR for diagnosis of human bocavirus infections and phylogenetic analysis. *J Clin Microbiol* 45: 2116–22.
106. Canducci F, Debiaggi M, Sampaolo M, et al. (2008) Two-year prospective study of single infections and co-infections by respiratory syncytial virus and viruses identified recently in infants with acute respiratory disease. *J Med Virol* 80: 716–23.
107. Bastien N, Chui N, Robinson JL, et al. (2007) Detection of human bocavirus in Canadian children in a 1-year study. *J Clin Microbiol* 45: 610–3.
108. Manning A, Russell V, Eastick K, et al. (2006) Epidemiological profile and clinical associations of human bocavirus and other human parvoviruses. *J Infect Dis* 194: 1283–90.
109. Arthur JL, Higgins GD, Davidson GP, Givney RC, Ratcliff RM (2009) A novel bocavirus associated with acute gastroenteritis in Australian children. *PLoS Pathog* 5: e1000391.
110. Villa L, Melon S, Suarez S, et al. (2008) Detection of human bocavirus in Asturias, Northern Spain. *Eur J Clin Microbiol Infect Dis* 27: 237–9.
111. Ligozzi M, Diani E, Lissandrini F, Mainardi R, Gibellini D (2017) Assessment of NS1 gene-specific real time quantitative TaqMan PCR for the detection of human bocavirus in respiratory samples. *Mol Cell Probes* 34: 53–5.
112. Kleines M, Scheithauer S, Rackowitz A, Ritter K, Hausler M (2007) High prevalence of human bocavirus detected in young children with severe acute lower respiratory tract disease by use of a standard PCR protocol and a novel real-time PCR protocol. *J Clin Microbiol* 45: 1032–4.
113. Christensen A, Nordbo SA, Krokstad S, Rognlien AG, Dollner H (2008) Human bocavirus commonly involved in multiple viral airway infections. *J Clin Virol* 41: 34–7.
114. Schenk T, Huck B, Forster J, et al. (2007) Human bocavirus DNA detected by quantitative real-time PCR in two children hospitalized for lower respiratory tract infection. *Eur J Clin Microbiol Infect Dis* 26: 147–9.
115. Kaida A, Kubo H, Takakura K, Iritani N (2010) Detection and quantitative analysis of human bocavirus associated with respiratory tract infection in Osaka City, Japan. *Microbiol Immunol* 54: 276–81.
116. Kantola K, Sadeghi M, Antikainen J, et al. (2010) Real-time quantitative PCR detection of four human bocaviruses. *J Clin Microbiol* 48: 4044–50.

SUMMARY

Human bocavirus (HBoV) was first detected in 2005 in nasopharyngeal aspirations of patients with respiratory tract symptoms. Since then, this single-stranded DNA virus has also been associated with gastrointestinal tract infection, mostly in infants and young children. However, the pathogenic role of HBoV in the gastrointestinal tract infection remains unclear. In the absence of *in vitro* culture system and animal models for the study of HBoV, the diagnosis of HBoV infection is based mainly on the detection of viral DNA in the clinical specimens using several molecular techniques. Overall, many aspects of HBoV infection are awaiting to be explored.

Molecular Epidemiology of Human Polyomaviruses

6

Dongyou Liu
Royal College of Pathologists of Australasia Quality Assurance Programs

Contents

6.1 INTRODUCTION

Polyomaviruses (in Greek, poly means many, multiple; oma means tumors) are a group of nonenveloped, double-stranded (ds) DNA viruses with a ~5 kb genome that commonly occur in animal and human populations and show notable differences in tissue tropism. Although these viruses seem harmless in immunocompetent hosts, they have the potential to induce tumor formation and cause other diseases in immunocompromised hosts (including neonates).

The first polyomavirus (i.e., murine polyomavirus or MPyV) was discovered in the 1950s as the transmissible agent for causing multiple parotid gland tumors in neonatal or immunosuppressed rodents. Following the identification of simian virus 40 (SV40) in 1960 from African green monkey kidney cells used for the production of polio- and adenovirus vaccine, the role of polyomaviruses in human diseases was confirmed with the isolation of polyomaviruses JC (JCPyV) and BK (BKPyV) in 1971 from patients displaying progressive multifocal leukoencephalopathy (PML) and polyomavirus-associated nephropathy, respectively [1,2]. Subsequent application of novel molecular techniques (e.g., rolling circle amplification and digital transcriptome subtraction that permit detection of rare organisms present at low levels) has since expanded the number of human polyomaviruses (HPyVs) to 13 [3–13].

Taxonomy. The family *Polyomaviridae* covers a large group of small, nonenveloped, dsDNA viruses that are organized on the basis of conserved amino acid blocks of the large T antigen (LTAg)-coding sequences into four genera: *Alphapolyomavirus, Betapolyomavirus, Gammapolyomavirus*, and *Deltapolyomavirus*. Of 102 polyomavirus species (whose LTAg coding sequences differ by >15%) identified to date, 13 are human pathogenic, including five in the genus *Alphapolyomavirus* (human polyomavirus 5, human polyomavirus 8, human polyomavirus 9, human polyomavirus 13, and human polyomavirus 14), four in the genus *Betapolyomavirus* (human polyomavirus 1, human polyomavirus 2, human polyomavirus 3, and human polyomavirus 4), and four in the genus *Deltapolyomavirus* (human polyomavirus 6, human polyomavirus 7, human polyomavirus 10, and human polyomavirus 11) (Table 6.1) [14]. Interestingly, Mexican polyomavirus was once considered a distinct virus but later found to be a variant of human polyomavirus 10 (MW polyomavirus). Furthermore, another human infecting polyomavirus (i.e., human polyomavirus 12 or HPyV12 belonging to the genus *Alphapolyomavirus*) was shown to be a variant of a nonhuman polyomavirus that naturally infects pygmy shrew (*Sorex araneus*) and thus renamed Sorex araneus polyomavirus (SaraPyV) [15,16].

Morphology. Mature virions of the family *Polyomaviridae* are particles of 40–45 nm in diameter covered by a capsid in the absence of a lipoprotein envelope. Displaying a T=7dextro (right-handed) icosahedral symmetry, the capsid is made up of 360 copies of the major capsid protein VP1 arranged in 72 pentamers (with each pentamer comprising five VP1 proteins), which interlink by the C-terminal arm and stabilize by calcium ions and disulfide bonds to form capsomers. A copy of the internal viral protein VP2 or VP3 is located in the cavity on the internal face of each pentamer (Figure 6.1), while a copy of VP4 (produced by bird polyomaviruses) connects between VP1 and the viral genome, which is a circular dsDNA situated beneath the capsid [17]. Consisting of 12% DNA and 88% protein (predominantly VP1), polyomavirus particles display a sedimentation coefficient of 240 S, while infectious particles and empty capsids show buoyant densities in CsCl of 1.34 and 1.29 g cm^{-3}, respectively. Expressed alone, VP1 may participate in the formation virus-like particles (VLPs) of 20–60 nm, which may package a lower level of histones than native virions.

Genome structure. Polyomaviruses possess a circular dsDNA genome of ~5,000 bp (ranging from 4,776 to 5,387 bp among human polyomaviruses) (Table 6.1), which consists of two coding regions, early and late, both initiating at the origin of DNA replication (ORI) from the noncoding control region (NCCR, which includes the transcription start sites and promoter/enhancer elements in addition to ORI) anticlockwise and clockwise, respectively (Figure 6.2) [18]. Expressed prior to the initiation of viral DNA replication, the early region encodes via alternative splicing the large T antigen (T-Ag or LTag) and small t antigen (t-Ag or sTag). Expressed after the onset of viral DNA replication, the late region encodes also via alternative splicing the capsid structural proteins VP1, VP2, and VP3 (which assemble in the nucleus to form the capsid) as well as the nonstructural agnoprotein (encoded by upstream of the start codon of VP1 of JCPyV, BKPyV, and SV40 only) involved in virus transcription, maturation, and egress. Functionally, among the early gene products, large T antigen (Tag) participates in cell cycle progression, inhibition of apoptosis, and viral replication, and small t antigen (tAg) contributes to cell cycle progression. Among the late gene products, VP1 (45 kDa) is involved in the capsid structure (external), viral attachment and entry, and antigenic determination, whereas VP2 (38 kDa) is involved in the capsid structure (internal) and viral uncoating during viral penetration, and VP3 (27 kDa) is involved in the capsid structure (internal) and viral uncoating during viral penetration. Some human polyomaviruses may possess another open reading frame (ORF) for either an alternative tumor antigen (ALTO) or a middle tumor antigen (MTAg), while avian polyomaviruses and primate polyomavirus SV40 may encode a late protein VP4 (derived from additional ORF located upstream of the VP2-encoding late mRNA) involved in the lysis of infected cells, genome packaging, and capsid formation [19].

Biology and epidemiology. As naked viruses, polyomaviruses are resistant to lipid solvents and stable at temperature of 50°C for 1 h but not so in the presence of 1 M $MgCl_2$.

Polyomaviruses demonstrate restricted host and cell specificity, which appears to be influenced by their divergence inside the hosts (ca. 10^{-8} substitutions per site per year), lineage duplications, and recombination.

Polyomavirus infection in permissive cells begins with the binding of the virion to a receptor on the outer cell membrane facilitating viral internalization. Specifically, BKPyV targets renal proximal tubular

TABLE 6.1 Molecular and clinical characteristics of human polyomaviruses

GENUS	*SPECIES/ALTERNATIVE NAME*	*ABBREVIATION*	*GENOME SIZE (bP)*	*GENOTYPE*	*ASSOCIATED DISEASE (TISSUE)*	*INITIAL DESCRIPTION*
Alphapolyomavirus	Human polyomavirus 5/Merkel cell polyomavirus	HPyV5/MCPyV/ MCV	5,387	Five genotypes [America/ Europe (Caucasian), Africa (sub-Saharan), Asia/Japan, Oceania, and South America (Amerindians)]	Merkel cell carcinoma (skin)	Feng H, et al. 2008 [5]
	Human polyomavirus 8/trichodysplasia spinulosa polyomavirus	HPyV8/TSPyV/TSV	5,232		Trichodysplasia spinulosa [facial follicular papules, keratotic protrusions (spicules or spines), alopecia of the eyelashes and brows] (skin/TS spicule)	van der Meijden E, et al. 2010 [6]
	Human polyomavirus 9	HPyV9	5,026		Unknown (serum)	Scuda N, et al. 2011 [8]
	Human polyomavirus 13/New Jersey polyomavirus	HPyV13/ NJPyV	5,108		Unknown (muscle)	Mishra N, et al. 2014 [12]
	Human polyomavirus 14/ LI polyomavirus	HPyV14/LIPyV	5,269		Unknown (skin)	Gheit T, et al. 2017 [13]
Betapolyomavirus	Human polyomavirus 1/BK polyomavirus/BK virus	HPyV1/BKPyV/ BKV	5,153	Four genotypes: I (Ia, Ib-1, Ib-2, Ic), II, III, IV (IVa-1, IVa-2, IVb-1, IVb-2, IVc-1, IVc-2)	BKPyV-associated nephropathy (BKVAN), hemorrhagic cystitis, ureteral stenosis (urine)	Gardner SD, et al. 1971 [2]
	Human polyomavirus 2/JC polyomavirus/JC virus	HPyV2/JCPyV/JCV	5,130	Seven genotypes: 1 (IA, IB), 2 (2A1, 2A2, 2B, 2D1, 2D2, 2E), 3 (3A, 3B), 4, 6, 7 (7A, 7B1, 7B2, 7C1, 7C2), 8 (8A, 8B)	Progressive multifocal leukoencephalopathy (PML; cognitive impairment, visual deficit, and motor dysfunction) (brain)	Padgett BL, et al. 1971 [1]

(*Continued*)

TABLE 6.1 (*Continued*) Molecular and clinical characteristics of human polyomaviruses

GENUS	*SPECIES/ALTERNATIVE NAME*	*ABBREVIATION*	*GENOME SIZE (BP)*	*GENOTYPE*	*ASSOCIATED DISEASE (TISSUE)*	*INITIAL DESCRIPTION*
	Human polyomavirus 3/KI polyomavirus/Karolinska Institute polyomavirus	HPyV3/KIPyV/KIV	5,040		Possible acute respiratory symptoms (nasopharynx)	Allander T, et al. 2007 [3]
	Human polyomavirus 4/ WU polyomavirus/ Washington University polyomavirus	HPyV4/WUPyV/ WUV	5,229		Possible acute respiratory symptoms (nasopharynx)	Gaynor AM, et al. 2007 [4]
Deltapolyomavirus	Human polyomavirus 6	HPyV6	4,926		Pruritic and dyskeratotic dermatosis (skin)	Schowalter RM, et al. 2010 [7]
	Human polyomavirus 7	HPyV7	4,952		Epithelial hyperplasia, pruritic rash (skin)	Schowalter RM, et al. 2010 [7]
	Human polyomavirus 10/MW polyomavirus/ Mexican polyomavirus	HPyV10/MWPyV/ MXPyV	4,927		Unknown (skin/anal condyloma)	Siebrasse EA, et al. 2012 [9]; Yu G, et al. 2012 [10]
	Human polyomavirus 11/STL polyomavirus	HPyV11/STLPyV	4,776		Unknown (stool)	Lim ES, et al. 2013 [11]

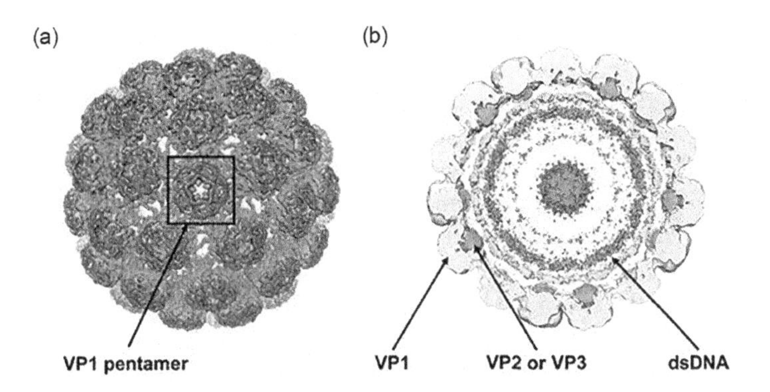

FIGURE 6.1 Cryo-electron microscopy structure of the BKPyV particles. (a) External view of BKPyV virion showing viral protein VP1 pentamers; (b) The unsharpened/unmasked virion map showing VP1, VP2 or VP3, and packaged dsDNA. (Photo credit: Helle F, Brochot E, Handala L, et al. *Viruses*. 2017;9(11):327.)

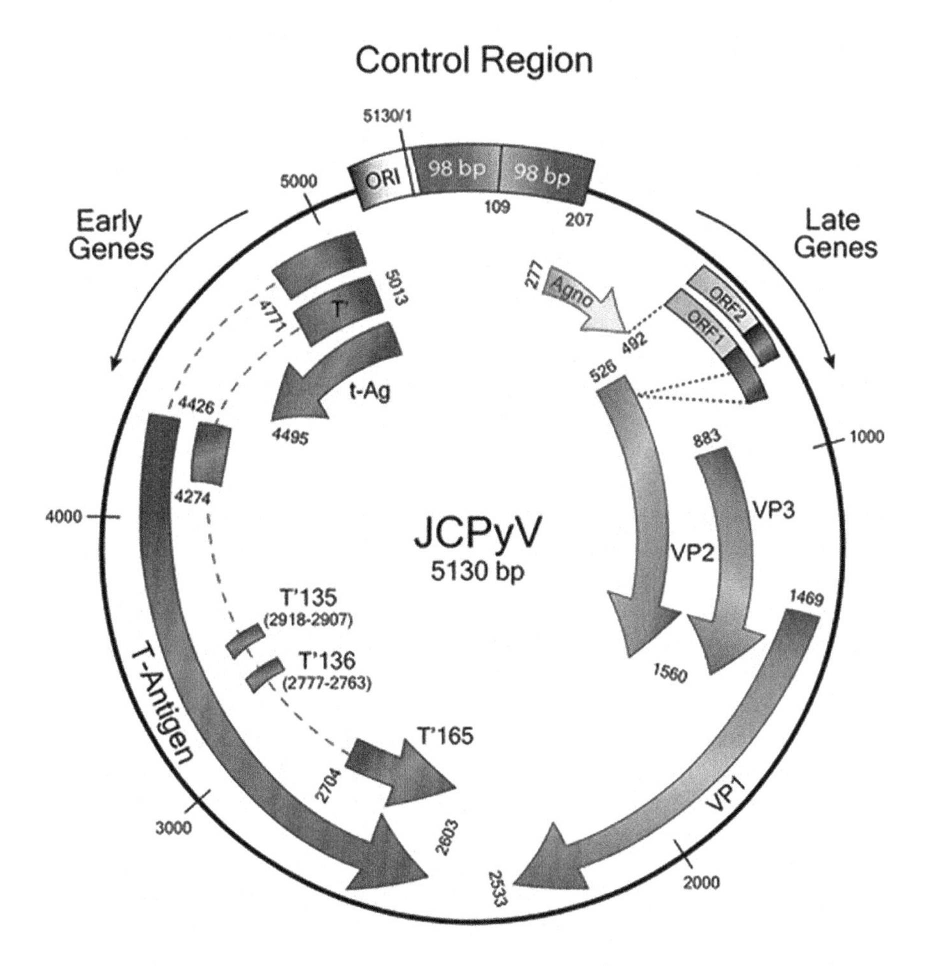

FIGURE 6.2 Schematic representation of JCPyV genome. Initiating from a common control region at the origin of DNA replication (ORI) anticlockwise and clockwise, respectively, the early genes encode large T-antigen and small t-antigen (T'135, T'136, and T'165), while the late genes encode capsid proteins VP-1, VP-2, and VP-3, and nonstructural agnoprotein. Additionally, ORF1 and ORF2 exist between Agno and VP1 genes. (Photo credit: Del Valle L, Piña-Oviedo S. *Front Oncol.* 2019;9:711.)

epithelial cells (through VP1 binding to cell receptors gangliosides GD1b and GT1b, with assistance from GD2 and GD3) and salivary gland cells; JCPyV infects renal cells (cell receptors GT1b, GD1b, and GD2) and glial cells (serotoninergic 5-HT_{2a} receptor); MCPyV is infective to dermal fibroblasts (cell receptor GT1b and possible co-receptor glycosaminoglycans); TSPyV invades the inner root sheath cells of the hair follicle; HPyV6 and HPyV7 recognize nonganglioside receptors. Alternatively, polyomavirus undertakes caveolae-mediated internalization (all human PyVs except JCPyV) or clathrin-mediated endocytosis (JCPyV), then escapes the endocytic compartment (ER), and enters the cytosol [20].

Once inside the host cell, polyomavirus virion decapsidizes in the cytosol and exposes the nuclear localization signals in VP2/VP3, which guide the virion into the nucleus. Transcription of the early genes gives rise to mRNAs encoding LTAg, STAg and alternative early proteins. LTAg binds the tumor suppressor proteins p53 and the retinoblastoma family members, STAg binds and inactivates cellular protein phosphatase 2A (PP2A), and middle T antigen binds PP2A, leading to the initiation of viral replication. Transcription of the late genes generates capsid proteins, which facilitate the assembly of the viral genome (including a package with histones H2A, H2B, H3, and H4) into viral particles. Using lysis-dependent or -independent mechanisms, viruses egress host cells. Interestingly, Merkel cells are nonpermissive for MCPyV and unable to sustain the viral life cycle, leading to MCPyV genome integration and transformation in these cells [20].

Primary infection appears to take place during early childhood and sustains throughout life as indicated by higher percentages of seropositivity in older people. Given that BKPyV, JCPyV, MWPyV, and STLPyV persist in the reno-urinary and gastrointestinal tracts and shed in urine and stool, they are likely transmitted via contaminated food and water. Furthermore, BKPyV, JCPyV, KIPyV, and WUPyV are often found in the tonsillar tissue and respiratory aspirates, suggesting an aerogenic route of infection. On the other hand, MCPyV, HPyV6, HPyV7, and TSPyV naturally inhabit the skin and are thus transmissible via contact of virus-bearing fluids and secretion with mucosal surfaces.

In recent years, some changes in polyomavirus disease epidemiology are being noticed as a consequence of medical intervention, including a decline of JCPyV-related progressive multifocal leukoencephalopathy (PML) in AIDS patients due to highly active anti-retroviral therapy and a reemergence of PML in *multiple sclerosis* patients following novel immunosuppressive therapies, as well as a more frequent occurrence of BKPyV-associated nephropathy in renal transplant patients treated with more potent immunosuppressive drugs.

Clinical features. Despite their common occurrence, human polyomavirus infections rarely induce clinically apparent diseases in the general population [21]. However, in individuals with immune suppression and dysfunction, human polyomaviruses reactivate and produce a range of clinical diseases, including polyomavirus-associated nephropathy (PyVAN) in 1%–10% of renal transplant patients, polyomavirus-associated hemorrhagic cystitis (PyVHC) in 5%–15% of allogeneic hematopoietic stem cell transplant (HSCT) recipients, and ureteral stenosis [due to human polyomavirus 1/BK polyomavirus (HPyV1/BKPyV)]; progressive multifocal leukoencephalopathy (PML; manifesting as cognitive impairment, visual deficit, and motor dysfunction) in HIV/AIDS-positive individuals [due to human polyomavirus 2/JC polyomavirus (HpyV2/JCPyV)]; Merkel cell carcinoma [a rare but aggressive skin cancer, strongly linked to human polyomavirus 5/Merkel cell polyomavirus (HPyV5/MCPyV)]; trichodysplasia spinulosa [facial follicular papules, keratotic protrusions (spicules or spines), and alopecia of the eyelashes and brows, due to human polyomavirus 8/trichodysplasia spinulosa polyomavirus (HPyV8/TSPyV)]; pruritic and dyskeratotic dermatosis [due to human polyomavirus 6 (HPyV6)]; epithelial hyperplasia, and pruritic rash [due to human polyomavirus 7 (HPyV7)]. The roles of other human polyomaviruses [human polyomavirus 9 (HPyV9), human polyomavirus 13/New Jersey polyomavirus (HPyV13/NJPyV), human polyomavirus 14/LI polyomavirus (HPyV14/LIPyV), human polyomavirus 3/KI polyomavirus (HPyV3/KIPyV), human polyomavirus 4/WU polyomavirus (HPyV4/WUPyV), human polyomavirus 10/MW polyomavirus/Mexican polyomavirus (HPyV10/MWPyV/MXPyV), human polyomavirus 11/STL polyomavirus (HPyV11/STLPyV)] in the causation of human diseases remain to be clarified (Table 6.1) [22–25].

Specifically, BKPyV often infects children of 3–4 years in age and lies dormant in the renal tubular epithelial cells, leading to seroprevalence of about 50% in children under 10 years of age and 60%–80%

in adulthood. Individuals with immune deficiency (e.g., AIDS and SOT) may develop BKPyV viruria and viremia, leading to PyVAN (in ~10% of renal transplant recipients), manifesting as tubulo-interstitial inflammation, premature renal failure without fever (1%–14% of cases), leukocytosis, hematuria or proteinuria, decreased kidney allograft function, and graft loss in at least 50% of kidney transplant patients. Ureteral stenosis may present with urinary obstruction and elevated serum creatinine levels without symptoms of pain or discomfort as a result of deinnervation in the transplanted kidney. Bone marrow transplant patients with hemorrhagic cystitis may show hematuria, dysuria, urgency of urination, increased frequency of urination, or suprapubic pain, along with severe bleeding and clot formation, and complications of urinary tract obstruction and renal failure within two months of transplant. Factors influencing the outcome of PyVAN range from recipient status (male gender, older age, low or undetectable BKPyV antibody, low or absent BKPyV-specific T-cell responses), donor status (recent BKPyV exposure, HLA mismatch, deceased donation), transplant-related issues (ischemia/reperfusion injury, ureteric stents, acute rejection, retransplantation after graft loss due to PyVAN), to therapy (cumulative corticosteroid exposure, tacrolimus or mycophenolate exposure, or mTOR inhibitor combinations) [17,26–28].

Progressive multifocal leukoencephalopathy (PML) often goes unnoticed or mistaken for a stroke initially, until significant impairment of cognition or motor function emerges (e.g., hemiparesis, loss of vision on one side, aphasia, and ataxia). Treatment of PML with aggressive antiretroviral therapy in patients with AIDS and rapid withdrawal of immunosuppressive therapy in patients with autoimmune diseases or organ transplants may result in an influx of lymphocytes into the brain and development of immune reconstitution inflammatory syndrome (IRIS), further exacerbating the neurological symptoms and causing massive swelling of the brain, herniation, and death of the patient. PML has *an estimated incidence of* 1.29 cases per 1,000 post-*solid organ transplant (SOT)* patient years, in comparison to 2.4 cases per 1,000 patient years in AIDS patients in the absence of combination antiretroviral therapy and 2.1–100 cases per 1,000 patient years in multiple sclerosis patients treated with natalizumab [29–31].

Merkel cell carcinoma (MCC) is a rare, rapidly expanding, neuroendocrine malignancy that initially appears as solitary purplish lesion on sun-exposed skin of elderly white and immunocompromised individuals (e.g., SOT and AIDS patients), with an incidence of 0.24–0.35 cases per 100,000 patient years and a mortality rate of 33%–46%. Detected in the majority of proven cases, MCPyV genome often undergoes chromosomal integration, leading to the expression of early viral gene regions (sTag), but not of late viral gene regions, and abrogation of the (cyto-)lytic viral replication cycle. Patients with higher antibody titer to MCPyV appear to have better clinical outcomes [32–35].

Trichodysplasia spinulosa is a rare proliferative-cytopathic skin disease of immunocompromised individuals, with an incidence of <1 case per 1,000 patient years post-transplant (mainly kidney transplant). Linked to a history of leukemia, chemotherapy, or transplantation, *trichodysplasia spinulosa* typically forms papules of thickened skin and small hair-like protrusions (called *spiculae*, which need to be distinguished from other digitate hyperkeratoses) in the forehead, nose, and ears, along with circumscribed alopecia (of the eyebrows), and has detectable PyV particles in inner root hair follicles.

Pathogenesis. Infections with human polyomaviruses may produce five pathologic patterns: (i) *cytopathic pattern* (the loss of infected cells through high-level virus replication, without significant inflammation, as exemplified by progressive multifocal leukoencephalopathy or PML due to JCV replication in the brain of patients with advanced AIDS), (ii) *immune-reconstitution inflammatory syndrome* (*IRIS*; a dominant inflammatory response to abundant antigen, typically after a brisk recovery of the cellular immune response, as exemplified by hemorrhagic cystitis associated with BKV following allogenic stem cell transplantation), (iii) *cytopathic inflammation* (high-level virus replication and a significant inflammatory response due to cytopathic lysis, necrosis, with infiltration of granulocytes and lymphocytes, as exemplified by nephropathy stage B in kidney allografts), (iv) *autoimmune pattern* (pathologic response to "self" triggered by viral antigens, as exemplified by DNA antibodies, anti-histone antibodies, or possibly other antinuclear antibodies induced by LTag complexed to DNA and nucleosomes), and (v) *oncogenic pattern* (viral early gene expression activating the host cells, without sufficient late gene expression to cause rapid host cell lysis, as exemplified by MCC due to chromosomal integration and truncating mutations in the LTag and/or VP1 that impairs MCPyV genome replication and late gene expression) [36–38].

Specifically, BKPyV has tropism for the uroepithelial cells of the genitourinary tract, and its reactivation may be stimulated by host factors (immunity, viral load, allogeneic immune response) and inflammation (tissue injury, leukocyte infiltration, release of pro-inflammatory cytokines), leading to progressive infection, renal tubular epithelial cell lysis, and renal dysfunction [39,40].

JCPyV has the ability to infect microvascular endothelial cells, cross into the brain, and cause lytic infection of the oligodendrocytes in the brain, resulting in the development of PML. Invariably, it takes several events for this to happen, including (i) host immune system impairment (e.g., following immunomodulatory drug treatment), (ii) viral NCCR rearrangement for increased viral transcription and replication, (iii) upregulation of DNA-binding factors for NCCR sequences, and (iv) viral migration across the blood–brain barrier [41].

MCC is attributed to the presence of clonally integrated MCPyV (HPyV5) or chronic ultraviolet light (UV) exposure, and the name Merkel cell carcinoma does not suggest its origin from Merkel cells but rather its similar appearance to Merkel cells (which are located in the basal layer of the epidermis and associated with sensory nerves). It is of interest to note that MCC from low UV exposure countries frequently contains MCPyV (MCPyV$^+$ MCC), while MCC from high UV exposure countries is often MCPyV negative (MCPyV$^-$ MCC) and have DNA mutations (e.g., inactivating mutations in *RB1* and *TP53*) bearing a UV signature instead [42,43].

Diagnosis. Given the clinical complexity of human polyomavirus infections, accurate diagnosis of these diseases requires input from physical examination, imaging study, biochemical test, histopathology, serology, and molecular procedures [44].

For BK-related PyVAN, electron microscopy reveals BKPyV in the nuclei of renal cells as crystalloid particles. Cytological detection of epithelial cells containing BK viral inclusions (decoy cells) in urine has a positive predictive value of 27% and a negative predictive value of 100% (Figure 6.3) [45]. This is followed by PCR to quantify the viral load from the urine and blood. A negative urinary PCR excludes BKPyV as a cause of disease. A positive serum/urine PCR ($>1 \times 10^4$ DNA copies/ml plasma or $>1 \times 10^7$ DNA copies/ml urine) has a positive predictive value of 50%–85% and a negative predictive value of 100%. Renal biopsy identifying the typical intranuclear viral inclusion bodies in the renal tubular epithelial cells provides ultimate confirmation of PyVAN (with false negatives of up to 30% due to the focality of the BKV in the renal tissue). Positive immunohistochemistry staining for SV40 T-antigen is pathognomonic for PyVAN [46].

For JCPyV-linked progressive multifocal leukoencephalopathy (PML), diagnosis relies on analysis of histological features and immunological marker expression profiles of the lesion. As an unusual demyelinating disease affecting multiple foci of the subcortical white matter, PML is characterized by perivascular cuffs of lymphocytes, enlarged oligodendrocytes with eosinophilic intranuclear inclusions (containing polyomavirus particles), and bizarre astrocytes, reminiscent of neoplastic cells (Figure 6.4) [47]. Magnetic

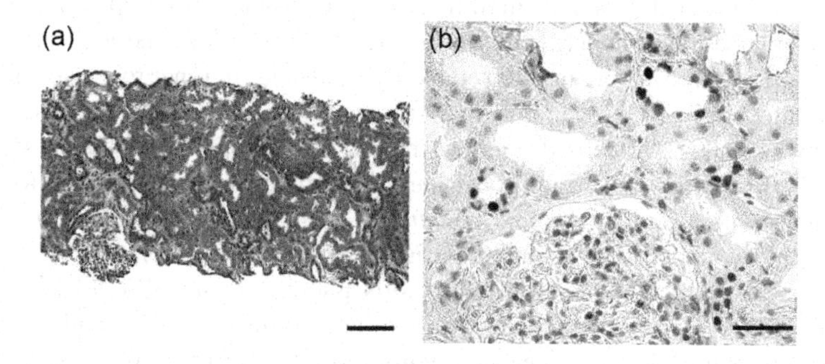

FIGURE 6.3 Histological analysis of renal allograft biopsy from the recipient at 12 weeks post-transplant. (a) HES (hematoxylin, eosin, and saffron)-stained section. Scale bar, 100 µm. (b) Immunohistochemistry staining of the same biopsy as in (a), BKPyV viral LTag expression (brown color) in tubular epithelial cells using cross-reacting monoclonal anti-SV40 LTag antibody Pab416 (Merck). Scale bar, 50 µm. (Photo credit: Lorentzen EM, Henriksen S, Kaur A, et al. *Virol J.* 2020;17(1):5.)

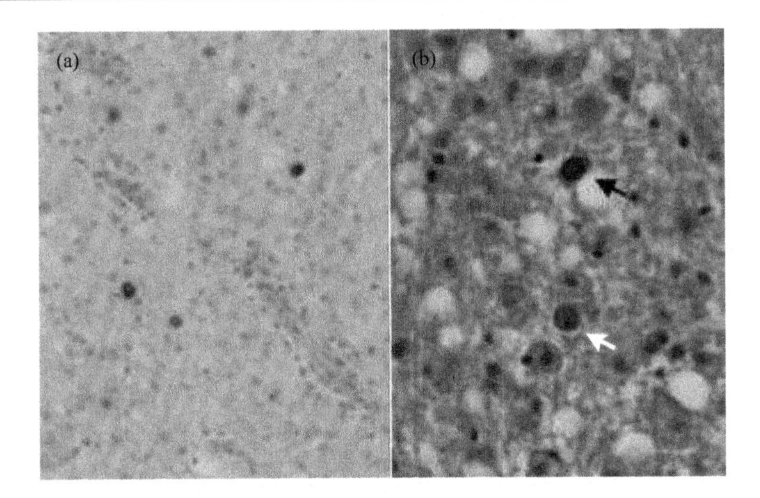

FIGURE 6.4 Development of demyelinating lesions in progressive multifocal leukoencephalopathy (PML). (a) Immunostaining for JCPyV capsid proteins reveals the presence of infected glial cells in the borders of expanding demyelinating lesions. (b) JCPyV-infected glial cells with full inclusions (black arrow) and cells with dot-shaped intranuclear structures (white arrow). (Photo credit: Ono D, Shishido-Hara Y, Mizutani S, et al. *Neuropathology*. 2019;39(4):294–306.)

resonance imaging (MRI) of the brain typically shows multiple high-signal-intensity lesions involving the uncinate fibers and the posterior parts of the brain but sparing the gray matter on T2-weighted and FLAIR sequences, hypointense lesions on T1-weighted images, and nonenhanced lesions (or variably enhanced lesions in the case of IRIS or AIDS) upon administration of contrast agent gadolinium. Because JCPyV viremia sometimes occurs in the absence of PML, PCR detection of virus DNA in the cerebrospinal fluid (CSF) or brain biopsy is necessary for the confirmation of PML.

For MCPyV-associated MCC, PCR detection of MCPyV DNA helps confirm the viral origin of MCC (80% of cases), since MCPyV DNA is not found in melanoma, basal cell carcinoma, uterine cervix, large bowel, ovary, breast, bone, and soft tissue tumors (Figure 6.5) [48]. Despite the fact that MCPyV DNA may occur in keratoacanthoma or squamous cell carcinoma, the viral load is two orders of magnitude higher in MCC than in keratoacanthoma or squamous cell carcinoma. MCC patients tend to have very high antibody titers to MCPyV in comparison with the general population, further supporting the role of MCPyV infection in MCC [49–51].

Treatment. Treatment strategy for BKPyV infection involves reducing immunosuppressive medications, monitoring serum creatinine and viral load periodically, and ancillary therapies (e.g., cidofovir, leflunomide, quinolones, IVIG) [52].

Treatment of JCV-related PML should center on restoration of the immune system and administration of antiviral drugs [e.g., mirtazapine alone or in combination with cytosine arabinoside (Ara-C)]. For example, adoptive transfer of JCV antigen-specific cytotoxic T cells (generated after *in vitro* stimulation with viral T and VP1 proteins) along with citalopram (a serotonin reuptake inhibitor) and CDV may help clear JCV from the CSF and improve clinical symptoms in hematopoietic cell transplant recipients with PML. However, PML still has a mortality rate of nearly 50% in patients with AIDS and of somewhat lower in patients with multiple sclerosis [41].

MCC typically presents with a solitary cutaneous or subcutaneous nodule in sun-exposed areas, its treatment is stage dependent (local or disseminated), and involves surgical resection of localized tumor in combination with radiotherapy or chemotherapy (especially for metastatic or refractory MCC) and immune-checkpoint inhibitors [53,54].

Treatment for trichodysplasia spinulosa consists of improving immune function (e.g., reducing maintenance immunosuppression in SOT recipients) and using topical antivirals (i.e., cidofovir 1%–3% cream).

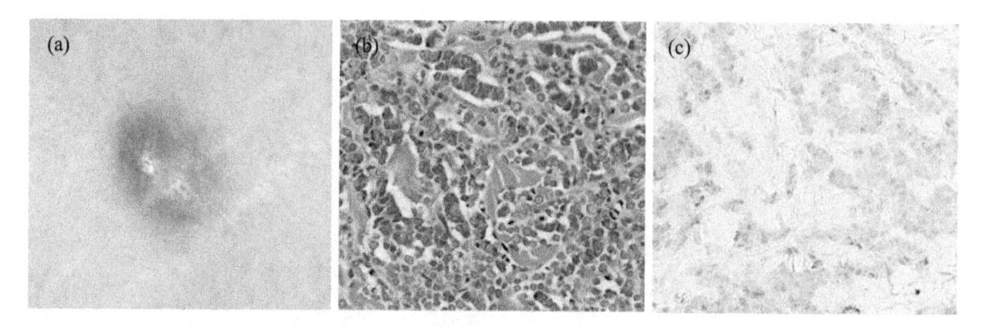

FIGURE 6.5 (a) Merkel cell carcinoma on the posterior surface of the left thigh during treatment with the oral JAK inhibitor, tofacitinib. (b) Histopathologic examination of the tumor. Hematoxylin-eosin staining reveals the proliferation of tumor cells within the dermis, with small, round basophilic cells arranged in cordlike structures. (c) Immunohistochemical examination shows tumor's positive reaction for MCPyV large T antigen. (Photo credit: Koike Y, Murayama N, Kuwatsuka Y, Utani A. *JAAD Case Rep.* 2017;3(6):498–500.)

6.2 MOLECULAR EPIDEMIOLOGY OF HUMAN POLYOMAVIRUSES

Serological approaches. Epidemiological surveys using VLP or VP1 capsomer-based ELISA reveal a widespread occurrence of human poliomaviruses in the general population. For example, 82%–99% and 39%–81% adults in USA, Australia, and Italy are serologically positive for BKPyV and JCPyV, respectively, while 60%–81% of adults in USA, Australia, and Italy harbor antibodies to MCPyV. However, a lower proportion of children have antibodies to BKPyV (50% in 2-year-olds and >90% in >10-year-olds), JCPyV, and MCPyV (10%). The findings suggest that human polyomavirus infection begins at early age and continues to expand in the adulthood. Nonetheless, the fact that polyomavirus-associated diseases are rare in the general population highlights the requirement of immune deficiency or SOT as a precondition for disease to emerge [55,56].

Molecular approaches. Based mainly on type-specific restriction of endonuclease sites of a 327-bp variable region of the gene coding for the major capsid protein VP1, BKPyV is separated into four subtypes (I, II, III, and IV), of which subtypes I and IV viruses make up 75% and 17.5% of cases, respectively, while subtypes II and III viruses account for 5.5% and 2% of cases, respectively. Further analysis of the nucleotide sequences (1,089 bp; nt 1564–2652) of the VP1 gene allows differentiation of subtype I into four subgroups [Ia (7%), Ib-1 (22%), Ib-2 (38%), and Ic (7%)] and subtype IV into six subgroups [IVa-1 (1.5%), IVa-2 (1.5%), IVb-1 (1.5%), IVb-2 (1%), IVc-1 (2.5%), and IVc-2 (10%)] (Table 6.1). Interestingly, subgroups Ia, Ib-1, Ib-2, and Ic predominate in Africa, southeast Asia, Europe, and northeast Asia, respectively. Furthermore, all subtypes of genotypes IV except IVc2 are commonly present in East Asia, while subtype IVc2 largely occurs in Europe, America, and Northeast Asia [57–59].

Similarly, the examination of the coding region polymorphisms of full-length genome sequences permits separation of JCPyV into seven genotypes (1–8, with genotype 5 being a variant of genotype 3), each containing multiple subtypes, that is, 1 [IA (European/European-American), IB(European/European-American)], 2 [2A1 (Asian/Native American), 2A2 (Asian/Native American), 2B (Asian/Eurasian), 2D1 (Asian/South Asian), 2D2 (Asian/South Asian), 2E (Western Pacific populations), 3 [3A (African/African-American), 3B (African/African-American)], 4 (European/European-American), 6 (African), 7 [7A (Asian), 7B1 (Asian), 7B2 (Asian), 7C1 (Asian/South Asian), 7C2 (Asian/South Asian)], 8 [8A (inhabitants of Papua New Guinea), and 8B (Western Pacific populations)] (Table 6.1). Further analysis suggests that JCPyV genotype 6 represents an ancestral type that gives rise to two evolutionary lineages, with one later yielding genotypes 1 and 4 and the other producing genotypes 2, 3, 7, and 8 [60,61].

Moreover, analysis of the 1,284-bp fragment (corresponding to the 1,468-bp fragment without the 144-bp noncoding region of fragment C) facilitates the identification of five major geographically related MCPyV genotypes from America/Europe (Caucasian), Africa (sub-Saharan), Asia/Japan, Oceania, and South America (Amerindians) [62].

6.3 FUTURE PERSPECTIVES

Human polyomaviruses are a group of small (40–45 nm), nonenveloped, icosahedral viruses with a circular dsDNA genome of ~5 kb. Transmitted via contaminated food/water (e.g., BKPyV, JCPyV) or skin contact (e.g., MCPyV), human polyomaviruses invade specific host cells and persist in two distinct nonpathological states, that is, latency or low-level replication. Any changes in host immune function (e.g., AIDS and SOT) may reactivate the viruses, leading to active viral replication and various clinical diseases (e.g., PVAN, PML, *MCC*). Recent application of molecular techniques has helped uncover many details about polyomavirus gene structures and regulation and enabled development of efficient tracking tools for polyomaviruses. Nonetheless, much is to be learnt about polyomaviruses, especially their oncogenic potential, before appropriate treatment, and prevention measures are available.

Except for MCPyV in MCC, none of the new HPyVs has yet to be implicated in human cancer. MCPyV has become well established as a major etiological agent for MCC through a precise application of molecular techniques such as the use of Southern blot to establish clonal integration, digital transcriptome subtraction to verify the expression of T-antigen, and RNA interference to show the essential nature of T-antigen expression. The clear association of MCPyV with Merkel cell cancer emphasizes the oncogenic potential of HPyVs, while the link of TSV to trichdysplasia spinulosa, a rare disease in immunosuppressed individuals, underlines the pathogenic ability of HPyVs to cause disease in populations with suppressed immune functions. Undoubtedly, with the help of increasingly sophisticated molecular techniques, it will be only a matter of time when the pathogenic mechanisms of human polyomaviruses are fully elucidated, and specific therapeutics for polyomavirus-induced diseases are developed.

REFERENCES

1. Padgett BL, Walker DL, ZuRhein GM, Eckroade RJ, Dessel BH. Cultivation of papova-like virus from human brain with progressive multifocal leucoencephalopathy. *Lancet.* 1971;1(7712):1257–60.
2. Gardner SD, Field AM, Coleman DV, Hulme B. New human papovavirus (B.K.) isolated from urine after renal transplantation. *Lancet.* 1971;1(7712):1253–7.
3. Allander T, Andreasson K, Gupta S, et al. Identification of a third human polyomavirus. *J Virol.* 2007;81(8): 4130–6.
4. Gaynor AM, Nissen MD, Whiley DM, et al. Identification of a novel polyomavirus from patients with acute respiratory tract infections. *PLoS Pathog.* 2007;3(5):e64.
5. Feng H, Shuda M, Chang Y, Moore PS. Clonal Integration of a polyomavirus in human Merkel cell carcinoma. *Science.* 2008;319(5866):1096–100.
6. van der Meijden E, Janssens RW, Lauber C, Bouwes Bavinck JN, Gorbalenya AE, Feltkamp MC. Discovery of a new human polyomavirus associated with trichodysplasia spinulosa in an immunocompromized patient. *PLoS Pathog.* 2010;6(7):e1001024.
7. Schowalter RM, Pastrana D V, Pumphrey KA, Moyer AL, Buck CB. Merkel cell polyomavirus and two previously unknown polyomaviruses are chronically shed from human skin. *Cell Host Microbe.* 2010;7(6):509–15.
8. Scuda N, Hofmann J, Calvignac-Spencer S, et al. A novel human polyomavirus closely related to the African green monkey-eerived lymphotropic polyomavirus. *J Virol.* 2011; 85(9):4586–90.

9. Siebrasse EA, Reyes A, Lim ES, et al. Identification of MW polyomavirus, a novel polyomavirus in human stool. *J Virol*. 2012;86(19):10321–6.
10. Yu G, Greninger AL, Isa P, et al. Discovery of a novel polyomavirus in acute diarrheal samples from children. *PLoS One*. 2012;7(11):e49449.
11. Lim ES, Reyes A, Antonio M, et al. Discovery of STL polyomavirus, a polyomavirus of ancestral recombinant origin that encodes a unique T antigen by alternative splicing. *Virology*. 2013;436(2):295–303.
12. Mishra N, Pereira M, Rhodes RH, et al. Identification of a novel polyomavirus in a pancreatic transplant recipient with retinal blindness and vasculitic myopathy. *J Infect Dis*. 2014;210(10):1595–9.
13. Gheit T, Dutta S, Oliver J, et al. Isolation and characterization of a novel putative human polyomavirus. *Virology*. 2017;506:45–54.
14. Polyomaviridae Study Group of the International Committee on Taxonomy of Viruses, Calvignac-Spencer S, Feltkamp MC, et al. A taxonomy update for the family Polyomaviridae. *Arch Virol*. 2016;161(6):1739–50.
15. Korup S, Rietscher J, Calvignac-Spencer S, et al. Identification of a novel human polyomavirus in organs of the gastrointestinal tract. *PLoS One*. 2013;8(3):1–7.
16. Gedvilaite A, Tryland M, Ulrich RG, et al. Novel polyomaviruses in shrews (*Soricidae*) with close similarity to human polyomavirus 12. *J Gen Virol*. 2017;98(12):3060–7.
17. Helle F, Brochot E, Handala L, et al. Biology of the BKPyV: an update. *Viruses*. 2017;9(11):327.
18. Del Valle L, Piña-Oviedo S. Human polyomavirus JCPyV and its role in progressive multifocal leukoencephalopathy and oncogenesis. *Front Oncol*. 2019;9:711.
19. Guidry JT, Scott RS. The interaction between human papillomavirus and other viruses. *Virus Res*. 2017; 231:139–47.
20. Bhattacharjee S, Chattaraj S. Entry, infection, replication, and egress of human polyomaviruses: an update. *Can J Microbiol*. 2017;63(3):193–211.
21. Hirsch HH, Babel N, Comoli P, et al. European perspective on human polyomavirus infection, replication and disease in solid organ transplantation. *Clin Microbiol Infect*. 2014;20 Suppl 7:74–88.
22. Bennett SM, Broekema NM, Imperiale MJ. BK polyomavirus: emerging pathogen. *Microbes Infect*. 2012;14(9):672–83.
23. White MK, Gordon J, Khalili K. The rapidly expanding family of human polyomaviruses: recent developments in understanding their life cycle and role in human pathology. *PLoS Pathog*. 2013;9(3):e1003206.
24. Burger-Calderon R, Webster-Cyriaque J. Human BK polyomavirus-The potential for head and neck malignancy and disease. *Cancers (Basel)*. 2015;7(3):1244–70.
25. Beckervordersandforth J, Pujari S, Rennspiess D, et al. Frequent detection of human polyomavirus 6 in keratoacanthomas. *Diagn Pathol*. 2016;11(1):58.
26. Gupta N, Lawrence RM, Nguyen C, Modica RF. Review article: BK virus in systemic lupus erythematosus. *Pediatr Rheumatol Online J*. 2015;13:34.
27. Papadimitriou JC, Randhawa P, Rinaldo CH, Drachenberg CB, Alexiev B, Hirsch HH. BK polyomavirus infection and renourinary tumorigenesis. *Am J Transplant*. 2016;16(2):398–406.
28. Levican J, Acevedo M, León O, Gaggero A, Aguayo F. Role of BK human polyomavirus in cancer. *Infect Agent Cancer*. 2018;13:12.
29. Delbue S, Comar M, Ferrante P. Review on the role of the human polyomavirus JC in the development of tumors. *Infect Agent Cancer*. 2017;12:10.
30. Nishiyama S, Misu T, Shishido-Hara Y, et al. Fingolimod-associated PML with mild IRIS in MS: a clinicopathologic study. *Neurol Neuroimmunol Neuroinflamm*. 2017;5(1):e415.
31. Paz SPC, Branco L, Pereira MAC, Spessotto C, Fragoso YD. Systematic review of the published data on the worldwide prevalence of John Cunningham virus in patients with multiple sclerosis and neuromyelitis optica. *Epidemiol Health*. 2018;40:e2018001.
32. Dalianis T, Hirsch HH. Human polyomaviruses in disease and cancer. *Virology*. 2013; 437(2):63–72.
33. Spurgeon ME, Lambert PF. Merkel cell polyomavirus: a newly discovered human virus with oncogenic potential. *Virology*. 2013;435(1):118–30.
34. Mulchan N, Cayton A, Asarian A, Xiao P. Merkel cell carcinoma: a case report and literature review. *J Surg Case Rep*. 2019;2019(11):rjz322.
35. Nguyen KD, Chamseddin BH, Cockerell CJ, Wang RC. The biology and clinical features of cutaneous polyomaviruses. *J Invest Dermatol*. 2019;139(2):285–92.
36. De Gascun CF, Carr MJ. Human polyomavirus reactivation: disease pathogenesis and treatment approaches. *Clin Dev Immunol*. 2013;2013:373579.

37. Barth H, Solis M, Kack-Kack W, Soulier E, Velay A, Fafi-Kremer S. In vitro and in vivo models for the study of human polyomavirus infection. *Viruses*. 2016;8(10):292.
38. Moens U, Macdonald A. Effect of the large and small T-antigens of human polyomaviruses on signaling pathways. *Int J Mol Sci*. 2019;20(16):3914.
39. El Hennawy HM. BK polyomavirus immune response with stress on BK-specific T cells. *Exp Clin Transplant*. 2018;16(4):376–85.
40. Alcendor DJ. BK polyomavirus virus glomerular tropism: implications for virus reactivation from latency and amplification during immunosuppression. *J Clin Med*. 2019;8(9):1477.
41. Bellizzi A, Anzivino E, Rodio DM, Palamara AT, Nencioni L, Pietropaolo V. New insights on human polyomavirus JC and pathogenesis of progressive multifocal leukoencephalopathy. *Clin Dev Immunol*. 2013;2013:839719.
42. Becker JC, Stang A, DeCaprio JA, et al. Merkel cell carcinoma. *Nat Rev Dis Primers*. 2017; 3:17077.
43. Robinson CG, Tan D, Yu SS. Recent advances in Merkel cell carcinoma. *F1000Res*. 2019; 8:F1000 Faculty Rev-1995.
44. Madhavan HN, Bagyalakshmi R, Revathy M, Aarthi P, Malathi J. Optimisation and analysis of polymerase chain reaction based DNA sequencing for genotyping polyoma virus in renal transplant patients: a report from South India. *Indian J Med Microbiol*. 2015;33 Suppl:37–42.
45. Lorentzen EM, Henriksen S, Kaur A, et al. Early fulminant BK polyomavirus-associated nephropathy in two kidney transplant patients with low neutralizing antibody titers receiving allografts from the same donor. *Virol J*. 2020;17(1):5.
46. Chong S, Antoni M, Macdonald A, Reeves M, Harber M, Magee CN. BK virus: current understanding of pathogenicity and clinical disease in transplantation. *Rev Med Virol*. 2019;29(4):e2044.
47. Ono D, Shishido-Hara Y, Mizutani S, et al. Development of demyelinating lesions in progressive multifocal leukoencephalopathy (PML): comparison of magnetic resonance images and neuropathology of post-mortem brain. *Neuropathology*. 2019;39(4):294–306.
48. Koike Y, Murayama N, Kuwatsuka Y, Utani A. A case of Merkel cell carcinoma development under treatment with a Janus kinase inhibitor. *JAAD Case Rep*. 2017;3(6):498–500.
49. Moens U, Rasheed K, Abdulsalam I, Sveinbjørnsson B. The role of Merkel cell polyomavirus and other human polyomaviruses in emerging hallmarks of cancer. *Viruses*. 2015; 7(4):1871-901.
50. DeCaprio JA. Merkel cell polyomavirus and Merkel cell carcinoma. *Philos Trans R Soc Lond B Biol Sci*. 2017;372(1732):20160276.
51. Prado JCM, Monezi TA, Amorim AT, Lino V, Paladino A, Boccardo E. Human polyomaviruses and cancer: an overview. *Clinics (Sao Paulo)*. 2018;73(suppl 1):e558s.
52. Ambalathingal GR, Francis RS, Smyth MJ, Smith C, Khanna R. BK polyomavirus: clinical aspects, immune regulation, and emerging therapies. *Clin Microbiol Rev*. 2017; 30(2):503–28.
53. Schadendorf D, Lebbé C, Zur Hausen A, et al. Merkel cell carcinoma: epidemiology, prognosis, therapy and unmet medical needs. *Eur J Cancer*. 2017;71:53–69.
54. Becker JC, Stang A, Hausen AZ, et al. Epidemiology, biology and therapy of Merkel cell carcinoma: conclusions from the EU project IMMOMEC. *Cancer Immunol Immunother*. 2018;67(3):341–51.
55. Viscidi RP, Rollison DE, Sondak VK, et al. Age-specific seroprevalence of Merkel cell polyomavirus, BK virus, and JC virus. *Clin Vaccine Immunol*. 2011;18(10):1737–43.
56. Kamminga S, van der Meijden E, Wunderink HF, Touzé A, Zaaijer HL, Feltkamp MCW. Development and evaluation of a broad bead-based multiplex immunoassay to measure IgG seroreactivity against human polyomaviruses. *J Clin Microbiol*. 2018;56(4):e01566–17.
57. Morel V, Martin E, François C, et al. A simple and reliable strategy for BK virus subtyping and subgrouping. *J Clin Microbiol*. 2017;55(4):1177–85.
58. Wunderink HF, de Brouwer CS, van der Meijden E, et al. Development and evaluation of a BK polyomavirus serotyping assay using Luminex technology. *J Clin Virol*. 2019;110:22–8.
59. Wunderink HF, De Brouwer CS, Gard L, et al. Source and relevance of the BK polyomavirus genotype for infection after kidney transplantation. *Open Forum Infect Dis*. 2019; 6(3):ofz078.
60. Ferenczy MW, Marshall LJ, Nelson CD, et al. Molecular biology, epidemiology, and pathogenesis of progressive multifocal leukoencephalopathy, the JC virus-induced demyelinating disease of the human brain. *Clin Microbiol Rev*. 2012;25(3):471–506.
61. Ryschkewitsch CF, Jensen PN, Major EO. Multiplex qPCR assay for ultra sensitive detection of JCV DNA with simultaneous identification of genotypes that discriminates non-virulent from virulent variants. *J Clin Virol*. 2013;57(3):243–8.
62. Martel-Jantin C, Filippone C, Tortevoye P, et al. Molecular epidemiology of Merkel cell polyomavirus: evidence for geographically related variant genotypes. *J Clin Microbiol*. 2014;52(5):1687–90.

SUMMARY

Human polyomaviruses (HPyVs) are a group of small (40–45 nm), nonenveloped, icosahedral viruses with a circular dsDNA genome of ~5 kb that commonly occur in animal and human populations. Transmitted via contaminated food/water (e.g., BKPyV, JCPyV) or skin contact (e.g., MCPyV), human polyomaviruses invade specific host cells and persist in two distinct nonpathological states, that is, latency or low-level replication. Any changes in host immune function (e.g., AIDS and SOT) may reactivate the viruses, leading to active viral replication and various clinical diseases (e.g., PVAN, PML, *MCC*). Except for MCPyV in MCC, none of the new HPyVs has yet to be implicated in human cancer. Recent application of molecular techniques has helped uncover many details about polyomavirus gene structures and regulation and enabled development of efficient tracking tools for polyomaviruses. Nonetheless, much is to be learnt about polyomaviruses, including their pathogenic mechanisms and oncogenic potential. Given the increasing sophistication of molecular techniques, it will not be long before the pathogenesis of polyomavirus infection is fully established and specific therapeutics for polyomavirus-induced diseases are discovered.

Genetics of Emerging Human Coronaviruses

7

Ming Wang
China Agricultural University

Dongyou Liu
Royal College of Pathologists of Australasia Quality Assurance Programs

Contents

7.1 INTRODUCTION

Although coronaviruses had been known to cause respiratory, enteric, and systemic infections in chicken and other animals from the 1930s, their role in human diseases was only established following the isolation of coronavirus strains OC43 and 229E from human respiratory samples by two independent groups in 1966 [1]. AS human coronaviruses (e.g., OC43 or 229E) are generally associated with mild respiratory symptoms (or common cold), they have not attracted much attention until 2003, when the sudden emergence of severe acute respiratory syndrome (SARS) involving a novel coronavirus (SARS-CoV) resulted in 8,096 cases and 774 deaths (9.5% mortality) [2]. This incidence expedited the research on human coronaviruses and led to the identification of two additional strains (NL63 and HKU1) in 2004 and 2005, respectively, which are implicated in common cold and/or pneumonia. After several years of relative inactivity, the world was awakened by the epidemics of Middle East respiratory syndrome (MERS) due to another new coronavirus (MERS-CoV) in 2012, which affected a total of 2,521 individuals with 919 deaths (36.4% mortality) [3]. In December 2019, a previously unknown coronavirus (SARS-CoV-2) was linked to a SARS-like disease (coronavirus disease of 2019 or COVID-19) that rapidly swept the world, producing 88,383,771 cumulative cases and 1,919,126 deaths (2.2% mortality) in 223 countries, areas or territories by January 11, 2021 [4].

Taxonomy. Falling under the order *Nidovirales* (which covers enveloped, nonsegmented, positive-sense, single-stranded RNA viruses characterized by their relatively large genomes; highly conserved genomic organization, with a large replicase gene preceding structural and accessory genes; expression of many nonstructural genes by ribosomal frameshifting; several unique or unusual enzymatic activities encoded within the large replicase–transcriptase polyprotein; and expression of downstream genes by synthesis of 3′ nested subgenomic mRNAs), the family Coronaviridae encompasses 2 subfamilies (Letovirinae and Orthocoronavirinae), 5 genera, 26 subgenera, and 47 species. While the subfamily Letovirinae comprises a single genus (Alphaletovirus), a single subgenus (Milecovirus), and a single species (Microhyla letovirus 1 or MLeV) affecting the ornamented pygmy frog (*Microhyla fissipes*), the subfamily Orthocoronavirinae is separated into four genera (Alphacoronavirus, Betacoronavirus, Gammacoronavirus, and Deltacoronavirus). In turn, the genus Alphacononavirus is divided into 14 subgenera, with human coronavirus 229E (HCoV-229E) found in the subgenus Duvinacovirus and human coronavirus NL63 (HCoV-NL63) in the subgenus Setracovirus. Furthermore, the genus Betacoronavirus is differentiated into five subgenera (Embecovirus, Hibecovirus, Merbecovirus, Nobecovirus, and Sarbecovirus), with human coronavirus OC43 (HCoV-OC43, also known as betacoronavirus 1 strain OC43) and human coronavirus HKU1 (HCoV-HKU1) found in the subgenus Embecovirus, Middle East respiratory syndrome-related virus (MERS-CoV) in the subgenus Merbecovirus, and severe acute respiratory syndrome-related coronavirus (SARS-CoV or SARS-CoV-1) and severe acute respiratory syndrome-related coronavirus 2 (SARS-CoV-2) in the subgenus Sarbecovirus (Table 7.1) [5]. Despite their classification in the same subgenus (Sarbecovirus) and their 79% sequence homology, SARS-CoV-2 is not a descendent of SARS-CoV and belongs to a different clade from SARS-CoV [6].

It is notable that of seven human coronaviruses identified to date, HCoV-229E, HCoV-NL63, HCoV-OC43, and HCoV-HKU1 are mainly implicated in nonfatal common cold (lower respiratory tract infection), croup/laryngotracheitis (HCoV-NL63), pneumonia (HCoV-HKU1), and severe lower respiratory tract infection (HCoV-OC43), while SARS-CoV, SARS-CoV-2, and MERS-CoV are responsible for severe respiratory diseases (SARS, COVID-19, and MERS), with fatality rates of 9.5%, 2.2%, and 36.4%, respectively (Table 7.1) [7,8].

Morphology. Coronaviruses are spherical or pleomorphic, enveloped, nonsegmented particles of ~120 nm in diameter harboring a linear, positive-sense, single-stranded (ss) RNA genome of ~30,000 nt (ranging from 27,317 to 30,738 nt in human coronaviruses). The virion possesses an envelope made up of envelope proteins (E) and membrane proteins (M) and decorated by two sets of surface projections (spikes), i.e., the long drumstick-shaped projections (~19 μm in length) composed of S proteins (which give coronaviruses their characteristic crown-like appearance), and the shorter club-shaped projections (~6 μm in length) composed of HE proteins [which are present in the genus Betacoronavirus, subgenus Embecovirus (e.g., HCoV-OC43 and HCoV-HKU1), but absent in the genus Betacoronavirus, subgenus Sarbecovirus (i.e., SARS-CoV and SARS-CoV-2), and subgenus Merbecovirus (e.g., MERS-CoV)]. Beneath the envelope is a helically symmetrical nucleocapsid formed by genomic RNA and phosphorylated nucleoprotein (N), which is buried inside phospholipid bilayers (Figure 7.1) [5].

Biology. Similar to other coronaviruses, human coronaviruses rely on host cells for replication. During host cell invasion, the S protein (as well as the HE protein in the case of HCoV-OC43 and HCoV-HKU1t) binds to the host cell receptor [e.g., angiotensin-converting enzyme 2 (ACE2) for SARS-CoV and SARS-CoV-2 and dipeptidyl peptidase 4 (DDP4) for MERS-CoV)] and mediates endocytosis of the virus into the host cell. Moreover, the S protein (S2) assists in the fusion of the virus membrane with the endosomal membrane and the release of ssRNA(+) genome into the cytoplasm. Subsequent synthesis from the genomic ssRNA(+), which serves as both the viral messenger RNA and genome, and proteolytic cleavage of replicase polyprotein into viral polymerase (RdRp) allows the generation of a double-stranded (ds)RNA genome. Transcription of the dsRNA genome yields viral mRNAs/new ssRNA(+) genomes, and translation of structural proteins from subgenomic mRNAs (sgRNAs) facilitates viral assembly at membranes of the endoplasmic reticulum (ER), the intermediate compartments, and/or the Golgi complex. Finally, new virions exit the host cells through budding and exocytosis [9,10].

Epidemiology. Orthocoronaviruses are infective to various animals (e.g., bats, rats, dogs, cats, piglets, whales, chicken, turkeys) and cause species-specific illness. It is generally believed that human coronaviruses originate from animal reservoirs (e.g., bats, rodents), and pass through intermediate hosts (e.g., civets

TABLE 7.1 Biological, clinical, and genetic characteristics of human coronaviruses

SPECIES	*ABBREVIATION*	*GENUS/SUBGENUS*	*ANIMAL RESERVOIR/ INTERMEDIATE HOST*	*GENOME (NT)*	*RECEPTOR*	*CLINICAL DISEASE*	*YEAR OF INITIAL IDENTIFICATION*
Human coronavirus 229E	HCoV-229E	Alphacoronavirus/ Duvinacovirus	Bats/camelids?	27,317	Aminopeptidase (APN)	Common cold (mild respiratory tract infection); incubation period 2–5 days; transmission via respiratory droplets and fomites; case fatality not available	1966
Human coronavirus NL63	HCoV-NL63	Alphacoronavirus/ Setracovirus	Bats/ unidentified	27,553	Angiotensin-converting enzyme 2 (ACE2)	Common cold (mild respiratory tract infection); incubation period 2–4 days; transmission via respiratory droplets and fomites; case fatality not available	2004
Human coronavirus OC43 (Betacoronavirus 1 strain OC43)	HCoV-OC43	Betacoronavirus/ Embecovirus	Rodents/bovines	30,738	9-*O*-acetylated sialic acid (9-*O*-Ac-Sia)	Common cold (mild respiratory tract infection); incubation period of 2–5 days; transmission via respiratory droplets and fomites; case fatality not available	1966
Human coronavirus HKU1	HCoV-HKU1	Betacoronavirus/ Embecovirus	Rodents/ unidentified	29,926	9-*O*-acetylated sialic acid (9-*O*-Ac-Sia)	Common cold (mild respiratory tract infection) and pneumonia; incubation period 2–4 days; transmission via respiratory droplets and fomites; case fatality not available	2005

(*Continued*)

TABLE 7.1 (*Continued*) Biological, clinical, and genetic characteristics of human coronaviruses

SPECIES	*ABBREVIATION*	*GENUS/SUBGENUS*	*ANIMAL RESERVOIR/ INTERMEDIATE HOST*	*GENOME (NT)*	*RECEPTOR*	*CLINICAL DISEASE*	*YEAR OF INITIAL IDENTIFICATION*
Severe acute respiratory syndrome-related coronavirus	SARS-CoV (SARS-CoV-1)	Betacoronavirus/ Sarbecovirus	Bats/palm civets	29,751	Angiotensin-converting enzyme 2 (ACE2)	Severe acute respiratory syndrome (SARS, also known as atypical pneumonia); incubation period 2–11 days; transmission via respiratory droplets, fomites, and fecal–oral; case fatality 9.5%	2003
Severe acute respiratory syndrome-related coronavirus-2	SARS-CoV-2	Betacoronavirus/ Sarbecovirus	Bats/pangolins?	29,903	Angiotensin-converting enzyme 2 (ACE2)	Coronavirus disease 2019 (COVID-19); incubation period 3–6 days; transmission via respiratory droplets, fomites, and fecal–oral; case fatality 2.2%	2019
Middle East respiratory syndrome-related coronavirus	MERS-CoV	Betacoronavirus/ Merbecovirus	Bats/dromedary camels	30,119	Dipeptidyl peptidase 4 (DDP4)	Middle East respiratory syndrome (MERS); incubation period 2–13 days; transmission via respiratory droplets and fomites; case fatality 36.4%	2012

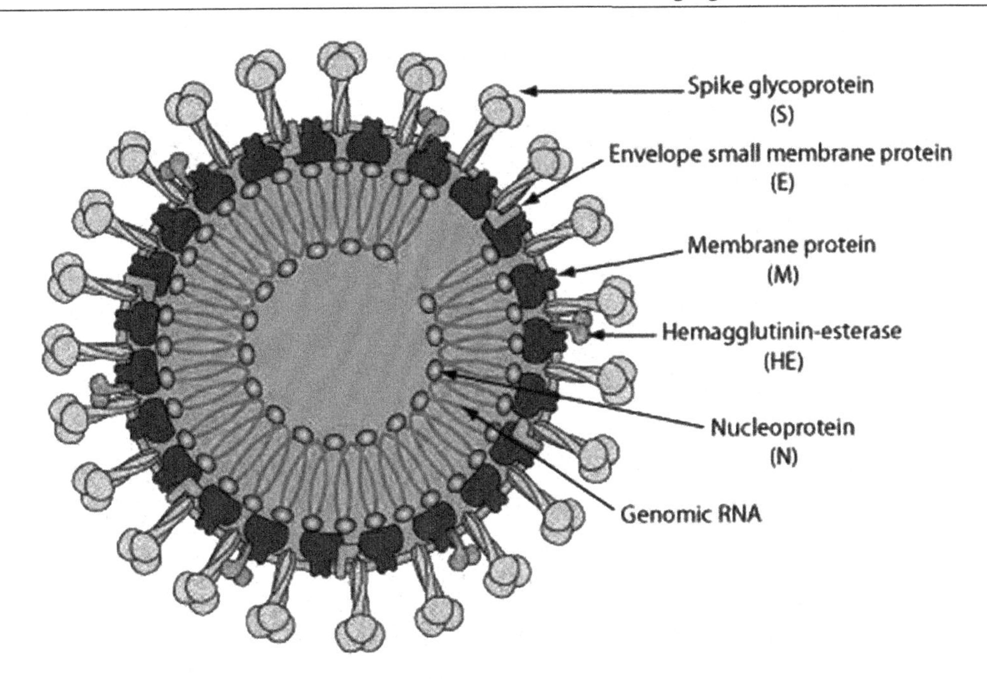

FIGURE 7.1 Schematic representation of betacoronavirus virion showing drumstick-shaped spike glycoprotein (S), club-shaped hemagglutinin-esterase [HE, which is present in the genus Betacoronavirus, subgenus Embecovirus (e.g., HcoV-HKU1 and HCoV-OC43), but absent in the genus Betacoronavirus, subgenus Sarbecovirus (i.e., SARS-CoV and SARS-CoV-2), and subgenus Merbecovirus (e.g, MERS-CoV)], membrane protein (M), envelope protein (E), and nucleocapsid composed of genomic RNA and phosphorylated nucleoprotein (N). (Photo credit: http://ruleof6ix.fieldofscience.com/2012/09/a-new-coronavirus-should-you-care.html.)

for SARS-CoV, dromedary camels for MERS-CoV, and possibly pangolins for SARS-CoV-2) before the establishment in humans [11–13]. Person-to-person spread of SARS-CoV, SARS-CoV-2, and MERS-CoV often takes place via respiratory droplets (coughs or sneezes), formites, and/or oral–fecal routes. Indeed, affected patients may shed SARS-CoV and SARS-CoV-2 in urine and stools typically between days 12–14 postinfection, and the shed viruses remain infectious for up to 4 days from patients with diarrhea [14].

The SARS epidemic in 2003 occurred predominately in Southeast Asia and had a total of 8,096 confirmed cases (including ~2,700 in China, ~1,000 in Hong Kong, ~220 in Taiwan, ~160 in Singapore, and a small number in 28 other countries), and 774 deaths (9.5% mortality). The MERS epidemic in 2012 generated 2,521 confirmed cases in 27 countries (mainly in the Middle East) and 919 deaths (36.4% mortality). However, the COVID-19 epidemic emerging in December 2019 has resulted in 88,383,771 cumulative cases and 1,919,126 deaths (2.2% mortality) over 223 countries, areas and territories by January 11, 2021.

Clinical features. SARS, COVID-19, and MERS are essentially respiratory diseases that present with fever, cough, dyspnea, and sore throat, as well as other symptoms, with fatality rates of 9.5%, 5.7%, and 36.4%, respectively (Table 7.2). Underlying conditions (e.g., *cardiovascular disease, chronic lung disease, diabetes, and malignancy) have a negative impact on disease outcomes. Common complications include intensive care unit admission, acute respiratory distress syndrome* (ARDS), *acute kidney injury* (AKI), *and deaths in hospitalized patients* (Table 7.2). MERS tends to have a more frequent development of ARDS and AKI than SARS and COVID-19, due possibly to the common occurrence of DDP4 receptor in tubules and glomeruli. A certain proportion of patients with SARS, COVID-19, and MERS may be asymptomatic [15].

Pathogenesis. While both SARS-CoV and SARS-CoV-2 recognize the same angiotensin-converting enzyme 2 (ACE2) receptor, which is an ectoenzyme anchored to the plasma membrane of the cells in the nasal mucosa, bronchus, lungs, heart, esophagus, kidneys, stomach, bladder, and ileum, they differ in their binding capacity [16,17]. Specifically, SARS-CoV contains mutations in the receptor-binding domain (RBD) of the S protein that strengthens its affinity with the ACE2 receptor. On the other hand,

TABLE 7.2 Clinical, radiographic, and laboratory findings of SARS, COVID-19, and MERS

DISEASE	*CLINICAL*	*RADIOGRAPHIC*	*LABORATORY*	*UNDERLYING CONDITIONS*	*COMPLICATIONS*
SARS	Fever (99%–100%), cough (57%–75%), dyspnea (40%–42%), sore throat (13%–25%), dizziness/confusion (4%–43%), diarrhea (23%–70%), nausea/vomiting (20%–35%)	Unilateral infiltrate (46%–54%), bilateral infiltrate (29%–45%), no finding (13%–25%)	Leukopenia (33.9%), lymphopenia (54%–70%), thrombocytopenia (44.8%), elevated aminotransferases (23%)	Cardiovascular disease (8%), chronic lung disease (1%–2%), diabetes (16%), malignancy (6%)	Intensive care unit admission (23%–34%), acute respiratory distress syndrome (20%), acute kidney injury (6.7%), deaths in hospitalized patients (3.6%–15.7%)
COVID-19	Fever (81%–91%), cough (48%–68%), dyspnea (19%–31%), sore throat (29%), dizziness/confusion (22%), diarrhea (16%), nausea/vomiting (6%)	Unilateral infiltrate (10%), bilateral infiltrate (84%–90%), no finding (14%)	Leukopenia (35%), lymphopenia (35%–72%), thrombocytopenia (12%), elevated aminotransferases (28%–35%)	Cardiovascular disease (10%–46%), chronic lung disease (1%–2%), diabetes (10%), malignancy (2%–4%)	Intensive care unit admission (24%), acute respiratory distress syndrome (18%–30%), acute kidney injury (3%), deaths in hospitalized patients (10%–11%)
MERS	Fever (81.7%–98%), cough (56.9%–83%), dyspnea (22%–72%), sore throat (9.1%–14%), dizziness/confusion (5.4%), diarrhea (19.4%–26%), nausea/vomiting (14%–21%)	Unilateral infiltrate (14.3%–62.6%), bilateral infiltrate (37.4%–75%), no finding (4.3%–30%)	Leukopenia (14%), lymphopenia (32%), thrombocytopenia (36%), elevated aminotransferases (11%–40%)	Cardiovascular disease (9.1%), chronic lung disease (10.2%), diabetes (18.8%), malignancy (15.5%)	Intensive care unit admission (53%–89%), acute respiratory distress syndrome (20%–30%), acute kidney injury (41%–50%), deaths in hospitalized patients (30%–40%)

SARS-CoV-2 has no amino acid substitutions in the RBD of the S protein but possesses six mutations in other regions of the RBD, increasing its efficiency for receptor binding. After entry into the host cells, SARS-CoV replicates in pneumocytes and enterocytes and downregulates the ACE2 receptor in the lung epithelium, leading to acute lung injury, acute respiratory disease syndrome (ARDS), and cytokine storm (as in COVID-19). ARDS is known to prevent enough oxygen from getting to the lungs and into the circulation and accounts for mortality of most respiratory disorders and acute lung injury [18]. MERS-CoV utilizes the dipeptidyl peptidase 4 (DPP4) receptor for entry into host cells (tubules and glomeruli) and behaves similarly to SARS-CoV and SARS-CoV-2 in its subsequent actions [19,20].

Diagnosis. Patients with SARS, COVID-19, and MERS may show radiographic (e.g., unilateral infiltrate, bilateral infiltrate, or no finding) and laboratory (leukopenia, lymphopenia, thrombocytopenia, elevated aminotransferases) findings (Table 7.2).

On CT, COVID-19 may show bilateral pulmonary parenchymal ground-glass, consolidative, or "crazy paving" pulmonary lesions, with a rounded shape and a peripheral distribution.

RT-PCR, alternative isothermal amplification, and CRISPR-Cas13a-based specific high sensitivity enzymatic reporter UnLOCKing (SHERLOCK) system may be utilized for specific and sensitive detection of SARS-CoV, SARS-CoV-2 (targeting ORF 1a, ORF 1b, S gene, N gene, M gene, RdRp gene, and 3′UTR), and MERS-CoV (targeting upE, ORF 1b, and ORF 1a) [21,22].

Furthermore, serological tests [e.g., enzyme-linked immunosorbent assay (ELISA), chemiluminescence assay (CLIA), immunofluorescence assay (IFA), western blot (WB), protein microarray, and neutralization] are useful for surveillance, prediction of disease outcome, and epidemiological investigation but not for early diagnosis of SARS, COVID-19, and MERS [23].

Treatment and prevention. Currently, no single specific antiviral therapy is available for SARS, COVID-19, and MERS. Various vaccine strategies (e.g., inactivated viruses, live-attenuated viruses, viral vector-based vaccines, subunit vaccines, recombinant proteins, and DNA vaccines) are being developed for control of these diseases [24].

Prevention of human coronavirus infections should focus on avoidance of the unprotected contact with both farm and wild animals, implementation of high vigilance, prompt screening and isolation of suspected cases, and development of vaccines [25,26].

7.2 GENETICS OF EMERGING HUMAN CORONAVIRUSES

Genome organization. Coronaviruses possess an enveloped, nonsegmented, polycistronic, positive-sense, single-stranded RNA (+ssRNA) genome of ~30,000 nt. Overall, SARS-CoV, SARS-CoV-2 and MERS-CoV share 82% identity in genome sequences, and >90 identity in essential enzyme and structural protein gene sequences. Organizationally, coronavirus genome is typically arranged in the order of the 5′ untranslated region (UTR); open reading frames ORF1a and ORF1b encoding nonstructural replicase proteins; genes encoding structural proteins spike (S), envelope (E), membrane (M), and nucleocapsid (N); accessory genes 3–9 encoding nonstructural accessory proteins; and the 3′ UTR (Figure 7.2) [27]. In the case of SARS CoV-2, a total of 11 genes with 11 open reading frames (ORFs) are found in the genome, that is, ORF1ab and ORF2 (spike protein); ORF3a and ORF4 (envelope protein); ORF5 (membrane protein); ORF6, ORF7a, ORF7b, ORF8, and ORF9 (nucleocapsid protein); and ORF10 (Table 7.3) [28,29]. It appears that SARS CoV, SARS CoV-2 and MERS-CoV differ mainly in ORF1a and the sequence of gene encoding spike (S) protein, which is involved in viral interaction with target cells.

The 5′-UTR contains a leader sequence and multiple stem loop structures, which act as mRNA for translation of the replicase polyproteins through frameshifting from ORF1a and ORF1b, which occupy two-thirds of the genome (about 20,000 nt). Furthermore, transcriptional regulatory sequences (TRSs) that are indispensable for the expression of each of these genes are at the beginning of each structural or accessory gene. The 3′ UTR includes a poly (A) tail and other RNA structures for replication and synthesis of viral RNA [30,31].

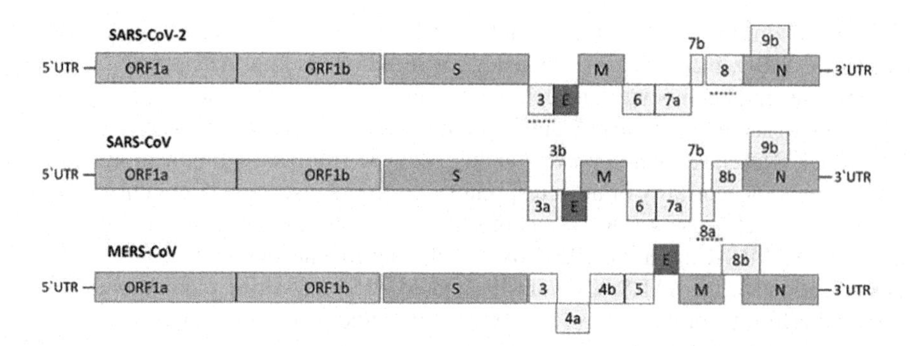

FIGURE 7.2 Genome organization of betacoronaviruses SARS-CoV-2, SARS-CoV, and MERS-CoV. The 29,903 nt ssRNA genome of SARS-CoV-2 contains the 5′-untranslated region (5′-UTR), open reading frames (orf) 1a/b (green box) encoding nonstructural proteins (nsp) for replication, genes encoding structural protein spike (S), envelope (E), membrane (M), and nucleocapsid (N), accessory proteins (orf 3, 6, 7a, 7b, 8, and 9b), and the 3′-untranslated region (3′-UTR). The red dotted lines indicate proteins with notable variations between SARS-CoV-2 and SARS-CoV. (Photo credit: Shereen MA, Khan S, Kazmi A, Bashir N, Siddique R. *J Adv Res.* 2020;24:91–8.)

TABLE 7.3 Open reading frames and encoded proteins in SARS CoV-2 genome

ORF NUMBER	*GENE*	*NUCLEOTIDE POSITIONS*	*PROTEIN SIZE (AA)*	*PROTEIN NAME*
1	ORF1ab	266–21,555	7,096	ORF1ab polyprotein
1	ORF1a	266–13,483	4,405	ORF1a polyprotein
2	ORF2 (S)	21,563–25,384	1,273	Spike (S) protein
3	ORF3a	25,393–26,220	275	ORF3a protein
4	ORF4 (E)	26,245–26,472	75	Envelope (E) protein
5	ORF5 (M)	26,523–27,191	222	Membrane (M) protein
6	ORF6	27,202–27,387	61	ORF6 protein
7	ORF7a	27,394–27,759	121	ORF7a protein
8	ORF7b	27,756–27,887	43	ORF7b protein
9	ORF8	27,894–28,259	121	ORF8 protein
10	ORF9 (N)	28,274–29,533	419	Nucleocapsid (N) protein
11	ORF10	29,558–29,674	38	ORF10 protein

Structural proteins. Encoded by the S, E, M, and N genes located downstream of ORF1b, structural proteins are involved in the formation of coronavirus particles [32].

The spike (S) protein (~150 kDa) is a glycoprotein, which functions as a class I fusion protein and mediates attachment to the host receptor (angiotensin-converting enzyme 2 or ACE2 in the case of SARS-CoV and SARS-CoV-2) and utilizes an N-terminal signal sequence to gain access to the ER. The S protein is cleaved by a host furin-like protease into two separate polypeptides S1 (which represents the large receptor-binding domain of the S protein) and S2 (which is the stalk of the spike projection) in SARS-CoV-2 but not in SARS-CoV (which lacks a furin cleavage site QTQTNSPRRARSVASQSIIA in the S protein).

The membrane (M) protein (25–30 kDa) has a small N-terminal glycosylated ectodomain and a much larger C-terminal endodomain extending 6–8 nm into the viral particle and interacts with the nucleocapsid (N) protein to encapsidate the RNA genome, facilitating viral assembly.

The envelope (E) protein (8–12 kDa) is a transmembrane protein with an N-terminal ectodomain and a C-terminal endodomain. Involved in the virus assembly and the release of the virus, the E protein oligomerizes and creates an ion channel, which is not required for viral replication but is critical for viral pathogenesis [33].

The nucleocapsid (N) protein (or nucleoprotein) has two separate domains, an N-terminal domain (NTD) and a C-terminal domain (CTD), both of which are capable of binding viral RNA. Through interactions with nsp3 and the M protein, the N protein helps tether the viral genome to the replicase–transcriptase complex (RTC) and packages the encapsidated genome into viral particles.

The hemagglutinin-esterase (HE, 48 kDa), which is present in the genus Betacoronavirus, subgenus Embecovirus (e.g., HcoV-HKU1 and HCoV-OC43) but absent in the genus Betacoronavirus, subgenus Sarbecovirus (i.e., SARS-CoV and SARS-CoV-2) and subgenus Merbecovirus (e.g., MERS-CoV), acts as a hemagglutinin for binding sialic acids on surface glycoproteins and shows acetyl-esterase activity, which together enhance S protein-mediated cell entry and virus spread through the mucosa.

Nonstructural proteins. Translation of ORF1a and ORF1b from coronavirus genomic mRNA generates polyproteins pp1a and pp1ab via –1 ribosomal frameshifting, which are then cleaved by viral chymotrypsin-like protease ($3CL^{pro}$) or main protease (M^{pro}) and one or two papain-like proteases into 16 nonstructural proteins (NSPs. or 15 in the genus Gammacoronavirus), which form the replication–transcription complex (RTC) in double-membrane vesicles (DMVs) involved in virus replication (Table 7.4) [6,32].

Accessory proteins. In SAR-CoV-2 genome, accessory proteins are encoded by ORF3a, ORF6, ORF7a, ORF7b, ORF8, and ORF10. These proteins are nonessential for replication in tissue culture, but some play important roles in viral pathogenesis. Specifically, the ORF3a protein is an ion channel protein that binds to

TABLE 7.4 Nonstructural proteins (NSPs) derived from ORF1ab polyprotein in SARS CoV-2 genome

NSP	*PROTEIN SIZE (aa)*	*PROPOSED FUNCTION*
NSP1	180	Also known as leader protein, which promotes host mRNA degradation selectively and through binding to the 40S ribosome of the host cell, inactivates translation, and inhibits interferon (IFN) signaling, leading to subdued host innate immune response
NSP2	638	Binds to prohibitin proteins (PHBs) 1 and 2, which are involved in cell cycle progression, migration, differentiation, apoptosis, and mitochondrial biogenesis, and thus plays a possible role in the disruption of host cell environment
NSP3	1945	As a papain-like proteinase protein (~200 kDa), NSP3 interacts with N protein via its Ubl1 and Ac domains; promotes cytokine expression via its ADRP activity; releases NSP1, NSP2, and NSP3 from the N-terminal region of polyproteins 1a and 1ab; blocks host innate immune response via its PLPro/deubiquitinase domain; and performs other unspecified functions via Ubl2, NAB, G2M, SUD, and Y domains
NSP4	500	Interacts with NSP3 (and possibly host proteins) for double-membrane vesicle (DMV) formation as well as for viral replication
NSP5	306	3C-like proteinase ($3CL^{pro}$) and main protease (M^{pro}) cleave at 11 sites of viral polyprotein to yield mature and intermediate nonstructural proteins (NSPs) and inhibit host IFN signaling
NSP6	290	Generates autophagosomes from endoplasmic reticulum (ER), which facilitate the assembly of replicase proteins; restricts autophagosome/lysosome expansion, thus preventing autophagosomes from delivering viral components for degradation in lysosomes; and induces DMV formation as transmembrane scaffold protein
NSP7	83	Forms hexadecameric complex with NSP8 and NSP12 to yield RNA polymerase activity of NSP8 and thus acts as processivity clamp for RNA polymerase
NSP8	198	Forms hexadecameric complex with NSP7 and NSP12 and acts as processivity clamp for RNA polymerase

(*Continued*)

TABLE 7.4 (*Continued*) Nonstructural proteins (NSPs) derived from ORF1ab polyprotein in SARS CoV-2 genome

NSP	*PROTEIN SIZE (aa)*	*PROPOSED FUNCTION*
NSP9	113	Interacts with the DEAD-box RNA helicase 5 (DDX5) cellular protein for viral replication
NSP10	139	Interacts with NSP14, which is an *S*-adenosylmethionine (SAM)-dependent (guanine-N7) methyl transferase (N7-MTase); stimulates the activity of NSP16, which is a 2′-*O*-methyltransferase (2-*O*-MT)
NSP11	13	Unknown; the first nine amino acids (sadaqsfln) in NSP11 are identical to the first nine in NSP12
NSP12	932	As primer-dependent RNA-dependent RNA polymerase (RdRp), NSP12 forms a complex with NSP7–NSP8 heterodimer and NSP8 monomer to confer processivity of NSP12; presence of NSP7 and NSP8 lowers the dissociation rate of NSP12 from RNA, leading to poor processivity in RNA synthesis
NSP13	601	RNA helicase unwinds duplex RNA; binds with NSP12 to increase helicase activity; functions as 5′ triphosphatase to introduce 5′-terminal cap of viral mRNA
NSP14	527	Guanine N7 methyltransferase adds 5′ cap to viral RNA; 3′ to 5′ exonuclease proofreads viral genome
NSP15	346	EndoRNAse/endoribonuclease degrades viral polyuridine sequences to evade host dsRNA sensors
NSP16	298	2′-*O*-ribose methyltransferase methylates the 2′-hydroxy group of adenine using *S*-adenosylmethionine as the methyl source and shields viral 5′-cap RNA from MDA5 recognition, exerting a negative impact on host innate immunity

TRAF3, which in turn activates ASC ubiquitination and leads to activation of caspase 1 and IL-1β maturation. The ORF6 protein interacts with NSP8, promoting RNA polymerase activity. The ORF7a protein is a type I transmembrane protein. The ORF7b protein is localized in the Golgi compartment. SARS-CoV-2 has a single ORF8 protein, while SARS CoV has ORF8a and ORF8b proteins. The ORF8b protein binds to the IRF association domain (IAD) of interferon regulatory factor 3 (IRF3), which in turn inactivates interferon signaling. A putative ORF9b is present, and the ORF10 protein (38 aa) in SARS-CoV-2 has unknown function [33].

Phylogenetic relationship. Genome-wide comparison indicates that SARS-CoV-2 shares 79.5% and 50% sequence identity to SARS-CoV and MERS-CoV, respectively. Considering that SARS-CoV-2 and SARS-CoV demonstrate 94.6% sequence identity in the seven conserved replicase domains in ORF1ab, both are classified in the subgenus Sarbecovirus (lineage B) and genus Betacoronavirus [34–36]. Nonetheless, SARS-CoV-2 is not a descendent of SARS-CoV and is in a different clade from SARS-CoV despite its 79.5% sequence homology with SARS-CoV (Figure 7.3) [37].

7.3 FUTURE PERSPECTIVES

The emergence of highly pathogenic human coronaviruses SARS-CoV, SARS-CoV-2, and MERS-CoV, which are responsible for severe respiratory diseases SARS, COVID-19, and MERS, in the last two decades provides a clear reminder on the existence and danger of zoonotic microorganisms that may break loose and create havoc in totally unprepared human populations. The latest pandemic linked to SARS-CoV-2 is especially a case in point, producing 88,383,771 cumulative infections and 1,919,126

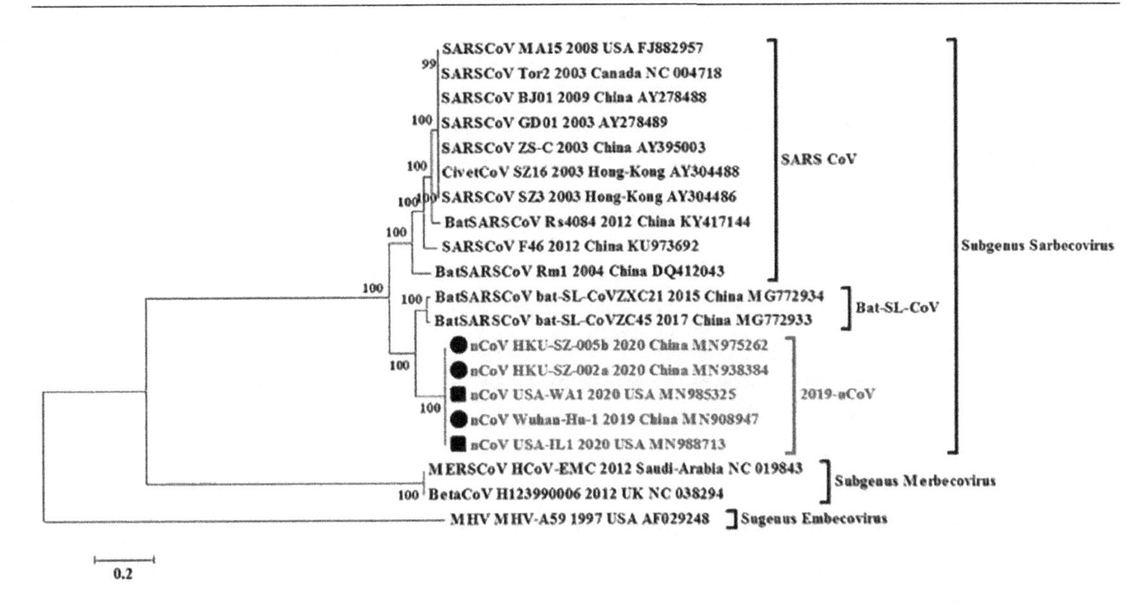

FIGURE 7.3 Phylogenetic relationship of SARS-CoV, SARS-CoV-2 (2019-nCoV), MERS-CoV, and betacoronaviruses of animal origin. The solid-black circles indicate SARS-CoV-2 isolates from China, and the solid-black squares indicate SARS-CoV-2 isolates from USA. (Photo credit: Malik YS, Sircar S, Bhat S, et al. *Vet Q.* 2020;40(1):68–76.)

deaths in 223 countries, areas and territories from December 2019 to January 11, 2021, with no ending on the horizon. Despite causing a lower mortality rate (2.2%), SARS-CoV-2 has a higher tendency to spread than SARS-CoV and MERS-CoV. The mechanism underlining the increased infectivity of SARS-CoV-2 is inadequately understood. This calls for further studies on the biology, epidemiology, immunology, and molecular pathogenesis of coronaviruses in general and SARS-CoV-2 in particular zoonotic with the goals to develop improved diagnostics, therapeutics, and vaccines for their control and eradication [38].

REFERENCES

1. Kahn JS; McIntosh K. History and recent advances in coronavirus discovery. *Pediatr Infect Dis J.* 2005; 24 (11): S223–7.
2. Song Z, Xu Y, Bao L, et al. From SARS to MERS, Thrusting coronaviruses into the spotlight. *Viruses.* 2019;11(1). pii: E59.
3. Al-Osail AM, Al-Wazzah MJ. The history and epidemiology of Middle East respiratory syndrome coronavirus. *Multidiscip Respir Med.* 2017;12:20.
4. Baek WK, Sohn SY, Mahgoub A, Hage R. A comprehensive review of severe acute respiratory syndrome coronavirus 2. *Cureus.* 2020;12(5):e7943.
5. Coronaviridae Study Group of the International Committee on Taxonomy of Viruses. The species severe acute respiratory syndrome-related coronavirus: Classifying 2019-nCoV and naming it SARS-CoV-2. *Nat Microbiol.* 2020;5(4):536–44.
6. Malik YS, Sircar S, Bhat S, et al. Emerging novel coronavirus (2019-nCoV)-current scenario, evolutionary perspective based on genome analysis and recent developments. *Vet Q.* 2020;40(1):68–76.
7. Brüssow H. The novel coronavirus - A snapshot of current knowledge. *Microb Biotechnol.* 2020;13(3):607–12.
8. Ciotti M, Angeletti S, Minieri M, et al. COVID-19 outbreak: An overview. *Chemotherapy.* 2020;1–9.
9. Fehr AR, Perlman S. Coronaviruses: An overview of their replication and pathogenesis. *Methods Mol Biol.* 2015;1282:1–23.

10. Chen Y, Liu Q, Guo D. Emerging coronaviruses: Genome structure, replication, and pathogenesis. *J Med Virol.* 2020;92(4):418–23.
11. Wang Y, Liu D, Shi W, et al. Origin and possible A47 genetic recombination of the Middle East respiratory syndrome coronavirus from the first imported case in China: Phylogenetics and coalescence analysis. *Virus Evol.* 2017;3(Suppl 1). pii: vew036.046.
12. Yusof MF, Queen K, Eltahir YM, et al. Diversity of Middle East respiratory syndrome coronaviruses in 109 dromedary camels based on full-genome sequencing, Abu Dhabi, United Arab Emirates. *Emerg Microbes Infect.* 2017;6(11):e101.
13. Sen S, Anand KB, Karade S, Gupta RM. Coronaviruses: Origin and evolution. *Med J Armed Forces India.* 2020;76(2):136–41.
14. Habibzadeh P, Stoneman EK. The novel coronavirus: A bird's eye view. *Int J Occup Environ Med.* 2020;11(2):65–71.
15. Rabaan AA, Al-Ahmed SH, Haque S, et al. SARS-CoV-2, SARS-CoV, and MERS-COV: A comparative overview. *Infez Med.* 2020;28(2):174–84.
16. Schäfer A, Baric RS, Ferris MT. Systems approaches to Coronavirus pathogenesis. *Curr Opin Virol.* 2014;6:61–9.
17. Gralinski LE, Baric RS. Molecular pathology of emerging coronavirus infections. *J Pathol.* 2015;235(2):185–95.
18. Jin Y, Yang H, Ji W, et al. Virology, Epidemiology, pathogenesis, and control of COVID-19. *Viruses.* 2020;12(4):372.
19. Letko M, Miazgowicz K, McMinn R, et al. Adaptive evolution of MERS-CoV to species variation in DPP4. *Cell Rep.* 2018;24(7):1730–7.
20. Rokni M, Ghasemi V, Tavakoli Z. Immune responses and pathogenesis of SARS-CoV-2 during an outbreak in Iran: Comparison with SARS and MERS. *Rev Med Virol.* 2020;30(3):e2107.
21. Loeffelholz MJ, Tang YW. Laboratory diagnosis of emerging human coronavirus infections - the state of the art. *Emerg Microbes Infect.* 2020;9(1):747–56.
22. Yan Y, Chang L, Wang L. Laboratory testing of SARS-CoV, MERS-CoV, and SARS-CoV-2 (2019-nCoV): Current status, challenges, and countermeasures. *Rev Med Virol.* 2020;30(3):e2106.
23. Hong KH, Lee SW, Kim TS, et al. Guidelines for laboratory diagnosis of coronavirus disease 2019 (COVID-19) in Korea. *Ann Lab Med.* 2020;40(5):351–60.
24. Ogimi C, Kim YJ, Martin ET, Huh HJ, Chiu CH, Englund JA. What's new with the old coronaviruses? *J Pediatric Infect Dis Soc.* 2020;9(2):210–7.
25. Xu J, Jia W, Wang P, et al. Antibodies and vaccines against Middle East respiratory syndrome coronavirus. *Emerg Microbes Infect.* 2019;8(1):841–56.
26. Ahn DG, Shin HJ, Kim MH, et al. Current status of epidemiology, diagnosis, therapeutics, and vaccines for novel coronavirus disease 2019 (COVID-19). *J Microbiol Biotechnol.* 2020;30(3):313–24.
27. Shereen MA, Khan S, Kazmi A, Bashir N, Siddique R. COVID-19 infection: Origin, transmission, and characteristics of human coronaviruses. *J Adv Res.* 2020;24:91–98.
28. Abdullahi IN, Emeribe AU, Mustapha JO, et al. Exploring the genetics, ecology of SARS-COV-2 and climatic factors as possible control strategies against COVID-19. *Infez Med.* 2020;28(2):166–73.
29. Uddin M, Mustafa F, Rizvi TA, et al. SARS-CoV-2/COVID-19: Viral genomics, epidemiology, vaccines, and therapeutic interventions. *Viruses.* 2020;12(5):E526.
30. Yang D, Leibowitz JL. The structure and functions of coronavirus genomic 3' and 5' ends. *Virus Res.* 2015;206:120–33.
31. Romano M, Ruggiero A, Squeglia F, Maga G, Berisio R. A structural view of SARS-CoV-2 RNA replication machinery: RNA synthesis, proofreading and final capping. *Cells.* 2020;9(5):E1267.
32. Yoshimoto FK. The proteins of severe acute respiratory syndrome coronavirus-2 (SARS CoV-2 or n-COV19), the cause of COVID-19. *Protein J.* 2020;39(3):198–216.
33. Schoeman D, Fielding BC. Coronavirus envelope protein: Current knowledge. *Virol J.* 2019;16(1):69.
34. Chafekar A, Fielding BC. MERS-CoV: Understanding the latest human coronavirus threat. *Viruses.* 2018;10(2). pii: E93.
35. Petrosillo N, Viceconte G, Ergonul O, Ippolito G, Petersen E. COVID-19, SARS and MERS: Are they closely related? *Clin Microbiol Infect.* 2020;26(6):729–34.
36. Ye ZW, Yuan S, Yuen KS, Fung SY, Chan CP, Jin DY. Zoonotic origins of human coronaviruses. *Int J Biol Sci.* 2020;16(10):1686–97.
37. Mousavizadeh L, Ghasemi S. Genotype and phenotype of COVID-19: Their roles in pathogenesis. *J Microbiol Immunol Infect.* 2020. doi:10.1016/j.jmii.2020.03.022.
38. Dhama K, Khan S, Tiwari R, et al. Coronavirus disease 2019–COVID-19. *Clin Microbiol Rev.* 2020;33:e00028–20.

SUMMARY

The family Coronaviridae comprises a large number of enveloped, nonsegmented, positive-sense, single-stranded RNA viruses with relatively large genomes and highly conserved genomic organization. While most coronaviruses cause respiratory, enteric, and systemic infections in chicken and other animals, seven are implicated in human respiratory diseases of varying severity (i.e., OC43, 229E, NL63, HKU1, SARS-CoV, MERS-CoV, and SARS-CoV-2). Notably, SARS-CoV (severe acute respiratory syndrome-related coronavirus) emerged in 2003, and resulted in 8,096 cases and 774 deaths (9.5% mortality); MERS-CoV (Middle East respiratory syndrome coronavirus) appeared in 2012 and affected a total of 2,521 individuals with 919 deaths (36.4% mortality); and SARS-CoV-2 (severe acute respiratory syndrome-related coronavirus 2) was linked to a SARS-like disease (COVID-19) in late 2019, producing 88,383,771 cumulative-cases and 1,919,126 deaths (2.2. mortality) in 223 countries, areas and territories by January 11, 2021. The emergence of highly pathogenic human coronaviruses SARS-CoV, SARS-CoV-2, and MERS in the last two decades provides a clear reminder on the existence and danger of zoonotic microorganisms that may break loose and create havoc in totally unprepared human populations. Further studies on the biology, epidemiology, immunology, and molecular pathogenesis of coronaviruses are urgently needed in order to develop accurate diagnostics and effective therapeutics for their control and eradication.

Molecular Mechanisms of Host Immune Responses to Rotavirus

8

Lijuan Yuan, Alissa Hendricks, and Ashwin Ramesh
Virginia Polytechnic Institute and State University

Contents

8.1 INTRODUCTION

Rotaviruses (RV) have been recognized as one of the leading viral etiological agents of acute gastroenteritis (AGE) in children <5 years of age in both developed and developing countries.[1] Despite the availability of RV vaccines, RV gastroenteritis is still responsible for ~215,000 deaths, 2.3 million hospitalizations, and 24 million outpatient visits among children <5 years of age annually.[2–4] RotaTeq™ (Merck), a pentavalent human-bovine reassortant oral vaccine, and Rotarix™ (GlaxoSmithKline), a monovalent attenuated human RV oral vaccine, are the two currently licensed, commercial RV vaccines that have been integrated into the national immunization programs of more than 100 countries.[5] These vaccines are highly efficacious (80%–90%) in high-income countries against severe RV AGE, but display only a moderate degree of protection (40%–60%) in low-income countries.[6–10] Improvement in the performance of these vaccines and development of alternative vaccine candidates are needed.[11] To date, the correlates of protective immunity against human rotavirus (HRV) diarrhea after vaccination have not been fully identified,[12] which hampers the development of more effective vaccines. Understanding the molecular mechanisms of host immune responses to RV infections will facilitate identification of the immune correlates and development of more effective vaccines.

8.1.1 Structure and Classification

RVs are classified in the *Rotavirus* genus within the family *Reoviridae*. Viruses belonging to this family are characterized by a segmented double-stranded RNA (dsRNA) genome encapsulated within a non-enveloped triple-layered protein (TLP) icosahedral capsid. Using electron cryomicroscopy, mature RV virions together with their protruding spike domains were measured to be around 100 nm in diameter.[13] The RV genome is made up of 11 segments of dsRNA that code for six viral structural proteins (VP1, VP2, VP3, VP4, VP6, and VP7) and six non-structural proteins (NSP1–6). All of the segments are monocistronic, excluding segment 11 that codes for NSP5 and NSP6.[14,15]

Contained within the TLP are the inner capsid layer (VP2) and the middle layer (VP6).[16,17] The outer capsid is comprised of the VP7 glycoproteins and the VP4 spike protein.[17,18] The VP4 protein, encoded by a single gene, gets cleaved by the digestive enzyme trypsin into two separate domains that maintain ionic interactions to form the stem VP5*, and the head VP8* that is responsible for attachment to host cellular receptors.[19,20] The inner capsid is made up of 120 molecules of VP2 proteins arranged as 60 dimers that protect the segmented dsRNA genome together with the RNA-dependent RNA polymerase (VP1)[21,22] and the mRNA capping enzyme (VP3) with methyltransferase and guanyltransferase activity.[23–25]

The *Rotavirus* genus currently contains eight species (groups) designated from A to H[26] and two tentative groups, RVI[27] and RVJ.[28] Strains belonging to RVA, RVB, and RVC groups are known to infect both humans and animals, whereas strains within RVD–RVH groups have only been known to infect animals. The two tentative groups RVI and RVJ have thus far only been isolated from animals. The traditional serological classification of RV groups was based on the divergence of antigens on VP6. Recently, a sequence-based classification method was developed to identify the genetic variability among VP6 proteins belonging to different RV species.[26] Within the RVA group, the G protein (Glycoprotein, VP7) and P protein (Protease-cleaved protein, VP4) are used to determine the serotypes, or genotypes, of RVA given their ability to segregate and therefore evolve independently.[15] The most prevalent RV infections in humans worldwide are caused by RVA, and therefore studies, if not specified, are of HRVs in this group.

8.1.2 Epidemiology of HRV

RV is highly contagious and transmitted through the fecal–oral route. It is commonly spread by contaminated food, water, hands, and fomites. A respiratory mode of transmission has also been proposed.[29] It has been found to cause disease in the young ones of many different animal species including, but not limited to, humans, cats, dogs, cattle, pigs, horses, rats, birds, and various exotic animal species such as giant pandas, raccoon dogs, emus, sugar gliders, salmon, and sea lions to name a few.[30]

The most common HRV types found to be endemic to the United States include G1P[8], G2P[4], G3P[8], G9P[8], and G12P[8]; although there are 8 predominant types in the United States, more than 40 types have been identified globally.[15,31] Since the introduction of vaccines, there has been a significant reduction in RV-associated diarrhea given the near 90% seroconversion rate and good vaccine coverage in high-income countries.[32,33]

HRV disease is most prevalent in the African and South- and Southeast-Asian countries.[34] A case study in 2013 analyzed the etiological causes of moderate to severe diarrhea in seven developing countries in sub-Saharan Africa and Southern Asia, and it found that the vast majority of the diarrhea cases within the first 2 years of life were caused by the HRV.[35] While the World Health Organization (WHO) advises that all countries should implement RV vaccines and more than 100 countries have licensed RotaTeq and/or Rotarix[5], the countries with the highest rates of diarrheal deaths have seen the lowest rate of RV vaccine seroconversions, averaging 43%.[32,36–38]

8.1.3 Pathogenesis of HRV

The traditional route of transmission (fecal–oral) of RV restricts infections to the mature non-dividing, differentiated enterocytes, located on the tip of villi,[39] as well as enteroendocrine cells, in the small intestine.[40,41] The main pathology of RV comes from damage caused by viral replication within the enterocytes causing malabsorptive diarrhea.[40–42] RV NSP4s are able to increase intracellular Ca^{2+} levels, which degrades tight junctions allowing paracellular water transportation. This can trigger the enteric nervous system's release of additional Ca^{2+} as well as Cl^-, contributing to the RV secretary diarrhea mechanism.[42,43] The classic histopathological markers of RV infections include villus blunting, crypt hyperplasia, and increased numbers of intraepithelial lymphocytes, but no significant inflammation.[40]

The RV utilizes integrins, sialic acids, and histo-blood group antigens (HBGAs) to trigger cellular entry of the viral particles.[44,45] The virus invades the enterocytes lining the middle and apical portions of villi, by direct membrane penetration or through endocytosis. Though there have been many advances in elucidating the entry pathway, it is still debated whether the virus utilizes endocytosis or direct entry through the cell membrane; evidence exists that the particles utilize both methods.[46] TLPs are released from non-polarized cells (MA104) by lysis,[47] but from epithelial cells (e.g., Caco-2) by a budding process that does not immediately kill the cells.[48] Through the activation of type I interferons (IFNs) and the production of virus-like small RNA (coded from NSP4), autophagy is initiated by RV, which could play a role in virus replication and the infection of additional cells.[49]

8.2 METHODOLOGIES

The molecular mechanisms of host immune responses have been investigated since the discovery of the RV to elucidate methods for the prevention of RV disease. As summarized in Table 8.1, the various methods discussed throughout this chapter have been used to determine initial virus recognition, transcriptional activity, protein localization, cytokine profile, antibody responses, B and T cell response, alongside many other properties.

TABLE 8.1 Key methods used to uncover the molecular mechanisms of host immune responses to RV

METHODS AND EXAMPLES	*REFERENCES*
FLOW CYTOMETRY (a laser-based assay that analyzes cells in a single-cell stream based on physical and chemical characteristics)	
• Determines the frequency of surface or intracellular receptors for a specific cell type	[50–53]
• Separates different cell types to determine the cells recruited at various times during infection	[50,54–57]
• Determines lymphocyte activation based on cell surface markers	[58]
Tetramer staining (based on flow cytometry–based assay, tetramer staining permits the *ex vivo* quantification and phenotypic characterization of T cells without T cell activation)	
• Characterizes antigen-specific T cells to particular T cell epitopes	[53]
POLYMERASE CHAIN REACTION (PCR) (PCR allows for the amplification of DNA; qPCR quantifies results and RT-PCR uses reverse transcriptase to analyze RNA)	
• Quantifies the mRNA levels downstream of transcription factors	[59–61]
• Determines the levels of viral RNA in different cells, tissues, and body fluids	[62]
• Compares the differences in mRNA in sorted cell types or different tissues	[50]
ENZYME-LINKED IMMUNOSORBENT ASSAY (ELISA) (a 96-well plate assay that uses enzyme-linked antibodies or antigens and color change to detect and quantify proteins)	
• Detects changes in cytokine production in biological fluids	[50,63,64]
• Quantifies antibody levels and their targets	[36–38,62,65–67]
ENZYME-LINKED IMMUNOSPOT (ELISPOT) ASSAY (modified from ELISA, ELISPOT enumerates antibody-producing plasma cells or cytokine-producing cells)	
• Detects effector and memory B cell responses to viral infection and vaccination	[68–70]
• Detects Th1, Th2, and regulatory T cell responses to infection and vaccination	[71]
Sf-9 cell immunocytochemical staining assay (a plate assay with recombinant baculovirus expressing foreign viral proteins as the capture antigen)	
• Measures viral-protein-specific serum antibody responses	[72,73]
WESTERN BLOT (a gel-based assay used to identify polypeptides; co-immunoprecipitation is a variation that the user will select for one peptide but stain for another to determine association)	
• Determines the presence and relative quantity of cytokines	[74]
• Elucidates the immune system antagonizing role of viral proteins	[74,75]
FLUORESCENCE MICROSCOPY or IMMUNOHISTOCHEMISTRY (use of fluorescent-tagged antibodies to visualize specific peptides)	
• Visualizes the localization, or co-localization, of peptides in culture or tissues	[63]
• Analyzes data from a reporter gene assay	[76]
REPORTER GENE ASSAY (integration of a plasmid containing a gene of interest with a fluorescent reporter tag)	
• Elucidates changes in the production or localization of a protein	[76]
SHORT INTERFERING RNA (siRNA) (a synthetic RNA duplex designed to specifically target a particular mRNA for degradation; it can be used for the inhibition of specific cellular processes)	
• Determines the role of individual proteins by creating a selective knockout	[59,77]

(*Continued*)

TABLE 8.1 (*Continued*) Key methods used to uncover the molecular mechanisms of host immune responses to RV

METHODS AND EXAMPLES	*REFERENCES*
SELCVTIVE ANTAGNOIST TREATMENTS (use of known inhibitors of intracellular processes to determine physiological requirements of certain pathways)	
• Brefeldin A to inhibit intracellular vesicle transportation	[78]
KNOCKOUT ANIMALS (animals bred or genetically engineered to lack a functioning gene of interest)	
• Determines the role of the gene or cell type of interest in controlling and regulating infection and illness	[79–83]
GNOTOBIOTIC (Gn) ANIMALS (raised under sterile conditions, devoid of maternal antibodies and all extraneous microbes, to minimize confounding variables)	
• Investigates the role of B cell, antibody and T cell responses in protective immunity	[68,70,84,85]
• Assesses the role of maternal antibodies	[86–89]
• Determines the role of the gut microbiome	[90–92]
• Elucidates the effects of probiotics and prebiotics	[52,93]

8.3 RECENT ADVANCES IN STUDIES OF HOST IMMUNE RESPONSES TO RV

8.3.1 Initial Virus Recognition of RV Infection

Germline encoded pattern recognition receptors (PRRs) are the first active line of defense against pathogens through the detection of pathogen-associated molecular patterns (PAMPs).[94] There are many PRRs that serve to recognize non-eukaryotic and non-self domains, and most notably in the defense against HRV, the recognition of dsRNA is by toll-like receptors (TLRs), RIG-I-like receptors (RLRs), and NOD-like receptors (NLRs). With TLRs, RLRs, and NLRs, epithelial cells and immune cells are capable of identifying initial infections and trigger antiviral pathways of the immune system.[95]

8.3.1.1 Toll-Like Receptors

Members of the TLR family are the major PRRs in cells. They are type I transmembrane proteins containing leucine-rich repeats that recognize viral and bacterial PAMPs in the extracellular environment (TLR1, TLR2, TLR4, TLR5, TLR6, and TLR11) or endolysosomes (TLR3, TLR7, TLR8, TLR9, and TLR10).[96] Multiple TLRs on mononuclear cells are upregulated after RV infection in humans and Gn pigs, including TLR2, TLR3, TLR4, TLR7, TLR8, and TLR9.[52,97] The presence of RV dsRNA is primarily detected by the intra-endosomal TLR3.[61,98] TLR7 signaling was found to activate dendritic and B cells after exposure to RV.[99] TLR2 has also been found to be activated by RV NSP4 in the extracellular space.[63]

A study in mice showed that TLR3-mediated production of type III IFN helps limit RV infectivity in adults; there is a strong inverse correlation between TLR3 activity and viral susceptibility in adults but not in neonates.[61] This study went on to test healthy human adult and child duodenal biopsies, finding that adults also had an increased TLR3 mRNA expression.[61] A higher level of TLR3 in adults relates to the viral epidemiology where adults are significantly less susceptible to HRV infection. However, adult mice remain susceptible to RV infection, indicating that TLR3-mediated signaling is not sufficient to protect against this pathogen. Through the use of RIG-I, MDA5, TLR3, and MAVS knockout mice and siRNA in IECs, it was shown that both RIG-I and MDA-5 were needed for IFN-β induction, while the knock-down of TLRs in the infected cells and animals did not affect the outcome of the infection.[59,100]

Downstream from TLR stimulation along the innate immune response signalling pathway is the adaptor myeloid differentiation primary response 88 (MyD88), which mediates signaling for all TLRs, except TLR3, and receptors for inflammasome cytokines IL-1 and IL-18.[101,102] The role of MyD88 signaling in the initial control of murine RV (EC strain) infection and the induction of RV-specific Ab responses were studied in MyD88 knockout mice.[83] MyD88 knockout mice shed significantly higher virus titers in feces at 2–4 days post-inoculation and had significantly higher copy numbers of the viral genome in the small intestine, colon, and peripheral blood than wild type mice. The loss of MyD88 also impaired humoral immunity to RV and shifted the IgG subtypes. These results indicate that MyD88-mediated TLR signaling contributes to the control of primary RV infection and the development of a properly polarized adaptive immunity to RV.[83]

8.3.1.2 RIG-1 Like Receptors

Two RLRs have been identified with the ability to detect dsRNA and are activated in the presence of replicating RVs. RIG-I is capable of detecting short- to medium-length dsRNA (from 21b to ~1kb), while melanoma differentiation associated protein 5 (MDA5) recognizes dsRNA of lengths >1kb.[59,103] Both of these proteins are able to activate very similar downstream cascades. Once activated, the RLRs will form a complex with the mitochondrial bound protein MAVS (mitochondrial-associated anti-viral signaling).[104] Alternatively, a third RLR can be activated, LGP2 (laboratory of genetics and physiology 2), which has been identified as a negative regulator of RIG-1/MDA5 in response to RV.[59]

Studies of the interactions between RV and the RLR pathway showed that RV modulates the pathway to decrease the production of IFN-β.[74] IECs infected with various strains of RV showed that the levels of MAVS proteins in the cells decreased as the levels of NSP3 and NSP1 increased throughout the progression of RV infection.[74] The addition of NSP1 to IECs drastically reduced cellular MAVS levels in a strain-independent manner. Further work on cell culture indicated that NSP1 is able to inhibit MAVS-mediated production of IFN-β in response to RV infection.[74]

Not only has it been shown that NSP1 can reduce the secretion of type I IFNs through the inhibition of MAVS in the cells, but it has also been shown to degrade RIG-1 in a proteasome-independent manner. This was determined through a multi-step analysis of the interactions of the two proteins. Western blots were used to show that the individual immunoprecipitation of NSP1 or RIG-1 from treated 293FT cells led to the coprecipitation of both proteins. When a truncated form of NSP1 that contained only the N-terminal domain was transfected into cells and translated, it was also sufficient to interact with RIG-1; this elucidated the specific domain of NSP1 that has evolved to interfere with the RLR. Moreover, through the use of a proteasome inhibitor, it was further established that NSP1 can directly degrade the protein even without the assistance of proteasomes.[75] These findings illustrate the evolution of RV with the innate immune system. It has been elucidated that RLRs are needed within the infected cells to initiate cytokine production, but for the virus' survival a non-structural protein, NSP1, has developed the ability to lessen the power of innate signaling throughout the course of infection.

8.3.1.3 NOD-Like Receptors

Although it was originally suspected in previous decades that NLRs played little to no role in the control of RV infection, recent studies have been able to uncover correlations and molecular interactions between specific NLRs and the RV genomic material. For other viral infections, it has been established in previous reviews that the pyrin and caspase domain containing NLR (NLRP and NLRC, respectively) subfamilies are able to respond to viral infection and activate programmed cell death pathways.[95,105] Using CRISPR-Cas9 technology, knockout mice were generated with diminished different sections of the NLR pathways, deducing the importance of the specific activation of Nlrp9b on IECs in the host defense against RV.[106] Nlrp9b deficiency results in increased susceptibility to RV infection. Increased RV susceptibility was also observed by eliminating Asc and Casp1, two proteins that are required for the formation of inflammasomes. Furthermore, upon establishing the NLR pathway activated by RV, its role in pathogenesis was analyzed. Using knockout mice and a mouse intestinal organoid model, Zhu et al. found that by the activation of Gsdmd by the Nlrp9b pathway pyroptosis, an inflammatory programmed cell death is activated to induce premature death of infected IECs, maintaining intestinal homeostasis, as shown through TUNEL staining of organoids that had been infected with RV *in vitro*.[106] The studies highlight an important anti-viral innate immune signaling pathway, the novel Nlrp9b NLR specifically expressed by IECs, in the host defense against RV infection.

8.3.2 Antiviral Cytokines and Their Associated Transcription Factors

In early innate responses of host cells to a viral infection, after the initiation of PRR cascades, antiviral genes are upregulated leading to the production and release of type I and III IFNs and pro-inflammatory cytokines through transcription factors such as NF-κB and IRF3 and 7. RVs have evolved multiple strategies to modulate IFN responses in infected cells[107], including the strategy to limit the expression of type I IFNs. NSP1 has also been known to bind to IRF3, 5, and 7 and mediate their degradation by proteasomes.[108,109] Figure 8.1 (adapted from the review by Arnold et al., *PLoS Pathog* 2013, with permission[110]) summarizes the interactions of RV with innate signaling pathways.

Many RV strains are able to inhibit the nuclear accumulation of NF-κB and the transcription factors STAT1 and STAT2.[76] RVs utilize a novel method for preventing the translocation of the STAT proteins.[111] Studies have found that the inhibition occurred after the STAT proteins were activated and bound to nuclear import proteins, without causing dysregulation to the import proteins or nuclear pores given that other transcription factors were unaffected. Although the viral mediator of this effect has not been identified, it would seem likely to be NSP1.[111]

To better understand the role of cytokines in RV infection, many animal studies have been conducted to evaluate the role of IFNs in controlling RV infections *in vivo*. While one study showed that the administration of INF-α was able to decrease the severity of RV infection in newborn calves, other studies found that knockout infant and adult mice without IFN-α/β receptors did not see a significant change in illness or infection.[79–81]

Type III IFNs have been found in recent years to play a substantial role in controlling enteric virus infection in intestinal epithelial cells. Engaging type III IFN receptors activates redundant pathways to type I; however it was also shown that mice without INF-λ receptors suffered from a much more severe infection than mice without type I (IFN-α/β) receptors.[79,80] One explanation for this differential effect is found in the fact that type I IFN receptors are nearly ubiquitous in nucleated cells, but type III IFN receptors are largely confined to epithelial cells.[112] Beyond these findings, it was also shown that there was an important synergistic effect generated by the combined production of INF-λ and IL-22. Both cytokines have receptors that are preferentially expressed by intestinal epithelial cells allowing for this effect to be critical for RV control.[113] Further studies are required to determine the exact nature of IFN-λ action against RV and whether IFN-λ is important for limiting RV infection in humans and other animals.

Though there are studies that suggest that type I and type III IFN play a critical role in RV infection clearance, it was later found that replication of the virus in intestinal epithelial cells was independent of either group IFN's function and an adaptive response was required to clear the infection.[114] What was found though is that these IFNs are able to control the rate of replication and severity of illness until an adaptive response can be generated, while type I IFNs were able to prevent the extra-intestinal spread of RV.[114] Altogether, these studies on IFNs offer an explanation to the early control of RV.

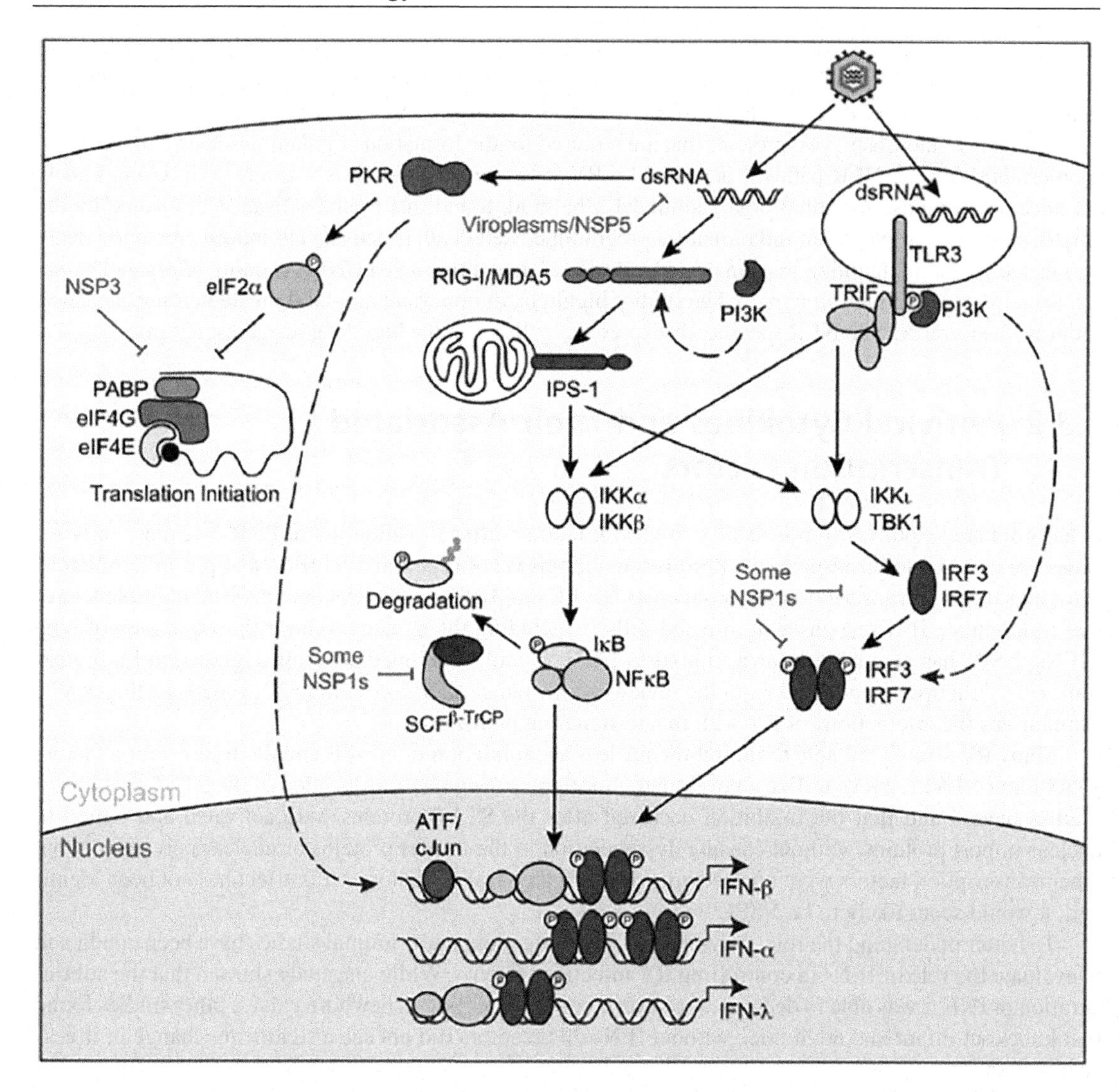

FIGURE 8.1 Rotavirus interactions with innate signaling pathways. Viral nucleic acids may be recognized in a host cell by membrane-bound TLR3 or cytoplasmic RLRs. When activated by nucleic-acid binding, RLRs recruit and activate the signaling adaptor molecule IPS-1 (IFN-β promoter stimulator 1, also known as MAVS), which recruits a signaling complex that activates latent cytoplasmic transcription factors such as IRF3 and NF-kB. TLR3 activation stimulates the recruitment of the adaptor TRIF, which acts as a platform for a variety of signaling molecules that also phosphorylate IRF3 or NF-kB. When signaled, the C-terminus of IRF3 is phosphorylated, causing a conformational change that leads to dimerization and nuclear translocation. NF-kB is held inactive by the inhibitor of NF-kB (IkB). NF-kB subsequently translocates to the nucleus. IRF3 and NF-kB bind to the IFN-β promoter in a cooperative manner with c-Jun/ATF-2 forming an enhanceosome complex initiating the transcription of IFN-β mRNA. Additional transcription factors, including IRF7, are induced by IFN-β and can also bind to the IFN-β promoter to enhance the transcription of IFN-β and IFN-α genes. RV can antagonize innate signaling pathways through several avenues, the primary one is the NSP1-induced degradation of IRF3 and IRF7. Some NSP1 proteins are also known to induce the degradation of β-TrCP. RV NSP3 can also impede antiviral responses by suppressing the translation of host mRNAs generated from IFN-stimulated genes. By sequestering viral RNAs within viroplasms, the virus can prevent their recognition by protein kinase R (PKR), RIG-I, MDA-5, and other sensors that upregulate antiviral responses. (Adapted from Arnold MM. et al. *PLoS Pathog* 2013; 9:e1003064, with permission from authors and publisher [110].)

8.3.3 Innate Immune Cell Activity

When the activation and recruitment pathways are activated, natural killer (NK) cells, neutrophils, and monocytes are recruited to the area. NK cells are recruited for the non-antigen specific destruction of infected tissues to prevent the spread of the intracellular virus. Neutrophils and monocytes aid in the clean-up process by degrading dead cells. There are only a few reports on NK cell responses in RV infection[115–119] and their role in controlling RV infection or disease has not been confirmed. Human peripheral blood mononuclear cells (PBMCs) stimulated with RV demonstrated significantly enhanced cytotoxic activity on K562 and LCL cells and treatment of the PBMCs with anti-NK cell marker CD16 monoclonal antibodies reduced the cytotoxic activity on these cells, suggesting that RV stimulates the activity of NK cells.[117] NK cells play an important role in murine experimental biliary atresia induced by RV infection[120,121]; however, an association between biliary atresia in human infants and RV infection has not been established. Future studies are needed to identify the role of NK cells in RV infection and disease. In this section, the role of macrophages (MØ) and dendritic cells (DC), and the potential role of newly identified innate lymphoid cells (ILC) in RV infection and immunity are discussed.

8.3.3.1 Macrophages

For defending against intestinal pathogens there are two groups of MØ that are found to play a role: intestinal MØ and blood monocytes/MØ. Intestinal MØ lack the antigen presentation and chemotaxis functions, although they can phagocytose apoptotic cells and be infected by RV to increase TLR and RLR activation, therefore boosting innate immune cell activation.[54,56,63,122] From studies using various mouse models, Gn pigs, and bone marrow derived MØ (BMDM), it has been established that the recruitment of CD14+ blood monocyte/MØ is needed for a robust response to RV infection.[54–56] CD14 is an important marker of MØ and DCs derived from monocytes, as it has been shown that CD14+ BMDMs have the ability to upregulate the TLR3-stimulated cytokine responses against dsRNA.[55] BMDMs are susceptible to infection by RV, and infection of BMDMs increases the production of IL-6 and IFN-β, determined via qPCR of transcripts.[54] Infection of migratory immune cells would offer an explanation for the observed RV antigens in tissues not known to support productive infections, including the cerebrospinal fluid.

8.3.3.2 Conventional Dendritic Cells

Conventional, or myeloid, DCs (cDCs) are known to be the subset of DCs critical for bridging the innate immune system to the adaptive by activating appropriate T helper (Th) cell populations. An *in vitro* study of PBMCs cultured with RV showed that viral infection was not sufficient for the maturation of immature cDCs, but with a minimal productive infection (shown by the presence of NSP4 in the supernatant), mature cDCs were capable of activating Th1 cells.[123] Another study showed that human DCs treated with the supernatant of RV-infected Caco-2 cells induced a poor Th1 response.[64] The ability of mature cDCs to stimulate Th1 activation when directly exposed to full RV particles is inhibited by the supernatant of RV-infected epithelial cells containing increased levels of TGF-β1.[64] Neutralization of TGF-β1 in the cell culture yielded a more potent increase in the activation of the Th1 response by cDCs.[64] These findings suggest that the induction of TGF-β1 by RV could be an immune evasion mechanism and may partially explain the weak RV-T cell response previously evidenced in humans. Given that RV infection is able to be cleared and adaptive immune memory cells are induced, the inhibition of cDCs by RV is able to be overcome by the host immune system.

8.3.3.3 Plasmacytoid DCs

Plasmacytoid DCs (pDCs) are crucial first responders to viral infection due to their ability to rapidly produce a large amount of type I IFN (IFN-α) upon recognition of virally encoded molecular patterns.

The robust IFN-α secretion limits viral replication, while simultaneously activating cells belonging to the adaptive immune system. During RV infection, pDCs are critical activators of B cells, and therefore critical in developing serum antibody responses.[58] Using purified human PBMCs co-cultured with RV, it was demonstrated that pDCs are necessary and sufficient for human B cell activation and that the B cell activation was mediated by IFN-α.[58] These *in vitro* findings were confirmed *in vivo*. After challenging wild type mice and knockout mice that are lacking functional pDCs and IFN receptors with murine RV, the number of RV-specific B cells and total serum antibody titers, as well as fecal IgA levels, were reduced in the knockout mice compared to the wild type, which led to increased RV fecal shedding.[58] There is also evidence that pDCs are necessary to stimulate RV-specific memory T cell responses.[116] pDC depletion from PBMCs was associated with a significant decrease in the frequency of RV-specific IFN-γ producing CD4+ and CD8+ T cells after incubation in the presence of rhesus rotavirus (RRV).

Two viral proteins VP4 and/or VP7 and the dsRNA genome are required for the initiation of the pDC IFNα response to RV; however, virus replication or activation by trypsin cleavage of VP4 is not required.[124] In the same study of pDC responses to RV by flow cytometry and Luminex, it was revealed that RRV can replicate in a small subset of human primary peripheral blood pDCs and it is independent of trypsin cleavage of VP4, which may explain the establishment of extraintestinal viremia and antigenemia after RRV infection and other RV strains.[124]

The intestinal pDCs in RV infection have not been well characterized. In human-RV infected Gn pigs, the frequencies of intestinal pDCs (SWC3lowCD4+) were significantly higher at 2 and 4 days post-infection than those of the non-infected controls, and the main IFN-α+ DC population detected after RV infection was intestinal pDCs followed by intestinal cDCs.[125] CCR9 is recognized to be a homing receptor for pDCs to the small intestine.[126] CD103 (integrin αEβ7) expression was also used as an intestinal homing marker for pDCs.[127] The frequencies of CD103+ pDCs in Gn pigs with vitamin A deficiency were found to be lower than those of vitamin A sufficient Gn pigs before RV challenge, and they were further reduced after challenge. The reduction in pDCs was associated with lowered RV-specific IgA antibody titers and increased virus shedding and diarrhea post-challenge.[127] Similarly, frequencies of intestinal pDCs were significantly lower in protein-deficient pigs than in protein-sufficient pigs, and those pigs developed more severe RV infection (diarrhea and shedding) after RV infection.[128] These data together indicate the importance of pDCs in RV protective immunity.

8.3.3.4 Innate Lymphoid Cells

ILCs are cells of the innate immune system that develop from lymphoid precursor cells instead of myeloid cells and play a similar role as helper T cells. It was not until 2011 that the cytokine-producing and lineage negative NK cell-like lymphoid cells were officially accepted as their own official cell type.[129,130] While NK cells are considered the innate counterparts of cytotoxic CD8$^+$ T cells, ILC1s, ILC2s, and ILC3s represent the innate counterparts of Th1, Th2, and Th17 CD4+ T cells. Various groups have begun to study the effect of ILCs on different pathogens. ILC1s and ILC3s are found to play a role in mucosal homeostasis by producing type I and II IFNs as well as in immune cell activation by producing cytokines GM-CSF (granulocyte-macrophage colony-stimulating factor) and IL-18.[131] As discussed above, type I IFN production is reduced during productive RV infection; however, using single cell analysis of mRNA levels, it was shown that there is a hematopoietic cell type within the intestinal tract of RV-infected mice, which produces a significant quantity of IFN-α/β.[77] It is possible that this hematopoietic cell type that is responsible for this early type I IFN response to RV infection is ILCs. In order to establish this connection, more research needs to be conducted in this area, and to date, there have been no published studies on the precise roles of ILCs in controlling any enteric viruses. However, there are studies on the role of ILCs in intestinal inflammatory diseases, including Crohn's disease and colitis, and it was found that aberrant ILC functions can increase the severity of the diseases through secretion of inflammatory cytokines and interaction with other immune and non-immune cells.[132] It is possible that ILCs play a critical role in either the development or control of RV disease and should be a target of future research.

8.3.4 Adaptive Immune Response

Even with a robust innate immune response, the proper clearance and control of a RV infection requires the adaptive, antigen-specific effector cells.[12] RV replicates in the enterocytes of the small intestine; therefore, local mucosal immunity is a critical factor in protection against RV diarrhea. At the site of an intestinal infection, cytotoxic CD8 T cells in gut associated lymphoid tissues (GALT) are activated through MHC class I molecules, which are ubiquitously expressed on all cells.[133] Once activated, CD8 T cells proceed to eliminate infected cells through directed killing and the release of various cytokines. Meanwhile, naïve CD4 T cells are activated by professional antigen-presenting cells (APC). MØ and DC present antigens to T cells through MHC class II. Upon activation, CD4 T cells can develop into different subsets of helper cells (Th1, Th2, Th3, Th17, and Th9) to assist in the clearance of viruses from infected tissues and follicular helper cells (Tfh) that stimulate B cell differentiation. Once B cells are activated, they go through a mutation and selection process to produce antigen-specific and therefore virus-specific antibodies to increase the precision and accuracy of the immune response.

8.3.4.1 T Cell Responses in RV Infection and Immunity

To understand the roles of CD8 and CD4 T cells during a RV infection, knockout mice that were either deficient in CD8 or CD4 T cells were utilized. A study showed that CD8 is required for the initial and rapid control of RV replication, while CD4 is required for the prevention of chronic RV shedding.[134] When CD8 cells were depleted in the CD4 T cell knockout mice, RV shedding increased dramatically, and when CD4 T cells were added to the knockout mice, antibody production increased to the point that RV shedding ceased, and thus resolution of RV infection in mice requires both CD4 and CD8 activities.[134]

Once naïve CD4 T cells are activated by professional APCs and they differentiate into Th cells, they need specific homing signals to migrate to the site of infection. An early study in humans showed that natural RV infection results in specific circulating memory CD4+ cells that mostly express the gut-homing receptor $\alpha_4\beta_7$ and this phenotype is not shared with memory cells elicited by intramuscular immunization.[135] The findings indicate that the expression of $\alpha_4\beta_7$ helps to segregate intestinal versus systemic immune response. In a more recent study in humans, circulating RV-specific CD4 T cells identified with a class II tetramer using flow cytometry display both $\alpha_4\beta_7$ and CCR9.[53]

Studies of T cell responses to human RV infection and vaccination in Gn pigs using intracellular staining and flow cytometry demonstrated that protection rates against diarrhea upon challenge with virulent human RV significantly correlated with the frequencies of virus-specific intestinal, but not systemic, IFN-γ producing CD4 and CD8 T cells at the time of challenge.[85] Using B cell-deficient Gn pigs, another study showed that after RV vaccination and challenge, the lack of B cell immune responses reduced the protection rate against RV diarrhea and shedding conferred by the vaccine, confirming the essential role of B cells in RV protective immunity.[82] When comparing between vaccinated and unvaccinated B cell-deficient Gn pigs post-challenge, the vaccinated pigs had significantly reduced diarrhea and virus shedding, suggesting that in the absence of B cells, vaccination stimulated the other arm of the adaptive immune response (T cells) that provided the observed partial protection. Further, depletion of CD8 cells in the vaccinated, B cell-deficient Gn pigs (by injecting anti-CD8 monoclonal antibodies) increased the severity of diarrhea and virus shedding and prolonged the virus shedding, indicating that CD8 T cells are largely responsible for the partial protection and are important for resolving RV infection.[82] In addition, several more Gn pig studies showed that the enhanced production of virus-specific IFN-γ producing CD4 and CD8 T cells in RV vaccinated Gn pigs by feeding either probiotics[136,137] or rice bran[93] is associated with increased protection against virulent human RV challenge induced diarrhea, further supporting the important protective role of effector T cells.

The regulatory T cell (Treg) responses to RV infections have been studied in mice and Gn pigs.[84,138] Murine RV (EC) infection in neonatal mice induced an expansion of the CD4+CD25+Foxp3+ Treg cell populations. When the Treg cells were depleted before RV infection, CD4 and CD8 T-cell proliferation and their IFN-γ secretion, as well as CD19+ B cell proliferation responses to RV were all enhanced in the mesenteric lymph nodes (MLNs) and spleen.[138] However, Treg cell depletion did not have an effect on diarrheal disease, virus shedding, and RV-specific IgA antibody response after RV infection. Intestinal and systemic CD4+CD25+Foxp3+ and CD4+CD25-Foxp3+ Treg cell responses were studied in Gn pigs after virulent human RV infection or attenuated RV vaccination.[84] Human RV infection reduced the frequencies and numbers of tissue-residing Treg cells and decreased the frequencies of IL-10 and TGF-β producing CD4+CD25-FoxP3+ Treg cells in the ileum, spleen and blood at post-inoculation day (PID) 28. The frequencies of IL-10 and TGF-β producing CD4+CD25-foxP3+ Treg cells in all sites at PID 28 were significantly, inversely correlated with the protection rate against RV-caused diarrhea.[84] Taken together, the studies of effector T cell and Treg cell responses in Gn pigs indicate that T cells are important players in RV infection and immunity. The dynamics of T cell responses after RV infection and vaccination are illustrated in Figure 8.2.

8.3.4.2 B Cells and Antibody Responses in RV Infection and Immunity

8.3.4.2.1 Intestinal Antibody-Secreting B Cells (ASCs)

After RV infection, robust B cell responses are generated in the intestinal lymphoid tissues. The role of effector and memory B cells in RV infection and immunity has been studied in mice, Gn pigs, and humans. Infection of mice with murine RV induces life-long immunity, characterized by large numbers of RV-specific ASCs and high levels of IgA in the intestine and in GALT.[57] Using ELISPOT assay, a

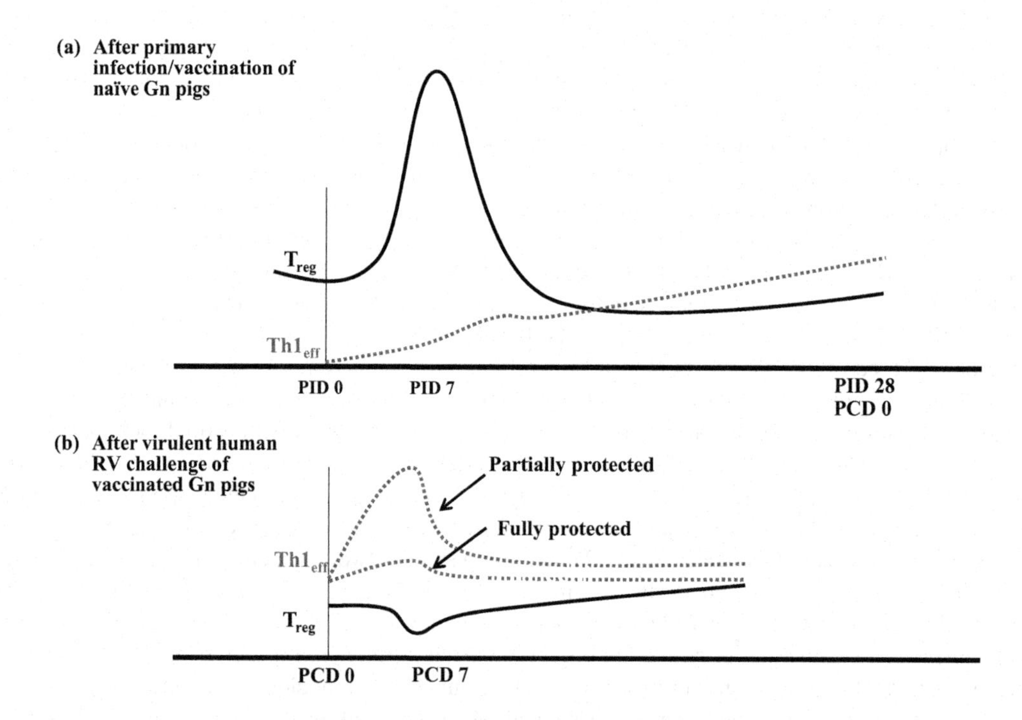

FIGURE 8.2 Model for the dynamics of effector T cell and CD4+CD25− Treg responses after RV primary infection/vaccination and challenge. During the acute phase of RV infection/vaccination (PID 1–7), virus-specific IFN-γ producing effector Th1 cell (Th1eff) responses are induced in the intestinal lymphoid tissue; meanwhile there is an expansion of CD25− Tregs at the same site. In the convalescent phase after RV infection/vaccination, the host resumes immune homeostasis and establishes the equilibrium between Th1eff cells and Tregs but at a different level, with increased numbers of Th1eff cells and decreased numbers of Tregs (a). After challenge, vaccinated and partially protected pigs develop anamnestic Th1eff cell responses that are stronger and faster than those after primary infection but the Treg levels remain similar to that at prechallenge (b). Fully protected pigs do not develop an anamnestic Th1eff cell immune response. (Adapted from Wen K. et al. *Immunology* 2012; 137; 160–71, with permission from authors and publisher [84].)

correlation between the increased frequency of RV-specific memory B cells in small intestinal lamina propria and enhanced protection against shedding was established.[139] Also using ELISPOT assay, studies of Gn pigs infected with human RV and vaccinated with various vaccines have consistently showed the correlation between intestinal IgA ASCs and memory B cells and the protection rate against RV diarrhea.[140] Further studies using adoptive transfer of different phenotypically defined memory B cells into B- and T-cell deficient mice chronically infected with RV showed that the memory B cell subset responsible for the secretory IgA response and long-term protective immunity to RV is the subset that expresses the intestinal homing receptor α4β7.[57] Retinoic acid (a metabolite of vitamin A) derived from DCs in GALT was identified to be the determining factor in the gut tropism of B cells by inducing high levels of α4β7.[141] Retinoic acid synergizes with IL-6 and IL-5, produced by the DCs to induce IgA secretion in mice and in cultured human cells. Vitamin A deficiency aggravated the course of RV infection in mice[142] and impaired intestinal IgA responses in rats,[143] and vitamin A supplementation restored IgA levels in malnourished mice.[144] Studies in Gn pigs showed that prenatal vitamin A deficiency impaired adaptive immune responses to RV infection and vaccination, leading to more severe diarrhea after virulent RV challenge, whereas vitamin A supplementation failed to compensate for the negative effects.[127,145,146]

8.3.4.2.2 Whole Virus Specific and Viral Protein Specific Antibodies

Serum and mucosal IgA and IgG antibodies have been extensively evaluated in children and in animal models after RV infection using ELISA with purified whole RV as the capture antigen. Serum virus-specific IgA titers greater than 1:800 were associated with reduced risk of RV reinfection and diarrhea in children after natural infection and were suggested to be a correlate of protection against RV infection and moderate-to-severe diarrhea.[147] Higher fecal IgA levels are also associated with protection against RV infection and illness in young children to a significant level.[65] In addition, serum IgA antibody levels have been shown to reflect the IgA antibody responses in the mucosal surface of the small intestine in humans[147] and animal models.[148]

Using either recombinant proteins as the capture antigen in ELISA or recombinant baculoviruses expressing individual RV proteins in Sf-9 cell immunocytochemical staining assays, the IgA antibodies were found mainly against the VP6 antigen, followed by homotypic VP4, NSP4, and VP7.[72,149] Antibodies to VP6 are non-neutralizing antibodies, and the levels of the antibodies specific to the viral protein have not been correlated with protection against natural infection. Analyses of immune responses to various RV proteins in infants and young children infected naturally with RV are often difficult, due to variation in age and physical conditions of infants and young children involved in such studies; the date of infection is often unknown, the dose and genotype of the infecting RV strain(s) are not well-defined, and the post-infection day for serum collection varies. In a study of serum samples from adult human volunteers orally challenged with virulent human RV, higher pre-challenge titers of IgG antibodies to homotypic VP7 and VP4 were correlated with increased resistance to RV infection.[150]

A recent study utilized RV-specific single cell-sorted intestinal B cells from human adults, barcode-based deep sequencing of antibody repertoires, monoclonal antibody expression, and serologic and functional characterization to identify the viral protein-specific antibodies that mediate heterotypic immunity to RV.[66] Stable triple layered RV particles labeled with (CDC-9 strain) labeled with Cy5 used in flow cytometry sorting provided the researchers the ability to collect RV-specific ASCs. The sorted ASCs then were used to create cDNA libraries for the Ig heavy and light chains (IGHC and IHLC) based on a system referred to as barcode-based deep sequencing. Single cell sequencing allowed for specific targeting of the V_H (heavy chain variable region) in ASCs and allowed for the amplification of IGHC/IGLC pairs with plate- and well-identifying barcodes.[151] With specific amplified sequences, mAb were sorted based on RV ELISA reactivity, and neutralization assays were conducted, identifying 31 IgG and 2 IgA, of which there were 10 mAb that were neutralizing.[66] This study went on to determine the protective efficacy of the different mAb. It was shown that the mAb to VP5*, the stalk region of the RV attachment protein VP4, were able to mediate heterotypic protective immunity. Heterotypic protective mAb against VP7 and VP8*, the cell-binding region of VP4, were also identified, but homotypic anti-VP7 and non-neutralizing VP8* responses are more common after RV infection in humans. This study demonstrated that humans can generate heterotypic neutralizing antibody responses to VP7, VP8*, and most often to VP5* after natural infection and the findings suggest a new target antigen VP5* for a broadly protective RV vaccine.[66]

8.4 FUTURE PERSPECTIVES

Even with the significant amount of knowledge that has been complied on the molecular mechanisms of host immune responses to RV infections, there are still areas of immune responses that have not had their role thoroughly divulged in RV immunity. The recent availability of a plasmid-based reverse genetics system[152] will help with the precise understanding of the role(s) of each viral protein and how they interact with specific aspects of the host immune system, which will aid in the development of more immunogenic and efficacious vaccines.

Based on our current knowledge, and the lack thereof, the following topics will require further research attention in the future: (i) the significance of the different innate immune cell responses in the resistance to RV infection and diseases; (ii) the identification of the immune correlates of protection against RV disease; (iii) the host genetic influence on RV infection and immunity; (iv) the role of gut microbiota and dietary metabolites in host responses to RV and immunity; and (v) transkingdom interactions of RV, human gut microflora, and host immune system.

Emerging evidence indicates that enteric viruses regulate, and are in turn regulated by, other microbes in the gut through a series of processes termed transkingdom interactions, which represents a changing paradigm in intestinal immunity to viral infection.[153] Early studies into the RV–microbiome interactions focused more on showing the effects of RV on the intestinal fauna.[154] However, as technology developed in the 2010s, more evidence has surfaced that the state of an individuals' microbiome can play a significant role in the host responses to the infection and illness caused by various gastrointestinal pathogens, including RV.[155,156] Changes in the microbiome of the gut have been shown to develop from malnutrition, diseases, and other environmental factors, which can lead to an increased susceptibility to RV infection and decreased efficacy of RV vaccination.[91] Furthermore, probiotics have been studied extensively for their preventive and therapeutic impact on RV infection and illness.[157]

One of the biggest road blocks for identifying the role of microbiome and probiotics in controlling and preventing enteric virus infection and illness is the lack of a reproducible environment. Our microbiome continuously changes from birth up to the moment life leaves our bodies, reconfiguring its metagenomic layout in response to daily variations in the environment, diet, and the physiological and immunological needs at various stages. To control the variables in the environment and diet, Gn animals stand out as essential model systems for studying the effects of microbiota on physiological and immunological responses to viral infections. Gn animals are completely devoid of all microbes and maternal antibodies and neonatal Gn pigs can be infected by HRV and they exhibit similar pathological changes as seen in humans.[91] By removing confounding variables, Gn animals have allowed researchers to study the implications of individual factors on RV vaccination, treatment, and immunity.[158]

The microbiome has been shown to promote the infectivity and pathogenesis of several enteric viruses, including poliovirus, reovirus, norovirus, and RV.[62,159,160] Utilizing both germ-free and antibiotic treatments to ablate the microbiota of mice, the impact of the microbiome as a whole on RV infectivity and diarrhea was determined.[62] Microbiota elimination by antibiotic treatment reduced RV infectivity in mice. In spite of a reduced viral load, microbiome ablated mice secreted a significantly increased magnitude of serum and fecal IgG and IgA antibodies for a longer duration and an increased number of small intestinal RV-specific IgA ASCs, compared to the control mice. The clinical relevance of the findings is that antibiotic administration before RV vaccination could potentially improve the seroconversion rates in low-resource countries and the vaccine efficacy.[62] Further studies are needed to identify the mechanisms of the impact of different types of human microbiomes on RV infection and the development of immune responses. Advancement in this area of research holds a great promise in improving the protective efficacy of enteric vaccines.

REFERENCES

1. Ngabo F, Tate JE, Gatera M, et al. Effect of pentavalent rotavirus vaccine introduction on hospital admissions for diarrhoea and rotavirus in children in Rwanda: a time-series analysis. *Lancet Glob Health* 2016;4:36.
2. Tate JE, Burton AH, Boschi-Pinto C, et al. 2008 estimate of worldwide rotavirus-associated mortality in children younger than 5 years before the introduction of universal rotavirus vaccination programmes: a systematic review and meta-analysis. *Lancet Infect Dis* 2012;12:136–41.
3. Parashar UD, Gibson CJ, Bresse JS, Glass RI. Rotavirus and severe childhood diarrhea. *Emerg Infect Dis* 2006;12:304–6.
4. Walker CL, Rudan I, Liu L, et al. Global burden of childhood pneumonia and diarrhoea. *Lancet* 2013;381:1405–16.
5. Data, statistics and graphics. 2018. (Accessed 4/15/2018, 2018, at http://www.who.int/immunization/monitoring_surveillance/data/en/.)
6. Gilmartin AA, Petri WA, Jr. Exploring the role of environmental enteropathy in malnutrition, infant development and oral vaccine response. *Philos Trans R Soc London Series B, Biol Sci* 2015;370: 20140143.
7. Zaman K, Dang DA, Victor JC, et al. Efficacy of pentavalent rotavirus vaccine against severe rotavirus gastroenteritis in infants in developing countries in Asia: a randomised, double-blind, placebo-controlled trial. *Lancet* 2010;376:615–23.
8. Glass RI, Parashar U, Patel M, Gentsch J, Jiang B. Rotavirus vaccines: successes and challenges. *J Infect* 2014;68(Suppl 1):S9–18.
9. Armah GE, Sow SO, Breiman RF, et al. Efficacy of pentavalent rotavirus vaccine against severe rotavirus gastroenteritis in infants in developing countries in sub-Saharan Africa: a randomised, double-blind, placebo-controlled trial. *Lancet* 2010;376:606–14.
10. Vesikari T. Rotavirus vaccination: a concise review. *Clin Microbiol Infect: The Official Publication of the European Society of Clinical Microbiology and Infectious Diseases* 2012;18(Suppl 5):57–63.
11. Velasquez DE, Parashar U, Jiang B. Decreased performance of live attenuated, oral rotavirus vaccines in low-income settings: causes and contributing factors. *Expert Rev Vaccines* 2018;17:145–61.
12. Desselberger U. Rotaviruses. *Virus Res* 2014;190:75–96.
13. Prasad BV, Wang GJ, Clerx JP, Chiu W. Three-dimensional structure of rotavirus. *J Mol Biol* 1988;199:269–75.
14. Mattion NM, Mitchell DB, Both GW, Estes MK. Expression of rotavirus proteins encoded by alternative open reading frames of genome segment 11. *Virology* 1991;181:295–304.
15. Matthijnssens J, Ciarlet M, Rahman M, et al. Recommendations for the classification of group A rotaviruses using all 11 genomic RNA segments. *Arch Virol* 2008;153:1621–9.
16. Estes MK, Greenberg HB. Rotaviruses. In: Knipe DM, Howley PM, eds. *Fields Virology*. 6th Edition. Philadelphia, PA: Wolters Kluwer Health/Lippincott Williams & Wilkins; 2013:1347–401.
17. McClain B, Settembre E, Temple BR, Bellamy AR, Harrison SC. X-ray crystal structure of the rotavirus inner capsid particle at 3.8 Å resolution. *J Mol Biol* 2010;397:587–99.
18. Settembre EC, Chen JZ, Dormitzer PR, Grigorieff N, Harrison SC. Atomic model of an infectious rotavirus particle. *EMBO J* 2011;30:408–16.
19. Clark S, Roth J, Clark M, Barnett B, Spendlove R. Trypsin enhancement of rotavirus infectivity: mechanism of enhancement. *J Virol* 1981;39:816–22.
20. Estes M, Graham D, Mason B. Proteolytic enhancement of rotavirus infectivity: molecular mechanisms. *J Virol* 1981;39:879–88.
21. Valenzuela S, Pizarro J, Sandino AM, et al. Photoaffinity labeling of rotavirus VP1 with 8-azido-ATP: identification of the viral RNA polymerase. *J Virol* 1991;65:3964–7.
22. Zeng CQ, Wentz MJ, Cohen J, Estes MK, Ramig RF. Characterization and replicase activity of double-layered and single-layered rotavirus-like particles expressed from baculovirus recombinants. *J Virol* 1996;70:2736–42.
23. Chen D, Luongo CL, Nibert ML, Patton JT. Rotavirus open cores catalyze 5′-capping and methylation of exogenous RNA: evidence that VP3 is a methyltransferase. *Virology* 1999;265:120–30.
24. Estrozi LF, Settembre EC, Goret G, et al. Location of the dsRNA-dependent polymerase, VP1, in rotavirus particles. *J Mol Biol* 2013;425:124–32.
25. Ogden KM, Snyder MJ, Dennis AF, Patton JT. Predicted structure and domain organization of rotavirus capping enzyme and innate immune antagonist VP3. *J Virol* 2014;88:9072–85.
26. Matthijnssens J, Otto PH, Ciarlet M, Desselberger U, Van Ranst M, Johne R. VP6-sequence-based cutoff values as a criterion for rotavirus species demarcation. *Arch Virol* 2012;157:1177–82.

27. Mihalov-Kovacs E, Gellert A, Marton S, et al. Candidate new rotavirus species in sheltered dogs, Hungary. *Emerg Infect Dis* 2015;21:660–3.
28. Banyai K, Kemenesi G, Budinski I, et al. Candidate new rotavirus species in Schreiber's bats, Serbia. *Infect Genet Evol* 2017;48:19–26.
29. Prince DS, Astry C, Vonderfecht S, Jakab G, Shen FM, Yolken RH. Aerosol transmission of experimental rotavirus infection. *Pediatr Infect Dis* 1986;5:218–22.
30. Ghosh S, Kobayashi N. Exotic rotaviruses in animals and rotaviruses in exotic animals. *Virusdisease* 2014;25:158–72.
31. Bowen MD, Mijatovic-Rustempasic S, Esona MD, et al. Rotavirus strain trends during the postlicensure vaccine era: United States, 2008–2013. *J Infect Dis* 2016;214:732–8.
32. Organization WH. Rotavirus vaccines: WHO position paper. *Wkly Epidemiol Rec* 2013;88:49–64.
33. Shah MP, Dahl RM, Parashar UD, Lopman BA. Annual changes in rotavirus hospitalization rates before and after rotavirus vaccine implementation in the United States. *PLoS One* 2018;13:e0191429.
34. Tate JE, Burton AH, Boschi-Pinto C, Parashar UD. Global, regional, and national estimates of rotavirus mortality in children <5 years of age, 2000–2013. *Clin Infect Dis* 2016;62(Suppl 2):S96–105.
35. Kotloff KL, Nataro JP, Blackwelder WC, et al. Burden and aetiology of diarrhoeal disease in infants and young children in developing countries (the Global Enteric Multicenter Study, GEMS): a prospective, case-control study. *Lancet* 2013;382:209–22.
36. Lee B, Carmolli M, Dickson DM, et al. Rotavirus-specific immunoglobulin A responses are impaired and serve as a suboptimal correlate of protection among infants in Bangladesh. *Clin Infect Dis* 2018;67(2), 186–192.
37. Kompithra RZ, Paul A, Manoharan D, et al. Immunogenicity of a three dose and five dose oral human rotavirus vaccine (RIX4414) schedule in south Indian infants. *Vaccine* 2014;32:A129–33.
38. Patel M, Glass RI, Jiang B, Santosham M, Lopman B, Parashar U. A systematic review of anti-rotavirus serum IgA antibody titer as a potential correlate of rotavirus vaccine efficacy. *J Infect Dis* 2013;208:284–94.
39. Osborne MP, Haddon SJ, Spencer AJ, Collins J. An electron microscopic investigation of time-related changes in the intestine of neonatal mice infected with murine rotavirus. *J Pediatr Gastroenterol Nutr* 1988;7:236–48.
40. Ramig RF. Pathogenesis of intestinal and systemic rotavirus infection. *J Virol* 2004;78:10213–20.
41. Lundgren O, Svensson L. Pathogenesis of rotavirus diarrhea. *Microbes Infect* 2001;3:1145–56.
42. Greenberg HB, Estes MK. Rotaviruses: from pathogenesis to vaccination. *Gastroenterology* 2009;136:1939–51.
43. Ball JM, Tian P, Zeng CQ, Morris AP, Estes MK. Age-dependent diarrhea induced by a rotaviral nonstructural glycoprotein. *Science* 1996;272:101–4.
44. Lopez S, Arias C. Early steps in rotavirus cell entry. Reoviruses: Entry, Assembly and Morphogenesis: Berlin, Heidelberg: Springer; CTMI 2006; 309:39–66.
45. Coulson BS. Expanding diversity of glycan receptor usage by rotaviruses. *Curr Opin Virol* 2015;15:90–6.
46. Ruiz MC, Leon T, Diaz Y, Michelangeli F. Molecular biology of rotavirus entry and replication. *Sci World J* 2009;9:1476–97.
47. McNulty MS, Curran WL, McFerran JB. The morphogenesis of a cytopathic bovine rotavirus in Madin-Darby bovine kidney cells. *J Gen Virol* 1976;33:503–8.
48. Gardet A, Breton M, Fontanges P, Trugnan G, Chwetzoff S. Rotavirus spike protein VP4 binds to and remodels actin bundles of the epithelial brush border into actin bodies. *J Virol* 2006;80:3947–56.
49. Zhou Y, Geng P, Liu Y, et al. Rotavirus-encoded virus-like small RNA triggers autophagy by targeting IGF1R via the PI3K/Akt/mTOR pathway. *Biochim Biophys Acta (BBA)-Mol Basis Dis* 2018;1864:60–8.
50. Saxena V, Shivakumar P, Sabla G, Mourya R, Chougnet C, Bezerra JA. Dendritic cells regulate natural killer cell activation and epithelial injury in experimental biliary atresia. *Sci Transl Med* 2011;3:102ra94–ra94.
51. Rojas OL, Narváez CF, Greenberg HB, Angel J, Franco MA. Characterization of rotavirus specific B cells and their relation with serological memory. *Virology* 2008;380:234–42.
52. Wen K, Azevedo MS, Gonzalez A, et al. Toll-like receptor and innate cytokine responses induced by lactobacilli colonization and human rotavirus infection in gnotobiotic pigs. *Vet Immunol Immunopathol* 2009;127:304–15.
53. Parra M, Herrera D, Calvo-Calle JM, et al. Circulating human rotavirus specific CD4 T cells identified with a class II tetramer express the intestinal homing receptors alpha4beta7 and CCR9. *Virology* 2014;452(3):191–201.
54. Di Fiore IJ, Holloway G, Coulson BS. Innate immune responses to rotavirus infection in macrophages depend on MAVS but involve neither the NLRP3 inflammasome nor JNK and p38 signaling pathways. *Virus Res* 2015;208:89–97.
55. Lee H-K, Dunzendorfer S, Soldau K, Tobias PS. Double-stranded RNA-mediated TLR3 activation is enhanced by CD14. *Immunity* 2006;24:153–63.
56. Zhang W, Wen K, Azevedo MS, et al. Lactic acid bacterial colonization and human rotavirus infection influence distribution and frequencies of monocytes/macrophages and dendritic cells in neonatal gnotobiotic pigs. *Vet Immunol Immunopathol* 2008;121:222–31.

57. Williams MB, Rose JR, Rott LS, Franco MA, Greenberg HB, Butcher EC. The memory B cell subset responsible for the secretory IgA response and protective humoral immunity to rotavirus expresses the intestinal homing receptor, alpha4beta7. *J Immunol* 1998;161:4227–35.
58. Deal EM, Lahl K, Narvaez CF, Butcher EC, Greenberg HB. Plasmacytoid dendritic cells promote rotavirus-induced human and murine B cell responses. *J Clin Invest* 2013;123:2464–74.
59. Broquet AH, Hirata Y, McAllister CS, Kagnoff MF. RIG-I/MDA5/MAVS are required to signal a protective IFN response in rotavirus-infected intestinal epithelium. *J Immunol* 2011;186:1618–26.
60. Xu J, Yang Y, Wang C, Jiang B. Rotavirus and coxsackievirus infection activated different profiles of toll-like receptors and chemokines in intestinal epithelial cells. *Inflamm Res* 2009;58:585–92.
61. Pott J, Stockinger S, Torow N, et al. Age-dependent TLR3 expression of the intestinal epithelium contributes to rotavirus susceptibility. *PLoS Pathog* 2012;8:e1002670.
62. Uchiyama R, Chassaing B, Zhang B, Gewirtz AT. Antibiotic treatment suppresses rotavirus infection and enhances specific humoral immunity. *J Infect Dis* 2014;210:171–82.
63. Ge Y, Mansell A, Ussher JE, et al. Rotavirus NSP4 triggers secretion of proinflammatory cytokines from macrophages via Toll-like receptor 2. *J Virol* 2013;87:11160–7.
64. Rodríguez L-S, Narváez CF, Rojas OL, Franco MA, Ángel J. Human myeloid dendritic cells treated with supernatants of rotavirus infected Caco-2 cells induce a poor Th1 response. *Cell Immunol* 2012;272:154–61.
65. Matson DO, O'Ryan ML, Herrera I, Pickering LK, Estes MK. Fecal antibody responses to symptomatic and asymptomatic rotavirus infections. *J Infect Dis* 1993;167:577–83.
66. Nair N, Feng N, Blum LK, et al. VP4-and VP7-specific antibodies mediate heterotypic immunity to rotavirus in humans. *Sci Transl Med* 2017;9(395). pii:eaam5434.
67. Günaydın G, Zhang R, Hammarström L, Marcotte H. Engineered Lactobacillus rhamnosus GG expressing IgG-binding domains of protein G: capture of hyperimmune bovine colostrum antibodies and protection against diarrhea in a mouse pup rotavirus infection model. *Vaccine* 2014;32:470–7.
68. Yuan L, Geyer A, Saif LJ. Short-term immunoglobulin A B-cell memory resides in intestinal lymphoid tissues but not in bone marrow of gnotobiotic pigs inoculated with Wa human rotavirus. *Immunology* 2001;103:188–98.
69. Yuan L, Kang SY, Ward LA, To TL, Saif LJ. Antibody-secreting cell responses and protective immunity assessed in gnotobiotic pigs inoculated orally or intramuscularly with inactivated human rotavirus. *J Virol* 1998;72:330–8.
70. Yuan L, Ward LA, Rosen BI, To TL, Saif LJ. Systematic and intestinal antibody-secreting cell responses and correlates of protective immunity to human rotavirus in a gnotobiotic pig model of disease. *J Virol* 1996;70:3075–83.
71. Azevedo MS, Yuan L, Pouly S, et al. Cytokine responses in gnotobiotic pigs after infection with virulent or attenuated human rotavirus. *J Virol* 2006;80:372–82.
72. Yuan L, Ishida S, Honma S, et al. Homotypic and heterotypic serum isotype-specific antibody responses to rotavirus nonstructural protein 4 and viral protein (VP) 4, VP6, and VP7 in infants who received selected live oral rotavirus vaccines. *J Infect Dis* 2004;189:1833–45.
73. Yuan L, Honma S, Ishida S, Yan XY, Kapikian AZ, Hoshino Y. Species-specific but not genotype-specific primary and secondary isotype-specific NSP4 antibody responses in gnotobiotic calves and piglets infected with homologous host bovine (NSP4[A]) or porcine (NSP4[B]) rotavirus. *Virology* 2004;330:92–104.
74. Nandi S, Chanda S, Bagchi P, Nayak MK, Bhowmick R, Chawla-Sarkar M. MAVS protein is attenuated by rotavirus nonstructural protein 1. *PLoS One* 2014;9:e92126.
75. Qin L, Ren L, Zhou Z, et al. Rotavirus nonstructural protein 1 antagonizes innate immune response by interacting with retinoic acid inducible gene I. *Virol J* 2011;8:526.
76. Holloway G, Truong TT, Coulson BS. Rotavirus antagonizes cellular antiviral responses by inhibiting the nuclear accumulation of STAT1, STAT2, and NF-κB. *J Virol* 2009;83:4942–51.
77. Sen A, Rothenberg ME, Mukherjee G, et al. Innate immune response to homologous rotavirus infection in the small intestinal villous epithelium at single-cell resolution. *Proc Natl Acad Sci USA* 2012;109:20667–72.
78. Bugarčić A, Taylor JA. Rotavirus nonstructural glycoprotein NSP4 is secreted from the apical surfaces of polarized epithelial cells. *J Virol* 2006;80:12343–9.
79. Angel J, Franco MA, Greenberg HB, Bass D. Lack of a role for type I and type II interferons in the resolution of rotavirus-induced diarrhea and infection in mice. *J Interferon Cytokine Res* 1999;19:655–9.
80. Pott J, Mahlakõiv T, Mordstein M, et al. IFN-λ determines the intestinal epithelial antiviral host defense. *Proc Natl Acad Sci USA* 2011;108:7944–9.
81. Schwers A, Broecke CV, Maenhoudt M, Beduin J, Werenne J, Pastoret P-P. Experimental rotavirus diarrhoea in colostrum-deprived newborn calves: assay of treatment by administration of bacterially produced human interferon (Hu-IFNα2). *Ann Rech Vét* 1985;16:213–8.

82. Wen K, Bui T, Weiss M, et al. B-cell-deficient and CD8 T-cell-depleted gnotobiotic pigs for the study of human rotavirus vaccine-induced protective immune responses. *Viral Immunol* 2016;29:112–27.
83. Uchiyama R, Chassaing B, Zhang B, Gewirtz AT. MyD88-mediated TLR signaling protects against acute rotavirus infection while inflammasome cytokines direct Ab response. *Innate Immun* 2015;21:416–28.
84. Wen K, Li G, Yang X, et al. CD4+ CD25- FoxP3+ regulatory cells are the predominant responding regulatory T cells after human rotavirus infection or vaccination in gnotobiotic pigs. *Immunology* 2012;137:160–71.
85. Yuan L, Wen K, Azevedo MS, Gonzalez AM, Zhang W, Saif LJ. Virus-specific intestinal IFN-gamma producing T cell responses induced by human rotavirus infection and vaccines are correlated with protection against rotavirus diarrhea in gnotobiotic pigs. *Vaccine* 2008;26:3322–31.
86. Nguyen TV, Yuan L, Azevedo MS, et al. Low titer maternal antibodies can both enhance and suppress B cell responses to a combined live attenuated human rotavirus and VLP-ISCOM vaccine. *Vaccine* 2006;24:2302–16.
87. Nguyen TV, Yuan L, Azevedo MS, et al. High titers of circulating maternal antibodies suppress effector and memory B-cell responses induced by an attenuated rotavirus priming and rotavirus-like particle-immunostimulating complex boosting vaccine regimen. *Clin Vaccine Immunol* 2006;13:475–85.
88. Parreno V, Hodgins DC, de Arriba L, et al. Serum and intestinal isotype antibody responses to Wa human rotavirus in gnotobiotic pigs are modulated by maternal antibodies. *J Gen Virol* 1999;80(Pt 6):1417–28.
89. Hodgins DC, Kang SY, de Arriba L, et al. Effects of maternal antibodies on protection and development of antibody responses to human rotavirus in gnotobiotic pigs. *J Virol* 1999;73:186–97.
90. Wang H, Gao K, Wen K, et al. Lactobacillus rhamnosus GG modulates innate signaling pathway and cytokine responses to rotavirus vaccine in intestinal mononuclear cells of gnotobiotic pigs transplanted with human gut microbiota. *BMC Microbiol* 2016;16:109.
91. Twitchell EL, Tin C, Wen K, et al. Modeling human enteric dysbiosis and rotavirus immunity in gnotobiotic pigs. *Gut Pathog* 2016;8:51.
92. Wen K, Tin C, Wang H, et al. Probiotic Lactobacillus rhamnosus GG enhanced Th1 cellular immunity but did not affect antibody responses in a human gut microbiota transplanted neonatal gnotobiotic pig model. *PLoS One* 2014;9:e94504.
93. Yang X, Wen K, Tin C, et al. Dietary rice bran protects against rotavirus diarrhea and promotes Th1-type immune responses to human rotavirus vaccine in gnotobiotic pigs. *Clin Vaccine Immunol* 2014;21:1396–403.
94. Takeuchi O, Akira S. Recognition of viruses by innate immunity. *Immunol Rev* 2007;220:214–24.
95. Jensen S, Thomsen AR. Sensing of RNA viruses: a review of innate immune receptors involved in recognizing RNA virus invasion. *J Virol* 2012;86:2900–10.
96. Newton K, Dixit VM. Signaling in innate immunity and inflammation. *Cold Spring Harb Perspect Biol* 2012;4:a006049.
97. Xu J, Yang Y, Sun J, et al. Expression of Toll-like receptors and their association with cytokine responses in peripheral blood mononuclear cells of children with acute rotavirus diarrhoea. *Clin Exp Immunol* 2006;144:376–81.
98. Gunaydin G, Nordgren J, Svensson L, Hammarstrom L. Mutations in toll-like receptor 3 are associated with elevated levels of rotavirus-specific IgG antibodies in IgA-deficient but not IgA-sufficient individuals. *Clin Vaccine Immunol* 2014;21:298–301.
99. Pane JA, Webster NL, Coulson BS. Rotavirus activates lymphocytes from non-obese diabetic mice by triggering toll-like receptor 7 signaling and interferon production in plasmacytoid dendritic cells. *PLoS Pathog* 2014;10:e1003998.
100. Sen A, Pruijssers AJ, Dermody TS, García-Sastre A, Greenberg HB. The early interferon response to rotavirus is regulated by PKR and depends on MAVS/IPS-1, RIG-I, MDA-5, and IRF3. *J Virol* 2011;85:3717–32.
101. Yamamoto M, Sato S, Mori K, et al. Cutting edge: a novel Toll/IL-1 receptor domain-containing adapter that preferentially activates the IFN-beta promoter in the Toll-like receptor signaling. *J Immunol* 2002;169:6668–72.
102. Takeda K, Akira S. TLR signaling pathways. *Semin Immunol* 2004;16:3–9.
103. Kato H, Takeuchi O, Mikamo-Satoh E, et al. Length-dependent recognition of double-stranded ribonucleic acids by retinoic acid-inducible gene-I and melanoma differentiation-associated gene 5. *J Exp Med* 2008;205:1601–10.
104. Borden EC, Sen GC, Uze G, et al. Interferons at age 50: past, current and future impact on biomedicine. *Nat Rev Drug Disc* 2007;6:975–90.
105. Kim YK, Shin J-S, Nahm MH. NOD-like receptors in infection, immunity, and diseases. *Yonsei Med J* 2016;57:5–14.
106. Zhu S, Ding S, Wang P, et al. Nlrp9b inflammasome restricts rotavirus infection in intestinal epithelial cells. *Nature* 2017;546:667.
107. Holloway G, Coulson BS. Innate cellular responses to rotavirus infection. *J Gen Virol* 2013;94:1151–60.

108. Barro M, Patton JT. Rotavirus nonstructural protein 1 subverts innate immune response by inducing degradation of IFN regulatory factor 3. *Proc Natl Acad Sci USA* 2005;102:4114–9.
109. Barro M, Patton JT. Rotavirus NSP1 inhibits expression of type I interferon by antagonizing the function of interferon regulatory factors IRF3, IRF5, and IRF7. *J Virol* 2007;81:4473–81.
110. Arnold MM, Sen A, Greenberg HB, Patton JT. The battle between rotavirus and its host for control of thc interferon signaling pathway. *PLoS Pathog* 2013;9:e1003064.
111. Holloway G, Dang VT, Jans DA, Coulson BS. Rotavirus inhibits IFN-induced STAT nuclear translocation by a mechanism that acts after STAT binding to importin-alpha. *J Gen Virol* 2014;95:1723–33.
112. Mahlakõiv T, Hernandez P, Gronke K, Diefenbach A, Staeheli P. Leukocyte-derived IFN-α/β and epithelial IFN-λ constitute a compartmentalized mucosal defense system that restricts enteric virus infections. *PLoS Pathog* 2015;11:e1004782.
113. Hernández PP, Mahlakõiv T, Yang I, et al. Interferon-λ and interleukin 22 act synergistically for the induction of interferon-stimulated genes and control of rotavirus infection. *Nat Immunol* 2015;16:698.
114. Lin J-D, Feng N, Sen A, et al. Distinct roles of type I and type III interferons in intestinal immunity to homologous and heterologous rotavirus infections. *PLoS Pathog* 2016;12:e1005600.
115. Comstock SS, Li M, Wang M, et al. Dietary human milk oligosaccharides but not prebiotic oligosaccharides increase circulating natural killer cell and mesenteric lymph node memory T cell populations in noninfected and rotavirus-infected neonatal piglets. *J Nutr* 2017;147:1041–7.
116. Mesa MC, Rodriguez LS, Franco MA, Angel J. Interaction of rotavirus with human peripheral blood mononuclear cells: plasmacytoid dendritic cells play a role in stimulating memory rotavirus specific T cells in vitro. *Virology* 2007;366:174–84.
117. Yasukawa M, Nakagomi O, Kobayashi Y. Rotavirus induces proliferative response and augments non-specific cytotoxic activity of lymphocytes in humans. Clin Exp Immunol 1990;80:49–55.
118. Myers TJ, Schat KA. Natural killer cell activity of chicken intraepithelial leukocytes against rotavirus-infected target cells. *Vet Immunol Immunopathol* 1990;26:157–70.
119. Vlasova AN, Shao L, Kandasamy S, et al. Escherichia coli Nissle 1917 protects gnotobiotic pigs against human rotavirus by modulating pDC and NK-cell responses. *Eur J Immunol* 2016;46:2426–37.
120. Miethke AG, Saxena V, Shivakumar P, Sabla GE, Simmons J, Chougnet CA. Post-natal paucity of regulatory T cells and control of NK cell activation in experimental biliary atresia. *J Hepatol* 2010;52:718–26.
121. Squires JE, Shivakumar P, Mourya R, Bessho K, Walters S, Bezerra JA. Natural killer cells promote long-term hepatobiliary inflammation in a low-dose rotavirus model of experimental biliary atresia. *PLoS One* 2015;10:e0127191.
122. Smith PD, Ochsenbauer-Jambor C, Smythies LE. Intestinal macrophages: unique effector cells of the innate immune system. *Immunol Rev* 2005;206:149–59.
123. Narváez CF, Angel J, Franco MA. Interaction of rotavirus with human myeloid dendritic cells. *J Virol* 2005;79:14526–35.
124. Deal EM, Jaimes MC, Crawford SE, Estes MK, Greenberg HB. Rotavirus structural proteins and dsRNA are required for the human primary plasmacytoid dendritic cell IFNalpha response. *PLoS Pathog* 2010;6:e1000931.
125. Gonzalez AM, Azevedo MS, Jung K, Vlasova A, Zhang W, Saif LJ. Innate immune responses to human rotavirus in the neonatal gnotobiotic piglet disease model. *Immunology* 2010;131:242–56.
126. Wendland M, Czeloth N, Mach N, et al. CCR9 is a homing receptor for plasmacytoid dendritic cells to the small intestine. *Proc Natl Acad Sci USA* 2007;104:6347–52.
127. Vlasova AN, Chattha KS, Kandasamy S, Siegismund CS, Saif LJ. Prenatally acquired vitamin A deficiency alters innate immune responses to human rotavirus in a gnotobiotic pig model. *J Immunol* 2013;190:4742–53.
128. Vlasova AN, Paim FC, Kandasamy S, et al. Protein malnutrition modifies innate immunity and gene expression by intestinal epithelial cells and human rotavirus infection in neonatal gnotobiotic pigs. *mSphere* 2017;2.
129. Eberl G, Colonna M, Di Santo JP, McKenzie AN. Innate lymphoid cells. Innate lymphoid cells: a new paradigm in immunology. *Science* 2015;348:aaa6566.
130. Spits H, Di Santo JP. The expanding family of innate lymphoid cells: regulators and effectors of immunity and tissue remodeling. *Nat Immunol* 2011;12:21–7.
131. Holmkvist P, Roepstorff K, Uronen-Hansson H, et al. A major population of mucosal memory CD4+ T cells, coexpressing IL-18Ralpha and DR3, display innate lymphocyte functionality. *Mucosal Immunol* 2015;8:545–58.
132. Geremia A, Arancibia-Carcamo CV. Innate Lymphoid Cells in Intestinal Inflammation. *Front Immunol* 2017;8:1296.
133. Lefrancois L. Cytotoxic T cells of the mucosal immune system. In: Mestecky J, Lamm ME, Strober W, Bienenstock J, McGhee JR, Mayer L, eds. *Mucosal Immunolgy*. Cambridge, Massachusetts: Elsevier Academic Press; 2005: 559–81.
134. McNeal MM, Rae MN, Ward RL. Evidence that resolution of rotavirus infection in mice is due to both CD4 and CD8 cell-dependent activities. *J Virol* 1997;71:8735–42.

135. Rott LS, Rose JR, Bass D, Williams MB, Greenberg HB, Butcher EC. Expression of mucosal homing receptor alpha4beta7 by circulating CD4+ cells with memory for intestinal rotavirus. *J Clin Invest* 1997;100:1204–8.
136. Wen K, Li G, Bui T, et al. High dose and low dose *Lactobacillus acidophilus* exerted differential immune modulating effects on T cell immune responses induced by an oral human rotavirus vaccine in gnotobiotic pigs. *Vaccine* 2012;30:1198–207.
137. Wen K, Liu F, Li G, et al. *Lactobacillus rhamnosus* GG dosage affects the adjuvanticity and protection against rotavirus diarrhea in gnotobiotic pigs. *J Pediatr Gastroenterol Nutr* 2015;60:834–43.
138. Kim B, Feng N, Narvaez CF, et al. The influence of CD4+ CD25+ Foxp3+ regulatory T cells on the immune response to rotavirus infection. *Vaccine* 2008;26:5601–11.
139. Moser CA, Offit PA. Distribution of rotavirus-specific memory B cells in gut-associated lymphoid tissue after primary immunization. *J Gen Virol* 2001;82:2271–4.
140. Yuan L, Saif LJ. Induction of mucosal immune responses and protection against enteric viruses: rotavirus infection of gnotobiotic pigs as a model. *Vet Immunol Immunopathol* 2002;87:147–60.
141. Mora JR, Iwata M, Eksteen B, et al. Generation of gut-homing IgA-secreting B cells by intestinal dendritic cells. *Science* 2006;314:1157–60.
142. Reifen R, Mor A, Nyska A. Vitamin A deficiency aggravates rotavirus infection in CD-1 mice through extensive involvement of the gut. *Int J Vitam Nutr Res* 2004;74:355–61.
143. Sirisinha S, Darip MD, Moongkarndi P, Ongsakul M, Lamb AJ. Impaired local immune response in vitamin A-deficient rats. *Clin Exp Immunol* 1980;40:127–35.
144. Nikawa T, Odahara K, Koizumi H, et al. Vitamin A prevents the decline in immunoglobulin A and Th2 cytokine levels in small intestinal mucosa of protein-malnourished mice. *J Nutr* 1999;129:934–41.
145. Kandasamy S, Chattha KS, Vlasova AN, Saif LJ. Prenatal vitamin A deficiency impairs adaptive immune responses to pentavalent rotavirus vaccine (RotaTeq(R)) in a neonatal gnotobiotic pig model. *Vaccine* 2014;32:816–24.
146. Chattha KS, Kandasamy S, Vlasova AN, Saif LJ. Vitamin A deficiency impairs adaptive B and T cell responses to a prototype monovalent attenuated human rotavirus vaccine and virulent human rotavirus challenge in a gnotobiotic piglet model. *PLoS One* 2013;8:e82966.
147. Velazquez FR, Matson DO, Guerrero ML, et al. Serum antibody as a marker of protection against natural rotavirus infection and disease. *J Infect Dis* 2000;182:1602–9.
148. To TL, Ward LA, Yuan L, Saif LJ. Serum and intestinal isotype antibody responses and correlates of protective immunity to human rotavirus in a gnotobiotic pig model of disease. *J Gen Virol* 1998;79:2661–72.
149. Lappalainen S, Blazevic V, Malm M, Vesikari T. Rotavirus vaccination and infection induce VP6-specific IgA responses. *J Med Virol* 2017;89:239–45.
150. Yuan L, Honma S, Kim I, Kapikian AZ, Hoshino Y. Resistance to rotavirus infection in adult volunteers challenged with a virulent G1P1A[8] virus correlated with serum immunoglobulin G antibodies to homotypic viral proteins 7 and 4. *J Infect Dis* 2009;200:1443–51.
151. Lu DR, Tan Y-C, Kongpachith S, et al. Identifying functional anti-*Staphylococcus aureus* antibodies by sequencing antibody repertoires of patient plasmablasts. *Clin Immunoly* 2014;152:77–89.
152. Kanai Y, Komoto S, Kawagishi T, et al. Entirely plasmid-based reverse genetics system for rotaviruses. *Proc Natl Acad Sci USA* 2017;114:2349–54.
153. Pfeiffer JK, Virgin HW. Viral immunity. Transkingdom control of viral infection and immunity in the mammalian intestine. *Science* 2016;351: aad5872.
154. Zhang M, Zhang M, Zhang C, et al. Pattern extraction of structural responses of gut microbiota to rotavirus infection via multivariate statistical analysis of clone library data. *FEMS Microbiol Ecol* 2009;70:177–85.
155. Leslie JL, Young VB. The rest of the story: the microbiome and gastrointestinal infections. *Curr Opin Microbiol* 2015;23:121–5.
156. Berger AK, Mainou BA. Interactions between enteric bacteria and eukaryotic viruses impact the outcome of infection. *Viruses* 2018;10:19.
157. Lei S, Twitchell E, Yuan L. Pathogenesis, immunity and the role of microbiome/probiotics in enteric virus infections in humans and animal models. In: Sun J, Dudeja PK, eds. *Mechanisms Underlying Host-Microbiome Interactions in Pathophysiology of Human Diseases.* 1st Edition. Cambridge, Massachusetts: Springer US; 2018:55–78.
158. Yuan LJ, Jobst PM, Weiss M. Gnotobiotic pigs: from establishing facility to modeling human infectious diseases. *Gnotobiotics* 2017:349–68.
159. Jones MK, Watanabe M, Zhu S, et al. Enteric bacteria promote human and mouse norovirus infection of B cells. *Science* 2014;346:755–9.
160. Kuss SK, Best GT, Etheredge CA, et al. Intestinal microbiota promote enteric virus replication and systemic pathogenesis. *Science* 2011;334:249–52.

SUMMARY

Rotavirus (RV) is a fecal–orally transmitted gastrointestinal pathogen that kills more than 200,000 children under the age of five worldwide each year and is most commonly spread by contaminated food, water, hands, and fomites. Even though nearly every healthy adult is immune to RV diarrhea, it is unclear how long-term immunity in humans is established. Currently, two vaccines RotaTeq™ and Rotarix™ are licensed in the United States and in >100 other countries throughout the world. While these vaccines provide a high level of protection against RV AGE in high-income countries, they appear to be less effective in low-resource countries where disease burden is the highest. Without antiviral drugs and with the existing vaccines falling short of expectation, further studies on the molecular mechanisms of host immune responses to RV are necessary to develop improved vaccines and therapeutic measures. Reviewing the latest findings on the initial recognition of RV infection, control of infection, and establishment of immune memory by the host immune system, perspectives on several aspects of RV infection that are fundamental to the elucidation of RV pathogenesis are presented.

Development of Vaccines for Avian Influenza Virus

9

Ioannis Sitaras and Erica Spackman
Agricultural Research Service, United States Department of Agriculture

Contents

9.1 INTRODUCTION

Avian influenza virus (AIV) is a type A influenza virus that originates from and is adapted to an avian species. With a worldwide distribution in wild birds, where the primary carriers are waterfowl and other aquatic birds, AIV can be transmitted to domestic poultry (most commonly chickens, ducks, and turkeys).

In gallinaceous birds (chickens and turkeys) two pathotypes, low pathogenic (LP) and highly pathogenic (HP), have been described.[1] The LP form causes no or mild disease in any avian species unless complicated by another agent, and it does not occur in meat. In contrast, the HP form causes severe, often 100% fatal, disease in chickens and turkeys. The severity of disease and the extent of mortality due to HPAIV in ducks are more variable. The HP form can be found in meat and eggs of chickens, turkeys, and ducks. Vaccine development to prevent disease and death from LPAIV and HPAIV in poultry and to eliminate HPAIV from meat and eggs is the focus of this chapter.

9.1.1 Avian Influenza Virus Biology

Belonging to the *Orthomyxoviridae* family, avian influenza (AI) viruses are single-stranded, negative-sense RNA viruses, with a genome composed of eight gene segments that encode for virus proteins necessary for a successful virus replication cycle (i.e., attachment and entry to cells, synthesis of RNA, virus release, etc.).[2] Table 9.1 presents an overview of the genes in AIVs, their sizes, and main protein functions.

AI viruses are classified according to their coat proteins, i.e., hemagglutinin (HA) and neuraminidase (NA), encoded by the respective genes (i.e., H5N1, H7N7, H7N9, H9N2, etc.). There exist 16 types of HA and 9 NA types, giving rise to 144 possible subtype combinations. As the immunologically most-important proteins, HA and NA are serotype determinants. In particular, HA is the most abundant surface protein and outnumbers NA by a 4:1 ratio.[3,4] The HA protein plays a pivotal role in the binding of the virus to cell-surface sialic-acid (SA) receptors, virulence, antigenicity, transmissibility, and host specificity.[5–8] Therefore, antibodies that inhibit the function of the HA protein are the most critical for vaccine efficacy.

Because of its abundance and importance, HA experiences the most intense selection pressure from the immune system, which makes it very prone to mutations. These mutations often alter the protein enough to escape neutralization by innate or acquired immunity and even breach host-range barriers. This type of altering of the HA protein is known as antigenic drift[4,5,9–11] and is further exacerbated by the

TABLE 9.1 Genes comprising the segmented genome of avian influenza viruses and their main protein functions

SEGMENT	*GENE(S)*	*GENE ABBREVIATION*	*GENE SIZE (KB)*	*MAIN PROTEIN FUNCTION*
1	Polymerase Basic 2	PB2	2.3	Component of viral polymerase
2	Polymerase Basic 1	PB1	2.3	Component of viral polymerase
	Polymerase Basic 1 F2	PB1-F2	0.27	Viral pathogenicity in some species (not expressed in all strains)
3	Polymerase Acidic	PA	2.2	Component of viral polymerase
4	Hemagglutinin	HA	1.7	Coat protein, host cell receptor binding and cell entry, major mediator of virulence in gallinaceous poultry.
5	Nucleoprotein	NP	1.5	Encapsidates RNA
6	Neuraminidase	NA	1.4	Coat protein, releases virus progeny by cleaving the newly budded virus from the host cell
7	Matrix 1	M1	0.8	Structural, virus assembly
	Matrix 2	M2	1.0	Ion channel that regulates pH for viral uncoating
8	Non-Structural 1	NS1	0.7	RNA transport, splicing, translation, host immune system modulation
	Non-Structural 2	NS2	0.8	Mediates export of vRNPs

error-prone RNA polymerase of influenza viruses, which is responsible for 10^6 more replication errors than the DNA polymerase).[4,9,12,13] As a consequence, antigenic drift leads to an infinite number of antigenically different influenza virus mutants that can be selected.

AIVs are categorized as low pathogenic and highly pathogenic (LPAIV and HPAIV, respectively), depending on their pathogenicity in gallinaceous birds (chickens and turkeys). The difference in disease severity between LPAIVs and HPAIVs is mainly because LPAIVs only replicate locally in the respiratory and intestinal tracts, while HPAIVs replicate systemically. The severity of clinical disease can range from subclinical to moderate symptoms with LPAIV infection, to acute disease and death with HPAIV infection. LPAIV infections are more common in poultry and wild birds and are associated with milder signs, which, however, can become exacerbated in the presence of other bacterial or viral infections.[14–16] Low pathogenic strains of the H5 and H7 subtypes are particularly monitored, since they have the potential to convert to HPAIVs through the accumulation of mutations during circulation in poultry. On the other hand, HPAIV infections are known to cause up to 100% mortality in chickens and turkeys.

9.1.2 Health and Economic Consequences of AIV Infections

The global spread of AIV infections carries huge financial consequences and is a threat to animal and human health. Estimates give the number of commercial poultry birds that die each year, either as a direct result of AI infections or as a result of stamping out policies, to hundreds of millions.[17] In addition to poultry health, certain AIVs can transmit to humans on rare occasions (Table 9.2). Historically, transmission of AIVs to humans appears to be limited to those with high levels of contact with infected poultry (e.g., poultry workers, cullers during outbreaks), and human-to-human transmission is extremely rare and limited to direct caregivers outside of a healthcare setting (i.e., immediate family members and where no infection control precautions were taken).[18–20] Two examples of AIVs with documented transmission potential to humans are the goose/Guangdong/1996 lineage H5Nx HPAIV (gs/GD/96 H5) and the Anhui/2013 lineage H7N9 LPAIV (Anhui/13 H7N9). From 2003 until March 2018, the World Health Organization (WHO) reported 860 cases of laboratory-confirmed human infection with gs/GD/96 H5 lineage strains, 454 of which were fatal, giving this virus a 53% mortality rate.[21] From 2013 until May 2018, Anhui/13 H7N9 was confirmed in 1,567 cases, of which 615 were fatal (39%

TABLE 9.2 Documented infections of humans with avian influenza virus as of April 25, 2018

SUBTYPE	PATHOTYPE	YEAR(S)	COUNTRY/REGION	CASES	DEATHS[B]
H7N9	LP and HP[a]	2013–present	China	1,625	623
H5N1	HP	1997–present	Asia, Africa	858	453
H7N7	HP	2003	Netherlands	89	1
H9N2	LP	1999–present	Asia, Egypt	28	0
H5N6	HP	2013–2016	China	14	6
H7N2	LP	2007	United Kingdom	4	0
H10N7	LP	2014	China	4	0
H10N8	LP	2013–2014	China	3	2
H7N2	LP	2002, 2016	United States	3	0
H7N3	LP and HP	2004	Canada	2	0
H7N3	LP	2006	United Kingdom	1	0
H6N1	LP	2013	Taiwan	1	0
H7N7	LP	1996	United Kingdom	1	0

Avian influenza virus (AIV) infection was confirmed by genetic analysis of the virus isolated from patients. Mild upper respiratory symptoms and conjunctivitis were the most common clinical presentations.

[a] LP, low pathogenicity; HP, highly pathogenic.

[b] Deaths were attributed to infection with AIV or complications arising from infection with AIV.

mortality rate).[22,23] Importantly, both LPAIVs and HPAIVs can infect humans, and the most common symptoms are conjunctivitis and mild upper respiratory disease; however, host and virus factors in combination are involved in cases of more severe disease.[24] Further details about human diseases caused by AIVs are available elsewhere.[24,25]

9.1.3 Consequences of AIV Infection to Meat and Egg Production and Safety

LPAIV infections are associated with drops in egg production (especially in turkeys), as well as reduced weight gain, which is particularly important in chickens and turkeys raised for meat. LPAIVs cause localized infection in the upper respiratory, intestinal, and hen reproductive tracts, do not spread into muscle, and are therefore not found in poultry meat.[14,26]

Besides massive production losses, HPAIVs pose some risk of infection through fresh and frozen raw poultry products, although this risk is rather low. HPAIVs have been recovered from chicken thigh and breast meat, as well as duck meat. In addition, HPAIVs have been detected in egg yolk and albumen of infected hens.[27] Human cases almost always involve handling live poultry or infected carcasses (e.g., during food preparation) or consuming infected/raw meat or eggs,[26,28–30] as AIVs are not heat stable and are rapidly inactivated at cooking and egg product pasteurization temperatures.[26,31,32]

9.2 VACCINATION AGAINST AVIAN INFLUENZA VIRUSES

9.2.1 General Considerations

When used properly, avian influenza vaccines can be critical in preserving the life of poultry, reducing the chance of human infections, and minimizing financial losses to the poultry industry.

Although vaccines against swine and equine influenza are routinely applied, vaccination of poultry is complicated by several factors. First, the variety of strains that can infect poultry are much more diverse than mammalian hosts, so vaccination is not possible until a specific strain is present in a local poultry population. Second, most AIV vaccines are inactivated[33] and require individual administration of vaccine to each animal. Indeed, multiple vaccinations are needed to maintain lifelong immunity in egg-laying birds. This makes vaccination in poultry economically nonviable in many cases and the extra withdrawal times prior to slaughter can add further cost to the farmer owing to feeding and housing the chickens and turkeys, which exceeds the sale value of the bird for meat. Third, vaccination for AIVs can affect international trade, as some countries do not want to trade with regions that vaccinate.[15] Also, surveillance programs must be modified during vaccination so that they can detect when vaccinated birds are infected with field viruses, but these programs are expensive (see Section 9.2.4). Another reason preventing countries from using vaccines to control the spread of AIV infections is the fear that vaccination may drive the selection of mutants capable of escaping vaccination-induced immunity[34–37] or may allow the silent spread of AIVs, since vaccinated animals may not show disease signs but still excrete virus into the environment, thus further complicating eradication efforts.[15,38,39] With prolonged vaccination, AIV vaccines need to be periodically updated to maintain an adequate antigenic match with the circulating field strains, much like the vaccines for human influenza. Surveillance programs to monitor vaccine efficacy also increase the cost of vaccination programs.

Despite the disadvantages listed above, vaccination against HPAI is used either as a routine or emergency measure in some countries when it seems economically viable. Strict biosecurity measures and culling of infected poultry, as well as pre-emptive culling of poultry at high risk of infection, are usually an early response to an outbreak with the goal of eradication (stamping-out). Such stamping out policies can have a devastating financial effect on the poultry industry and carry ethical dilemmas because all the poultry is destroyed and the meat and eggs are frequently discarded.[40,41]

In the field, the success of vaccines for controlling AIV outbreaks has been variable; vaccination programs can fail at the flock and/or regional levels. Recent reports indicate that the reasons why vaccination programs have failed did not necessarily have to do with the vaccines themselves, but rather with mistakes in vaccine administration and vaccination strategies. Such mistakes include "cutting" vaccine doses to reduce costs (resulting in poor levels of immunity and inadequate herd immunity), having poorly trained personnel administer the vaccine, and inadequate vaccination coverage (i.e., vaccination not applied to an epidemiologically relevant geographic area or population).[33–37,40–50] Other factors, such as poor management or immunosuppression in the host, may also contribute to vaccination failures.[51]

9.2.2 Aims of an Avian Influenza Vaccination Program

In order for any avian influenza vaccination program to be considered successful, it must meet as many of the following aims as possible: (i) greatly reduce or eliminate morbidity and mortality, (ii) significantly reduce virus shedding from infected birds, as well as the duration of shedding by inducing sufficiently high levels of immunity in a large enough percentage of the vaccinated population (herd immunity), and (iii) stop transmission of the virus to other birds.[14,38,52] Vaccination-induced immunity should be relatively long-lasting (i.e., at least 6 months), and should be the result of a minimum number of vaccinations, in order to keep the cost of the vaccination campaign as low as possible.[52] Vaccination programs should establish criteria for when to end vaccination or the minimum specific goals of vaccination and metrics for measuring success should be developed.

9.2.3 Challenges of an Avian Influenza Vaccination Program

Vaccination is impeded by high labor costs and in many cases is associated with trade embargos, thereby further damaging the poultry industry and adding to the financial losses.[15] Although evidence to the contrary has been reported,[53] such trade embargos are in place because there is the belief that vaccination can mask symptoms of infection, thus allowing the silent spread of the virus and increasing chances of transmission. This is particularly true regarding AIV of the H5 and H7 subtypes, which are notifiable to the OIE.

Selection of a vaccine-seed strain is based on the HA protein and must be the result of careful evaluation of circulating field strains through surveillance programs (see Section 9.2.6.1); otherwise, the vaccine may not protect adequately. Vaccines must be of the same HA subtype as the field virus, but there is antigenic variation within subtypes, so, for example, no one H5 strain will provide protection against all H5 subtype strains. Importantly, the exact matching of the vaccine seed strain to currently circulating strains is not the only factor in protection since highly immunogenic HA proteins of the same subtype as the field strain can protect against moderately antigenically divergent strains.[41,54–57]

Finally, vaccination as a means of controlling avian influenza infections should always be used in conjunction with biosecurity measures, not in spite of them. Such measures most importantly include culling of infected animals, pre-emptive culling, restriction of movement within the outbreak zones, implementation of rapid diagnosis and surveillance programs, and effective disinfection of premises with infected animals.[14,46,48,52,58,59]

9.2.4 Avian Influenza Vaccination Programs

There are three types of vaccination programs against AIV depending on the situation: routine, preventative, and emergency.

Routine vaccination is applied in countries where AIVs are already endemic, to contain the spread of disease, minimize poultry losses as well as implications to human health, and reduce the financial impact on the poultry industry. Local conditions may prevent the eradication of the infection by means of usual biosecurity measures such as control of movement, culling of infected poultry, and pre-emptive culling. Routine vaccination may not necessarily be applied *ad infinitum*, since conceivably it may eventually lead to a situation where the spread of infection can be reduced to such an extent that stamping out and surveillance could eradicate the infection. Due to antigenic drift, vaccine-seed strains need to be periodically evaluated with the current field strains. As protection efficacy deteriorates it will often be apparent in the field as flocks which are well vaccinated will start to present clinical disease.

Preventative vaccination programs are initiated before the virus is present in a poultry population but when there is a potential and credible threat of introduction of an AIV from wild/migratory birds or domestic birds in the region (e.g., when an AIV is already introduced into one production sector (e.g., table-egg layers, broiler chickens, turkeys, breeders) and needs to be prevented from posing a threat to another production sector). Preventative vaccination may also be used for zoo, ornamental, and pet birds if they are considered to be at risk. Usually, preventative vaccination programs come to an end when the perceived risk of infection no longer applies.

Emergency vaccination is usually employed during the early stages of an outbreak and the virus is in the population (e.g., within a control zone). Often there are epidemiological reasons to believe that biosecurity measures alone (including culling of infected poultry and pre-emptive culling) will be unable to contain the infection. This is mostly due to parameters such as high poultry density, local restrictions, etc. Oftentimes, emergency vaccination is used in the form of ring vaccination (i.e., vaccination of flocks within an area surrounding an outbreak incident). The radius of ring vaccination should be determined from epidemiological data and mathematical model projections that calculate the spread and containment of the infection by taking into consideration a multitude of parameters such poultry density, wind intensity and direction, transmissibility of the virus, etc. It is advisable to use ring vaccination in the framework of a Differentiating Infected from Vaccinated Animals (DIVA) strategy.

9.2.5 Differentiating Infected from Vaccinated (DIVA) Strategy

During vaccination against AIV, the potential exists for vaccinated animals to acquire AIV infection (vaccines do not provide sterilizing immunity), which may go unnoticed, due to the lack of clinical signs as a consequence of vaccine protection. Also, for surveillance programs there is a need to determine whether an antibody-positive bird was exposed to field virus in addition to being vaccinated. Therefore, being able to differentiate vaccinated from infected animals (DIVA) is very important to know as soon as possible whether the vaccine has failed to stop the spread of infection. The simplest way to differentiate is to include unvaccinated sentinel birds within a flock. If a virus is present and actively spreading, these sentinel birds will be the first to show symptoms of avian influenza infection.[38,60] Another way to differentiate vaccinated from infected birds is to use an inactivated whole-virus vaccine that has the same HA as the targeted circulating field strain, but a different NA. Serological testing of birds can then show whether they have antibodies against the vaccine strain NA, or the field strain NA, in which case they are infected. The same applies when using vectored vaccines, since vaccinated birds will only exhibit antibodies to the HA, but not to the NA or internal proteins such as the matrix or nucleoprotein.[38,61–63] Some drawbacks are that DIVA strategies can only be successful when sensitive and accurate diagnostic tests are available for antibodies to the differentiating NA subtype. Importantly, because of the vaccine, they will often not mount a strong antibody response to the NA of the field virus, so high sensitivity of tests is critical. Additionally, DIVA strategies add cost to vaccination programs.

9.2.6 Avian Influenza Vaccines

9.2.6.1 *Inactivated Whole-Virus Vaccines*

Inactivated whole-virus AIV vaccines are by far the most widely applied worldwide, accounting for about 95% of all vaccines used.[33] Inactivated vaccines are inexpensive and highly effective, although multiple doses may be needed for life-long protection in long lived birds, such as table-egg layers and breeders. The main disadvantage of inactivated vaccines is the required parenteral route of administration, which is labor-intensive and demands the employment of qualified and trained personnel.[14,15,38,52,58] Withdrawal times are often required by food hygiene regulations for mineral oil adjuvanted vaccines in birds for human consumption and in broiler chickens, which can be longer than the life span of the bird.

Inactivated vaccines are produced by selecting a seed strain (i.e., antigen) that is ideally highly immunogenic and has broad antigenic reactivity among strains within the subtype. Immunogenicity can be determined by *in vivo* studies with the vaccine, and the antigenic reactivity is often determined by hemagglutination inhibition (HI) assay. In the HI assay, serum with the prospective vaccine is tested for its ability to neutralize the hemagglutinating function of the HA protein. However, *in vivo* protection needs to be demonstrated. For safety purposes, vaccine seed strains are almost always LP strains or HP strains that have been engineered to be LP by reverse genetics (RG) if natural LP strains are not available. Using LP strains also eliminates the need for high biosecurity facilities and the regulatory oversight of HP strains for vaccine production.

Once the seed strain is selected, a stock of antigen is grown, often in specific pathogen free (SPF) or specific antibody negative (SAN) embryonating chicken eggs. Therefore, an additional requirement for the seed strain is that it should be able to replicate to high titers in eggs. Since the development of RG methods, the HA from strains that do not replicate well can be put in a "backbone" of a virus that does replicate adequately well.[64] What constitutes "adequately well" is not well defined since the necessary dose depends on potency. Prior to application in the field, safety and efficacy testing is also conducted on the new vaccine. If it is necessary to use a new strain for adequate protection (i.e., existing vaccines are not protective against a new AIV strain), the entire process of producing an inactivated vaccine will take at least 6–9 months after the decision to vaccinate is made (Figure 9.1).

9.2.6.2 *Vectored Vaccines*

Vectored vaccines are the second most commonly utilized vaccines against AIV, accounting for approximately 4.5% of vaccines used. They are produced by inserting the HA of interest into a live viral vector. Vectors that have been used include the following: fowl pox virus (FPV), herpes virus of turkeys (HVT), non-virulent avian paramyxovirus type 1 (APMV-1), and for ducks, attenuated duck enteritis virus.[15,38,52,58] Fowl pox virus vectored vaccines containing an H5 HA have been licensed for use in many countries,

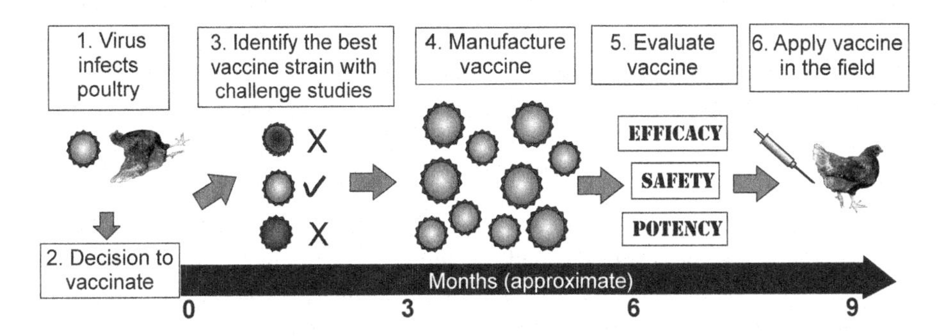

FIGURE 9.1 An overview of the steps in AIV vaccine development process and an approximate time-line based on inactivated vaccines..

and an AMPV-1 viral vector utilizing the La Sota strain and expressing an H5 HA has been licensed in China.[38,65–67] Recombinant vaccines have the advantage that DIVA strategies targeting antibodies to the NA protein can be employed, since birds that have been vaccinated should only have antibodies to the HA. Recombinant APMV-1 vaccines have the added advantage that they can be mass administered by spray vaccination. However, pre-existing immunity to the vectors (maternal or active) causes failure of the vaccine to induce immunity.[52,58] The use of APMV-1 vectored AIV vaccines has been largely discontinued because pre-existing immunity is so widespread in chickens. HVT vectored vaccines are not affected by maternal immunity[68] and can be mass applied when administered *in ovo*. Similar to the inactivated vaccines, multiple vaccinations are necessary for adequate immunity for longer lived birds.

9.2.6.3 Adjuvants Used in Avian Influenza Vaccines

Adjuvants are compounds that are added to vaccines to boost the level of immune response and lengthen its duration. For AI vaccines, only inactivated whole-virus vaccines and virus-like particle vaccines require the use of adjuvants because they are not able to replicate, and thus cannot stimulate the immune response sufficiently by themselves. Adjuvants should ideally be safe, stable, cost-effective, and should not cause localized inflammatory reactions and pain.[69]

Currently, almost all inactivated whole-virus vaccines use different sorts of oil emulsion adjuvants (i.e., water-in-oil, oil-in-water, and water-in-polymer). A multitude of different types of adjuvants exist (e.g., mineral oil, vegetable oil, synthetic polymer and ester, mineral nanoparticles, seaweed-derived, calcium phosphate, and de-acylated chitin).[69,70] Data suggest that mineral oil– and vegetable oil–based adjuvants are optimal in chickens.[69] Data for adjuvant efficacy in other avian species are lacking.

9.2.7 Evaluating Vaccine Efficacy

Numerous metrics are employed to evaluate AI vaccines in poultry, many of which are common for any vaccine (e.g., safety, potency, and efficacy). Although the antibody to the challenge virus has positive predictive value for protection,[71] the most accurate method for determining efficacy is *in vivo* challenge studies. It should also be noted that the protection observed in laboratory testing often exceeds what can be accomplished in the field where birds are stressed more and exposed to other infectious agents.

Potency is a quantitative measure of efficacy relative to antigen mass (i.e., vaccine dose) and is frequently expressed as a 50% protective dose. For example, the dose at which 100% of birds are protected against a challenge dose of 10^6 50% egg infectious doses (EID_{50}). Determining the potency requires challenge studies in the target species using difference doses of the vaccine with a uniform dose of the challenge virus. A dose of 10^6 EID_{50} is common for HPAIV since it represents a high challenge dose. The potency of an inactivated AIV vaccine may vary by species. Finally, potency testing should be conducted with vaccines prepared with the same adjuvant that will be utilized in the final product.

In addition to potency, commercial inactivated vaccines are tested for safety and efficacy. The exact procedures can vary somewhat by country when trials are being conducted for licensure, but the scientific principles are the same. Safety testing is mainly for assuring that the virus is properly inactivated (usually with formalin). Inactivation is determined with a bioassay, typically replication in embryonating chicken eggs (ECE).

Vaccines are most often tested for efficacy with challenge studies in chickens. Although there is some variation in protection among species, inactivated vaccines tend to provide similar levels of protection in chicken, turkeys, and ducks. For efficacy testing, SPF or SAN chickens are vaccinated with the test vaccine using the appropriate dose based on potency testing. Young chickens are typically used for logistical reasons (smaller and less expensive to buy and care for), and in real world applications, most birds will be first vaccinated when they are young when other vaccines are applied.

After vaccination, the chickens are held, normally for 3–4 weeks, before they are challenged to allow immunity to develop. Serum should be collected within 3 days prior to the challenge so antibody levels can be evaluated at the time of challenge. Antibody levels are determined by hemagglutination inhibition (HI) assay and should be tested with both the vaccine virus and the challenge virus if they are different strains. It is possible to use virus neutralization assay, but HI assay is faster and less expensive. Other antibody tests used with AIV (e.g., agar gel immunodiffusion [AGID] assay and commercial ELISA) are not suitable because they do not measure protective antibodies. Both AGID and commercial ELISA target antibodies to the nucleoprotein, and therefore will not detect any antibody to vectored vaccines, and although they may detect antibodies to inactivated whole-virus vaccines, those are not protective antibodies.

In addition to selecting the correct assay to measure protective antibodies it should be noted that antibodies only have a positive predictive value.[71] Generally, birds with titers of over 40 to the challenge virus are expected to be protected. However, birds with lower or even no detectable titers may be protected.[72] Here protection is defined as prevention of disease and death. Sometimes reduction in quantities of virus shed is used as a metric as well; a 100-fold reduction in shed titers is the typical goal.

The chickens are challenged with a high dose, usually 10^6 EID_{50} per bird, of the virus being evaluated. A high dose is used to assure a high margin of protection with a high level of field challenge. The virus can be administered in several ways, intra-tracheally, intra-ocularly, intra-nasally, and intra-choanally. All of which mimic the natural respiratory route of exposure to some extent.

The metrics for protection that are evaluated after challenge are the following: morbidity, mortality, and virus shed (how much virus is excreted) from the oral route and cloacal route. Morbidity and mortality are generally evaluated daily, while the timing of sample collection (i.e., swabs) for virus shed is somewhat variable. Most procedures include some combination of 1, 2, 3, and 4 days post-challenge (or all four time points). Less commonly, samples for virus shed may be collected at later time points (5–14 days). All of these data are combined to measure efficacy against a given strain.

9.3 RECENT ADVANCES IN AVIAN INFLUENZA VACCINES

Influenza is the proof-of-concept agent for new vaccine technology. Although most novel technologies are intended for use in humans, they sometimes have the potential for use in animal or avian species. Since the field of avian influenza vaccines is so rapidly changing the reader should consult recent literature for the state-of-the-art developments. In many cases, it has not been determined if these novel vaccine platforms can be manufactured cheaply enough to be economically viable for poultry.

9.3.1 Nucleic Acid Vaccines

Nucleic acid vaccines can be mRNA-based, DNA-based, or RNA replicon particle-based. All are able to induce both cellular and humoral immune responses. mRNA vaccines are comprised of a vector carrying the information of the antigen of interest, 5′ and 3′ untranslated regions, CAP structure, and poly-A tail. mRNA vaccines contain only the sequence of the nucleic acid of the antigen of interest and they do not require an adjuvant. DNA-based vaccines rely on bacterial plasmids that express the protein of interest and are used to transfect cells. While they are able to encode multiple genes of interest, they are poorly transported into target cells, and there is always some danger of genomic integration into the host cell.

Finally, RNA replicon particle-based vaccines are positive or negative single-stranded or double-stranded RNA viruses that are engineered to express the antigen of interest. They are mainly based on alphaviruses, against which birds have reduced host immunity, and they cannot integrate into the host genome like DNA-based vaccines.[14] One advantage is that they can induce both humoral and cell mediated immunity.[73] The alphavirus-based replicon has performed well in laboratory studies in chickens and turkeys,[74,75] and has been provisionally licensed in the US, but has not been used in the field yet.

9.3.2 Cytokine Adjuvants

A relatively recent development is the potential use of cytokines as adjuvants for avian influenza vaccines. A number of cytokine genes have been discovered in chickens and preliminary testing with interleukin 1β (IL-1β), as well as interferon-α (IFN-α) and interferon-γ (IFN-γ) showed that their use led to increased antibody responses.[76] Whether cytokines will be used as vaccine adjuvants remains to be seen, since more testing needs to be done.

9.4 FUTURE PERSPECTIVES OF AVIAN INFLUENZA VACCINES AND VACCINATION

There are critical needs to both improve vaccines and find better correlates of protection that will reduce or even eliminate the need for *in vivo* challenge studies. Although the vaccines are generally efficacious, the seed strain needs to be antigenically matched and immunogenic. Adjuvants or methods to enhance and broaden immunity would increase the lifespan of the vaccine and reduce the frequency of seed strain updates. Mineral oil– and vegetable oil–based adjuvants are very common and work very well in chickens.[69] Inactivated vaccines only induce humoral immunity, and therefore modifying the vaccine to stimulate innate immunity and cellular immunity would be beneficial. One of the advantages of vectored vaccines and the alphavirus replicon vaccines is that they do induce cellular immunity.[73] Adjuvants and vaccine platforms that improve mucosal immunity would be very beneficial as well.

Enhancement of immunity should also improve the duration of immunity. There are a few studies that have evaluated the duration of immunity beyond a single vaccination and a few weeks,[77,78] but in the field it is clear that repeated vaccinations are required for adequate lifelong immunity. Extending immunity would improve control by substantially increasing population immunity and would reduce the cost of vaccination. Ideally vaccines should be able to be mass applied, some vectored vaccines can be (e.g., HVT can be applied *in ovo*), but the inactivated vaccines must be used for booster vaccinations because of immunity to the vector. Developing a vaccine which does not have interference from prior immunity, but which can be mass applied would be very helpful.

Another critical need for HPAIV vaccines is improved DIVA strategies. Current methods lack sensitivity and are expensive. As discussed earlier, currently the most reliable way to detect whether there is a field virus challenge in a vaccinated population is to use sentinel birds.

All of these objectives, such as enhanced and broad immunity, mass application, and compatibility with a DIVA strategy, must be accomplished in an economically viable manner. Costs must be sufficiently low for farmers and poultry producers to afford the vaccine, while allowing for adequate income for commercial production.

Another critical need is to develop a better way to test vaccines without the *in vivo* challenge model currently being utilized. The current method is expensive and takes several months to complete, and because HPAIV requires a high biosafety level, the labs that can conduct the testing are limited. As described earlier, antibody titers to the challenge virus only provide a positive prediction of immunity (mostly against death and clinical disease). Unfortunately, cases where birds are protected in the absence of detectable antibodies cannot be predicted with *in vitro* assays. Along these lines, vaccines need to be tested in individual species, although the results for chickens are often valid and extrapolated to other species. Correlates of protection should ideally be applicable to turkeys and domestic ducks as well as chickens.

Also, one goal of vaccination is to reduce virus shed. A reduction of titers by 100 fold is the goal. Not all vaccines reduce shed titers even if they protect 100% against morbidity and mortality, but there is currently no way to determine how a vaccine will affect virus shed titers except the *in vivo* challenge test. As vaccine technology improves, one goal will be to provide vaccines that are more consistent in reducing virus shed.

REFERENCES

1. Swayne DE, Suarez DL, Sims LD. Influenza. In: Swayne D, ed. *Diseases of Poultry*. 13th ed. Ames, IA: Blackwell; 2013:181–218.
2. Shaw ML, Palese, P. Orthomyxoviridae: The viruses and their replication. In: Knippe DM, Howley PM, ed. *Fields Virology*. 6th ed. Philadelphia, PA: Lippincott Williams & Wilkins; 2013.
3. Bouvier NM, Palese P. The biology of influenza viruses. *Vaccine* 2008;26 Suppl 4:D49–53.
4. Webster RG, Bean WJ, Gorman OT, Chambers TM, Kawaoka Y. Evolution and ecology of influenza A viruses. *Microbiol Rev* 1992;56:152–79.
5. Hampson AW. Influenza virus antigens and 'antigenic drift'. *Perspect Med Virol* 2002;7:49–85.
6. Webster RG, Rott R. Influenza virus A pathogenicity: the pivotal role of hemagglutinin. *Cell* 1987;50:665–6.
7. Steinhauer DA. Role of hemagglutinin cleavage for the pathogenicity of influenza virus. *Virology* 1999; 258:1–20.
8. Rott R. The pathogenic determinant of influenza virus. *Vet Microbiol* 1992;33:303–10.
9. Nelson MI, Holmes EC. The evolution of epidemic influenza. *Nat Rev Genet* 2007;8:196–205.
10. Gerhard W, Webster RG. Antigenic drift in influenza A viruses. I. Selection and characterization of antigenic variants of A/PR/8/34 (HON1) influenza virus with monoclonal antibodies. *J Exp Med* 1978;148:383–92.
11. Lee CW, Saif YM. Avian influenza virus. *Comp Immunol Microbiol Infect Dis* 2009;32:301–10.
12. Steinhauer DA, Holland JJ. Rapid evolution of RNA viruses. *Ann Rev Microbiol* 1987;41:409–33.
13. Holland J, Spindler K, Horodyski F, Grabau E, Nichol S, VandePol S. Rapid evolution of RNA genomes. *Science* 1982;215:1577–85.
14. Rahn J, Hoffmann D, Harder TC, Beer M. Vaccines against influenza A viruses in poultry and swine: status and future developments. *Vaccine* 2015;33:2414–24.
15. Spackman E, Pantin-Jackwood MJ. Practical aspects of vaccination of poultry against avian influenza virus. *Vet J* 2014;202:408–15.
16. Alexander DJ. A review of avian influenza in different bird species. *Vet Microbiol* 2000;74:3–13.
17. Otte J, Hinrichs J, Rushton J, Roland-Holst D, Zilberman D. Impacts of avian influenza virus on animal production in developing countries. *CAB Rev: Perspect Agr Vet Sci Nutr Nat Resour* 2008;3:1–18.
18. Peiris JS, de Jong MD, Guan Y. Avian influenza virus (H5N1): a threat to human health. *Clin Microbiol Rev* 2007;20:243–67.
19. Beigel JH, Farrar J, Han AM, et al. Avian influenza A (H5N1) infection in humans. *New Engl J Med* 2005;353:1374–85.
20. Cinatl J, Jr., Michaelis M, Doerr HW. The threat of avian influenza A (H5N1). Part I: epidemiologic concerns and virulence determinants. *Med Microbiol Immunol* 2007;196:181–90.
21. Cumulative number of confirmed human cases for avian influenza A (H5N1) reported to WHO, 2003–2018. 2018. (Accessed 01 June, 2018, at http://www.who.int/influenza/human_animal_interface/2018_03_02_tableH5N1.pdf?ua=1.)
22. Asian lineage avian influenza a (H7N9) virus. 2018. (Accessed 01 June, 2018, at https://www.cdc.gov/flu/avianflu/h7n9-virus.htm.)
23. H7N9 situation update. 2018. (Accessed 01 June, 2018, at http://www.fao.org/ag/againfo/programmes/en/empres/h7n9/situation_update.html.)
24. Cox NJ, Trock SC, Uyeki TM. Public health implication of animal influenza viruses. In: Swayne DE, ed. *Animal Influenza Virus*. 2nd ed. Ames, IA: Blackwell; 2017:92–132.
25. Spackman E. Avian influenza virus. In: Liu D, ed. *Handbook of Foodborne Diseases*. Boca Raton, FL: CRC Press; 2018:.
26. Chmielewski RA, Beck JR, Swayne DE. Thermal inactivation of avian influenza virus and Newcastle disease virus in a fat-free egg product. *J Food Prot* 2011;74:1161–8.
27. Cappucci DT, Jr., Johnson DC, Brugh M, et al. Isolation of avian influenza virus (subtype H5N2) from chicken eggs during a natural outbreak. *Avian Dis* 1985;29:1195–200.
28. Beato MS, Capua I, Alexander DJ. Avian influenza viruses in poultry products: a review. *Avian Pathol* 2009;38:193–200.
29. Harder TC, Teuffert J, Starick E, et al. Highly pathogenic avian influenza virus (H5N1) in frozen duck carcasses, Germany, 2007. *Emerg Infect Dis* 2009;15:272–9.
30. Tumpey TM, Suarez DL, Perkins LE, et al. Characterization of a highly pathogenic H5N1 avian influenza A virus isolated from duck meat. *J Virol* 2002;76:6344–55.

31. Chmielewski RA, Beck JR, Swayne DE. Evaluation of the U.S. Department of Agriculture's egg pasteurization processes on the inactivation of high-pathogenicity avian influenza virus and velogenic Newcastle disease virus in processed egg products. *J Food Prot* 2013;76:640–5.
32. Thomas C, King DJ, Swayne DE. Thermal inactivation of avian influenza and Newcastle disease viruses in chicken meat. *J Food Prot* 2008;71:1214–22.
33. Swayne DE, Pavade G, Hamilton K, Vallat B, Miyagishima K. Assessment of national strategies for control of high-pathogenicity avian influenza and low-pathogenicity notifiable avian influenza in poultry, with emphasis on vaccines and vaccination. *Rev Sci Tech* 2011;30:839–70.
34. Smith GJ, Fan XH, Wang J, et al. Emergence and predominance of an H5N1 influenza variant in China. *Proc Natl Acad Sci USA* 2006;103:16936–41.
35. Cattoli G, Fusaro A, Monne I, et al. Evidence for differing evolutionary dynamics of A/H5N1 viruses among countries applying or not applying avian influenza vaccination in poultry. *Vaccine* 2011;29:9368–75.
36. Balish AL, Davis CT, Saad MD, et al. Antigenic and genetic diversity of highly pathogenic avian influenza A (H5N1) viruses isolated in Egypt. *Avian Dis* 2010;54:329–34.
37. Lee CW, Senne DA, Suarez DL. Effect of vaccine use in the evolution of Mexican lineage H5N2 avian influenza virus. *J Virol* 2004;78:8372–81.
38. Capua I, Alexander DJ. Avian influenza vaccines and vaccination in birds. *Vaccine* 2008;26 Suppl 4:D70–3.
39. Webster RG, Peiris M, Chen H, Guan Y. H5N1 outbreaks and enzootic influenza. *Emerg Infect Dis* 2006;12:3–8.
40. Sitaras I, Kalthoff D, Beer M, Peeters B, de Jong MC. Immune escape mutants of Highly Pathogenic Avian Influenza H5N1 selected using polyclonal sera: identification of key amino acids in the HA protein. *PLoS One* 2014;9:e84628.
41. Sitaras I, Rousou X, Kalthoff D, Beer M, Peeters B, de Jong MC. Role of vaccination-induced immunity and antigenic distance in the transmission dynamics of highly pathogenic avian influenza H5N1. *J R Soc Interface* 2016;13:20150976.
42. Cattoli G, Milani A, Temperton N, et al. Antigenic drift in H5N1 avian influenza virus in poultry is driven by mutations in major antigenic sites of the hemagglutinin molecule analogous to those for human influenza virus. *J Virol* 2011;85:8718–24.
43. Peyre M, Samaha H, Makonnen YJ, et al. Avian influenza vaccination in Egypt: limitations of the current strategy. *J Mol Genet Med* 2009;3:198–204.
44. Domenech J, Dauphin G, Rushton J, et al. Experiences with vaccination in countries endemically infected with highly pathogenic avian influenza: the Food and Agriculture Organization perspective. *Rev Sci Tech* 2009;28:293–305.
45. Soares Magalhaes RJ, Pfeiffer DU, Otte J. Evaluating the control of HPAIV H5N1 in Vietnam: virus transmission within infected flocks reported before and after vaccination. *BMC Vet Res* 2010;6:31.
46. Swayne DE. Impact of vaccines and vaccination on global control of avian influenza. *Avian Dis* 2012; 56:818–28.
47. Swayne DE. The role of vaccines and vaccination in high pathogenicity avian influenza control and eradication. *Expert Rev Vaccines* 2012;11:877–80.
48. Pavade G, Awada L, Hamilton K, Swayne DE. The influence of economic indicators, poultry density and the performance of veterinary services on the control of high-pathogenicity avian influenza in poultry. *Rev Sci Tech* 2011;30:661–71.
49. Grund C, Abdelwhab ESM, Arafa AS, et al. Highly pathogenic avian influenza virus H5N1 from Egypt escapes vaccine-induced immunity but confers clinical protection against a heterologous clade 2.2.1 Egyptian isolate. *Vaccine* 2011;29:5567–73.
50. FAO. H5N1 HPAI Global Overview January–March 2012.
51. Spackman E, Stephens CB, Pantin-Jackwood MJ. The effect of infectious bursal disease virus-induced immunosuppression on vaccination against highly pathogenic avian influenza virus. *Avian Dis* 2018;62:36–44.
52. Swayne DE. Avian influenza vaccines and therapies for poultry. *Comp Immunol Microbiol Infect Dis* 2009;32:351–63.
53. Ellis TM, Leung CY, Chow MK, et al. Vaccination of chickens against H5N1 avian influenza in the face of an outbreak interrupts virus transmission. *Avian Pathol* 2004;33:405–12.
54. Sitaras I, Rousou X, Peeters B, de Jong MCM. Mutations in the haemagglutinin protein and their effect in transmission of highly pathogenic avian influenza (HPAI) H5N1 virus in sub-optimally vaccinated chickens. *Vaccine* 2016;34:5512–8.
55. Abbas MA, Spackman E, Fouchier R, et al. H7 avian influenza virus vaccines protect chickens against challenge with antigenically diverse isolates. *Vaccine* 2011;29:7424–9.

56. Spackman E, Swayne DE, Pantin-Jackwood MJ, et al. Variation in protection of four divergent avian influenza virus vaccine seed strains against eight clade 2.2.1 and 2.2.1.1. Egyptian H5N1 high pathogenicity variants in poultry. *Influenza Other Resp Viruses* 2014;8: 654–62.
57. Swayne DE, Perdue ML, Beck JR, Garcia M, Suarez DL. Vaccines protect chickens against H5 highly pathogenic avian influenza in the face of genetic changes in field viruses over multiple years. *Vet Microbiol* 2000;74:165–72.
58. Swayne DE, Spackman E, Pantin-Jackwood M. Success factors for avian influenza vaccine use in poultry and potential impact at the wild bird-agricultural interface. *EcoHealth* 2014;11:94–108.
59. Bunn D, Beltran-Alcrudo D, Cardona C. Integrating surveillance and biosecurity activities to achieve efficiencies in national avian influenza programs. *Prev Vet Med* 2011;98:292–4.
60. Uttenthal A, Parida S, Rasmussen TB, Paton DJ, Haas B, Dundon WG. Strategies for differentiating infection in vaccinated animals (DIVA) for foot-and-mouth disease, classical swine fever and avian influenza. *Expert Rev Vaccines* 2010;9:73–87.
61. Capua I, Terregino C, Cattoli G, Mutinelli F, Rodriguez JF. Development of a DIVA (Differentiating Infected from Vaccinated Animals) strategy using a vaccine containing a heterologous neuraminidase for the control of avian influenza. *Avian Pathol* 2003;32:47–55.
62. Suarez DL. Overview of avian influenza DIVA test strategies. *Biologicals* 2005;33:221–6.
63. Suarez DL. DIVA vaccination strategies for avian influenza virus. *Avian Dis* 2012;56:836–44.
64. Kandeil A, Moatasim Y, Gomaa MR, et al. Generation of a reassortant avian influenza virus H5N2 vaccine strain capable of protecting chickens against infection with Egyptian H5N1 and H9N2 viruses. *Vaccine* 2015;34:218–224.
65. Beard CW, Schnitzlein WM, Tripathy DN. Protection of chickens against highly pathogenic avian influenza virus (H5N2) by recombinant fowlpox viruses. *Avian Dis* 1991;35:356–9.
66. Qiao CL, Yu KZ, Jiang YP, et al. Protection of chickens against highly lethal H5N1 and H7N1 avian influenza viruses with a recombinant fowlpox virus co-expressing H5 haemagglutinin and N1 neuraminidase genes. *Avian Pathol* 2003;32:25–32.
67. Ge J, Deng G, Wen Z, et al. Newcastle disease virus-based live attenuated vaccine completely protects chickens and mice from lethal challenge of homologous and heterologous H5N1 avian influenza viruses. *J Virol* 2007;81:150–8.
68. Rauw F, Palya V, Gardin Y, et al. Efficacy of rHVT-AI vector vaccine in broilers with passive immunity against challenge with two antigenically divergent Egyptian clade 2.2.1 HPAI H5N1 strains. *Avian Dis* 2012;56: 913–22.
69. Lone NA, Spackman E, Kapczynski D. Immunologic evaluation of 10 different adjuvants for use in vaccines for chickens against highly pathogenic avian influenza virus. *Vaccine* 2017;35:3401–8.
70. Burakova Y, Madera R, McVey S, Schlup JR, Shi J. Adjuvants for animal vaccines. *Viral Immunol* 2018;31:11–22.
71. Swayne DE, Suarez DL, Spackman E, et al. Antibody titer has positive predictive value for vaccine protection against challenge with natural antigenic-drift variants of H5N1 high-pathogenicity avian influenza viruses from Indonesia. *J Virol* 2015;89:3746–62.
72. Swayne DE, Kapczynski DR. Vaccines and vaccination for avian influenza in poultry. In: Swayne DE, ed. *Animal Influenza*. 2nd ed. Ames, IA: Wiley-Blackwell; 2017:378–438.
73. Hubby B, Talarico T, Maughan M, et al. Development and preclinical evaluation of an alphavirus replicon vaccine for influenza. *Vaccine* 2007;25:8180–9.
74. Kapczynski DR, Sylte MJ, Killian ML, Torchetti MK, Chrzastek K, Suarez DL. Protection of commercial turkeys following inactivated or recombinant H5 vaccine application against the 2015U.S. H5N2 clade 2.3.4.4 highly pathogenic avian influenza virus. *Vet Immunol Immunopathol* 2017;191:74–9.
75. Bertran K, Balzli C, Lee DH, Suarez DL, Kapczynski DR, Swayne DE. Protection of White Leghorn chickens by U.S. emergency H5 vaccination against clade 2.3.4.4 H5N2 high pathogenicity avian influenza virus. *Vaccine* 2017;35:6336–44.
76. Asif M, Jenkins KA, Hilton LS, Kimpton WG, Bean AG, Lowenthal JW. Cytokines as adjuvants for avian vaccines. *Immunol Cell Biol* 2004;82:638–43.
77. Santos JJS, Obadan AO, Garcia SC, et al. Short- and long-term protective efficacy against clade 2.3.4.4 H5N2 highly pathogenic avian influenza virus following prime-boost vaccination in turkeys. *Vaccine* 2017;35:5637–43.
78. Abdelwhab EM, Grund C, Aly MM, Beer M, Harder TC, Hafez HM. Multiple dose vaccination with heterologous H5N2 vaccine: immune response and protection against variant clade 2.2.1 highly pathogenic avian influenza H5N1 in broiler breeder chickens. *Vaccine* 2011;29:6219–25.

SUMMARY

Avian influenza virus (AIV) is of critical concern for food security; therefore, controlling the spread of AI infections is of paramount importance. Vaccination is a common tool to reduce production losses, halt spread in poultry, and reduce risk to public health by decreasing the environmental load of AIV. The most widely used vaccine types are inactivated whole-virus vaccines and live vectored vaccines. Other AIV vaccines, such as RNA particles, are also available, but are not widely used in the field. As the epidemiology of AIV has changed in recent years, leading to frequent transmission to domestic birds, the use of vaccine has increased and will probably continue to do so. This chapter reviews the technical aspects of vaccine development for AIV, including the basics of current AIV vaccine production and technology, the approaches for vaccine evaluation and testing, and the need for further improvements in vaccines.

Genetic Manipulation of Human Adenovirus 10

Si Qin
Hunan Agricultural University

Dongyou Liu
Royal College of Pathologists of Australasia Quality Assurance Programs

Contents

10.1 INTRODUCTION

Harboring a linear, double-stranded (ds) DNA genome without envelope, human adenovirus (HAdV) was first isolated from adenoid tissue in 1953 and subsequently confirmed as the cause of mostly mild, self-limiting illnesses involving respiratory, gastrointestinal, ophthalmologic, genitourinary, and neurologic systems in susceptible populations (e.g., young children, crowded communities, schools, and military training camps as well as immunocompromised individuals) throughout the world.

Following the realization of its abilities to infect a diversity of replicating and quiescent cells at high efficiency even without certain genes, to grow to high titers, and to induce little genotoxicity *in vivo*, HAdV has been exploited extensively as a vector for vaccine development and gene therapy.

Taxonomy. Classified in the order Rowavirales, the family Adenoviridae encompasses 80 species of non-enveloped, icosahedral, dsDNA viruses that are organized under five genera: *Atadenovirus* (8 species), *Aviadenovirus* (15 species), *Ichtadenovirus* (1 species), *Mastadenovirus* (50 species, including 7 human-infecting species), and *Siadenovirus* (6 species). Of the seven human-infecting species (human mastadenoviruses A–G) in the genus *Mastadenovirus*, 103 serotypes are identified, including 90 affecting humans, 1 affecting bovines, and 12 affecting simians. Specifically, human mastadenovirus A (with high oncogenicity in rodents and low G+C% in genome) contains 3 human-infecting serotypes (HAdV-12, HAdV-18, and HAdV-31), human mastadenovirus B includes 14 human-infecting serotypes and 1 simian-infecting serotype (HAdV-3, HAdV-7, HAdV-11, HAdV-14, HAdV-16, HAdV-21, HAdV-34, HAdV-35,

HAdV-50, HAdV-55, HAdV-76, HAdV-77, HAdV-78, HAdV-79, and SAdV-21), human mastadenovirus C (with the ability to recombine) comprises 5 human-infecting serotypes, 1 bovine-infecting serotype, and 1 simian-infecting serotype (BAdV-9, HAdV-1, HAdV-2, HAdV-5, HAdV-6, HAdV-89, and SAdV-31), human mastadenovirus D consists of 64 human-infecting serotypes (HAdV-8, HAdV-9, HAdV-10, HAdV-13, HAdV-15, HAdV-17, HAdV-19, HAdV-20, HAdV-22, HAdV-23, HAdV-24, HAdV-25, HAdV-26, HAdV-27, HAdV-28, HAdV-29, HAdV-30, HAdV-32, HAdV-33, HAdV-36, HAdV-37, HAdV-38, HAdV-39, HAdV-42, HAdV-43, HAdV-44, HAdV-45, HAdV-46, HAdV-47, HAdV-48, HAdV-49, HAdV-51, HAdV-53, HAdV-54, HAdV-55, HAdV-56, HAdV-57, HAdV-58, HAdV-59, HAdV-60, HAdV-61, HAdV-62, HAdV-63, HAdV-64, HAdV-65, HAdV-66, HAdV-67, HAdV-68, HAdV-69, HAdV-70, HAdV-71, HAdV-72, HAdV-73, HAdV-74, HAdV-75, HAdV-80, HAdV-81, HAdV-82, HAdV-83, HAdV-84, HAdV-85, HAdV-86, HAdV-87, HAdV-88, and HAdV-90), human mastadenovirus E has 1 human-infecting serotype and 4 simian-infecting serotypes (HAdV-4, SAdV-22, SAdV-23, SAdV-24, and SAdV-25), human mastadenovirus F (with restricted growth capacity) contains 2 human-infecting serotypes (HAdV-40 and HAdV-41), and human mastadenovirus G has 1 human-infecting serotype and 5 simian-infecting serotypes (HAdV-52, SAdV-2, SAdV-7, SAdV-11, SAdV-12, and SAdV-15) [1,2]. Interestingly, adenoviruses most commonly used in gene therapy belong to human mastadenoviruses B (e.g., serotypes 3 and 11), C (especially serotypes 2 and 5 with rare occurrence in the general population), and D [3].

Morphology. Adenovirus is a non-enveloped particle of 70–90 nm in diameter with an icosahedral capsid. The capsid is composed of 240 hexons (capsomers, each containing three identical polypeptides II) of 8–10 nm in diameter and 12 penton bases (vertex capsomers, each containing five polypeptides III), from which a fiber or spike (each containing three polypeptides IV that interlink to form a distal knob) of 9–77.5 nm in length protrudes to give the virion its characteristic morphology (Figure 10.1). A combination of penton base (polypeptides III) and fiber (polypeptides IV) makes up so-called penton (III and IV), which is instrumental for receptor binding and internalization. In addition, 12 copies of polypeptide IX are located in the center of each facet formed by 9 hexons in the genus *Mastadenovirus*; two monomers of IIIa sit beneath the vertex region; multiple proteins VI form a ring under the hexons; polypeptide VIII is present in the inner surface of the hexon capsid; other polypeptides (monomers of IIIa, trimers of IX, and multimers of VI) interact with hexons to produce a continuous protein shell; and four polypeptides [V (which is found only in the genus *Mastadenovirus*), VII, X (also known as mu or μ), and terminal protein] complex with the linear dsDNA genome forming a virion core (Figure 10.1) [2,4].

Genome. Adenovirus genome comprises a linear, non-segmented, dsDNA of 26,163–48,395 bp, which is flanked on both ends by hair-pin-like, inverted terminal repeats (ITRs) of 36–371 bp. ITRs act as a self-primer (origin of replication) to promote primase-independent DNA synthesis and facilitate integration into the host genome. Covalently linked to the 5′ end of each DNA strand is ~200 bp sequence encoding a terminal protein (TP), which functions as the packaging signal for proper viral transcript packaging. With a G+C content of 33.6%–66.9%, the adenovirus genome includes early and late gene regions that generate about 40 different polypeptides, mostly via complex splicing mechanisms. Superficially, the early gene region (E1 to E4, expressed before DNA replication) encodes non-structural proteins involved in activating transcription of other viral regions and altering the cellular environment to promote viral production; namely, the early protein E1 (E1A and E1B) induces mitogenic activity in the host cell and stimulates expression of other viral genes; E2 (E2A and E2B) mediates viral DNA replication; E3 subverts host defense mechanisms, and E4 alters host cell signaling (Table 10.1). The early (intermediate) gene region produces IX (pIX) and IVa2, which influence hexon protein interactions and help package viral DNA into the immature virion, respectively. The late gene region (expressed after DNA replication) encodes structural proteins after activation of the major late promoter, including late proteins L1–L5 for virion assembly and late proteins VA RNA I and II for inhibiting activation of the host interferon response, impeding host micro-RNA processing and thus influencing host gene expression (Table 10.1; Figure 10.2) [4–6].

Biology. HAdV is capable of either lytic infection in epithelial cells or latent infection in lymphoid cells.

HAdV lytic infection (so-called viral reproduction cycle) takes place after viral entry and replication inside human epithelial cells. Viral entry involves binding of the fiber knob domain to host cell receptor [e.g., coxsackie and adenovirus receptor (CAR), blood coagulation factor X (FX), and heparan sulfate

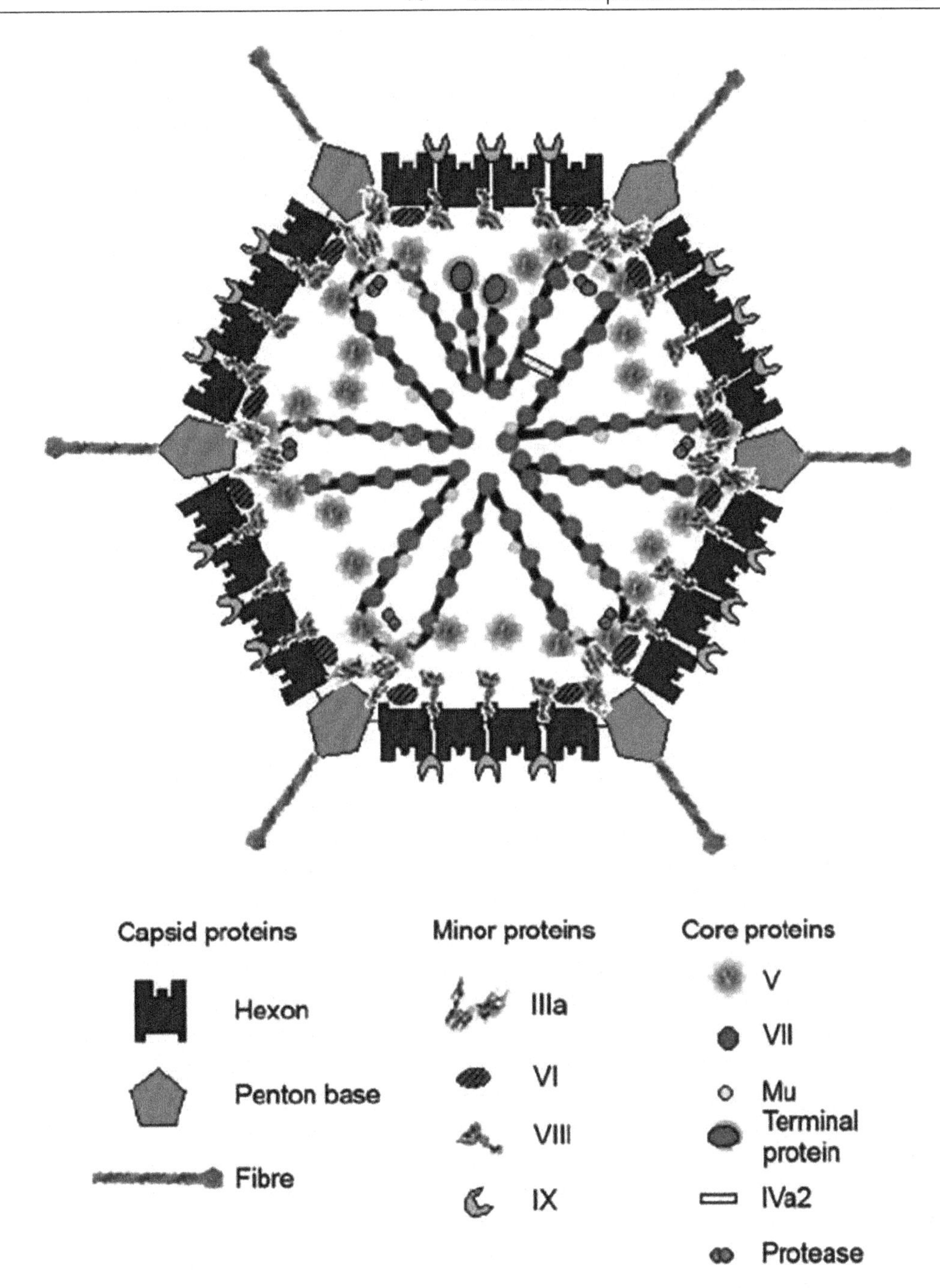

FIGURE 10.1 Schematic representation of adenovirus particle. The icosahedral capsid is composed of hexon, penton base, and knobbed fiber. Core proteins include V, VII, Mu, terminal protein, IVa, and protease. Minor proteins are IIIa, V, VIII, and IX. (Photo credit: Pettersson U. *Ups J Med Sci.* 2019;124(2):83–93.)

proteoglycans (HSPG)] for species A, C, E, and F and alternative receptors [e.g., membrane cofactor protein (MCP orCD46), desmoglein-2, sialic acid, and GD1a glycan for species B and D] and subsequent interaction via arginine-glycine-aspartic acid (RGD) motifs in the penton base with host cellular integrins (e.g., $\alpha v\beta 3$, $\alpha v\beta 5$, $\alpha v\beta 1$, $\alpha 3\beta 1$, and $\alpha 5\beta 1$), leading to cytoskeleton alteration and formation of clathrin-coated vesicles (which later become endosomes) harboring the virion [7–9]. After undergoing endosomal

TABLE 10.1 Predicted structural (S) and non-structural (NS) viral proteins from human adenovirus serotype 2 genome

PROTEIN	*FUNCTION*	*PRESENCE*
E1A	NS	Mastadenoviruses only
E1B	NS	Mastadenoviruses only
E2A	NS; DBP (DNA-binding protein)	All adenoviruses
E2B	NS; DNA polymerase	All adenoviruses
E2B	S; Term, pTP (terminal protein)	All adenoviruses
E3	NS	Mastadenoviruses only
E4	NS	Mastadenoviruses and atadenoviruses only
E4	NS; dUTPase	Some mastadenoviruses and aviadenoviruses only
L1	NS; scaffold protein	All adenoviruses
L1	S (pIIIa); p-protein	All adenoviruses
L2	S (III); penton base	All adenoviruses
L2	S (pVII); major core	All adenoviruses
L2	S (V); minor core	Mastadenoviruses only
L2	S (pX); X/μ	All adenoviruses
L3	S (pVI)	All adenoviruses
L3	S (II); hexon	All adenoviruses
L3	S; protease	All adenoviruses
L4	NS	All adenoviruses
L4	NS; p-protein	All adenoviruses
L4	S (pVIII)	All adenoviruses
L5	S (IV); fiber	All adenoviruses
Intermediate	S (IX)	Mastadenoviruses only
Intermediate	S (IVa2)	All adenoviruses

acidification, the virion disassembles and releases the protein VII–viral DNA complex into the cytoplasm, which then translocates along microtubules to the nucleus for viral replication. In general, adenovirus does not integrate its genome into host cell chromosomes and usually replicates as linear, extra-chromosomal DNA elements, which are packaged into newly assembled capsids in the nucleus. Following transportation of newly assembled capsids to the cytoplasm and by inhibition of host macromolecular synthesis, the virus induces cell lysis and death and is ready to infect the next cell [10].

HAdV latent infection occurs after the lytic infection, with the virus persisting in lymphoid organs (e.g., adenoids, tonsils, or Peyer's patches) for years. Upon certain stimulus, the latent virus may re-activate, re-infect, and replicate in epithelial cells, producing clinical disease.

Epidemiology. HAdV particles have a molecular weight of 150–180 MDa and a buoyant density in CsCl of 1.31–1.36 g cm^{-3}, are stable on storage in the frozen state, and are insensitive to mild acids, lipid solvents (due to its lack of an envelope), and ethanol, but are inactivated by heat (56°C for >10 min), formaldehyde, and bleach. Therefore, HAdV can remain viable in the environment (including summer camps, playgrounds, dormitories, and school grounds) for long periods.

Respiratory route (cough or sneeze) represents the most common mode of HAdV transmission, although fecal-oral contamination of food or water (e.g., ineffective chlorine treatment of public swimming pools) and direct contact of contaminated fomites may also play a part in its transmission. In addition, exposure to cervical canal secretions during birth and in solid organ transplants (especially the liver and kidney) is another potential means of HAdV transmission.

HAdV serotypes 1, 2, 3, 4, 7, 8, 14, 21, and 41 are linked to outbreaks in many parts of the world. In particular, HAdV serotypes 7 and 14 appear to be highly virulent and accounted for nearly 20% of

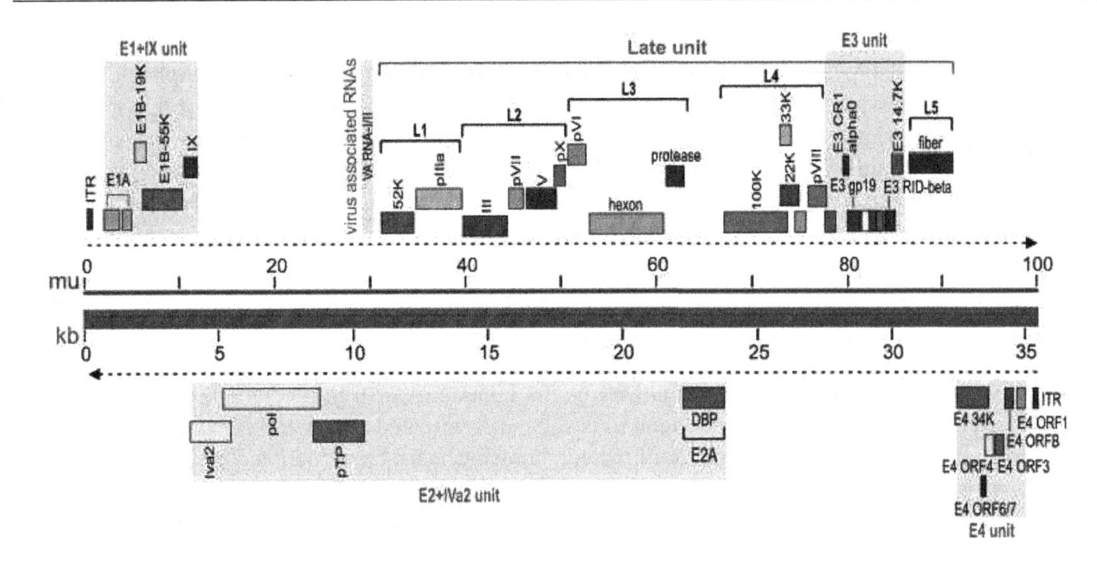

FIGURE 10.2 Genome organization of human adenovirus serotype 5 (Ad5). Composed of a 36 kb linear dsDNA, with map units (mu) showing as red line, the Ad5 genome consists of early units (E1 to E4) and late units (L1 to L5). The genes listed in the upper part are transcribed from left to right, while those in the lower bottom part are transcribed from right to left, as indicated by the dotted arrows. The E1 gene products (E1A and E1B) participate in viral replication; the E2 gene products (E2A and E2B) provide the machinery for viral DNA replication and transcription of late genes, which mostly encode structural proteins for virus packaging; the E3 proteins modulate host immune response and are not essential for viral production *in vitro*; and the E4 proteins alter host cell signaling. (Photo credit: Lee CS, Bishop ES, Zhang R, et al. *Genes Dis.* 2017;4(2):43–63.)

all adenoviral infections during the last global survey. Notably, >80% of confirmed cases involve children<4 years old (due to lack of humoral immunity). Individuals with compromised immune functions have a great risk of disseminated disease and mortality in comparison with an immunocompetent person. Furthermore, adenoviral infections presenting as acute respiratory distress syndrome (ARDS) are observed with high frequency in Malaysia, South Korea, and China [11].

Clinical features. With an incubation period of 4–8 days, HAdV infections typically affect the upper or lower respiratory tracts, conjunctiva, or gastrointestinal (GI) tract (Table 10.2), with clinical symptoms ranging from respiratory tract infections (fever, pharyngitis, tonsillitis, rhinitis, cough, sore

TABLE 10.2 Association of human mastadenovirus serotypes with clinical infections and diseases

SPECIES	*SEROTYPES*	*CLINICAL INFECTIONS AND DISEASES*
Human mastadenovirus A	12, 18, 31	Gastrointestinal, respiratory, urinary, cryptic enteric infections; meningoencephalitis; obesity
Human mastadenovirus B	3, 7, 11, 14, 16, 21, 34, 35, 50, 55, 76–79	Conjunctivitis; gastrointestinal, respiratory, urinary infections; pneumonia; meningoencephalitis; cystitis
Human mastadenovirus C	1, 2, 5, 6, 89	Respiratory, gastrointestinal infections; pneumonia; hepatitis; obesity
Human mastadenovirus D	8–10, 13, 15, 17, 19, 20, 22–30, 32, 33, 36–39, 42–49, 51, 53–75, 80–88, 90	Conjunctivitis; gastrointestinal infection; meningoencephalitis; obesity
Human mastadenovirus E	4	Conjunctivitis; respiratory infection; pneumonia
Human mastadenovirus F	40, 41	Gastrointestinal infection; infantile diarrhea
Human mastadenovirus G	52	Gastrointestinal infection

throat, dyspnea, and pneumonia in newborns and infants, due mainly to serotypes 1–5, 7, 14, and 21), keratoconjunctivitis (shipyard conjunctivitis, pharyngoconjunctival fever, pink eye, photophobia, foreign body sensation, excessive tearing, ocular and periorbital pain, decreased visual acuity, due mainly to serotypes 8, 19, and 37 and occasionally to serotypes 3, 4, 7, 11, and 14), GI infections (diarrhea and acute gastroenteritis due mainly to serotypes 40 and 41), hemorrhagic cystitis (fever, hypertension, proteinuria associated with glomerulonephritis, due mainly to serotypes 11 and 21), urinary tract infections (dysuria, hematuria, hemorrhagic cystitis, and renal allograft dysfunction, due mainly to serotypes 11, 34, 35, 3, 7, and 21), to ARDS [12–14].

In immunocompromised patients [stem cell or solid organ transplantation, severe combined immunodeficiency (SCID) syndrome, and human immunodeficiency virus (HIV)], meningoencephalitis, interstitial/tubulointerstitial nephritis (due mainly to serotypes 11, 34, and 35), and gastroenteritis are often observed, while patients undergoing chemotherapy and bone marrow transplantation, or having severe lymphopenia or graft-versus-host disease tend to develop disseminated adenoviral infections [13].

Pulmonary sequelae of HAdV infections include bronchiectasis, bronchiolitis obliterans, and hyperlucent lung, while extrapulmonary complications consist of meningoencephalitis, hemorrhagic colitis, hepatitis, cholecystitis, pancreatitis, myocarditis, nephritis, neutropenia, disseminated intravascular coagulation, and hemophagocytic lymphohistiocytosis.

Pathogenesis. HAdVs have the ability to recognize specific receptors on the epithelial cells of the GI tract, respiratory tract, eyes, and urinary bladder, gain entry, and replicate in these cells utilizing host cellular machineries (so-called lytic infection). In the meanwhile, HAdVs are capable of inhibiting host cell protein synthesis and subverting host immune responses. Furthermore, HAdVs can remain latent in lymphoid cells (so-called latent infection) and reactivate when conditions become favorable [15,16].

Diagnosis. Diagnosis of HAdV infection involves clinical evaluation, imaging (chest X-ray and CT), laboratory culture/viral isolation (nasal swab, blood, urine, or stool), serological assays [enzyme-linked immunosorbent assay (ELISA) and latex agglutination test (LAT)], and molecular tests (targeting fiber or hexon genes).

Treatment and prevention. Most HAdV patients have mild and self-limiting illnesses and recover without medical care. HAdV patients with lingering symptoms may benefit from bronchodilator medication (to open the airways), oral rehydration or increased fluid intake, and rest. Immunocompromised individuals with severe HAdV infections require antiviral drugs (e.g., ribavirin and cidofovir) [17].

As HAdVs can remain viable on objects and surfaces (e.g., doorknobs, towels, and medical instruments) for 3–8 weeks, prevention of HAdV infection and outbreak requires implementation of measures (e.g., frequent handwashing, sanitizing surfaces, staying at home when ill, avoiding close contact with people who are sick, and covering nose and mouth when sneezing or coughing) to disrupt viral transmission.

10.2 GENETIC MANIPULATION OF HUMAN ADENOVIRUS

Characteristics of adenoviral vectors. As HAdV tends to cause minor, self-limiting illnesses in humans and has the ability to infect cells containing specific receptors even in the absence of certain early or later genes, it has been exploited for delivering genes of interest into mammalian cells for high-level transgene expression and gene therapy.

Compared to other widely used vectors, adenovirus (e.g., serotype 5) stands out by its capability to transduce efficiently a broad range of dividing and non-dividing cells that express primary adenovirus receptor and secondary integrin receptors, its ease of manipulation, its potential to propagate to stable high-titers ($>10^{13}$ viral particles/mL), and its low genotoxicity *in vivo* due to its episomal location (Table 10.3). Its

TABLE 10.3 Key features of commonly used vectors for gene delivery

VECTOR	*GENOME*	*INSERT SIZE*	*VECTOR GENOME IN HOST CELL*	*ADVANTAGES*	*DISADVANTAGES*
Adenovirus	Double-stranded DNA, 36,000 bp (serotype 5)	5,000–36,000 bp	Episomal	Efficient transduction; high titer (1×10^{13} viral particles/mL)	Potently inflammatory
Adeno-associated virus	Single-stranded DNA, 8,500 nt	Up to 5,000 nt	Episomal (90%) or integrated site-specifically (<10%)	Non-pathogenic; non-inflammatory; safe transgene delivery	Small packaging capacity; requirement of helper AdV for replication; difficult to produce pure viral stocks
Lentivirus	Single-stranded RNA, 8,000 nt	Up to 9,000 nt	Integrated	Efficient transduction	Potentially oncogenic
Herpes simplex virus 1	Double-stranded DNA, 150,000 bp	Up to 50,000 bp	Episomal	Large packaging capacity; strong tropism for neuronal cells	Inflammatory; transient gene expression in non-neuronal cells
Baculovirus	Double-stranded DNA, 80,000–180,000 bp	Limit unknown	Episomal or integrated	Large cargo size; high level of gene expression	Inflammatory; limited mammalian host range
Plasmid	Double-stranded DNA, 2,000–100,000 bp	Up to 15,000 bp	Episomal or integrated	Broad host cell range; easy production; high level of gene expression	Low transduction efficiency, low specificity

perceived drawback of being potently inflammatory is circumvented by the development of gutless adenoviral vectors (see below) [11].

Indeed, a recent survey of the registered gene therapy clinical trials indicates that adenovirus predominates vector systems used (21%), followed by retrovirus (18.6%), naked plasmid DNA (17.2%), adeno-associated virus (7.2%), lentivirus (6%), vaccinia virus (5.2%), herpes simplex virus (3.7%), poxvirus (2.9%), and others (18%) [6,11].

Adenoviral vectors deficient in E1. Removal (or replacement) of the early non-structural E1 (E1a and E1b) gene produces an adenoviral vector that is incapable of replication in most cell lines. With a capacity of introducing an insert of 5.2 kb, this first-generation adenoviral vector facilitates short-term transgene expression *in vitro* and *in vivo*, as residual viral gene expression leads to immune-mediated loss of the vector and transduced cells *in vivo*.

Adenoviral vectors deficient in E1–E4. Deletion of E2, E3, and E4 on top of E1 deficiency represents the second-generation adenoviral vector that prolongs transgene expression and increases cloning capacity. However, adenoviral vector lacking E1–E4 still triggers immune responses *in vivo* and reduces the yields of transduced cells. Furthermore, removal of E1–E4 often causes an overall reduction in the vector genome size, which somehow decreases the heat stability of the vector.

Gutless adenovirus vectors [also known as helper-dependent adenoviruses (HD-Ad) or high-capacity adenoviruses (HC-Ad)]. Considering the fact that adenoviral vectors lacking E1 or E1–E4 are potently inflammatory and induce unwanted cellular immune responses (including the risk of anaphylactic shock or death after repeated use), a gutless adenovirus vector is engineered to keep only the 5′ and 3′ ITRs and the packaging signal (Ψ), which permit genome replication and final packaging into the capsid, respectively. As this third-generation adenoviral vector does not encode any viral proteins, it elicits no host immune responses and has the capacity to accommodate an insert of up to 36 kb. Nonetheless, it depends on a helper adenovirus for replicative and packaging functions *in trans*. The helper adenovirus provides the necessary proteins *in trans* for the packaging of a gutless adenovirus vector, but it is not packaged along with the desired adenovirus vector. This is attributed to the fact that the helper adenovirus has its packaging sequences flanked by *lox*P recognition sites, which are sufficiently excised by Cre recombinase, and therefore, the helper adenovirus DNA remains unpackaged. Alternatively, an FLP/FRT system may be used to select against the helper adenovirus more efficiently (about 100% for FLP recombinase versus 80% for Cre recombinase) (Figure 10.3) [6].

Use of adenoviral vectors for gene delivery. In view of their obvious merits, adenoviral vectors are a widely applied therapeutic platform for vaccine development and disease treatment.

As adenoviral vectors can stimulate robust cytotoxic T-cell response as well as humoral response, leading to efficient destruction of virus-infected cells and intracellular pathogens, they are utilized in the delivery of vaccine molecules against human viral (e.g., HIV, Ebola, and influenza), bacterial (e.g., tuberculosis), and protozoan (e.g., malaria) pathogens [18–23].

Another area in which adenoviral vectors find ready application relates to anticancer therapies since cytotoxic T-cell response induced by adenoviral vectors is potent against cancerous cells. Specifically, replication-defective adenoviruses allow direct delivery of immune-related genes/epitopes into tumor cells that elicit a local antitumoral immune response. Furthermore, replication-competent adenoviruses (or oncolytic adenoviruses) are employed to replicate in and lyse cancerous cells during the lytic life cycle of the virus [24,25]. In addition, adenoviral delivery or overexpression of tumor-suppressor, cytotoxic/suicide, or anti-PD1/PD-L1 immune checkpoint genes may trigger cancerous cell cycle arrest and apoptosis [26–28].

Adenoviruses are also evaluated in gene therapies for cardiovascular diseases, neurodegenerative disorders, diabetes, metabolic diseases, cystic fibrosis, angina pectoris, OTC deficiency, and so on.

Moreover, adenoviruses are shown to mediate delivery of the clustered regularly interspaced short palindromic repeats (CRISPR)/CRISPR-associated protein (Cas) 9 system into human hepatocytes for efficient genome editing [29].

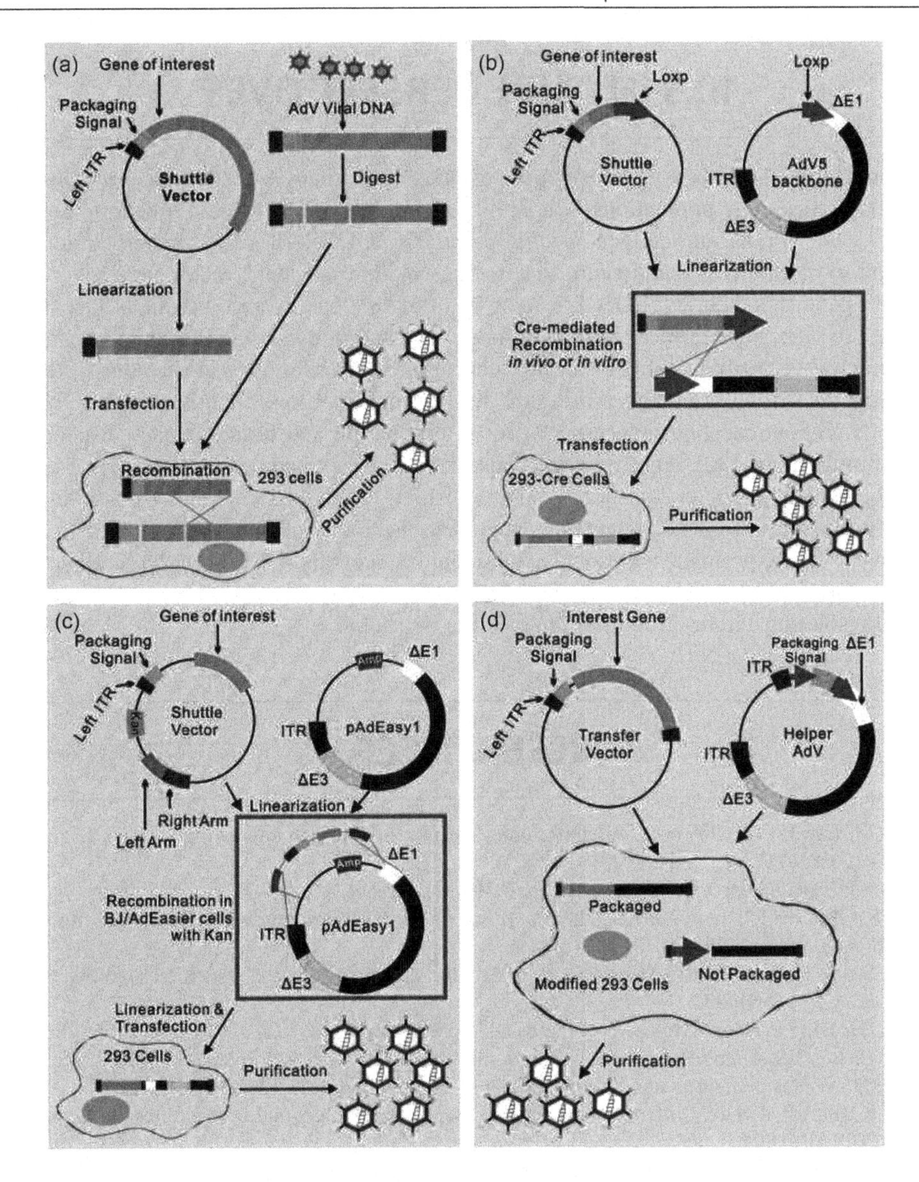

FIGURE 10.3 Four ways to generate and produce adenovirus vectors for gene delivery. (a) Recombination in human embryonic kidney HEK-293 cells. The gene of interest (GOI) is first cloned into a shuttle vector, which contains 5′-ITR, a packaging signal, and homologous regions to the adenoviral genome. Adenoviruses are generated in HEK-293 cells through recombination between shuttle vector and adenoviral backbone vector, which is unable to produce virus by its self. (b) Cre/LoxP-mediated recombination. The GOI is cloned into a shuttle vector that contains the LoxP site(s). Cre recombinase-mediated recombination occurs with a LoxP-containing adenoviral backbone vector *in vitro* or 293-Cre cells, leading to the generation of adenoviruses. (c) AdEasy system. The GOI is subcloned into a shuttle vector that contains 5′-ITR and packaging signal, as well as a kanamycin-containing bacterial replication unit flanked by homologous arms. Recombinant adenoviral plasmids are generated through homologous recombination between the linearized shuttle vector and ampicillin-resistant adenoviral backbone vector, such as pAdEasy1, in bacterial strain BJ5183 under kanamycin selection. The resultant adenoviral plasmids are linearized and used for adenovirus production in HEK-293 cells. (d) Use of helper adenovirus for production of HC-AdV (or HD-AdV, or Gutless AdV). The GOI is cloned into a transfer vector that contains both ITRs and packaging signal only. Adenoviruses are generated with a helper adenovirus, which is not packaged due to the deletion of packaging signal in the modified HEK-293 cells, usually through Cre/LoxP or FLP/FRT excision system. (Photo credit: Lee CS, Bishop ES, Zhang R, et al. *Genes Dis.* 2017;4(2):43–63.)

10.3 FUTURE PERSPECTIVES

Human adenovirus (HAdV) is a non-enveloped, dsDNA virus that tends to induce mild, self-limiting illnesses among susceptible populations. The ability of HAdV to infect both replicating and quiescent cells, to elicit robust transgene antigen-specific cellular (i.e. CD8+ T cells) and/or humoral immune responses, and to cause little genotoxicity in vivo, has made it an ideal vector for delivering vaccine molecules and gene therapeutics. Indeed, a number of HAdV-based, anti-infectious disease vaccines have advanced to later stage clinical trials. On the other hand, notwithstanding their enormous potential, few adenoviral vector-based anticancer products are available for clinical use. One of the lingering issues relates to the difficulty in determining the host immune response that is adequate for lysing the cancerous cells without causing unnecessary host inflammation and tissue damage, especially during secondary administration [30]. Another issue concerns the low transduction efficiency of cancer cells, which lack the specific receptor for adenoviral vector [31]. A third issue is about possible contamination with helper adenovirus when a gutless adenovirus vector is utilized. Although self-inactivating helper adenovirus offers promise, additional assessment is necessary. Clearly, widespread application of adenovirus-vectored infectious disease vaccines and anti-cancer and other disease therapies is dependent on successful resolution and/or circumvention of these issues [32,33].

REFERENCES

1. Ismail AM, Lee JS, Lee JY, et al. Adenoviromics: Mining the human adenovirus species D genome. *Front Microbiol.* 2018;9:2178. (correction 2018;9: 3005).
2. Pettersson U. Encounters with adenovirus. *Ups J Med Sci.* 2019;124(2):83–93.
3. Khanal S, Ghimire P, Dhamoon AS. The repertoire of adenovirus in human disease: The innocuous to the deadly. *Biomedicines.* 2018;6(1):30.
4. Saha B, Wong CM, Parks RJ. The adenovirus genome contributes to the structural stability of the virion. *Viruses.* 2014;6(9):3563–83.
5. Sohn SY, Hearing P. Adenovirus early proteins and host sumoylation. *mBio.* 2016;7(5):e01154–16.
6. Lee CS, Bishop ES, Zhang R, et al. Adenovirus-mediated gene delivery: Potential applications for gene and cell-based therapies in the new era of personalized medicine. *Genes Dis.* 2017;4(2):43–63.
7. Kosulin K. Intestinal HAdV infection: Tissue specificity, persistence, and implications for antiviral therapy. *Viruses.* 2019;11(9):804.
8. Pennington MR, Saha A, Painter DF, et al. Disparate entry of adenoviruses dictates differential innate immune responses on the ocular surface. *Microorganisms.* 2019;7(9):351.
9. Stasiak AC, Stehle T. Human adenovirus binding to host cell receptors: A structural view. *Med Microbiol Immunol.* 2020;209(3):325–33.
10. Radke JR, Cook JL. Human adenovirus infections: Update and consideration of mechanisms of viral persistence. *Curr Opin Infect Dis.* 2018;31(3):251–6.
11. Crenshaw BJ, Jones LB, Bell CR, Kumar S, Matthews QL. Perspective on adenoviruses: Epidemiology, pathogenicity, and gene therapy. *Biomedicines.* 2019;7(3):61.
12. Ponterio E, Gnessi L. Adenovirus 36 and obesity: An overview. *Viruses.* 2015;7(7):3719–40.
13. Lynch JP 3rd, Kajon AE. Adenovirus: Epidemiology, global spread of novel serotypes, and advances in treatment and prevention. *Semin Respir Crit Care Med.* 2016;37(4):586–602.
14. Labib BA, Minhas BK, Chigbu DI. Management of adenoviral keratoconjunctivitis: Challenges and solutions. *Clin Ophthalmol.* 2020;14:837–52.
15. Cook J, Radke J. Mechanisms of pathogenesis of emerging adenoviruses. *F1000Res.* 2017;6:90.
16. Prusinkiewicz MA, Mymryk JS. Metabolic reprogramming of the host cell by human adenovirus infection. *Viruses.* 2019;11(2):141.
17. Gonzalez G, Yawata N, Aoki K, Kitaichi N. Challenges in management of epidemic keratoconjunctivitis with emerging recombinant human adenoviruses. *J Clin Virol.* 2019;112:1–9.

18. Hartman ZC, Appledorn DM, Amalfitano A. Adenovirus vector induced innate immune responses: Impact upon efficacy and toxicity in gene therapy and vaccine applications. *Virus Res.* 2008;132(1–2):1–14.
19. Afkhami S, Yao Y, Xing Z. Methods and clinical development of adenovirus-vectored vaccines against mucosal pathogens. *Mol Ther Methods Clin Dev.* 2016;3:16030.
20. Knippertz I, Deinzer A, Dörrie J, Schaft N, Nettelbeck DM, Steinkasserer A. Transcriptional targeting of mature dendritic cells with adenoviral vectors via a modular promoter system for antigen expression and functional manipulation. *J Immunol Res.* 2016;2016:6078473.
21. Gilbert SC, Warimwe GM. Rapid development of vaccines against emerging pathogens: The replication-deficient simian adenovirus platform technology. *Vaccine.* 2017; 35(35A):4461–4.
22. Chen S, Tian X. Vaccine development for human mastadenovirus. *J Thorac Dis.* 2018;10(Suppl 19):S2280–94.
23. Coughlan L. Factors which contribute to the immunogenicity of non-replicating adenoviral vectored vaccines. *Front Immunol.* 2020;11:909.
24. Uusi-Kerttula H, Hulin-Curtis S, Davies J, Parker AL. Oncolytic adenovirus: Strategies and insights for vector design and immuno-oncolytic applications. *Viruses.* 2015;7(11):6009–6042.
25. Gao J, Zhang W, Ehrhardt A. Expanding the spectrum of adenoviral vectors for cancer therapy. *Cancers (Basel).* 2020;12(5):E1139.
26. Kuroki LM, Jin X, Dmitriev IP, et al. Adenovirus platform enhances transduction efficiency of human mesenchymal stem cells: An opportunity for cellular carriers of targeted TRAIL-based TR3 biologics in ovarian cancer. *PLoS One.* 2017;12(12):e0190125.
27. Baker AT, Aguirre-Hernández C, Halldén G, Parker AL. Designer oncolytic adenovirus: Coming of age. *Cancers (Basel)* 2018; 10(6): 201.
28. Martin NT, Wrede C, Niemann J, et al. Targeting polysialic acid-abundant cancers using oncolytic adenoviruses with fibers fused to active bacteriophage borne endosialidase. *Biomaterials.* 2018;158:86–94.
29. Tsukamoto T, Sakai E, Iizuka S, Taracena-Gándara M, Sakurai F, Mizuguchi H. Generation of the adenovirus vector-mediated CRISPR/Cpf1 system and the application for primary human hepatocytes prepared from humanized mice with chimeric liver. *Biol Pharm Bull.* 2018;41(7):1089–95.
30. Atasheva S, Shayakhmetov DM. Adenovirus sensing by the immune system. *Curr Opin Virol.* 2016;21:109–13.
31. Beatty MS, Curiel DT. Chapter two--Adenovirus strategies for tissue-specific targeting. *Adv Cancer Res.* 2012;115:39–67
32. Chira S, Jackson CS, Oprea I, et al. Progresses towards safe and efficient gene therapy vectors. *Oncotarget.* 2015;6(31):30675–703.
33. Alonso-Padilla J, Papp T, Kaján GL, et al. Development of novel adenoviral vectors to overcome challenges observed with HAdV-5-based constructs. *Mol Ther.* 2016;24(1):6–16.

SUMMARY

Human adenovirus (HAdV) is a non-enveloped, double-stranded DNA virus implicated in mostly mild, self-limiting illnesses among susceptible populations (e.g., young children, crowded communities, schools, and military training camps, as well as immunocompromised individuals) worldwide. Owing to its ability to infect a diversity of replicating and quiescent cells at high efficiency even in the absence of certain genes, to grow to high-titers, and to induce little genotoxicity *in vivo*, HAdV represents an ideal vector for vaccine development and gene therapy. To realize its full potential as a vector for cancer vaccine, however, several issues need to be addressed, including (i) the difficulty in determining host immune response that is adequate for lysing the cancerous cells without causing unnecessary host inflammation and tissue damage, especially during secondary administration; (ii) the low transduction efficiency of cancer cells which lack the specific receptor for adenoviral vector; and (iii) possible contamination with helper adenovirus when a gutless adenovirus vector is utilized. Successful application of adenoviral vectors for the treatment of other diseases faces similar challenges.

SECTION II

Molecular Analysis and Manipulation of Foodborne Bacteria

Molecular Mechanisms of *Listeria monocytogenes* Stress Responses

11

Alberto Alía, Alicia Rodríguez, María J. Andrade, and Juan J. Córdoba
University of Extremadura

Contents

11.1 INTRODUCTION

Listeria monocytogenes is a Gram-positive facultative intracellular bacterium responsible for a severe foodborne infection known as listeriosis.[1] As a ubiquitous organism that can be found in fresh produce, animals, food processing equipment and premises, this bacterium is more likely to cause death than other foodborne bacterial pathogens.[2]

Realizing the potential risk of *L. monocytogenes* and other foodborne pathogens to human health,[3] many different preserving methods have been applied in the food industry to process safe and high-quality products, including heat treatments, refrigeration, acidification, hygienization, desiccation, and addition of salts or sugars.[4,5] However, some of the traditional preserving methods are not applicable to certain foods or fail to completely eliminate this and other foodborne pathogens. Therefore, alternative technologies such as ultrasound under pressure at non-lethal (manosonication, MS) and lethal temperatures (manothermosonication, MTS), pulsed electric field (PEF), high hydrostatic pressure (HHP), and UV-light are being used.[6] Mechanistically, all of the above methods aim to cause stress to and eventually kill *L. monocytogenes*, being cold and heat, acidic and alkaline, osmotic and drying, and oxidative stresses. However, *L. monocytogenes* demonstrates a remarkable capacity to withstand these stresses, and recover quickly after the treatments. The adaptive response of this food pathogen to the aforementioned stresses is of significant concern to the food industry because these control strategies may just not fail to prevent the survival and development of *L. monocytogenes*, they may even bring out stress-resistant strains with high pathogenic potential.[4,7] Exposure of the pathogen to a mild type of stress can cause cross-protection against other stress factors, modify the virulence, and give rise to more persistent strains. The ability of *L. monocytogenes* to survive and multiply under a wide range of environmental-stress conditions normally encountered in foods is determined by the induction of the resistance mechanisms that the bacterium possesses. Several transcriptional regulators important for stress response and virulence gene expression have been identified in this microorganism.[5]

This chapter gives an overview of the molecular mechanisms, which *L. monocytogenes* uses to tolerate the different types of stresses that are encountered in foods and food environments throughout the processing and storage either in the industry or in the market.

11.2 GENES INVOLVED IN *L. MONOCYTOGENES* STRESS TOLERANCE AND VIRULENCE

The ability of *L. monocytogenes* to respond rapidly to adverse or changing environmental conditions is mediated by different associations between various alternative σ factors, which guide RNA polymerase to specific promoters to activate their transcription, enabling the bacterium to survive in rapidly changing conditions. These σ factors are key components of the complex regulatory networks of *L. monocytogenes* including multiplex transcriptional regulators of stress-response and virulence genes, regulation of genes encoding other regulators, and regulation of small RNAs.[8,9]. One of these factors is σ^B, encoded by the *sigB* gene, which has provided enough evidence that it contributes to the ability of *L. monocytogenes* to multiply and survive under non-host-associated environmental stress conditions such as acid, osmotic, oxidative, cold, nutrient limitation, and energy stresses (Table 11.1).[8] Previous studies have shown that σ^B regulates the expression of some virulence and stress-related genes that are necessary for survival in the host, either during gastrointestinal passage or during intracellular survival and growth. Furthermore, σ^B also contributes to transcription of the *prfA* gene, which encodes the global *L. monocytogenes* virulence gene regulator, PrfA.[8]

TABLE 11.1 *Listeria monocytogenes* σ^B contribution to stress resistance (Chaturongakul et al.)[8]

STRESS CONDITION	*Σ^B-DEPENDENT GENES*	*PHENOTYPIC DATA SUPPORTING Σ^B CONTRIBUTIONS*
Acid	*gadA*, *gadB*, and *gadC*	Survival of the Δ*sigB* strain is lower than the wild type when logarithmic-phase cells are exposed to pH 2.5 for 1 h
Osmotic	*betL*, *gbuA*, *bilEAB*, and *opuCABCD*	Reduction of the ability to use carnitine and betaine as osmoprotectans in the Δ*sigB* strain
Oxidative	*lmo0669* and *lmo1433*	Survival of the Δ*sigB* strain is lower than the wild type when stationary-phase cells are exposed to 13 mM CHP for 15 min
Nutrient availability (energy)	*opuCA* and carbon metabolism genes (e.g. *ldh*)	Viability of the Δ*sigB* strain is impaired when glucose is limited
Cold or freezing	*ltrC*	The σ^B factor is not required during the growth of *L. monocytogenes* in rich media at 4°C or for survival after freeze–thaw cycles, but it contributes to the growth in minimal media and in meats at low temperatures
High hydrostatic pressure	n.d.	Survival of the Δ*sigB* strain is more limited than the wild type at 350 Mpa after 28 min
Arsenate	*lmo2230* (encodes an arsenate reductase protein)	n.d.

Δ*sigB:* absence of the *sigB* gene.
n.d.: not determined.

Another σ factor is σ^L, also known as σ^{54} or RpoN, which is encoded by the *sigL* gene and plays a role in the intracellular replication and stress response of this microorganism, contributing to osmotolerance and influencing on its susceptibility to the antibacterial peptide mesentericin Y105.[8,10,11] On the other hand, the σ^C factor or extracytoplasmic function σ factors (encoded by the *sigC* gene) seem to be activated upon heat stress.[8] Furthermore, the σ factors regulate small RNAs controlling their transcription such as the control of the expression of binding proteins, and chaperones such as *Hfq*, which play an important role in stress response in the presence of salt and ethanol.[12]

Several σ factors are involved in the virulence and stress-response genes. However, the stress-related σ^B, encoded by the *sigB* gene, is the only one linked to virulence.[13] The emerging consensus is that, while σ^B may contribute to modulate *prfA*-dependent expression, it does not play a significant influence on it during invasive infection, its involvement being restricted to the gastrointestinal phase before entry.[13–16] PrfA is the major regulator of the virulence genes in *L. monocytogenes*.[17] Once within the host, *L. monocytogenes* parasitizes macrophages and active epithelial cells, hepatocytes, fibroblasts, and cells of the endothelium and nervous tissue where the entry is mediated via two listerial surface-associated proteins (internalins InlA and InlB). After that, other virulence proteins aid in the survival of this pathogen by promoting vacuole escape (by activation of pore-forming toxin listeriolysin O or LLO, phospholipases PlcA and PlcB, and protease Mpl), cytosolic replication (sugar-phosphate permease Hpt), and actin-based cell-to-cell spreading to adjacent cells (actin-polymerizing protein ActA, InlC), where the cycle starts again.[13] PrfA regulon is selectively activated during host cell infection where the genes encoding these virulence factors are coordinately expressed under the positive control of PrfA. This protein comes from the cAMP receptor protein (Crp)/fumarate nitrate reductase regulator (Fnr) family of bacterial transcription factors that can bind DNA as dimers on specific sites in gene promoters and activate transcription.[13,17] PrfA was initially identified for its ability to stimulate the expression of the *hly* gene or encoding the pore-forming toxin LLO. The *hly* gene is co-expressed together with a PrfA-dependent virulence cluster encompassing the *prfA* gene itself, as well as the *plcA* and *plcB* genes encoding phospholipases and the *mpl* gene which encodes a metalloprotease and ActA, the protein that promotes actin comet tail formation by recruiting the cellular actin nucleator complex Arp2/3.[18–24] PrfA also activates other genes like the *inlA* and *inlB* genes encoding the internalin family proteins InlA and InlB, the *inlC* gene, encoding a short secreted internalin-like protein important for cell-to-cell spread and the dampening of inflammatory response, the *bsh* gene, encoding a bile-salt hydrolase participating in the intestinal and hepatic phases of infection, and the *uhpT* gene, encoding a hexose phosphate transporter.[21,25,26]

Apart from the *sigB* and *prfA* genes, there are other transcriptional regulators that contribute to an increase of gene expression during infection favoring virulence of *L. monocytogenes*. The two-component system VirR/S controls the expression of 17 genes, including its own operon, contributes to the modification of bacterial surface components, and/or provides resistance to antimicrobial peptides.[27] The *dlt* gene encodes proteins involved in D-alanylation of lipoteichoic acids and it has been shown to be required for virulence[28]; the *lmo1695* gene is required for lysinylation of phospholipids in listerial membranes and confers resistance to cationic antimicrobial peptides[29]; the *lmo1696* gene encodes a putative glycopeptide antibiotic resistance protein of the VanZ family[30]; and finally, the ABC transporter AnrAB is involved in resistance to the bacteriocin nisin and several other antibiotics.[31] Furthermore, the nutrient-responsive regulator CodY, which coordinates the synthesis of branched-chain amino-acids (BCAA), stimulates the expression of virulence genes in response to the low availability of BCAA.

In addition to the action of the transcriptional regulators, which take part in the control of virulence gene expression in *L. monocytogenes*, other studies have shown a large variety of RNA-mediated regulatory mechanisms of bacterial gene expression.[17] One of them is the 5′-untranslated regions (UTRs), which are involved in the control of the expression of several *L. monocytogenes* virulence genes at the post-transcriptional level. Thus, e.g. the 5′-UTR of the *actA*, *hly*, *inlA*, and *iap* genes are required for the optimal expression of these virulence genes.[17] Furthermore, riboswitches consist in short RNA sequences that make conformational changes upon ligand binding and adjust the expression of certain operons according to the availability of specific metabolites. Some of these operons are *pdu*, *cob*, and *eut*, which have been proposed to confer a metabolic advantage to pathogenic species in competition with the gut flora in the host.[17] Finally, the *Listeria* genome encodes many non-coding RNA (ncRNAs) that share a partial complementarity with targets located elsewhere in the genome and can regulate them. To date, more than 150 putative acting sRNAs, named Rli, have been annotated and it is suggested that they have a regulatory function in virulence; however, the functions of most *Listeria* sRNAs remain uncharacterized.[15,32–34]

11.3 MOLECULAR MECHANISMS UNDERLYING *L. MONOCYTOGENES* STRESS TOLERANCE RESPONSE TO ADVERSE CONDITIONS IN FOODS AND FOOD ENVIRONMENTS

L. monocytogenes are exposed to numerous stress factors such as heat, freezing, acid, detergents, dehydration, osmolites, and oxidation in foods and food environments throughout the processing and storage in the industry. The microorganism may normally confront more than one type of stress at the same time to survive in foods. In this section, the mechanisms used for *L. monocytogenes* to tolerate adverse conditions and persist in foods are in-depth described. Table 11.2 shows an overview of the main mechanisms utilized by *L. monocytogenes* to tolerate stress.

TABLE 11.2 Overview of the main mechanisms of adaptation of *Listeria monocytogenes* to tolerate stress

TYPE OF STRESS	*MECHANISM OF ADAPTATION*	*REFERENCES*
Cold stress response	Cold shock proteins (Csps), cold acclimation proteins (Caps), and DNA helicases (*lmo0866*, *lmo1722*, and *lmo1450)* are important for *L. monocytogenes* growth	[36,37]
	Csps participate in the control of replication, transcription, and translation favoring adaptation to a cold environment	[38]
	The *LisR*, *lmo1172*, and *lmo1060* gene activation permits tolerance of *L. monocytogenes* to cold stress	[35]
	lhkA, *yycJ*, and *yycF* genes are transcriptionally active in response to cold stress	[35]
	Increased transcription of the *groEL*, *clpP*, and *clpB* genes	[35]
	Increased expression of the *flp* and *trxB* genes associated with the growth of *L. monocytogenes* at low temperatures	[35]
	Increased transcription of the *hisJ*, *trpG*, *cysS*, and *aroA* genes involved in amino acid synthesis as a consequence of the Caps	[35]
	Activation of the *ltrA*, *ltrB*, *and ltrC* genes is essential for the growth of *L. monocytogenes* (<4°C)	[41]
	flaA encodes the *L. monocytogenes* major flagellin protein FLaA, increasing flagellum production at low temperatures	[42]
Hot stress response	Heat shock proteins (Hsps) which contain highly conserved chaperones and ATP-bounded proteases	[5]
	The GroES, GroEL, and DnaK chaperone proteins help protein aggregation at high temperatures	[46]
	Activation of the *CtsR* gene, which controls the *ClpP*, *ClpE*, and *ClpC* genes that remove damaged or misfolded proteins	[47]
	The *fri* gene encoding a single ferritin gene is strongly overexpressed after heat shock or chemical stress	[50,51,127]
Acidic stress response	The induction of the σ^B, *prfA*, *hrcA*, and *ctsR* genes increases the activation of the fatty acid biosynthetic genes	[57]
	The induction of dehydrogenases (*GuaB*, *PDuQ*, and *lmo0560*), reductases (*YegT*), and respiratory enzymes is related to adaptation to acid stress consisting in accelerating electron transfer through enhanced oxidation–reduction potential	[5,56]
	Respiratory enzymes together with the F_0F_1 ATPase action can contribute to proton efflux events	[53,58]
	The promoting of the *atpC* and *atpD* genes encoding H^+ ATPase, which are regulated by the *prfA* genes, is linked to the onset of the acid tolerance response	[5]
	Enzymes encoded by the *arcA*, *arcB*, *arcC*, and *arcD* genes are involved in the conversion of arginine to ornithine and transfer the ornithine outside the cell in exchange for arginine	[60]
	The glutamate decarboxylase acid resistance system including the *gadA*, *gadB*, and *gadC* genes is related to acid stress tolerance	[59,61–64,73]
	The activation of *lmo0038* gene and the genes encoding LisRK and HtrA plays an important acid stress adaptation role in *L. monocytogenes*	
	Changes in expression of the stress-related genes (*gad2* and *sigB*) of *L. monocytogenes* are produced in response to acid stress	[65]

(*Continued*)

TABLE 11.2 (*Continued*) Overview of the main mechanisms of adaptation of *Listeria monocytogenes* to tolerate stress

TYPE OF STRESS	*MECHANISM OF ADAPTATION*	*REFERENCES*
Alkaline stress response	Activation of the stress-protective chaperones DnaK and GroEL is stimulated by alkaline stress conditions	[67]
Osmotic stress response	The *treA* gene is involved in the trehalose metabolism and is related to osmotic stress resistance	[68]
	The accumulation of glycine betaine, an important osmoprotectan, is mediated by the *BetL* and *Gbu* genes	[76]
	The *OpuC* gene has a direct role in regulating carnitine production, an important osmoprotectant component	[77]
	The biosynthesis of proline is regulated by the *ProBA* gene and gone up under osmotic stress	[74]
	Activation of the Kdp system, which allows K^+ import, makes easy a rapid response to osmotic stress	[69,72]
	The *lmo2085* and *lmo1078* genes encode proteins, which modify cell wall structure in response to osmotic stress	[79,80]
	The *lmo0501* gene is inhibited under osmotic stress conditions	[81]
	Activation of the *lstC* gene is needed to high salt tolerance adaptation	[82]
	The genes encoding the proteases HtrA and ClpP and the guanosine tetra/penta-phosphate synthetase RelA are essential for osmotic stress adaptation	[72]
	The virulence *iap* gene plays an important role in salt adaptation	[83,84]
	The *sigB* gene, which controls various genes, is also implied in osmotic stress tolerance	[8]
Desiccation	Microbial biofilm formation and the increase of extrapolymeric substances	[71,72]
	Accumulation of osmolytes such as betaine, carnitine, proline, and trehalose	[68,69]
Oxidative stress response	Activation of the *sigB* and *perR* genes, which both encode stress regulator proteins; the *sod, kat,* and *fri* genes, which are involved in oxidation resistance of the microorganism, and the *recA* implied in the DNA repair are some essential genes for the persistence and growth under oxidative stress	[92,93,95,98]
	The *kat* gene works together with superoxide dismutases encoded by the *sod* gene to prevent oxidative attack	[93]
	A regulator encoded by the *sigB* gene regulates the expression of *sod, kat, fri,* and other oxidation resistance genes under oxidative stress conditions	[96,97]
	The *fri* gene is important in the protection of *L. monocytogenes* cells from hydrogen peroxide	[98]
	The *perR* gene has been reported to be associated with the ability to tolerate oxidative stress	[99]
	The *recA* gene contributes to DNA repair after cell impairment under stress conditions	[95,96,100]

11.3.1 Cold and Hot Stress Responses

11.3.1.1 Cold Stress Response

L. monocytogenes is well known for its capacity to survive and multiply under a broad range of environmental-stress conditions, including the ability to grow at temperatures as low as –0.4°C.[35] The characterization of molecular mechanisms that helps *L. monocytogenes* to survive and grow at low temperatures and develop adaptation to cryotolerance is necessary to search for strategies to reduce or prevent *L. monocytogenes* growth in refrigerated ready-to-eat (RTE) foods.[35,36]

Cold stress resistance is a biological property mediated through many molecular response mechanisms. The reactions occurring in bacteria against stress are the form of long-term adaptation or immediate (shock) response. If a decrease of temperatures occurs suddenly, the bacterium will suffer a cold shock stress, which requires appropriate cold shock adaptation molecular response mechanisms; once at low temperatures, the microorganism must acclimatize to the cold stress environment and adapt its growth at low temperatures.[37] Typical molecular cold stress adaptation mechanisms documented in food-borne pathogens including *L. monocytogenes* are (i) modulation of nucleic acid structures (DNA and RNA) to relieve constraints associated with increased stability; (ii) maintenance of structural integrity in cell membranes; (iii) uptake of compatible solutes; (iv) non-specific stress response mechanisms; and (v) production of various cold stress proteins known as cold shock proteins (Csps) and cold acclimation proteins (Caps).[37] It is important to underline that DNA helicases and Csps are important for the growth of *L. monocytogenes*.[5] These Csps are members of small and highly conserved chaperones, which play a role in the control of replication, transcription, and translation[38] and could favor the cold adaptation of *L. monocytogenes*.[5,39]

On the other hand, it has been described that the absence of the *lisR*, *lmo1172*, and *lmo1060* genes reduces the tolerance of the bacterium to cold stress, while the *lhkA*, *yycJ*, and *yycF* genes are transcriptionally active in response to such stress.[35] Liu et al.[40] reported an enhanced expression of the *yycJ* gene in response to cold stress, which encodes a putative member of the two-component signal transduction system in *L. monocytogenes*. Similarly, the *lhkA* gene, which encodes a histidine kinase sensor, was also induced at low temperatures.[37] In addition, the activation of the *psr* gene under cold stress conditions provokes the regulation of its own expression as well as of penicillin-binding protein 5 while its deletion leads to various cell-wall alteration of the bacterium.[37] Liu et al.[40] also detected increased transcription of the *groEL*, *clpP*, and *clpB* genes during cold acclimation of *L. monocytogenes*. These authors also observed an induction of the transcription of both *flp* and *trxB* genes, which encode Flp and thioredoxin reductase, that are associated with growth at low temperatures and a risen transcription of the *hisJ*, *trpG*, *cysS*, and aroA genes involved in amino acid synthesis that encode amino acid biosynthetic enzymes due to a Caps. Zheng and Kathariou[41] describe in their study that the inactivation of three low temperature-requirement genes (*ltrA*, *ltrB*, and *ltrC)* seems to be essential for growth at low temperature (4°C) but dispensable for growth at 8°C or higher temperatures.[41]

Other studies carried out by Liu et al.[42] demonstrate that an increase of flagellum production during the growth of this bacterium at low temperatures occurs. The *flaA* gene encodes the *L. monocytogenes* major flagellin protein FLaA.[43] It has been proven that the expression of such gene was higher when *L. monocytogenes* is grown at 25°C and 10°C than at 37°C.[43] This increment of flagellum production is unclear; however, motility is associated with microbial biofilm formation in other microorganisms having been described as an adaptation strategy of bacteria against cold stress environments.[37] Moreover, the *lmo0866*, *lmo1722*, and *lmo1450* genes encoding helicases identified in *L. monocytogenes* take part in the cold stress response transcriptome.[44] Finally, Azizoglu and Kathariou[36] have shown the necessity of the *lmo0866* gene for *L. monocytogenes* to survive and grow under low-temperature conditions.

11.3.1.2 Hot Stress Response

L. monocytogenes is a food-associated pathogen, which possesses a relatively high heat resistance being able to survive in some cases during pasteurization in dairy products.[5,45] In addition, the exposure to heat treatment induces the microorganism to activate cellular mechanisms to maintain homeostasis and repair the effect of heat treatment with removal of degraded proteins, renovation of metabolic functions, and production of new proteins.[5] This response is related to the rise in the production of heat shock proteins (Hsps), which contain highly conserved chaperones and ATP-bounded proteases. The protection of the cells is provided by the *GroES*, *GroEL*, and *DnaK* genes involved in their corresponding chaperone proteins, by helping protein aggregation at high temperatures.[46] The gene repressor *CtsR* controls the expression of the *ClpP, ClpE,* and *ClpC* genes encoding ATP-bounded proteases and removes damaged or misfolded proteins.[46] Hu et al.[47] showed that the activation of the *ClpB* gene makes the precipitated proteins active again. Furthermore, it has been reported that sensitivity to heat stress increases in *L. monocytogenes* cells in the absence of the *fri* gene, which encodes a single ferritin protein.[48] *L. monocytogenes* ferritin is a major cold shock protein, which is also strongly overexpressed after heat shock or chemical stresses.[49–51] Thus, the above proteins are responsible for the response of *L. monocytogenes* to hot stress.

11.3.2 Acidic and Alkaline Stress Responses

11.3.2.1 Acidic Stress Response

Acidification is frequently used in the food industry to control pathogenic microorganisms such as *L. monocytogenes* in foods.[52] The use of organic acids as food preservatives[53] and the compounds produced by lactic acid bacteria used as protective cultures[54] may cause an acidic environment in foods that *L. monocytogenes* needs to face to survive. In addition, it is noteworthy to mention that *L. monocytogenes* also needs to withstand stress provoked by inorganic acids present in the gastrointestinal tracts of human hosts during the infection process. The survival and adaptation of this pathogen in relation to this stress are essential for systemic infection.[5,55] Different mechanisms of adaptation to acidic stress have been described including stimulation in the activity of DNA repair enzymes, regulators, and oxidative stress defenses, improvement in the amino acid catabolism and proton pump, and modifications in the composition of the cell membrane.[56]

The nature of the organic acid used plays a role in the adaptation mechanism of this pathogen to acidic stress. Thus, Bowman et al.[57] have discovered that the use of sodium diacetate at pH 5.0 promotes changes in the expression of a wide range of genes (σ^B, *prfA*, *hrcA*, and *ctsR*) and increases activation of oxidative stress defenses, DNA repair, intermediary metabolism, cell wall modification, and cofactor and fatty acid biosynthetic genes. Other studies in which *L. monocytogenes* faced hydrochloric acid have demonstrated that enzyme dehydrogenases and reductases, osmolyte transport, protein folding and repair, general stress resistance, flagella synthesis, and metabolism are all associated with acid stress responses in this bacterium.[58,59]

Other adaptation mechanisms used by *L. monocytogenes* to confront acid stress consists in accelerating electron transfer through an enhanced oxidation–reduction potential.[5] The induction of genes involved in dehydrogenases (*guaB*, *pduQ*, and *lmo0560*), reductases (*yegT*), and respiratory enzymes is related to adaptation to acid stress in this bacterium.[56] The aforementioned enzymes together with the F_0F_1 ATPase action can contribute to proton efflux events.[53,58]

H^+ ATPase, encoded by the *atpC* and *atpD* genes which are regulated by the *prfA* gene, is linked to the onset of the acid tolerance response.[5] The *L. monocytogenes* arginine deiminase, involved in the conversion of arginine to ornithine and the transfer of ornithine outside the cell in exchange for arginine because of the action of the enzymes encoded by the *arcA*, *arcB*, *arcC*, and *arcD* genes,[60] is also associated with the resistance of the bacterium of acid stress conditions. Moreover, the glutamate decarboxylase acid resistance system including the *gadA*, *gadB*, and *gadC* genes and the *lmo0038* gene, encoding putative peptidylarginase deiminase family proteins LisRK and HtrA, plays also an important acid stress adaptation role in *L. monocytogenes*.[59,61–64]

Recently, Makariti et al.[65] have demonstrated changes in the expression of the stress-related genes (*gad2* and *sigB*) of *L. monocytogenes* in response to acid stress.

11.3.2.2 Alkaline Stress Response

Most of the detergents and sanitizing agents used in food processing environments may provoke alkaline stress conditions that *L. monocytogenes* should withstand to survive.[56] Various mechanisms can be utilized by this bacterium during alkaline stress adaptation that include metabolic changes to increase acid generation and the stimulation of transporters and enzymes and cell surface modifications both associated with proton retention.[5,66] Moreover, it has been found out that 390 gene transcripts in *L. monocytogenes* cells are expressed in different grades in relation to alkaline stress. Such genes take part in general stress response, solute transportation, and various metabolic systems.[67] In addition, the activation of the genes involved in stress-protective chaperones DnaK and GroEL is also stimulated by alkaline stress conditions,[67] increasing the synthesis of different proteins. However, to date, little is still known about the mechanism of alkaline stress adaptation in this food-borne pathogen.

11.3.3 Water Availability Stress-Related Responses

One of the strategies of the food industry to ensure the safety and quality of its products relies on the limitation of the water availability for microorganisms. Two different water-related stresses can be distinguished: desiccation (drying or matrix effect) and osmotic stress. This is of significant concern to the food industry since part of the reason for this pathogen to survive in foods and food processing plants is due to its ability to withstand matric and osmotic stress associated with desiccation or external osmolytes. This bacterium has evolved several complex interplaying and different molecular mechanisms to respond to either desiccation or osmotic stress encountered in food and food environments, which is described below.

11.3.3.1 Osmotic Stress Response

In the food industry, the high levels of sugar and sodium salts are used as food preservatives by their dehydration effect since they increase intracellular solute concentrations and may disrupt many biological functions. *L. monocytogenes* may encounter osmotic stress during a shift to a hyperosmotic solution or due to dehydration. In this bacterium, several proteins have been shown to be linked to salt and sugar stress response.[68–70] These proteins are involved in the production of compatible solutes, changes of the cell wall, and regulatory processes, some of them being general stress proteins too.[71]

One of the main osmotic stress adaptation strategies is the accumulation of compatible solutes.[5,72–74] The major osmoprotectants described are trehalose, glycine betaine, carnitine, and amino acids. The compatible solute trehalose is important for the survival of *L. monocytogenes* under adverse environmental conditions.[72,75] The phosphotrehalase gene *treA* (*lmo1254*) is involved in the trehalose metabolism and this is related to stress resistance.[68] The accumulation of glycine betaine is mediated by the *betL* and *gbu* genes,[76] while the *opuC* gene has a direct role in regulating carnitine production.[77] Regarding amino acids, proline is an important osmoprotectant.[78] The biosynthesis of this amino acid is regulated by the *ProBA* gene and gone up under osmotic stress.[74]

Other mechanisms to adaptation of osmotic stress used for *L. monocytogenes* are detailed below. One of them consists in the activation of the Kdp system, which allows K+ import, making easy a rapid response to such stress.[72,78] In addition, two genes that encode proteins modifying cell wall structure in response to osmotic stress have been reported: Lmo2085, a putative peptidoglycan bound protein, and Lmo1078, a putative UDP-glucose phosphorylase catalyzing the formation of UDP-glucose.[79,80] Regarding the regulatory proteins involved in osmotic stress response, the *lmo0501* gene has shown its activity reduced under such environmental stress conditions.[81] The *lstC* gene of *L. monocytogenes* is also implicated in high salt tolerance.[82] In addition, some general stress response proteins are essential for osmotic stress adaptation in *L. monocytogenes*. Some of them are the serine protease HtrA, the protease ClpP, and the guanosine tetra/penta-phosphate synthetase RelA.[72] It is important to underline that the virulence *iap* gene also plays an important role in salt adaptation.[83,84] As it has been described before, many of the mentioned genes are under the control of the alternative sigma factor σ^B, which controls various genes implied in stress tolerance, metabolism, transport, cell envelope, and virulence.[8]

11.3.3.2 Desiccation

The importance and frequency of desiccation stress provoked by the drying process (relative humidity variations) of foods or by daily cleaning–disinfection procedures on surfaces of equipment in the food industry[71] require particular attention to understand how *L. monocytogenes* can adapt, tolerate, and survive it. Gram-positive bacteria, such as *L. monocytogenes*, due to their thicker peptidoglycans, are more resistant to dryness.[85]

Although not much is known about the mechanisms associated with desiccation adaptation in *L. monocytogenes*, different cellular mechanisms toward this stress tolerance have been described. Firstly, it must be underlined that it has been proven that the microbial biofilm formation increased the tolerance of bacteria in general, and *L. monocytogenes*, in particular, to desiccation stress.[71,72] Such biofilms are mainly formed by extracellular polymeric substances, which form a water-rich gel around bacterial cells. Although it is stated that the increase in extrapolymeric substances by *L. monocytogenes* cells in biofilms is one of the main mechanisms of adaptation against desiccation,[86,87] no information about the molecular basis of this mechanism of adaptation against desiccation has been reported.

Another adaptation strategy of this pathogen against desiccation consists in the accumulation of osmolytes, such as betaine, carnitine, proline, and trehalose.[68,69] The molecular mechanisms utilized by this bacterium against drying stress are the same described against osmotic stress in Section 11.2.3.2.

The desiccation stress response may also be associated with the presence of HSP or DNA damage repair. HSP is able to overcome the loss of water through the building of hydrogen bonds to other molecules.[88] The mechanism used to confront heat stress (detailed in Section 11.1.2) may also be used by this microorganism to withstand desiccation. Regarding DNA, it has been hypothesized that cells under desiccation use DNA to stabilize proteins. Therefore, DNA itself could also be a molecular shield against drying stress.[89]

11.3.4 Oxidative Stress Response

L. monocytogenes can get exposed to atmospheric changes-based oxidative stress along with chemical agents like detergents and disinfectants.[5] Reactive oxygen species (ROS) like superoxide, hydroxyl radicals, and hydrogen peroxide can be produced and accumulated in the microorganism.[57,72] To resist the exposure to oxidative agents, *L. monocytogenes* cells need to activate the protein, membrane, and nucleic acid damage repair mechanism.[5] For this reason, bacterial ROS detoxification systems including superoxide dismutase (Sod), catalase (Kat), and alkyl hydroperoxidase (AhpCF), which participate in oxidative stress protection of *L. monocytogenes*,[5,90] are activated against this type of stress. There are some key genes needed for the persistence and growth of *L. monocytogenes* under oxidative stress conditions, including the *sigB* and *perR* genes, both encoding stress regulator proteins, the *sod, kat,* and *fri* genes, involved in oxidation resistance, and the *recA* implicated in DNA repair.[91–96] It is reported that the catalase encoded by the *kat* gene works together with (Sods) encoded by the *sod* gene to prevent an oxidative attack in *L. monocytogenes*.[93] A regulator encoded by the *sigB* gene regulates the expression of the *sod, kat,* and *fri* genes, and other oxidation resistance genes in *L. monocytogenes* under oxidative stress environments.[96,97] The *fri* gene encoding ferritin-like protein is important in the protection of *L. monocytogenes* cells from hydrogen peroxide[98] and the *perR* gene has been also reported to be associated with the ability to tolerate such type of stress.[99] Finally, the *recA* gene contributes to DNA repair after cell impairment under stress conditions.[95,96,100] It is noteworthy to underline that oxidative stress is considered to be a secondary stress, as a result of bacterial exposure to acid, heat, high salt concentrations, and toxic agents such as antibiotics.[101] The secondary oxidative stress response may have a significant impact on food safety since cells are exposed to multiple harsh conditions in food processing environments.

11.4 EFFECTS OF FOOD PROCESSING AND TREATMENTS ON *L. MONOCYTOGENES* STRESS TOLERANCE

Processing and treatments of raw materials and pre- and elaborated foods to eliminate contamination and/or to avoid the growth of *L. monocytogenes* may provoke different types of stresses that affect survival, virulence, and tolerance response of this microorganism against them.

Dairy products such as cheeses and meat products offer a suitable environment for the survival and growth of this bacterium, allowing this pathogen to display tolerance responses that can favor its presence in elaborated foods and persistence in their processing plants.[7,102,103] In cheeses and dairy products, their composition, moisture content, pH, content of salt, and ripening conditions favor the existence of acid, osmotic, drying, oxidative, heat, and cold stress conditions that *L. monocytogenes* needs to be confronted to survive.[104] In the case of meat products, it has been demonstrated that this pathogen responds to the conditions encountered during the fermentation and ripening of sausages activating acidic and osmotic stress mechanisms.[103] This bacterium can overcome the inhibitory effect of low pH by developing adaptive mechanisms, among which the F_0-F_1 ATPase, the glutamate decarboxylase (GAD) system, the arginine deiminase system, and sigma factors and regulators may be highlighted.[105] To cope with osmotic, drying, and cold stress caused by the composition and processing of these products, this bacterium mainly synthesizes or utilizes compatible solutes to balance the intracellular and extracellular environments.[7,69] To withstand the exposure to oxidative agents, this bacterium needs to activate protein, membrane, and nucleic acid damage repair mechanisms[7,90]; whilst to tolerate heat stress, *L. monocytogenes* synthesizes Hsps.[106] All these molecular mechanisms are detailed in Section 11.3 of this chapter. In addition, it has been proven that an increase in the expression of the stress-related *sigB* gene implies activation of the virulence-related genes in *L. monocytogenes*.[102]

Some of the traditional preserving methods such as heat treatments, refrigeration, acidification, hygienization, desiccation, and addition of salts or sugars may cause different types of stresses to *L. monocytogenes*. Nonetheless, the extraordinary adaptability of this bacterium often renders these treatments ineffective, and helps select stress-resistant strains that thrive on conditions associated with such treatments.[4,7]

Additionally, some of the aforementioned methods are not applicable to the hygienization of certain foods and are unable to completely inhibit the growth of this pathogen. Application of ultrasound under pressure at non-lethal (MS) and lethal temperatures (MTS), PEF, HHP, E-beam treatment, and UV-light may be helpful in this regard.[6] Indeed, both MS and MTS aid the control of *L. monocytogenes* growth in juices and other foods depending on pH and medium composition[107–111]; however, no studies have been conducted to evaluate the influence of these non-thermal technologies on changes in virulence of *L. monocytogenes* and stress tolerance by this bacterium.

PEF inhibits the growth of *L. monocytogenes* in fruit juices and milk.[112,113] However, the mechanism by which this technique inactivates this microorganism is not completely understood.[3] Some studies have reported that application of PEF may induce DNA and membrane integrity damages to *L. monocytogenes* cells and generation of toxic compounds,[3,114] but the molecular mechanism of *L. monocytogenes* to adapt and survive this treatment remains unclear.

HHP, also known as pascalization, is a non-thermal pasteurization consisting of treatments above 100 MPa.[115] This technique has been successfully applied to reduce or control the growth of *L. monocytogenes* in some fruit and vegetable juices, jams and smoothies,[115–119] cheeses,[120] cooked chicken,[121] and RTE-cooked ham.[122] Many authors highlight that the inactivation of *L. monocytogenes* by HHP is a multi-target process.[6,115,123] It is clear that the membrane is harmed by HHP, but additional damaging events seem to be necessary to prevent the survival of this bacterium in food, including such as an increase of solute loss during pressurization, protein coagulation, key enzyme inactivation, conformational changes in the ribosome, and alteration of recovery mechanisms.[123] It has been also hypothesized that HHP may affect cytoplasmic or membrane enzymes, disturb cellular metabolism, and induce the generation of ROS provoking stress to *L. monocytogenes* cells.[6] Although this method is increasingly used, its effect on the expression and regulation of stress-related genes in *L. monocytogenes* cells, which survive this treatment, has not been studied yet. UV has been traditionally employed for air, surface, and water decontamination, but recently this methodology is being successfully applied in the food industry in liquid foods and the surfaces of solid foods contaminated with *L. monocytogenes*.[6,124] Although scarce studies have been performed about the influence of UV on the growth of *L. monocytogenes* in foods, it is known that the short-wave UV light region (UV-C) from 200 to 280 nm is the most germicidal and its range of maximum effectiveness (260–265 nm) corresponds to the peak of maximum DNA absorption. Thus, UV mainly affects the integrity of genetic material as well as proteins.[125]

So far, no studies have been conducted to evaluate the molecular mechanism of *L. monocytogenes* to confront the adverse conditions associated with the use of this technology. Recently, Horita et al.[126] have proposed a combined strategy consisting of reformulation, active packaging, and the use of non-thermal post-packaging decontamination technology such as HHP, PEF, E-Beam, and UV-light to control *L. monocytogenes* in cooked RTE meat products. Although these non-thermal technologies are increasingly applied to control *L. monocytogenes* in foods, little is still known about stress responses induced by these treatments in surviving strains.

11.5 CONCLUSIONS AND FUTURE PERSPECTIVES

L. monocytogenes has evolved different molecular mechanisms to tolerate adverse conditions and withstand various types of stress, which are linked to the processing or storage of foods, and the use of traditional preserving methods (e.g. dehydration or utilization of salts and sugars) or new methods (e.g. PEF and HHP, among others) to inhibit the growth of this pathogen. Understanding the mechanisms of *L. monocytogenes* stress tolerance may help design new and efficient strategies for controlling its survival and growth in foods and for reducing listeriosis risk in consumers.

Although important advances have been made in understanding *L. monocytogenes* stress response using broth models, there is a need to conduct additional studies in food substrates under actual food processing conditions. Furthermore, while the σ^B factor takes part in *L. monocytogenes* stress response and virulence, the involvement of other stress-related genes on the virulence of this bacterium remains inadequately defined. Similarly, transcriptional regulators are active participants of *L. monocytogenes* response and virulence under acidic, osmotic, heat, and cold stresses, but their role in dealing with alkaline, desiccation, and oxidative stresses requires clarification. Moreover, studies on *L. monocytogenes* tolerance of hygienization, dehydration, oxidative agents, and non-thermal technologies used in the food industries are required.

Studies of *L. monocytogenes* stress response and virulence under food processing conditions may also benefit from the use of MS-based proteomic approach, which not only provides additional details on the mechanisms of adaptation and survival of this pathogen under adverse conditions, but also helps in the search for novel ways to block the expression of proteins that regulate stress adaptation and render *L. monocytogenes* resistant to stress stimulation.

ACKNOWLEDGMENTS

This work is funded through projects RTA-2013-00070-C03-03 and RTA2017-00027-C03-03 from the Instituto Nacional de Investigación y Tecnología Agraria y Agroalimentaria (INIA), project IB16149 from the Junta de Extremadura, and project GR15108 from FEDER. A. Alía is supported by a pre-doctoral fellowship from the Junta de Extremadura-Consejería de Economía E Infraestructura-, Fondo Social Europeo-, "Una manera de hacer Europa" (PD16023).

REFERENCES

1. Chlebicz A, Śliżewska K. Campylobacteriosis, salmonellosis, yersiniosis, and listeriosis as zoonotic foodborne pathogens: a review. *Int J Environ Res Public Health.* 2018;15(5):1–28.
2. Buchanan RL, Gorris LGM, Hayman MM, Jackson TC, Whiting RC. A review of *Listeria monocytogenes*: an update on outbreaks, virulence, dose-response, ecology, and risk assessments. *Int J Food Microbiol.* 2017;221:37–53.
3. Rajkovic A, Smigic N, Devlieghere F. Contemporary strategies in combating microbial contamination in food chain. *Int J Food Microbiol.* 2010;141:S29–S42.
4. Beales N. Adaptation of microorganisms to cold temperatures, weak acid preservatives, low pH, and osmotic stress: a review. *Compr Rev Food Sci Food Saf.* 2004;3(1):1–20.
5. Kocaman N, Sarimehmetoğlu B. Stress responses of *Listeria monocytogenes. Ankara Üniv Vet Fak Derg.* 2016;63:421–7.
6. Cebrián G, Mañas P, Condón S. Comparative resistance of bacterial foodborne pathogens to non-thermal technologies for food preservation. *Front Microbiol.* 2016;7:734.
7. Melo J, Andrew PW, Faleiro ML. *Listeria monocytogenes* in cheese and the dairy environment remains a food safety challenge: the role of stress responses. *Food Res Int.* 2015;67:75–90.
8. Chaturongakul S, Raengpradub S, Wiedmann M, Boor KJ. Modulation of stress and virulence in *Listeria monocytogenes. Trends Microbiol.* 2008;16:388–96.
9. Zhang Q, Feng Y, Deng L, et al. SigB plays a major role in *Listeria monocytogenes* tolerance to bile stress. *Int J Food Microbiol.* 2011;145(1):238–43.
10. Robichon D, Gouin E, Debarbouille M, Cossart P, Cenatiempo Y, Hechard Y. The *rpoN* (sigma54) gene from *Listeria monocytogenes* is involved in resistance to mesentericin Y105, an antibacterial peptide from *Leuconostoc mesenteroides. J Bacteriol.* 1997;179(23):7591–94.
11. Okada Y, Okada N, Makino SI, Asakura H, Yamamoto S, Igimi S. The sigma factor *RpoN* (σ^{54}) is involved in osmotolerance in *Listeria monocytogenes. FEMS Microbiol Lett.* 2006;263:54–60.
12. Christiansen JK, Larsen MH, Ingmer H, Søgaard-Andersen L, Kallipolitis BH. The RNA-binding protein Hfq of *Listeria monocytogenes*: role in stress tolerance and virulence. *J Bacteriol.* 2004;186(11):3355–62.
13. de las Heras A, Cain RJ, Bielecka MK, Vazquez-Boland JA. Regulation of *Listeria* virulence: PrfA master and commander. *Curr Opin Microbiol.* 2011;14(2):118–27.
14. Garner MR, Njaa BL, Wiedmann M, Boor KJ. Sigma B contributes to *Listeria monocytogenes* gastrointestinal infection but not to systemic spread in the guinea pig infection model. *Infect Immun.* 2006;74:876–86.
15. Toledo-Arana A, Dussurget O, Nikitas G, et al. The *Listeria* transcriptional landscape from saprophytism to virulence. *Nature.* 2009;459(7249):950–6.
16. Ollinger J, Bowen B, Wiedmann M, Boor KJ, Bergholz TM. *Listeria monocytogenes* σ^B modulates PrfA-mediated virulence factor expression. *Infect Immun.* 2009;77(5):2113–24.
17. Lebreton A, Cossart P. RNA- and protein-mediated control of *Listeria monocytogenes* virulence gene expression. *RNA Biol.* 2017;14(5):460–70.
18. Leimeister-Wachter M, Haffner C, Domann E, Goebel W, Chakraborty T. Identification of a gene that positively regulates expression of listeriolysin, the major virulence factor of *Listeria monocytogenes. Proc Natl Acad Sci.* 1990;87(21):8336–40.

19. Körner H, Sofia HJ, Zumft WG. Phylogeny of the bacterial superfamily of Crp-Fnr transcription regulators: exploiting the metabolic spectrum by controlling alternative gene programs. *FEMS Microbiol Rev.* 2003;27:559–92.
20. Hamon MA, Ribet D, Stavru F, Cossart P. Listeriolysin O: the swiss army knife of *Listeria. Trends Microbiol.* 2012;20(8):360–8.
21. Kreft J, Vázquez-Boland JA. Regulation of virulence genes in *Listeria. Int J Med Microbiol.* 2001;291:145–57.
22. Vazquez-Boland JA, Kocks C, Dramsi S, et al. Nucleotide-sequence of the lecithinase operon of *Listeria monocytogenes* and possible role of lecithinase in cell-to-cell spread. *Infect Immun.* 1992;60(1):219–30.
23. Poyart C, Abachin E, Razafimanantsoa I, Berche P. The zinc metalloprotease of *Listeria monocytogenes* is required for maturation of phosphatidylcholine phospholipase C: direct evidence obtained by gene complementation. *Infect Immun.* 1993;61:1576–80.
24. Kocks C, Gouin E, Tabouret M, Berche P, Ohayon H, Cossart P. *L. monocytogenes*-induced actin assembly requires the actA gene product, a surface protein. *Cell.* 1992;68:521–31.
25. Pizarro-cerda J, Ku A. Entry of *Listeria monocytogenes* in mammalian epithelial cells: an updated view. *Cold Spring Harb Perspect Med.* 2012; 2(11):1–18.
26. Dussurget O, Cabanes D, Dehoux P, et al. *Listeria monocytogenes* bile salt hydrolase is a PrfA-regulated virulence factor involved in the intestinal and hepatic phases of listeriosis. *Mol Microbiol.* 2002;45:1095–106.
27. Kang J, Wiedmann M, Boor KJ, Bergholz TM. VirR-mediated resistance of *Listeria monocytogenes* against food antimicrobials and cross-protection induced by exposure to organic acid salts. *Appl Environ Microbiol.* 2015;81(13):4553–62.
28. Abachin E, Poyart C, Pellegrini E, et al. Formation of D-alanyl-lipoteichoic acid is required for adhesion and virulence of *Listeria monocytogenes.* Mol Microbiol. 2002;43(1):1–14.
29. Thedieck K, Hain T, Mohamed W, et al. The MprF protein is required for lysinylation of phospholipids in listerial membranes and confers resistance to cationic antimicrobial peptides (CAMPs) on *Listeria monocytogenes. Mol Microbiol.* 2006;62:1325–39.
30. Arthur M, Courvalin, P. Genetics of glycopeptide resistance in enterococci. *Antimicrob Agents Chemother.* 1993;37(8):1563–71.
31. Collins B, Curtis N, Cotter PD, Hill C, Ross RP. The ABC transporter AnrAB contributes to the innate resistance of *Listeria monocytogenes* to nisin, bacitracin, and various β-lactam antibiotics. *Antimicrob Agents Chemother.* 2010;54:4416–23.
32. Schultze T, Izar B, Qing X, Mannala GK, Hain T. Current status of antisense RNA-mediated gene regulation in *Listeria monocytogenes. Front Cell Infect Microbiol.* 2014;4:1–6.
33. Sesto N, Koutero M, Cossart P. Bacterial and cellular RNAs at work during *Listeria* infection. *Future Microbiol.* 2014;9:1025–37.
34. Mraheil MA, Billion A, Mohamed W, et al. The intracellular sRNA transcriptome of *Listeria monocytogenes* during growth in macrophages. *Nucleic Acids Res.* 2011;39:4235–48.
35. Chan CY, Hu Y, Chaturongakul S, et al. Contributions of two-component regulatory systems, alternative σ factors, and negative regulators to *Listeria monocytogenes* cold adaptation and cold growth. *J Food Prot.* 2008;71(2):420–5.
36. Azizoglu RO, Kathariou S. Inactivation of a cold-induced putative RNA helicase gene of *Listeria monocytogenes* is accompanied by failure to grow at low temperatures but does not affect freeze-thaw tolerance. *J Food Prot.* 2010;73:1474–9.
37. Tasara T, Stephan R. Cold stress tolerance of *Listeria monocytogenes*: a review of molecular adaptive mechanisms and food safety implications. *J Food Prot.* 2006;69(6):1473–84.
38. Samara A, Koutsoumanis KP. Effect of treating lettuce surfaces with acidulants on the behaviour of Listeria monocytogenes during storage at 5°C and 20°C and subsequent exposure to simulated gastric fluid. *Int J Food Microbiol.* 2009;19:1–7.
39. Schmid B, Klumpp J, Raimann E, Loessner MJ, Stephan R, Tasara T. Role of cold shock proteins in growth of *Listeria monocytogenes* under cold and osmotic stress conditions. *Appl Environ Microbiol.* 2009;2009:1621–7.
40. Liu X, Basu U, Miller P, McMullen LM. Stress response and adaptation of *Listeria monocytogenes* 08-5923 exposed to a sublethal dose of Carnocyclin A. *Appl Environ Microbiol.* 2014;80:3835–41.
41. Zheng W, Kathariou S. Transposon-induced mutants of *Listeria monocytogenes* incapable of growth at low temperature (4°C). *FEMS Microbiol Lett.* 1994;121:287–92.
42. Liu S, Graham JE, Bigelow L, Morse PD, Wilkinson BJ. Identification of *Listeria monocytogenes* genes expressed in response to growth at low temperature. *Appl Environ Microbiol.* 2002;68:1697–705.
43. Dons L, Rasmussen OF, Olsen JE. Cloning and characterization of a gene encoding flagellin of *Listeria monocytogenes. Mol Microbiol.* 1992;6:2919–29.

44. Chan YC, Raengpradub S, Boor KJ, Wiedmann M. Microarray-based characterization of the *Listeria monocytogenes* cold regulon in log- and stationary-phase cells. *Appl Environ Microbiol.* 2007;73(20):6484–98.
45. Sergelidis D, Abrahim A. Adaptive response of *Listeria monocytogenes* to heat and its impact on food safety. *Food Control.* 2009;20(1):1–10.
46. Hu Y, Oliver HF, Raengpradub S, et al. Transcriptomic and phenotypic analyses suggest a network between the transcriptional regulators HrcA and σ^B in *Listeria monocytogenes. Appl Environ Microbiol.* 2007;73:7981–91.
47. Hu Y, Raengpradub S, Schwab U, et al. Phenotypic and transcriptomic analyses demonstrate interactions between the transcriptional regulators QsR and sigma B in *Listeria monocytogenes. Appl Environ Microbiol.* 2007;73(24):7967–80.
48. Dussurget O, Dumas E, Archambaud C, et al. *Listeria monocytogenes* ferritin protects against multiple stresses and is required for virulence. *FEMS Microbiol Lett.* 2005;250(2):253–61.
49. Hèbraud M, Guzzo J. The main cold shock protein of *Listeria monocytogenes* belongs to the family of ferritin-like proteins. *FEMS Microbiol Lett.* 2000;190(1):29–34.
50. Phan-Thank L, Gormon T. Analysis of heat and cold shock proteins in *Listeria* by two-dimensional electrophoresis. *Electrophoresis.* 1995;16:444–50.
51. Phan-Thank L, Gormon T. Stress proteins in *Listeria monocytogenes. Electrophoresis.* 1997;18:1464–71.
52. Jones TH, Vail KM, McMullen LM. Filament formation by foodborne bacteria under sublethal stress. *Int J Food Microbiol.* 2013;165(2):97–110.
53. Skandamis PN, Gounadaki AS, Geornaras I, Sofos JN. Adaptive acid tolerance response of *Listeria monocytogenes* strains under planktonic and immobilized growth conditions. *Int J Food Microbiol.* 2012;159(2):160–6.
54. Buchanan RL, Edelson SG. Culturing enterohemorrhagic *Escherichia coli* in the presence and absence of glucose as a simple means of evaluating the acid tolerance of stationary-phase cells. *Appl Environ Microbiol.* 1996;62(11):4009–13.
55. Rychli K, Grunert T, Ciolacu L, et al. Exoproteome analysis reveals higher abundance of proteins linked to alkaline stress in persistent *Listeria monocytogenes* strains. *Int J Food Microbiol.* 2016;218:17–26.
56. Soni KA, Nannapaneni R, Tasara T. The contribution of transcriptomic and proteomic analysis in elucidating stress adaptation responses of *Listeria monocytogenes. Foodborne Pathog Dis.* 2011;8(8):843–52.
57. Bowman JP, Chang KJL, Pinfold T, Ross T. Transcriptomic and phenotypic responses of *Listeria monocytogenes* strains possessing different growth efficiencies under acidic conditions. *Appl Environ Microbiol.* 2010;76:4836–50.
58. Phan-Thanh L, Jänsch L. Elucidation of mechanisms of acid stress in *Listeria monocytogenes* by proteomic analysis. *Microb Proteomics Funct Biol Whole Org.* 2005:49:75–88.
59. Wemekamp-Kamphuis HH, Wouters JA, De Leeuw PPLA, Hain T, Chakraborty T, Abee T. Identification of sigma factor σ^B-controlled genes and their impact on acid stress, high hydrostatic pressure, and freeze survival in *Listeria monocytogenes* EGD-e. *Appl Environ Microbiol.* 2004;70:3457–66.
60. Ryan S, Begley M, Gahan CGM, Hill C. Molecular characterization of the arginine deiminase system in *Listeria monocytogenes*: regulation and role in acid tolerance. *Environ Microbiol.* 2009;11(2):432–45.
61. Cotter PD, Gahan CG, Hill C. Analysis of the role of the *Listeria monocytogenes* F_0F_1-ATPase operon in the acid tolerance response. *Int J Food Microbiol.* 2000;60(2–3):137–46.
62. Cotter PD, Emerson N, Gahan CGM, Hill C. Identification and disruption of lisRK, a genetic locus encoding a two-component signal transduction system involved in stress tolerance and virulence in *Listeria monocytogenes. J Bacteriol.* 1999;181(21):6840–3.
63. Chen J, Jiang L, Chen Q, et al. *lmo0038* is involved in acid and heat stress responses and specific for *Listeria monocytogenes* lineages I and II, *and Listeria ivanovii. Foodborne Pathog Dis.* 2009;6(3):365–76.
64. Wonderling LD, Wilkinson BJ, Bayles DO. The *htrA* gene *of Listeria monocytogenes* 10403S is essential for optimal growth under stress conditions. *Appl Environ Microbiol.* 2004;70(4):1935–43.
65. Makariti IP, Printezi A, Kapetanakou AE, Zeaki N, Skandamis PN. Investigating boundaries of survival, growth and expression of genes associated with stress and virulence of *Listeria monocytogenes* in response to acid and osmotic stress. *Food Microbiol.* 2015;45(PB):231–44.
66. Gardan R, Cossart P, Labadie J. Identification of *Listeria monocytogenes* genes involved in salt and alkaline-pH tolerance. *Appl Environ Microbiol.* 2003;69(6):3137–43.
67. Giotis ES, Muthaiyan A, Natesan S, Wilkinson BJ, Blair IS, McDowell DA. Transcriptome analysis of alkali shock and alkali adaptation in *Listeria monocytogenes* 10403S. *Foodborne Pathog Dis.* 2010;7(10):1147–57.
68. Ells TC, Truelstrup Hansen L. Increased thermal and osmotic stress resistance in *Listeria monocytogenes* 568 grown in the presence of trehalose due to inactivation of the phosphotrehalase-encoding gene *treA. Appl Environ Microbiol.* 2011;77(19):6841–51.
69. Sleator RD, Gahan CGM, Hill C. A postgenomic appraisal of osmotolerance in *Listeria monocytogenes. Appl Environ Microbiol.* 2003;69:1–9.

70. Bae D, Liu C, Zhang T, Jones M, Peterson SN, Want C. Global gene expression of *Listeria monocytogenes* to salt stress. *J Food Prot.* 2012;75(5):906–12.
71. Esbelin J, Santos T, Hébraud M. Desiccation: an environmental and food industry stress that bacteria commonly face. *Food Microbiol.* 2018;69:82–8.
72. Burgess CM, Gianotti A, Gruzdev N, et al. The response of foodborne pathogens to osmotic and desiccation stresses in the food chain. *Int J Food Microbiol.* 2016;221:37–53.
73. Sleator RD, Hill C. A novel role for the LisRK two-component regulatory system in listerial osmotolerance. *Clin Microbiol Infect.* 2005;11(8):599–601.
74. Sleator RD, Wouters J, Gahan CGM, Abee T, Hill C. Analysis of the role of *OpuC*, an osmolyte transport system, in salt tolerance and virulence potential of *Listeria monocytogenes. Appl Environ Microbiol.* 2001;67:2692–8.
75. Elbein AD, Pan YT, Pastuszak I, Carroll D. New insights on trehalose: a multifunctional molecule. *Glycobiology.* 2003;13(4):17–27.
76. Angelidis AS, Smith GM. Three transporters mediate uptake of glycine betaine and carnitine by *Listeria monocytogenes* in response to hyperosmotic stress. *Appl Environ Microbiol.* 2003;69(2):1013–22.
77. Fraser KR, Harvie D, Coote PJ, O'Byrne CP. Identification and characterization of an ATP binding cassette L-carnitine transporter in *Listeria monocytogenes. Appl Environ Microbiol.* 2000;66(11):4696–704.
78. Sleator RD, Hill C. Bacterial osmoadaptation: the role of osmolytes in bacterial stress and virulence. *FEMS Microbiol Rev.* 2002;26(1):49–71.
79. Chassaing D, Auvray F. The *lmo1078* gene encoding a putative UDP-glucose pyrophosphorylase is involved in growth of *Listeria monocytogenes* at low temperature. *FEMS Microbiol Lett.* 2007;275(1):31–7.
80. Utratna M, Shaw I, Starr E, O'Byrne CP. Rapid, transient, and proportional activation of σ^B in response to osmotic stress in *Listeria monocytogenes. Appl Environ Microbiol.* 2011;77(21):7841–5.
81. Michel E, Stephan R, Tasara T. The *lmo0501* gene coding for a putative transcription activator protein in *Listeria monocytogenes* promotes growth under cold, osmotic and acid stress conditions. *Food Microbiol.* 2011;28(7):1261–5.
82. Burall LS, Simpson AC, Chou L, Laksanalamai P, Datta AR. A novel gene, *lstC*, of *Listeria monocytogenes* is implicated in high salt tolerance. *Food Microbiol.* 2015;48:72–82.
83. Burall LS, Laksanalamai P, Datta AR. *Listeria monocytogenes* mutants with altered growth phenotypes at refrigeration temperature and high salt concentrations. *Appl Environ Microbiol.* 2012;78(4):1265–72.
84. Olesen I, Vogensen FK, Jespersen L. Gene transcription and virulence potential of *Listeria monocytogenes* strains after exposure to acidic and NaCl stress. *Foodborne Pathog Dis.* 2009;6(6):669–80.
85. Bale MJ, Bennett PM, Beringer JE, Hinton M. The survival of bacteria exposed to desiccation on surfaces associated with farm buildings. *J Appl Bacteriol.* 1993;75(6):519–28.
86. Harmsen M, Lappann M, Knøchel S, Molin S. Role of extracellular DNA during biofilm formation by *Listeria monocytogenes. Appl Environ Microbiol.* 2010;76(7):2271–9.
87. Hingston PA, Stea EC, Knøchel S, Hansen T. Role of initial contamination levels, biofilm maturity and presence of salt and fat on desiccation survival of *Listeria monocytogenes* on stainless steel surfaces. *Food Microbiol.* 2013;36(1):46–56.
88. França MB, Panek AD, Eleutherio ECA. Oxidative stress and its effects during dehydration. *Comp Biochem Physiol Part A Mol Integr Physiol.* 2007;146(4):621–31.
89. García-Fontana C, Narváez-Reinaldo JJ, Castillo F, González-López J, Luque I, Manzanera M. A new physiological role for the DNA molecule as a protector against drying stress in desiccation-tolerant microorganisms. *Front Microbiol.* 2016;7:1–12.
90. Lungu B, Ricke SC, Johnson MG. Growth, survival, proliferation and pathogenesis of *Listeria monocytogenes* under low oxygen or anaerobic conditions: a review. *Anaerobe.* 2009;15:7–17.
91. Bredeche MF, Ehrlich SD, Michel B. Viability of rep recA mutants depends on their capacity to cope with spontaneous oxidative damage and on the *Dnak* chaperone protein. *J Bacteriol.* 2001;183:2165–71.
92. Fiorini F, Stefanini S, Valenti P, Chiancone E, De Biase D. Transcription of the *Listeria monocytogenes fri* gene is growth-phase dependent and is repressed directly by Fur, the ferric uptake regulator. Gene. 2008;410(1):113–21.
93. Fisher CW, Lee D, Dodge BA, Hamman KM, Robbins JB, Martin SE. Influence of catalase and superoxide dismutase on ozone inactivation of *Listeria monocytogenes. Appl Environ Microbiol.* 2000;66:1405–9.
94. Rea R, Hill C, Gahan CGM. *Listeria monocytogenes* PerR mutants display a small-colony phenotype, increased sensitivity to hydrogen peroxide, and significantly reduced murine virulence. *Appl Environ Microbiol.* 2005;71(12):8314–22.
95. Van Der Veen S, Abee T. Importance of *SigB* for *Listeria monocytogenes* static and continuous-flow biofilm formation and disinfectant resistance. *Appl Environ Microbiol.* 2010;76:7854–60.

96. Huang Y, Morvay AA, Shi X, Suo Y, Shi C, Knøchel S. Comparison of oxidative stress response and biofilm formation of *Listeria monocytogenes* serotypes 4b and 1/2a. *Food Control.* 2018;85:416–22.
97. Dieuleveux V, Van Der Pyl D, Chataud J, Gueguen M. Purification and characterization of anti-*Listeria* compounds produced by *Geotrichum candidum. Appl Environ Microbiol.* 1998;64(2):800–3.
98. Olsen KN, Larsen MH, Gahan CGM, et al. The Dps-like protein Fri of *Listeria monocytogenes* promotes stress tolerance and intracellular multiplication in macrophage-like cells. *Microbiology.* 2005;151:925–33.
99. Brenot A, King KY, Caparon MG. The PerR regulon in peroxide resistance and virulence of *Streptococcus pyogenes. Mol Microbiol.* 2005;55:221–34.
100. Hasset DJ, Cohen MS. Bacterial adaptation to oxidative stress: implication for pathogenesis and interaction with phagocytic cells. *Faseb J Off Publ Fed Am Soc Exp Biol.* 1989;3:2574–82.
101. Mols M, Abee T. Primary and secondary oxidative stress in *Bacillus. Environ Microbiol.* 2011;13(6):1387–94.
102. Rantsiou K, Mataragas M, Alessandria V, Cocolin L. Expression of virulence genes of *Listeria monocytogenes* in food. *J Food Saf.* 2012;32(2):161–8.
103. Mataragas M, Rovetto F, Bellio A, et al. Differential gene expression profiling of *Listeria monocytogenes* in Cacciatore and Felino salami to reveal potential stress resistance biomarkers. *Food Microbiol.* 2015;46:408–17.
104. D'Amico DJ, Druart MJ, Donnelly CW. 60-Day aging requirement does not ensure safety of surface-mold-ripened soft cheeses manufactured from raw or pasteurized milk when *Listeria monocytogenes* is introduced as a postprocessing contaminant. *J Food Prot.* 2008;71(8):1563–71.
105. Faleiro ML, Andrew PW, Power D. Stress response of *Listeria monocytogenes* isolated from cheese and other foods. *Int J Food Microbiol.* 2003;84(2):207–16.
106. Van der Veen S, Hain T, Wouters JA, et al. The heat-shock response of *Listeria monocytogenes* comprises genes involved in heat shock, cell division, cell wall synthesis, and the SOS response. *Microbiology.* 2007; 53(10):3593–607.
107. Baumann AR, Martin SE, Feng H. Power ultrasound treatment of *Listeria monocytogenes* in apple cider. *J Food Prot.* 2005;68(11):2333–40.
108. Mañas P, Pagán R, Raso J. Predicting lethal effect of ultrasonic waves under pressure treatments on *Listeria monocytogenes* ATCC 15313 by power measurements. *J Food Sci.* 2000;65(4):663–7.
109. Pagán R, Mañas P, Raso J, Condón S. Bacterial resistance to ultrasonic waves under pressure at nonlethal (manosonication) and lethal (manothermosonication) temperatures. *Appl Environ Microbiol.* 1999;65(1):297–300.
110. Piyasena P, Mohareb E, McKellar RC. Inactivation of microbes using ultrasound: a review. *Int J Food Microbiol.* 2003;87(3):207–16.
111. Guzel BH, Arroyo C, Condón S, Pagán R, Bayindirli A, Alpas H. Inactivation of *Listeria monocytogenes* and *Escherichia coli* by ultrasonic waves under pressure at nonlethal (manosonication) and lethal temperatures (manothermosonication) in acidic fruit juices. *Food Bioprocess Technol.* 2014;7(6):1701–12.
112. Fleischman GJ, Ravishankar S, Balasubramaniam VM. The inactivation of *Listeria monocytogenes* by pulsed electric field (PEF) treatment in a static chamber. *Food Microbiol.* 2004;21(1):91–5.
113. Mosqueda-Melgar J, Raybaudi-Massilia RM, Martín-Belloso O. Influence of treatment time and pulse frequency on *Salmonella enteritidis, Escherichia coli* and *Listeria monocytogenes* populations inoculated in melon and watermelon juices treated by pulsed electric fields. *Int J Food Microbiol.* 2007;117(2):192–200.
114. García D, Gómez N, Mañas P, Condón S, Raso J, Pagán R. Occurrence of sublethal injury after pulsed electric fields depending on the micro-organism, the treatment medium pH and the intensity of the treatment investigated. *J Appl Microbiol.* 2005;99(1):94–104.
115. Daher D, Le Gourrierec S, Pérez-Lamela C. Effect of high pressure processing on the microbial inactivation in fruit preparations and other vegetable based beverages. *Agriculture.* 2017;7(9):72.
116. Scolari G, Zacconi C, Busconi M, Lambri M. Effect of the combined treatments of high hydrostatic pressure and temperature on *Zygosaccharomyces bailii* and *Listeria monocytogenes* in smoothies. *Food Control.* 2015;47:166–74.
117. Erkmen O, Dogan C. Effects of ultra high hydrostatic pressure on *Listeria monocytogenes* and natural flora in broth, milk and fruit juices. Int J Food Sci Technol. 2004;39(1):91–7.
118. Préstamo G, Sanz PD, Fonberg-Broczek M, Arroyo G. High pressure response of fruit jams contaminated with *Listeria monocytogenes. Lett Appl Microbiol.* 1999;28(4):313–6.
119. Shahbaz HM, Yoo S, Seo B, et al. Combination of TiO_2-UV photocatalysis and high hydrostatic pressure to inactivate bacterial pathogens and yeast in commercial apple juice. *Food Bioprocess Technol.* 2016;9(1):182–90.
120. Evrendilek GA, Koca N, Harper JW, Balasubramaniam VM. High-pressure processing of Turkish white cheese for microbial inactivation. *J Food Prot.* 2008;71(1):102–8.
121. Patterson MF, Mackle A, Linton M. Effect of high pressure, in combination with antilisterial agents, on the growth of *Listeria monocytogenes* during extended storage of cooked chicken. *Food Microbiol.* 2011;28(8):1505–8.

122. Marcos B, Aymerich T, Monfort JM, Garriga M. High-pressure processing and antimicrobial biodegradable packaging to control *Listeria monocytogenes* during storage of cooked ham. *Food Microbiol.* 2008;25(1):177–82.
123. Mañas P, Pagán R. Microbial inactivation by new technologies of food preservation. *J Appl Microbiol.* 2005;98(6):1387–99.
124. Gayán E, Serrano MJ, Pagán R, Álvarez I, Condón S. Environmental and biological factors influencing the UV-C resistance of *Listeria monocytogenes.* Food Microbiol. 2015;46:246–53.
125. Gayán E, Condón S, Álvarez I. Biological aspects in food preservation by ultraviolet light: a review. *Food Bioprocess Technol.* 2014;7(1):1–20.
126. Horita CN, Baptista RC, Caturla MYR, Lorenzo JM, Barba FJ, Sant'Ana AS. Combining reformulation, active packaging and non-thermal post-packaging decontamination technologies to increase the microbiological quality and safety of cooked ready-to-eat meat products. *Trends Food Sci Technol.* 2018;72:45–61.

SUMMARY

Listeria monocytogenes is a foodborne bacterial pathogen that has evolved various mechanisms to withstand the different types of stress (e.g. acidic, alkaline, osmotic, heat, cold, desiccation, and oxidative stress) associated with food and food-processing environments. As *L. monocytogenes* stress response plays a key role in its survival in guts and on surface of food processing plants, proper understanding of the molecular mechanisms of its stress tolerance is vital for developing effective measures for its control in foods and its prevention in human infection. While many important findings on *L. monocytogenes* stress response have been made on broth models, there is a lack of such data on food substrates. In addition, the use of MS-based proteomic analyses may be necessary to uncover further insights on *L. monocytogenes* stress tolerance in foods and its effect on virulence.

Molecular Pathogenesis of *Aeromonas* Infection

12

Elena Mendoza-Barberá, Susana Merino, and Juan M. Tomás
University of Barcelona

Contents

12.1 INTRODUCTION

The genus *Aeromonas* encompasses ubiquitous, water-borne bacteria naturally found in several water sources and also in different types of food.[1] Although *Aeromonas* was initially placed in the family *Vibrionaceae*, currently it is considered a member of the family *Aeromonadaceae*.[2] Being Gram-negative, facultatively anaerobic, and chemo-organotrophic bacteria, with an optimal growing temperature of about 22°C–28°C, members of the *Aeromonadaceae* family are generally motile by polar flagellation, and capable of reducing nitrates to nitrites and catabolizing glucose and several carbohydrates to acids and often gases. Apart from the genera *Aeromonas*, *Tolumonas*, and *Oceanimonas*,[3] some authors also include the genera *Oceanisphaera* and *Zobellella* in the family *Aeromonadaceae*.[4]

One of the most significant advances in the taxonomic analyses was the introduction and application of genotypic methods. DNA hybridization studies revealed the existence of multiple hybridization groups (HGs) within each of the recognized mesophilic species (*A. hydrophila*, *A. sobria*, and *A. caviae*).[5] HGs may be either defined species or reference strains representing unassigned species, and the term "phenospecies" is used to refer to a single heterogeneous species containing multiple HGs.[6] Although there are a few more validly published species names in the genus *Aeromonas*, the second edition of *Bergey's Manual of Systematic Bacteriology* only recognizes the following 17[3]: *A. hydrophila* (HG1), *A. bestiarum* (HG2), *A. salmonicida* (HG3), *A. caviae* (HG4), *A. media* (HG5), *A. eucrenophila* (HG6), *A. sobria* (HG7), *A. veronii* (bv. *sobria* (HG8) and bv. *veronii* (HG10)), *A. jandaei* (HG9), *A. schubertii* (HG12), *A. trota* (HG14), *A. allosaccharophila* (HG15), *A. encheleia* (HG16), *A. popoffii* (HG17), *A. culicicola*, *A. simiae*, and *A. molluscorum*. The relative ease of assessing high-quality draft genomes these days provides opportunities for determining taxonomic relationships and discovering new taxa. This is particularly relevant in the greatly diverse *Aeromonas* genus, as it has been shown that the virulence of *Aeromonas* depends not only on the infection route and animal used as a model organism, but also on the bacterial strain.[7]

Aeromonas strains are predominantly pathogenic to poikilothermic animals including amphibians, fish, and reptiles, although they can also be associated with infections of birds and mammals. For instance, *A. hydrophila* and *A. jandaei* cause aeromoniasis in eels,[8] *A. salmonicida* induces systemic furunculosis in salmonid,[9] and *A. hydrophila* and *A. veronii* are responsible for similar diseases in carp, perch, salmon, and other fish.[10] In humans, aeromonads are implicated in several intestinal and extraintestinal diseases and syndromes ranging from relatively mild illnesses (e.g., acute gastroenteritis or superficial wound infections) to more complicated pathologies (e.g., respiratory tract or eye infections, and even life-threatening conditions including septicemia, necrotizing fasciitis, meningitis, or myonecrosis).[11–14]

Microbiological infection is a consequence of an interaction between pathogen and host. *Aeromonas* utilizes flagella, pili, and adhesins for attachment and entry into host cells,[15] siderophores and outer-membrane proteins (OMPs) for multiplication in host tissue, and capsule, S-layer, lipopolysaccharide, and porins for resistance to host defense mechanisms. Furthermore, *Aeromonas* employs type II, III, IV, and VI secretion systems[7,16–18] for the production of several extracellular proteins (e.g., cytotoxic and cytotonic enterotoxins, hemolysins, and various hydrolytic enzymes) that cause damage to host cells and eventual cell death.[15] In this regard, a significant proportion of the *A. hydrophila* isolates from chlorinated and non-chlorinated water contains genes related to cytotoxic activity,[19] whose expressions appear to be influenced by the prevailing environmental temperature.[20]

Despite the enterotoxic potential of some *Aeromonas* strains, there has been a lack of large epidemic outbreaks reported or adequate animal model for simulating the pathological conditions associated with disparate human infection outcome. Members of the *Aeromonas* genus, however, could pose a serious public health risk as many strains can grow and produce exotoxins at low temperatures,[21] as well as in the presence of high salt concentrations, which are widely used to extend the shelf-life and preserve the quality of fresh produce, seafood, and meat products.[22] In addition, the role of aeromonads as important human pathogens in natural disasters has been reinforced by studies in the aftermath of Hurricane Katrina in New Orleans[23] and the tsunami that struck Thailand in December 2004,[16] in which *Aeromonas* species were isolated in high quantity from floodwater samples or skin and soft tissue infections of the survivors, respectively.

12.2 SURFACE POLYSACCHARIDES

12.2.1 Capsule

The capsule is a structure that usually covers the outer membrane of the bacterial cell. It is highly hydrated (with approximately 95% water) and made up of repetitions of monosaccharides that are linked to each other by glycosidic bonds forming homo- or heteropolymers. Although aeromonads are generally described as non-capsulated, capsules have been identified in both motile and non-motile *Aeromonas*. In *A. hydrophila* AH-3 (serogroups O:11 and O:34), a capsule composed of D-glucose, L-rhamnose, D-mannose, D-mannuronic acid, and acetic acid heteropolymers is found.[24] In *A. salmonicida*, which is also capsulated, the capsule includes N-acetylmannosamine but lacks acetic acid.[25]

Because of its exposure to the environment, the capsule is thought to play a role in the colonization and virulence potential of many pathogenic bacteria. Indeed, the capsule has been shown to prevent phagocytosis, facilitate adherence, enhance interactions with other bacteria and host tissue, increase resistance to the complement system, and act as a barrier against hydrophobic toxins.[24,26,27] Specific studies have demonstrated that the capsule increases the invasiveness of both psychrophilic and mesophilic aeromonads in fish cell lines[28–31] and protects the bacterium from host immune responses by preventing complement-mediated killing and macrophage opsonization.[25,32,33]

A gene cluster for capsule production has been identified in *A. hydrophila* PPD134/91 (serogroup O:18).[34] It contains 13 genes organized in three regions. Genes from regions I and III are involved in capsule export, whereas region II genes are associated with capsule biosynthesis.[34]

12.2.2 Lipopolysaccharide (LPS)

LPS is a surface glycoconjugate unique to Gram-negative bacteria and a key elicitor of innate immune responses, ranging from local inflammation to disseminated sepsis. It is a major component of the outer leaflet of the outer membrane and consists of three domains linked to each other: the extremely variable O-specific polysaccharide (or O-antigen), the core oligosaccharide (OS), and the conserved and toxic lipid A component.[35]

The O-antigen, which is the most surface-exposed LPS moiety, mediates pathogenicity by protecting invading bacteria from serum complement killing and phagocytosis.[36] It is usually attached to a terminal residue of the outer core, and composed of OS polymers of repeating subunits of 1–6 sugars. The individual chains of these O-antigens vary in length (of up to 40 repeated units), and constitute the hydrophilic domain of the LPS molecule, as well as a major antigenic determinant of the Gram-negative cell wall. The chemical composition, structure, and antigenicity of the O-antigens differ widely among Gram-negative bacteria, giving rise to a large number of O-antigen groups or serogroups.[37] A total of 96 O-antigen serogroups have been identified in *Aeromonas* strains,[38,39] and there are still more to be discovered as many strains identified to date do not fit into an existing serogroup. Serogroup O:11 is responsible for severe infections in humans such as septicemia, meningitis, and peritonitis, and serogroup O:34 (the most common in mesophilic *Aeromonas*) is associated with wound infections in humans and outbreaks of septicemia in fish.[40] *Aeromonas* LPS molecules are mainly high heterogeneous mixtures of smooth LPS (S-LPS) with a varying proportion of rough LPS (R-LPS), which lacks the O-specific chain. These molecules show marked dissimilarities in the ability to induce oxidative burst in human granulocytes[41] and to activate the host complement system.[42] Also, in serogroups O:13, O:33, O:34, and O:44, S-LPS is abundant at 20°C and of high osmolarity, whereas R-LPS is common at 37°C and of low osmolarity, and this LPS thermoregulation appears to be linked to colonization.[29,43] When strains of *A. hydrophila* O:34 were grown at 20°C, the presence of their S-LPS was correlated with higher virulence in fish and mice than their R-LPS counterparts grown at higher temperatures.[44] S-LPS also helps protect bacteria from the bactericide effects of the non-immune serum thus avoiding cell lysis, since the complement component C3b binds to the long O-antigen chains far away from the membrane and is therefore unable to form the complement attack complex.[45]

Although O-antigens are highly variable as their chemical composition is very diverse, based on the determination or inference of some aeromonad O-antigen structures or compositions by chemical analysis or by bioinformatic analysis of the genome sequences. This is the case of *A. hydrophila* O:34,[46] *A. salmonicida* subsp. *salmonicida*,[47] the O:11-antigen of *A. hydrophila* LL1,[48] or *A. caviae* ATCC 15468.[49] In addition, the genes involved in O-antigen biosynthesis have been described in *A. hydrophila* strains PPD134/91 (serogroup O:18)[34] and AH-3 (serogroup O:34).[50] Like other clusters involved in polysaccharide biosynthesis, three classes of genes exist: genes involved in the biosynthesis of activated sugars, genes that encode glycosyltransferases, and genes whose products are necessary for the O-antigen translocation and polymerization. LPS biosynthesis begins at the cytosolic membrane, and the molecule is then transferred to the outer membrane, where it becomes surface-exposed. The O-antigen is synthesized as a lipid-linked glycan intermediate (in bacteria, undecaprenyl phosphate or Und-P) by a process that is remarkably similar to the biogenesis of lipid-linked OSs for protein N-glycosylation.[51] While other strains rely on WecA to transfer N-acetylglucosamine-1-phosphate from UDP-N-acetylglucosamine to Und-P, *A. hydrophila* AH-3 utilizes WecP instead, which transfers N-acetylgalactosamine.[52] The differences in substrate specificity (and also in membrane topology) between WecP and WecA suggest their distinct phylogenetic branches/origins. After this initial transfer reaction, specific glycosyltransferases move particular sugars onto the Und-P to create an O-antigen unit.

The most prominent activity of LPS is its ability to induce the inflammatory host response. In particular, the endotoxic properties of LPS are dependent on the release of lipid A from lysed bacteria, which can provoke a major systemic inflammation known as septic or endotoxic shock.[53] Lipid A is a highly conserved structure covalently linked to the polysaccharide complex. Representing the lipid component of LPS, lipid A makes up the hydrophobic, membrane-anchoring region. Its biological activity appears to depend on its conformation, which is determined by the glucosamine disaccharide, the PO_4 groups, the acyl chains, and the 3-*deoxy*-D-*manno*-octulosonic acid (Kdo)-containing inner core. In *Aeromonas* spp., like in other Gram-negative bacteria, lipid A induces B cell polyclonal activation and response to immunoglobulin M, provoking leukopenia, septic shock, hemorrhagic necrosis of tumors, diarrhea, or even death.[54,55]

Structurally, lipid A consists of a phosphorylated N-acetylglucosamine (NAG) dimer with six or seven saturated fatty acids attached. Some of these fatty acids are attached directly to the NAG dimer while others are esterified to the 3-hydroxy fatty acids characteristically present.

It has been shown that the lipid A components of *A. salmonicida* subsp. *salmonicida* contain three major molecules differing in their acylation patterns (tetra-, penta-, or hexa-acylated lipid A species).[56] The tetra-acylated lipid A structure contains a 3-(dodecanoyloxy)tetradecanoic acid at N-2′, a 3-hydroxytetradecanoic acid at N-2, and a 3-hydroxytetradecanoic acid at O-3. The penta-acylated lipid A molecule has a similar fatty acid distribution pattern but, additionally, carries a 3-hydroxytetradecanoic acid at O-3′. Lastly, in the hexa-acylated lipid A structure, a 3-hydroxytetradecanoic acid molecule at O-3′ is esterified with a secondary 9-hexadecenoic acid.

The last LPS domain, the core OS, is structurally divided into two regions of different sugar composition, as determined in *A. hydrophila* AH-3 (O:34): the inner and the outer core.[57] The inner core of *A. hydrophila* is highly similar to that of *A. salmonicida* A450,[58] suggesting that it is a conserved structure among different species. It is attached to the lipid A at the 6′ position of one NAG, and contains Kdo or a derivative residue (3-*glycero*-D-*talo*-octulosonic acid). The outer core, in contrast, provides an attachment site to the O-polysaccharide and shows more structural diversity than those of *A. salmonicida* subsp. *salmonicida* A450[58] and *A. hydrophila* AH-3.[59] The complete lipid A-core OS unit is translocated to the periplasmic face of the inner membrane by the MsbA transporter, which is a member of the glyco ATP-binding cassette (ABC) transporters superfamily requiring ATP hydrolysis.[60]

At the genetic level, three gene clusters associated with LPS core biosynthesis have been isolated in *A. hydrophila* AH-3 (O:34), whose regions 2 and 3 contain identical genes to *A. salmonicida* A450.[59] Furthermore, the complete genomics and proteomics of *A. salmonicida* have been established, showing a unique or prevalent LPS core type.[61]

12.2.3 Surface α-Glucan

A. hydrophila AH-3 α-glucan (D-glucose residues connected by α-1,4 linkages and sometimes branched by α-1,6 linkages) is a surface polysaccharide exported via WecP (also used by the O:34-antigen LPS) and ligated to the surface through the O:34-antigen polysaccharide ligase WaaL.[62] Despite the common use of the export and ligation systems, experimental studies demonstrate that α-glucan and O:34-antigen LPS are independent polysaccharides, as mutants lacking either one of the polysaccharides or both can be found.[62] In *Aeromonas*, α-glucan production may not have a significant role in cell adhesion but does seem to participate in biofilm formation.[62] In this regard, the role of *Aeromonas* surface α-glucan polysaccharide could be similar to that of some *Escherichia coli* exopolysaccharides, which are integral elements of biofilms and hold together the different protein, lipid, and polysaccharide components of these layers.[63]

12.2.4 S-Layers

Surface layers or S-layers are bi-dimensional crystalline protein arrays produced by a broad range of bacteria to form the outermost cell envelope and are composed of a single protein of 40–200 kDa, depending upon bacterial species,[64] that can be glycosylated in Gram-negative bacteria.[65] Once secreted, S-layer proteins auto-assemble to form a 5–10 nm thick paracrystalline structure that contains pores of identical size (2–8 nm in diameter) and morphology.[66] S-layers constitute one of the most abundant protein moieties in the cell and have been associated with several possible functions that relate to pathogenicity. Because of its exposition on the cell surface, it plays a major role in diverse biological functions such as adhesion, protection against complement and attack by phagocytes, antigenic properties, anchoring site for hydrolytic exoenzymes, bacteriophage receptor, and others.[67]

In *Aeromonas*, the S-layer is composed of a unique protein (VapA, sometimes called AshA in mesophilic aeromonads) that forms a tetragonal array complex on the surface of the strains that contain the *vapA* gene, covering the entire bacterial cell and constituting the predominant surface antigen.[68] Although the first finding of an S-layer (initially called A-layer) in *Aeromonas* came from *A. salmonicida*,[69] it has also been described in strains of *A. hydrophila* and *A. veronii* bv. *sobria* of LPS serogroups O:11, O:14, and O:81.[70,71] All these layers are morphologically similar but differ at the genetic and functional levels, probably carrying out different roles in pathogenicity.[72] Indeed, it has been shown that the presence of the S-layer in *A. hydrophila* (O:11) increases its capacity of adherence, contributing to the colonization of the intestinal mucosa and generating a major resistance to opsonophagocytosis.[69] The S-layer of *A. salmonicida*, on the other hand, has been shown to promote association with extracellular matrix proteins and macrophages, increasing protection against proteases and oxidative killing.[32,68]

All the S-layers of *Aeromonas* strains tested to date have an LPS molecule that contains O-antigen polysaccharides of homogeneous chain lengths, suggesting the possible implication of LPS in linking the

S-layer to the bacterial cell surface. In this regard, mutants of *A. salmonicida* lacking the S-LPS O-antigen have been found to be unable to assemble a functional S-layer and, in addition, mutations resulting in partial loss of the LPS O-antigen also lead to S-layer disruption.[73]

12.3 SECRETION SYSTEMS

To transport proteins across the cytoplasmic membrane to the cell surface or extracellular space, Gram-negative bacteria have developed six major secretion systems (Types I–VI). These mechanisms can be Sec-dependent (Types II and V) or Sec-independent (Types I, III, IV, and VI),[74] although an alternative Sec-independent pathway known as twin-arginine translocation system (Tat-system), which recognizes proteins containing two identical arginine residues in the signal sequence, is also employed to transport already folded proteins across the inner membrane.[75] Proteins secreted via the Sec-dependent pathway contain an N-terminal signal peptide and use the Sec translocase for transport across the cytoplasmic membrane, while Sec-independent pathways export proteins from the cytoplasm to the extracellular environment in one step, without the involvement of periplasmic intermediates. Of the six secretion systems molecularly characterized in Gram-negative bacteria, types II, III, IV, and VI have been described in *Aeromonas*.

12.3.1 Type II Secretion System

The type II secretion system (T2SS), conserved in most Gram-negative bacteria, is a multiprotein complex made of four subassemblies: the outer-membrane complex, the inner-membrane platform, the secretion ATPase, and the pseudopilus.[76] The outer-membrane complex serves as the channel through which folded periplasmic T2SS substrates are translocated. Since it is located in the outer membrane, proteins secreted through this channel must be first delivered to the periplasm via the Sec or Tat secretion pathways and therefore have a Sec- or Tat-type cleavable signal sequence at their N-termini.[76] This complex is composed of a multimeric protein called the secretin, which has a long N-terminus believed to extend to the periplasm to make contact with other T2SS proteins in the inner membrane.[77] The inner-membrane platform is embedded in the inner membrane but extends into the periplasm to coordinate the export of substrates by contacting the secretin, the pseudopilus, and the ATPase.[76] It is composed of multiple copies of at least four proteins, and it plays a critical role in the secretion process. In addition, it contains some structural proteins of the T2SS such as ExeA and ExeB, which have been found in *A. hydrophila*[78] and *A. salmonicida*.[79] The T2SS pseudopilus is structurally similar to proteins that comprise type IV pili on bacterial cell surfaces,[80] and the ATPase (located in the cytoplasm) is the energy-providing protein that powers the system.[76]

T2SS is present in all known members of *A. hydrophila*, and essential for the extracellular secretion of many virulence factors[16,18] such as the aerolysin-related cytotoxic enterotoxin Act, in *A. hydrophila* SSU, which is perhaps the most potent known virulence factor.[81]

12.3.2 Type III Secretion System

The type III secretion system (T3SS) is considered an important virulence factor because it transports bacterial proteins, frequently involved in pathogenicity, directly from the bacterial cytoplasm into the cytosol of host cells.[82,83] Once the effector proteins are inside the host cell, they normally disrupt the cell cytoskeleton and induce apoptosis.[82,84] Although the T3SS is considered a Sec-independent system, the assembly of the secretion apparatus probably requires the Sec machinery, as several components have the N-terminal signal sequences characteristic of proteins secreted through this pathway.[85]

The T3SS is a complex multicomponent system formed by three different types of proteins: structural components called injectisomes, secretion substrates known as effectors, and molecular chaperones that assist and protect structural and effector proteins during transport. The structural components of these systems are typically encoded in a few operons, which can be found either in pathogenicity islands in the bacterial chromosome or on plasmids. Because they are commonly subjected to horizontal gene transfer, bacteria that are evolutionary distinct may have closely related systems and vice-versa.[86]

The T3SS, sometimes called injectisome, comprises about 20 different proteins that assemble to form a needle-like structure that can be also broken down into three main components: a base complex or basal body, the needle component, and the translocon.[87] The base complex contains cytoplasmic components and spans the inner and outer membrane, forming a socket-like structure consisting of several rings with a center rod. A filament called the needle is encased by this socket-like structure and extends from it into the extracellular space. This filament has an inner hollow core that is wide enough to permit an unfolded effector to traverse.[88] The T3SS tip complex, on the outer end of the needle, is critical for sensing contact with host cells and regulating the secretion of effectors, and also necessary for the insertion of the translocon into the host cell membranes.[89] The T3SS translocon, in turn, is needed for the passage of effectors through the host cell membranes, but not for the secretion of effectors outside the bacterium.[90] Translocons are thought to assemble upon contact with host cells and form a pore that is essential for effector delivery.[83] More recently, however, an alternative two-step model of translocation of Type 3 effectors has been proposed. In this model, the effectors and translocon components are secreted before host cell contact and remain associated with the bacteria, perhaps in lipid vesicles. After contact with host cells, the translocon and tip proteins then form a pore through which the effectors pass.[91]

A functioning T3SS has been described in *A. salmonicida*[92] and *A. hydrophila* strains AH-1,[93] AH-3,[82] and SSU,[94] and appears to be similar to that of *Yersinia*.[82] Furthermore, five effector toxins have been identified in *Aeromonas*: AexT, AopP, AopO, and AopH in *A. salmonicida*[95–97] (AexT in *A. hydrophila* AH-3 as well[98]) and AexT-like (AexU) in *A. hydrophila* SSU.[99] AexT is a bifunctional toxin, homologous to the also bifunctional effectors ExoT/ExoS of *Pseudomonas aeruginosa*, showing ADP-ribosyltransferase and GTPase-acting protein activities. AopP belongs to the YopJ family, a group of T3SS effectors that interfere with signaling pathways of mitogen-activated protein kinases (MAPK) and/or the nuclear factor kappa B (NF-κB). Lastly, AopO (a serine/threonine kinase) and AopH (a tyrosine phosphatase) are homologs to the *Yersinia enterocolitica* effectors AopO and YopH, respectively.

12.3.3 Type IV Secretion System

The type IV secretion system (T4SS) delivers effector proteins or other macromolecules directly into the cytosols of eukaryotic target cells, to aid bacterial colonization and survival within host cells or tissues.[100,101] However, it is also capable of performing conjugal genetic transfer between bacteria by a contact-dependent process,[101] which plays a crucial role in the spread of antibiotic-resistance genes.[102] T4SS conjugal transfer has been shown in a strain of *A. caviae* isolated from a hospital effluent[103] and in the strain CECT 5761T of *A. culicicola* (a synonym of *A. veronii*) isolated from the midgut of the mosquito *Culex quinquefasciatus*.[102] In both cases, the T4SS genes are encoded in plasmids. Interestingly, sequence analysis of the *A. culicicola* T4SS showed it to be more similar to that of *E. coli* than to that of *A. caviae*, suggesting that *A. culicicola* might have acquired the T4SS from *E. coli*, which inhabits the mosquito midgut as well.

12.3.4 Type VI Secretion System

The type VI secretion system (T6SS) was first identified as a new secretion system in *P. aeruginosa*[104] and *Vibrio cholerae*[104,105] in 2006, and appears to constitute a phage-tail-spike-like injectisome with the potential to introduce effector proteins directly into the cytoplasm of host cells.[17]

T6SS is fairly well conserved in a wide range of Gram-negative bacterial species, where it translocate proteins into a variety of recipient cells, including eukaryotic cell targets and, more commonly, other bacteria. In addition, T6SS is capable of transporting effector proteins from one bacterium to another in a contact-dependent manner, which is believed to play a role in bacterial communication and interactions in the environment.[106]

T6SS is very large, with up to 21 proteins encoded within a contiguous gene cluster.[106] It has been proposed that some of these proteins, which are structural components of the T6SS apparatus, may also serve as effector proteins. That is the case of VgrG (valine-glycine repeat protein G) and Hcp (hemolysin-coregulated protein), whose encoding genes (*hcp* and *vgrG*) are often found in the proximity of genes that code for putative effector proteins[107] and Hcp has been shown to function as an antimicrobial pore-forming protein when secreted.[108] In addition, the genes of *A. hydrophila* SSU *vasH* (coding for the transcriptional regulator VasH) and *vasK* (coding for the helical transmembrane protein VasK) are essential for the expression of the T6SS genes, as *vasH* and *vasK* mutants result in decreased antiphagocytic activity and attenuated virulence.[17] However, T6SS is not crucial for *A. hydrophila* virulence, as some members of a recently described hypervirulent *A. hydrophila* pathotype of freshwater fishes have a complete T6SS while others retain only a few core components.[16,109] Other T6SS effector proteins have also been identified, many of them directed against the bacterial cell wall and membrane, supporting a role for T6SS in promoting interspecies bacterial competition.[106]

12.4 IRON-BINDING SYSTEMS

Iron is an essential element for the metabolic processes of almost all living bacteria. However, iron molecules are hardly accessible *in vivo* because most of them are bound to specialized iron-binding proteins such as hemoglobin, transferrin, lactoferrin, or ferritin.[110,111] Because of the low availability of iron in the environment, bacteria have developed a series of mechanisms to obtain protein-associated iron from their hosts. Iron acquisition is recognized as one of the key steps in the survival of bacterial pathogens within their hosts, and it has been shown to significantly contribute to virulence.[111,112]

Generally, iron-binding systems can be divided into two categories: siderophore-dependent and siderophore-independent.[113] Siderophores are low-molecular-weight chelators that microbes secrete into the environment under iron-limiting conditions. These peptides need specific cell membrane-bound receptors as well as a cell-associated system to incorporate the metal into the bacterial metabolism. Under iron limiting conditions, mesophilic *Aeromonas* can produce either enterobactin or a set of four bis-catecholate siderophores named amonabactin, but never both. In this genus, therefore, biosynthesis of these two molecules is encoded by two distinct gene groups: the *amo* genes, in strains that produce amonabactin, and the *aeb* genes (aeromonad enterobactin biosynthesis), in enterobactin hydroxypyroridone-producing strains.[114] In contrast to enterobactin, which is found in different Gram-negative bacteria, amonabactin is only known in *Aeromonas* spp.[115] In particular, the *A. hydrophila* receptor for this molecule has been shown to have low specificity, enabling the transport of an extraordinary wide range of siderophores with different chelating groups like catecholate, hydroxamate, and hydroxypyroridone.[116] Once iron is bound to the siderophore, the iron–siderophore complex associates with a siderophore receptor located on the outer membrane of the cell, and it is transported into the cytoplasm in an ATP-dependent manner. Then, in the cytoplasm, iron is dissociated from the iron–siderophore complex and is used for cellular processes.[110]

Some microbes, in addition, use siderophore-independent mechanisms to obtain host-produced iron-containing molecules, such as transferrin or heme, through specialized receptors. For instance, *A. salmonicida* has been shown to express three iron-regulated outer membrane receptors, which can bind to specific host heme-binding proteins without the intervention of siderophores, under iron-limited conditions.[117] In Gram-negative bacteria, heme uptake by siderophore-independent mechanisms usually involves not only outer membrane receptors but also a TonB-dependent internalization system by which heme iron is translocated into the cytoplasm, with the aid of ExbB and ExbD accessory proteins.[118,119] The energy generation system required for the transport of ferri-siderophores, and other iron sources specifically bound to the outer membrane receptors, is driven by ATP hydrolysis and implicates an ABC.[118] In this regard, in *A. hydrophila*, the TonB2 energy transduction system has been recently found to be implied in catecholamine growth promotion.[120]

The expression of genes involved in iron acquisition is tightly regulated by the ferric uptake regulator protein Fur, which acts as an iron-responsive DNA-binding repressor.[110,121] In this regard, iron-regulated proteins of *A. salmonicida* have been found to be protective antigens for fish and are good candidates for the improvement of vaccines.[122]

12.5 ADHESINS

Adhesins are the proteins responsible for the first contact between a host and a pathogen, which is a critical step in the infection process. In *Aeromonas*, two types of adhesins have been described: those associated with filamentous structures and those associated with proteins of the outer membrane.

12.5.1 Filamentous Adhesins: Fimbriae/Pili

Fimbriae/pili are filamentous extracellular structures formed by subunits known as pilin that allow bacteria to adhere to different surfaces. Although pili are often described as adhesive organelles, they have been implicated in other functions, such as phage binding, DNA transfer, biofilm formation, cell aggregation, host cell invasion, and twitching motility.[123] The pili of Gram-negative bacteria are divided into four main groups based on their assembly pathway: pili assembled by the chaperone-usher pathway, type IV pili, pili assembled by the extracellular nucleation/precipitation pathway, and pili assembled by the alternative chaperon-usher pathway.[123] Furthermore, the pili can also be classified according to their structural differences. In this sense, two distinct pili types have been described in *Aeromonas* strains collected from environmental samples: short, rigid (S/R) pili that are found in high numbers on the bacterial cell and long, wavy (L/W) pili, which are found in smaller numbers.[124]

S/R pili have a length of 0.6 to 2 µm and share common epitopes in some mesophilic *Aeromonas*[125] as well as in the psychrophilic species *A. salmonicida* subsp. *salmonicida*.[79] In *A. salmonicida*, S/R pili appear to play a role in the initial stages of colonization of Atlantic salmon, as *A. salmonicida* lacking the S/R pilus operon shows a reduced ability to adhere to the host's gastrointestinal tract and, once adhered, did not show any higher ability to invade the host than the wild type.[126] S/R pili predominate in aeromonads with elevated pili numbers and, in some clinical strains, they can be induced under certain environmental conditions.[127] On the other hand, L/W pili are large, fine, and flexible adhesins, also considered hemagglutinins, predominant in strains with a small number of pili. They measure 4–7 nm in length and are composed of pilins of between 19 and 23 KDa in size. Amino acid sequence analysis indicates that they correspond to type IV pili, which are important structures for adhesion to epithelial cells and are involved in biofilm formation and twitching motility.[128]

Two different type-IV pili have been described in gastroenteritis-associated *Aeromonas* species: the bundle-forming pili (Bfp) and the type IV pili (Tap). Bfp pili are involved in adhesion to intestinal cells and are considered major colonization factors in mesophilic *Aeromonas* spp. In this genus, the first isolated Bfp was that of *A. veronii* bv. *sobria*,[129] which has been shown by genetic means to be a member of the MSHA (mannose-sensitive hemagglutinin) pilus family.[130] Bfp is required for *A. veronii* adherence and biofilm formation, with both major and minor pilin proteins involved in this process.[130] Tap differs from Bfp in N-terminal sequences and molecular weights and exhibits the highest homology with the type IV pili of *Pseudomonas* and pathogenic *Neisseria* species.[128] Out of the four Tap pili identified in *A. hydrophila* (TapABCD), TapD has been shown to be essential for the secretion of many virulence factors, such as proteases, hemolysin, and DNase, contributing to type II secretion.[128,130]

In addition to the Bfp and Tap pili, a third type IV pilus has also been identified in *A. salmonicida* subsp. *salmonicida*, named Flp due to its homology with the Flp system of *Actinobacillus actinomycetemcomitans*.[131] The Flp pilus is widespread among bacteria and, although in *A. salmonicida* it does not appear to be involved in virulence, it has been shown to contribute to biofilm formation and autoagglutination.[131,132]

12.5.2 Non-Filamentous Adhesins: Outer-Membrane Proteins

Besides filamentous and other macromolecule adhesins, such as S-layers or polysaccharides, several OMPs have been identified in *Aeromonas* spp. as monomeric or non-filamentous adhesins. OMPs facilitate adherence of non-piliated aeromonads to host cells,[133] acting as adhesion enhancers for *Aeromonas* during colonization.[134]

Several OMPs annotated as potential invasins in the genome of *A. hydrophila* ATCC 7966T[125] are predicted to contain bacterial immunoglobulin (BIG)-like domains. In this regard, the intimin protein of *E. coli*[135] and the invasin protein of *Yersinia* spp.,[136] both BIG domain-containing OMPs, have been shown to be involved in adherence and invasion of host tissues, respectively. However, *Aeromonas* BIG-like proteins AigA and AigB show most sequence homology not to these proteins but to the BIG-domain-containing Lig lipoproteins of pathogenic *Leptospira* spp.[137,138] These lipoproteins, located on the bacterial cell surface, bind some extracellular matrix proteins such as fibronectin, collagen, and laminin and are essential for the colonization and pathogenesis of virulent *Leptospira*.[138] This suggests that *Aeromonas* Lig-like proteins may have a role in pathogenesis, which is supported by the fact that the production of Lig proteins in *Leptospira* spp. is characteristic of extremely pathogenic strains.[137]

As OMPs are located on the bacterial cell surface and extremely immunogenic, they represent potential vaccine candidates. For instance, the OmpW protein, widespread among bacteria, demonstrates vaccine potential against *V. cholerae*[139] and that of *A. hydrophila* is highly immunogenic in common carp.[140] The OmpG protein, which is highly conserved in aeromonads, provides some protection against *A. hydrophila* and *A. sobria* in European eels.[141]

12.6 SECRETED PATHOGENIC FACTORS

12.6.1 Exotoxins

A wide range of exotoxins are secreted by members of the *Aeromonas* genus, and many of them are considered to be important virulence factors. Not all *Aeromonas* strains produce all known toxins. However, some strains may have toxin-coding genes that they do not express or may express only under certain growth conditions.[142] Among other exotoxins, both cytotonic and cytotoxic enterotoxins have been described in *Aeromonas*.[84]

Cytotonic enterotoxins stimulate cyclic adenosine monophosphate (cAMP)-mediated sequences of events in cells. Although the mechanisms of action are similar to those of the choleric toxin,[143] the cytotonic enterotoxins produced by *Aeromonas* spp. show variable reactivity to the choleric antitoxin.[144] Cytotonic enterotoxins are divided into two groups: heat-labile, without cross-reactivity with the choleric antitoxin, and heat-stable, which react with the choleric antitoxin. Examples of these groups are the heat-labile cytotoxic enterotoxin Alt from *A. hydrophila* SSU, which shows sequence similarity to the C-terminus of *A. hydrophila* phospholipase C, and the heat-stable cytotonic enterotoxin Ast. Both enterotoxins have been found to increase the cAMP levels in the intestinal mucosa cells of rats.[145]

The other types of enterotoxins, the cytotoxic or cytolytic enterotoxins, are generally isolated from patients suffering from diarrhea.[145] These toxins are secreted as inactive precursors that undergo processing at both the N- and C-terminal ends to achieve the biological activity. Once active, these toxins bind to a glycoprotein on the surface of the target cell and after oligomerization, they form pores in the host's cell membrane causing cell death.[146] The pore-forming cytotoxic enterotoxin Act, from *A. hydrophila* SSU,[147] has been shown to play an important role in *Aeromonas* infections,[148] through involvement in several biological activities such as hemolysis, cytotoxicity, enterotoxicity, and lethality.[94,149] This enterotoxin shows significant homology with aerolysin, although some dissimilarities at the amino acid level could lead to different protein folding patterns and result in differential neutralization of these toxins by specific monoclonal antibodies.[150] Act is therefore an aerolysin-related toxin, which interacts with target cell membranes through the 3′-OH end of the membrane-constituent cholesterol molecules.[147] This interaction leads to Act activation, subsequent oligomerization, and pore formation.[151,152] In addition, Act has been shown to upregulate the production of proinflammatory cytokines, stimulating the production of iNOS (inducible nitric oxide synthase) which, in turn, through nitric oxide production, is an essential element of antimicrobial immunity and host-induced tissue damage. In *A. hydrophila*, Act activation also leads to the upregulation of genes encoding cyclooxygenase-2 (COX-2) and antiapoptotic protein Bcl-2, which prevent a possible massive apoptosis induced by the inflammatory host response. Moreover, it activates the arachidonic acid (AA) metabolism in macrophages by inducing group V secretory phospholipase A2 (sPLA2) in the membrane of these cells, where concentrations of AA are limited, leading to eicosanoid production and subsequent cAMP synthesis.[149]

Besides cytotoxic hemolytic enterotoxins, *Aeromonas* spp. strains produce at least two other classes of hemolysins without enterotoxic properties: α-hemolysins and β-hemolysins. The former are synthesized in the stationary growth phase and lead to reversible cytotoxic effects and incomplete erythrocytes lysis, while the latter are usually synthesized in the exponential growth phase. They are both thermostable and pore-forming toxins that lead to osmotic lysis and complete destruction of erythrocytes.[153]

12.6.2 Other Extracellular Enzymes

Aeromonas spp. secretes a wide range of extracellular enzymes, including proteases, metalloproteases, lipases, and collagenases, that are involved in pathogenesis. Although their role in pathogenicity is still to be determined in many cases, they show enormous potential to adapt to environmental changes. Extracellular proteases underscore the metabolic versatility that allows *Aeromonas* to persist in different habitats, facilitating ecological interactions with other organisms. In general, proteases contribute to pathogenicity by promoting invasion via direct damage of host tissue or by proteolytic activation of toxins.[22] In addition, they aid in the establishment of infection by overcoming the initial host defenses through inactivation of the complement system or provision of nutrients for cell proliferation.[154] In *Aeromonas* spp., three different types of proteases have been identified: a temperature-labile serine protease and two metalloproteases; both are temperature-stable but respectively sensitive and insensitive to EDTA (ethylenediaminetetraacetic acid).[155] Furthermore, different aminopeptidases have been described with distinctive specific activities, including the extracellular activation of aerolysin, a β-hemolysin precursor.[155]

Many bacteria also produce lipases or triacylglycerol hydrolases. Lipases, besides other diverse functions, act as pathogenic factors by interacting with human leukocytes or by affecting several immune system functions through the generation of free fatty acids.[156] In *A. hydrophila*, different lipases have been described such as Ah65 lipase/acyltransferase, H3, Lip, and Apl1, which demonstrate phospholipase C activity.[155,157] Particularly in species of serogroup O:34, phospholipase A1 and phospholipase C have been described as having lecithinase and cytotoxic activities, playing a role as a virulence factor.[158]

Finally, other extracellular enzymes such as glycerophospholipid-cholesterol acyltransferases (GCAT) are involved in *Aeromonas* pathogenesis. Isolated from *A. hydrophila* and *A. salmonicida*, these enzymes have been shown to digest and eventually lyse erythrocyte membranes.[155]

12.7 FLAGELLA

Bacterial flagella are complex systems that allow bacteria to move toward nutrients and favorable environments and away from resource-restricted locations (a process known as chemotaxis). Out of the different possible bacterial movements, swarming (for movement over surfaces, mainly through lateral flagella) and swimming (for motility in liquid media and linked to polar flagella) are correlated with the presence of flagella. In bacteria with a single polar flagellum, such as *Aeromonas*, the counter-clockwise rotation propels the cell forward in a run, whereas the clockwise rotation propels the cell backward with a concomitant random reorientation.[159] In general, the prokaryotic flagellum is composed of three main structural components: the basal body, (embedded in the bacterial surface), the filament and the hook, both of which are located extracellularly. The entire system is driven by a protein engine (the motor) located at the base of the basal body, on the inner cell membrane.[160]

12.7.1 Polar Flagella

Polar flagella allow members of the *Aeromonas* genus to be motile in liquid environments. Although this structure is constitutively expressed, some psychrophilic aeromonads show a temperature-regulated expression of polar flagellum. Such is the case of *A. salmonicida*, previously defined as aflagellated and non-motile, which has been shown to express an unsheathed polar flagellum at extremely low frequencies when grown at 30°C or 37°C, but at much higher frequencies (with subsequently increased movement) when grown at higher temperatures.[161] The aeromonad polar flagellum has two flagellin subunits (FlaA and FlaB) that are approximately 32 kDa in size and are exported as unfolded monomers via the T3SS. Following secretion, flagellin monomers are folded, and assemble into filaments assisted by the filament-capping protein FlaH.[162–165] Functional flagellin proteins comprise three main domains: a conserved N-terminal domain, required for export of the protein via the T3SS; a variable central D2/D3 antigenic domain, which is exposed to the outer surface and can be modified by glycosylation; and a very well conserved C-terminal domain that is required for recognition by the flagellin-specific chaperone FlaJ.[166]

Other notable components of the aeromonad polar flagellum are the flagellar motor components. Structurally, the polar flagellar motor is divided into two main parts: the rotor (the rotatory component) and the stator (the stationary component). The rotor is composed of proteins FliM, FliN, and FliG, which constitute the C ring structure at the base of the basal body, and the stator consists of membrane-embedded proteins surrounding the MS-ring that form proton or sodium ion channels and couple the flow of ions to flagellar rotation.[167] In the proton-driven motor of *E. coli* and *S. enterica* sv. *Typhimurium* (the most studied models of flagellum within Gram-negative bacteria), the stator is composed of two integral membrane proteins, MotA and MotB.[167,168] In *Aeromonas* spp., however, the polar flagella stator is more complex than those in *E. coli* and *Salmonella* and shows higher similarity to those in *Vibrio alginolyticus* and *Vibrio parahaemolyticus*.[169] Similar to the systems of these two species, the stator complex of *Aeromonas* is driven by sodium ions rather than protons and requires four proteins: PomA, PomB, MotX, and MotY. PomA and PomB are paralogs of the proton-driven proteins MotA and MotB, respectively,[170,171] and although MotX and MotY do not have paralogous proteins in *E. coli*, they form the T-ring component of the basal body in *Vibrio* spp.[172] Besides these proteins, mesophilic *Aeromonas* spp. such as *A. hydrophila* have two additional polar stator proteins named $PomA_2$ and $PomB_2$. The arranged PomA-B and PomA$_2$-B$_2$ sets are sodium-coupled stator complexes, albeit with different sodium sensitivities, and have been shown to be redundant since neither of them is required by itself for full motility.[171]

12.7.2 Lateral Flagella

Some aeromonads such as *A. hydrophila* and *A. caviae* express a single polar flagellum when grown in a liquid medium but also express a completely distinct system of peritrichous flagella when grown on solid or viscous media.[173] It has been shown that approximately 60% of mesophilic *Aeromonas* species possess dual flagellar systems, expressing these peritrichous lateral flagella for movement across a solid surface.[174,175] Lateral flagella filaments on aeromonads are composed of repeating lateral flagellin units known as LafA, which form unsheathed flagella filaments generated through the T3SS, in the same way as the polar flagella. Swarming motility over solid surfaces mediated by the lateral flagella is powered by the stator complex, which comprises proteins LafT and LafU, and it is thought to be proton-driven, unlike the sodium-driven polar flagella stator.[159,171] Most of the lateral flagella-containing mesophilic aeromonads possess a single lateral flagellin gene (*lafA*). However, the genome of *A. caviae* Sch3, as well as those of a few other aeromonad strains, harbors two lateral flagellin genes (*lafA1* and *lafA2*). In addition, the lateral flagella filament of *A. caviae* Sch3 comprises two distinct lateral flagellins known as LafA1 and LafA2.[175]

12.7.2 Flagellin Glycosylation

Protein glycosylation, the most abundant polypeptide chain modification in nature, was long thought to exist only in eukaryotes. The first bacterial and archaeal glycoproteins were discovered on the S-layers of the archaeon *Halobacterium salinarum*[176] and of two hyperthermophilic *Clostridium* species.[177] The glycosylation process generally entails the covalent attachment of glycans either to the amide nitrogen of Asn residues (*N*-glycosylation, first described in Campylobacter *jejuni*[178]) or to the hydroxyl oxygen of Ser, Thr, or Tyr residues (*O*-glycosylation, which occurs in all three domains of life[179]). In Gram-negative bacteria, the functionality of glycosylated flagellins is diverse. In some species, like *C. jejuni*, glycosylation is necessary for filament assembly[180] and plays an essential role in the intestine colonization process.[181] In other species, like *P. aeruginosa*, the absence of glycosylation affect neither flagellar assembly nor motility, but it is needed for its proinflammatory mechanisms.[182]

The structural characterization of the flagellins of some Gram-negative bacteria such as *P. aeruginosa*, *Helicobacter pylori*, *Clostridium botulinum*, and *C. jejuni*/*C. coli*, which have been shown to be glycosylated,[183] led to the identification of pseudaminic acid, a sugar of nine carbon atoms similar to *N*-acetylneuraminic acid or sialic acid. In aeromonads, O-glycosylation was first described in *A. caviae*, when it was observed that the polar flagellins FlaA and FlaB of this species showed aberrant migration as analyzed by SDS-PAGE.[165] Later studies have found that the purified flagellins FlaA and FlaB of *A. caviae* Sch3 are O-glycosylated at six or seven sites, respectively, with a non-ulosonate sugar derivate of 373 Da linked to the Ser or Thr residues of the flagellin central D2/D3 domain, predicted to be surface-exposed.[184] Glycosylation is required for flagellar assembly,[185] although different glycosylation patterns have been described within *Aeromonas* species. For instance, while *A. hydrophila* AH-3 has one polar and one lateral flagellin, both of them glycosylated, *A. hydrophila* AH-1 has one polar and two lateral flagellins, and only the polar flagellum is glycosylated.[186] In *A. hydrophila* AH-3, glycosylation of the lateral flagellin occurs at three flagellin sites and always by the addition of a single pseudaminic acid derivative of 376 Da, which is also used to modify the polar flagellin.[187,188] Moreover, a lipid carrier protein (WecX) is implied in this process.[189]

At the genetic level, the proteins encoded by the genes *flmA*, *flmB*, *neuA*, *flmD*, and *neuB* are needed not only for O-antigen LPS biosynthesis but also for flagellin glycosylation.[184] In addition, there is strong evidence on the involvement of motility accessory factor (Maf) proteins in the transfer of activated pseudaminic acid onto the flagellin monomers. For instance, the genes coding for Maf proteins are present in the flagellar glycosylation loci of bacteria that glycosylate their flagellins with non-ulosonic acids,[190] and genes coding for putative Maf proteins are required for flagellin glycosylation and formation of a fully functional flagellum in *A. caviae* and *A. hydrophila*.[187,190,191] Interestingly, a gene homologous to the *Campylobacter* Maf family of genes (named *maf5*) has been identified in the lateral flagella structural locus of *A. hydrophila*. The product of this gene is required for lateral flagella production,[162] similar to what it has been shown for Maf proteins of the polar system, whose mutants cause loss of the polar flagellin.[163,164]

12.7.3 Genetic Organization of Flagellar Systems

The genes involved in the biosynthesis of the polar flagella are similarly arranged in different species such as mesophilic *Aeromonas*, *Vibrio* spp., and *P. aeruginosa*.[164,169] In *A. hydrophila*, the polar flagellum and its regulatory proteins are encoded by more than 50 genes distributed in five chromosomal regions, although most of them are contained in just two loci (Figure 12.1). Region 1 comprises three gene clusters and contains genes coding for chemotaxis factors (*cheVR*) and structural proteins (*flgBCDEFGHIJKL*). It is similar to region 1 of *V. parahaemolyticus*, although the latter has three additional flagellin genes (*flaCDE*). Region 2 shows similar organization in *Vibrio* and *Aeromonas*, although two more genes (*flaF*, encoding a flagellin, and *flaI*, coding for a putative chaperone) can be found in the *Vibrio* genome. However, the *fli* genes of *Vibrio* are contained in region 2 while those of *Aeromonas* are located in region 3. *Aeromonas* region 2 also contains the *maf-1* gene, which is absent in *Vibrio* but has been reported in *H. pylori*, *Clostridium acetobutylicum*, and *C. jejuni*.[192] Interestingly, region 2 has been shown to be identical in *A. hydrophila* AH-3 and *A. caviae* Sch3, and it is also highly conserved among other *Aeromonas* species.[164,165] The aeromonad region 3 contains genes coding for proteins involved in flagellar biogenesis such as *fliEF*, *fliGHIJKLM*, *fliOPQR*, *flhAB*, and *flhFG*. In addition, it contains the *fliA* (σ28 factor), *pomAB* (which encode orthologs of the MotA and MotB motor proteins of *Pseudomonas*), and genes involved in chemotaxis. It shows similar organization to region 2 in *V. parahaemolyticus*, except for the genes *pomA* and *pomB*, which are absent in *Vibrio*. The next region, region 4, includes a gene that encodes the sodium-driven motor protein MotX in both *Vibrio* and *Aeromonas*. This protein, together with PomAB encoded in region 3, constitutes the motor of the polar flagella. Similarly, in *Pseudomonas*, this region contains genes that code for the motor proteins MotCD. Lastly, *Aeromonas* region 5 contains the master regulatory genes (*flrABC*). In *Vibrio* and *Pseudomonas*, instead, it contains a gene that codes for the motor protein MotY, which is found in a different region and with different behavior in *Aeromonas*.[170]

The complete set of genes involved in the formation of a functional and inducible lateral flagellar system has been described in two bacterial species: *V. parahaemolyticus* [193] and *A. hydrophila*.[162] The lateral flagella of these species are encoded by 38 genes arranged in a single chromosomal region in *A. hydrophila* but distributed in two discontinuous regions in *V. parahaemolyticus* (Figure 12.2). Besides the structural genes of region 1, *V. parahaemolyticus* contains the $motY_L$ gene in region 2. This gene, which encodes an outer-membrane motor protein similar to *MotYp* in *V. alginolyticus*,[194] is absent in *Aeromonas*. On the other hand, the *A. hydrophila* lateral flagella region contains the *maf5* gene, possibly involved in lateral flagellin glycosylation, which is absent in *V. parahaemolyticus*.[162–164]

Since the flagella biosynthesis requires the expression of many genes, it is metabolically very expensive and is therefore highly regulated by a hierarchical transcription cascade that involves transcriptional and post-translational mechanisms. The regulatory flagellar hierarchy has been studied in *A. hydrophila* AH-3 for both polar[195] and lateral[196] flagella. Like in other bacteria with polar flagellation such as *V. cholerae*[197] and *P. aeruginosa*,[198] the polar flagellum synthesis of *A. hydrophila* involves four transcriptional levels (gene classes I–IV), where each level is needed for the activation of the next one (Figure 12.3). Class I genes express proteins FlrA (in *A. hydrophila* and *V. cholerae*) and FleQ (in *P. aeruginosa*), which bind to σ^{54} factor and activate class II genes. FlrA is a major regulator of the polar flagellar system and *flrA* mutants have been shown indeed to be unable to produce polar flagella in *A. hydrophila* AH-3.[162] Class II

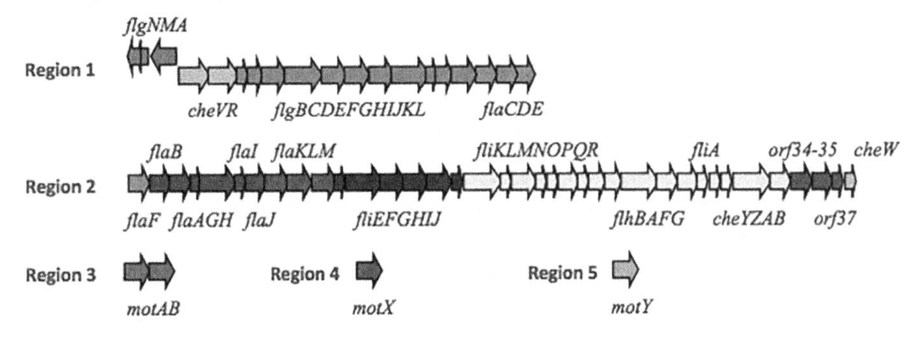

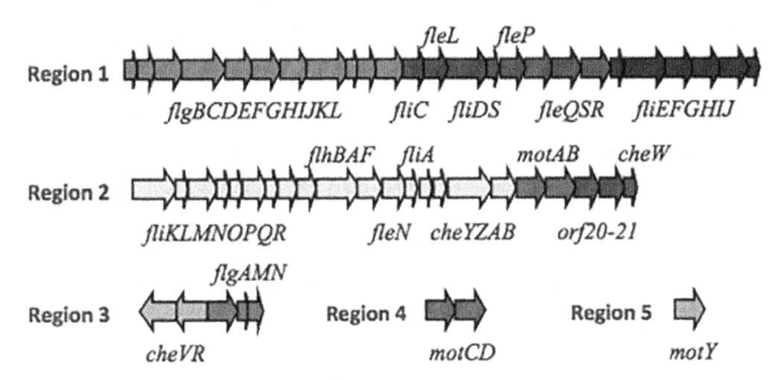

FIGURE 12.1 Genetic organization of polar flagellar genes in *A. hydrophila*, *V. parahaemolyticus*, and *P. aeruginosa*.[164,169]

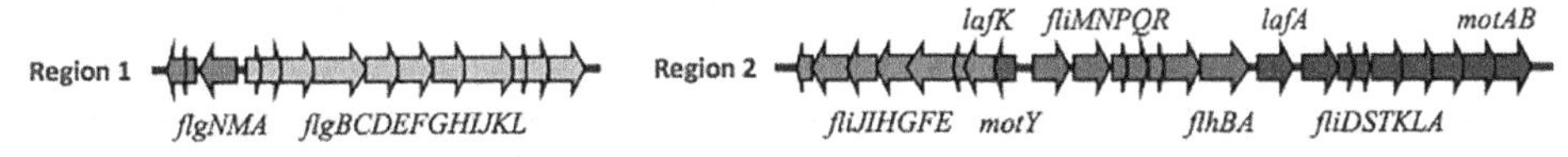

FIGURE 12.2 Genetic organization of lateral flagellar genes in *A. hydrophila* and *V. parahaemolyticus*.[162,193]

Aeromonas hydrophila

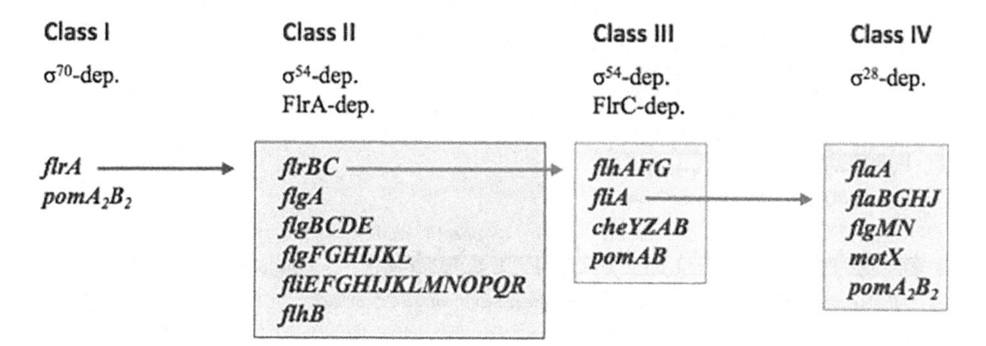

Pseudomonas aeruginosa

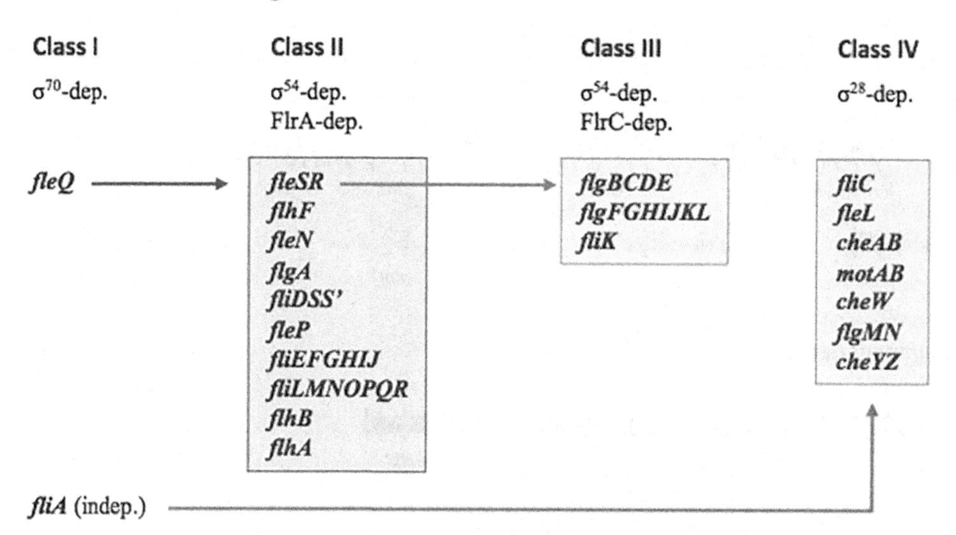

Vibrio cholerae

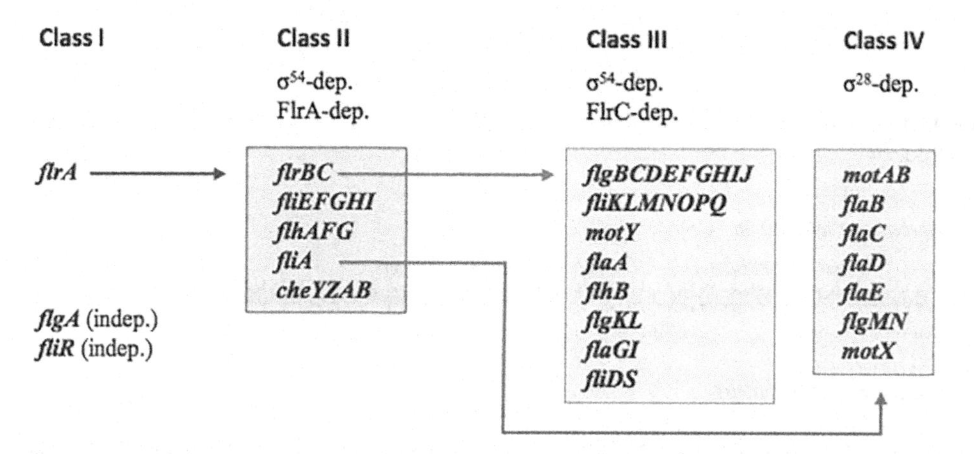

FIGURE 12.3 Comparison of *A. hydrophila, P. aeruginosa,* and *V. cholerae* polar flagellum gene transcription hierarchies.[197–199]

genes include most of those that code for flagellar structural proteins and also the genes that encode the enhancer-binding protein FlrC and its associated kinase FlrB. FlrBC proteins of both *A. hydrophila* and *V. cholerae*, and their similar counterparts in *P. aeruginosa* (FleSR), are associated with σ^{54} factor and are required for the expression of class III genes. This third class of genes includes those that code for the σ^{28} factor which, in turn, activates the late (class IV) flagellar genes. Class IV genes code for several filament proteins such as flagellins (*flaA* and *flaB*), the filament cap (*flaH*), and the flagellin-specific chaperone (*flaJ*) and also for the motor components and the antisigma factor FlgM.[199] FlgM, which is transcribed directly from class II genes, negatively regulates class IV promoters by binding to σ^{28} (FliA) and inhibiting its function. Once the structure of the basal body is complete, FlgM is exported along with the hook and basal body components and therefore frees σ^{28} to initiate the transcription of the class IV promoters.[200]

The *Aeromonas* lateral flagellar system, on the other hand, is regulated by LafK. This transcriptional regulator, which shows similarity to polar FlrA and FleQ proteins, is σ^{70}-dependent and associates with σ^{54} factor to activate the expression of class II genes. This second class of genes includes the gene coding for the σ^{28} factor, which in turn activates the late (class III) flagellar genes. Interestingly, it has been observed that the loss of the polar flagellar regulator FlaK can be compensated by the lateral regulator LafK, in *V. parahaemolyticus*.[201] However, in *A. hydrophila*, LafK is not able to compensate for the disruption of the polar flagellar regulator FlrA, despite the sequence similarity between these two proteins. In this regard, it has been suggested that FlrA and LafK may bind different DNA-binding sites and therefore fail to compensate for each other.[162,199]

In contrast to the constitutively expressed polar flagella genes, those required for lateral flagella biosynthesis are expressed only under certain conditions, mainly when bacteria are grown on solid or highly viscous media.[202] However, this system has an initial constitutive expression that can be explained by the fact that class I genes are σ^{70}-dependent, which is transcribed in all liquid, viscous, and solid media, but translated only in viscous or solid environment.[196]

12.7.4 Colonization, Motility, and Virulence Factor

Depending on the environmental conditions, bacteria can move freely or remain in the same place to form colony groups and colonize surfaces. As a group, bacteria can optimize growth and survival by the presence of different cell types that can perform specialized functions, therefore having better access to nutrients and better defense mechanisms for protection. In addition, bacteria in colonies secrete polysaccharides to form biofilms, which enhance adhesion, survival, and movement.

The flagellum structure is critical for successful colonization as, once bacteria have reached the host's mucosa, it becomes necessary for motility, adhesion, and invasion. The production of functional flagella is therefore a key point for *Aeromonas* pathogenesis, as the virulent role of motility has been reported in several pathogenic bacteria. Some examples include *H. pylori* and *P. aeruginosa*, where motility is crucial for the infection of stomach and lungs, respectively[203,204]; *V. cholerae*, where motility is necessary to colonize the intestinal mucosa[205]; *Proteus mirabilis*, in which swarming is associated with important urinary tract infections[206]; or *Y. enterocolitica*, where motility is associated with the invasion of epithelial cells.[207] Furthermore, flagella have been reported to act like adhesins, most probably via their flagellin D2/D3 domain.[208] For instance, the flagellin of *P. aeruginosa* can bind to the lungs mucine,[209] and some enteropathogenic strains of *E. coli* adhere to the intestinal mucosa via a flagellum-dependent mechanism.[210] Particularly in *Aeromonas*, motility and flagella are associated with biofilm formation, which generally goes along with persistent infections. In this regard, the polar[165] and lateral[175] flagella of *A. caviae*, and also glycosylation of the *A. hydrophila* polar flagella,[211] have been shown to be essential for adherence to HEp-2 cells. Glycosylation has been also shown to play a role in the proinflammatory action of flagellin in *P. aeruginosa*, as IL-8 release from A549 cells stimulated with non-glycosylated flagellin is significantly reduced when compared to wild-type flagellin,[182] and similar pathogenic mechanisms have been observed for *Aeromonas* polar[164] and lateral[162] flagella. In this regard, flagellin has also been shown to be a direct target of the innate immune system through TLR-5, leading to an increased production of IL-6 in mammals.[212] Also, in *E. coli*, *S. enterica* sv. *Typhimurium*, and *P. aeruginosa*, flagellins have been shown to stimulate the secretion of IL-8 in epithelial cells, attracting neutrophils and macrophages to the infection site.[208]

12.8 CONCLUSION

Aeromonas is a complex genus of Gram-negative bacteria that includes 36 recognized species, of which 19 are implicated in human infections, and 4 (*A. caviae, A. dhakensis, A. veronii, A. hydrophila*) account for approximately 95% of clinical isolations. Human aeromoniasis typically manifests as gastroenteritis, wound infection, bacteremia/sepsis, and other infections, depending largely on the immune status of affected individuals. Molecular pathogenesis of aeromonads lies in their production and use of surface polysaccharides (capsule, LPS, surface α-glucan, and S-layer), secretion systems, iron-binding systems, adhesins (pili/fimbriae and and outer membrane proteins), secreted pathogenic factors (exotoxins and other extracellular enzymes), and flagella (both polar and lateral) for motility, adherence, invasion, colonization, protection, and immune evasion. Extensive past efforts have uncovered many valuable insights on this emerging pathogen, which underline the development of improved identification and typing techniques. Given the current lack of effective vaccines against pathogenic Aeromonas spp., further research on the host-parasite interactions using state-of-art technologies is urgently required.

REFERENCES

1. Janda, J. M. & Abbott, S. L. The genus *Aeromonas*: taxonomy, pathogenicity, and infection. *Clin. Microbiol. Rev.* 23, 35–73 (2010).
2. Colwell, R. R., Macdonell, M. T. & De Ley, J. Proposal to recognize the family *Aeromonadaceae* fam. nov. *Int. J. Syst. Bacteriol.* 36, 473–7 (1986).
3. Martin-Carnahan, A. & Joseph, S. W. Order XII. *Aeromonadales.* in *Bergey's Manual® of Systematic Bacteriology* (eds. Brenner, D. J., Krieg, N. R. & Staley, J. T.) 2 Part B, 556–78 (Springer US, New York, 2005).
4. Huys, G. The family *Aeromonadaceae.* in *The Prokaryotes: Gammaproteobacteria* (eds. Rosenberg, E., DeLong, E. F., Lory, S., Stackebrandt, E. & Thompson, F.) 27–57 (Springer, Berlin, 2014).
5. Popoff, M. Y., Coynault, C., Kiredjian, M. & Lemelin, M. Polynucleotide sequence relatedness among motile *Aeromonas* species. *Curr. Microbiol.* 5, 109–14 (1981).
6. Carnahan, A. M. *Aeromonas* taxonomy: a sea of change. *Med. Microbiol. Lett.* 2, 206–11 (1993).
7. Yu, H. B. et al. Identification and characterization of putative virulence genes and gene clusters in *Aeromonas hydrophila* PPD134/91. *Appl. Environ. Microbiol.* 71, 4469–77 (2005).
8. Austin, B. et al. Characterization of atypical *Aeromonas salmonicida* by different methods. *Syst. Appl. Microbiol.* 21, 50–64 (1998).
9. Janda, J. M. Recent advances in the study of the taxonomy, pathogenicity, and infectious syndromes associated with the genus *Aeromonas*. *Clin. Microbiol. Rev.* 4, 397–410 (1991).
10. Joseph, S. W. & Carnahan, A. The isolation, identification, and systematics of the motile *Aeromonas* species. *Annu. Rev. Fish Dis.* 4, 315–43 (1994).
11. Janda, J. M. & Abbott, S. L. Evolving concepts regarding the genus *Aeromonas*: an expanding panorama of species, disease presentations, and unanswered questions. *Clin. Infect. Dis.* 27, 332–44 (1998).
12. Janda, J. M. & Abbot, S. L. The genus *Aeromonas.* in *Human Pathogens* (eds. Austin, B., Altwegg, M., Gosling, P. J. & Josheph, S.) 151–73 (John Wiley & Sons Ltd., Chichester, 1996).

13. Chen, P.-L., Lamy, B. & Ko, W.-C. *Aeromonas dhakensis*, an increasingly recognized human pathogen. *Front. Microbiol.* 7, 793 (2016).
14. Teunis, P. & Figueras, M. J. Reassessment of the enteropathogenicity of mesophilic *Aeromonas* species. *Front. Microbiol.* 7, 1395 (2016).
15. Gavín, R., Merino, S. & Tomas, J. M. Molecular mechanisms of bacterial pathogenesis from an emerging pathogen: *Aeromonas* spp. *Recent Res. Dev. Infection Immun.* 1, 337–54 (2003).
16. Pang, M. et al. Novel insights into the pathogenicity of epidemic *Aeromonas hydrophila* ST251 clones from comparative genomics. *Sci. Rep.* 5, 9833 (2015).
17. Suarez, G. et al. Molecular characterization of a functional type VI secretion system from a clinical isolate of *Aeromonas hydrophila*. *Microb. Pathog.* 44, 344–61 (2008).
18. Tseng, T.-T., Tyler, B. M. & Setubal, J. C. Protein secretion systems in bacterial-host associations, and their description in the Gene Ontology. *BMC Microbiol.* 9, Suppl 1, S2 (2009).
19. Ørmen, Ø & Østensvik, Ø. The occurrence of aerolysin-positive *Aeromonas* spp. and their cytotoxicity in Norwegian water sources. *J. Appl. Microbiol.* 90, 797–802 (2001).
20. Mateos, D., Anguita, J., Naharro, G. & Paniagua, C. Influence of growth temperature on the production of extracellular virulence factors and pathogenicity of environmental and human strains of *Aeromonas hydrophila*. *J. Appl. Bacteriol.* 74, 111–8 (1993).
21. Majeed, K. N., Egan, A. F. & Rae, I. C. Mac. Production of exotoxins by *Aeromonas* spp. at 5°C. *J. Appl. Bacteriol.* 69, 332–7 (1990).
22. Kirov, S. M. *Aeromonas* and *Plesiomonas* species. in *Food Microbiology: Fundamentals and Frontiers* (eds. Doyle, M. P., Beuchat, L. R. & Montville, T. J.) 265–87 (ASM Press, Washington, DC, 1997).
23. Presley, S. M. et al. Assessment of pathogens and toxicants in New Orleans, LA following Hurricane Katrina. *Environ. Sci. Technol.* 40, 468–74 (2006).
24. Martínez, M. J. et al. The presence of capsular polysaccharide in mesophilic *Aeromonas hydrophila* serotypes O:11 and O:34. *FEMS Microbiol. Lett.* 128, 69–73 (1995).
25. Garrote, A., Bonet, R., Merino, S., Simon-Pujol, M. D. & Congregado, F. Occurrence of a capsule in *Aeromonas salmonicida*. *FEMS Microbiol. Lett.* 74, 127–31 (1992).
26. Merino, S. & Tomás, J. M. Bacterial capsules and evasion of immune responses. in Pettis G. (ed) *eLS* 1–10 (John Wiley & Sons Ltd, New York, 2015).
27. Merino, S. et al. The role of the capsular polysaccharide of *Aeromonas hydrophila* serogroup O:34 in the adherence to and invasion of fish cell lines. *Res. Microbiol.* 148, 625–31 (1997).
28. Merino, S. et al. The role of the capsular polysaccharide of *Aeromonas salmonicida* in the adherence and invasion of fish cell lines. *FEMS Microbiol. Lett.* 142, 185–9 (1996).
29. Merino, S. et al. Mesophilic *Aeromonas* sp. serogroup O:11 resistance to complement-mediated killing. *Infect. Immun.* 64, 5302–9 (1996).
30. Merino, S., Rubires, X., Aguillar, A., Guillot, J. F. & Tomás, J. M. The role of the O-antigen lipopolysaccharide on the colonization in vivo of the germfree chicken gut by *Aeromonas hydrophila* serogroup O:34. *Microb. Pathog.* 20, 325–33 (1996).
31. Merino, S., Rubires, X., Aguilar, A. & Tomás, J. M. The role of flagella and motility in the adherence and invasion to fish cell lines by *Aeromonas hydrophila* serogroup O:34 strains. *FEMS Microbiol. Lett.* 151, 213–7 (2006).
32. Garduño, R. A., Thornton, J. C. & Kay, W. W. *Aeromonas salmonicida* grown in vivo. *Infect. Immun.* 61, 3854–62 (1993).
33. Merino, S., Albertí, S. & Tomás, J. M. *Aeromonas salmonicida* resistance to complement-mediated killing. *Infect. Immun.* 62, 5483–90 (1994).
34. Zhang, Y. L., Arakawa, E. & Leung, K. Y. Novel *Aeromonas hydrophila* PPD134/91 genes involved in O-antigen and capsule biosynthesis. *Infect. Immun.* 70, 2326–35 (2002).
35. Nikaido, H. Outer membrane. in *Escherichia coli and Salmonella: Cellular and Molecular Biology* (eds. Neidhardt, F. C., Curtiss III, R. & Neidhardt, R.) 29–47 (ASM Press, Washington, DC, 1996).
36. Joiner, K. A. Complement evasion by bacteria and parasites. *Annu. Rev. Microbiol.* 42, 201–30 (1988).
37. Raetz, C. R. H. Bacterial lipopolysaccharides: a remarkable family of bioactive macroamphiphiles. in *Escherichia coli and Salmonella: Cellular and Molecular Biology* (eds. Neidhardt, F. C., Curtiss III, R. & Ingraham, J. L.) 1, 1035–63 (ASM Press, Washington, DC, 1996).
38. Sakazaki, R. & Shimada, T. O-serogrouping scheme for mesophilic *Aeromonas* strains. *Jpn. J. Med. Sci. Biol.* 37, 247–55 (1984).
39. Thomas, L. V., Gross, R. J., Cheasty, T. & Rowe, B. Extended serogrouping scheme for motile, mesophilic *Aeromonas* species. *J. Clin. Microbiol.* 28, 980–4 (1990).

40. Janda, J. M., Abbott, S. L., Khashe, S., Kellogg, G. H. & Shimada, T. Further studies on biochemical characteristics and serologic properties of the genus *Aeromonas*. *J. Clin. Microbiol.* 34, 1930–3 (1996).
41. Kapp, A., Freudenberg, M. A. & Galanos, C. Induction of human granulocyte chemiluminescence by bacterial lipopolysaccharides. *Infect. Immun.* 55, 758–61 (1987).
42. Freudenberg, M. A. & Galanos, C. Metabolism of LPS in vivo. in *Bacterial Endotoxic Lipopolysaccharides, Immunopharmacology and Pathophysiology* (eds. Ryan, J. L. & Morrison, D. C.) 2, 275–94 (CRC Press, Boca Raton, FL, 1992).
43. Aguilar, A., Merino, S., Rubires, X. & Tomas, J. M. Influence of osmolarity on lipopolysaccharides and virulence of *Aeromonas hydrophila* serotype O:34 strains grown at 37 degrees C. *Infect. Immun.* 65, 1245–50 (1997).
44. Merino, S., Camprubí, S. & Tomás, J. M. Effect of growth temperature on outer membrane components and virulence of *Aeromonas hydrophila* strains of serotype O:34. *Infect. Immun.* 60, 4343–9 (1992).
45. Albertí, S. et al. Analysis of complement C3 deposition and degradation on *Klebsiella pneumoniae. Infect. Immun.* 64, 4726–32 (1996).
46. Knirel, Y. A., Shashkov, A. S., Senchenkova, S. N., Merino, S. & Tomás, J. M. Structure of the O-polysaccharide of *Aeromonas hydrophila* O:34; a case of random O-acetylation of 6-deoxy-L-talose. *Carbohydr. Res.* 337, 1381–6 (2002).
47. Wang, Z. et al. Structural and serological characterization of the O-chain polysaccharide of *Aeromonas salmonicida* strains A449, 80204 and 80204-1. *Carbohydr. Res.* 340, 693–700 (2005).
48. Dooley, J. S., Lallier, R., Shaw, D. H. & Trust, T. J. Electrophoretic and immunochemical analyses of the lipopolysaccharides from various strains of *Aeromonas hydrophila. J. Bacteriol.* 164, 263–9 (1985).
49. Wang, Z., Liu, X., Li, J. & Altman, E. Structural characterization of the O-chain polysaccharide of *Aeromonas caviae* ATCC 15468 lipopolysaccharide. *Carbohydr. Res.* 343, 483–8 (2008).
50. Jimenez, N. et al. The *Aeromonas hydrophila* wb*O34 gene cluster: genetics and temperature regulation. *J. Bacteriol.* 190, 4198–209 (2008).
51. Valvano, M. O Antigen biosynthesis. in *Comprehensive Natural Products II: Chemistry and Biology* (eds. Mander, L. & Liu, H.-W.) 6, 297–314 (Elsevier, Oxford, 2010).
52. Merino, S. et al. A UDP-HexNAc:polyprenol-P GalNAc-1-P transferase (WecP) representing a new subgroup of the enzyme family. *J. Bacteriol.* 193, 1943–52 (2011).
53. Lüderitz, O. et al. Chemical structure and biological activities of lipid A's from various bacterial families. *Naturwissenschaften* 65, 578–85 (1978).
54. Morrison, D. C. Bacterial endotoxins and pathogenesis. *Rev. Infect. Dis.* 5, Suppl 4, S733–47 (1983).
55. Merino, S., Rubires, X., Knochel, S. & Tomas, J. M. Emerging pathogens: *Aeromonas* spp. *Int. J. Food Microbiol.* 28, 157–68 (1995).
56. Wang, Z., Li, J. & Altman, E. Structural characterization of the lipid A region of *Aeromonas salmonicida* subsp. salmonicida lipopolysaccharide. *Carbohydr. Res.* 341, 2816–25 (2006).
57. Knirel, Y. A., Vinogradov, E., Jimenez, N., Merino, S. & Tomás, J. M. Structural studies on the R-type lipopolysaccharide of *Aeromonas hydrophila. Carbohydr. Res.* 339, 787–93 (2004).
58. Wang, Z., Li, J., Vinogradov, E. & Altman, E. Structural studies of the core region of *Aeromonas salmonicida* subsp. salmonicida lipopolysaccharide. *Carbohydr. Res.* 341, 109–17 (2006).
59. Jimenez, N. et al. Molecular analysis of three *Aeromonas hydrophila* AH-3 (serotype O34) lipopolysaccharide core biosynthesis gene clusters. *J. Bacteriol.* 190, 3176–84 (2008).
60. Cuthbertson, L., Kos, V. & Whitfield, C. ABC transporters involved in export of cell surface glycoconjugates. *Microbiol. Mol. Biol. Rev.* 74, 341–62 (2010).
61. Jimenez, N. et al. Genetics and proteomics of *Aeromonas salmonicida* lipopolysaccharide core biosynthesis. *J. Bacteriol.* 191, 2228–36 (2009).
62. Merino, S. et al. *Aeromonas* surface glucan attached through the O-antigen ligase represents a new way to obtain UDP-glucose. *PLoS One* 7, e35707 (2012).
63. Sutherland, I. W. Biofilm exopolysaccharides: a strong and sticky framework. *Microbiology* 147, 3–9 (2001).
64. Sleytr, U. B. & Beveridge, T. J. Bacterial S-layers. *Trends Microbiol.* 7, 253–60 (1999).
65. Messner, P. Prokaryotic glycoproteins: unexplored but important. *J. Bacteriol.* 186, 2517–9 (2004).
66. Fagan, R. P. & Fairweather, N. F. Biogenesis and functions of bacterial S-layers. *Nat. Rev. Microbiol.* 12, 211–22 (2014).
67. Beveridge, T. J. et al. Functions of S-layers. *FEMS Microbiol. Rev.* 20, 99–149 (1997).
68. Chu, S. et al. Structure of the tetragonal surface virulence array protein and gene of *Aeromonas salmonicida. J. Biol. Chem.* 266, 15258–65 (1991).

69. Kay, W. W. et al. Purification and disposition of a surface protein associated with virulence of *Aeromonas salmonicida*. *J. Bacteriol.* 147, 1077–84 (1981).
70. Kokka, R. P., Vedros, N. A. & Janda, J. M. Electrophoretic analysis of the surface components of autoagglutinating surface array protein-positive and surface array protein-negative *Aeromonas hydrophila* and *Aeromonas sobria*. *J. Clin. Microbiol.* 28, 2240–7 (1990).
71. Esteve, C. et al. Pathogenic *Aeromonas hydrophila* serogroup O:14 and O:81 strains with an S layer. *Appl. Environ. Microbiol.* 70, 5898–904 (2004).
72. Noonan, B. & Trust, T. J. The synthesis, secretion and role in virulence of the paracrystalline surface protein layers of *Aeromonas salmonicida* and *A. hydrophila*. *FEMS Microbiol. Lett.* 154, 1–7 (1997).
73. Griffiths, S. G. & Lynch, W. H. Characterization of *Aeromonas salmonicida* variants with altered cell surfaces and their use in studying surface protein assembly. *Arch. Microbiol.* 154, 308–12 (1990).
74. Kostakioti, M., Newman, C. L., Thanassi, D. G. & Stathopoulos, C. Mechanisms of protein export across the bacterial outer membrane. *J. Bacteriol.* 187, 4306–14 (2005).
75. Palmer, T. & Berks, B. C. Moving folded proteins across the bacterial cell membrane. *Microbiology* 149, 547–56 (2003).
76. Korotkov, K. V., Sandkvist, M. & Hol, W. G. J. The type II secretion system: biogenesis, molecular architecture and mechanism. *Nat. Rev. Microbiol.* 10, 336–51 (2012).
77. Korotkov, K. V., Gonen, T. & Hol, W. G. J. Secretins: dynamic channels for protein transport across membranes. *Trends Biochem. Sci.* 36, 433–43 (2011).
78. Li, G., Miller, A., Bull, H. & Howard, S. P. Assembly of the type II secretion system: identification of ExeA residues critical for peptidoglycan binding and secretin multimerization. *J. Bacteriol.* 193, 197–204 (2011).
79. Reith, M. E. et al. The genome of *Aeromonas salmonicida* subsp. salmonicida A449: insights into the evolution of a fish pathogen. *BMC Genomics* 9, 427 (2008).
80. Sauvonnet, N., Vignon, G., Pugsley, A. P. & Gounon, P. Pilus formation and protein secretion by the same machinery in Escherichia coli. *EMBO J.* 19, 2221–8 (2000).
81. Galindo, C. L. et al. *Aeromonas hydrophila* cytotoxic enterotoxin activates mitogen-activated protein kinases and induces apoptosis in murine macrophages and human intestinal epithelial cells. *J. Biol. Chem.* 279, 37597–612 (2004).
82. Vilches, S. et al. Complete type III secretion system of a mesophilic *Aeromonas hydrophila* strain. *Appl. Environ. Microbiol.* 70, 6914–9 (2004).
83. Galán, J. E. & Collmer, A. Type III secretion machines: bacterial devices for protein delivery into host cells. *Science* 284, 1322–8 (1999).
84. Austin, B. & Austin, D. A. *Bacterial Fish Pathogens: Disease of Farmed and Wild Fish.* (Springer, Dordrecht, 2007).
85. Hueck, C. J. Type III protein secretion systems in bacterial pathogens of animals and plants. *Microbiol. Mol. Biol. Rev.* 62, 379–433 (1998).
86. Gophna, U., Ron, E. Z. & Graur, D. Bacterial type III secretion systems are ancient and evolved by multiple horizontal-transfer events. *Gene* 312, 151–63 (2003).
87. Abrusci, P., McDowell, M. A., Lea, S. M. & Johnson, S. Building a secreting nanomachine: a structural overview of the T3SS. *Curr. Opin. Struct. Biol.* 25, 111–7 (2014).
88. Demers, J.-P. et al. High-resolution structure of the Shigella type-III secretion needle by solid-state NMR and cryo-electron microscopy. *Nat. Commun.* 5, 4976 (2014).
89. Picking, W. L. et al. IpaD of Shigella flexneri is independently required for regulation of Ipa protein secretion and efficient insertion of IpaB and IpaC into host membranes. *Infect. Immun.* 73, 1432–40 (2005).
90. Håkansson, S. et al. The YopB protein of *Yersinia pseudotuberculosis* is essential for the translocation of Yop effector proteins across the target cell plasma membrane and displays a contact-dependent membrane disrupting activity. *EMBO J.* 15, 5812–23 (1996).
91. Edgren, T., Forsberg, Å., Rosqvist, R. & Wolf-Watz, H. Type III secretion in *Yersinia*: injectisome or not? *PLoS Pathog.* 8, e1002669 (2012).
92. Burr, S. E., Stuber, K., Wahli, T. & Frey, J. Evidence for a type III secretion system in *Aeromonas salmonicida* subsp. salmonicida. *J. Bacteriol.* 184, 5966–70 (2002).
93. Yu, H. B. et al. A type III secretion system is required for *Aeromonas hydrophila* AH-1 pathogenesis. *Infect. Immun.* 72, 1248–56 (2004).
94. Sha, J. et al. The type III secretion system and cytotoxic enterotoxin alter the virulence of *Aeromonas hydrophila*. *Infect. Immun.* 73, 6446–57 (2005).
95. Burr, S. E., Stuber, K. & Frey, J. The ADP-ribosylating toxin, AexT, from *Aeromonas salmonicida* subsp. salmonicida is translocated via a type III secretion pathway. *J. Bacteriol.* 185, 6583–91 (2003).

96. Fehr, D. et al. AopP, a type III effector protein of *Aeromonas salmonicida*, inhibits the NF-κB signalling pathway. *Microbiology* 152, 2809–18 (2006).
97. Fehr, D. et al. *Aeromonas* exoenzyme T of *Aeromonas salmonicida* is a bifunctional protein that targets the host cytoskeleton. *J. Biol. Chem.* 282, 28843–52 (2007).
98. Vilches, S. et al. *Aeromonas hydrophila* AH-3 AexT is an ADP-ribosylating toxin secreted through the type III secretion system. *Microb. Pathog.* 44, 1–12 (2008).
99. Sha, J. et al. Further characterization of a type III secretion system (T3SS) and of a new effector protein from a clinical isolate of *Aeromonas hydrophila* – Part I. *Microb. Pathog.* 43, 127–46 (2007).
100. Backert, S. & Meyer, T. F. Type IV secretion systems and their effectors in bacterial pathogenesis. *Curr. Opin. Microbiol.* 9, 207–17 (2006).
101. Cascales, E. & Christie, P. J. The versatile bacterial type IV secretion systems. *Nat. Rev. Microbiol.* 1, 137–49 (2003).
102. Rangrez, A. Y. et al. Detection of conjugation related type four secretion machinery in *Aeromonas culicicola*. *PLoS One* 1, e115 (2006).
103. Rhodes, G. et al. Complete nucleotide sequence of the conjugative tetracycline resistance plasmid pFBAOT6, a member of a group of IncU plasmids with global ubiquity. *Appl. Environ. Microbiol.* 70, 7497–510 (2004).
104. Mougous, J. D. et al. A virulence locus of *Pseudomonas aeruginosa* encodes a protein secretion apparatus. *Science* 312, 1526–30 (2006).
105. Pukatzki, S. et al. Identification of a conserved bacterial protein secretion system in *Vibrio cholerae* using the Dictyostelium host model system. *Proc. Natl. Acad. Sci. U. S. A.* 103, 1528–33 (2006).
106. Russell, A. B., Peterson, S. B. & Mougous, J. D. Type VI secretion system effectors: poisons with a purpose. *Nat. Rev. Microbiol.* 12, 137–48 (2014).
107. De Maayer, P. et al. Comparative genomics of the Type VI secretion systems of *Pantoea* and *Erwinia* species reveals the presence of putative effector islands that may be translocated by the VgrG and Hcp proteins. *BMC Genomics* 12, 576 (2011).
108. Bingle, L. E., Bailey, C. M. & Pallen, M. J. Type VI secretion: a beginner's guide. *Curr. Opin. Microbiol.* 11, 3–8 (2008).
109. Rasmussen-Ivey, C. R. et al. Classification of a hypervirulent *Aeromonas hydrophila* pathotype responsible for epidemic outbreaks in warm-water fishes. *Front. Microbiol.* 7, 1615 (2016).
110. Crosa, J. H. Genetics and molecular biology of siderophore-mediated iron transport in bacteria. *Microbiol. Rev.* 53, 517–30 (1989).
111. Litwin, C. M. & Calderwood, S. B. Role of iron in regulation of virulence genes. *Clin. Microbiol. Rev.* 6, 137–49 (1993).
112. Braun, V. Bacterial iron transport related to virulence. in *Concepts in Bacterial Virulence* 12, 210–33 (Karger, Basel, 2004).
113. Byers, B. R., Massad, G., Barghouthi, S. & Arceneaux, J. E. Iron acquisition and virulence in the motile aeromonads: siderophore-dependent and- independent systems. *Experientia* 47, 416–8 (1991).
114. Massad, G., Arceneaux, J. E. & Byers, B. R. Diversity of siderophore genes encoding biosynthesis of 2,3-dihydroxybenzoic acid in *Aeromonas* spp. *Biometals* 7, 227–36 (1994).
115. Telford, J. R. & Raymond, K. N. Coordination chemistry of the amonabactins, bis(catecholate) siderophores from *Aeromonas hydrophila*. *Inorg. Chem.* 37, 4578–83 (1998).
116. Stintzi, A. & Raymond, K. N. Amonabactin-mediated iron acquisition from transferrin and lactoferrin by *Aeromonas hydrophila*: direct measurement of individual microscopic rate constants. *J. Biol. Inorg. Chem.* 5, 57–66 (2000).
117. Ebanks, R. O., Dacanay, A., Goguen, M., Pinto, D. M. & Ross, N. W. Differential proteomic analysis of *Aeromonas salmonicida* outer membrane proteins in response to low iron and in vivo growth conditions. *Proteomics* 4, 1074–85 (2004).
118. Stojiljkovic, I. & Perkins-Balding, D. Processing of heme and heme-containing proteins by bacteria. *DNA Cell Biol.* 21, 281–95 (2002).
119. Postle, K. & Larsen, R. A. TonB-dependent energy transduction between outer and cytoplasmic membranes. *BioMetals* 20, 453–65 (2007).
120. Dong, Y. et al. Catecholamine-stimulated growth of *Aeromonas hydrophila* requires the TonB2 energy transduction system but is independent of the amonabactin siderophore. *Front. Cell. Infect. Microbiol.* 6, 183 (2016).
121. de Lorenzo, V., Wee, S., Herrero, M. & Neilands, J. B. Operator sequences of the aerobactin operon of plasmid ColV-K30 binding the ferric uptake regulation (fur) repressor. *J. Bacteriol.* 169, 2624–30 (1987).

122. Hirst, I. D. & Ellis, A. E. Iron-regulated outer membrane proteins of *Aeromonas salmonicida* are important protective antigens in Atlantic salmon against furunculosis. *Fish Shellfish Immunol.* 4, 29–45 (1994).
123. Proft, T. & Baker, E. N. Pili in Gram-negative and Gram-positive bacteria – structure, assembly and their role in disease. *Cell. Mol. Life Sci.* 66, 613–35 (2009).
124. Ho, A. S., Mietzner, T. A., Smith, A. J. & Schoolnik, G. K. The pili of *Aeromonas hydrophila*: identification of an environmentally regulated 'mini pilin'. *J. Exp. Med.* 172, 795–806 (1990).
125. Seshadri, R. et al. Genome sequence of *Aeromonas hydrophila* ATCC 7966T: jack of all trades. *J. Bacteriol.* 188, 8272–82 (2006).
126. Dacanay, A., Boyd, J., Fast, M., Knickle, L. & Reith, M. *Aeromonas salmonicida* type I pilus system contributes to host colonization but not invasion. *Dis. Aquat. Organ.* 88, 199–206 (2010).
127. Kirov, S. M., Jacobs, I., Hayward, L. J. & Hapin, R. H. Electron microscopic examination of factors influencing the expression of filamentous surface structures on clinical and environmental isolates of *Aeromonas veronii* biotype sobria. *Microbiol. Immunol.* 39, 329–38 (1995).
128. Pepe, C. M., Eklund, M. W. & Strom, M. S. Cloning of an *Aeromonas hydrophila* type IV pilus biogenesis gene cluster: complementation of pilus assembly functions and characterization of a type IV leader peptidase/N-methyltransferase required for extracellular protein secretion. *Mol. Microbiol.* 19, 857–69 (1996).
129. Kirov, S. M. & Sanderson, K. Characterization of a type IV bundle-forming pilus (SFP) from a gastroenteritis-associated strain of *Aeromonas veroniibiovar* sobria. *Microb. Pathog.* 21, 23–34 (1996).
130. Hadi, N. et al. Bundle-forming pilus locus of *Aeromonas veronii* bv. sobria. *Infect. Immun.* 80, 1351–60 (2012).
131. Boyd, J. M. et al. Contribution of type IV pili to the virulence of *Aeromonas salmonicida* subsp. salmonicida in Atlantic salmon (*Salmo salar* L.). *Infect. Immun.* 76, 1445–55 (2008).
132. Planet, P. J., Kachlany, S. C., Fine, D. H., DeSalle, R. & Figurski, D. H. The widespread colonization island of *Actinobacillus actinomycetemcomitans*. *Nat. Genet.* 34, 193–8 (2003).
133. Rocha-De-Souza, C. M. et al. Identification of a 43-kDa outer-membrane protein as an adhesin in *Aeromonas caviae*. *J. Med. Microbiol.* 50, 313–9 (2001).
134. Torres, A. G., Vazquez-Juarez, R. C., Tutt, C. B. & Garcia-Gallegos, J. G. Pathoadaptive mutation that mediates adherence of shiga toxin-producing *Escherichia coli* O111. *Infect. Immun.* 73, 4766–76 (2005).
135. Luo, Y. et al. Crystal structure of enteropathogenic *Escherichia coli* intimin–receptor complex. *Nature* 405, 1073–7 (2000).
136. Hamburger, Z. A., Brown, M. S., Isberg, R. R. & Bjorkman, P. J. Crystal structure of invasin: a bacterial integrin-binding protein. *Science* 286, 291–5 (1999).
137. Cerqueira, G. M. et al. Distribution of the leptospiral immunoglobulin-like (lig) genes in pathogenic *Leptospira* species and application of ligB to typing leptospiral isolates. *J. Med. Microbiol.* 58, 1173–81 (2009).
138. Choy, H. A. et al. Physiological osmotic induction of *Leptospira interrogans* adhesion: LigA and LigB bind extracellular matrix proteins and fibrinogen. *Infect. Immun.* 75, 2441–50 (2007).
139. Das, M., Chopra, A. K., Cantu, J. M. & Peterson, J. W. Antisera to selected outer membrane proteins of *Vibrio cholerae* protect against challenge with homologous and heterologous strains of *V. cholerae*. *FEMS Immunol. Med. Microbiol.* 22, 303–8 (1998).
140. Maiti, B., Shetty, M., Shekar, M., Karunasagar, I. & Karunasagar, I. Evaluation of two outer membrane proteins, Aha1 and OmpW of *Aeromonas hydrophila* as vaccine candidate for common carp. *Vet. Immunol. Immunopathol.* 149, 298–301 (2012).
141. Guan, R., Xiong, J., Huang, W. & Guo, S. Enhancement of protective immunity in European eel (*Anguilla anguilla*) against *Aeromonas hydrophila* and *Aeromonas sobria* by a recombinant *Aeromonas* outer membrane protein. *Acta Biochim. Biophys. Sin.* 43, 79–88 (2011).
142. Sha, J., Lu, M. & Chopra, A. K. Regulation of the cytotoxic enterotoxin gene in *Aeromonas hydrophila*: characterization of an iron uptake regulator. *Infect. Immun.* 69, 6370–81 (2001).
143. Cahill, M. M. Virulence factors in motile *Aeromonas* species. *J. Appl. Bacteriol.* 69, 1–16 (1990).
144. Potomski, J., Burke, V., Robinson, J., Fumarola, D. & Miragliotta, G. *Aeromonas* cytotonic enterotoxin cross reactive with cholera toxin. *J. Med. Microbiol.* 23, 179–86 (1987).
145. Chopra, A. K. & Houston, C. W. Enterotoxins in Aeromonas-associated gastroenteritis. *Microbes Infect.* 1, 1129–37 (1999).
146. Asao, T., Kinoshita, Y., Kozaki, S., Uemura, T. & Sakaguchi, G. Purification and some properties of *Aeromonas hydrophila* hemolysin. *Infect. Immun.* 46, 122–7 (1984).
147. Ferguson, M. R. et al. Hyperproduction, purification, and mechanism of action of the cytotoxic enterotoxin produced by *Aeromonas hydrophila*. *Infect. Immun.* 65, 4299–308 (1997).
148. Xu, X. J. et al. Role of a cytotoxic enterotoxin in *Aeromonas*-mediated infections: development of transposon and isogenic mutants. *Infect. Immun.* 66, 3501–9 (1998).

149. Chopra, A. K. et al. The cytotoxic enterotoxin of *Aeromonas hydrophila* induces proinflammatory cytokine production and activates arachidonic acid metabolism in macrophages. *Infect. Immun.* 68, 2808–18 (2000).
150. Buckley, J. T. & Howard, S. P. The cytotoxic enterotoxin of *Aeromonas hydrophila* is aerolysin. *Infect. Immun.* 67, 466–7 (1999).
151. Sha, J., Kozlova, E. V & Chopra, A. K. Role of various enterotoxins in *Aeromonas hydrophila*-induced gastroenteritis: generation of enterotoxin gene-deficient mutants and evaluation of their enterotoxic activity. *Infect. Immun.* 70, 1924–35 (2002).
152. Parker, M. W., van der Goot, F. G. & Buckley, J. T. Aerolysin – the ins and outs of a model channel-forming toxin. *Mol. Microbiol.* 19, 205–12 (1996).
153. Thelestam, M. & Ljungh, A. Membrane-damaging and cytotoxic effects on human fibroblasts of alpha- and beta-hemolysins from *Aeromonas hydrophila. Infect. Immun.* 34, 949–56 (1981).
154. Leung, K. Y. & Stevenson, R. M. Tn5-induced protease-deficient strains of *Aeromonas hydrophila* with reduced virulence for fish. *Infect. Immun.* 56, 2639–44 (1988).
155. Pemberton, J. M., Kidd, S. P. & Schmidt, R. Secreted enzymes of *Aeromonas. FEMS Microbiol. Lett.* 152, 1–10 (1997).
156. Stehr, F., Kretschmar, M., Kröger, C., Hube, B. & Schäfer, W. Microbial lipases as virulence factors. *J. Mol. Catal. B Enzym.* 22, 347–55 (2003).
157. Chuang, Y. C., Chiou, S. F., Su, J. H., Wu, M. L. & Chang, M. C. Molecular analysis and expression of the extracellular lipase of *Aeromonas hydrophila* MCC-2. *Microbiology* 143, 803–12 (1997).
158. Merino, S. et al. Cloning, sequencing, and role in virulence of two phospholipases (A1 and C) from mesophilic *Aeromonas* sp. serogroup O:34. *Infect. Immun.* 67, 4008–13 (1999).
159. Merino, S. & Tomás, J. M. Lateral flagella systems. in *Pili and Flagella: Current Research and Future Trends* (ed. Jarrell, K. F.) 173–90 (Caister Academic Press, Norwich, 2009).
160. Terashima, H., Kojima, S. & Homma, M. Flagellar motility in bacteria: structure and function of flagellar motor. *Int. Rev. Cell Mol. Biol.* 270, 39–85 (2008).
161. Umelo, E. & Trust, T. J. Identification and molecular characterization of two tandemly located flagellin genes from *Aeromonas salmonicida* A449. *J. Bacteriol.* 179, 5292–9 (1997).
162. Canals, R. et al. Analysis of the lateral flagellar gene system of *Aeromonas hydrophila* AH-3. *J. Bacteriol.* 188, 852–62 (2006).
163. Canals, R. et al. The UDP N-acetylgalactosamine 4-epimerase gene is essential for mesophilic *Aeromonas hydrophila* serotype O34 virulence. *Infect. Immun.* 74, 537–48 (2006).
164. Canals, R. et al. Polar flagellum biogenesis in *Aeromonas hydrophila. J. Bacteriol.* 188, 3166 (2006).
165. Rabaan, A. A., Gryllos, I., Tomás, J. M. & Shaw, J. G. Motility and the polar flagellum are required for *Aeromonas caviae* adherence to HEp-2 cells. *Infect. Immun.* 69, 4257–67 (2001).
166. Smith, K. D. et al. Toll-like receptor 5 recognizes a conserved site on flagellin required for protofilament formation and bacterial motility. *Nat. Immunol.* 4, 1247–53 (2003).
167. Berg, H. C. The rotary motor of bacterial flagella. *Annu. Rev. Biochem.* 72, 19–54 (2003).
168. Macnab, R. M. How bacteria assemble flagella. *Annu. Rev. Microbiol.* 57, 77–100 (2003).
169. McCarter, L. L. Polar flagellar motility of the *Vibrionaceae. Microbiol. Mol. Biol. Rev.* 65, 445–62 (2001).
170. Molero, R., Wilhelms, M., Infanzon, B., Tomas, J. M. & Merino, S. *Aeromonas hydrophila* motY is essential for polar flagellum function, and requires coordinate expression of motX and Pom proteins. *Microbiology* 157, 2772–84 (2011).
171. Wilhelms, M. et al. Two redundant sodium-driven stator motor proteins are involved in *Aeromonas hydrophila* polar flagellum rotation. *J. Bacteriol.* 191, 2206–17 (2009).
172. Terashima, H., Fukuoka, H., Yakushi, T., Kojima, S. & Homma, M. The *Vibrio* motor proteins, MotX and MotY, are associated with the basal body of Na^+-driven flagella and required for stator formation. *Mol. Microbiol.* 62, 1170–80 (2006).
173. Shimada, T., Sakazaki, R. & Suzuki, K. Peritrichous flagella in mesophilic strains of *Aeromonas. Jpn. J. Med. Sci. Biol.* 38, 141–5 (1985).
174. Kirov, S. M. et al. Lateral flagella and swarming motility in *Aeromonas* species. *J. Bacteriol.* 184, 547–55 (2002).
175. Gavín, R. et al. Lateral flagella of *Aeromonas* species are essential for epithelial cell adherence and biofilm formation. *Mol. Microbiol.* 43, 383–97 (2002).
176. Mescher, M. F. & Strominger, J. L. Purification and characterization of a prokaryotic glycoprotein from the cell envelope of *Halobacterium salinarium. J. Biol. Chem.* 251, 2005–14 (1976).
177. Sleytr, U. B. & Thorne, K. J. Chemical characterization of the regularly arranged surface layers of *Clostridium thermosaccharolyticum* and *Clostridium thermohydrosulfuricum. J. Bacteriol.* 126, 377–83 (1976).

178. Szymanski, C. M., Yao, R., Ewing, C. P., Trust, T. J. & Guerry, P. Evidence for a system of general protein glycosylation in *Campylobacter jejuni*. *Mol. Microbiol.* 32, 1022–30 (1999).
179. Logan, S. M. Flagellar glycosylation – a new component of the motility repertoire? *Microbiology* 152, 1249–62 (2006).
180. Guerry, P. et al. Changes in flagellin glycosylation affect *Campylobacter* autoagglutination and virulence. *Mol. Microbiol.* 60, 299–311 (2006).
181. Szymanski, C. M., Logan, S. M., Linton, D. & Wren, B. W. *Campylobacter* – a tale of two protein glycosylation systems. *Trends Microbiol.* 11, 233–8 (2003).
182. Verma, A., Arora, S. K., Kuravi, S. K. & Ramphal, R. Roles of specific amino acids in the N terminus of *Pseudomonas aeruginosa* flagellin and of flagellin glycosylation in the innate immune response. *Infect. Immun.* 73, 8237–46 (2005).
183. Nothaft, H. & Szymanski, C. M. Protein glycosylation in bacteria: sweeter than ever. *Nat. Rev. Microbiol.* 8, 765–78 (2010).
184. Tabei, S. M. B. et al. An *Aeromonas caviae* genomic island is required for both O-antigen lipopolysaccharide biosynthesis and flagellin glycosylation. *J. Bacteriol.* 191, 2851–63 (2009).
185. Gryllos, I., Shaw, J. G., Gavín, R., Merino, S. & Tomás, J. M. Role of flm locus in mesophilic Aeromonas species adherence. *Infect. Immun.* 69, 65–74 (2001).
186. Fulton, K. M., Mendoza-Barberá, E., Twine, S. M., Tomas, J. & Merino, S. Polar glycosylated and lateral non-glycosylated flagella from *Aeromonas hydrophila* strain AH-1 (Serotype O11). *Int. J. Mol. Sci.* 16, 28255–69 (2015).
187. Canals, R. et al. Non-structural flagella genes affecting both polar and lateral flagella-mediated motility in *Aeromonas hydrophila*. *Microbiology* 153, 1165–75 (2007).
188. Wilhelms, M., Fulton, K. M., Twine, S. M., Tomás, J. M. & Merino, S. Differential glycosylation of polar and lateral flagellins in *Aeromonas hydrophila* AH-3. *J. Biol. Chem.* 287, 27851–62 (2012).
189. Merino, S. et al. *Aeromonas hydrophila* flagella glycosylation: involvement of a lipid carrier. *PLoS One* 9, e89630 (2014).
190. Parker, J. L., Day-Williams, M. J., Tomas, J. M., Stafford, G. P. & Shaw, J. G. Identification of a putative glycosyltransferase responsible for the transfer of pseudaminic acid onto the polar flagellin of *Aeromonas caviae* Sch3N. *Microbiologyopen* 1, 149–60 (2012).
191. Parker, J. L. et al. Maf-dependent bacterial flagellin glycosylation occurs before chaperone binding and flagellar T3SS export. *Mol. Microbiol.* 92, 258–72 (2014).
192. Karlyshev, A. V., Linton, D., Gregson, N. A. & Wren, B. W. A novel paralogous gene family involved in phase-variable flagella-mediated motility in *Campylobacter jejuni*. *Microbiology* 148, 473–80 (2002).
193. Stewart, B. J. & McCarter, L. L. Lateral flagellar gene system of *Vibrio parahaemolyticus*. *J. Bacteriol.* 185, 4508–18 (2003).
194. Okabe, M., Yakushi, T., Kojima, M. & Homma, M. MotX and MotY, specific components of the sodium-driven flagellar motor, colocalize to the outer membrane in *Vibrio alginolyticus*. *Mol. Microbiol.* 46, 125–34 (2002).
195. Raetz, C. R. H. & Whitfield, C. Lipopolysaccharide endotoxins. *Annu. Rev. Biochem.* 71, 635–700 (2002).
196. Wilhelms, M., Gonzalez, V., Tomás, J. M. & Merino, S. *Aeromonas hydrophila* lateral flagellar gene transcriptional hierarchy. *J. Bacteriol.* 195, 1436–45 (2013).
197. Syed, K. A. et al. The *Vibrio cholerae* flagellar regulatory hierarchy controls expression of virulence factors. *J. Bacteriol.* 191, 6555–70 (2009).
198. Dasgupta, N. et al. A four-tiered transcriptional regulatory circuit controls flagellar biogenesis in *Pseudomonas aeruginosa*. *Mol. Microbiol.* 50, 809–24 (2003).
199. Wilhelms, M., Molero, R., Shaw, J. G., Tomás, J. M. & Merino, S. Transcriptional hierarchy of *Aeromonas hydrophila* polar-flagellum genes. *J. Bacteriol.* 193, 5179–90 (2011).
200. Chilcott, G. S. & Hughes, K. T. Coupling of flagellar gene expression to flagellar assembly in *Salmonella enterica* serovar typhimurium and *Escherichia coli*. *Microbiol. Mol. Biol. Rev.* 64, 694–708 (2000).
201. Kim, Y.-K. & McCarter, L. L. Cross-regulation in *Vibrio parahaemolyticus*: compensatory activation of polar flagellar genes by the lateral flagellar regulator LafK. *J. Bacteriol.* 186, 4014–8 (2004).
202. Merino, S., Shaw, J. G. & Tomás, J. M. Bacterial lateral flagella: an inducible flagella system. *FEMS Microbiol. Lett.* 263, 127–35 (2006).
203. Kao, C.-Y., Sheu, B.-S. & Wu, J.-J. *Helicobacter pylori* infection: an overview of bacterial virulence factors and pathogenesis. *Biomed. J.* 39, 14–23 (2016).
204. Feldman, M. et al. Role of flagella in pathogenesis of *Pseudomonas aeruginosa* pulmonary infection. *Infect. Immun.* 66, 43–51 (1998).
205. Almagro-Moreno, S., Pruss, K. & Taylor, R. K. Intestinal colonization dynamics of *Vibrio cholerae*. *PLoS Pathog.* 11, e1004787 (2015).

206. Schaffer, J. N. & Pearson, M. M. *Proteus mirabilis* and urinary tract infections. *Microbiol. Spectr.* 3, UTI-0017-2013 (2015).
207. Young, G. M., Badger, J. L. & Miller, V. L. Motility is required to initiate host cell invasion by *Yersinia enterocolitica. Infect. Immun.* 68, 4323–6 (2000).
208. Ramos, H. C., Rumbo, M. & Sirard, J.-C. Bacterial flagellins: mediators of pathogenicity and host immune responses in mucosa. *Trends Microbiol.* 12, 509–17 (2004).
209. Lillehoj, E. P., Kim, B. T. & Kim, K. C. Identification of *Pseudomonas aeruginosa* flagellin as an adhesin for Mucl mucin. *Am. J. Physiol. Cell. Mol. Physiol.* 282, L751–6 (2002).
210. Girón, J. A., Torres, A. G., Freer, E. & Kaper, J. B. The flagella of enteropathogenic *Escherichia coli* mediate adherence to epithelial cells. *Mol. Microbiol.* 44, 361–79 (2002).
211. Merino, S., Wilhelms, M. & Tomás, J. M. Role of *Aeromonas hydrophila* flagella glycosylation in adhesion to Hep-2 cells, biofilm formation and immune stimulation. *Int. J. Mol. Sci.* 15, 21935–46 (2014).
212. Hayashi, F. et al. The innate immune response to bacterial flagellin is mediated by Toll-like receptor 5. *Nature* 410, 1099–103 (2001).

SUMMARY

Aeromonas species are ubiquitous, water-borne bacteria that are found in different water sources and various types of food. Being mostly infective to poikilothermic animals, *Aeromonas* spp. are also important pathogens of birds and mammals, including humans, in which they cause several intestinal and extraintestinal diseases. Among pathogenic factors reported from *Aeromonas* spp. are cell-surface polysaccharides (e.g., capsules, lipopolysaccharides, surface α-glucan, or S-layers), iron-binding systems for survival within their hosts (using siderophore-dependent and independent mechanisms), or exotoxins and other extracellular enzymes secreted through several secretion systems. Besides expressing colonization factors (e.g., fimbria, pili, and some non-filamentous adhesins), the motile members of *Aeromonas* can swim in liquid media with the help of a polar flagellum (which can also be glycosylated). More than half of the species have another lateral flagellar system that is expressed in viscous or solid media for swarming over surfaces. Both flagellar systems are associated with persistent infections and, along with some other above-mentioned factors, have been shown to significantly contribute to *Aeromonas* pathogenicity and virulence.

Molecular Identification of *Campylobacter* spp.

13

Dongyou Liu
Royal College of Pathologists of Australasia Quality Assurance Programs

Contents

13.1 INTRODUCTION

While *Campylobacter*-induced symptoms were first observed from affected children by Theodor Escherich in 1886, the causal agent has remained elusive for many decades, due mainly to the inability to isolate the culprit agent and also to the resemblance of this agent to other related microorganisms. With the establishment of the *Campylobacter* genus in 1963, the clinical relevance of this Gram-negative, spiral-shaped, non-spore-forming bacterium in the causation and development of diarrheal diseases (campylobacteriosis) was finally recognized in the 1970s. Subsequent studies have expanded the coverage of the genus *Campylobacter* to 33 species, of which *C. jejuni* and *C. coli* are responsible for almost 90% and 5% of clinical cases of human campylobacteriosis, respectively. Representing one of four key global causes of diarrheal diseases, *Campylobacter* spp. also play a role in immune mimicry, with serious consequences (e.g., Guillain–Barré syndrome or GBS and reactive arthritis or REA). For many years, diagnosis of human campylobacteriosis relied heavily on laboratory cultivation of *Campylobacter* bacteria followed by biochemical and serological characterization. Recent development and application of molecular techniques herald a game-change in the identification, typing, and phylogenetic analysis of *Campylobacter* bacteria and offer unprecedented opportunity to reveal the underlying mechanisms of *Campylobacter* pathogenicity and to design innovative countermeasures against campylobacteriosis.

TABLE 13.1 *Campylobacter* spp. implicated in human diseases

SPECIES	*NATURAL HOSTS*	*HUMAN DISEASES*
C. jejuni	Turkeys, chicken, waterfowl, sheep, cattle, dogs, cats, pigs	Gastroenteritis, bacteremia, meningitis, appendicitis, myocarditis, Guillain–Barré syndrome, reactive arthritis; 90% of cases
C. coli	Pigs, chicken, turkeys, sheep, cattle, dogs, cats	Gastroenteritis, abortion, bacteremia; 5% of cases
C. concisus	Humans	Inflammatory bowel disease, diarrhea, Barrett's esophagus, periodontal disease; rare
C. curvus	Humans	Gastroenteritis, abscess; rare
C. gracilis	Humans	Periodontal disease; rare
C. hominis	Humans	Septicemia; rare
C. rectus	Humans	Periodontal diseases, IBD; rare
C. showae	Humans	IBD; rare
C. ureolyticus	Humans	IBD, gastroenteritis, genital tract disease; rare
C. fetus subsp. *fetus*	Sheep, goats, cattle	Bacteremia, abortion, meningitis, abscess, gastroenteritis; rare
C. fetus subsp. *testudinum*	Sheep, goat, cattle	Leukemia; rare
C. hyointestinalis	Pigs, cattle, birds, tortoises, shellfish	Gastroenteritis; rare
C. insulaenigrae	Marine mammals	Gastroenteritis, bacteremia; rare
C. lari	Gulls, chicken	Bacteremia; rare
C. sputorum	Pigs	Gastroenteritis, abscess; rare
C. upsaliensis	Dogs, cats	Gastroenteritis, bacteremia; rare

Taxonomy. Forming one of the four genera (i.e., *Campylobacter*, *Arcobacter*, *Dehalospirilum*, and *Sulfurospirilum*) in the family *Campylobacteraceae*, order Campylobacterales, class Epsilonproteobacteria, and phylum Proteobacteria, the genus *Campylobacter* encompasses 33 species (and 13 subspecies) of Gram-negative bacteria (including *C. avium*, *C. blaseri*, *C. canadensis*, *C. coli*, *C. concisus*, *C. corcagiensis*, *C. cuniculorum*, *C. curvus*, *C. fetus*, *C. geochelonis*, *C. gracilis*, *C. helveticus*, *C. hepaticus*, *C. hominis*, *C. hyointestinalis*, *C. iguanorium*, *C. insulaenigrae*, *C. jejuni*, *C. lanienae*, *C. lari*, *C. mucosalis*, *C. ornithocola*, *C. peloridis*, *C. pinnipediorum*, *C. portucalensis*, *C. rectus*, *C. showae*, *C. sputorum*, *C. subantarcticus*, *C. troglodytis*, *C. upsaliensis*, *C. ureolyticus*, and *C. volucris*). Notably, several former members of the genus *Campylobacter* have been relocated to the genus *Helicobacter*, family *Helicobacteraceae* (i.e., *C. fennelliae* and *C. pyloris*), and the genus *Arcobacter*, family *Campylobacteraceae* (i.e., *C. butzleri*, *C. cryaerophila*, and *C. nitrofigilis*).

Among *Campylobacter* spp., *C. jejuni* comprises two subspecies: *C. jejuni* subsp. *jejuni* (simply called *C. jejuni)* and *C. jejuni* subsp. *doylei*, of which the former is much more common than the latter. Following the structural diversity of capsule polysaccharides (CPS) on the bacterial surface, *C. jejuni* is distinguished into 47 serotypes [1]. Clinically, *C. jejuni* is responsible for causing approximately 90% of human campylobacteriosis cases (mostly gastroenteritis and sometimes GBS, as a result of molecular mimicry between *C. jejuni* sialylated lipooligosaccharides and human nerve gangliosides). Furthermore, *C. coli* is implicated in about 5% of human campylobacteriosis cases [2]. Other *Campylobacter* species may be occasionally associated with human diseases (Table 13.1) [3,4].

Morphology. Morphologically, *Campylobacter* (meaning "curved rod" in Greek and Latin) is a spiral-shaped, "S"-shaped, or curved, non-spore-forming rod of 0.5–5 μm in length and 0.2–0.8 μm in width. Assisted by a single polar flagellum present at one or both ends of the cell, *Campylobacter* moves in a corkscrew-like motion (Figure 13.1) [5]. Electron microscopy reveals the presence of outer membrane,

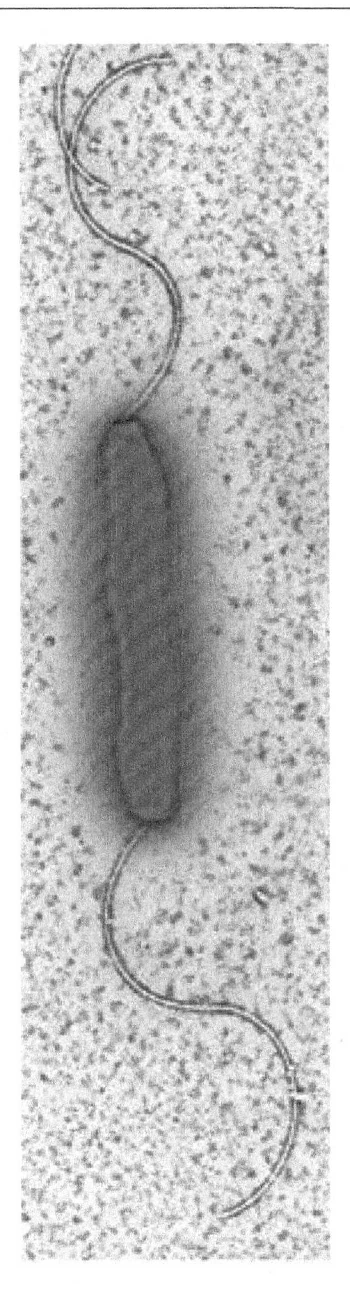

FIGURE 13.1 Electron micrograph of erythromycin-resistant *Campylobacter jejuni* strain 68-ER, a descendant of *C. jejuni* NCTC11168 resulting from *in vitro* step-wise selection by erythromycin, showing long, spiral, and complete flagella filaments on two sides. Photo credit: Hao H, Fang X, Han J, et al. *Front Microbiol.* 2017;8:729.

inner membrane, chemoreceptors, ribosomes, storage granules, nucleoid, and other structures within *C. jejuni* bacteria (Figure 13.2) [6].

Genome. The complete genome sequences of >100 *Campylobacter* species or strains are currently available in the NCBI database [7–9]. Specifically, *C. jejuni* strain BfR-CA-14430 from chicken meat possesses a chromosome of 1,645,980 bp, 1,665 coding sequences, and a plasmid of 41,772 bp that encodes for 46 genes, with an overall GC content of 30.4%. Notable genome features include the existence of genes encoding cytolethal distending toxins (*CdtA*, *CdtB*, and *CdtC*), fibronectin-binding

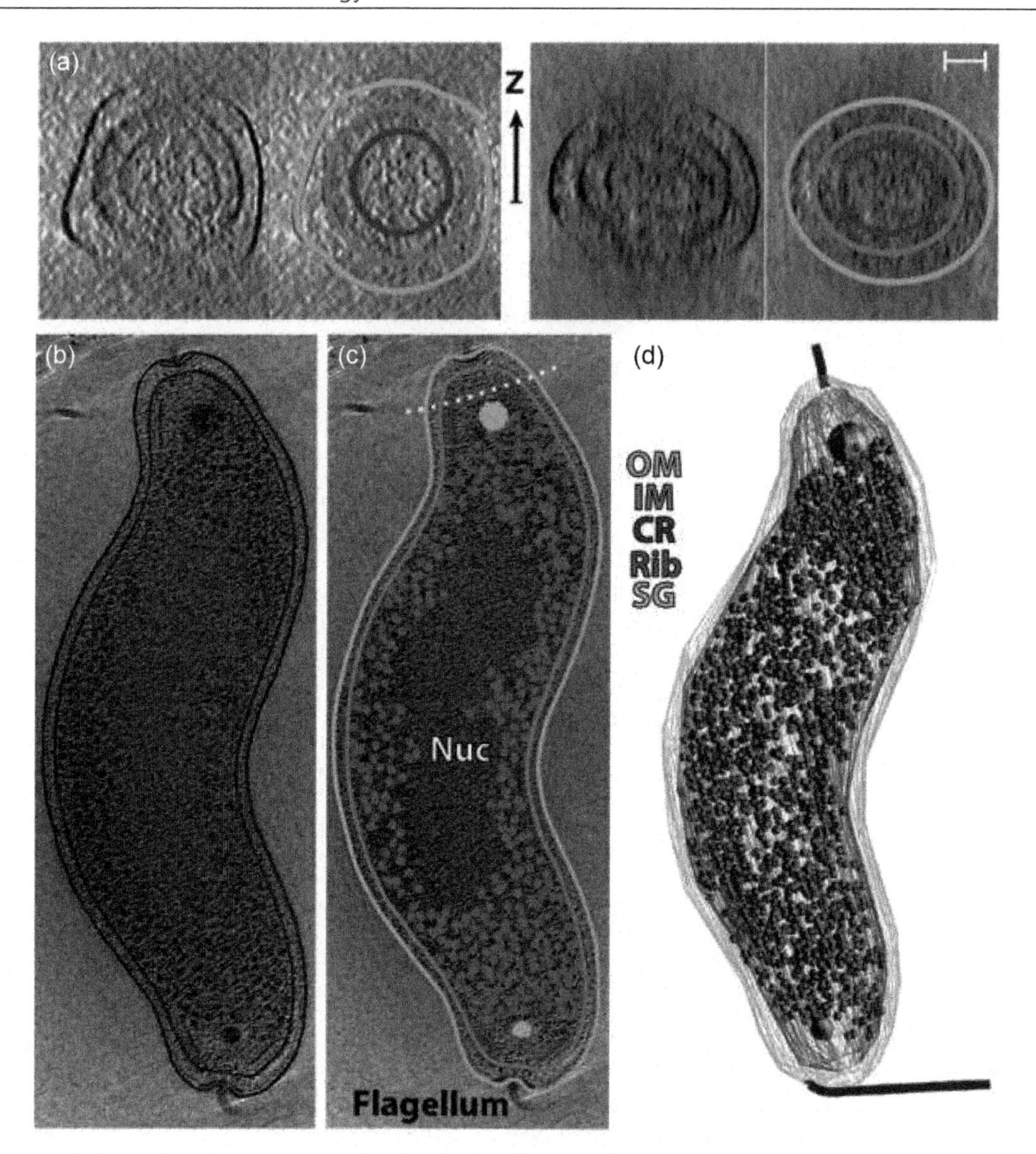

FIGURE 13.2 Electron micrographs revealing the ultrastructure of *Campylobacter jejuni*. (a–c) Duplicate images on the right highlight outer membrane (OM), inner membrane (IM), chemoreceptors (CR), ribosomes (Rib), storage granules (SG), and nucleoid (Nuc). (A) Two tomographic slices through a *C. jejuni* cell to show how the chemoreceptor arrays form a complete collar (see dotted line in [b] for approximate location), with the arrows indicating the direction of the electron beam along which resolution is attenuated. (b) Tomographic slice of a *C. jejuni* cell. (c) A model highlighting the various features of (b), noting the central and polar ribosome exclusion zones (REZ), and the variability in the thickness of the periplasm. (d) 3D model of *C. jejuni* based on the full tomograms from (b), displaying two membranes, flagella, storage granules, and ribosomes. The left model (c) has two poles, flagella, and polar REZs. Scale bar: 100 nm across a–d. (Photo credit: Müller A, Beeby M, McDowall AW, Chow J, Jensen GJ, Clemons WM Jr. *Microbiologyopen*. 2014;3(5):702–10.)

protein (CadF), *Campylobacter* invasion antigens (CiaB and CiaC), nonfunctional α-2,3-sialyltransferase (CstIII), *N*-acetylneuraminic acid biosynthesis proteins (NeuA1, NeuB1, and NeuC1, which are linked to GBS onset), capsule biosynthesis (*kpsC* and *kpsF*), and flagellar protein glycosylation (*pseA-I*). Among the genes carried in the 41.7 kb plasmid are *tetO* (for tetracycline resistance), *virB2-11*, and *virD4* (for a putative type IV secretion system or T4SS, as well as for conjugative DNA transfer between *Campylobacter* strains). The presence of a beta-lactam resistance gene (bla_{OXA-61}, equivalent to Cj0299 in *C. jejuni* strain NCTC 11168), plasmid-carried tetracycline resistance gene (*tetO*), and a point mutation in the DNA gyrase (*gyrA*, i.e., T86I)) in this strain helps explain for its apparent resistance to ampicillin, tetracycline, and (fluoro-)quinolones [10]. Similarly, *C. coli* strain YH502 from retail chicken harbors a chromosome of 1,718,974 bp and a mega-plasmid (pCOS502) of 125.9 kb, with an overall GC content of 31.2%, 1,931 coding sequences, and 53 non-coding RNAs. In addition to several virulence genes [e.g., *cdtA*, *cdtB*, and *cdtC* (cytolethal distending toxins)] in the chromosome, type VI secretion system [T6SS, as indicated by the presence of *vgrG* (valine-glycine repeats) and *hcp* (hemolysin correlated protein)] and antimicrobial resistance genes (beta-lactams, fluoroquinolones, and aminoglycoside) are found in the plasmid [11].

Biology. *Campylobacter* is essentially a microaerophilic and thermophilic bacterium with the ability to grow in 10% of CO_2 and 5% of O_2, at a temperature of 30°C–46°C (optimally 40°C–42°C) and at a water activity (a_w) above 0.987 (optimally 0.997). While *C. jejuni* survives for >4 h at 27°C, it does not seem to multiply outside animal hosts or in food during storage. Under unfavorable conditions, *Campylobacter* forms a biofilm on abiotic surfaces, which ensures a supply of nutrients and provides mechanical protection.

After ingestion of undercooked meat (particularly poultry), contaminated milk, or water by a host, *C. jejuni* passes through the stomach, reaches and crosses the mucus layer of the small intestine with the help of its polar flagellum, adheres and invades the intestinal epithelial cells, and then enters into the blood circulation and migrates to other parts of the body.

While most *Campylobacter* spp. reside in the gastrointestinal tract of domestic animals, wildlife, and human as commensal organisms, some can induce animal and human diseases. Interestingly, *C. jejuni* (naturally occurring in the avian intestine) and *C. coli* occur in chicken and cattle at a ratio of 9:1, but at a reversed ratio in pigs. Although *C. jejuni* and *C. coli* are considered animal-hosted species, they may cause acute bacterial gastroenteritis as well as GBS in humans via contaminated food (poultry 30%, cattle 20%–30%, and game) or water, with a minimal dose of 500–800 and an optimal dose of 9×10^4 *C. jejuni* bacteria to establish human infection [12]. Epidemiological surveys suggest that 60%–80% and 98% of chicken meat in the EU and in the United States, respectively, are contaminated with *Campylobacter* bacteria. *Campylobacter*-infected humans may also shed bacteria in stool for a period of 2–3 weeks (ranging from 3 days to several months), providing another potential source of contamination. The fact that *C. jejuni* is capable of surviving significantly longer in feces than *C. coli* may in part explain for a more common involvement of *C. jejuni* (90%) than *C. coli* (5%) in human campylobacteriosis [13]. Notable animal pathogens include *C. fetus* (the causative agent of bovine and ovine abortions), *C. hepaticus* (involved in spotty liver disease in chicken), *C. avium* (chicken and turkeys), *C. canadensis* (whooping crane), *C. cuniculorum* (rabbits), *C. subantarcticus* (albatross chicks and gentoo penguins), *C. troglodytis* (chimpanzees), and *C. volucris* (gulls) [14].

Epidemiology. *Campylobacter* spp. occur in a wide range of warm-blooded animals (e.g., poultry, cattle, pigs, sheep, ostriches, cats, and dogs) as well as shellfish [12,15]. As the most common bacterial cause of human gastroenteritis in the world, human campylobacteriosis is frequently encountered in Africa, Asia, the Middle East, North America, and Europe. The annual incidences of foodborne campylobacteriosis are estimated at 550 million globally (of which 220 million are under the age of 5 years), including 9 million (71 cases per 100,000) in the European Union and 2 million (14 cases per 100,000) in the USA [16].

Clinical features. Human campylobacteriosis has an incubation period of 2–3 days (range 1–10 days). Although most patients infected with *Campylobacter* (typically *C. jejuni* and *C. coli*) are asymptomatic, some may present with a range of symptoms, from mild, non-inflammatory, self-limiting diarrhea (three or more loose or liquid stools per day, lasting 24–48 h), abdominal pain, fever, headache, nausea, and/or vomiting, to severe inflammatory bloody diarrhea.

In children, *C. jejuni* and *C. coli* infections often result in diarrhea (bloody stools in 50% of cases), abdominal pain, vomiting, fever, dehydration, seizures, meningismus, and encephalopathy as well as bacteremia (primarily in immunocompromised individuals) [17].

In pregnant women, *C. jejuni* infection may lead to abortion, premature labor, and/or neonatal septicemia and meningitis.

Major complications linked to human campylobacteriosis include GBS (a polio-like form of paralysis characterized by a progressive symmetrical weakness in the limbs, with or without hyporeflexia, and possible respiratory and severe neurological dysfunction; with 30%–40% of cases preceded by *C. jejuni* infection), REA (painful inflammation of the knees and ankles, occurring about 11 days after the onset of diarrhea and lasting for 21 days or more without long-term sequelae), and irritable bowel syndrome. Other sequelae range from bacteremia, bursitis, osteitis, soft tissue infections; cholecystitis, hepatitis, pancreatitis, peritonitis (in patients with peritoneal dialysis), septic pseudoaneurysm, pericarditis, myocarditis; erythema nodosum, glomerulonephritis, hemolytic anemia, IgA nephropathy, intestinal perforation, to postinfectious inflammatory bowel disease (IBD) [18].

Pathogenesis. A diversity of mechanisms are employed by *Campylobacter* spp. to invade host intestinal cells/tissues and survive within the host environment. These include the use of polar flagellum for motility, generation of toxins and other proteins for adhesion and invasion, secretion of toxins and other molecules for invasion and host interaction, and evolution of drug resistance. Clinical manifestations of campylobacteriosis are largely the outcomes of the interactions between *Campylobacter* bacteria and the host.

(i) **Motility**. Composed of multimers of flagellin proteins (FlaA and FlaB), *Campylobacter* polar flagellum is critical for moving through the mucus layer, adhering to intestinal epithelial cells, secreting and transporting invasive proteins (e.g., *Campylobacter* invasive antigens or *Cia*) to host cells via secretion systems, and forming biofilms for survival in the host [19].

(ii) **Invasion**. *Campylobacter* generates various molecules that aid in its invasion and colonization of intestinal epithelium (columnar epithelial cells) and compromise intestinal permeability. These include adhesins [*Campylobacter* adhesion to fibronectin (CadF), major outer membrane protein (MOMP), periplasmic binding protein (PEB1), P95, *jejuni* lipoprotein A (JlpA), *Campylobacter* autotransporter protein A (CapA), fibronectin-like protein A (FlpA)], toxins [e.g., zonula occludens toxin (Zot), cytolethal distending toxin (Cdt)], invasion proteins [*Campylobacter* invasion antigens C and D (CiaC, CiaD), hemolytic phospholipase A_2 (PLA_2), serine protease high-temperature requirement A (HtrA)], virulence factors [e.g., *VirB11*, *IcbA*, *PldA*, *WlaN*, *IamA*, and *Cgt*], and chemotactic factors [20,21].

Encoded by prophages, Zot is a known virulence factor that causes prolonged damage to the intestinal epithelium and thus increases intestinal permeability, through activation of intestinal epithelial cells for production of proinflammatory cytokines (e.g., TNF-α and IL-8), chemokine CXCL16, and TLR3, upregulation of macrophage response to enteric bacterial species, and induction of cell apoptosis. Consisting of subunits CdtA (for enzymatic reaction), CdtB, and CdtC (both for binding to the membrane of a eukaryotic cell), Cdt halts the eukaryotic cell during the G2/M phase and interrupts cell transition into the mitosis phase, leading to cellular death. In addition, PLA_2 is a membrane protein that exhibits cytolytic effects on host cells and causes tissue destruction related to intestinal inflammation (Figure 13.3) [22–24].

(iii) **Immune interaction**. *Campylobacter* flagellins are remarkably unprovocative to the host immune system, permitting its adhesion to and invasion of host intestinal cells in the early stage of infection without hindrance by host innate immunity. However, as the infection progresses, *Campylobacter* produces various toxins and other molecules that help its establishment inside the host and at the same time provoke the host immune system. For example, proinflammatory cytokines [e.g., interleukin (IL)-8 and tumor necrosis factor (TNF)-α in THP-1 macrophages; IL-8 and cyclooxygenase (COX)-2 in HT-29 cells] elicited by *Campylobacter* infection lead to upregulated surface expression of lipopolysaccharides receptors [e.g., Toll-like receptor (TLR) 4 and myeloid differentiation factor 2] in HT-29 cells, activation of neutrophil adherence molecule CD11b and oxidative burst response in neutrophils, and assembly of IFI16 inflammasome

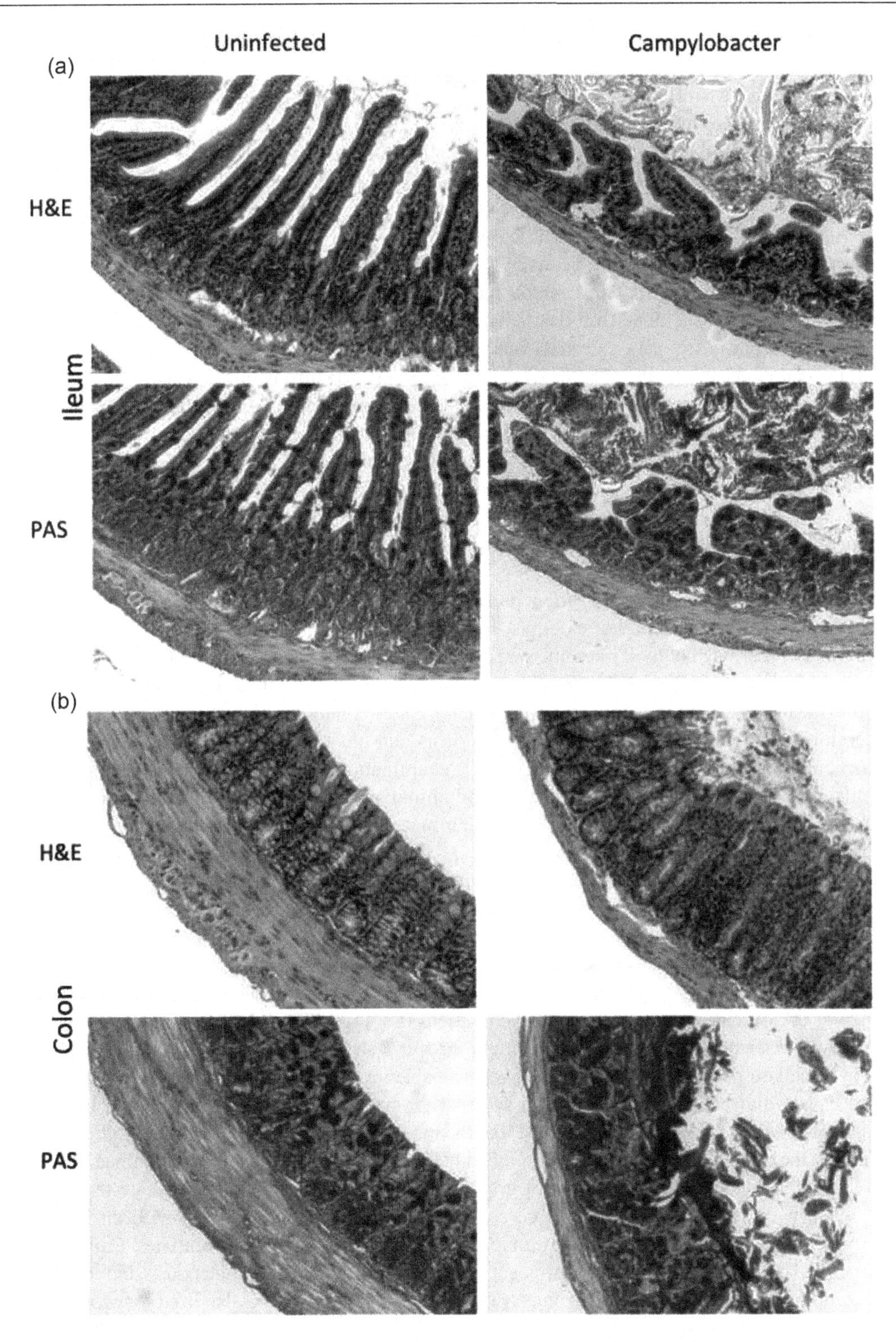

FIGURE 13.3 Intestinal sections from dZD fed mice stained for H&E or PAS, showing the structural damages caused by *Campylobacter* bacteria. Panel A: ileum. Panel B: colon. (Photo credit: Giallourou N, Medlock GL, Bolick DT, et al. *PLoS Pathog*. 2018;14(3):e1007083.)

(which promotes caspase-1 activation and subsequent production of proinflammatory cytokines IL-1β and 1L-18) in both intestinal epithelial cells and macrophages.

Structurally similar to GM1 ganglioside on peripheral nerve myelin and GQ1b ganglioside on cranial nerve myelin, *C. jejuni* lipopolysaccharide induces cross-reacting antibodies, and subsequent neural damage, leading to GBS (an acute inflammatory immune-mediated polyradiculoneuropathy affecting the myelin sheath and related Schwann-cell components and manifesting as a progressive symmetrical weakness in the limbs, with or without hyporeflexia) and Miller Fisher syndrome (MFS, a clinical variant of GBS defined by acute-onset ophthalmoparesis, areflexia, and ataxia), respectively [25].

(iv) **Antibiotic resistance**. *Campylobacter* has the ability to undergo genetic changes (e.g., point mutations, rearrangements, or deletions; lateral or horizontal gene transfer, acquisition of genetic variation from an external source) and builds up resistance to antibiotics, including fluoroquinolones (e.g., ciprofloxacin), macrolides (e.g., erythromycin), aminoglycosides (e.g., gentamycin, kanamycin, and streptomycin), tetracyclines, and β-lactams (e.g., penicillins and cephalosporins), enhancing its survival in host alimentary tract. Indeed, continuous horizontal gene transfer through reassortment of existing alleles underlines the emergence of a large number of sequence types (STs) (genotypes; 6,942 to date) in *Campylobacter.* Conserved substitutions (e.g., A2075G) in domain V of the 23S rRNA gene represent another mechanism for macrolide resistance in *C. jejuni* and *C. coli* [26–31].

Diagnosis. *Campylobacter* infection often presents with nonspecific clinical symptoms, which may resemble those of other diseases. For example, bloody stools and vomiting without fever in infants with *Campylobacter* enteritis may look like intussusception; severe right lower quadrant pain without diarrhea in older children with acute *Campylobacter* ileocecitis may appear similar to appendicitis; and colitis and bloody diarrhea in patients with severe *Campylobacter* enteritis may give an impression of IBD.

Therefore, accurate diagnosis of campylobacteriosis requires input from imaging study, *in vitro* culture, biochemical tests, serological assays, histologic examination, and molecular procedures.

While imaging study helps rule out intra-abdominal processes (e.g., intussusception), histologic examination facilitates differentiation of acute inflammatory changes in *Campylobacter* infection from chronic changes associated with IBD.

Stool culture facilitates isolation and subsequent morphological and biochemical identification of *Campylobacter* spp., which grow well on media containing selective antibiotics [Butzler (sheep blood agar to 10% with bacitracin, novobiocin, colistin, cephalothin, and actidione), Blaser media (agar-blood sheep to 10% with vancomycin, trimethoprim, polymyxin B, cephalothin, and amphotericin B), Skirrow media (horse blood agar lysate to 7% with vancomycin, polymyxin B, and trimethoprim), Preston medium (cefoperazone), or CASA agar], under microaerobic conditions (5% O_2, 10% CO_2, 85% N_2, and H_2), at 42°C for 48h. *Campylobacter* typically appears as a comma- or spiral-shaped Gram-negative bacillus and displays oxidase and catalase production. Blood culture is invaluable for infants younger than 3 months of age with fever and bloody diarrhea, patients with signs of sepsis or systemic manifestations, and immunocompromised patients. Nonetheless, it should be noted that the media used for *Campylobacter* culture may specifically enhance the growth and isolation of *C. jejuni* and *C. coli* and select against the less defined species [32].

Biochemical tests include assessment of growth requirement for H_2, indoxyl acetate test, hippurate test (*C. jejuni* hydrolyzes hippurate while *C. coli* does not), arylsulfatase test, growth on MacConkey agar, production of H_2S, and *l*-alanine aminopeptidase activity (which differentiates *Campylobacter, Helicobacter, Arcobacter,* and other Gram-negative bacteria). Additionally, matrix-assisted laser desorption ionization-time of flight mass spectrometry (MALDI-TOF MS) may be also used for protein composition analysis of *Campylobacter* cells.

Serologic assays [e.g., enzyme-linked immunosorbent assays (ELISA), flow cytometry, and quantitative immunofluorescence] allow detection of specific antibodies in recent *Campylobacter* infection in patients with REA or GBS who have negative stool studies but are ineffective for diagnosing acute *Campylobacter* infection.

Molecular procedures are increasingly applied for identification, typing, and phylogenetic analysis of *Campylobacter* spp. because of their extreme sensitivity, exquisite specificity, and speedy turnaround (see Section 13.2).

Treatment. For most patients with mild, self-limited *Campylobacter* enteritis, supportive care is adequate. For patients with acute infection, oral or intravenous rehydration is necessary. Zinc supplementation may help shorten the duration of diarrhea in children 6 months to 5 years of age with zinc deficiency or malnutrition. Antibiotics (azithromycin 10 mg $kg^{-1} day^{-1}$ for 3 days or erythromycin 40 mg $kg^{-1} day^{-1}$ in 4 doses for 5 days) may be prescribed to patients with severe disease (e.g., bloody stools, high fever, extra-intestinal infection, worsening, relapsing, or prolonged symptoms exceeding >1 week), elderly, pregnant, or immunocompromised individuals. This usually leads to the elimination of *Campylobacter* from the stool within 2–3 days.

Prevention. Prevention of *Campylobacter* infection involves avoiding exposure to animals (particularly poultry, which is implicated in 50%–80% of all human cases), proper cooking of chicken meat, hand hygiene and thorough cleansing of cutting boards and utensils, drinking of pasteurized milk and chlorinated water, and excluding individuals with diarrhea from food handling and patient/child care [33]. Furthermore, bacteriocins, bacteriophages, and probiotics may be utilized for control of *Campylobacter* infection in chicken [34].

13.2 MOLECULAR IDENTIFICATION OF *CAMPYLOBACTER* SPP.

13.2.1 Overview

Principles. Conventional techniques (e.g., microscopy, biochemical tests, and serological assays) assess phenotypic properties of *Campylobacter* bacteria, which may change with external conditions, yielding inconsistent test results. Furthermore, *C. jejuni* and *C. coli* are not always culturable on selective agar plates and indistinguishable based on cellular or biochemical characteristics. By contrast, molecular techniques focus on genotypic characteristics of *Campylobacter* bacteria, with the diagnostic decision based on the presence or absence of unique genes, the size differences of shared genes, and the variations in repetitive elements, restriction enzyme sites, or GC contents in the genomes. Since genotypic features stay constant independent of bacterial cultural parameters and viability, they offer enormous potential for improving the specificity, sensitivity, and speed of laboratory detection and identification of *Campylobacter* bacteria.

Gene targets. Variations in GC contents, restriction enzyme sites, random primer sites, and repetitive elements in *Campylobacter* genomes form the basis of early generation molecular techniques (e.g., DNA–DNA hybridization, PFGE, RAPD, and MLVA). Although molecular techniques targeting genomic differences have proven useful in helping determine the species/subspecies status of or phylogenetic relationship among *Campylobacter* strains, they lack desired convenience and reproducibility. The subsequent focus on ribosomal RNA (16S, 23S, and 5S) and related ITS/IGS, as well as housekeeping and other genes (e.g., *flaA* and *flaB*), shared among *Campylobacter* species/subspecies has greatly improved the specificity and reproducibility of molecular diagnosis of campylobacteriosis. The characterization and application of various species-specific genes have further streamlined molecular identification, strain typing, and phylogenetic analysis of *Campylobacter* bacteria.

Nucleic acid amplification. Apart from ribosomal RNA (16S, 23S, and 5S) and related ITS/IGS, most shared and unique genes in *Campylobacter* exist in a single copy. Given that early molecular procedures (e.g., PFGE and FISH) did not involve nucleic acid amplification, they were relatively insensitive. The advent of polymerase chain reaction (PCR), which utilizes oligonucleotides to amplify specific gene regions *in vitro*, has revolutionized the molecular diagnosis of microbial diseases including campylobacteriosis. Subsequent refinements led to the development of multiplex PCR (which detects multiple targets in a single reaction), nested PCR (which involves two rounds of amplification on a gene target), reverse transcription PCR (RT-PCR; which targets RNA instead of DNA as the amplification template), real-time PCR (which amplifies and detects simultaneously), quantitative PCR (qPCR, which measures fluorescence intensity during amplification to quantify nucleic acid templates), and digital PCR (dPCR, which separates a sample into a large number of small reaction chambers and determines the absolute copy number of a gene target by the resulting number of positive versus negative reactions as calculated using a Poisson statistical algorithm, generating significantly more precise outcomes than qPCR) [35–41]. Apart from PCR, other occasionally applied nucleic acid amplification procedures consist of ligase chain reaction, nucleic acid sequence-based amplification, loop-mediated isothermal amplification (LAMP), and so on [42].

Detection platforms. Agarose gel electrophoresis represents a straightforward, inexpensive, and highly popular platform for detecting non-amplified and amplified nucleic acids. However, as this approach involves significant manual handling, it increases the risk of amplicon cross-contamination. The introduction of intercalating fluorescent dyes that accumulate in each cycle with an intensity directly proportional to the amount of target template DNA and that are detectable with specialized laser-imaging systems enables real-time detection of amplified nucleic acids, making agarose gel electrophoresis essentially obsolete. Since real-time detection platform requires minimal manual handling, it not only eliminates the risk of amplicon cross-contamination, but also provides instant result availability [43].

Miniaturization and point-of-care devices. The increasing automation of real-time detection platform has resulted in sophisticated machines with skyrocketing costs that are beyond the capability of many laboratories in economically disadvantaged countries. There is an obvious need for cost-effective molecular testing platforms with uncompromised sensitivity, specificity, and speed. Miniaturization and point-of-care devices represent some promising efforts toward this goal. Indeed, the use of biosensors (based on nucleic acids, aptamers, proteins, antibodies, and whole cells) and other new technologies in miniaturized and point-of-care devices dramatically simplifies the detection of *Campylobacter* [44]. For example, a combination of an organic light-emitting diode biosensor together with a DNA probe attached to a glass slide and an Alexa Fluor fluorophore-labeled secondary DNA probe permits rapid detection of 0.37 ng μl^{-1} DNA and 1.5×10^1 CFU g^{-1} of *Campylobacter* [45].

13.2.2 Species-Specific Identification

Species identity of *Campylobacter* bacteria can be ascertained through the presence of unique genes, size/nucleic acid base differences in the shared genes, and variations in GC content, restriction enzyme sites, and repetitive elements in the genomes. Notably, the relatively conserved 16S rRNA gene allows *Campylobacter* genus-specific detection, while the species- and strain-specific intervening sequences in the 23S rRNA gene and the internal transcribed spacer (ITS) region facilitate differentiation among *Campylobacter* species and strains. Indeed, using oligonucleotide primers derived from shared and specific genes [e.g., 16S rRNA, 23S rRNA, ITS, flagellin (*flaA* and *flaB*), heat shock proteins (*hsp60*), hippuricase (*hipO*), cytolethal distending toxin (*cdt*), aspartokinase (*asp*), lipid A acyltransferase (*lpxA*), ATP-binding protein (*cje0832*), VS1, and cytochrome c oxidase (*ccoN*)] in PCR, *Campylobacter* spp. have been successfully detected and quantified from food, milk, vegetable, stool, and environmental water samples, with the detection limit approaching to ten bacteria or 2 CFU and testing time of <4 h, which compare favorably to *in vitro* culture and microscopy [46,47]. Commercial kits are also available for rapid detection of *Campylobacter* and other gastrointestinal bacterial pathogens (e.g., BD MAX™ Enteric Bacterial Panel PCR; RIDAGENE Bacterial Stool Panel; BioFire FilmArray) [48].

13.2.3 Strain Typing

The ability to type *Campylobacter* strains is important for the identification of novel bacterial strains, the discrimination of closely related isolates, the investigation of the bacterial organization at the genome level, and the tracking of infection patterns and routes of transmission [49].

Several phenotypic procedures (e.g., biotyping, Penner serotyping, and multilocus enzyme electrophoresis) have been utilized in the early studies of *Campylobacter* strains. Biotyping separates *Campylobacter* strains according to their metabolic activities (e.g., colonial morphology, environmental tolerances, and biochemical reactions); Penner serotyping divides *C. jejuni* strains into 47 serotypes based on the structural diversity of surface-located CPS; and multilocus enzyme electrophoresis differentiates *Campylobacter* strains by the varying electrophoretic mobility of constitutive enzymes under nondenaturing conditions [50].

Genotypic methods that are applied to *Campylobacter* strains include pulse-field gel electrophoresis (PFGE), multilocus sequence typing (MLST), ribotyping, flagellin typing, amplified fragment length polymorphism (AFLP), and multiplex PCR-restriction fragment length polymorphism (RFLP). Of these, PFGE and MLST are commonly utilized. PFGE is a variant of agarose-gel electrophoresis that uses restriction enzymes to digest chromosomal DNA (and proteases and RNases to remove unwanted proteins and RNA) into 8–25 large DNA fragments of 40–600 kb size, and alternating currents to separate these fragments. MLST involves amplification of several (typically seven) housekeeping or virulence-associated gene fragments of ~400 bp, digestion with restriction enzymes, separation by agarose gel electrophoresis, and subsequent classification into allelic profiles or STs [50,51]. Moreover, PCR amplification and sequencing analysis of the strain-specific sequences in the ITS region also allow accurate typing of *Campylobacter* strains.

The recent application of next-generation sequencing enables rapid whole-genome sequence (WGS) analysis of *Campylobacter* strains, which offers a most discriminatory typing tool in comparison with serotyping, PFGE, AFLP, and PCR-based methods [52,53]. In addition, comparative genomic fingerprinting allows the detection of *Campylobacter* genes with high variability among the different species of bacterial clusters that have been previously identified by microarray comparative genomic hybridization.

13.2.4 Phylogenetic Analysis

Phylogenetic analysis helps estimate the relationships among the species or strains and usually involves the following steps: (i) selection of appropriate markers, (ii) amplification, sequencing, and assembly of the target sequence, (iii) alignment of the target sequence with homologous sequences from a national database such as GenBank using an alignment program (e.g., Clustal W algorithm), (iv) reconstruction of the phylogenetic tree from the aligned sequences using a program (e.g., maximum likelihood or neighbor-joining method), and (vi) evaluation of the phylogenetic tree. Fortunately, there exists an integrated program (i.e., Molecular Evolutionary Genetic Analysis X or MEGA X) that takes care of the last three steps in a single environment, and that also caters to other programs for particular steps if desired.

The most widely used markers for phylogenetic analysis of *Campylobacter* species and strains include 16S rRNA and housekeeping genes (e.g., *hsp60*). In a recent report, Silva et al. compared the full-length 16S rRNA gene sequence from FMV-PI01 isolate with other 16S rRNA gene sequences deposited in the GenBank using BLASTN algorithm; aligned the 1513 nucleotide positions of 16S rRNA gene sequences with Clustal W algorithm and trimmed positions with missing data; reconstructed the phylogenetic tree by the neighbor-joining method; and estimated stability of grouping by bootstrap analysis, set for 1,000 replications (all carried out with MEGA X software). The results confirmed the identity of FMV-PI01 isolate as a novel species (*C. portucalensis*) within the genus *Campylobacter* (Figure 13.4) [54].

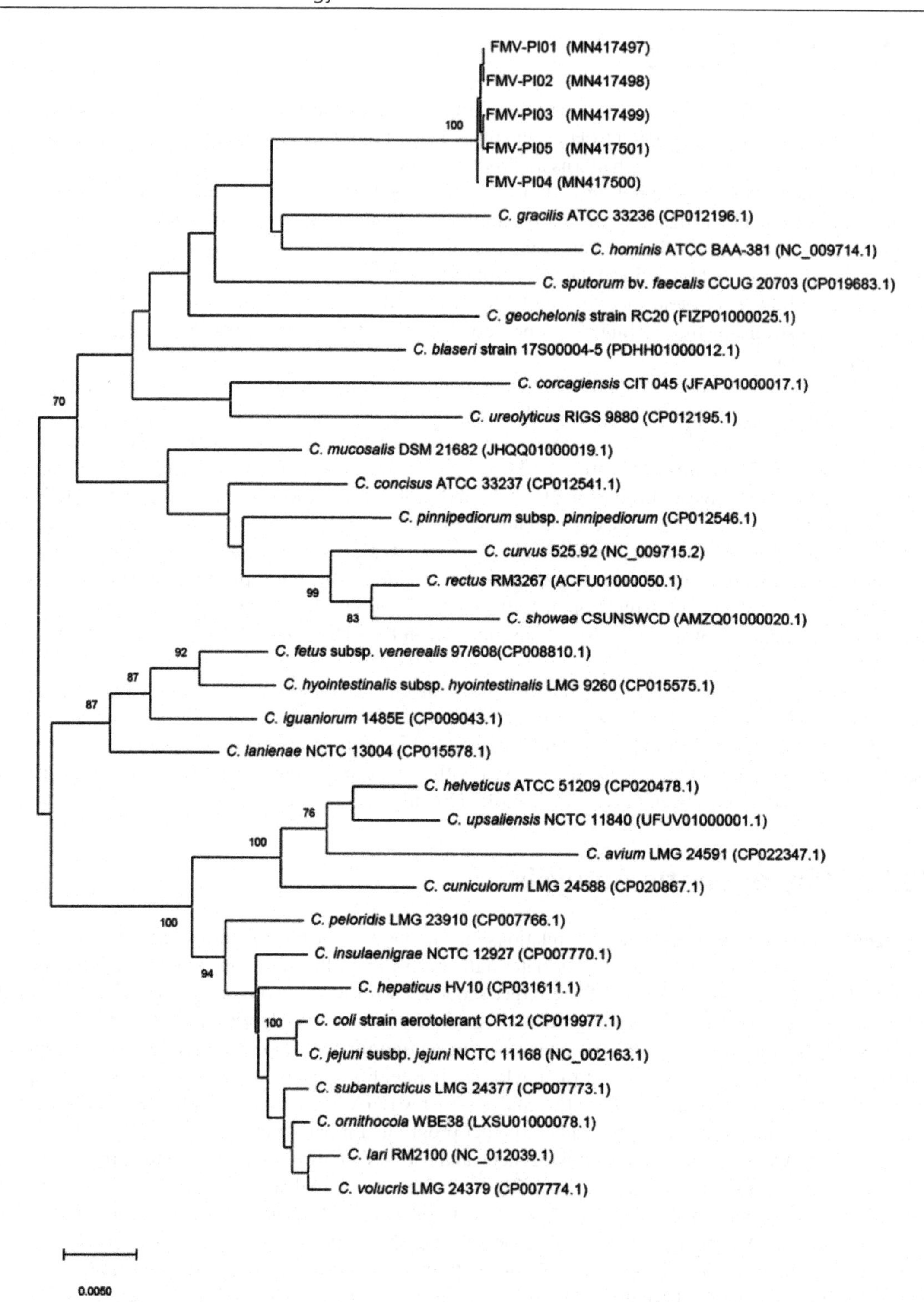

FIGURE 13.4 Phylogenetic tree based on 16S rRNA gene sequences of *Campylobacter* species, reconstructed by the neighbor-joining method. Bootstrap values (%) obtained from 1,000 simulations are indicated at the nodes. Bootstrap values lower than 70% are not shown. Bar: 0.0050 substitutions per site. (Photo credit: Silva MF, Pereira G, Carneiro C, et al. *PLoS One*. 2020;15(1):e0227500.)

13.4 FUTURE PERSPECTIVES

Campylobacter is an important foodborne pathogen that is capable of causing not only severe diarrhea, but also extra-gastrointestinal diseases (e.g., bacteremia, GBS, and REA) in humans. Prompt diagnosis of campylobacteriosis plays a critical role in helping initiate early countermeasures that prevent the disease from progressing into a state of no return. Demonstrating superior sensitivity, specificity, and speed, molecular procedures have done in a relatively short period much more than conventional laboratory diagnostic tests over a century in attaining the goals of rapid and precise identification, strain typing, and phylogenetic analysis of *Campylobacter* bacteria. Despite these remarkable achievements, there is still room for improvement in miniaturization and point-of-care application of molecular procedures. Additionally, creative use of molecular procedures will yield further insights into the mechanisms of *Campylobacter* pathogenicity and contribute to the development of effective therapies for clinical diseases associated with this zoonotic bacterium.

REFERENCES

1. Pike BL, Guerry P, Poly F. Global distribution of *Campylobacter jejuni* Penner serotypes: A systematic review. *PLoS One*. 2013;8(6):e67375.
2. Skarp-de Haan CP, Culebro A, Schott T, et al. Comparative genomics of unintrogressed *Campylobacter coli* clades 2 and 3. *BMC Genomics*. 2014;15: 129.
3. Akutko K, Matusiewicz K. *Campylobacter concisus* as the etiologic agent of gastrointestinal diseases. *Adv Clin Exp Med*. 2017;26(1):149–54.
4. Liu F, Ma R, Wang Y, Zhang L. The clinical importance of *Campylobacter concisus* and other human hosted *Campylobacter* species. *Front Cell Infect Microbiol*. 2018;8:243.
5. Hao H, Fang X, Han J, et al. *Cj0440c* affects flagella formation and *in vivo* colonization of erythromycin-susceptible and- resistant *Campylobacter jejuni*. *Front Microbiol*. 2017;8:729.
6. Müller A, Beeby M, McDowall AW, Chow J, Jensen GJ, Clemons WM Jr. Ultrastructure and complex polar architecture of the human pathogen *Campylobacter jejuni*. *Microbiologyopen*. 2014;3(5):702–10.
7. Thomas DK, Lone AG, Selinger LB, et al. Comparative variation within the genome of *Campylobacter jejuni* NCTC 11168 in human and murine hosts. *PLoS One*. 2014;9(2):e88229.
8. Redondo N, Carroll A, McNamara E. Molecular characterization of *Campylobacter* causing human clinical infection using whole-genome sequencing: Virulence, antimicrobial resistance and phylogeny in Ireland. *PLoS One*. 2019;14(7):e0219088.
9. He Y, Reed S, Strobaugh TP Jr. Complete genome sequence and annotation of *Campylobacter jejuni* YH003, isolated from retail chicken. *Microbiol Resour Announc*. 2020;9(4):e01307–19.
10. Epping L, Golz JC, Knüver MT, et al. Comparison of different technologies for the decipherment of the whole genome sequence of *Campylobacter jejuni* BfR-CA-14430. *Gut Pathog*. 2019;11: 59.
11. Ghatak S, He Y, Reed S, Strobaugh T Jr, Irwin P. Whole genome sequencing and analysis of *Campylobacter coli* YH502 from retail chicken reveals a plasmid-borne type VI secretion system. *Genom Data*. 2017;11:128–31.
12. Cody AJ, Colles FM, Sheppard SK, Maiden MC. Where does *Campylobacter* come from? A molecular odyssey. *Adv Exp Med Biol*. 2010;659:47–56.
13. Sheppard SK, Maiden MC. The evolution of *Campylobacter jejuni* and *Campylobacter coli*. *Cold Spring Harb Perspect Biol*. 2015;7(8):a018119.
14. Chlebicz A, Śliżewska K. Campylobacteriosis, salmonellosis, yersiniosis, and listeriosis as zoonotic foodborne diseases: A review. *Int J Environ Res Public Health*. 2018;15(5):E863.
15. Mohammadpour H, Berizi E, Hosseinzadeh S, Majlesi M, Zare M. The prevalence of *Campylobacter* spp. in vegetables, fruits, and fresh produce: A systematic review and meta-analysis. *Gut Pathog*. 2018;10:41.
16. Kaakoush NO, Castaño-Rodríguez N, Mitchell HM, Man SM. Global epidemiology of *Campylobacter* infection. *Clin Microbiol Rev*. 2015;28(3):687–720.
17. Same RG, Tamma PD. *Campylobacter* infections in children. *Pediatr Rev*. 2018;39(11):533–41.

18. El-Salhy M, Mazzawi T, Gundersen D, Hatlebakk JG, Hausken T. Changes in the symptom pattern and the densities of large-intestinal endocrine cells following *Campylobacter* infection in irritable bowel syndrome: A case report. *BMC Res Notes*. 2013;6:391.
19. Agnetti J, Seth-Smith HMB, Ursich S, et al. Clinical impact of the type VI secretion system on virulence of *Campylobacter* species during infection. *BMC Infect Dis*. 2019;19(1):237.
20. Boehm M, Simson D, Escher U, et al. Function of serine protease HtrA in the lifecycle of the foodborne pathogen *Campylobacter jejuni*. *Eur J Microbiol Immunol*. 2018;8(3):70–7.
21. Liu J, Parrish JR, Hines J, Mansfield L, Finley RL Jr. A proteome-wide screen of *Campylobacter jejuni* using protein microarrays identifies novel and conformational antigens. *PLoS One*. 2019;14(1):e0210351.
22. Kumar A, Gangaiah D, Torrelles JB, Rajashekara G. Polyphosphate and associated enzymes as global regulators of stress response and virulence in *Campylobacter jejuni*. *World J Gastroenterol*. 2016;22(33):7402–14.
23. Lai CK, Chen YA, Lin CJ, et al. Molecular mechanisms and potential clinical applications of *Campylobacter jejuni* cytolethal distending toxin. *Front Cell Infect Microbiol*. 2016;6:9.
24. Giallourou N, Medlock GL, Bolick DT, et al. A novel mouse model of *Campylobacter jejuni* enteropathy and diarrhea. *PLoS Pathog*. 2018;14(3):e1007083.
25. Rodríguez Y, Rojas M, Pacheco Y, et al. Guillain–Barré syndrome, transverse myelitis and infectious diseases. *Cell Mol Immunol*. 2018;15(6):547–62.
26. Hao H, Ren N, Han J, et al. Virulence and genomic feature of multidrug resistant *Campylobacter jejuni* isolated from broiler chicken. *Front Microbiol*. 2016;7:1605.
27. Bolinger H, Kathariou S. The current state of macrolide resistance in *Campylobacter* spp.: Trends and impacts of resistance mechanisms. *Appl Environ Microbiol*. 2017;83(12):e00416–17.
28. Yao H, Liu D, Wang Y, Zhang Q, Shen Z. High prevalence and predominance of the aph(2″)-If gene conferring aminoglycoside resistance in *Campylobacter*. *Antimicrob Agents Chemother*. 2017;61(5):e00112–17.
29. Elhadidy M, Miller WG, Arguello H, et al. Genetic basis and clonal population structure of antibiotic resistance in *Campylobacter jejuni* isolated from broiler carcasses in Belgium. *Front Microbiol*. 2018;9:1014.
30. Keske Ş, Zabun B, Aksoy K, Can F, Palaoğlu E, Ergönül Ö. Rapid molecular detection of gastrointestinal pathogens and its role in antimicrobial stewardship. *J Clin Microbiol*. 2018;56(5):e00148-18.
31. Otigbu AC, Clarke AM, Fri J, Akanbi EO, Njom HA. Antibiotic sensitivity profiling and virulence potential of *Campylobacter jejuni* isolates from estuarine water in the Eastern Cape Province, South Africa. *Int J Environ Res Public Health*. 2018;15(5):925.
32. Perry JD. A decade of development of chromogenic culture media for clinical microbiology in an era of molecular diagnostics. *Clin Microbiol Rev*. 2017;30(2):449–79.
33. Facciolà A, Riso R, Avventuroso E, Visalli G, Delia SA, Laganà P. *Campylobacter*: From microbiology to prevention. *J Prev Med Hyg*. 2017;58(2):E79–92.
34. Aprea G, Zocchi L, Di Fabio M, De Santis S, Prencipe VA, Migliorati G. The applications of bacteriophages and their lysins as biocontrol agents against the foodborne pathogens *Listeria monocytogenes* and *Campylobacter*: An updated look. *Vet Ital*. 2018;54(4):293–303.
35. Bonilauri P, Bardasi L, Leonelli R, et al. Detection of food hazards in foods: Comparison of real time polymerase chain reaction and cultural methods. *Ital J Food Saf*. 2016;5(1):5641.
36. Andersen SC, Fachmann MSR, Kiil K, Møller Nielsen E, Hoorfar J. Gene-based pathogen detection: Can we use qPCR to predict the outcome of diagnostic metagenomics? *Genes*. 2017;8(11):332.
37. Andersen SC, Kiil K, Harder CB, et al. Towards diagnostic metagenomics of *Campylobacter* in fecal samples. *BMC Microbiol*. 2017;17(1):133.
38. Jäckel C, Hammerl JA, Rau J, Hertwig S. A multiplex real-time PCR for the detection and differentiation of *Campylobacter* phages. *PLoS One*. 2017;12(12):e0190240.
39. Magana M, Chatzipanagiotou S, Burriel AR, Ioannidis A. Inquiring into the gaps of *Campylobacter* surveillance methods. *Vet Sci*. 2017;4(3):36.
40. Papić B, Pate M, Henigman U, et al. New approaches on quantification of *Campylobacter jejuni* in poultry samples: The use of digital PCR and real-time PCR against the ISO standard plate count method. *Front Microbiol*. 2017;8:331.
41. Ricke SC, Feye KM, Chaney WE, Shi Z, Pavlidis H, Yang Y. Developments in rapid detection methods for the detection of foodborne *Campylobacter* in the United States. *Front Microbiol*. 2019;9:3280.
42. Quyen TL, Nordentoft S, Vinayaka AC, et al. A sensitive, specific and simple loop mediated isothermal amplification method for rapid detection of *Campylobacter* spp. in broiler production. *Front Microbiol*. 2019;10:2443.
43. Mirski T, Bartoszcze M, Bielawska-Drózd A, et al. Microarrays – New possibilities for detecting biological factors hazardous for humans and animals, and for use in environmental protection. *Ann Agric Environ Med*. 2016;23(1):30–6.

44. Shams S, Bakhshi B, Tohidi Moghadam T, Behmanesh M. A sensitive gold-nanorods-based nanobiosensor for specific detection of *Campylobacter jejuni* and *Campylobacter coli*. *J Nanobiotechnol.* 2019;17(1):43.
45. Vidic J, Manzano M, Chang CM, Jaffrezic-Renault N. Advanced biosensors for detection of pathogens related to livestock and poultry. *Vet Res.* 2017;48(1):11.
46. Iraola G, Pérez R, Betancor L, et al. A novel real-time PCR assay for quantitative detection of *Campylobacter fetus* based on ribosomal sequences. *BMC Vet Res.* 2016;12(1):286.
47. Lv R, Wang K, Feng J, Heeney DD, Liu D, Lu X. Detection and quantification of viable but non-culturable *Campylobacter jejuni*. *Front Microbiol.* 2020;10:2920.
48. Gueudet T, Paolini MC, Buissonnière A, et al. How to interpret a positive *Campylobacter* PCR result using the BD MAX™ system in the absence of positive culture?. *J Clin Med.* 2019;8(12):2138.
49. Good L, Miller WG, Niedermeyer J, et al. Strain-specific differences in survival of *Campylobacter* spp. in naturally contaminated turkey feces and water. *Appl Environ Microbiol.* 2019;85(22):e01579-19.
50. Eberle KN, Kiess AS. Phenotypic and genotypic methods for typing *Campylobacter jejuni* and *Campylobacter coli* in poultry. *Poult Sci.* 2012;91(1):255–64.
51. Kiatsomphob S, Taniguchi T, Tarigan E, Latt KM, Jeon B, Misawa N. Aerotolerance and multilocus sequence typing among *Campylobacter jejuni* strains isolated from humans, broiler chickens, and cattle in Miyazaki Prefecture, Japan. *J Vet Med Sci.* 2019;81(8):1144–51.
52. Jansen van Rensburg MJ, Swift C, Cody AJ, Jenkins C, Maiden MC. Exploiting bacterial whole-genome sequencing data for evaluation of diagnostic assays: *Campylobacter* species identification as a case study. *J Clin Microbiol.* 2016;54(12):2882–90.
53. Llarena AK, Taboada E, Rossi M. Whole-genome sequencing in epidemiology of *Campylobacter jejuni* infections. *J Clin Microbiol.* 2017;55(5):1269–75.
54. Silva MF, Pereira G, Carneiro C, et al. *Campylobacter portucalensis* sp. nov., a new species of *Campylobacter* isolated from the preputial mucosa of bulls. *PLoS One.* 2020;15(1):e0227500.

SUMMARY

The genus *Campylobacter* covers 33 Gram-negative, spiral-shaped, non-spore-forming species that inhabit the gastrointestinal tract of domestic animals, wildlife, and human as commensal organisms. However, some *Campylobacter* spp. can cause diseases in animal and human hosts. Among human-pathogenic *Campylobacter* species, *C. jejuni* and *C. coli* are responsible for almost 90% and 5% of clinical cases of human campylobacteriosis, respectively. Manifesting as severe diarrhea as well as extra-gastrointestinal diseases (e.g., bacteremia, Guillain–Barré syndrome, and reactive arthritis), human campylobacteriosis may have serious consequences if not diagnosed and treated promptly. Conventional diagnosis of human campylobacteriosis relies heavily on laboratory cultivation of *Campylobacter* bacteria followed by biochemical and serological characterization. Recent application of molecular techniques has greatly improved the performance of *Campylobacter* identification, typing, and phylogenetic analysis and helped reveal new insights into the mechanisms of *Campylobacter* pathogenicity, based on which innovative therapies against these zoonotic bacteria will be developed.

Molecular Epidemiology of *Salmonella*

14

S.I. Smith
Nigerian Institute of Medical Research (NIMR)

A. Seriki and A. Ajayi
University of Lagos

Contents

14.1 INTRODUCTION

Salmonella spp. have been continually implicated in periodic disease outbreaks, with non-typhoidal *Salmonella* (NTS) serovars causing 93.8 million cases of gastroenteritis and 155,000 deaths worldwide yearly.[1,2] Classified in the family Enterobacteriaceae, the *Salmonella* genus comprises Gram-negative, facultatively anaerobic, peritrichously flagellated rods that affect a vast host range, including humans, insects, ruminants, birds, reptiles, amphibians, and even fishes. To date, over 2,500 serotypes have been identified in the genus, which seems to keep expanding unabatedly.[1,3,4] Most often, human *Salmonella* infections are acquired through fecal-oral route via contaminated food or water and by direct contact with animals.[5] However, insects, plants, and algae are also capable of accommodating *Salmonella* and thus might be involved in the dissemination of the pathogen.[6]

The nomenclature of *Salmonella* has evolved over time. In the past, *Salmonella* was usually christened by the original place of isolation such as *Salmonella* Lagos, *Salmonella* GoldCoast, *Salmonella*

Budapest, and so on. However, this system was replaced by the classification based on the susceptibility of *Salmonella* to different arrays of bacteriophages (so-called phage typing), which has indeed been useful. On a broader note, *Salmonella* serotypes can be grouped into typhoidal *Salmonella* (TS) and non-typhoidal *Salmonella* (NTS) based on associated clinical syndromes.[7,8] The White–Kauffmann–Le Minor classification scheme employs the flagella antigen (H) and somatic antigen (O) and separates over 2,500 known *Salmonella* serotypes into various groups. Unique antigenic formulas are derived for each serovar such as 4,12:eh:1,7 for *Salmonella* Kaapstad where 4,12 are O antigen, eh phase 1 H antigen, and 1,7 are phase 2 H antigen. Currently, *Salmonella* is divided into two major species *Salmonella enterica* and *Salmonella bongori*, of which *S. enterica* is further distinguished into six subspecies (*S. enterica* subsp *enterica* I, *S. enterica* subsp *salamae* II, *S. enterica* subsp *arizonae* IIIa, *S. enterica* subsp *diarizonae* IIIb, *S. enterica* subsp *houtenae* IV, *S. enterica* subsp *indica* VI).[9]

Serovars that belong to *S. enterica* subsp *enterica* I are extremely diverse and have mainly been associated with human disease and other warm-blooded animals while members of other subspecies have been isolated from cold-blooded animals and the environment.[10,11] Despite their vast host range, *Salmonella* serovars exhibit host specificity and adaptation. *S. Typhi*, *S. Gallinarum*, and *S. Abortusovis* are host specific infecting only humans, fowl, and sheep, respectively. Host-adapted serovars such as *S. Dublin* and *S. Choleraesuis* primarily cause systemic disease in both cattle and pigs, respectively; however, humans and other animals could be infected. Some other serovars such as *S. Typhimurium* and *S. Enteritidis* are referred to as generalist because they cause disease in a broader range of hosts.[1,3]

Diseases elicited by *Salmonella* in humans are characterized by clinical patterns, which consist of four groups (i.e., enteric fever, gastroenteritis, bacteremia, and septicemia and convalescent lifetime carrier state).[7,8] The time frame and outcome of the infection are dependent on infective dose, the genetic makeup of the infecting organism and host, and also the immune status of the host.[12] Transmission of *Salmonella* is mostly foodborne via consumption of contaminated food products from poultry, bovine, pigs, vegetables, and fruits as well as contaminated drinking water. Nosocomial exposure and direct contact with infected animals may also play an accessory role in its transmission.[6,11,13]

14.2 ISOLATION AND IDENTIFICATION OF *SALMONELLA*

Taking into consideration the large distribution of *Salmonella* serovars, proper isolation and identification of the organism is key to effective epidemiological surveillance to track, control, and forestall outbreaks. Investigations have in the past relied on traditional methods that entail culture, biochemical characterization, phage typing, and serology. These procedures require a considerable amount of time, thus having a minimum of 3 days for a negative result and 5 days to confirm a positive result.[14]

14.2.1 Culture Isolation

Since *Salmonella* is a non-fastidious bacterium, the use of standard culture methods allows for its isolation and colony morphology assessment from food, clinical, or environmental samples (Figure 14.1). However, an optimal isolation method is still lacking as current procedures are often labor intensive and/or time consuming, and false-positive and false-negative results are rather common. Blood culture for example can only detect about 45% to 70% of patients with typhoid fever.[15–17] This is not to say that the traditional culture method has lost its place of relevance. To successfully isolate and detect *Salmonella* using culture media, the choice of a suitable (representative) sampling procedure combined with a sensitive culture method will give a desirable result.[18] Routinely, a combination of enrichment

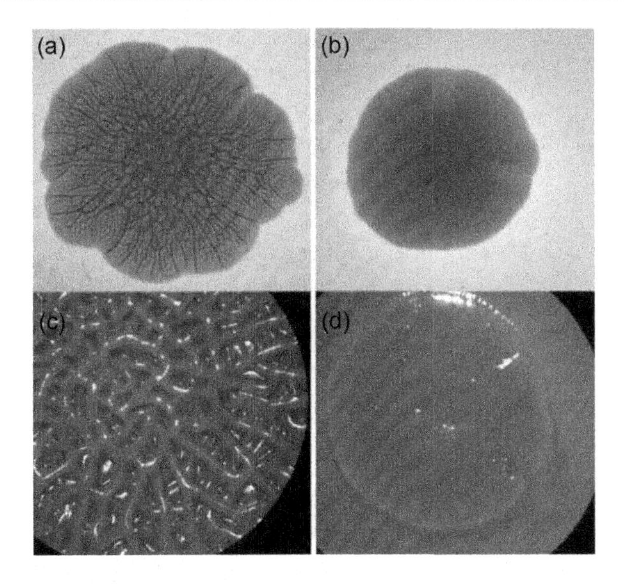

FIGURE 14.1 *Salmonella* ser. Agona culture on Congo red (CR) agar after incubation at 20°C, showing rdar (red, dry, and rough) morphology of *Salmonella* ser. Agona strain 71-3 close up (a) and through a magnifying glass (c); and bdar (brown, dry, and rough) morphology of *Salmonella* ser. Agona strain 1454-1 close up (b) and through a magnifying glass (d). (Photo credit: Vestby LK, Møretrø T, Ballance S, Langsrud S, Nesse LL. *BMC Vet Res*. 2009;5:43.)

and plating on two selective media is carried out (Figure 14.2). Pre-enrichment is done using a non-selective medium (buffered peptone water). This is important because it facilitates the multiplication of an otherwise small number of *Salmonella*, which would have probably been killed by the toxicity of selective enrichment media such as selenite F broth. Also it may help resuscitate salmonellae that have been exposed to adverse conditions such as freezing, heating, exposure to organic acids, and the likes.[19] After pre-enrichment, it is subjected to selective enrichment using selenite F broth, Rappaport Vassiliadis, or Tetrathionate broth (Müller-Kauffmann). Then a selective plating out is done on selective solid media such as brilliant green agar (BGA), *Salmonella Shigella* agar (SSA), bismuth sulfite (BS) agar, and xylose lysine desoxycholate (XLD) agar. Isolates obtained from culture are further characterized by a biochemical test or analytical profile index 20 E (API 20 E) kit and serotyping using the White–Kauffmann–Le Minor scheme.[20] To circumvent the various challenges of trying to isolate *Salmonella* from different samples, several molecular methods have been developed for detection and typing of *Salmonella* serotypes for extensive epidemiological surveys. Also important to note is that DNA-based methods of investigating *Salmonella* are not only rapid but also give information on the genetic and genomic determinants that confer traits of virulence, drug resistance, and host adaptation to the target organism rather than the phenotypic traits.[21]

14.2.2 PCR Detection

Molecular methods that demonstrate higher specificity and sensitivity than traditional culture methods are increasingly finding relevance in the identification and typing of *Salmonella*. The invention of polymerase chain reaction (PCR) in the early 1980s by Kary Mullis signaled a wave of extraordinary change in the fields of biological sciences and medicine.[22] Researchers have since latched on this technique to detect *Salmonella* and even characterize serotype-specific serovars from various samples including food, blood, stool, and water. Several genes in the genome of *Salmonella* such as pathogenicity determinants and metabolism have been explored to this end. Freitas et al.[23] deployed a multiplex

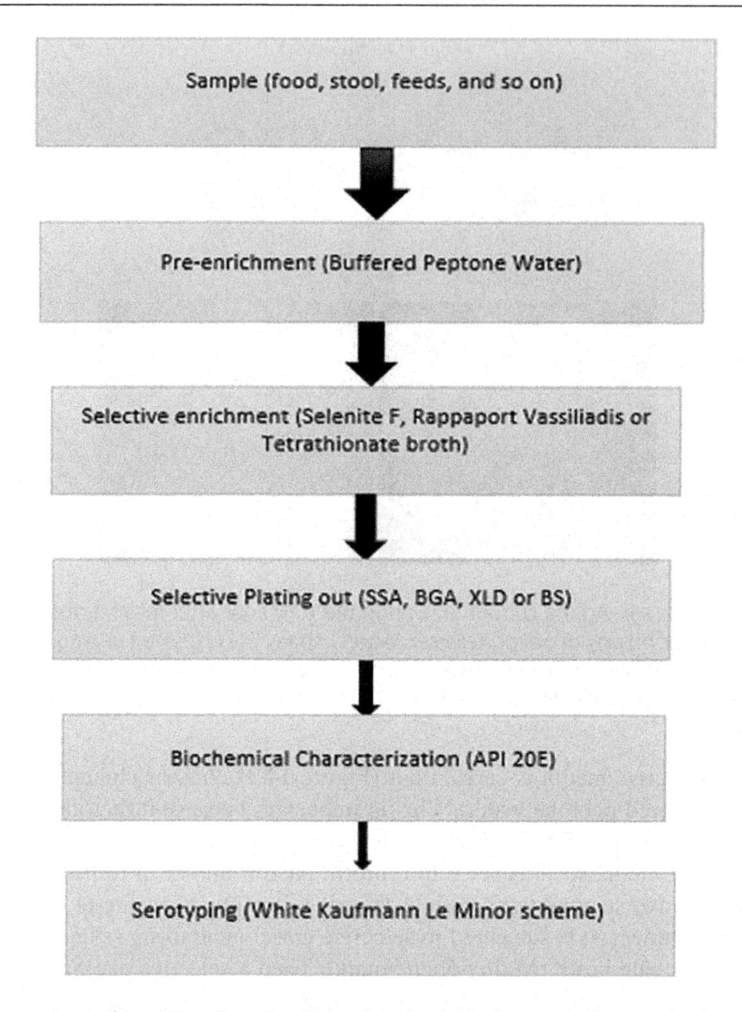

FIGURE 14.2 Flow chart of traditional methods used in the isolation and characterization of *Salmonella*.

PCR in the detection of *Salmonella* Enteritidis, Typhi, and Typhimurium from poultry meat in Brazil, achieving a *Salmonella* genus-specific detection in 2.74% of samples. Ali et al.[24] applied an ultra-rapid and simple multiplex PCR technique with a combination of primers targeting the *tyv* (*rfbE*), *prt* (*rfbS*), and *invA* genes for the detection of *S. Typhi*, producing a highly specific band of 784 bp. In a similar study, Thong et al.[16] detected *S. Typhi* and *S. Paratyphi* from stool, blood, and food samples using multiplex PCR. Not only has PCR been used in detecting TS, but it has also been used in detecting NTS. Smith et al.[25] utilized a PCR assay for detection of both TS and NTS species from snack and food commonly sold in Lagos Nigeria. After comparative genomics analysis, Li et al.[26] developed a novel Hexa-plex PCR for the detection and serotyping of *Salmonella* spp. including *S. Typhimurium*, *S. Enteritidis*, *S. Agona*, *S. Choleraesuis*, and *S. Pullorum* simultaneously. Other variants of PCR have also been employed in the detection of *Salmonella*. Zhang et al.[27] used a generic and differential FRET-PCR targeting the tetrathionate reductase response regulator gene to successfully detect seven plasmids that contained partial sequences of *S. bongori* and six *S. enterica* subspecies. Trinucleotide repeat sequence (TRS) PCR reported by Majchrzak et al.[28] provided a potential tool for inter-serovar discrimination of *S. Enteritidis*, *S. Typhimurium*, *S. Infantis*, *S. Virchow*, *S. Hardar*, *S. Newport*, and *S. Anatum*. Tennant et al.[29] described a *clyA*-based real-time PCR with the capacity to detect TS and paratyphoidal *Salmonella* in the blood, which was extremely fast in comparison with blood culture. *Salmonella invA* gene produces invasion protein, which, as a virulence determinant, facilitates attachment and invasion

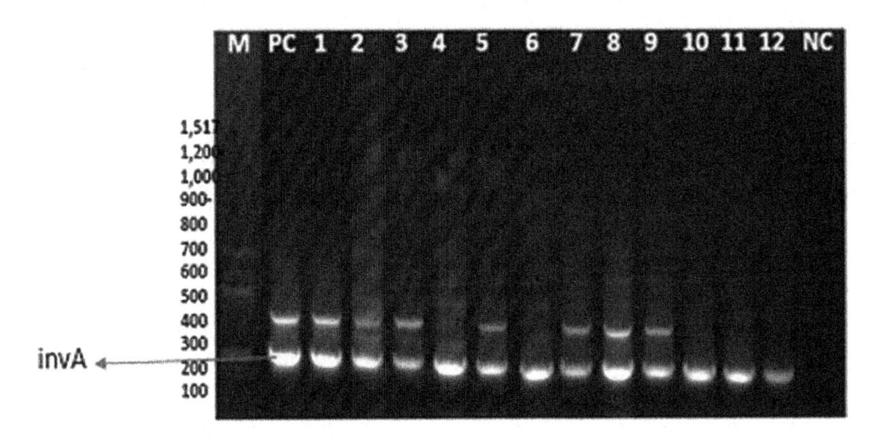

FIGURE 14.3 Multiplex PCR detection of the *invA* gene in *Salmonella* spp.

of M-cells. Because *invA* is located in the chromosome and is specific to *Salmonella* spp., it has been proposed as a gene for the molecular identification or detection of *Salmonella*.[30] Fazl et al.[31] and Ezzat et al.[32] reported 100% detection of *invA* gene in *Salmonella* spp. by PCR. Figure 14.3 presents a gel image of amplified *invA* gene products from *Salmonella* spp. using multiplex PCR.

The White–Kaufmann–Le Minor scheme widely used for serotyping *Salmonella* spp. is complex and requires over 250 typing sera and 350 different antigens,[2] which in reality are expensive and beyond the reach of non-reference laboratories across the world, most especially sub-Saharan Africa. Several attempts to serotype *Salmonella* isolates using PCR have yielded satisfactory results. Shah et al.[33] using serotype-specific *rfbS* sequence related to the O- and H-antigens were able to distinguish *S. Gallinarum* from *S. Pullorum* of poultry origin and other *Salmonella* serotypes in less than 3 h. Ranieri et al.[2] correctly identified serogroups O:4, O:7, O:8, O:9, O:3,10, O:13, and O:18 using PCR. Lim and Thong[34] also reported an H-typing multiplex PCR that identified flagella antigen 'a', 'b', or 'd'. Tennant et al.[35] used PCR to differentiate diphasic and monophasic *S. Typhimurium* from other O serogroup B, H:i serovars (e.g., *S. Enteritidis*, *S. Dublin*, and *S. Stanleyville*), with 100% sensitivity and specificity.

14.2.3 Molecular Typing

Molecular methods for typing of *Salmonella* are of epidemiological significance during foodborne salmonellosis outbreaks, as they help link the disease to foods or food ingredients that contain culprit organisms.[36] Among the widely applied molecular methods are pulsed-field gel electrophoresis (PFGE), random amplified polymorphic DNA (RAPD), ribotyping, enterobacterial repetitive intergenic consensus (ERIC) PCR, repetitive sequence-based (rep) PCR, multiple locus variable number tandem repeat analysis (MLVA), multilocus sequence typing (MLST), and whole-genome sequencing (WGS). Based on a PFGE study, Pang et al.[37] suggested the existence of a major worldwide clone of *S. Enteritidis*. A comparison of 107 food and poultry isolates from Germany with 124 human isolates from Taiwan revealed that the X353N3 pattern was common. Modarressi and Thong[38] utilized PFGE to type 88 *Salmonella* isolates from chicken, beef, and street food in Malaysia into 61 distinct pulsed-field profiles. Ifeanyi et al.[39] determined the genetic diversity of *Salmonella* serovars isolated from children with gastroenteritis in Abuja Nigeria. Considered as the gold standard for typing of outbreaks isolates, PFGE is nonetheless laborious and time consuming. Hyeon et al.[40] examined three methods (PFGE, rep-PCR, and MLST) for subtyping *S. Enteritidis* isolates obtained from food and human sources and showed that rep-PCR had a better resolution of discrimination with a Simpson's diversity index of 0.82 than that of PFGE, which is 0.71. Given the genetic diversity of *Salmonella* serovars, the value of molecular methods in the tracking of widespread outbreaks cannot be overemphasized. In Germany between 2013 and 2014, two consecutive large outbreaks of *S. Muenchen* were linked to pig farms. *S. Muenchen* isolated from stool of patients, environmental surfaces, and food shared indistinguishable PFGE patterns indicating clonal relatedness, thus linking the human disease outbreak to the pig farms.[41]

Synergistic application of molecular typing methods enables harnessing robust epidemiological information about *Salmonella.* After a comparative study involving PFGE, ERIC-PCR, MLVA, and MLST, Campioni et al.[42] noted that PFGE and ERIC-PCR were more efficient than MLVA for the typing of 188 *S. Enteritidis* strain from different sources isolated over a period of 24 years in Brazil. However, as a reliable, rapid, and reproducible typing method, MLVA has been employed in the investigation of outbreaks and appeared to provide additional epidemiological information and evolutionary relationships on the strains concerned.[43] Boland et al.[44] typed monophasic *Salmonella* 4[5]:i strains associated with human infection in Belgium between 2008–2011 using MLVA and identified 55 different profiles with a Simpson's diversity index of 0.942. Kjeldsen et al.[45] successfully divided 272 *S. Dublin* isolates from an epidemiologically confirmed outbreak into 103 types using MLVA. Compared to the PFGE "gold standard," MLVA displays a significantly enhanced discriminatory power. Using MLST analysis, Hughes et al.[46] identified a predominant strain of *S. Typhimurium* circulating among garden bird populations in the United Kingdom that was rarely detected in other host species.

14.3 MOLECULAR EPIDEMIOLOGY OF *SALMONELLA*: RECENT ADVANCES

Salmonella evolves steadily at the genomic level, leading to an increased number of virulence strains and multidrug-resistance (MDR).[47] MDR in *S. enterica* bacteria was defined as resistance to chloramphenicol, trimethoprim-sulfamethoxazole (TMP-SMX), and ampicillin[48,49,] and the emergence of MDR in TS and NTS is a public health concern.[50] Resistance to trimethoprim, chloramphenicol, and ampicillin by MDR strains is encoded by plasmid, which is linked to the H1 incompatibility group (incH1).[51–53] On the other hand, resistance to fluoroquinolones in *S. Typhi* is either mediated by chromosomal mutations or plasmid.[54] The treatment of choice for infection with MDR *S. enterica* serovar Typhi (*S. Typhi*) is fluoroquinolones and third-generation cephalosporins,[49] although fluoroquinolones-resistant *S. Typhi* has been found in different studies.[55,56]

MDR *Salmonella* strains are prevalent in the Asian continent, Africa, and Latin America,[57] while the incidence of MDR *S. Typhi* H58 clone is high in the Indian subcontinent, Southern Asia, Kenya, and East Africa.[58] Horizontal gene transfer is a significant factor in the spread of antimicrobial drug-resistance genes, mostly when these genes are associated with transposons, plasmids, and integrons.[59] Resistance of *S. Enterica* to antibiotics is to some extent due to genes carried on integrons and several *S. Enterica* integron gene cassette arrays were reported,[59] which played a role in horizontal gene transfer in the spread of *S. Enterica* MDR globally. Reports by Centers for Disease Control and Prevention in 2004 stated that the most common *S. Enterica* MDR pattern of *S. Enterica* was resistance to ampicillin, chloramphenicol, streptomycin, sulfamethoxazole, and tetracycline.[60]

The emergence of antimicrobial resistance in *Salmonella* strains, particularly the MDR *Salmonella* serovars, has a major economic impact. Rapid, specific, and sensitive methods (e.g., PCR) are needed for the detection of *Salmonella* serovars.[61] Alternative molecular methods that supplement PCR include loop-mediated isothermal amplification (LAMP) and other nucleic acid amplification procedures (e.g., DNA and WGS).[62–64] LAMP amplifies DNA with high specificity, rapidity, and efficiency.[62] Comparative genomic analyses,[64] determination of clonality by PFGE,[65] and microarray-based assays offer additional approaches for *Salmonella* investigations.[66,67] The DNA microarray technology is noted for its rapidity, flexibility, sensitivity, and specificity and allows high-throughput analysis for detection and identification of pathogenic microorganisms.[68] Biosensors such as nano- and micro-scaled and aptamers represent recent advances for detection of *Salmonella*.[69] Made up of three elements including a biological capture molecule (probes and antibodies), biosensors provide a method for converting capture molecule–target interactions into a signal and an output.[61] In addition, metagenomics permit a culture-independent analysis of the genetic materials of all the microbial DNA in a sample[70] and can be used to analyze the microorganisms within a sample and to detect rare and novel pathogens.[71]

With advances in next-generation sequencing, there is now a drive toward the increasing adoption of WGS as a primary method for isolate characterization. WGS has the potential to replace multiple methods that are in current usage, as it generates high-resolution genomic data rapidly and at a reduced cost.[72] MLST and the *Salmonella in silico* typing resource (SISTR) are examples of the *in silico* prediction tools employed in WGS.[73]

SISTR is a bioinformatics platform for rapid *in silico* analysis and provides serovar prediction using a genoserotyping approach for subtyping *Salmonella*.[72,73] MLST based on PCR determines the nucleotide sequences of a series of housekeeping genes and facilitates the discovery of the clonal lineages and evolutionary pathways of organisms.[74,75] Typically, MLST detects or differentiates strains into a sequence of seven housekeeping genes namely, *purE*, *thr*A, *dna*N, *aro*C, *his*D, *hem*D, and *suc*A for *Salmonella*.[76] However, the utility of these methods may be compromised when only a small subset of serotypes is available.[77]

14.4 FUTURE PERSPECTIVES

Excessive or indiscriminate use of antibiotics is the main factor underscoring the development of resistance; hence a routine review of therapeutic policies, such as the control of over-prescription of antimicrobials in humans and the regulation of antimicrobials for use in animals, is required. In addition, determining the patterns of antimicrobial resistance of enteric fever isolates is required for adequate prevention of *Salmonella* infection. The application of rapid, robust, and cost-effective technologies such as PCR, LAMP, DNA sequencing, DNA microarray technology, biosensors, and metagenomics improves the diagnosis of *Salmonella*. Also, a composite reference standard (CRS) for the diagnosis of *Salmonella* serovar bacteria is required for a reliable diagnosis.[78] The development of a number of potential invasive non-typhoidal *Salmonella* (iNTS) vaccines, which include subunit-based, live-attenuated, and recombinant antigen-based substances and a multivalent vaccine, is needed to target different serovars.[79,80]

While the molecular tools currently deployed in the epidemiological surveillance of *Salmonella* seem to give desirable results, their full potentials are yet to be harnessed most especially in developing countries where expertise and financial competence are lacking. Newer methods of serotyping as discussed above could in the future supplement and replace the White–Kaufmann–Le Minor scheme for *Salmonella*.

REFERENCES

1. Hoelzer K, Switt AIM, Wiedmann M. Animal contact as a source of human nontyphoidal salmonellosis. *Vet Res* 2011;42:34.
2. Ranieri LM, Shi C, Switt AIM, den Balkker HC, Wiedmann M. Comparison of typing methods with a new procedure based on sequence characterization for *Salmonella* serovar prediction. *J Clin Microbiol* 2013;51(6):1786–97.
3. Evangelopoulou G, Kritas S, Govaris A, Burriel AR. Animal salmonellosis: A brief review of host adaptation and host specificity of *Salmonella* spp. *Vet World* 2013;6(10):703–8.

4. Eng S-K, Pusparajah P, Ab Mutalib N-S, Ser H-L, Chan K-G, Lee L-H. *Salmonella*: A review on pathogenesis, epidemiology and antibiotic resistance. *Front Life Sci* 2015;8(3):284–93.
5. Adagbada AO, Coker AO, Smith SI, Adesida SA. The prevalence and plasmid profile of non-typhoidal salmonellosis in children in Lagos metropolis, South-western Nigeria. *Pan Afri Med J* 2014; 19:359.
6. Levantesi C, Bonadonna L, Briancesco R., Grohmann E, Toze S, Tandoi V. *Salmonella* in surface and drinking water: Occurrence and water-mediated transmission. *Food Res Int* 2012;45:587–602.
7. Pui CF, Wong WC, Chai LC, et al. *Salmonella*: A foodborne pathogen. *Int Food Res J* 2011;18:465–73.
8. Smith SI, Seriki A, Ajayi A. Typhoidal and non-typhoidal *Salmonella* infections in Africa. *Eur J Clin Microbiol Infect Dis* 2016;35(12):1913–22.
9. Grimont PAD, Weill FX. *Antigenic Formulae of the Salmonella Serovars*, Ninth edn., World Health Organization collaborating center for reference and research on *Salmonella* Institute Pasteur, Paris, France. 2007.
10. Brenner FW, Villar RG, Angulo FJ, Tauxe R, Swaminathan B. *Salmonella* nomenclature. *J Clin Microbiol* 2000;38(7):2465–7.
11. Haeusler GM, Curtis N. Non-typhoidal *Salmonella* in children: Microbiology, epidemiology and treatment. In: *Hot Topics in Infection and Immunity in Children IX*, Curtis N, Finn A, Pollard AJ (eds.) Advances in experimental medicine and biology 764. Springer Science+Business Media, New York. 2013; pp. 13–26.
12. Akyala AI, Alsam S. Extended spectrum beta lactamase producing strain of *Salmonella* species – A systematic review. *J Microbiol Res* 2015;5(2):57–70.
13. Malorny B, Lofstrom C, Wagner M, Kramer N, Hoorfar J. Enumeration of *Salmonella* bacteria in food and feed samples by real-time PCR for quantitative microbial risk assessment. *Appl Environ Microbiol* 2008;74(5):1299–304.
14. Gelinski JMCN, Martin G, Destro MT, Landgraf M, Franco BDGM. Rapid detection of *Salmonella* in foods using a combination of SPRINT, MSRV and *Salmonella* latex test. *Braz J Pharm Sci* 2002;38(3):315–21.
15. Love BC, Rostagno MH. Comparison of five culture methods for *Salmonella* isolation from swine fecal samples of known infection status. *J Vet Diagn Invest* 2008;20:620–4.
16. Thong KL, Tec CSJ, Chua KH. Development and evaluation of a multiplex polymerase chain reaction for the detection of *Salmonella* species. *Trop Biomed* 2014;31(4):680–97.
17. Ogunremi D, Nadin-Davis S, Dupcas AA, et al. Evaluation of a multiplex PCR assay for the identification of *Salmonella* serovar Enteritidis and Typhimurium using retail and abattoir samples. *J Food Prot* 2017;80(2):295–301.
18. Carrique-Mas JJ, Barnes S, McLaren I, Davies R. Comparison of three plating media for the isolation of *Salmonella* from poultry environmental samples in Great Britain using ISO 6579:2002 (Annex D). *J Appl Microbiol* 2008;107:1976–83.
19. World Organisation for Animal Health (OIE). *Manual of Diagnostic Tests and Vaccines for Terrestrial Animals*. OIE, Paris. 2012; Available at: https://www.oie.int/doc/ged/D12009.PDF (accessed on 3 March 2018).
20. Global Salm-Surv, Rene S (ed.) *A global Salmonella Surveillance and Laboratory Support Project of the World Health Organization. Laboratory Protocols Level 1 Training Course: Isolation of Salmonella*, 4th edn., Hendriksen, Peabody, MA. 2003; pp. 3–17.
21. Lauri A, Castiglioni B, Mariani P. Comprehensive analysis of *Salmonella* sequence polymorphisms and development of a LDR-UA assay for the detection and characterization of selected serotypes. *Appl Microbiol Biotechnol* 2011;91:189–210.
22. Vynck M, Trypsteen W, Thas O, Vandekerchhove L, De Spiegelaere W. The future of digital polymerase chain reaction in virology. *Mol Diagn Ther* 2016;20(5):437–47.
23. Freitas CG, Santana AP, da Silva PHC, et al. PCR multiplex for detection of *Salmonella* Enteritidis, Typhi and Typhimurium and occurrence in poultry meat. *Int J Food Microbiol* 2010;139:15–22.
24. Ali K, Zeynab A, Zahra S, Akbar K, Saeid M. Development of an ultra-rapid and simple multiplex polymerase reaction technique for detection of *Salmonella* typhi. *Saudi Med J* 2006;27(8):1134–8.
25. Smith S, Opere B, Fowora M, et al. Molecular characterization of *Salmonella* spp. directly from snack and food commonly sold in Lagos, Nigeria. *Res Note* 2012;43(3):718–22.
26. Li R, Wang Y, Shen J, Wu C. Development of a novel hexa-plex PCR method for identification and serotyping of *Salmonella* species. *Foodborne Pathog Dis* 2014;11(1):75–7.
27. Zhang J, Wei L, Kelly P, et al. Detection of *Salmonella* spp. using a generic and differential FRET-PCR. *PLoS One* 2013;8(10):e76053.
28. Majchrzak M, Krzyzanowska A, Kubiak AB, et al. TRS-based PCR as a potential tool for inter-serovar discrimination of *Salmonella* Enteritidis, *S.* Typhimurium, *S.* Infantis, *S.* Virchow, *S.* Hadar, *S.* Newport and *S.* Anatum. *Mol Biol Res* 2014;41:7121–32.

29. Tennant SM, Toema D, Amar F, et al. Detection of typhoidal and paratyphoidal *Salmonella* in blood by real-time polymerase chain reaction. *Clin Infect Dis* 2015;61(S4):S241–50.
30. Smith SI, Fowora MA, Atiba A, et al. Molecular detection of some virulence genes in *Salmonella* spp. isolated from food samples in Lagos Nigeria. *Anim Vet Sci* 2015;3(1):22–7.
31. Fazl AA, Salchi TZ, Jamshidian M, Amini K, Jangjou AH. Molecular detection of *inv*A, *ssa*P, *sse*C and *pip*B genes in *Salmonella* Typhimurium isolated from human and poultry in Iran. *Afri J Microbiol Res* 2013;7(13):1104–8.
32. Ezzat ME, Shabana II, Esawy AM, Elsotohy ME. Detection of virulence gene in *Salmonella* serovars isolated from broilers. *Anim Vet Sci* 2014;2(6):189–93.
33. Shah DH, Park J-H, Cho M-R, Kim M-C, Chae J-S. Allele specific PCR method based on *rfbs* sequence for distinguishing *Salmonella* Gallinarium from *Salmonella* Pullorum: Serotype specific *rfbs* sequence polymorphism. *J Microbiol Methods* 2004; 60:169–77.
34. Lim BK, Thong KL. Application of PCR based serogrouping of selected *Salmonella* serotypes in Malaysia. *J Infect Dev Ctries* 2009; 3(6):420–8.
35. Tennant SM, Diallo S, Levy H, et al. Identification by PCR of non-typhoidal *Salmonella enterica* serovars associated with invasive infections among febrile patients in Mali. *PLoS Negl Trop Dis* 2010;4(3):e621.
36. Ngoi ST, Tec CSJ, Chai LC, Thong KL. Overview of molecular typing tool for the characterization of *Salmonella enterica* in Malaysia. *Biomed Environ Sci* 2015;28(10):751–64.
37. Pang J-C, Chiu T-H, Helmuth R, Schroeter A, Guerra B, Tsen H-T. A pulsed field gel electrophoresis (PFGE) study that suggests a major worldwide clone of *Salmonella enterica* serovar Enteritidis. *Int J Food Microbiol* 2007;116:305–12.
38. Modarressi S, Thong KL. Isolation and molecular subtyping of *Salmonella enterica* from chicken, beef and street foods in Malaysia. *Sci Res Essays* 2010;5(18):2713–20.
39. Ifeanyi CIC, Bassey BE, Ikeneche NF, Al-Gallas N. Molecular characterization and antibiotic resistance of *Salmonella* in children with acute gastroenteritis in Abuja, Nigeria. *J Infect Dis Dev Ctries* 2014;8(6):712–9.
40. Hyeon J-Y, Chon J-W, Park J-H, et al. A comparison of subtyping methods for differentiating *Salmonella enterica* serovar Enteritidis isolated from food and humans. *Osong Public Health Res Perspect* 2013;4(1):27–33.
41. Schielke A, Rabsch W, Prager R, et al. Two consecutive large outbreaks of *Salmonella* Muenchen linked to pig farming in Germany, 2013 to 2014. Is something missing in our regulatory framework. *Euro Surveill* 2017;22(18):30528.
42. Campioni F, Pitondo-Silva A, Bergamini AMM, Falcao JP. Comparison of four molecular methods to type *Salmonella* Enteritidis strains. *APMIS* 2015; 123(5):422–6.
43. Peter T, Bertrand S, Björkman JT, et al. Multi-laboratory validation study of multilocus variable number tandem repeat analysis (MLVA) for *Salmonella enterica* serovar Enteritidis. *Euro Surveill* 2017;22(9):30477.
44. Boland C, Bertrand S, Matheus W, Dierick K, Wattiau P. Molecular typing of monophasic *Salmonella* 4,[5]:i strains isolated in Belgium (2008–2011). *Vet Microbiol* 2013;168:447–50.
45. Kjeldsen MK, Torpdahl M, Campos J, Pedersen K, Nielsen EM. Multiple locus variable number tandem repeat analysis of *Salmonella enterica* subsp. enterica serovar Dublin. *J Appl Microbiol* 2014; 116(4):1044–54.
46. Hughes LA, Wigley P, Bennett M, Chantrey J, Williams N. Multi-locus sequence typing of *Salmonella enterica* serovar Typhimurium isolates from wild birds in northern England suggest host-adapted strain. *Lett Appl Microbiol* 2010; 51:477–9.
47. Hurley D, McCusker MP, Fanning S, Martins M. *Salmonella*–host interactions – Modulation of the host innate immune system. *Front Immunol* 2014;5:481–91.
48. Wasfy MO, Frenck R, Ismail TF, Hoda M, Malone JL, Mahoney FJ. Trends of multiple-drug resistance among *Salmonella* serotype typhi isolates during a 14-year period in Egypt. *Clin Infect Dis* 2002;35(10): 1265–8.
49. Rahman BA, Wasfy MO, Maksoud MA, Hanna N, Dueger E, House B. Multi-drug resistance and reduced susceptibility to ciprofloxacin among *Salmonella enterica* serovar typhi isolates from the Middle East and Central Asia. *New Microbe New Infect* 2014;2:88–92.
50. Fàbrega A, Vila J. *Salmonella enterica* serovar Typhimurium skills to succeed in the host: Virulence and regulation. *Clin Microbiol Rev* 2013;26(2):308–41.
51. Threlfall EJ. Antimicrobial drug resistance in *Salmonella*: Problems and perspectives in food- and water-borne infections. *FEMS Microbiol Rev* 2002;26:141–8.
52. Roumagnac P, Weill FX, Dolecek, C, et al. Evolutionary history of *Salmonella* typhi. *Science* 2006;314:1301–4.
53. Holt KE, Phan MD, Baker S, et al. Emergence of a globally dominant IncHI1 plasmid type associated with multiple drug resistant typhoid. *PLoS Negl Trop Dis* 2011;5:e1245.
54. Slayton RB, Date KA, Mintz ED. Vaccination for typhoid fever in sub-Saharan Africa. *Hum Vaccin Immunother* 2013;9(4):903–6.

55. Shrestha KL, Pant ND, Bhandari R, Khatri S, Shrestha B, Lekhak B. Reemergence of the susceptibility of the *Salmonella* spp. isolated from blood samples to conventional first line antibiotics. *Antimicrob Resist Infect Control* 2016;5:22.
56. Veeraraghavan B, Anandan S, Sethuvel DPM, Puratchiveeran N, Walia K, Ragupathi NKD. Molecular characterization of intermediate susceptible typhoidal *Salmonella* to ciprofloxacin, and its impact. *Mol Diagn Ther* 2016;20:213–9.
57. Nagshetty K, Channappa ST, Gaddad SM. Antimicrobial susceptibility of *Salmonella* typhi in India. *J Infect Dev Ctries* 2010;4(2):70–3.
58. Kariuki S, Revathi G, Kiiru J, et al. Typhoid in Kenya is associated with a dominant multidrug-resistant *Salmonella enterica* serovar typhi haplotype that is also widespread in Southeast Asia. *J Clin Microbiol* 2010;48:2171–6.
59. Krauland MG, Marsh JW, Paterson DL, Harrison LH. Integron-mediated multidrug resistance in a global collection of nontyphoidal *Salmonella* enteric isolates. *Emerg Infect Dis* 2009;15(3):388–96.
60. Centers for Disease Control and Prevention. *National Antimicrobial Resistance Monitoring System for Enteric Bacteria (NARMS): Human Isolates Final Report, 2004.* The Centers;Atlanta, GA. 2007.
61. Priyanka B, Rajashekhar KP, Dwarakanath S. A review on detection methods used for food borne pathogens. *Indian J Med Res* 2016;144(3):327–38.
62. Wong YP, Othman S, Lau YL, Radu S, Chee HY. Loop-mediated isothermal amplification (LAMP): A versatile technique for detection of micro-organisms. *J Appl Microbiol* 2017;124: 626–43.
63. Chiu L-H, Chiu C-H, Horn Y-M, et al. Characterization of 13 multi-drug resistant *Salmonella* serovars from different broiler chickens associated with those of human isolates. *BMC Microbiol* 2010;10:86.
64. Tasmin R, Hasan NA, Grim CJ, et al. Genotypic and phenotypic characterization of multidrug resistant *Salmonella* Typhimurium and *Salmonella* Kentucky strains recovered from chicken carcasses. *PLoS One* 2017;12(5):e0176938.
65. Yan L, Hongyu Z, Jian S, et al. Characterization of multidrug-resistant *Salmonella enterica* serovars Indiana and Enteritidis from chickens in Eastern China. *PLoS One* 2014;9(5):e96050.
66. Wattiau P, Weijers T, Andreoli P, et al. Evaluation of the Premi test *Salmonella*, a commercial low-density DNA microarray system intended for routine identification and typing of *Salmonella enterica. Int J Food Microbiol* 2008;123(3):293–8.
67. Braun SD, Ziegler A, Methner U, et al. Fast DNA serotyping and antimicrobial resistance gene determination of *Salmonella enterica* with an oligonucleotide microarray-based assay, *PLoS One* 2012;7:e46489.
68. Reza R, Payam B, Raheleh R. DNA microarray for rapid detection and identification of food and water borne bacteria: From dry to wet lab. *Open Microbiol J* 2017;11:330–8.
69. Pashazadeh P, Mokhtarzadeh A, Hasanzadeh M, Hejazi M, Hashemi M, de la Guardia M. Nano-materials for use in sensing of *Salmonella* infections: Recent advances. *Biosens Bioelectron* 2017;87:1050–64.
70. Ronholm J, Nasheri N, Petronella N, Pagotto F. Navigating microbiological food safety in the era of whole genome sequencing. *Clin Microbiol Rev* 2016;29:837–57.
71. Lan H, Li B. Recent and the latest developments in rapid and efficient detection of *Salmonella* in food and water. *Adv Tech Biol Med* 2017;5:244.
72. Yoshida CE, Kruczkiewicz P, Laing CR, et al. The *Salmonella in silico* typing resource (SISTR): An open web-accessible tool for rapidly typing and subtyping draft *Salmonella* genome assemblies. *PLoS One* 2016;11(1):e0147101.
73. Robertson J, Yoshida C, Kruczkiewicz P, et al. Comprehensive assessment of the quality of Salmonella whole genome sequence data available in public sequence databases using the *Salmonella in silico* typing resource (SISTR). *Microb Genom* 2018;4:e000151.
74. Ranjbar R, Elhaghi P, Shokoohizadeh L. Multilocus sequence typing of the clinical isolates of *Salmonella enterica* serovar Typhimurium in Tehran hospitals. *Iran J Med Sci* 2017;42(5):443–8.
75. Alikhan N-F, Zhou Z, Sergeant MJ, Achtman M. A genomic overview of the population structure of *Salmonella. PLoS Genet* 2018;14(4):e1007261.
76. Yang F, Jiang Y, Yang L, et al. Molecular and conventional analysis of acute diarrheal isolates identifies epidemiological trends, antibiotic resistance and virulence profiles of common enteropathogens in Shanghai. *Front Microbiol* 2018;9:164.
77. Smith S, Braun S, Akintimehin F, et al. Serogenotyping and antimicrobial susceptibility testing of *Salmonella* spp. isolated from retail meat samples in Lagos, Nigeria. *Mol Cell Probes* 2016;30:189–94.
78. Maheshwari V, Kaore NM, Ramnani VK, Sarda S. A comparative evaluation of different diagnostic modalities in the diagnosis of typhoid fever using a composite reference standard: A tertiary hospital based study in Central India. *J Clin Diagn Res* 2016;10(10):DC01–4.

79. Tennant, SM, Calman AM, Simona R, Martin LB, Khan MI. Nontyphoidal *Salmonella* disease: Current status of vaccine research and development. *Vaccine* 2016;34:2907–10.
80. Andrea H, Ursula P, Im J, Baker S, Meyer CG, Marks F. Current perspectives on invasive nontyphoidal *Salmonella* disease. *Curr Opin Infect Dis* 2017;30:498–503.

SUMMARY

Advances in molecular biology have helped speed up the detection and diagnosis of pathogens in food, clinical, and environmental samples, contributing to better decision-making during outbreaks and more effective control and management. By giving a rich insight into the molecular epidemiology of *Salmonella*, a known foodborne pathogen, this chapter highlights both culture and culture-independent methods used in the surveillance of salmonellosis and also explores recent developments in typing, serotyping, and PCR techniques employed in the study of *Salmonella*.

Genomic and Transcriptomic Analyses of *Vibrio* Infection

15

Haoran An
Tsinghua University

Dongyou Liu
Royal College of Pathologists of Australasia Quality Assurance Programs

Contents

15.1 INTRODUCTION

The genus *Vibrio* covers a large group of Gram-negative, non-spore-forming, facultatively aerobic, curved rods that occupy diverse ecological niches, including humans, animals, aquaculture, and the environment. Of over 70 species identified to date, three (i.e., *V. cholerae*, *V. parahaemolyticus*, and *V. vulnificus*) are frequent human food-/water-borne pathogens, several (e.g., *V. alginolyticus*, *V. fluvialis*, *V. furnissii*, *V. metschnikovii*, and *V. hollisae*) are occasionally involved in human infections, and a few (e.g., *V. anguillarum*, *V. salmonicida*, and *V. harveyi*) cause significant problems in aquaculture.

As the cause of cholera, *V. cholerae* is notorious for its role in at least eight large cholera epidemics in the history of mankind, leaving millions of casualties in its wake [1]. Initially described by Filippo Pacini (who called it vibrion for its motility) in 1854 from the intestinal mucosa of fatal victims in Florence, Italy,

V. cholerae was not isolated until 1883, when Robert Koch successfully obtained pure cultures of this comma-shaped bacillus on gelatin plates from victims of a cholera outbreak in India. Despite continuing intervention efforts, food-/water-borne *V. cholerae* infection is still responsible for nearly 4 million cases of acute diarrhea (or cholera) and >100,000 deaths per year in Africa, Asia, Middle East, and South and Central America [2].

V. parahaemolyticus was identified in the 1950s from human gastroenteritis cases in Japan due to consumption of raw and undercooked shellfish. The emergence of pandemic serovar O3:K6 strain/clone in Calcutta, India, in 1966, has increased the prominence of *V. parahaemolyticus* as the cause of seafood-associated gastroenteritis, particularly in the United States [3].

V. vulnificus (formerly *Beneckea vulnifica*) was isolated in 1976 in Atlanta from blood samples of patients who presented with sepsis after consumption of raw oysters. To date, *V. vulnificus* outbreaks are regularly reported in places along the Gulf Coast (e.g., New Orleans and Florida) [4].

Taxonomy. The genus *Vibrio* is classified taxonomically in the family *Vibrionaceae* [consisting of the genera *Vibrio* (Genus I), *Photobacterium* (Genus II), and *Salinivibrio* (Genus III)], order *Vibrionales*, class *Gammaproteobacteria*, and domain *Bacteria*. To date, >70 species are recognized in the genus *Vibrio*, of which three (*V. cholerae*, *V. parahaemolyticus*, and *V. vulnificus*) are regularly implicated in human food-/water-borne gastroenteritis, five (*V. alginolyticus*, *V. fluvialis*, *V. furnissii*, *V. metschnikovii*, and *V. hollisae*) are infrequent human pathogens, and three (*V. anguillarum*, *V. salmonicida*, and *V. harveyi*) are important causes of aquatic animal diseases [5].

Analysis of the sugar composition of the somatic "O" antigen (located in surface lipopolysaccharide) helps differentiate *V. cholerae* into 206 serotypes, of which serotypes O1 and O139 "Bengal" are linked to epidemic cholera worldwide and in South-East Asia, respectively. Serotype O1 strains are further separated based on biochemical differences and bacteriophage susceptibility into Classical and El Tor (hemolysin-producing) biotypes/phenotypes. Additionally, strains in both Classical and El Tor biotypes/phenotypes are subdivided into serotypes Ogawa (containing O antigens A and B), Inaba (containing O antigens A and C), and an unstable intermediate serotype Hikojima (containing O antigens A, B, and C). Furthermore, serotype O1 is distinguished into toxigenic or non-toxigenic strains according to its toxin-producing capability. O139 "Bengal" differs from O1 El Tor biotype in surface polysaccharide antigen. O2-O138 and O140-O206 (so-called non-O1 and non-O139) strains are incapable of agglutinating with "O" antiserum, although they may cause mild diarrhea and extraintestinal infections, but not epidemics. Application of ribotyping techniques allows identification of more than six ribotypes in Classical biotype, at least five ribotypes in El Tor biotype, and three ribotypes in O139 biotype.

Combined use of somatic (O) and capsular (K) antigens permits separation of *V. parahaemolyticus* into 13 O serotypes and 71 K serotypes. Notably, pandemic strain belonging to O3:K6 clone harbors a bacteriophage with several unique open reading frames involved in the regulation of virulence-associated genes. Furthermore, *V. parahaemolyticus* may be distinguished into various ribotypes, sequence types, and patterns depending on the methods used.

Similarly, *V. vulnificus* may be differentiated into one flagellar (H) serotype and seven somatic (O) serotypes, as well as three biotypes. Specifically, biotype 1 (positive for indole and ornithine decarboxylation reactions) occurs in salt or brackish water throughout the world and causes a spectrum of diseases in humans (with a mortality rate of >50%). Biotype 2 (negative for indole and ornithine decarboxylation reactions) is present in saltwater in Eastern and Western Europe and infects primarily eel (*Anguilla*) and occasionally humans. Biotype 3 (negative for o-nitrophenol production) is a hybrid of biotypes 1 and 2 and is responsible for causing serious infections requiring amputation (a mortality rate of <8%) in individuals handling freshwater fish (*Tilapia*) in Israel.

Morphology and biology. *Vibrio* spp. are Gram-negative, non-spore-forming, facultatively anaerobic, curved rods of approximately 1.4–2.6 µm in length and 0.5–0.8 µm in width. They utilize a single, sheathed, polar flagellum (*V. cholerae* and *V. vulnificus*) or multiple thin flagella projecting in all directions (*V. parahaemolyticus*) for motility and have respiratory (oxygen-utilizing) and fermentative metabolisms

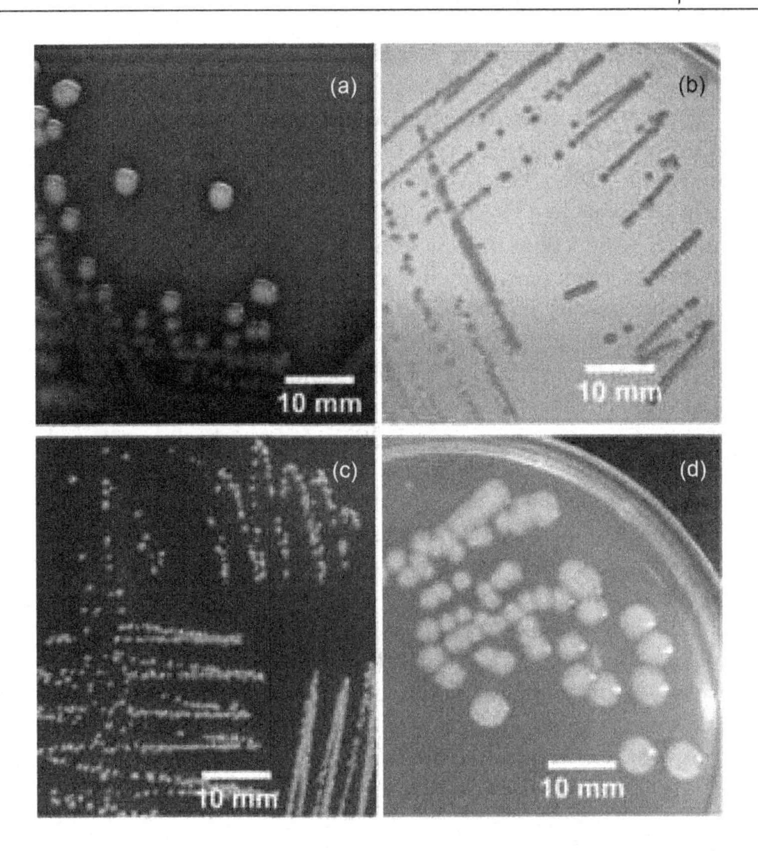

FIGURE 15.1 *Vibrio cholerae* colonies formed on blood agar (a), MacConkey agar (b), TCBS agar (c), and LB agar (d) after 24 h of cultivation. (Photo credit: Huang J, Chen Y, Chen J, et al. *PeerJ*. 2019;7:e7959.)

for producing oxidase and catalase and reducing nitrate and fermenting glucose without generating gas. Most *Vibrio* spp. grow well in standard growth media (e.g., nutrient agar/broth or media supplemented with required concentrations of NaCl (0.5%–3%) and/or a mixture of salts (preferably sea salts) (Figure 15.1) [6]. *V. cholerae* differs from other *Vibrio* spp. in that its growth is enhanced by addition of 1% NaCl and unaffected in nutrient broth without added NaCl.

Epidemiology. *Vibrio* spp. are free-living organisms with the ability to survive in the aquatic environment (fresh, estuarine, and marine waters), aquatic animals (e.g., copepods, chironomids, and fish), and plants [7]. However, a small number of *Vibrio* spp. have also adopted a lifestyle that involves humans or marine animals (e.g., fish and eels).

Human infection with *Vibrio* organisms results mainly from ingestion of raw or poorly cooked seafood, drinking of contaminated freshwater, or exposure of warm coastal waters to cut and bruises on the skin. Vectors for *V. cholerae* include zooplankton (e.g., copepods), chironomid insects, and cyanobacteria. While human *V. cholerae* infections occur throughout the world, with a total of 589,854 cases (and 7,816 deaths) notified to WHO from 58 countries in the year of 2011 alone, those caused by *V. parahaemolyticus* and *V. vulnificus* are mainly found in the southeastern USA, where raw oysters are frequently consumed or handled. Furthermore, although *V. parahaemolyticus* is 3–5 times more commonly detected than *V. vulnificus*, the former leads to fewer hospitalizations and deaths than the latter.

Clinical features. Human infections with *Vibrio* spp. may be asymptomatic or display (i) gastroenteritis (diarrhea 100%, abdominal cramps 89%, nausea 76%, vomiting 55%, fever 47%, bloody stools 29%, headache 24%, and myalgia 24%, after an incubation period of 12–52 h or average 19 h); (ii) skin and soft tissue (or wound) infection (swelling 100%, pain 100%, erythema 100%, hemorrhagic bullae 30%–50%, soft-tissue necrosis 30%–50%, and gangrene <10%, usually 3–24h after handling contaminated crabs, lobsters, or mussels; injury by fishhook within fish; or stepping on seashells, crustaceans, or stingrays); and (iii) septicemia (hypotension 100%, tachycardia 80%–90%, fever >90%, multiple hemorrhagic bullae 80%–100%, shock 50%–70%, multiple organ dysfunction 30%–50%, hypothermia <10%, and acute respiratory distress syndrome <5%, about 12–48h following the consumption of raw seafood or exposure of broken skin to warm seawater).

V. cholerae O1 and O139 infections in humans have an incubation period of 2h to 5days. While 75% of patients show no obvious symptoms, they discharge bacteria in feces for 7–14days after infection, which are infective to other people. The remaining 25% of patients may develop mild (18%), moderate (5%), and acute (2%) diseases. The acute disease (or cholera) is associated with watery diarrhea (rice water stool appearing gray and containing flecks of mucus, resulting in loss of >1 l of water and salts per h), vomiting, intense thirst, muscle cramps, loss of body weight, loss of normal skin turgor, dry mucous membranes, sunken eyes, lethargy, anuria, weak pulse, kidney failure, shock, coma, and death (especially in malnourished children or HIV-infected individuals) [8]. In patients who survive the acute disease, symptoms often subside in 3–6days, and bacteria are cleared in two weeks. *V. cholerae* non-O1 and non-O139 strains are non-choleragenic and involved mainly in sporadic, mild (non-choleragenic) gastroenteritis and invasive extraintestinal disease (septicemia).

V. parahaemolyticus infection in humans may manifest as gastroenteritis and septicemia (25% of cases). Similarly, human infection with *V. vulnificus* tends to induce acute gastroenteritis (nausea, vomiting, and abdominal pain), septicemia (fever, chills, obtundation, lethargy, or disorientation, thrombocytopenia), cutaneous lesions (e.g., necrotizing fasciitis, cellulitis, bullae, and ecchymoses in the extremities), other symptoms (peritonitis, pneumonia, endometritis, meningitis, septic arthritis, osteomyelitis, endophthalmitis, and keratitis), shock, and death (>50%).

Pathogenesis. *V. cholerae* possesses several virulence-associated genes [e.g., cholera toxin (CT) and toxin co-regulated pilus (TCP)], which are involved in host invasion and disruption of host fluid balance, leading to dehydration, hypokalemia (loss of potassium), metabolic acidosis (bicarbonate loss), and renal failure [9,10]. Similarly, *V. parahaemolyticus* produces thermostable direct hemolysin (TDH) and TDH-related hemolysin (TRH) for invasion and establishment in a host. Specifically, TDH acts as a porin in the enterocyte plasma membrane and facilitates the influx of multiple ionic species, culminating in cell swelling and death due to osmotic imbalance [11]. *V. vulnificus* employs its pili and flagella for attachment and invasion, its membrane proteins (OmpU and IlpA), hemolysin (VvhA), and metalloproteases (VvpE and VvpM) for hemolysis, cell apoptosis, and tissue necrosis, and its capsular polysaccharide (CPS, which is encoded by four essential genes *wcvA*, *wcvF*, *wcvI*, and *orf4*) for interaction with the host immune system, leading to cellular damage, cytotoxicity, and systemic disease (e.g., bullous cutaneous lesions).

Diagnosis. Conventional diagnosis of vibriosis involves isolation and subsequent identification of *Vibrio* organisms from stools and other specimens (Figure 15.1). Recent application of nucleic acid amplification techniques (e.g., PCR and emerging whole genome sequencing technology) allows improved identification and typing of *Vibrio* spp.

Treatment and prevention. Rapid fluid and electrolyte replacement (e.g., WHO/UNICEF ORS standard sachet, rice-based oral rehydration solution, and isotonic fluids) and antibiotics (e.g., tetracycline) represent viable treatment options for *V. cholerae* infection (cholera). Antibiotics (e.g., tetracycline, chloramphenicol, ciprofloxacin, sulfamethoxazole/trimethoprim, and doxycycline) along with plenty of fluids are also recommended for *V. parahaemolyticus* patients with prolonged diarrhea. Antibiotics (e.g., doxycycline and ceftazidime or cefotaxime for adults, trimethoprim-sulfamethoxazole and an aminoglycoside for children) help reduce the fatality rate associated with *V. vulnificus* infection, while surgery (e.g., aggressive surgical debridement or amputation to remove necrotic tissues) is considered for patients with severe soft tissue infection causing thrombosis of the blood vessels supplying the infected area.

Prevention of *Vibrio* infections centers on avoidance of eating raw oysters or shellfish harvested from warm salt and brackish water, personal hygiene, water treatment, emergency responses, and immunization.

15.2 GENOMIC AND TRANSCRIPTOMIC ANALYSES OF *VIBRIO* INFECTION: CURRENT STATUS

15.2.1 Overview of Genomic and Transcriptomic Techniques

Compared to **genetics** that focuses on individual genes and their roles in inheritance, **genomics** examines the structure, function, evolution, mapping, and editing of the genome. In other words, genomics studies the collective characterization, quantification, and interactions of all genes in an organism. Based on specific purposes, genomics may be divided into **structural genomics** (which uses genomic sequencing and modeling to investigate the three-dimensional structure of every protein encoded by a genome rather than one particular protein), **functional genomics** (which utilizes microarrays and bioinformatics to conduct genome-wide instead of gene-by-gene analysis of the functions of genes, RNA transcripts, and protein products, particularly the patterns of gene expression under various conditions), **epigenomics** (which relies on genomic high-throughput assays to analyze the complete set of reversible epigenetic modifications on DNA or histones that affect gene expression without alteration in DNA sequence), and **metagenomics** (which employs 16S rRNA gene sequencing, Sanger sequencing, and massively parallel pyrosequencing to examine microbial diversity on genetic material recovered directly from environmental specimens).

Transcriptomics studies transcriptome, a word first used in the 1990s to refer to the sum of all RNA transcripts in an organism, to find out whether a cellular process is active or dormant. It not only enables large-scale identification of transcriptional start sites, alternative promoter usage, and novel splicing alterations, but also facilitates disease profiling such as disease-associated single nucleotide polymorphisms (SNPs), allele-specific expression, and gene fusions.

Early techniques for investigating RNA transcripts include conversion of mRNA transcripts to complementary DNA (cDNA) using reverse transcriptase, northern blot (a low-sensitivity and low-throughput technique for displaying the size and amounts of small RNA), and serial analysis of gene expression (SAGE, which involves Sanger sequencing of concatenated random transcript fragments and subsequent quantification by matching the fragments to known genes). However, these techniques capture only a tiny subsection of a transcriptome and have been largely superseded by high-throughput techniques (e.g., cDNA microarray, RNA-seq, and quantitative PCR) that cover entire transcripts.

Microarray is a high-throughput technique that measures the abundances of a defined set of transcripts via hybridization to an array of complementary probes. Initially, expressed sequence tags (EST, a short nucleotide sequence generated from a single RNA transcript) library provided sequence information for microarray. Further improvements in microarray gene coverage and sensitivity enable detection of low-abundance transcripts.

RNA-seq (also referred to as massively parallel cDNA sequencing) is a high-throughput technique that involves reverse transcription of RNA and sequencing analysis of the resulting cDNA. It represents an unbiased method to assess transcript abundance under a variety of conditions and reveals details about genes and transcripts (e.g., alternative gene spliced transcripts; allele-specific expression identification; high sequence similarity between alternatively spliced isoforms) without a priori knowledge about the genomic features. Used on microbial pathogens, RNA-seq facilitates quantification of gene expression changes, identification of novel virulence factors, prediction of antibiotic resistance, and elucidation of host–pathogen immune interactions [12].

Reverse transcriptase quantitative PCR (RT-qPCR) is a fast, accurate, sensitive, and highly reproducible method for quantification of mRNA (limited to amplicons of <300 bp, usually near the 3′ end of the coding region, but not the 3′UTR) that helps validate transcriptomic results obtained from microarray and RNA-seq.

15.2.2 Genomic Analysis of Foodborne *Vibrio* spp.

The use of genomic techniques helps uncover critical insights on the phylogenetic relationships, virulence mechanisms, and other aspects of *Vibrio* spp. (Figure 15.2) [13,14].

V. cholerae. *V. cholerae* strain O1 biovar El Tor N16961 (pandemic strain) has a 4.03 Mb genome that is located in two chromosomes, with chromosome I consisting of 2.96 Mb with 2,690 genes, 37 pseudogenes, and 2,534 proteins; and chromosome II containing 1.07 Mb with 1,003 genes, 29 pseudogenes, and 970 proteins. While the genes involved in essential cell functions (e.g., DNA replication, transcription, translation, and cell-wall biosynthesis) and pathogenicity (e.g., toxins, surface antigens, and adhesins) are found in the large chromosome, the genes implicated in adaptation to environmental changes (e.g., an integron island, host addiction genes, and other hypothetic genes) are present in the small chromosome. *V. cholerae* possesses two virulent genome islands (GIs), that is, TCP (or VPI) and CTX–ϕ, which are closely connected to cholera disease. The TCP gene cluster (~41 kb) encodes a type IV-like pilus that participates in intestinal colonization during infection and serves as a receptor for CTX–ϕ temperate phage acquisition. The CTX–ϕ (6.9 kb) is a filamentous phage containing genes for enterotoxin CTXA/B subunits. Furthermore, many of the core genes (~1,500) are well-conserved in both O1 and non-O1 serotypes. Because of their relative conservation and large gene numbers, the core genome is more powerful in determining phylogenetic reconstruction among closely related taxa than 16S/23S rRNA and other essential housekeeping genes [15].

V. parahaemolyticus. The genome of *V. parahaemolyticus* strain RIMD 2210633 (pandemic strain) is 5.17 Mb in size, with chromosome I possessing 3.29 Mb, 3,222 genes, and 3079 proteins; and chromosome II containing 1.88 Mb, 1769 genes, and 1752 proteins. Interestingly, chromosome II (1.88 Mb) in *V. parahaemolyticus* is larger than those in *V. cholerae* (1.07 Mb) and *V. vulnificus* (1.39 Mb) [16].

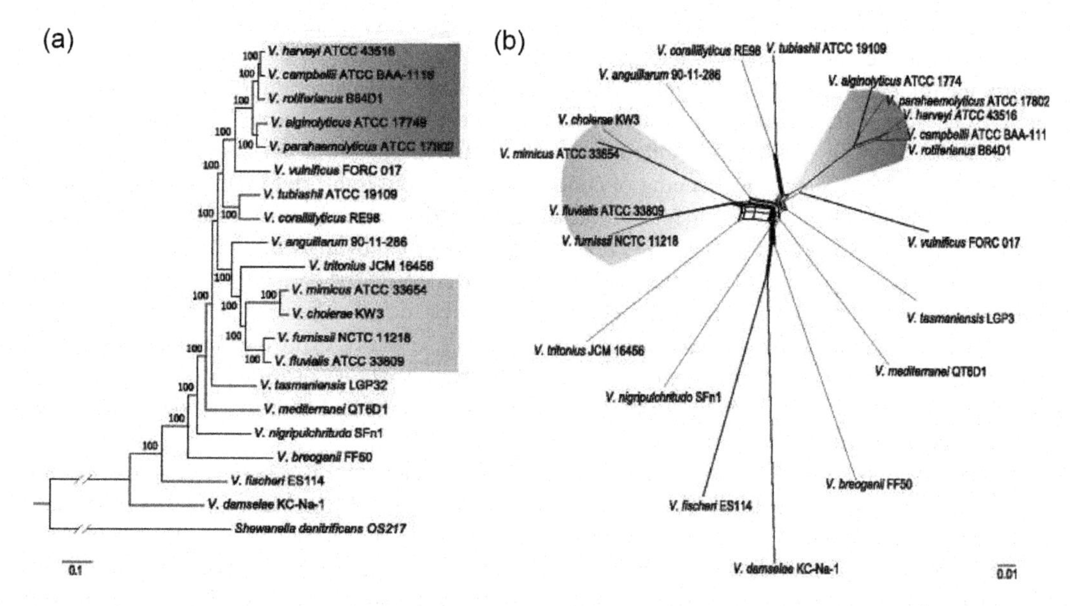

FIGURE 15.2 Phylogenetic relationships of the 20 vibrios with complete genomes. The Harveyi clade and the Cholerae clade are highlighted in red and yellow respectively. (a) Core-genome tree reconstructed by RAxML software; the tree is rooted using *S. denitrificans* OS217; the number at each node denotes the bootstrap value based on 1,000 replicates, and the scale bar indicates the number of substitutions per site. (b) Phylogenetic network reconstructed by SplitsTree4 software. (Photo credit: Lin H, Yu M, Wang X, Zhang XH. *BMC Genomics*. 2018;19(1):135.)

V. vulnificus. Sequencing analysis of *V. vulnificus* strain ASM221513vl reveals a 4.78 Mb genome organized in two chromosomes: chromosome I measures 3.39 Mb with 3,347 genes, 141 pseudogenes, and 3,047 protein; and chromosome II measures 1.39 Mb with 1,462 genes, 141 pseudogenes, and 1,390 proteins. In addition, a plasmid of 0.007 Mb with 63 genes is also found in this strain [17].

Given that vibrios experience changing environmental conditions, splitting of the genome into two replicons appears to favor rapid DNA replication, as evidenced by the doubling time of only 8–9 min in *V. parahaemolyticus*. In all *Vibrio* species, chromosome II includes a large and hypervariable segment (called the superintegron; spanning >100 kb), which comprises >200 open reading frames (with flanking sites and site-specific recombinases/integrases to entrap and acquire gene cassettes and promoters into a larger array of genes) and exists as individual mobile gene cassettes (MGC).

15.3.3 Transcriptomic Analysis of Foodborne *Vibrio* spp.

The ability of human-pathogenic *Vibrio* spp. to switch between aquatic environment and host gastrointestinal tract is possibly reflected by their seemingly effortless modulation of transcriptomic patterns for survival and proliferation. Detection of the transcriptomic profile of *Vibrio* spp. by microarray and RNA-seq is important for understanding their basic biological function and host–pathogen interactions [18].

V. cholerae. Microarray and RNA-seq analyses of *V. cholerae* have so far revealed transcriptomic changes involving 3,834 genes under 145 different experimental conditions, including the presence of serine hydroxamate, bile, and stress as well as gene deletion backgrounds (e.g., *rpoN*, *rpoH*, *cgtA*, *cpxR*, and *nqrA*) (see http://bioinfo.life.hust.edu.cn/mrvc) [19].

DNA methyltransferases (MTases) catalyze the covalent attachment of methyl moieties to specific nucleotides in the genome, a process known as DNA methylation, which represents a fundamental mechanism for epigenetic regulation in all domains of life. In bacteria, DNA MTases generate three modified DNA bases: 6-methyladenine (6mA), 4-methylcytosine (4mC), and 5-methylcytosine (5mC), which are protected from digestion by a co-transcribed cognate restriction enzyme. Some MTase genes that are not accompanied by a cognate restriction enzyme are called 'orphan' MTases, which are known to regulate diverse host cell processes. Using bisulfite sequencing, transcriptomics, and transposon insertion site sequencing, the orphan 5mC methyltransferase (VchM) in *V. cholerae* was characterized in relation to its DNA targets, genetic interactions, and the gene networks that it regulates. While VchM is required for optimal *V. cholerae* growth *in vitro* and during infection, it appears to be dispensable in the usually essential σ^E cell envelope stress pathway [20].

V. parahaemolyticus. Sun et al. [21] conducted a transcriptomic study of *V. parahaemolyticus* CHN25 response to artificial gastric fluid (AGF) stress using microarray followed by qRT-PCR and noted that 547 genes displayed increasing transcriptional levels (change ≥2.0-fold) and 663 genes had decreasing transcriptional levels (change ≤0.5-fold) in logarithmic growth phase (LGP) cells after exposure to AGF (pH 4.9) for 30 min. Interestingly, 11.6% of the upregulated genes relate to sugar transport, nitrogen metabolism, energy production, and protein biosynthesis, whereas 14.0% of the downregulated genes are ATP-binding cassette (ABC) transporters and flagellar biosynthesis genes. However, only 52 genes were upregulated and 108 genes were downregulated in stationary growth phase (SGP) cells after exposure to AGF (pH 4.9) for 30 min. It is clear from this transcriptome profile that *V. parahaemolyticus* employs distinct molecular strategies for dealing with acid stress. In a separate study, Song et al [22] examined the effect of benzyl isothiocyanate (BITC) treatment on *V. parahaemolyticus* using RNA-seq and qRT-PCR and identified 195 upregulated genes and 137 downregulated genes, suggesting that BITC may represent an effective control strategy for inhibiting *V. parahaemolyticus* growing in foods [23].

V. vulnificus. Through comparative transcriptomic analyses of *V. vulnificus* strains CMCP6 and YJ016 in either artificial seawater or human serum, Price and Gibas [12] observed that genomes with short reads are vulnerable to false positives when using a heterologous genome for RNA-Seq read alignment. However, these false positives can be largely avoided by using longer reads (>200 bp) with more depth.

15.3 FUTURE PERSPECTIVES

Human-pathogenic *Vibrio* spp. (*V. cholerae*, *V. parahaemolyticus*, and *V. vulnificus*) continue to be implicated in food- and water-borne diseases in many parts of the world despite the availability of antibiotics [24–26]. This suggests the current failure to keep *Vibrio* bacteria away from human food and water supplies and the inability to build up protective immunity in humans against *Vibrio* infection. While the former requires additional investigations into the ecology and epidemiology of *Vibrio* spp., the latter hinges on a better understanding on the molecular pathogenesis of *Vibrio* infection, for which genomic and transcriptomic approaches have unquestionably an indispensable role to play. Needless to say, genomic and transcriptomic studies of human-pathogenic *Vibrio* spp. over the past decade have uncovered many critical insights into the molecular mechanisms of *Vibrio* infection [27–36]. However, since *Vibrio* spp. are still able to outsmart humans in such an audacious way, it is obvious that they are miles ahead in terms of their ingenuity to exploit human host and their capability to thwart any human immune defense. This calls for further genomic and transcriptomic analysis of *Vibrio* spp, which will reveal vital clues on how these bacteria manage to avoid destruction by the human immune system and contribute to the development of innovative intervention measures against *Vibrio* infection.

REFERENCES

1. Orata FD, Keim PS, Boucher Y. The 2010 cholera outbreak in Haiti: how science solved a controversy. *PLoS Pathog.* 2014;10(4):e1003967.
2. Robins WP, Mekalanos JJ. Genomic science in understanding cholera outbreaks and evolution of *Vibrio cholerae* as a human pathogen. *Curr Top Microbiol Immunol.* 2014;379:211–29.
3. Espejo RT, García K, Plaza N. Insight into the origin and evolution of the *Vibrio parahaemolyticus* pandemic strain. *Front Microbiol.* 2017;8:1397.
4. Heng SP, Letchumanan V, Deng CY, et al. *Vibrio vulnificus*: An environmental and clinical burden. *Front Microbiol.* 2017;8:997.
5. Thompson FL, Iida T, Swings J. Biodiversity of vibrios. *Microbiol Mol Biol Rev.* 2004;68(3):403–31.
6. Huang J, Chen Y, Chen J, et al. Exploration of the effects of a *degS* mutant on the growth of *Vibrio cholerae* and the global regulatory function of *degS* by RNA sequencing. *PeerJ.* 2019;7:e7959.
7. Xie T, Pang R, Wu Q, et al. Cold tolerance regulated by the pyruvate metabolism in *Vibrio parahaemolyticus. Front Microbiol.* 2019;10:178.
8. Chen WD, Lai LJ, Hsu WH, Huang TY. *Vibrio cholerae* non-O1 – the first reported case of keratitis in a healthy patient. *BMC Infect Dis.* 2019;19(1):916.
9. Ceccarelli D, Hasan NA, Huq A, Colwell RR. Distribution and dynamics of epidemic and pandemic *Vibrio parahaemolyticus* virulence factors. *Front Cell Infect Microbiol.* 2013;3:97.
10. Pennetzdorfer N, Lembke M, Pressler K, Matson JS, Reidl J, Schild S. Regulated proteolysis in *Vibrio cholerae* allowing rapid adaptation to stress conditions. *Front Cell Infect Microbiol.* 2019;9:214.
11. Rivera-Cancel G, Orth K. Biochemical basis for activation of virulence genes by bile salts in *Vibrio parahaemolyticus. Gut Microbes.* 2017;8(4):366–73.

12. Price A, Gibas C. The quantitative impact of read mapping to non-native reference genomes in comparative RNA-Seq studies. *PLoS One*. 2017;12(7):e0180904.
13. Lin H, Yu M, Wang X, Zhang XH. Comparative genomic analysis reveals the evolution and environmental adaptation strategies of vibrios. *BMC Genomics*. 2018;19(1):135.
14. Ramamurthy T, Mutreja A, Weill FX, Das B, Ghosh A, Nair GB. Revisiting the global epidemiology of cholera in conjuction with the genomics of *Vibrio cholerae. Front Public Health*. 2019;7:203.
15. Schoolnik GK, Yildiz FH. The complete genome sequence of *Vibrio cholerae*: a tale of two chromosomes and of two lifestyles. *Genome Biol*. 2000;1(3):Reviews1016.
16. Chung HY, Lee B, Na EJ, et al. Potential survival and pathogenesis of a novel strain, *Vibrio parahaemolyticus* FORC_022, isolated from a soy sauce marinated crab by genome and transcriptome analyses. *Front Microbiol*. 2018;9:1504.
17. Pan J, Sun Y, Yao W, Mao H, Zhang Y, Zhu M. Complete genome sequence of the *Vibrio vulnificus* strain VV2014DJH, a human-pathogenic bacterium isolated from a death case in China. *Gut Pathog*. 2017;9:67.
18. Rahaman MH, Islam T, Colwell RR, Alam M. Molecular tools in understanding the evolution of *Vibrio cholerae. Front Microbiol*. 2015;6:1040.
19. Zhang Z, Chen G, Hu J, et al. Mr.Vc: a database of microarray and RNA-seq of *Vibrio cholerae. Database (Oxford)*. 2019;2019:baz069.
20. Chao MC, Zhu S, Kimura S, et al. A cytosine methyltransferase modulates the cell envelope stress response in the cholera pathogen. *PLoS Genet*. 2015;11(11):e1005666.
21. Sun X, Liu T, Peng X, Chen L. Insights into *Vibrio parahaemolyticus* CHN25 response to artificial gastric fluid stress by transcriptomic analysis. *Int J Mol Sci*. 2014;15(12):22539–62.
22. Song J, Hou HM, Wu HY, et al. Transcriptomic analysis of *Vibrio parahaemolyticus* reveals different virulence gene expression in response to benzyl isothiocyanate. *Molecules*. 2019;24(4):E761.
23. Pang R, Xie T, Wu Q, et al. Comparative genomic analysis reveals the potential risk of *Vibrio parahaemolyticus* isolated from ready-to-eat foods in China. *Front Microbiol*. 2019;10:186.
24. Boyd EF, Carpenter MR, Chowdhury N, et al. Post-genomic analysis of members of the family *Vibrionaceae*. Microbiol Spectr. 2015;3(5).
25. Chowdhury FR, Nur Z, Hassan N, von Seidlein L, Dunachie S. Pandemics, pathogenicity and changing molecular epidemiology of cholera in the era of global warming. *Ann Clin Microbiol Antimicrob*. 2017;16(1):10.
26. diCenzo GC, Finan TM. The divided bacterial genome: Structure, function, and evolution. *Microbiol Mol Biol Rev*. 2017;81(3):e00019-17.
27. Kim IH, Kim BS, Lee KS, Kim IJ, Son JS, Kim KS. Identification of virulence factors in *Vibrio vulnificus* by comparative transcriptomic analyses between clinical and environmental isolates using cDNA microarray. *J Microbiol Biotechnol*. 2011;21(12):1228–35.
28. Kim IH, Son JS, Wen Y, et al. Transcriptomic analysis of genes modulated by cyclo (L-phenylalanine-L-proline) in *Vibrio vulnificus*. *J Microbiol Biotechnol*. 2013;23(12):1791–801.
29. Kim D, Na EJ, Kim S, et al. Transcriptomic identification and biochemical characterization of HmpA, a nitric oxide dioxygenase, essential for pathogenesis of *Vibrio vulnificus. Front Microbiol*. 2019;10:2208.
30. Nydam SD, Shah DH, Call DR. Transcriptome analysis of *Vibrio parahaemolyticus* in type III secretion system 1 inducing conditions. *Front Cell Infect Microbiol*. 2014;4:1.
31. Williams TC, Blackman ER, Morrison SS, Gibas CJ, Oliver JD. Transcriptome sequencing reveals the virulence and environmental genetic programs of *Vibrio vulnificus* exposed to host and estuarine conditions. *PLoS One*. 2014;9(12):e114376.
32. Urmersbach S, Aho T, Alter T, Hassan SS, Autio R, Huehn S. Changes in global gene expression of *Vibrio parahaemolyticus* induced by cold- and heat-stress. *BMC Microbiol*. 2015;15:229.
33. Lambert B, Dassanayake M, Oh DH, Garrett SB, Lee SY, Pettis GS. A novel phase variant of the cholera pathogen shows stress-adaptive cryptic transcriptomic signatures. *BMC Genomics*. 2016;17(1):914.
34. Zhu C, Sun B, Liu T, et al. Genomic and transcriptomic analyses reveal distinct biological functions for cold shock proteins (VpaCspA and VpaCspD) in *Vibrio parahaemolyticus* CHN25 during low-temperature survival. *BMC Genomics*. 2017;18(1):436.
35. Sepúlveda-Cisternas I, Lozano Aguirre L, Fuentes Flores A, Vásquez Solis de Ovando I, García-Angulo VA. Transcriptomics reveals a cross-modulatory effect between riboflavin and iron and outlines responses to riboflavin biosynthesis and uptake in *Vibrio cholerae. Sci Rep*. 2018;8(1):3149.
36. Saul-McBeth J, Matson JS. A periplasmic antimicrobial peptide-binding protein is required for stress survival in *Vibrio cholerae. Front Microbiol*. 2019;10:161.

SUMMARY

The genus *Vibrio* encompasses >70 Gram-negative, non-spore-forming, curved rod species that occupy a diversity of ecological niches, including humans, animals, aquaculture, and the environment. Of these, three (i.e., *V. cholerae*, *V. parahaemolyticus*, and *V. vulnificus*) are frequent human food-/water-borne pathogens, causing acute diarrhea (cholera), seafood-associated gastroenteritis, and seafood-related sepsis, respectively. Despite the availability of antibiotics, these human-pathogenic *Vibrio* spp. continue to pose significant health risks in many parts of the world. This highlights the need to keep *Vibrio* bacteria away from human food and water supplies and to build up protective immunity in humans against *Vibrio* infection. Therefore, further investigations into the ecology and epidemiology of *Vibrio* spp. and improved understanding on the molecular pathogenesis of *Vibrio* infection are critical. There is no doubt that the application of genomic and transcriptomic approaches will help reveal valuable insights into pathogen–host interactions and contribute to the development of effective control measures against *Vibrio* spp.

Genetic Manipulation of *Escherichia coli* 16

Xi He
Hunan Agricultural University

Dongyou Liu
Royal College of Pathologists of Australasia Quality Assurance Programs

Contents

16.1 INTRODUCTION

Escherichia coli is a Gram-negative, rod-shaped bacterium that constitutes 0.1%–5% of gut microbiota in warm-blooded mammals and humans. First described from infant gut by Theodor Escherich in 1884 and previously referred to as *Bacterium coli* and *Bacillus coli*, *E. coli* represents one of six species (i.e., *E. albetii, E. coli, E. fergusonii, E. hermanii, E. marmotae,* and *E. vulneris*) in the genus *Escherichia.* Apart from non-pathogenic strains (e.g., laboratory strain K12) that are widely utilized as laboratory cloning host and for industrial production of proteins, *E. coli* also includes several pathogenic strains or pathotypes (e.g., O157:H4) implicated in human intestinal (e.g., diarrhea, dysentery, and enteritis) and extraintestinal diseases (urinary tract infection, pneumonia, neonatal meningitis, and sepsis).

Taxonomy. Classified taxonomically in the family Enterobacteriaceae, order Enterobacterales, class Gammaproteobacteria, and phylum Proteobacteria, the genus *Escherichia* currently comprises six valid species (i.e., *E. albetii, E. coli, E. fergusonii, E. hermanii, E. marmotae,* and *E. vulneris*), with *E. coli* as the type species. Most *E. coli* strains colonize the gastrointestinal tract of warm-blooded animals including humans as harmless commensal microorganisms. However, a small number of strains show the potential to cause both intestinal and extraintestinal infections in the human population.

Through their interactions with eukaryotic cells, adhesion/colonization mechanisms, toxin/virulence factor production, and clinical disease profiles, pathogenic *E. coli* strains are distinguishable into 13 pathotypes, with 10 involved in intestinal infections [i.e., enterotoxigenic *E. coli* (ETEC), enteropathogenic *E. coli* (EPEC), enterohemorrhagic *E. coli* (EHEC), enteroaggregative *E. coli* (EAEC), Shiga-toxin-producing enteroaggregative *E. coli* (STEAEC), enteroinvasive *E. coli* (EIEC), diffusely adhering *E. coli* (DAEC), cell detaching *E. coli* (CDEC), necrotoxic *E. coli* (NTEC), and adherent invasive *E. coli* (AIEC)] and 3 implicated in extraintestinal infections [i.e., septicemia causing *E. coli* (SCEC), neonatal meningitis causing *E. coli* (NMEC), and uropathogenic *E. coli* (UPEC)] (Table 16.1) [1].

TABLE 16.1 Pathogenic mechanisms and clinical characteristics of *Escherichia coli* pathotypes

PATHOTYPE	*PATHOGENIC MECHANISMS*	*CLINICAL CHARACTERISTICS*
Enterotoxigenic *E. coli* (ETEC)	Non-T3SS-dependent pathotype; intestinal colonization and elaboration of diarrheagenic enterotoxin(s), which activate adenylyl and guanylate cyclase leading to formation of cAMP and cGMP and stimulate water and electrolyte secretion by intestinal endothelial cells; major molecular diagnostic markers *elt* and *est*	Sudden onset of watery diarrhea (mild, self-limiting, or severe cholera-like), vomiting, dry mouth, rapid pulse, lethargy, decreased skin turgor, decreased blood pressure, muscle cramps, and shock due to progressive loss of fluids (dehydration) and electrolytes (sodium, potassium, chloride, and bicarbonate)
Enteroaggregative *E. coli* (EAEC)	Non-T3SS-dependent pathotype; formation of aggregative adherence (AA) pattern (prominent, 'stacked brick' auto-agglutination of the bacterial cells to each other) during adhesion to HEp-2 cells; individuals with single-nucleotide polymorphisms (SNPs) in the IL-8 gene promoter and lactoferrin gene show higher susceptibility; major molecular diagnostic markers *aggR*, *aatA*, and *aaiC*	Watery diarrhea with or without blood and mucus, abdominal pain, nausea, vomiting, and low-grade fever; persistent diarrhea in malnourished children and immunocompromised individuals (specifically HIV-infected patients)
Shiga toxin-producing enteroaggregative *E. coli* (STEAEC)	Non-T3SS-dependent pathotype; an EAEC strain (O104:H4) with EHEC phenotypes (Stx production and strong cell adherence)	Hemolytic-uremic syndrome (HUS), with a mortality rate of 1%
Diffusely adherent *E. coli* (DAEC)	Non-T3SS-dependent pathotype; exhibition of diffuse adherence (DA) to epithelial cells due to the production of adhesins; major molecular diagnostic marker Afa/Dr adhesins	Watery diarrhea
Adherent-invasive *E. coli* (AIEC)	Non-T3SS-dependent pathotype; adhesion to intestinal epithelial cells through type 1 pili (especially FimH adhesin variant) that interacts with host glycoprotein CEACAM6; replication in macrophages; transient AIEC colonization eliciting intestinal inflammation and altering microbiota composition, leading to Crohn's disease (CD)	Fever, fatigue, abdominal pain, cramping, nausea, vomiting, bloody stool, mouth sores, reduced appetite, weight loss, perianal disease, diarrhea

(*Continued*)

TABLE 16.1 (*Continued*) Pathogenic mechanisms and clinical characteristics of *Escherichia coli* pathotypes

PATHOTYPE	*PATHOGENIC MECHANISMS*	*CLINICAL CHARACTERISTICS*
Enteropathogenic *E. coli* (EPEC)	T3SS-dependent pathotype; induction of an attaching and effacing (A/E) lesion (or pedestal-like structure) on the lumenal surfaces of host small intestine, without the production of Shiga toxins; major molecular diagnostic marker *eae*	Watery diarrhea, vomiting, fever, malaise, and dehydration; bloody diarrhea after the colonization of the mid-distal small intestine (ileum)
Enterohemorrhagic *E. coli* (EHEC)	T3SS-dependent pathotype; formation of attaching and effacing (A/E) lesion in the cecum and ascending colon, and production of Shiga toxins (Stx1 and/or Stx2), leading to edema, erythema (redness), hemorrhage, erosion, occasionally long ulcer-like lesion, and marked narrowing of the luminal space; the most prevalent serotype is O157:H7; major molecular diagnostic markers *stx1* and *stx2*	Hemorrhagic colitis (afebrile bloody colitis or bloody stools with ulcerations of the bowel), sudden onset of abdominal pain, severe cramps, diarrhea within 24 h; HUS (acute renal failure, hemolytic anemia, and thrombocytopenia) due to O157:H7 infection
Enteroinvasive *E. coli* (EIEC)	T3SS-dependent pathotype; epithelial invasion of the large bowel leading to inflammation and ulceration of the mucosa (resembling *Shigella* spp.); major molecular diagnostic marker *ipaH*	Invasive inflammatory colitis with a watery diarrhea syndrome (symptoms of bacillary dysentery); scanty dysenteric stools containing blood and mucus, fever, and severe cramps in severe cases
Septicemia causing *E. coli* (SCEC)	Adhesion through non-fimH mechanisms and invasion of kidney epithelial cells and entry into the bloodstream	Sudden high fever with chills, nausea, vomiting, diarrhea, abdominal pain, hypotension, confusion, anxiety, tachypnea (short of breath), tachycardia (rapid heart rate), and uremia
Neonatal meningitis causing *E. coli* (NMEC)	Ability to survive in blood and invade meninges (cerebral microvascular endothelial cells) to cause meningitis in infants; possession of *kpsII*, K1, *neuC*, *iucC*, *sitA*, and *vat* genes	Fever, failure to thrive, neurologic signs, jaundice, decreased feeding, periods of apnea, and listlessness in neonates; irritability, lethargy, vomiting, lack of appetite, and seizures in infants of <1 month of age; neck rigidity, tense fontanels, and fever in infants of >4 months of age; headache, vomiting, confusion, lethargy, seizures, and fever in older children
Uropathogenic *E. coli* (UPEC)	Colonization of the periurethral and vaginal areas and the urethra; ascending into the bladder lumen and growing as planktonic cells in urine; adherence to the surface and interaction with the bladder and epithelium; biofilm formation; invasion and replication in the urothelium; kidney colonization and host tissue damage	Urinary tract infection (UTI) including cystitis and pyelonephritis (90% of all cases)

Moreover, serological examination of somatic (O), flagellar (H), capsular (K), and fimbrial (F) antigens separates *E. coli* strains into >700 serogroups (serotypes, e.g., O157:H7), which often belong to more than one pathotype and may correlate with certain clinical syndromes. Genotyping analyses (including DNA/DNA hybridization and 16S rRNA sequence alignment) further differentiates *E. coli* strains into six phylogenetic groups (A, B1, B2, D, E, and Shigella), which suggests a close genetic relationship between *Shigella* and *E. coli*. Multi-locus sequence typing (MLST) also helps distinguish *E. coli* strains into sequence types (STs), which allow typing of *E. coli* strains that do not express O- or H-antigens in vitro or that autoagglutinate [2].

Morphology. *E. coli* is a Gram-negative, rod-shaped, motile, non-spore-forming, facultatively anaerobic bacillus of 1–2 μm in length and 0.35–0.6 μm in width. The bacterium typically forms smooth, circular, low-convex colonies of about 3–4 mm in diameter with an entire edge on sheep blood agar (Figure 16.1); non-spreading black colonies with a characteristic greenish-black metallic sheen on EMB agar; and deep red colonies on MacConkey agar. Structurally, *E. coli* possesses fimbriae (pili, including 100–1,000 common fimbriae and 1–6 conjugative fimbriae or sex pili), 5–10 flagella (5–10 μm in length), capsule, outer membrane, periplasm and cell wall, cytoplasmic membrane, and cytoplasm. The overall compositions of *E. coli* include about 55% protein, 25% nucleic acids, 9% lipids, 6% cell wall, 2.5% glycogen, and 3% other metabolites.

Genome. *E. coli* strain MG1655 (a derivative of the commensal K-12 strain) harbors a circular genome of 4.63 Mb length with 4,288 protein-coding genes. Some *E. coli* pathotypes contain larger genomes [e.g., enterohemorrhagic *E. coli* strain O157:H7 Sakai (5.50 Mb), enteroaggregative *E. coli* strain O42 (5.36 Mb), and UPEC isolates 536 (4.94 Mb), UTI89 (5.07 Mb), and CFT073 (5.23 Mb)] in addition to 1–5 plasmids [3,4].

Biology. *E. coli* tolerates temperatures of 8°–48° (optimal 39°) and pH 6–8, but does not grow in media with >0.65 M NaCl. *E. coli* reduces nitrates to nitrites and generates succinate, ethanol, acetate,

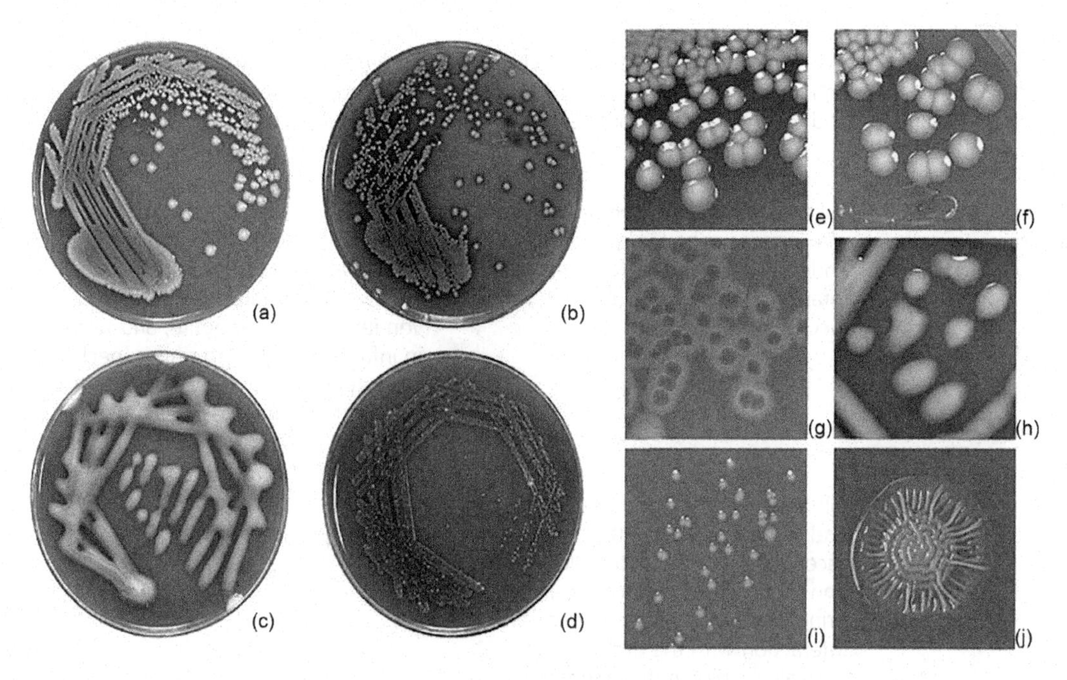

FIGURE 16.1 *Escherichia coli* colonies on sheep blood agar after 24 h at 37°C. Most strains produce smooth, circular, low-convex colonies of about 3–4 mm diameter with entire edge (a and b). Partial digestion of erythrocytes may cause discoloration of agar under colonies (b and g) and in their vicinity (e and f). Isolates from the urinary tract are often beta-hemolytic and may occasionally form mucoid colonies (c and h), small-colony variants (or dwarf colonies) (d and i), or rough colonies (j). (Photo credit: https://commons.wikimedia.org/wiki/File:Escherichia_coli_on_agar.jpg.)

and carbon dioxide. Most *E. coli* strains are positive for catalase, but negative for oxidase, citrate, urease, and hydrogen sulfide. Furthermore, *E. coli* is positive for indole production and the methyl red test. *E. coli* grows well in rich-nutrient broths and readily adapts in the mammalian intestine, including the colonic mucosa (EIEC, EHEC, and EAEC) and small intestine (ETEC and EPEC). In the intestinal mucosa, *E. coli* typically forms a complex biofilm in association with other microbes (Figure 16.2) [5] and reaches population densities of 10^6–10^9 cells per g of feces. Far from a pure parasite, *E. coli* produces vitamin K and vitamin B12 for mammalian hosts, consumes oxygen that enters the gut to enhance the survival of its anaerobic neighbors, and competes with and limits other pathogens (e.g., *Staphylococcus aureus*, which is linked to asthma, obesity, and diabetes) for the benefit of host health.

Epidemiology. Forming part of the natural flora of animals including humans as well as birds, reptiles, and fish, *E. coli* frequently contaminates soil, water source, and plants. Human infection with *E. coli* (e.g., O157:H7 and O121:H19, which have a very low infectious dose of <50 organisms) usually occurs through consumption of contaminated food products (e.g., undercooked meat, unpasteurized milk, and unclean salad leaves) and unfiltered drinking water, or through direct person-to-person spread. Pathotypes ETEC, EPEC, and EAEC represent the major causes of infantile diarrhea in the developing countries, while pathotypes EHEC, EAEC, and STEAEC are associated with food poisoning outbreaks in the developed countries. Based on data received from 27 countries in the Americas, Europe, and Western-Pacific, it is estimated that STEC alone caused 2,801,000 acute illnesses annually, 3,890 cases of hemolytic-uremic syndrome (HUS), and 230 deaths between 1998 and 2017.

Clinical features. Human infection with *E. coli* pathotypes may manifest as intestinal infection (watery diarrhea or dysentery, abdominal cramping, usually without fever, nausea, vomiting, and dehydration), urinary tract infection [UTI and occasionally HUS, which is characterized by hemolytic anemia (destruction of red blood cells), thrombocytopenia (low platelet count), and acute kidney failure], blood infection (septicemia), and meningitis. Complications may include pneumonia and intra-abdominal infection (cholecystitis, cholangitis, and peritonitis) [6].

Pathogenesis. *E. coli* pathotypes EHEC, EPEC, and EIEC utilize a type three secretion system (T3SS) (encoded by LEE, LEE, and pINV, respectively) to deliver effector proteins directly into the eukaryotic host cell for intimate attachment and attaching and effacing lesion, remodeling actin,

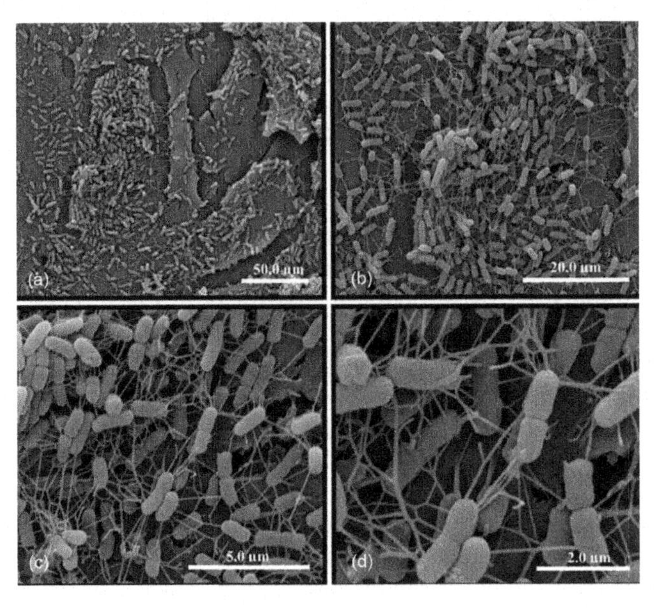

FIGURE 16.2 Scanning electron micrograph of human sepsis-causing *Escherichia coli* (SCEC) after 3 h of incubation with Vero cells. (a) Bacterial adhesion to the cell surface; (b) microcolony formation; (c) microcolony details; (d) detailed network of the extracellular matrix surrounding microcolonies on the cell surface. (Photo credit: Conceição RA, Ludovico MS, Andrade CG, Yano T. Braz *J Med Biol Res*. 2012; 45(5): 417–24.)

disrupting gut integrity, manipulating host immune response, balancing apoptosis and maintaining bacterial survival, and inhibiting phagocytosis. On the other hand, *E. coli* pathotypes ETEC, EAEC, STEAEC, DAEC, and AIEC are non-T3SS-dependent and employ pili and fimbriae for adherence and secrete toxins to facilitate entry into host cells [7].

Diagnosis. Initial laboratory diagnosis of *E. coli* infection involves complete blood count (to rule out leukocytosis, hemolysis, and thrombocytopenia), biochemical tests (to rule out dehydration, electrolyte disturbance, and uremia), and in vitro culture (sorbitol MacConkey agar to differentiate non-pathogenic *E. coli* from pathogenic *E. coli* O157; HEp-2, HeLa, Y1 adrenal and Chinese hamster ovarian cells to detect *E. coli* pathotypes). Commercially available enzyme-linked immunosorbent assays (ELISA) enable the detection of Shiga toxins in hemorrhagic stool samples. Molecular techniques (e.g., PCR) targeting toxin and other genes are used to further confirm the diagnosis [8].

Treatment and prevention. Most patients with *E. coli* infection recover in a week or so with supportive care (e.g., fluid, electrolyte, and nutrition supplements). Patients with HUS (usually 2 weeks after *E. coli* infection) require both supportive care and antibiotic treatment. Prevention of *E. coli* infection involves personal hygiene; proper cleaning, processing, and cooking of fruits, vegetables, and meat; avoidance of touching/petting farm animals; and provision of chlorinated drinking water.

16.2 GENETIC MANIPULATION OF *E. COLI*

E. coli is noted for its fast growth on chemically defined and relatively inexpensive media, its extensively characterized omic (genomic, transcriptomic, proteomic, and metabolomic) features, its genetic manipulability, its capacity to undergo bacterial conjugation, its efficiency in transferring DNA molecules into other cells, its tendency to not aggregate, its industrial scalability, and its sound biosafety. These features make non-pathogenic *E. coli* strains (e.g., K12) an ideal means for gene cloning, transferring large DNA fragments, laboratory and industrial production of proteins and biochemicals (e.g., biofuels, amino acids, sugar alcohol, and biopolymers), and other molecular applications [9–13].

Tools for genetic manipulation of *E. coli*. Plasmids and transposons represent the most common tools for genetic manipulation of *E. coli* and indeed other bacteria [14,15].

Plasmids are self-replicating DNA that provide additional features to the host. In contrast to host chromosomal DNA, plasmids generally exist as an extrachromosomal element in bacteria, although some may integrate into the host genome. Structurally, a cloning plasmid typically contains the origin of replication, regulatory elements to control expression and transcription termination (e.g., promoter, start codon, and stop codon), reporter genes [e.g., green fluorescent protein (GFP), LacZ], a multiple cloning site (MCS) for insertion of the target gene, and antibiotic selection markers (e.g., Amp^r, Kan^r, Cm^r, and Tet^r) (Figure 16.3) [16]. From the early versions (e.g., pMB1/ColE1/pBR322) in the 1970s with relatively simple structure and low capacity (producing 5–20 copies per cell), plasmids have evolved to include a few additional features and show improved yield (e.g., up to 700 copies per cell in pUC series). Plasmids not only allow for manipulation of *E. coli,* but also enable cloning and sequencing, mutant generation, protein expression, and many other applications in molecular biology. Considering that plasmids (of 2–100 kb in length) generally take up insert DNA of up to 15 kb well, but not larger fragments, bacterial artificial chromosome (BAC) is engineered to clone DNA sequences of 150–350 kb in *E. coli* and other bacteria.

Transposons (Tn, also known as jumping genes) are mobile genetic elements that contain genes related to transposition and self-propagation only and have the ability to transpose or move independently from one chromosomal location to another in the host DNA and induce mutations under the control of host factors (so-called directed mutation). Present in virtually all living organisms, transposons usually recognize specific target sites in the host chromosome. A transposon-mediated mutation within a gene may lead to its inactivation, while that in the upstream regulatory region may activate (and in rare cases inactivate) its expression.

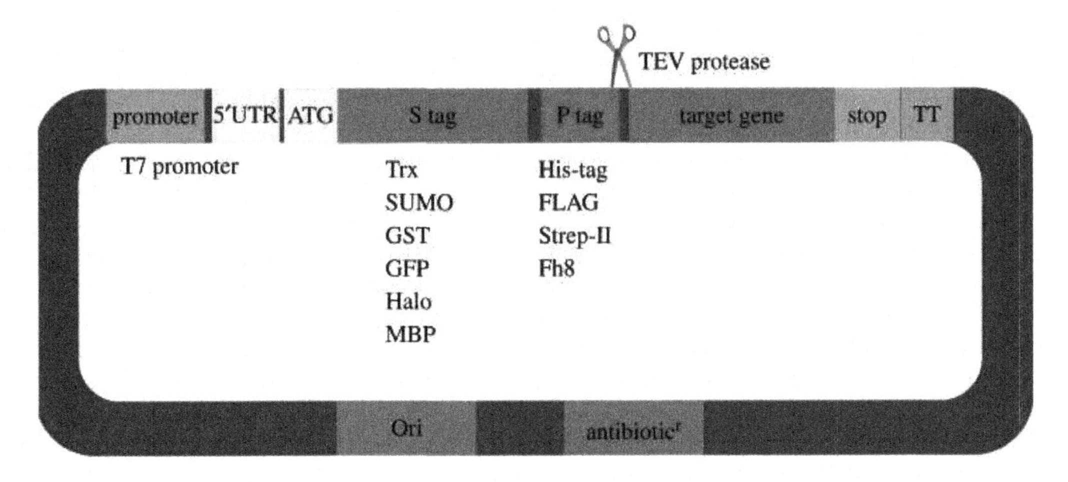

FIGURE 16.3 An exemplary plasmid vector for high-throughput expression of cytoplasmic proteins in *Escherichia coli*. Note: T7 promoter (which controls protein expression), 5′-UTR (5′-untranslated region), ATG (start/initiation codon), S tags (solubility and translation initiation tag or reporter protein), P tag (purification tag), TEV protease (for removing tags), target gene (gene of interest for protein expression), stop (stop codon), TT (transcriptional terminator), antibiotic (selection marker), and Ori (origin of replication). Use of this plasmid in *E. coli* DE3 host (which has a chromosomal copy of T7 phage RNA polymerase gene under the control of the *lac* promoter derivative *lac*UV5) permits accumulation of ~50% of total cellular proteins. (Photo credit: Jia B, Jeon CO. *Open Biol.* 2016;6(8):160196.)

Of three general classes of transposons, class I (compound/composite/non-replicative) transposons consist of Tn5 (kanamycin resistant), Tn7 (trimethoprim, streptothricin, spectinomycin, or streptomycin resistant), Tn9 (chloramphenicol resistant), and Tn10 (tetracycline resistant), which utilize transposases to catalyze 3′ and 5′ ends of the donor DNA, go through strand transfer, and insert itself into the target DNA (so-called the cut-and-paste method or conservative transposition); class II (complex/replicative) transposons, represented by Tn3 (ampicillin resistant), nick the 3′ end of the donor DNA, fuse with the target DNA through strand transfer (instead of cleaving the 5′ end of the donor DNA), form a cointegrate DNA structure, and insert into the target DNA through site-specific recombination (so-called the copy-and-paste method); and class III comprises bacteriophage Mu and related phages.

In bacteria, there exist small transposons [or insertion sequence (IS) elements] that encode transposase for catalyzing the hopping event, whereas in eukaryotes, other types of transposons (e.g., retrotransposons) are present. Compared to transposons that carry genes not directly related to insertion functions, IS elements only carry genes for insertion. Interestingly, class I transposons are flanked by IS elements, which allow their movement independent of the rest of the transposons.

Apart from plasmids and transposons, the clustered regularly interspaced short palindromic repeats-associated system (CRISPR-Cas system) has been recently developed for targeted gene editing in both prokaryotes (e.g., *E. coli*) and eukaryotes (e.g., *Saccharomyces cerevisiae*, plants, *Drosophila*, and human cell lines) [17–19]. The use of this strategy allows precise and efficient genome modifications (including gene deletion and insertion) in *E. coli* [17].

Genetic manipulation of *E. coli* for molecular cloning and gene transfer. Non-pathogenic *E. coli* strains (e.g., K12) are not only safe to handle and easy to grow, but also receptive to foreign DNA introduction and efficient for gene transfer to other bacteria via conjugation. Not surprisingly, *E. coli* offers a preferred host for molecular cloning and gene transfer.

Typically, molecular cloning involves restriction enzyme digestion of insert and plasmid DNA, ligation of digested DNA containing cohesive ends with T4 ligase, and transfer into *E. coli* cells via electroporation or rendered competent by chemical treatment (chemically competent). After growth in selective media, *E. coli* cells harboring foreign DNA are identified for further verification. Further refinements over the years have enabled T-A cloning as well as ligation-free cloning.

Like other bacteria, *E. coli* is capable of transferring DNA contained in plasmid or transposon to other bacteria by direct cell-to-cell contact or by a bridge-like connection between two cells (so-called conjugation).

Genetic manipulation of *E. coli* for the production of proteins and biochemicals. *E. coli* has been utilized widely in both research laboratories and pharmaceutical industries for in vitro production of proteins and biochemicals. In vitro expression by *E. coli* of a large number of specific proteins allows their characterization [20–22]. Large-scale production by *E. coli* of various biochemicals [from biofuels (e.g., hydrogen, bioethanol, 1-propanol, 1-butanol, 3-methyl-1-butanol, isopropanol, isobutanol, L-lactic acid, D-lactic acid, and succinic acid), amino acids (e.g., L-threonine, L-valine, L-phenylalanine, L-tryptophan, and L-tyrosine), sugar alcohols (e.g., xylitol and mannitol), to diols and polymers (e.g., PHA, taxadiene, echinomycin, anthracyclines, CoQ10)] supports environmentally friendly biotech manufacturing [23–30].

Genetic manipulation of *E. coli* for biosensors. *E. coli* biosensors using plasmid or chromosomal constructs provide cost-effective and reasonably sensitive approaches for measuring cellular processes and detecting environmental traits or hazards. The sensitivity of analytical methods is higher and more accurate, but biosensors are a good alternative for the fast detection of hazards. Also, they can be coupled with the controlled production of metabolites of commercial importance [31].

Genetic manipulation of *E. coli* for tetracycline degradation. *E. coli* containing a tetracycline enzyme (TetX monooxygenase) gene from *Bacteroides fragilis* has been shown to degrade tetracycline commonly used in poultry, cattle, and fisheries to prevent bacterial infections and to boost growth [32]. Biological degradation of tetracycline residues in wastewater provides a sustainable way to reduce antibiotic buildup and resistance.

16.3 FUTURE PERSPECTIVES

E. coli is a Gram-negative bacterial species that encompasses both non-pathogenic and pathogenic strains. While non-pathogenic strains form part of commensal microorganisms in the mammalian gut, pathogenic strains are implicated in intestinal as well as extraintestinal diseases in human populations. In light of their ability to grow well on chemically defined media and their ease of genetic manipulation, *E. coli* non-pathogenic strains have been used extensively for molecular cloning, gene transfer, and in vitro production of proteins and biochemicals. There is no doubt that with continuing advances in molecular technologies, our understanding and appreciation of *E. coli* as a pathogen and also an experimental tool will expand dramatically. This in turn will further enhance its value and versatility for laboratory and industrial applications in the years to come.

REFERENCES

1. Robins-Browne RM, Holt KE, Ingle DJ, Hocking DM, Yang J, Tauschek M. Are *Escherichia coli* pathotypes still relevant in the era of whole-genome sequencing? *Front Cell Infect Microbiol.* 2016;6:141.
2. Fratamico PM, DebRoy C, Liu Y, Needleman DS, Baranzoni GM, Feng P. Advances in molecular serotyping and subtyping of *Escherichia coli. Front Microbiol.* 2016;7: 644.
3. Lukjancenko O, Wassenaar TM, Ussery DW. Comparison of 61 sequenced *Escherichia coli* genomes. *Microb Ecol.* 2010;60(4):708–20.
4. Kurokawa M, Ying BW. Experimental challenges for reduced genomes: The cell model *Escherichia coli. Microorganisms.* 2019;8(1). pii: E3.
5. Conceição RA, Ludovico MS, Andrade CG, Yano T. Human sepsis-associated *Escherichia coli* (SEPEC) is able to adhere to and invade kidney epithelial cells in culture. *Braz J Med Biol Res.* 2012;45(5):417–24.

6. Wasiński B. Extra-intestinal pathogenic *Escherichia coli* - threat connected with food-borne infections. *Ann Agric Environ Med.* 2019;26(4):532–7.
7. Clements A, Young JC, Constantinou N, Frankel G. Infection strategies of enteric pathogenic *Escherichia coli. Gut Microbes.* 2012;3(2):71–87.
8. Peña-Gonzalez A, Soto-Girón MJ, Smith S, et al. Metagenomic signatures of gut infections caused by different *Escherichia coli* pathotypes. *Appl Environ Microbiol.* 2019; 85(24):e01820–19.
9. Yan MY, Yan HQ, Ren GX, Zhao JP, Guo XP, Sun YC. CRISPR-Cas12a-assisted recombineering in bacteria. *Appl Environ Microbiol.* 2017;83(17):e00947–17.
10. Loureiro A, da Silva GJ. CRISPR-Cas: Converting a bacterial defence mechanism into a state-of-the-art genetic manipulation tool. *Antibiotics (Basel).* 2019;8(1):18.
11. Massip C, Branchu P, Bossuet-Greif N, et al. Deciphering the interplay between the genotoxic and probiotic activities of *Escherichia coli* Nissle 1917. *PLoS Pathog.* 2019; 15(9):e1008029.
12. Shiriaeva A, Fedorov I, Vyhovskyi D, Severinov K. Detection of CRISPR adaptation. *Biochem Soc Trans.* 2020;48(1):257–69.
13. Yu X, Lin C, Yu J, Qi Q, Wang Q. Bioengineered *Escherichia coli* Nissle 1917 for tumour-targeting therapy. *Microb Biotechnol.* 2020;13(3):629–36.
14. Goodall ECA, Robinson A, Johnston IG, et al. The essential genome of *Escherichia coli* K-12. *mBio.* 2018;9(1):e02096–17.
15. Nyerges Á, Bálint B, Cseklye J, Nagy I, Pál C, Fehér T. CRISPR-interference-based modulation of mobile genetic elements in bacteria. *Synth Biol (Oxf).* 2019;4(1):ysz008.
16. Jia B, Jeon CO. High-throughput recombinant protein expression in *Escherichia coli*: Current status and future perspectives. *Open Biol.* 2016;6(8). pii: 160196.
17. Jiang Y, Chen B, Duan C, Sun B, Yang J, Yang S. Multigene editing in the *Escherichia coli* genome via the CRISPR-Cas9 system. *Appl Environ Microbiol.* 2015;81(7):2506–14. [correction: 2016;82(12):3693].
18. Hook CD, Samsonov VV, Ublinskaya AA, et al. A novel approach for *Escherichia coli* genome editing combining in vivo cloning and targeted long-length chromosomal insertion. *J Microbiol Methods.* 2016;130:83–91
19. Puttamreddy S, Minion FC. Transposon mutagenesis of foodborne pathogenic *Escherichia coli. Methods Mol Biol.* 2019;2016:73–80.
20. Zhang J, Quan C, Wang C, Wu H, Li Z, Ye Q. Systematic manipulation of glutathione metabolism in *Escherichia coli* for improved glutathione production. *Microb Cell Fact.* 2016;15:38.
21. Smolskaya S, Andreev YA. Site-specific incorporation of unnatural amino acids into *Escherichia coli* recombinant protein: Methodology development and recent achievement. *Biomolecules.* 2019;9(7):255.
22. Smolskaya S, Logashina YA, Andreev YA. *Escherichia coli* extract-based cell-free expression system as an alternative for difficult-to-obtain protein biosynthesis. *Int J Mol Sci.* 2020;21(3):928.
23. Chen X, Zou L, Tian K, et al. Metabolic engineering of *Escherichia coli*: A sustainable industrial platform for bio-based chemical production. *Biotechnol Adv.* 2013; 31:1200–23.
24. Mahalik S, Sharma AK, Mukherjee KJ. Genome engineering for improved recombinant protein expression in *Escherichia coli. Microb Cell Fact.* 2014;13:177.
25. Przystałowska H, Lipiński D, Słomski R. Biotechnological conversion of glycerol from biofuels to 1,3-propanediol using Escherichia coli. *Acta Biochim Pol.* 2015;62(1):23–34.
26. Egoburo DE, Diaz Peña R, Alvarez DS, Godoy MS, Mezzina MP, Pettinari MJ. Microbial cell factories à *la carte*: Elimination of global regulators Cra and ArcA generates metabolic backgrounds suitable for the synthesis of bioproducts in *Escherichia coli. Appl Environ Microbiol.* 2018;84(19). pii: e01337–18.
27. Li M, Nian R, Xian M, Zhang H. Metabolic engineering for the production of isoprene and isopentenol by *Escherichia coli. Appl Microbiol Biotechnol.* 2018;102(18):7725–38.
28. Cummings M, Peters AD, Whitehead GFS, et al. Assembling a plug-and-play production line for combinatorial biosynthesis of aromatic polyketides in *Escherichia coli. PLoS Biol.* 2019;17(7):e3000347.
29. de la Calle, M.E., Cabrera, G., Cantero, D. et al. A genetically engineered *Escherichia coli* strain overexpressing the nitroreductase NfsB is capable of producing the herbicide D-DIBOA with 100% molar yield. *Microb Cell Fact.* 2019;18:86.
30. Kopp J, Slouka C, Spadiut O, Herwig C. The rocky road from fed-batch to continuous processing with *E. coli. Front Bioeng Biotechnol.* 2019;7: 328.
31. Razmi N, Hasanzadeh M, Willander M, Nur O. Recent progress on the electrochemical biosensing of *Escherichia coli* O157:H7: Material and methods overview. *Biosensors (Basel).* 2020;10(5):54.
32. Mu Z, Zou Z, Yang Y, et al. A genetically engineered *Escherichia coli* that senses and degrades tetracycline antibiotic residue. *Synth Syst Biotechnol.* 2018;3(3):196–203.

SUMMARY

Escherichia coli is a Gram-negative, rod-shaped bacterium that makes up 0.1%–5% of gut microbiota in warm-blooded mammals and humans. Apart from non-pathogenic strains that form symbiotic relationships with mammalian hosts, *E. coli* also includes several pathogenic strains or pathotypes implicated in human intestinal (e.g., diarrhea, dysentery, and enteritis) and extraintestinal diseases (urinary tract infection, pneumonia, neonatal meningitis, and sepsis). Given their ability to grow well on chemically defined media and their ease of genetic manipulation, *E. coli* non-pathogenic strains have been used extensively for molecular cloning, gene transfer, and in vitro production of proteins and biochemicals. There is no doubt that with continuing advances in molecular technologies, our understanding and appreciation of *E. coli* as a pathogen and also an experimental tool will expand dramatically. This in turn will render this bacterium even more versatile for laboratory and industrial applications in the years to come.

SECTION III

Molecular Analysis and Manipulation of Foodborne Fungi

Molecular Pathogenesis of *Encephalitozoon* Infection

17

Xinming Tang
Chinese Academy of Agricultural Sciences

Xun Suo
China Agricultural University

Dongyou Liu
Royal College of Pathologists of Australasia Quality Assurance Programs

Contents

17.1 INTRODUCTION

Encephalitozoon is a microsporidian genus that covers a small number of spore-forming, obligate intracellular parasites of mammals and wildlife [1]. As an accidental host, humans (particularly those with suppressed immune functions) may acquire the infection through ingestion/drinking of food/water contaminated with *Encephalitozoon* spores. Like other microsporidians, *Encephalitozoon* had long been a low-profile organism attracting little attention. With the emergence of HIV/AIDS in the later 1980s and a rapidly aging world population, infection with *Encephalitozoon* has become not so infrequent after all. Accounting for about 10% of clinical human microsporidiosis cases, *Encephalitozoon* is second only to *Enterocytozoon* (whose type species *Enterocytozoon bieneusi* is responsible for nearly 90% of human microsporidiosis cases) in disease-causing capacity. Through establishment of in vitro and in vivo models and application of molecular techniques in the past decades, our understanding of *Encephalitozoon* biology, epidemiology, immunology, and pathogenesis has much improved. However, there is still a long way to go before *Encephalitozoon* infection is completely under our control.

Taxonomy. Microsporidia are a unique group of organisms that show characteristics of both eukaryotes (protozoa/fungi) and prokaryotes. First, microsporidia contain a nucleus, mitotic spindle-separated chromosome, an intracytoplasmatic membrane system (cytoskeleton), and polyadenylation on mRNA, typical of eukaryotes. Second, microsporidia lack visible mitochondria but appear to have remnants of mitochondria in the cytoplasm and include homologous sequences encoding mitochondria in the genome, suggestive of protozoa. Third, microsporidia possess chitin and trehalose, fungus-like cell cycles, and sequence homology in the α- and β-tubulin as well as Hsp70 (a heat shock protein or chaperonin) genes to fungi. Fourth, microsporidia have a small genome size and simplified ribosome structure (presence of 16S, 23S, and 5S rRNA, but absence of 5.8S rRNA), indicative of prokaryotes.

Not surprisingly, the taxonomy of microsporidia has been a topic of considerable debate and uncertainty, changing from protozoa to fungi. Lately, there is another attempt to move microsporidia back to protozoa, on the ground that microsporidia are vegetatively wall-less and typically phagotrophs similar to protozoa, whereas fungi are vegetatively walled osmotrophs. Until the dust settles, it is natural to consider microsporidia as fungi for the time being.

To date, about 1,500 microsporidian species have been identified and classified into over 200 genera. Human-infecting microsporidians consist of 17 species belonging to 8 genera, including those causing nonintestinal infections [i.e., *Anncaliia* (e.g., *Anncaliia algerae*, synonyms *Brachiola algerae* and *Nosema algerae; Anncaliia vesicularum*), *Nosema* (e.g., *Nosema ocularum; Nosema connori*, synonym *Brachiola connori*), *Vittaforma* (e.g., *Vittaforma corneae*, synonym *Nosema corneum*), *Microsporidium* (e.g., *Microsporidium ceylonensis*; *Microsporidium africanum*), *Encephalitozoon* (e.g., *Encephalitozoon cuniculi*; *Encephalitozoon hellem*), *Pleistophora* (e.g., *Pleistophora ronneafiei*), and *Trachipleistophora* (*Trachipleistophora hominis*; *Trachipleistophora anthropophthera*)], intestinal infection [*Enterocytozoon* (i.e., *Enterocytozoon bieneusi*)], and both intestinal and disseminated infections in AIDS patients [*Encephalitozoon* (*Septata*) *intestinalis*].

Falling under the family Unikaryonidae, suborder Apansporoblastina (membrane absent), phylum Microsporidia, and kingdom Fungi, the genus comprises six recognized species [i.e., *E. cuniculi*, *Encephalitozoon hellem*, *Encephalitozoon intestinalis* (formerly *Septata intestinalis*), *Encephalitozoon romaleae*, *Encephalitozoon lacertae*, and *Encephalitozoon pogonae*] in addition to several unassigned species. Out of these, *E. cuniculi*, *E. hellem*, and *E. intestinalis* are implicated in human infections (often involving people who have HIV/AIDS, undertake organ transplantation, or suffer CD8+ T-lymphocyte deficiency) [1].

E. cuniculi occurs in >20 mammalian species (including rabbits, rodents, carnivores, birds, monkeys, and so on) throughout the world; *E. hellem* mainly affects birds; and *E. intestinalis* shows the sporadic presence in wild animals [2,3]. Based on immunological and genotyping analyses, *E. cuniculi* is differentiated into four genotypes: ECI (rabbit, containing three ITS 5′-GTTT-3′ repeats), ECII (mouse, containing two ITS 5′-GTTT-3′ repeats), ECIII (dog, containing four ITS 5′-GTTT-3′ repeats), and ECIV (human, containing five ITS 5′-GTTT-3′ repeats); *E. hellem* consists of two genotypes (seven subtypes: 1A, 1B, 1C, 2A, 2B, 2C, and 2D); and *E. intestinalis* includes one genotype (two subtypes: 1A and 1B). Humans are susceptible to infection with any of the four *E. cuniculi* genotypes as well as *E. hellem* and *E. intestinalis* (which accounts for more human cases than the other two *Encephalitozoon* species and is second only to *Enterocytozoon bieneusi* in its capacity of causing microsporidiosis).

Having nearly identical life cycle to *E. hellem* and *E. cuniculi*, and forming a sister taxon to *E. hellem* on the basis of SSU rRNA gene analysis, *E. romaleae* occurs in the grasshopper (*Romalea microptera*) in Florida. *Encephalitozoon lacerate* is found in the European wall lizard (*Podarcis muralis*) in France. Closely related to *E. cuniculi* and *E. lacerate*, *Encephalitozoon pogonae* was previously reported as an unidentified microsporidian species or isolate of *E. cuniculi* affecting bearded dragons (*Pogona vitticeps*). Specifically, *E. pogonae* resides primarily in macrophages within foci of granulomatous inflammation of different organs of bearded dragons.

Morphology. Microsporidia are characterized by the formation of spores of various sizes (1–20 µm) depending on species. *Encephalitozoon* spores are relatively small (of 1–4 µm in diameter) (Figure 17.1c) and are covered by a thick wall of three layers: an electron-dense proteinaceous outer layer known as exospore, an electron-lucent chitinous middle layer known as endospore, and a fibrous inner layer known as the plasma membrane. Beneath the plasma membrane is the sporoplasm (cytoplasm), which contains nucleus, ribosomes, polaroplasts, and extrusion apparatus (composed of a posterior coiled polar filament/polar tubule and an anterior straight portion with an anchoring disc) (Figure 17.1b). In response to environmental stimuli (e.g., variations in pH, calcium, or other ion concentration) or in contact with a suitable host cell, the spore undergoes germination, with a long polar filament extruding from the anterior end of the spore and elongating into a hollow tube (a polar tube of 50–100 µm in length and 0.1–0.15 µm in diameter, with its diameter increasing to 0.4 µm during sporoplasm passage and its length shortening by 5%–10% after sporoplasm passage). Sporoplasm along with its nucleus flows through the polar tube and forms a droplet at its distal end. The droplet may be covered by a new membrane formed with polaroplast and released at the tip of the polar tube sometime later. If the polar tube is next to a cell, it may puncture the cell membrane and eject the sporoplasm into the cytoplasm (Figure 17.1a) [4,5]. The extrusion apparatus (polar tube) is unique to microsporidia and helps differentiate from all other organisms [6].

Genome. Comparison of whole-genome sequences from three human-infecting *Encephalitozoon* species (*E. cuniculi*, *E. hellem*, and *E. intestinalis*) and nonhuman pathogenic *E. romaleae* reveals a notable reduction in genome sizes from 2.9 Mb (*E. cuniculi*) and 2.5 Mb (*E. hellem* and *E. romaleae*) to 2.3 Mb (*E. intestinalis*), including massive gene losses (from 2,093 to 1,848 predicted ORF) and intergenic region shortenings (to average just over 100 bp) [7–12]. A major reason for the genome size differences is

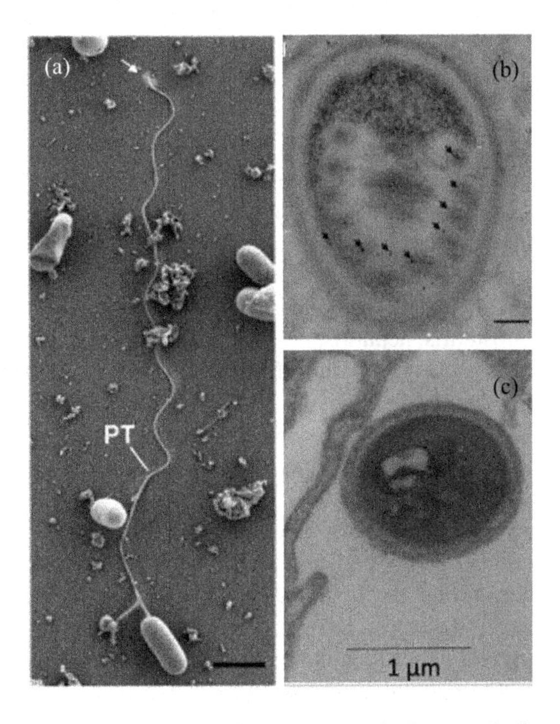

FIGURE 17.1 (a) The polar tube (PT) germinated from an *Encephalitozoon hellem* spore stained with EhPTP4 at the end (white arrow); (b) the polar tubes of an *E. hellem* spore highlighted by 10-nm colloidal gold particles (black arrows) (scale bar, 200 nm); (c) *Encephalitozoon cuniculi* spore attaching to a macrophage during infection. (Photo credits: (a & c) Han B, Polonais V, Sugi T, et al. *PLoS Pathog*. 2017;13(4):e1006341; (b) Pereira A, Alvares-Saraiva AM, Konno FTC, et al. *PLoS Negl Trop Dis*. 2019;13(9):e0007674.)

attributable to the presence of large subtelomeric regions in *E. cuniculi* and their absence in *E. intestinalis* and *Encephalitozoon* spp. [13]. Additionally, variations in genome sizes and gene contents also occur among *E. cuniculi* genotypes ECI, ECII, and ECIII. For example, the *E. cuniculi* ECI gene ECU06_0740 is absent from *E. cuniculi* ECII/ECIII genomes as well as *E. hellem* and *E. romaleae* genomes, but present in the *E. intestinalis* genome [14,15].

Life cycle and epidemiology. Like other microsporidia, *Encephalitozoon* undergoes three developmental phases in its life cycle: (i) an infective phase (involving polar tube extrusion and possible injection of the sporoplasm into the cytoplasm of the host cell); (ii) a proliferative phase (or merogony); and (iii) a spore-forming phase (or sporogony) [16]. Upon encountering a suitable host cell (e.g., enterocyte), a long polar filament extrudes from the spore and elongates into a hollow tube (polar tube), through which microsporidial sporoplasm along with its nucleus flows out, forms a droplet, which may be released from the tip of the polar tube sometime later and moves into the host cell. If a cell is close by, the polar tube may pierce the cell membrane and eject the sporoplasm into the cytoplasm [17]. The droplet (or sporoplasm) then undergoes repeated asexual divisions by binary or multiple fissions and transforms into meronts (a process termed merogony). Enclosed by plasma membrane and being larger than mature spores, meronts are rounded, irregular, or elongated simple cells with little differentiated cytoplasm and are located at the periphery of the membrane-bound parasitophorous vacuoles [18]. Meronts further develop into sporonts (a process termed sporogony) having one or two nuclei (if cell division is in progress) and a characteristic dense surface coat, which later becomes the exospore (outer layer) of the spore wall. Through binary or multiple fissions, sporonts multiply and divide into sporoblasts, which then mature into environmentally resistant, infective spores (of 1–3 µm in size) following synthesis of spore organelles.

The resulting spores are infective to adjacent host cells. When discharged with feces, sputum, or urine, these spores can survive for over a year in the environment and are highly resistant to chlorine disinfection, although they are sensitive to ultraviolet-sterilization (at 6 mJ cm^{-2}). Furthermore, the polar tube tolerates 1%–3% sodium dodecyl sulfate (SDS), 5–8 N H_2SO_4, 1–2 N HCl, chloroform, 1% guanidine HCl, 0.1 M proteinase K, and 8–10 M urea but dissolves in 50% 2-mercaptoethanol (2-ME) or 1% dithiothreitol (DTT).

Human infection with *Encephalitozoon* results largely from ingestion/drinking of food and water contaminated with spores. Furthermore, inhalation of airborne particles containing spores from disturbed excrement may represent another possible route of infection. In addition, human-to-human transmission via smear (keratoconjunctivitis), organ transplantation, or sexual contact may also occur.

Clinical features. *E. cuniculi* and *E. hellem* typically cause a range of nonintestinal diseases in immunocompromised individuals, including keratoconjunctivitis, sinusitis, bronchiolitis, pneumonia, nephritis, urethritis, cystitis, prostatitis, hepatitis, peritonitis, gastroenteritis, cholangitis prostatic abscess, tongue ulcer, cerebral infection, and renal failure. Most patients develop bilateral conjunctival inflammation and bilateral punctate epithelial keratopathy, leading to decreased visual acuity and occasionally corneal ulcers. It is notable that *E. cuniculi* affects nearly all organ systems and may induce a spectrum of diseases from no symptoms to severe disease, while *E. hellem* often infects the keratoconjunctiva, urinary tract, nasal sinuses, and bronchial system [1].

On the other hand, *E. intestinalis* is associated with both intestinal and disseminated diseases in immunocompromised persons (e.g., AIDS patients), leading to enteritis with diarrhea, nausea, vomiting, malabsorption, weight loss, perforation, peritonitis cholangitis, and cholecystitis. In particular, sclerosing cholangitis-type lesions induce progressive, irregular obstruction and dilation of the intra- and extrahepatic bile ducts, resulting in right upper-quadrant pain and an increasing alkaline phosphatase level [1].

Diagnosis. Diagnosis of microsporidiosis is possible using microscopic, serologic, and molecular techniques.

In combination with various stains (e.g., chromotrope 2R stain, modified trichrome stain, Luna calcofluor stain, chitin-binding fluorochrome, and so on), microscopic examination of feces, urine, cell culture, and tissue biopsy allows visualization of microsporidian spores and polar tubes [19].

Serologic techniques [e.g., indirect immunofluorescent-antibody testing (IFAT), enzyme-linked immunosorbent assay (ELISA), direct agglutination test (DAT), and Western blotting] are useful for species-specific detection of *Encephalitozoon* (especially *E. cuniculi*) in serum and other specimens. However, these tests are incapable of distinguishing between active and past infections.

Molecular techniques such as PCR and sequencing permit detailed analysis of *Encephalitozoon* rRNA, internal transcribed spacer (ITS), polar tube protein (PTP), spore wall protein 1 (SWP-1), and other genes and facilitate sensitive and rapid detection, speciation, and genotyping of *Encephalitozoon* spp. [20].

Treatment and prevention. *Encephalitozoon* infections may be effectively treated and controlled with albendazole (an inhibitor of microtubule assembly) or fumagillin (a natural product of *Aspergillus fumigatus*). However, systematic administration of fumagillin is linked to adverse effects (e.g., thrombocytopenia, neutropenia, and hyperlipidemia) in humans. Furthermore, as benzimidazoles have been used for treating parasitic organisms over the past 30 years, *Encephalitozoon* resistance to these drugs is not uncommon. As protease inhibitor (antiretroviral) therapy helps restore immune competence, it may be considered for HIV patients with microsporidiosis [21,22].

Prevention of *Encephalitozoon* infections involves avoidance of contaminated food and water, proper hygiene, and disinfection [23].

17.2 MOLECULAR PATHOGENESIS OF *ENCEPHALITOZOON* INFECTION: RECENT DEVELOPMENTS

Encephalitozoon spp. are obligate intracellular parasites that have evolved various mechanisms for infecting a diversity of animal hosts and to invade, replicate, and mature in many types of host cells (e.g., enterocytes, monocytes, and macrophages). It is even more remarkable that they can take advantage of host cell machineries for their growth and replication and persist inside host cells without getting eliminated by the host immune network. This suggests their ability to evade host immune surveillance through reduced provocation to and active subjugation of the host immune network. Experimental research conducted over the past decades has uncovered invaluable clues to the molecular basis of *Encephalitozoon* pathogenicity.

Host cell invasion. *Encephalitozoon* spores possess extrusion apparatus in the form of coiled polar filament/polar tubule and anterior anchoring disc. When a host cell is nearby, the coiled polar filament (polar tubule) extrudes from the spore (driven by changing osmotic pressure within the spore), inverts, elongates, and forms a hollow polar tube, allowing the sporoplasm as well as the nucleus to migrate and release from the tip of the polar tube and form a new cell. Mediated by the spore wall protein EnP1, the new cell binds to glycosaminoglycans on the plasma membrane and subsequently enters into the cytoplasm. Additionally, some spores may employ their polar tube to pierce through the host cell plasma membrane and inject the sporoplasm as well as the nucleus directly into the cytoplasm of the host cell and transform into a new parasite.

To date, five spore wall proteins are found in *Encephalitozoon* spp., including two exospore proteins [spore wall protein 1 (SWP1; absent in meronts; present in early sporonts) and SWP2 (present in mature spores)] and three endospore proteins [endospore protein 1 (Enp1), Enp2, and *E. cuniculi* chitin deacetylase-like protein (EcCDA)]. Enp1 is probably an adherence ligand allowing the parasite to attach to host cells and potentially modulate infection. Furthermore, three proline-rich polar tube proteins are identified in *Encephalitozoon* spp. [polar tube protein 1 (PTP1), PTP2, and PTP3]. Spore wall proteins are involved in the maintenance of *Encephalitozoon* spores, while polar tube proteins contribute to the normal function of *Encephalitozoon* polar tube [4,24].

Additionally, *Encephalitozoon* spores may enter the host cell via phagocytosis, even in nonphagocytic cells, with the formation of phagocytic vacuoles along the way, that allow spores to hatch and deposit sporoplasms, thus avoiding digestion.

Immune response and evasion. Host resistance to *Encephalitozoon* infection relies on a robust production of IL-12 by macrophages or dendritic cells, which in turn enhances the production of Th1 cytokines (e.g., IFN-γ by γδ T cells) in the circulation and tissues. The availability of a large quantity of IFN-γ upregulates class I molecules and stimulates the proliferation of antigen-specific $CD8^+$ T-cells, leading to lysis of the infected cells by a perforin-dependent mechanism and efficient control of parasite replication [25–28].

However, *Encephalitozoon* spp. have acquired several mechanisms to evade host immune killing. For instance, *E. intestinalis* shows a capacity to inhibit dendritic cell differentiation through an IL6-dependent mechanism. Upon exposure to high loads of *E. intestinalis* spores, dendritic cell cultures upregulate the surface expression of MHC class II and costimulatory molecules CD86 and CD40 as well as the secretion of IL-6, resulting in the formation of immature infected and mature bystander dendritic cells. This IL-6-dependent inhibition of dendritic cell differentiation and maturation appears to provide a means for *E. intestinalis* spores to evade host immune attack [29]. Furthermore, *E. cuniculi* spores demonstrate an inhibitory effect on staurosporine-induced apoptosis while suppressing pro-apoptosis genes and upregulating anti-apoptosis genes in human macrophage-differentiated cell line THP1, offering another potential way to evade the host immune system [30]. Although host humoral response is largely ineffective against microsporidian infection, B-1 cells (a subtype of B cells involved in the secretion of natural antibodies and IL-10 as well as other cytokines) may act as antigen-presenting cells and mediate M1 macrophage profile that contributes to the amelioration of microbicidal functions and disruption/reduction of *E. cuniculi* mechanisms involved in the evasion of host immune response [5].

It should be noted that the host immune response to *Encephalitozoon* infection may sometimes bring out undesirable consequences and cause various clinical symptoms. For example, invasion of enterocytes by *Encephalitozoon* spores elicits the production of certain cytokines (e.g., IL8), which activate resident phagocytes and attract phagocytes from blood to into lamina propria. The activated leukocytes then secrete prostaglandins, leukotrienes, platelet-activating factor, histamine, serotonin, and adenosine, which induce neurotransmitter-mediated intestinal secretion of chloride and water and inhibit the absorption of epithelial cells. Additionally, the activated mast cells produce proteases and oxidants, which hinder epithelial cell growth and induce villus atrophy and crypt hyperplasia. Altogether, these changes contribute to malabsorption of vitamin B_{12}, D-xylose, and fat and result in osmotic diarrhea, necrosis, and disruption of other tissues and organs [31,32].

17.3 FUTURE PERSPECTIVES

Encephalitozoon is a remarkable intracellular parasite that demonstrates the capacity to infect a diverse range of animals (including humans) and grow and replicate in a broad spectrum of host cells. In addition, *Encephalitozoon* forms spores for extended survival under harsh conditions and employs its polar tube for efficient entry into host cells. What is more, *Encephalitozoon* displays enormous ingenuity to somehow evade host immune surveillance and blunt host immune defense. Compared to what *Encephalitozoon* can do and respond, we have much to learn. For example, we do not know precisely why *E. cuniculi* and *E. hellem* mainly cause nonintestinal diseases while *E. intestinalis* induces both intestinal and systematic diseases. We still do not fully understand what *Encephalitozoon* does to evade the host immune system and thwart/divert host immune attack. We also have no clue about the best approach to treat and prevent *Encephalitozoon* infection [33]. Until we find answers to these and other pertinent questions, there is no room for complacency.

REFERENCES

1. Mathis A, Weber R, Deplazes P. Zoonotic potential of the microsporidia. *Clin Microbiol Rev.* 2005;18(3):423–45.
2. Hinney B, Sak B, Joachim A, Kváč M. More than a rabbit's tale - *Encephalitozoon* spp. in wild mammals and birds. *Int J Parasitol Parasites Wildl.* 2016;5(1):76–87.
3. Vergneau-Grosset C, Larrat S. Microsporidiosis in vertebrate companion exotic animals. *J Fungi (Basel).* 2015;2(1). pii: E3.
4. Han B, Polonais V, Sugi T, et al. The role of microsporidian polar tube protein 4 (PTP4) in host cell infection. *PLoS Pathog.* 2017;13(4):e1006341.
5. Pereira A, Alvares-Saraiva AM, Konno FTC, et al. B-1 cell-mediated modulation of M1 macrophage profile ameliorates microbicidal functions and disrupt the evasion mechanisms of *Encephalitozoon cuniculi. PLoS Negl Trop Dis.* 2019;13(9):e0007674.
6. Sibley LD. Invasion and intracellular survival by protozoan parasites. *Immunol Rev.* 2011;240(1):72–91.
7. Katinka MD, Duprat S, Cornillot E, et al. Genome sequence and gene compaction of the eukaryote parasite *Encephalitozoon cuniculi. Nature.* 2001;414(6862):450–3.
8. Corradi N, Pombert JF, Farinelli L, Didier ES, Keeling PJ. The complete sequence of the smallest known nuclear genome from the microsporidian *Encephalitozoon intestinalis. Nat Commun.* 2010;1: 77.
9. Peyretaillade E, Parisot N, Polonais V, et al. Annotation of microsporidian genomes using transcriptional signals. *Nat Commun.* 2012;3:1137.
10. Pombert JF, Selman M, Burki F, et al. Gain and loss of multiple functionally related, horizontally transferred genes in the reduced genomes of two microsporidian parasites. *Proc Natl Acad Sci USA.* 2012;109(31):12638–43.
11. Selman M, Sak B, Kváč M, Farinelli L, Weiss LM, Corradi N. Extremely reduced levels of heterozygosity in the vertebrate pathogen *Encephalitozoon cuniculi. Eukaryot Cell.* 2013;12(4):496–502.
12. Xiang H, Zhang R, Butler RR 3rd, et al. Comparative analysis of codon usage bias patterns in microsporidian genomes. *PLoS One.* 2015;10(6):e0129223.
13. Dia N, Lavie L, Faye N, et al. Subtelomere organization in the genome of the microsporidian *Encephalitozoon cuniculi*: patterns of repeated sequences and physicochemical signatures. *BMC Genomics.* 2016;17:34.
14. Pombert JF, Xu J, Smith DR, et al. Complete genome sequences from three genetically distinct strains reveal high intraspecies genetic diversity in the microsporidian *Encephalitozoon cuniculi. Eukaryot Cell.* 2013;12(4):503–11.
15. Pelin A, Moteshareie H, Sak B, et al. The genome of an *Encephalitozoon cuniculi* type III strain reveals insights into the genetic diversity and mode of reproduction of a ubiquitous vertebrate pathogen. *Heredity (Edinb).* 2016;116(5):458–65.
16. Izquierdo F, Moura H, Bornay-Llinares FJ, et al. Production and characterization of monoclonal antibodies against *Encephalitozoon intestinalis* and *Encephalitozoon* sp. spores and their developmental stages. *Parasit Vectors.* 2017;10(1):560.
17. Heinz E, Hacker C, Dean P, et al. Plasma membrane-located purine nucleotide transport proteins are key components for host exploitation by microsporidian intracellular parasites. *PLoS Pathog.* 2014;10(12):e1004547.
18. Lee SC, Heitman J. Dynamics of parasitophorous vacuoles formed by the microsporidian pathogen *Encephalitozoon cuniculi. Fungal Genet Biol.* 2017;107:20–23.
19. Visvesvara GS. In vitro cultivation of microsporidia of clinical importance. *Clin Microbiol Rev.* 2002;15(3):401–13.
20. Robertson LJ, Clark CG, Debenham JJ, et al. Are molecular tools clarifying or confusing our understanding of the public health threat from zoonotic enteric protozoa in wildlife? *Int J Parasitol Parasites Wildl.* 2019;9:323–41.
21. Anane S, Attouchi H. Microsporidiosis: epidemiology, clinical data and therapy. *Gastroenterol Clin Biol.* 2010;34(8–9):450–64.
22. Kotkova M, Sak B, Kvetonova D, Kvac M. Latent microsporidiosis caused by *Encephalitozoon cuniculi* in immunocompetent hosts: a murine model demonstrating the ineffectiveness of the immune system and treatment with albendazole. *PLoS One.* 2013;8(4):e60941.
23. Furuya K. Spore-forming microsporidian encephalitozoon: current understanding of infection and prevention in Japan. *Jpn J Infect Dis.* 2009;62(6):413–22.
24. Xu Y, Weiss LM. The microsporidian polar tube: a highly specialised invasion organelle. *Int J Parasitol.* 2005;35(9):941–53.

25. Khan IA, Moretto M, Weiss LM. Immune response to *Encephalitozoon cuniculi* infection. *Microbes Infect.* 2001;3(5):401–5.
26. Ghosh K, Weiss LM. T cell response and persistence of the microsporidia. *FEMS Microbiol Rev.* 2012;36(3):748–60.
27. Moretto MM, Harrow DI, Hawley TS, Khan IA. Interleukin-12-producing CD103+ CD11b- CD8+ dendritic cells are responsible for eliciting gut intraepithelial lymphocyte response against *Encephalitozoon cuniculi. Infect Immun.* 2015;83(12):4719–30.
28. Sak B, Kotková M, Hlásková L, Kváč M. Limited effect of adaptive immune response to control encephalitozoonosis. *Parasite Immunol.* 2017;39(12): e12496.
29. Bernal CE, Zorro MM, Sierra J, et al. Encephalitozoon intestinalis inhibits dendritic cell differentiation through an IL-6-dependent mechanism. *Front Cell Infect Microbiol.* 2016;6:4.
30. Sokolova YY, Bowers LC, Alvarez X, Didier ES. Encephalitozoon cuniculi and Vittaforma corneae (Phylum Microsporidia) inhibit staurosporine-induced apoptosis in human THP-1 macrophages in vitro. *Parasitology.* 2019;146(5):569–79.
31. Texier C, Brosson D, El Alaoui H, Méténier G, Vivarès CP. Post-genomics of microsporidia, with emphasis on a model of minimal eukaryotic proteome: a review. *Folia Parasitol (Praha).* 2005;52(1–2):15–22.
32. Li Z, Hao Y, Wang L, Xiang H, Zhou Z. Genome-wide identification and comprehensive analyses of the kinomes in four pathogenic microsporidia species. *PLoS One.* 2014;9(12):e115890.
33. García-Torres I, De la Mora-De la Mora I, Hernández-Alcántara G, et al. First characterization of a microsporidial triosephosphate isomerase and the biochemical mechanisms of its inactivation to propose a new druggable target. *Sci Rep.* 2018;8(1):8591.

SUMMARY

Encephalitozoon is a microsporidian genus that covers several spore-forming, obligate intracellular parasites of mammals and wildlife. Being an accidental host, humans may acquire the infection through ingestion/drinking of food/water contaminated with *Encephalitozoon* spores, leading to intestinal and/or systematic diseases. Accounting for about 10% of clinical human microsporidiosis cases, *Encephalitozoon* is second only to *Enterocytozoon bieneusi* (which is responsible for nearly 90% of human microsporidiosis cases) in disease-causing capacity. Through establishment of in vitro and in vivo models and application of molecular techniques in the past decades, our understanding of *Encephalitozoon* biology, epidemiology, immunology, and pathogenesis has much improved. Nonetheless, until today, we know little about why *E. cuniculi* and *E. hellem* mainly cause nonintestinal diseases whereas *E. intestinalis* induces both intestinal and systematic diseases. Similarly, we still do not fully understand what *Encephalitozoon* does to evade the host immune system and thwart/divert host immune attack. Only until we find answers to these and other pertinent questions, we may be able to design effective countermeasures against *Encephalitozoon* infection.

18 Molecular Detection and Quantification of Aflatoxin-Producing Molds

María J. Andrade, Alicia Rodríguez, Juan J. Córdoba, and Mar Rodríguez
University of Extremadura

Contents

18.1 INTRODUCTION

Aflatoxins are toxic metabolites produced mainly by *Aspergillus flavus* and *Aspergillus parasiticus* in foods. *A. flavus* usually produces only aflatoxins B_1 and B_2 and not all of its strains isolated from the natural habitats are aflatoxin-producers at least under laboratory conditions. In contrast, most strains of *A. parasiticus* are known to produce aflatoxins B_1, B_2, G_1, and G_2.[1]

As aflatoxin contamination occurs in a variety of foods, the availability of rapid and reliable procedures to detect aflatoxin-producing molds as well as mycotoxins in food commodities is important.[2] In addition, prompt detection of aflatoxigenic molds in the foodstuffs helps prevent aflatoxins from entering the food chain. Traditional methods for the detection of aflatoxins as well as aflatoxigenic mold strains rely on culture techniques, chemical analyses, enzyme-linked immunosorbent assay (ELISA), and chromatographic techniques. Despite their relatively high sensitivity and specificity, these techniques are often slow, tedious, and dependent on mycological expertise. The recent development of molecular methods has greatly improved the detection and quantification of aflatoxigenic molds in foods. The most common molecular methods are PCR-based [e.g., conventional PCR, real-time PCR (qPCR), reverse transcription PCR (RT-PCR), and reverse transcription real-time PCR (RT-qPCR)], although the emergence of loop-mediated isothermal amplification (LAMP) of DNA and DNA biosensors provides some useful alternatives to PCR procedures. Apart from PCR, LAMP, and DNA biosensors that target nucleic acids of aflatoxigenic molds, metabolic non-destructive techniques (e.g., biosensors, color imaging, electronic nose, fluorescence, infrared spectroscopy, or hyperspectral imaging (HSI) have been also developed for detecting aflatoxigenic molds and aflatoxins in foods.

This chapter gives an overview of the genomic and metabolic non-destructive methods described in the literature for detecting and quantifying aflatoxigenic molds as well as aflatoxins present in foods or produced by isolates of aflatoxigenic molds.

18.2 DETECTION AND QUANTIFICATION OF AFLATOXIN-PRODUCING MOLDS IN FOODS BY GENOMIC METHODS

A range of genomic methods has been developed for the detection and quantification of aflatoxigenic molds in feed and food commodities. The success of these techniques depends largely on the availability of adequate DNA or RNA. Therefore, both the sample preparation procedures and the genomic assays currently available in the literature are discussed in this section.

18.2.1 Mold DNA and RNA Extraction Methods

Extraction of nucleic acids is of utmost importance, especially when working with food products, due to its influence on the sensitivity of the following used method. Food components, such as fats, polysaccharides, polyphenols, and other secondary compounds, could inhibit the activity of the polymerase in the PCR reactions with the subsequent reduction of sensitivity.[3] Concretely, quality and yield of the obtained nucleic acid are two critical factors, being necessary to adapt the extraction procedure to the food matrix.[4] Furthermore, the mold cell wall is a complex structure composed of chitin, glucans, lipids, and other polymers that are extremely resistant, which is a major challenge when extracting mold nucleic acids.[5]

Regarding DNA, several protocols resulting in suitable purity and quantitative yields have been described for different contaminated food commodities. Remarkably, no single method of DNA extraction is appropriate for all the food matrices. Some of the methods described in the literature are in-house developed protocols using lysis buffers, such as cetyltrimethylammonium bromide (CTAB) or sodium dodecyl sulfate (SDS), followed by treatment with chloroform, phenol, or phenol-chloroform and subsequent precipitation with isopropanol or ethanol. Before the addition of the lysis buffer, a physical disruption treatment based on grinding the sample in a mortar and pestle together with liquid nitrogen is commonly used. Glass beads have been also used for the disruption of mold material.[6,7] During the extraction, the use of digestion enzymes, such as Proteinase K, lyticase, and RNAse, has been extensively proposed too. These laboratory extraction methods have been reported for direct DNA extraction from different foods including corn,[8] figs,[9,] and several other feed commodities.[10]

Moreover, some commercial DNA extraction kits have been developed to extract DNA from aflatoxigenic molds. For instance, Chen et al.[11] used the DNeasy Plant Mini Kit from Qiagen after grinding artificially contaminated peanut kernels in a mortar with a lysis buffer and liquid nitrogen. Mayer et al.[12] also used such a kit after grinding artificially infected pepper, paprika, and maize kernels. Sardiñas et al.[13] optimized a proper protocol using polyvinylpyrrolidone into the DNeasy Plant Mini Kit buffers to eliminate the polyphenols, which could interfere in PCR reactions.

In addition, the combination of laboratory extraction methods with a commercial DNA kit extraction has been described as an appropriate extraction method for the following detection of aflatoxigenic molds by PCR[14] and qPCR.[15,16] Both used the EZNA® Fungal DNA kit (Omega Bio-tek, Inc.) for purifying the extracted DNA.

Some authors have also included an enrichment step before DNA extraction to dilute the PCR inhibitors of the matrix with the subsequent improvement of the sensitivity of the method.[8] The drawback of this procedure is the expense of the time needed for the detection.[4]

Concerning the extraction and purification of aflatoxigenic mold RNA, there are few available methods to be used in routine food analysis, since the procedure is notoriously difficult due to its co-purification and co-precipitation with polysaccharides, proteins, and polyphenols, which generally compromise the isolation of RNA.[17] Furthermore, changes in the gene expression may occur throughout the handling of samples and RNA isolation.[4] Until now, only methods to extract RNA from culture media inoculated with aflatoxigenic molds have been reported. These methods include different extraction buffers, physical grinding methods, and purification using commercial kits or organic solvents. A classical method consisting of pulverization of the mold mycelia with liquid nitrogen in a mortar and pestle before using an extraction buffer composed of NaCl, EDTA, Tris-HCl and SDS, phenol:chloroform:isoamyl alcohol, and LiCl for RNA precipitation was described by Sweeney et al.[18] A method combining TRIZOL® (Sigma Aldrich®) as the extraction buffer and an amalgamator for the physical disruption of the mycelia has been also reported for aflatoxigenic molds.[19]

Furthermore, different commercial kits have been used in the literature, such as the Spectrum Plant Total RNA Kit protocol from Sigma Aldrich®[20,21] and the RNeasy kit from Qiagen,[22,23] after the physical disruption of mycelia. Leite et al.[24] developed a suitable RNA extraction method combining physical grinding by using beat-beating with the RNeasy kit. The use of a mechanical beat-beating method has several advantages over the traditional methods such as the reduction in time, multiple sample extraction, a reduction of the risk of cross-contamination, and the avoidance of using liquid nitrogen, which is frequently problematic and hazardous.[24]

After obtaining the mold RNA, a treatment with the RNase-Free DNase is commonly carried out to eliminate possible trace amounts of contaminating DNA and to ensure the successful application of the following selected RT-PCR method.

18.2.2 Detection and Quantification by DNA-Based Methods

DNA-based methods are currently considered good tools for the detection and quantification of toxigenic molds including the aflatoxigenic ones because of their high sensitivity, specificity, and rapidness. These methods have been mainly based on conventional PCR and qPCR. The latter procedures are the most valuable since they combine the sensitivity of conventional PCR with the quantification of specific DNA targets.[5] Additionally, qPCR is faster and less prone to cross-contamination than conventional PCR and has potential for automation.[25] By virtue of the abovementioned advantages of the qPCR, several protocols based on both SYBR® Green and TaqMan® methodologies have been set up in the last years for detecting and quantifying aflatoxigenic molds. Nevertheless, it is necessary to remark that there is no specific PCR for any one of the biologically produced aflatoxins.[26] Recently, LAMP has been considered an appropriate alternative to PCR-based methods, and also few DNA biosensors are available for aflatoxigenic molds.

18.2.2.1 PCR

One of the most important aspects to be taken into account when developing PCR-based methods is the target DNA region of the organism of interest.[5] Because genes involved in the aflatoxin biosynthetic pathway appear to be exclusively present in molds potentially producing aflatoxins, such genes have been used in several PCR and qPCR methods for their sensitive and accurate detection (Table 18.1). The regulation of the biosynthesis of aflatoxins is a complex process involving several interconnecting networks.[27] As many as 30 genes are potentially involved in aflatoxin biosynthesis.[28] The aflatoxin pathway genes are organized within a DNA cluster. Two genes, *aflR* and *aflS* (formerly *aflJ*), located in this gene cluster are the pathway-specific regulatory genes.[27–29] Other physically unrelated genes, *veA* and *laeA*, have shown to have a key role in the production of aflatoxins.[28,30,31] Most of the available PCR methods are based on the aflatoxin regulatory (*aflR*) gene and the aflatoxin pathway genes *aflD* (formerly *nor-1*), *aflM* (formerly *ver-1*), and *aflP* (formerly *omtA*). Chang et al.[32] stated that the coordinate transcription of the latter three genes is activated by the *aflR* gene.

TABLE 18.1 Genomic methods for the detection and quantification of the main aflatoxigenic molds related to food

MOLD	• *TARGET GENE*[a]	*FOOD MATRIX*	*DETECTION LIMIT*	*REFERENCES*
PCR				
• *A. flavus, A. parasiticus*	• *aflD, aflM, aflP*	• Corn	• 10^2 spores g^{-1}	• [8]
• *A. flavus*	• *aflD, aflM*	• Fig	• -	• [9]
• *A. flavus, A. parasiticus*	• *aflP*	• Cooked meat products, fig, almond, pistachio	• 2–3 log cfu g^{-1}	• [14]
• *A. flavus*	• ITS	• Wheat flour	• 10^2 spores g^{-1}	• [38]
• **Nested PCR**				
• *A. flavus*	• *aflR*	• Groundnut, maize	• 2 log cfu g^{-1}	• [34]
• **Multiplex PCR**				
• *A. flavus, A. parasiticus*	• *aflD, aflM, aflP*	• -	• -	• [35]
• *A. flavus, A. parasiticus*	• *aflR, aflD, aflM, aflP*	• -	• -	• [37]
• *A. parasiticus*	• *aflR, aflD, aflM, aflP*	• Peanut kernels	• -	• [11]
• **Real-Time PCR**				
• *A. flavus*	• *aflD*	• Pepper, paprika, maize kernel	• -	• [12]

(*Continued*)

TABLE 18.1 (*Continued*) Genomic methods for the detection and quantification of the main aflatoxigenic molds related to food

MOLD	• *TARGET GENE*[a]	*FOOD MATRIX*	*DETECTION LIMIT*	*REFERENCES*
• **Real-Time PCR**				
• *A. flavus, A. parasiticus*	• *aflP*	• Wheat, raisin, nuts (almond, peanut, walnut), spices (black pepper, paprika, oregano), ripened meat products, ripened cheese	• 1–2 log cfu g^{-1}	• [16]
• *A. flavus, A. parasiticus*	• *aflD*	• Peanut	• -	• [7]
• *A. flavus, A. parasiticus*	• ITS2 rDNA	• Wheat flour	• 10^2 spores mL^{-1}	• [40]
• **RT-PCR**[b]				
• *A. parasiticus*	• *aflQ, aflR*	• -	• -	• [18]
• *A. flavus, A. parasiticus*	• *aflD, aflG, aflH, aflI, aflK, aflM, aflO, aflP, aflQ, aflS, aflR*	• -	• -	• [47]
• *A. flavus*	• *aflD, aflS, aflQ, aflR*	• -	• -	• [19]
• **RT-qPCR**[b]				
• *A. flavus*	• *aflD, aflS, aflO, aflR*	• -	• -	• [22]
• *A. flavus*	• *aflR*	• -	• -	• [21]
• *A. flavus, A. parasiticus*	• *aflP, aflR, aflS*	• -	• -	• [20]
• *A. flavus*	• *aflD, aflR*	• -	• -	• [48]
• **LAMP**				
• Aflatoxin producers in *Aspergillus* section *Flavi*	• *aflD*	• Rice, maize, raisin, fig, hazelnut, almond, paprika, ginger	• 211 spores/reaction (after disruption)	• [43]
• *A. flavus, A. parasiticus A. nomius*	• *amy1, acl1*	• Peanut, Brazil nut, maize	• $10–10^5$ spores g^{-1}	• [45]

• [a] *aflD*: formerly *nor-1*; *aflM*: formerly *ver-1*; *aflP*: formerly *omtA* or *omt1*; *aflS*: formerly *aflJ*; and *aflQ:* formerly *ord1*.
[b] The housekeeping gene *β-tubulin* was used as an endogenous control.

These procedures have been mainly developed for the detection of aflatoxigenic mold species belonging to *Aspergillus*, mainly *A. parasiticus* and *A. flavus* (Table 18.1), the major species of concern in food commodities. Both species share a 90%–99% identity in aflatoxin-clustered genes,[3,33] which allows their detection with the same PCR system (primers/reaction).[34]

Manonmani et al.[34] proposed a nested PCR method based on the *aflR* gene for evaluating the presence of aflatoxigenic molds in foods. They assayed the specificity of the method with pure and mixed mold cultures, achieving only amplification for *A. flavus* and *A. parasiticus*. For the former species, the developed method showed a low sensitivity in groundnuts and maize (Table 18.1). Luque et al.[14] described the application of a PCR based on the *aflP* gene able to detect aflatoxigenic *A. flavus* and *A. parasiticus* in a wide range of food products (Table 18.1).

It is noteworthy that most of the authors have used more than one gene, which code for different stages in the aflatoxin biosynthetic pathway strengthening the validity of a particular strain's identification as aflatoxigenic.[4,8] Concretely, Shapira et al.[8] used PCR based on three genes involved in aflatoxin biosynthesis to selectively distinguish *A. parasiticus* and *A. flavus* from other molds. The PCR showed a high sensitivity after a 24 h enrichment step of artificially contaminated corn (Table 18.1). The same three genes were targeted by the method developed by Färber et al.[9] to be applied to detect aflatoxigenic *A. flavus* in contaminated figs (Table 18.1). Furthermore, Geisen et al.[35] set up a multiplex PCR method able to detect aflatoxin-producing molds by targeting three aflatoxin biosynthetic genes as target sequences at once (Table 18.1) with the subsequent savings of reagents, time, and effort of analysis.[5] A triplet banding pattern was obtained with aflatoxin-producing strains of *A. flavus* and *A. parasiticus* and also with *A. versicolor* producer of sterigmatocystin. This PCR result is not surprising, as sterigmatocystin is a toxic intermediate of aflatoxins biosynthesis and the same enzymes should be needed for its biosynthesis. In relation to this, Paterson et al.[36] considered that any primers designed for aflatoxin species in food should be carefully tested with numerous exclusively sterigmatocystin-producing molds. However, the multiplex PCR reaction developed by Geisen et al.[35] appeared to distinguish aflatoxigenic *A. parasiticus* and *A. flavus* from most of the non-aflatoxigenic food-related mold strains. Similarly, Chen et al. and Criseo et al.[11,37] developed PCR methods using sets of primers for four genes (Table 18.1) able to detect aflatoxigenic strains of *A. parasiticus* and *A. flavus*. While Chen et al.[11] used the primers previously reported by Geisen et al.[35] for the *aflD* and *aflM* genes and those described by Shapira et al.[8] for the *aflP* and *aflR* genes, Criseo et al.[37] used the three primers set described by Geisen et al.[35] for the *aflD, aflM,* and *aflP* genes and that described by Shapira et al.[8] for the *aflR* gene. Additionally, the quadruplex PCR set up by Chen et al.[11] was successfully applied to detect *A. parasiticus* in artificially contaminated peanut kernels. Regarding differentiation of aflatoxigenic from non-aflatoxigenic strains, it was not always possible when using both quadruplex PCR systems, which is in accordance with the results obtained by Geisen et al.[35].

Because the PCR-based methods reported to detect aflatoxigenic mold strains targeting aflatoxin biosynthetic genes do not discriminate between *A. parasiticus* and *A. flavus*, González-Salgado et al.[38] developed a PCR protocol based on the multicopy internal transcribed region of the rDNA unit (ITS1-5.8S-ITS2 rDNA) for the specific detection of *A. flavus*. Discrimination between *A. flavus* and *A. parasiticus* is vital because they have different secondary metabolite profiles. The developed PCR allowed to detect a low level of spores after an incubation of 16 h (Table 18.1). However, it has to be emphasized that the detection of *A. parasiticus* and *A. flavus* using genes different from those involved in the aflatoxin biosynthesis is no guarantee of their aflatoxigenicity.[26]

18.2.2.2 Real-Time PCR

Several qPCR methods mainly based on the *aflD* gene for detecting aflatoxigenic *A. flavus* and *A. parasiticus* from different food products are found in the literature. Mayer et al.[12] outlined a TaqMan® qPCR method based on the *aflD* gene to detect aflatoxigenic *A. flavus* in pepper, paprika, and maize kernels. They obtained a good correlation between the quantification of the gene of the aflatoxin biosynthetic pathway and the cfu of the mold. The same gene was used by Passone et al.[7] for developing a qPCR system to detect and quantify *A. flavus* and *A. parasiticus* in stored peanuts (Table 18.1). No statistical correlation between aflatoxin load and the *aflD* gene was obtained by Iheanacho et al.[39] when applying a TaqMan® qPCR to feed samples. Rodríguez et al.[16] set up two reliable qPCR methods based on SYBR® Green and TaqMan® technologies for quantifying aflatoxin-producing molds in a wide range of food commodities, including cereal, raisins, nuts, spices, and ripened meat and dairy products (Table 18.1).

Apart from the genes involved in the aflatoxin biosynthetic pathway, the multicopy ITS2 rDNA has been used as a target for the detection and quantification of *A. flavus* and *A. parasiticus* by two qPCR protocols.[40] A low detection limit of the methods was obtained after a 16 h incubation of artificially contaminated wheat flour (Table 18.1).

18.2.2.3 Loop-Mediated Isothermal Amplification

The LAMP technique consists of a DNA-based technology that enables rapid, specific, and sensitive diagnosis of microorganisms with a minimum of technical apparatus and sample processing effort.[41] This method can detect the presence of target nucleic acids in a sample as evidence for the presence of the organism by which it has been produced.[41,42] DNA amplification is done under isothermal conditions using a DNA polymerase and a set of four specially designed primers with six distinct sites in the target DNA, which enable highly specific amplification of such nucleic acid.[43,44] It is not therefore dependent on sophisticated laboratory equipment needed for other genomic detection technologies.[41]

Only a few LAMP methods have been developed for the detection of aflatoxigenic mold strains. In a recent study, an *aflD* gene-specific LAMP assay was developed for the detection of aflatoxin-producing *Aspergillus* spp.[43]. The usefulness of such assay was demonstrated by analyzing naturally contaminated rice, maize, hazelnuts, almonds, figs, raisins, and powdered paprika and ginger. Furthermore, good correlations between a LAMP signal and the presence of *Aspergillus* section *Flavi* and aflatoxins in rice samples were reported. The great advantage of this assay was that spores could be directly used in the analysis and that no DNA needs to be isolated. These authors used the *aflD* gene because it was the only gene for which sequences were available in GenBank from a substantial number of different mold species. This LAMP assay was slightly less sensitive when compared to the assay previously published by Luo et al.[45] (Table 18.1). The latter authors used the three sets of LAMP primers developed by Luo et al.[6] targeting the *acl1* gene of *A. flavus* and the *amy1* gene of *A. nomius* and *A. parasiticus*. The *acl1* gene is a housekeeping gene coding for ATP citrate lyase subunit 1. The *amy1* gene coding for alpha-amylase is indirectly connected to aflatoxin biosynthesis in *Aspergillus* spp.[46]. Luo et al.[6,45] described an alternative rapid and reliable method for the detection of the aflatoxigenic species *A. flavus*, *A. nomius*, and *A. parasiticus* to estimate contamination levels in commodities, such as Brazil nuts, peanuts, maize, or green coffee beans.

18.2.3 RNA-Based Methods

RNA can be used to evaluate the expression of genes involved in the aflatoxin biosynthesis. Such gene expression and, subsequently, the aflatoxin production can vary depending on environmental and nutritional factors, which has not been fully elucidated yet. Therefore, the presence or lack of mRNAs may permit direct differentiation between mycotoxin-producing and non-producing strains.[4] Because the mycotoxin production always follows the gene transcription, when the window between gene activation and mycotoxin production is long enough, predictions could be made and, consequently, some preventive or corrective actions might be taken in the food industry for preventing aflatoxin accumulation in foods.[21] Although the Northern hybridization analyses are reliable techniques for the detection of mRNAs in aflatoxigenic molds, its sensitivity cannot be sufficient for the detection of low levels of gene transcription.[18] Nonetheless, RT-PCR has been demonstrated to be a more sensitive, quicker, simpler, and safer alternative compared to the Northern hybridization analysis.[18] Thus, RT-PCR methods have aimed to detect the expression of the genes involved in the aflatoxin biosynthesis for evaluating how environmental or stressful conditions affect the production of aflatoxins. Sweeney et al.[18] developed an RT-PCR method to investigate the effect of different physiological factors on the aflatoxin gene expression in *A. parasiticus*. Two primer pairs were designed for the *aflQ* and *aflR* genes (Table 18.1). The procedure allowed to correlate the aflatoxin production with the detection of the transcription of the two analyzed genes in culture media. Scherm et al.[47] differentiated aflatoxin-producing from aflatoxin-non-producing strains of *A. flavus* and *A. parasiticus* by analyzing the expression of nine structural genes and two regulatory genes of the aflatoxin B_1 biosynthetic pathway (Table 18.1). Three (*aflD*, *aflO*, and *aflP*) of the 11 evaluated genes were identified as the most consistently correlated with the ability to produce aflatoxins.

Regarding the multiplex RT-PCR method, Degola et al.[19] developed a quintuplex RT-PCR procedure for the detection of aflatoxigenic strains of *A. flavus*. The detection of the expression of the *aflD*, *aflS*, *aflQ*, and *aflR* genes was carried out (Table 18.1).

RT-qPCR has been also reported as highly sensitive to quantify small changes in gene expression. Concretely, the expression of genes involved in aflatoxin biosynthesis in response to variation of temperature, water activity, and CO_2 exposure has been evaluated. Thus, Gallo et al.[22] studied the effect of different temperature and water activity values on the gene expression of *A. flavus* on an almond-based medium. They focused on the expression of the two regulatory genes, *aflR* and *aflS*, and two structural genes, *aflD* and *aflO*, involved in the early and late stages of the aflatoxin biosynthesis pathway, respectively (Table 18.1). They reported that temperature seems to act as a crucial aspect influencing aflatoxin production, which is strictly correlated to the stimulation of *aflD* and *aflO* gene expression, but not to that of aflatoxin regulatory genes, whose functional products are most likely subordinated to other regulatory processes acting at the post-translational level. Likewise, Bernáldez et al.[21] considered that the regulatory *aflR* gene alone is not a good indicator for aflatoxin B_1 production of *A. flavus* in a maize agar (Table 18.1). In contrast, the expression of *aflR* and *aflS* genes was reported to have a strong correlation with the aflatoxin production of *A. flavus* and *A. parasiticus* in a dry-cured ham-based medium[20] (Table 18.1). Medina et al.[48] developed an RT-qPCR based on the TaqMan® methodology to evaluate the impact of interactions between water activity, temperature, and CO_2 exposure on the expression of the *aflD* and aflR genes as well as growth and aflatoxin B_1 production on a conducive medium (Table 18.1). On the other hand, Baquião et al.[23] reported the existence of correlation between the ratio of *aflS/aflR* expression and the biosynthesis of aflatoxin and cyclopiazonic acid (CPA) in *A. flavus* strains that exclusively produced aflatoxin B_1 or CPA or that produced both mycotoxins. These authors used the *β-tubulin* and *calmodulin* as endogenous control genes.

Other RNA-based methods, such as microarrays and RNA-sequencing, can be found in the literature to evaluate the impact of environmental factors on the stimulation of genes involved in aflatoxin biosynthesis. However, there are still few studies dealing with such an issue. Concretely, Schmidt-Heydt et al.[1] used a toxin gene cluster microarray to evaluate the regulation of the aflatoxin cluster genes of *A. flavus* in relation to changing temperature and water activity environmental conditions. A correlation between the ratio of *aflR/aflJ* expression and the increase of aflatoxin B_1 biosynthesis was obtained. Regarding the RNA-sequencing, it was used by Medina et al.[49] to study the impact of different abiotic factors on the transcriptome and aflatoxin biosynthesis of *A. flavus.*

18.2.4 DNA Biosensors

DNA biosensors, based on nucleic acid recognition processes, allow to carry out rapid, simple, and inexpensive analyses. However, scarce works based on DNA biosensors are found in the literature for aflatoxigenic molds. Tombelli et al.[10] developed a DNA-based piezoelectric biosensor specific for the detection of a previously generated amplicon of the *aflD* gene of *A. flavus* and *A. parasiticus.* The biosensor was successfully applied on PCR products obtained from samples derived from four different animal feeds, such as barley, maize, and soybean flour, and a complete corn feed for pigs, previously colonized by *A. flavus* or *A. parasiticus.* The detection limit was established for 0.04 μM of the amplicon. Similarly, Sedighi-Khavidak et al.[50] developed an impedimetric electrochemical DNA sensor using gold nanoparticles on the glassy carbon electrode for the sensitive detection of the *aflD* gene of *A. flavus* in pistachio. Its detection limit was established for 0.55 nm.

18.3 DETECTION AND QUANTIFICATION OF AFLATOXIN-PRODUCING MOLDS IN FOODS BY METABOLIC NON-DESTRUCTIVE METHODS

Non-destructive techniques for testing internal or superficial contamination, without interfering in any way with the presentation of the foods, has been developed. These procedures have been developed to overcome the disadvantages of conventional invasive methods such as chromatographic or genomic methods.

Non-destructive methods are suitable for testing both mycotoxin-producing molds and mycotoxins. They have recently become a major area of interest[51] because there is an urgent requirement for the development of rapid, robust, and cost-effective alternative technologies for real-time monitoring of aflatoxins.[52]

Techniques such as biosensors, color imaging, electronic nose, fluorescence, infrared spectroscopy, HSI, and so on have been applied for aflatoxigenic molds and/or aflatoxin detection in foods.

18.3.1 Biosensors

Biosensors provide new analytical tools that are sensitive, fast, and portable. The receptors developed for mycotoxin detection are biomolecules and synthetic chemicals. Being small and neutral molecules, mycotoxins can be better detected through labels like enzymes, nanoparticles or quantum dots, redox molecules, and fluorophores.[53] Several types of biosensors have been described to detect aflatoxins, including optical, electrochemical, and piezoelectric biosensors.

Optical immunoassays have been used to detect aflatoxins combining optical measurement with antibody–antigen interaction. The used techniques are evanescent wave biosensors, surface plasmon resonance, colorimetry, chemiluminescence, and fluorescent assays.[52] Several authors have described suitable optical immunoassay biosensors for evaluating aflatoxin contamination in milk, rice, and cereals (Table 18.2).

TABLE 18.2 Non-destructive methods for the detection of aflatoxigenic molds and aflatoxins (AFs) in foods

METHOD[a]	• *OBJECT*[b] *(MOLD/AFs)*	*DETECTION LEVEL*	*FOOD MATRIX*	*REFERENCES*
Evanescent wave (OB)	• AF M_1	• 0.045 ng mL^{-1}	• Milk	• [66]
• **Surface plasmon-enhanced (OB)**	• AF M_1	• 0.018 ng mL^{-1}	• Milk	• [67]
• **Colorimetry (OB)**	• AFs	• 0.011 ng mL^{-1}	• Rice	• [68]
• **Fluorescence (OB)**	• AF B_1	• 0.04 ng mL^{-1}	• Rice	• [69]
• **Amperometry (EB)**	• AF M_1	• 0.01 ng mL^{-1}	• Milk	• [70]
• **Impedimetry (EB)**	• AF B_1	• 5 ng mL^{-1}	• Rice	• [71]
• **Voltammetry (EB)**	• AF B_1	• 0,0035 ng mL^{-1}	• Corn	• [72]
• **Potentiometry (EB)**	• AF B_1	• 0.087 ng g^{-1}	• Peanut	• [73]
• **Piezoelectric biosensor**	• AF B_1	• 0.01 ng mL^{-1}	• Milk	• [74]
• **Color imaging**	• *A. flavus* *A. flavus* AF B_1	• 89% 75% 0.003 ng g^{-1}	• Maize Maize Groundnut	• [55] [75,76]
• **Electronic nose**	• Total AFs *A. flavus* *Aspergillus* sp.	• 100% 100% 86.84%	• Maize Wheat Peanut	• [56] [77] [78]
• **Fluorescence spectroscopy**	• Total AFs AF B_1, AF B_2 *A. flavus*	• 0.012 ng g^{-1} 0.014 ng g^{-1} 100%	• Nut, pistachio, cashew Peanut Pistachio	• [79] [80] [81]
• **Near-infrared spectroscopy**	• Total AFs and *Aspergillus* sp. Total AFs *A. parasiticus* *A. flavus*	• 100% 96.9% 100% 98%	• Fig Rice Maize -	• [82] [60] [83] [55]
• **Mid-infrared spectroscopy**	• Total AFs *Aspergillus* sp. AF B1	• 90.6% 100% 0.008 ng g^{-1}	• Rice Peanut Peanuts	• [60] [84] [85]

(Continued)

TABLE 18.2 (*Continued*) Non-destructive methods for the detection of aflatoxigenic molds and aflatoxins (AFs) in foods

METHOD[a]	• *OBJECT*[b] *(MOLD/AFs)*	*DETECTION LEVEL*	*FOOD MATRIX*	*REFERENCES*
• **Hyperspectral imaging**	• AF B_1	• 82.5%	• Maize	• [61]
	AF B_1	98%	Maize	[86]
	A. flavus	90%–100%	Pea, lentil, bean	[87]
	Aflatoxigenic molds	94%–97%	Peanut	[88]
• **Thermal imaging**	• *A. flavus*	• 92%–100%	• Pistachio	• [64]
	Aspergillus sp.	96%–100%	Wheat	[65]

• [a] OB: optical biosensor; EB: electrochemical biosensor.
[b] AF B_1: aflatoxin B_1; AF M_1: aflatoxin M_1; AF B_2: aflatoxin B_2.

Electrochemical sensing is based on changes in current, potential, or impedance according to the used sensing biological element. The electrochemical biosensors employed to determine aflatoxins can be voltammetric, amperometric, conductometric, potentiometric, and impedimetric in nature.[52] Electrochemical biosensors have been reported for being used when investigating aflatoxin contamination in milk, rice, peanuts, and corns (Table 18.2).

Piezoelectric biosensors rely on the piezoelectric effect and are able to exchange mechanical input to an electric output or vice versa. These biosensors are suitable for affinity-based sensors, especially for immunosensors. Various piezoelectric biosensors have been designed for the onset recognition of aflatoxins in foods, using anti-aflatoxin B_1 on a gold-coated quartz crystal.[52] The piezoelectric biosensor has been used for the detection of aflatoxin contamination in milk (Table 18.2).

18.3.2 Color Imaging

A color imaging system essentially consists of a sample-holding platform, a digital camera for capturing the image, an image capture board for digitizing the image, a light source for proper illumination, and computer hardware and software to process the images. Machine vision systems can provide rapid and accurate information about external quality aspects of food grains,[54] including damage caused by aflatoxigenic molds.[55] This technique is mainly useful for food grains, like for the detection of *A. flavus* contamination in maize and aflatoxin B_1 in groundnuts (Table 18.2).

18.3.3 Electronic Nose

The electronic nose system consists of an array of electronic chemical gas sensors with different selectivity, a signal collection unit that converts the sensor signal to a readable format, and software analysis of the data to produce characteristic output related to the encountered odor. The interpretation of the signal is accomplished using multivariate techniques such as pattern recognition algorithms, discriminant functions, cluster analysis, and artificial neural networks.[51]

The underlying hypothesis for developing electronic nose-based sensors for food safety/quality evaluation is that the growth and metabolism of mycotoxin-producing molds are associated with the production of off-flavors and changes in the volatile compound composition potentially detectable by the instrument.[56] The electronic nose has been reported by several authors to be used in cereals and peanuts for aflatoxins and *A. flavus* contamination (Table 18.2).

18.3.4 Spectroscopy Techniques

The measurement techniques based on fluorescence spectroscopy (FS), near-infrared spectroscopy (NIRS), mid-infrared spectroscopy (MIRS), and HSI have provided interesting and promising results for the detection of aflatoxins and mold contamination in several varieties of foods. Concretely, these rapid and non-destructive techniques have been used for the detection of aflatoxins as well as aflatoxigenic molds in different varieties of agricultural products, such as corn, rice, wheat, peanuts, almond kernels, pistachio nuts, dried figs, chili peppers, and so on.[57,58]

18.3.4.1 Fluorescence Spectroscopy

FS is the emission of light after absorption of ultraviolet, visible, or infrared light of a fluorescent molecule or substructure, called fluorophore. The fluorophore absorbs energy in the form of light at a specific wavelength and liberates energy in the form of emission of light at a longer wavelength.

Aflatoxins B and aflatoxins G can emit fluorescence in the bright-blue (425–480 nm) and blue-green (480–500 nm) spectral range, respectively, which makes possible the detection of aflatoxin contamination by their fluorescence characteristics.[57] These aflatoxins with fluorescence properties thus can only be identified if no, or very low, background fluorescent constituents are present.[59] However, nuts and dried fruits contain various fluorescent constituents that render the aflatoxin determination difficult.[58] This technique has been used for the detection of aflatoxigenic molds in different agricultural products[57] and aflatoxin contamination (Table 18.2).

18.3.4.2 Near-Infrared Spectroscopy

The near-infrared region (NIR) in the electromagnetic spectrum is defined to be from 780 to 2,526 nm, which is located between the red band of the visible light and the mid-infrared regions. The NIRS detection technique depends on the interaction between incident light and molecules in samples, the result of molecular overtones, and combined vibrations associated mainly with C–H, O–H, and N–H functional groups.[57,58]

NIRS is the most widely applied spectroscopic technique for determining the physicochemical quality of agricultural products. NIRS is also a promising approach for the non-destructive detection of toxigenic molds and aflatoxin contamination in a wide variety of agricultural products due to its high detection efficiency and current low cost.[58] Concretely, the NIRS has been used for the detection of aflatoxin-producing molds and aflatoxin contamination in cereals and dried fruits (Table 18.2).

18.3.4.3 Mid-Infrared Spectroscopy

The mid-infrared region (MIR) addresses vibrations of substances based on the radiation from 2,526 to 25,000 nm, which is sufficiently energetic for the MIR range to excite molecular vibrations to higher energy levels. The main difference between NIR and MIR is that absorption in the MIR range corresponds to the fundamental frequency of molecular vibrations, whereas absorption in NIR corresponds to overtones and combinations of vibrations.[60] MIRS has a few advantages because it gathers additional information from vibrations of all functional groups, including C–O and C=O, not only the C–H, O–H, and N–H groups that are observed with NIRS.[58] MIRS has been proven to be a very promising tool for the detection of aflatoxigenic molds and aflatoxins in various agricultural products (Table 18.2).

18.3.4.4 Hyperspectral Imaging

HSI is a technique that integrates spectroscopic and imaging techniques to provide both spectral and spatial information of the tested sample simultaneously. The collected data are arranged into a three-dimensional data cube, commonly called a hypercube, with two spatial dimensions and one spectral dimension, which are made up of hundreds of contiguous wavebands for each spatial position of the tested sample.[51,57] The basis of HSI is that all biological materials reflect, scatter, absorb, and emit electromagnetic energy in a distinctive pattern at specific wavelengths.

The spectral wavelength range can be divided into visual near-infrared hyperspectral imaging (VIS-NIR HSI) (400–1000 nm) and short-wave infrared hyperspectral imaging (SWIR HSI) (1,000–2,500 nm).[61] Aflatoxin B_1 discriminant models, using controlled samples in a controlled environment, were built by the VIS-NIR HSI and SWIR HSI spectra, respectively, and their accuracies were 98% and 88%.[62] But the VIS-NIR HSI method encounters both false-positive and false-negative detection, because the measured bright greenish-yellow fluorescence could be also emitted by other intermediate metabolites, i.e., kojic acid and oxidized form of kojic acid. Thus, the VIS-NIR HIS for aflatoxin approach is only used as a presumptive test method.[61]

HSI has found wide application in the detection, identification, and discrimination of aflatoxigenic molds and aflatoxins in cereals, legumes, peanuts, and so on (Table 18.2). The main limiting factors of the use of HIS include the high cost of equipment and the lengthy time needed for pre-processing of the data and classification.[51]

18.3.5 Thermal Imaging

Thermal imaging (TI) is a non-invasive analytical tool suitable for the food industry. The basic principle of TI is that all materials emit infrared irradiation, which is a band of invisible light found on the electromagnetic spectrum with wavelengths of 0.75–100 μm.[63] Therefore, the technique utilizes the radiation to produce a pseudo image of the thermal distribution of the body surface.[51]

NIR imaging systems also exist, but they measure diffuse reflected radiation rather than emitted radiation. Thermography, which measures a large number of point temperatures of the target surface, is a powerful tool for visualizing and analyzing targets with thermal gradients.[63]

The main limitations of such a technique consist of the difficulty of the image to be accurately interpreted when based upon certain objects with erratic temperatures and the potential influence of the thermal interference from the ambient environment on the test results.[63]

TI has been applied by some researchers for identifying mold infection in cereals and pistachio[64,65] (Table 18.2).

18.4 CONCLUSIONS AND FUTURE PERSPECTIVES

Different molecular methods are available for the detection and quantification of aflatoxigenic molds in foods. These procedures essentially include conventional PCR, real-time PCR (qPCR), reverse transcription PCR (RT-PCR), and reverse transcription real-time PCR (RT-qPCR). Although PCR and qPCR methods are able to detect such molds, they do not assure the aflatoxin production. Therefore, the generation of the expected amplification product by PCR or qPCR does not necessarily imply that the analyzed strain can produce aflatoxins because these techniques do not allow to obtain information about the expression of the genes. Thus, RT-PCR and RT-qPCR seem to be the future genomic methods to detect aflatoxigenic molds since they can be used to prevent mycotoxin production at the post-transcriptional level of gene expression. The main downside of RT-PCR and RT-qPCR to be overcome consists of their development directly in the food matrix because only *in vitro* procedures have been reported. Other RNA-based methods, such as RNA sequencing and microarrays, are emerging techniques, which could give valuable information for minimizing the presence of aflatoxins in food.

LAMP has been also described as an alternative to overcome the limitations of PCR-based methods due to its ease and rapidness. Additionally, a promising future is predicted for the DNA biosensors due to their numerous valuable potential functions for aflatoxigenic evaluation in food. Nevertheless, they have the limitation of detecting the gene expression or the diagnostic between alive and dead cells; however, detection of mRNA as a target in the future can solve this problem on DNA biosensors. DNA-based biosensors also exhibit several advantages with regard to classical techniques, such as electrophoresis analysis of amplified DNA, which is not automatable and provides no sequence confirmation.

The main limitation of most of the genomic techniques revised in this chapter consists of the required tedious sample preparation, which leads to the destruction of the food products. To overcome such an issue, non-destructive analytical tools based on metabolomics have been recently described. In this sense, metabolic non-destructive methods, such as biosensors, color imaging, electronic nose, fluorescence, infrared spectroscopy, or HSI, are currently promising techniques for both detecting aflatoxigenic molds and mycotoxin production.

The application of proteomic approaches has been lately suggested as another powerful strategy to understand the aflatoxigenic character of molds present in foods. However, there are very limited proteomic studies on aflatoxin-producing molds. Performing new studies based on proteomics should be of great interest from the food safety point of view.

ACKNOWLEDGMENTS

This work is supported by the Spanish Ministry of Economy and Competitiveness, Government of Extremadura, and FEDER (AGL2013-45729-P, AGL2016-80209-P, GR15108).

REFERENCES

1. Schmidt-Heydt M, Abdel-Hadi A, Magan N, Geisen R. Complex regulation of the aflatoxin biosynthesis gene cluster of *Aspergillus flavus* in relation to various combinations of water activity and temperature. *Int J Food Microbiol* 2009;135(3):231–7.
2. Edwards SG, O'Callaghan J, Dobson ADW. PCR-based detection and quantification of mycotoxigenic fungi. *Mycol Res* 2002;106(9):1005–25.
3. Perrone G, Gallo A, Susca A. *Aspergillus.* In: Liu D, editor. *Molecular Detection of Foodborne Pathogens.* Boca Raton, FL: CRC Press (Taylor and Francis); 2010. pp. 529–48.
4. Konietzny U, Greiner R. The application of PCR in the detection of mycotoxigenic fungi in foods. *Brazilian J Microbiol* 2003;34(4):283–300.
5. Hayat A, Paniel N, Rhouati A, Marty JL, Barthelmebs L. Recent advances in ochratoxin A-producing fungi detection based on PCR methods and ochratoxin A analysis in food matrices. *Food Control* 2012;26(2):401–15.
6. Luo J, Vogel RF, Niessen L. Development and application of a loop-mediated isothermal amplification assay for rapid identification of aflatoxigenic molds and their detection in food samples. *Int J Food Microbiol* 2012;159(3):214–24.
7. Passone MA, Rosso LC, Ciancio A, Etcheverry M. Detection and quantification of *Aspergillus* section *Flavi* spp. in stored peanuts by real-time PCR of *nor-1* gene, and effects of storage conditions on aflatoxin production. *Int J Food Microbiol* 2010;138(3): 276–81.
8. Shapira R, Paster N, Eyal O, Menasherov M, Mett A, Salomon R. Detection of aflatoxigenic molds in grains by PCR. *Appl Environ Microbiol* 1996;62(9):3270–3.

9. Färber P, Geisen R, Holzapfel WH. Detection of aflatoxinogenic fungi in figs by a PCR reaction. *Int J Food Microbiol* 1997;36(2–3):215–20.
10. Tombelli S, Mascini M, Scherm B, Battacone G, Migheli Q. DNA biosensors for the detection of aflatoxin producing *Aspergillus flavus* and *A. parasiticus*. *Monatshefte fur Chemie* 2009;140(8): 901–7.
11. Chen RS, Tsay JG, Huang YF, Chiou RYY. Polymerase chain reaction-mediated characterization of molds belonging to the *Aspergillus flavus* group and detection of *Aspergillus parasiticus* in peanut kernels by a multiplex polymerase chain reaction. *J Food Prot* 2002;65(5):840–4.
12. Mayer Z, Bagnara A, Färber P, Geisen R. Quantification of the copy number of *nor*-1, a gene of the aflatoxin biosynthetic pathway by real-time PCR, and its correlation to the cfu of *Aspergillus flavus* in foods. *Int J Food Microbiol* 2003;82(2):143–51.
13. Sardiñas N, Gil-Serna J, Santos L, et al. Detection of potentially mycotoxigenic *Aspergillus* species in *Capsicum* powder by a highly sensitive PCR-based detection method. *Food Control* 2011;22:1363–6.
14. Luque MI, Rodríguez A, Andrade MJ, Martín A, Córdoba JJ. Development of a PCR protocol to detect aflatoxigenic molds in food products. *J Food Prot* 2012;75(1):85–94.
15. Rodríguez A, Luque MI, Andrade MJ, Rodríguez M, Asensio MA, Córdoba JJ. Development of real-time PCR methods to quantify patulin-producing molds in food products. *Food Microbiol* 2011;28(6):1190–9.
16. Rodríguez A, Rodríguez M, Luque MI, Martín A, Córdoba JJ. Real-time PCR assays for detection and quantification of aflatoxin-producing molds in foods. *Food Microbiol* 2012;31(1):89–99.
17. Rodríguez A, Rodríguez M, Andrade MJ, Córdoba JJ. Detection of filamentous fungi in foods. *Curr Opin Food Sci* 2015;5:36–42.
18. Sweeney MJ, Pàmies P, Dobson ADW. The use of reverse transcription-polymerase chain reaction (RT-PCR) for monitoring aflatoxin production in *Aspergillus parasiticus* 439. *Int J Food Microbiol* 2000;56(1):97–103.
19. Degola F, Berni E, Spotti E, Ferrero I, Restivo FM. Facing the problem of "false positives": re-assessment and improvement of a multiplex RT-PCR procedure for the diagnosis of *A. flavus* mycotoxin producers. *Int J Food Microbiol* 2009;129(3): 300–5.
20. Peromingo B, Rodríguez M, Delgado J, Andrade MJ, Rodríguez A. Gene expression as a good indicator of aflatoxin contamination in dry-cured ham. *Food Microbiol* 2017;67:31–40.
21. Bernáldez V, Córdoba JJ, Magan N, Peromingo B, Rodríguez A. The influence of ecophysiological factors on growth, *aflR* gene expression and aflatoxin B_1 production by a type strain of *Aspergillus flavus*. *LWT - Food Sci Technol* 2017;83:283–91.
22. Gallo A, Solfrizzo M, Epifani F, Panzarini G, Perrone G. Effect of temperature and water activity on gene expression and aflatoxin biosynthesis in *Aspergillus flavus* on almond medium. *Int J Food Microbiol* 2016;217:162–9.
23. Baquião AC, Lopes EL, Corrêa B. Molecular and mycotoxigenic biodiversity of *Aspergillus flavus* isolated from Brazil nuts. *Food Res Int* 2016;89:266–71.
24. Leite GM, Magan N, Medina Á. Comparison of different bead-beating RNA extraction strategies: An optimized method for filamentous fungi. *J Microbiol Methods* 2012;88(3):413–8.
25. Hernández M, Hansen F, Cook N, Rodríguez-Lázaro D. Real-Time PCR methods for detection of foodborne bacterial pathogens in meat and meat products. In: Toldrá F, editor. *Safety of Meat and Processed Meat*. New York: Springer; 2009. pp. 427–46.
26. Levin RE. PCR detection of aflatoxin producing fungi and its limitations. *Int J Food Microbiol* 2012; 156(1):1–6.
27. Georgianna DR, Payne GA. Genetic regulation of aflatoxin biosynthesis: from gene to genome. *Fungal Genet Biol* 2009;46(2):113–25.
28. Yu J. Current understanding on aflatoxin biosynthesis and future perspective in reducing aflatoxin contamination. *Toxins (Basel)* 2012;4(1):1024–57.
29. Amare MG, Keller NP. Molecular mechanisms of *Aspergillus flavus* secondary metabolism and development. Fungal Genet Biol 2014;66:11–8.
30. Cary JW, Harris-Coward PY, Ehrlich KC, et al. NsdC and NsdD affect *Aspergillus flavus* morphogenesis and aflatoxin production. *Eukaryot Cell* 2012;11(9):1104–11.
31. Amaike S, Keller NP. Distinct roles for VeA and LaeA in development and pathogenesis of *Aspergillus flavus*. *Eukaryot Cell* 2009;8(7):1051–60.
32. Chang PK, Ehrlich KC, Yu J, Bhatnagar D, Cleveland TE. Increased expression of *Aspergillus parasiticus aflR*, encoding a sequence-specific DNA-binding protein, relieves nitrate inhibition of aflatoxin biosynthesis. *Appl Environ Microbiol* 1995;61(6):2372–7.
33. Yu J, Chang P, Ehrlich KC, et al. Clustered pathway genes in aflatoxin biosynthesis. *Appl Environ Microbiol* 2004;70(3):1253–62.

34. Manonmani HK, Anand S, Chandrashekar A, Rati ER. Detection of aflatoxigenic fungi in selected food commodities by PCR. *Process Biochem* 2005;40(8):2859–64.
35. Geisen R. Multiplex polymerase chain reaction for the detection of potential aflatoxin and sterigmatocystin producing fungi. *Syst Appl Microbiol* 1996;19(3):388–92.
36. Paterson RRM. Identification and quantification of mycotoxigenic fungi by PCR. *Process Biochem* 2006;41(7):1467–74.
37. Criseo G, Bagnara A, Bisignano G. Differentiation of aflatoxin-producing and non-producing strains of *Aspergillus flavus* group. *Lett Appl Microbiol* 2001;33(4):291–5.
38. González-Salgado A, González-Jaén T, Vázquez C, Patiño B. Highly sensitive PCR-based detection method specific for *Aspergillus flavus* in wheat flour. *Food Addit Contam Part A* 2008;25(6):758–64.
39. Iheanacho HE, Dutton MF, Steenkamp PA, et al. Real time PCR of Nor~1 (*aflD*) gene of aflatoxin producing fungi and its correlative quantization to aflatoxin levels in South African compound feeds. *J Microbiol Methods* 2014;97(1):63–7.
40. Sardiñas N, Vázquez C, Gil-Serna J, González-Jaén MT, Patiño B. Specific detection and quantification of *Aspergillus flavus* and *Aspergillus parasiticus* in wheat flour by SYBR® Green quantitative PCR. *Int J Food Microbiol* 2011;145(1):121–5.
41. Niessen L. The application of loop-mediated isothermal amplification (LAMP) assays for the rapid diagnosis of food-borne mycotoxigenic fungi. *Curr Opin Food Sci* 2018;23:11–22.
42. Niessen L. Current state and future perspectives of loop-mediated isothermal amplification (LAMP)-based diagnosis of filamentous fungi and yeasts. *Appl Microbiol Biotechnol* 2015;99(2):553–74.
43. Niessen L, Bechtner J, Fodil S, Taniwaki MH, Vogel RF. LAMP-based group specific detection of aflatoxin producers within *Aspergillus* section *Flavi* in food raw materials, spices, and dried fruit using neutral red for visible-light signal detection detection. *Int J Food Microbiol* 2018;266:241–50.
44. Notomi T, Okayama H, Masubuchi H, et al. Loop-mediated isothermal amplification of DNA. *Nucleic Acids Res* 2000;28(12):E63.
45. Luo J, Vogel RF, Niessen L. Rapid detection of aflatoxin producing fungi in food by real-time quantitative loop-mediated isothermal amplification. *Food Microbiol* 2014;44:142–8.
46. Fakhoury AM, Woloshuk CP. *Amy1*, the alpha-amylase gene of *Aspergillus flavus*: involvement in aflatoxin biosynthesis in maize kernels. *Phytopathology* 1999;89(10):908–14.
47. Scherm B, Palomba M, Serra D, Marcello A, Migheli Q. Detection of transcripts of the aflatoxin genes *aflD*, *aflO*, and *aflP* by reverse transcription-polymerase chain reaction allows differentiation of aflatoxin-producing and non-producing isolates of *Aspergillus flavus* and *Aspergillus parasiticus*. *Int J Food Microbiol* 2005;98(2):201–10.
48. Medina Á, Rodríguez A, Sultan Y, Magan N. Climate change factors and *Aspergillus flavus*: effects on gene expression, growth and aflatoxin production. *World Mycotoxin J* 2015;8(2):171–9.
49. Medina A, Gilbert MK, Mack BM, et al. Interactions between water activity and temperature on the *Aspergillus flavus* transcriptome and aflatoxin B_1 production. *Int J Food Microbiol* 2017;256:36–44.
50. Sedighi-Khavidak S, Mazloum-Ardakani M, Rabbani Khorasgani M, Emtiazi G, Hosseinzadeh L. Detection of *aflD* gene in contaminated pistachio with *Aspergillus flavus* by DNA based electrochemical biosensor. *Int J Food Prop* 2017;20(1):S119–30.
51. Orina I, Manley M, Williams PJ. Non-destructive techniques for the detection of fungal infection in cereal grains. *Food Res Int* 2017;100:74–86.
52. Eivazzadeh-Keihan R, Pashazadeh P, Hejazi M, de la Guardia M, Mokhtarzadeh A. Recent advances in Nanomaterial-mediated Bio and immune sensors for detection of aflatoxin in food products. *Trends Anal Chem* 2017;87:112–28.
53. Chauhan R, Singh J, Sachdev T, Basu T, Malhotra BD. Recent advances in mycotoxins detection. *Biosens Bioelectron* 2016;81:532–45.
54. Vithu P, Moses JA. Machine vision system for food grain quality evaluation: A review. *Trends Food Sci Technol* 2016;56:13–20.
55. Tallada JG, Wicklow DT, Pearson TC, Armstrong PR. Detection of fungus-infected corn kernels using near-infrared reflectance spectroscopy and color imaging. *Trans ASABE* 2011;54(3):1151–8.
56. Campagnoli A, Dell'Orto V, Savoini G, Cheli F. Screening cereals quality by electronic nose: the example of mycotoxins naturally contaminated maize and durum wheat. *AIP Conf Proc* 2009;1137:507–10.
57. Tao F, Yao H, Hruska Z, Burger LW, Rajasekaran K, Bhatnagar D. Recent development of optical methods in rapid and non-destructive detection of aflatoxin and fungal contamination in agricultural products. *Trends Anal Chem* 2018;100:65–81.
58. Wu Q, Xie L, Xu H. Determination of toxigenic fungi and aflatoxins in nuts and dried fruits using imaging and spectroscopic techniques. *Food Chem* 2018;252:228–42.

59. Smeesters L, Meulebroeck W, Raeymaekers S, Thienpont H. Optical detection of aflatoxins in maize using one- and two-photon induced fluorescence spectroscopy. *Food Control* 2015;51:408416.
60. Shen F, Wu Q, Shao X, Zhang Q. Non-destructive and rapid evaluation of aflatoxins in brown rice by using near-infrared and mid-infrared spectroscopic techniques. *J Food Sci Technol* 2018;55(3):1175–84.
61. Chu X, Wang W, Yoon SC, Ni X, Heitschmidt GW. Detection of aflatoxin B_1 (AFB_1) in individual maize kernels using short wave infrared (SWIR) hyperspectral imaging. *Biosyst Eng* 2017;157(17):13–23.
62. Wang W, Ni X, Lawrence KC, Yoon SC, Heitschmidt GW, Feldner P. Feasibility of detecting Aflatoxin B_1 in single maize kernels using hyperspectral imaging. *J Food Eng* 2015;166:182–92.
63. Chen Q, Zhang C, Zhao J, Ouyang Q. Recent advances in emerging imaging techniques for non-destructive detection of food quality and safety. *Trends Anal Chem* 2013;52:261–74.
64. Kheiralipour K, Ahmadi H, Rajabipour A, Rafiee S, Javan-Nikkhah M, Jayas DS. Development of a new threshold based classification model for analyzing thermal imaging data to detect fungal infection of pistachio kernel. *Agric Res* 2013;2(2):127–31.
65. Chelladurai V, Jayas DS, White NDG. Thermal imaging for detecting fungal infection in stored wheat. *J Stored Prod Res* 2010;46(3):174–9.
66. Guo H, Zhou X, Zhang Y, Song B, Zhang J, Shi H. Highly sensitive and simultaneous detection of melamine and aflatoxin M1 in milk products by multiplexed planar waveguide fluorescence immunosensor (MPWFI). *Food Chem* 2016;197:359–66.
67. Karczmarczyk A, Dubiak-Szepietowska M, Vorobii M, Rodriguez-Emmenegger C, Dostálek J, Feller KH. Sensitive and rapid detection of aflatoxin M1 in milk utilizing enhanced SPR and p(HEMA) brushes. *Biosens Bioelectron* 2016;81:159–65.
68. Du B, Wang P, Xiao C, et al. Antibody-free colorimetric determination of total aflatoxins by mercury(II)-mediated aggregation of lysine-functionalized gold nanoparticles. *Microchim Acta* 2016;183(4):1493–500.
69. Xu W, Xiong Y, Lai W, Xu Y, Li C, Xie M. A homogeneous immunosensor for AFB_1 detection based on FRET between different-sized quantum dots. *Biosens Bioelectron* 2014;56:144–50.
70. Paniel N, Radoi A, Marty J-L. Development of an electrochemical biosensor for the detection of aflatoxin M_1 in milk. *Sensors* 2010;10(10):9439–48.
71. Li Z, Ye Z, Fu Y, Xiong Y, Li Y. A portable electrochemical immunosensor for rapid detection of trace aflatoxin B_1 in rice. *Anal Methods* 2016;8(3):548–53.
72. Zhang X, Li CR, Wang WC, et al. A novel electrochemical immunosensor for highly sensitive detection of aflatoxin B_1 in corn using single-walled carbon nanotubes/chitosan. *Food Chem* 2016;192:197–202.
73. Li Q, Lv S, Lu M, Lin Z, Tang D. Potentiometric competitive immunoassay for determination of aflatoxin B_1 in food by using antibody-labeled gold nanoparticles. *Microchim Acta* 2016;183(10):2815–22.
74. Jin X, Jin X, Chen L, Jang J, Shen G, Yu R. Piezoelectric immunosensor with gold nanoparticles enhanced competitive immunoreaction technique for quantification of aflatoxin B_1. *Biosens Bioelectron* 2009;24:2580–5.
75. Pearson TC, Wicklow DT. Detection of corn kernels infected by fungi. *Trans ASABE* 2006;49(4):1235–45.
76. Kumar V, Bagwan NB, Koradia VG, Padavi RD. Colour sorting - an effective tool to remove aflatoxin contaminated kernels in groundnut. *Indian Phytopathol* 2010;63(4):449–51.
77. de Lacy Costello BPJ, Ewen RJ, Gunson H, Ratcliffe NM, Sivanand PS, Spencer-Phillips PTN. A prototype sensor system for the early detection of microbially linked spoilage in stored wheat grain. *Meas Sci Technol* 2003;14(4):397–409.
78. Shen F, Wu Q, Liu P, Jiang X, Fang Y, Cao C. Detection of *Aspergillus* spp. contamination levels in peanuts by near infrared spectroscopy and electronic nose. *Food Control* 2018;93:1–8.
79. Lunadei L, Ruiz-Garcia L, Bodria L, Guidetti R. Image-based screening for the identification of bright greenish yellow fluorescence on pistachio nuts and cashews. *Food Bioprocess Technol* 2013;6(5):1261–8.
80. Sajjadi SM, Abdollahi H, Rahmanian R, Bagheri L. Quantifying aflatoxins in peanuts using fluorescence spectroscopy coupled with multi-way methods: resurrecting second-order advantage in excitation-emission matrices with rank overlap problem. *Spectrochim Acta - Part A Mol Biomol Spectrosc* 2016;156:63–9.
81. Hadavi E. Several physical properties of aflatoxin-contaminated pistachio nuts: Application of BGY fluorescence for separation of aflatoxin-contaminated nuts. *Food Addit Contam* 2005;22(11):1144–53.
82. Durmuş E, Güneş A, Kalkan H. Detection of aflatoxin and surface mould contaminated figs by using Fourier transform near-infrared reflectance spectroscopy. *J Sci Food Agric* 2017;97(1):317–23.
83. Falade T, Sultanbawa Y, Fletcher M, Fox G. Near Infrared Spectrometry for rapid non-invasive modelling of *Aspergillus* -contaminated maturing kernels of maize *(Zea mays L.). Agriculture* 2017;7:77.
84. Kaya-Celiker H, Mallikarjunan PK, Kaaya A. Mid-infrared spectroscopy for discrimination and classification of *Aspergillus* spp. contamination in peanuts. *Food Control* 2015;52:103–11.
85. Sieger M, Kos G, Sulyok M, Godejohann M, Krska R, Mizaikoff B. Portable infrared laser spectroscopy for on-site mycotoxin analysis. *Sci Rep* 2017;7:1–6.

86. Kimuli D, Wang W, Lawrence KC, Yoon SC, Ni X, Heitschmidt GW. Utilisation of visible/near-infrared hyperspectral images to classify aflatoxin B_1 contaminated maize kernels. *Biosyst Eng* 2018;166: 150–60.
87. Karuppiah K, Senthilkumar T, Jayas DS, White NDG. Detection of fungal infection in five different pulses using near-infrared hyperspectral imaging. *J Stored Prod Res* 2016;65:13–8.
88. Qiao X, Jiang J, Qi X, Guo H, Yuan D. Utilization of spectral-spatial characteristics in shortwave infrared hyperspectral images to classify and identify fungi-contaminated peanuts. *Food Chem* 2017;220:393–9.

SUMMARY

Aflatoxins are toxic metabolites produced mainly by *Aspergillus flavus* and *Aspergillus parasiticus* in foods. As foods containing aflatoxins are linked to foodborne illnesses in the human population, it is important to be able to detect aflatoxin-producing molds as well as mycotoxins in the food commodities. This helps prevent aflatoxins from entering the food chain. Traditional methods for the detection of aflatoxins and aflatoxigenic mold strains rely on culture techniques, chemical analyses, enzyme-linked immunosorbent assay (ELISA), and chromatographic techniques. Despite their relatively high sensitivity and specificity, these techniques are often slow, tedious, and dependent on mycological expertise. The recent development of molecular methods has greatly improved the detection and quantification of aflatoxigenic molds in foods. This chapter gives an overview of the genomic and metabolic non-destructive methods for detecting and quantifying aflatoxigenic molds as well as aflatoxins present in foods or produced by isolates of aflatoxigenic molds.

Molecular Identification and Subtyping of Toxigenic and Pathogenic *Penicillium* and *Talaromyces*

19

Josué Delgado
University of Málaga

Elena Bermúdez, Miguel A. Asensio, and Félix Núñez
University of Extremadura

Contents

19.1 INTRODUCTION

Being one of the most common fungal genera, *Penicillium* (of family Aspergillaceae, order Eurotiales, class Eurotiomycetes, subphylum Pezizomycotina, phylum Ascomycota, subkingdom Dikarya, kingdom Fungi) covers a large number of species that are associated with organic matter in nature and widespread in food commodities.[1–3] Recent application of molecular techniques and extrolite profiling has helped redefine the genus *Penicillium* into *Penicillium* sensu stricto and *Talaromyces*, with the latter containing all species of the former *Penicillium* subgenus *Biverticillium*.[4,5] First described as a sexual state of *Penicillium*, the genus *Talaromyces* is currently taxonomically classified in the family Trichocomaceae, order Eurotiales.

Macroscopically, colonies of penicillia are usually greenish, sometimes white, mostly consisting of a dense felt of conidiophores, which may be velvety, lanose, funiculose, or fasciculate, or sinnematous. Microscopically, conidiophores are hyaline, smooth- or rough-walled, single or bundled, consisting of a single stipe terminating either in a whorl of phialides or in a penicillus, which is composed of branches and metulae (with cells between metulae and stipe being referred to as branches). The branching pattern can be biverticillate, terverticillate, or quaterverticillate. Conidia are produced in long chains, divergent or in columns, globose, ellipsoidal, cylindrical or fusiform, hyaline or greenish, smooth- or rough-walled.[6]

The genus *Penicillium* comprises fast-growing fungi, which produce high numbers of exogenous dry-walled spores easily disseminated by air.[7] *Penicillium* species grow in a range of temperature (–3°C to 40°C), pH (2.1–10), and water activity (a_w 0.8–0.99),[8] permitting their development in a variety of environments. Most foodborne *Penicillium* species are psychrotolerant, mainly mesophilic with optimum growth temperature around 25°C, but some are hardy with the ability to grow at 37°C.[7]

Members of the *Penicillium* genus are capable of growing on foods and produce a large number of exometabolites (or secondary metabolites) during morphological and chemical differentiation under the environmental conditions that are rather more restrictive than those allowing optimal growth. Endometabolites (primary metabolites) are generated by most microorganisms, while exometabolites, exoproteins, and exopolysaccharides seem to be species-specific,[1] which may be exploited for molecular identification. Because of their intrinsic toxicity, some exometabolites are commonly referred to as mycotoxins, with *Penicillium* forming one of the most prevalent mycotoxin-producing fungal genera in foods (Table 19.1). Several mycotoxins produced by *Penicillium* species are classified by the International Agency for Research on Cancer (IARC) as group 3 (not classifiable as to its carcinogenicity to humans) including citrinin, patulin, or penicillic acid, or group 2B (possibly carcinogenic to humans) including ochratoxin A (OTA), sterigmatocystin, or griseofulvin.[9] The latter group has obvious detrimental effects on humans (Table 19.1).

The main interest in the *Penicillium* pathology stems from its ability to produce mycotoxins, although some species are considered as opportunistic human pathogens. The first human infection was reported in 1973, with *Penicillium marneffei* (renamed *Talaromyces marneffei*) being the culprit agent.[29] Thereafter, several cases involving mainly immunocompromised individuals have been reported, of which 80% are caused by *T. marneffei*.[30] While *T. marneffei* infection appears rare elsewhere, it is endemic in some parts of East and Southeast Asia. From 668 cases of penicilliosis marneffei reviewed in mainland China between 1984 and 2009, 99.4% were reported in the southern part of China.[31] *T. marneffei* can infect organs such as lung, liver, and skin, leading to fever, weight loss, and anemia, and has most commonly been isolated from skin, blood, and bone marrow (Figure 19.1).[30]

Other *Penicillium* species may be also implicated in human infections, particularly in immunodeficient individuals or those with chronic diseases. For example, *Penicillium piceum* was responsible for pulmonary nodule and adjacent rib osteomyelitis in a patient with X-linked chronic granulomatous disease.[32] *Penicillium purpurogenum, Penicillium chrysogenum, Penicillium decumbens, Penicillium janthinellum, Penicillium brevicompactum, Penicillium citrinum*, and *Penicillium lilanicum* have also been reported to infect different anatomic locations, such as lung, esophagus, brain, or urinary

TABLE 19.1 Main mycotoxins produced by *Penicillium* spp. and their effects on human health

TOXIN	*PRODUCER*	*EFFECT*	*REFERENCES*
Citreoviridin	*P. islandicum* *P. citreonigrum*	Neurotoxic	[7]
Citrinin	*P. expansum* *P. viridicatum* *P. verrucosum*	Nephrotoxic	[7,10]
Cyclopiazonic acid	*P. commume* *P. camemberti* *P. griseofulvum*	Neurotoxic and mutagenic	[7,11,12]
Griseofulvin	*P. canescens* *P. concentricum* *P. griseofulvum* *P. janczewskii* *P. raistrickii*	Hepatotoxic	[13,14]
Mycophenolic acid	*P. brevicompactum* *P. roqueforti*	Immunosuppressive	[15–20]
Ochratoxin A	*P. nordicum* *P. verrucosum*	Nephrotoxic	[21,22]
Patulin	*P. concentricum* *P. coprobium* *P. dipodomyicola* *P. glandicola* *P. vulpinum* *P. roqueforti* *P. expansum* *P. brevicompactum* *P. carneum* *P. paneum* *P. clavigerum* *P. gladioli* *P. griseofulvum* *P. sclerotigneum*	Mitochondrial toxicity, mutagenic, neurotoxic, immunotoxic, genotoxic	[7,19,23,24]
Penicillic acid	*P. roqueforti*	Cytotoxic	[19,20]
Penitrem A	*P. crustosum*	Tremors and convulsions	[25]
Roquefortine C	*P. roqueforti* *P. crustosum* *P. carneum* *P. chrysogenum* *P. expansum* *P. paneum*	Neurotoxic	[7,19,20,25]
Secalonic acid	*P. oxalicum*	Hepatotoxic	[26]
Sterigmatocystin	*P. commune*	Mutagenic and teratogenic	[27]
Verrucosidin	*P. aurantiogriseum* *P. polonicum*	Genotoxic, Tremorgenic	[13,20,28]

tract.[33] *P. chrysogenum* has been associated with intestinal invasion and disseminated disease in immunosuppressed patients, due to either human immunodeficiency virus or post-transplantation immunosuppressant medications.[34] This species may also cause invasive pulmonary mycosis in transplant patients.[35] Other *Penicillium* species have been described in endocarditis, peritonitis, and endophthalmitis.[33] However, the infection by *Penicillium* spp. is extremely uncommon overall.

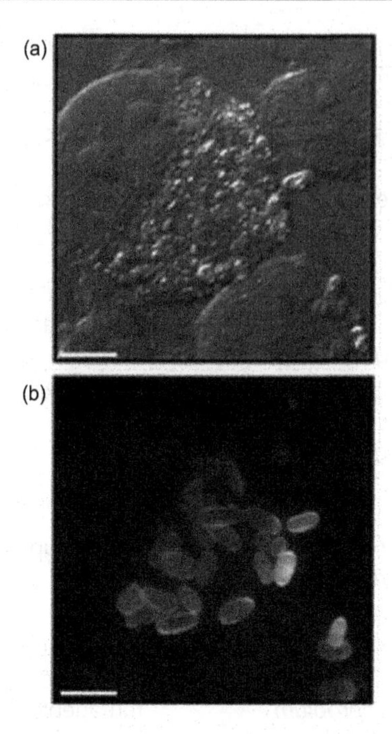

FIGURE 19.1 Murine macrophage J774 infected with *Penicillium marneffei* conidia for 24 h contains numerous yeast cells dividing by fission as examined using differential interference contrast (a) and stained by fluorescent brightener 28 (calcofluor) (b). (Photo credit: Boyce KJ, McLauchlan A, Schreider L, *PLoS Pathog.* 2015; 11(3): e1004790.)

Another implication for human health is the ability of *Penicillium* to provoke allergies. Serine proteases are considered the major allergenic proteins in many *Penicillium* species,[36] followed by extracellular polysaccharides.[37] An association between culturable airborne *Penicillium* and outdoor concentration of culturable fungi has been found, but not for other fungi such as *Cladosporium*, *Alternaria*, or *Aspergillus*. In addition, outdoor *Penicillium* exposure is primarily associated with increased asthma symptoms in children, even in those who showed a negative skin test against *Penicillium* antigens.[38] However, in adults, *Penicillium*, *Cladosporium*, and *Aspergillus* showed a very close relationship for reactivity against fungal antigens.[39]

Penicillium-related allergy is commonly linked to occupational exposure, leading to respiratory and skin symptoms as well as immunoglobulin E-mediated asthma[40] and chronic recurrent dermatitis due to contact with *Penicillium* from salami skin.[41] Also, a type of hypersensitivity pneumonitis, called salami brusher's disease, was described in patients working in a salami factory cleaning the white mold growing on salami surface using a manual wire brush.[42] The symptoms were caused by an immunologic reaction to inhaled *Penicillium* from salami skin.

19.2 METHODOLOGIES FOR DETECTION AND IDENTIFICATION OF TOXIGENIC AND PATHOGENIC *PENICILLIA* AND *TALAROMYCES* SPP.

Diagnostic and epidemiological investigations of pathologies caused by foodborne *Penicillium* spp. require the detection and identification of the responsible mold. Traditional methods for fungal identification are primarily based on cultivation and morphological and physiological assessments, which are relatively inexpensive but laborious and time-consuming, taking several days or even weeks, and require substantial technical and observational skills. Because of the difficulty in interpreting morphological observations, the results obtained can be inconclusive or inaccurate.

New generation methods utilize a polyphasic approach that combines morphological, chemical, and molecular characteristics of *Penicillium* for identification. The most common chemotaxonomic markers include proteins, lipids, and carbohydrates. For phylogenetic analysis, short standardized DNA sequences are often used.[43] Because the nucleic acids contained in conidia provide all the required information, molecular methods allow the characterization of molds and their toxicological potential even without visible mycelium. This reduces the need for specialized taxonomic expertise in the identification of penicillia. Gene sequencing is now considered the gold standard for *Penicillium* identification at the species level.[44]

19.2.1 Nucleic Acid-Based Methods

Several nucleic acid-based methods have been designed to detect, identify, and quantify the presence of *Penicillium* in clinical or toxicological specimens, including PCR-based and loop-mediated isothermal amplification (LAMP)-based methodologies. PCR-based methods (e.g., nested PCR, PCR-enzyme immunoassay, and PCR-based hybridization) as well as PCR typing techniques [e.g., random amplified polymorphic DNA (RAPD) profiling, amplified fragment length polymorphism (AFLP) fingerprinting, and microsatellite markers] have been applied for routine clinical diagnostic and epidemiological investigations on *Penicillium* spp.[45] For accurate identification of pathogenic or toxigenic molds, judicious selection of the target sequence is critically important. The internal transcribed spacer (ITS) region of the fungal ribosomal RNA (rRNA) gene is of high taxonomic value, particularly the two variable ITS1 and ITS2 spacer subregions flanking the highly conserved 5.8S nuclear rRNA gene (Figure 19.2). Unfortunately, consensus is lacking in relation to the DNA targets, probes, and the PCR equipment used for fungal identification in a routine laboratory setting.[46]

Mycotoxins result from a multistep synthesis pathway that requires a variable number of enzymatic reactions. Generally, genes responsible for mycotoxins biosynthesis are organized in clusters comprising open reading frames (ORF), but only some of these ORF have been characterized. Mycotoxin-related genes are only present in molds with the potential to produce the corresponding mycotoxin and may be exploited for species-independent quantitation of mycotoxins-producing fungi. However, given that a mycotoxin can be produced by different mold species and genera, that different biosynthetic pathways can be utilized in a single species to produce a mycotoxin, and that not every member within a mycotoxigenic species carries the complete set of genes to produce the corresponding mycotoxin, the absence or presence of genes related to mycotoxin production does not always allow the identification of molds.

19.2.1.1 Conventional PCR and Real-Time qPCR

PCR-based techniques are useful for early, rapid, accurate, and sensitive detection of toxigenic and pathogenic molds in different matrices. The conventional PCR technique is suitable for determining the presence of a toxigenic species in foods. Real-time qPCR, using both probe-based and DNA intercalating dyes methodologies, is a powerful tool for detecting and quantifying DNA from a broad-spectrum of mold species from a variety of sources. qPCR also dramatically reduces the risk of carryover contamination by environmental amplicons and the potential for false-positive results.[46] Therefore, qPCR is most often used

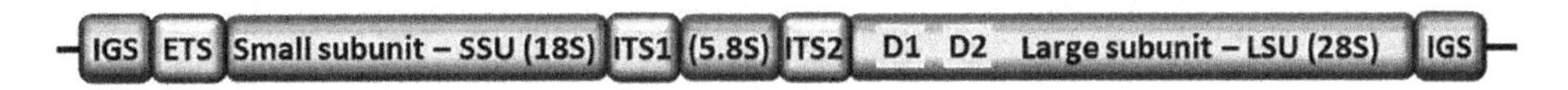

FIGURE 19.2 Diagram of the main fungal ribosomal RNA regions used for taxonomical studies. IGS: intergenic spacer; ETS: external transcribed spacer; ITS: internal transcribed spacer.

to determine the presence of mycotoxigenic molds. Multiplex PCR methods allow simultaneous amplification of several target sequences in a single PCR, thus simplifying the detection procedure, decreasing the amount of reagents as well as time and cost of analysis.

Table 19.2 summarizes commonly used PCR methods for the detection and identification of pathogenic or toxigenic species from the genus *Penicillium*.

19.2.1.2 Loop-Mediated Isothermal Amplification (LAMP) Assay

LAMP is a simple, rapid, robust, highly specific, and sensitive method for amplification of DNA from pathogenic organisms at a constant temperature without the need for thermal denaturation. The amplified DNA products may be visualized via turbidity by magnesium pyrophosphate precipitation, gel electrophoresis, fluorescence using DNA intercalating dyes, color change by complexometric dyes attaching divalent

TABLE 19.2 Selective PCR-based methods used to investigate *Penicillium* spp. and related *Talaromyces marneffei*

MOLD	*MYCOTOXIN*	*TARGET*	*MATRICES*	*DETECTION LIMIT*	*REFERENCES*
Conventional PCR and real-time qPCR					
P. aurantiogriseum	Ochratoxin A	*otanps*	Cooked ham	1–10 conidia g^{-1}	[47]
	Verrucosidin	SVr1 probe	Dry-fermented sausage	10^1 cfu g^{-1}	[48]
P. camemberti	Cyclopiazonic acid	*dmaT*	Dry-cured cheese	10^1 cfu g^{-1}	[49]
P. commune	Aflatoxins	*omt-1*	Dry-cured meats and cheese	2.4×10^3 cfu g^{-1}	[50]
			Black pepper	10^2 cfu g^{-1}	[51]
	Cyclopiazonic acid	*dmaT*	Dry-cured ham	10^1 cfu g^{-1}	[49]
	Sterigmatocystin	*fluG*	Peanut, paprika, dry-cured ham, cheese	10^1–10^2 cfu g^{-1}	[27]
	Patulin	*idh*	Cooked ham	10 conidia g^{-1}	[52]
P. crustosum	Penitrems	β-tubulin	Wine and must	1 fg DNA	[53]
P. expansum	Patulin	*patF*	Apple	0.1 ng DNA	[54]
	Patulin	*idh*	Meat products and peach	10 conidia g^{-1}	[52]
	Patulin	β-tubulin	Wine and must	1 fg DNA	[53]
P. melanoconidium	Verrucosidin	SVr1 probe	Dry-cured ham	10^1 cfu g^{-1}	[48]
P. nordicum	Ochratoxin A	*otanps*	Ripened cheese	1–10 conidia g^{-1}	[47]
P. polonicum	Verrucosidin	SVr1 probe	Dry-ripened cheese	10^1 cfu g^{-1}	[48]
P. verrucosum	Ochratoxin A	rRNA	Soybeans	pg DNA level	[55]
T. marneffei	Systemic mycosis	*MP1*	Clinical isolates, and plasma	100	[56]
		5.8S rRNA	Blood	10 cells	[57]
P. chrysogenum	-	β-tubulin	Wine and must	1 fg DNA	[105]

(Continued)

TABLE 19.2 (*Continued*) Selective PCR-based methods used to investigate *Penicillium* spp. and related *Talaromyces marneffei*

MOLD	*MYCOTOXIN*	*TARGET*	*MATRICES*	*DETECTION LIMIT*	*REFERENCES*
Conventional and quantitative multiplex PCR					
P. expansum *P. griseofulvum* *P. nordicum* *P. verrucosum*	Patulin Patulin Ochratoxin A Ochratoxin A	*idh* *otanps*	Fruits, peanut, wheat, spices, dry-meat products, cheese	10^1–10^3 cfu g^{-1} 10^1–10^3 cfu g^{-1} 10^1 cfu g^{-1} 10^1 cfu/g	[58]
P. aurantiogriseum *P. chrysogenum* *P. citrinum* *P. commune* *P. spinulosum*	-	ITS	Grain	1 pg-10 ng DNA	[59]
P. verrucosum *P. viridicatum*	Ochratoxin A	*otanps*	Grain	10^3 conidia ml^{-1}	[60]
Penicillium sp.	Ochratoxin A	*pks*	Maize	10^3 conidia g^{-1}	[61]
P. expansum *P. nordicum*	Patulin Ochratoxin A	*idh* *otanps*	Apple, peanut, wheat, paprika, dry-fermented sausage	10^3–10^4 cfu g^{-1}	[62]

metal ions, or lateral flow assay using fluorescein isothiocyanate-labeled DNA probes.[63] Sample preparation for LAMP is straightforward, without the need for an extensive DNA purification,[64] as its sensitivity is unaffected by the presence of non-target DNA in samples and its amplification process is not inhibited by substances that are known to interfere with PCR (e.g., blood, serum, plasma, or heparin).[65]

LAMP has been successfully used for the specific detection of both toxigenic and pathogenic penicillia (Table 19.3). Similar to other molecular methods, rRNA or genes related to the biosynthesis of mycotoxins are usually selected as LAMP targets.

19.2.1.3 DNA Microarrays

DNA microarrays provide a rapid method for the detection and identification of filamentous fungi. In this assay, the target fungal DNA is first amplified and the PCR products are then labeled and used for hybridization to the probes on the array. As the hybridization results can be evaluated by naked eyes,

TABLE 19.3 Selective loop-mediated isothermal amplification (LAMP) protocols used to investigate toxigenic *Penicillium* spp. and pathogenic *Talaromyces marneffei*

MOULD	*TARGET*	*DETECTION LIMIT:*	*REFERENCES*
P. expansum	Ribosomal RNA large subunit	-	[66]
T. marneffei	Ribosomal RNA large subunit	-	[66]
	ITS	2 genome copies	[65]
P. nordicum	*otapks*	100 fg DNA 10^2–10^5 conidia	[67]
P. oxalicum	Gene coding for protein PDE_07106	100 pg genomic DNA	[68]

DNA microarrays facilitate rapid fungal identification without expensive equipment and highly specialized technical skills.

The simultaneous use of probes directed to the ITS region and β-tubulin gene allows the detection of fungal species belonging to *Aspergillus*, *Eurotium*, and *Penicillium* with 97.4% concordance rate between microarray and DNA sequencing results at the species or genus levels.[69] Aoyama et al.[70] designed a DNA microarray for the detection of 142 fungal strains, including 28 penicillia. The D2 region of LSU rRNA was used for genus-specific probes (Figure 19.1), and *Penicillium* species were distinguished at the subgenus level into 10 groups, with 2–3 probes designed for each group, based on similarity in the nucleotide sequence of the target region.

An oligonucleotide microarray exploiting the sequence variations of the elongation factor 1-alpha (EF-1α) coding regions, the ITS regions of the rRNA, and genes leading to toxin production was developed to identify potentially mycotoxigenic fungi from maize-derived foods, including eight penicillia.[71] A barcode oligonucleotide array based on sequences of *COX1* gene from 60 species of *Penicillium* subgenus *Penicillium* and 12 allied species was designed to detect species belonging to this subgenus in ecological surveys, epidemiological studies, and food spoilage.[72] However, the high sequence similarity of the *COX1* gene among *Penicillium* species makes it difficult to find taxon-specific probes for many species, and the array has limited utility for several clades.[72]

An oligonucleotide microarray based on specific probes complementary to ITS1 and ITS2 subregions proved to be highly discriminative, leading to unequivocal identification of species from pathogenic fungi responsible for invasive and superficial mycoses, including *T. marneffei*.[73]

19.2.1.4 PCR-Based Subtyping Methods

PCR-based subtyping methods exploit the polymorphisms of fungal DNA for studies on the population genetics and molecular epidemiology of foodborne penicillia. Based on gel electrophoresis of PCR product patterns with or without prior restriction digestion, these methods do not reveal sufficient phylogenetic details for reliable classification and identification of fungi and are of limited use to distinguish closely related strains on a population level.[74]

PCR DNA fingerprinting methods. PCR-based DNA fingerprinting methods (e.g., RAPD and AFLP) are very informative techniques for assessing the genetic diversity of microorganisms without prior sequence information of genomic DNA. Based on PCR amplification of DNA polymorphisms with short primers of 8–15 nts in length, these techniques are capable of generating a characteristic banding fingerprint for individual *Penicillium* species.

Through the use of arbitrary primers that can simultaneously anneal to multiple sites in the whole genome under low stringent conditions, RAPD permits molecular diagnosis of many fungal species, estimation of genetic variations among closely related species, and analysis of the intraspecific diversity of molds. Indeed, RAPD banding patterns allowed distinguishing different strains of *T. maffernei*.[75] The main limitations of this technique are related to the intralaboratory or interlaboratory reproducibility that could be affected by slight variations in the temperature or calibration of the thermocycler. Therefore, fingerprinting could be regarded as an intermediate step between massive morphological observation and focused gene sequencing.[44]

In PCR-AFLP, DNA is first digested with two restriction enzymes and subsequently amplified under stringent conditions to yield a characteristic subset of restriction fragments. AFLP analysis has a good discriminatory power for differentiating isolates belonging to the same species.[76] Examination of 185 OTA-producing isolates of *Penicillium verrucosum* revealed 155 different AFLP patterns indicating a high genetic diversity in this species.[77] However, AFLP is time-consuming, not always repeatable between laboratories, and the polymorphism due to different alleles from a single locus may lead to markers with non-Mendelian behavior.[78]

RAPD and AFLP analyses are also helpful in differentiating foodborne ochratoxin-producing penicillia into *Penicillium nordicum* and *P. verrucosum*.[79] Both RAPD and AFLP profiles were useful for typing *Penicillium commune* and evaluating the distribution of mold strains as well as possible contamination points in the processing environment of cheese dairies.[76] For *P. commune*, AFLP fingerprinting matched with morphotyping and showed higher discriminatory power than RAPD.[76] Restriction fragment length polymorphism (RFLP) combines a PCR (which amplifies a specific region of the genome) and restriction enzyme digestion to produce fragments that form distinct electrophoretic band patterns and represents a sensitive, rapid, and inexpensive method to detect polymorphisms within and between restriction sites.

PCR-RFLP analysis of the ITS region has been successfully used to differentiate 11 mold genera, 22 species among the genus *Penicillium*,[80,] and 12 of 13 species of biverticillate penicillia.[81] This technique can facilitate rapid and easy identification of penicillia without sequencing,[82] particularly when multiple isolates have to be authenticated before analysis for phylogenetic assessment or population genetics.[81] In addition, given that the PCR-RFLP profile could correlate with metabolic activity, this technique may be used to discriminate between toxigenic and non-toxigenic strains of *Penicillium aurantiogriseum*.[83]

Multilocus sequence typing (MLST). MLST targets tandem repeat polymorphisms within housekeeping loci detecting genetic variation produced by neutral mutations and genetic relatedness between isolates. This technique provides an accurate and reliable DNA sequence-based identification system for the Trichocomaceae.[84] MLST of eight nuclear gene fragments of *T. marneffei* (*Aba*A, *Cpe*A, *Stl*A, *Icl*1, *PAA*, *NGS*, *LNS*, *MP1*, and *COX1*) demonstrated a high degree of discriminatory power and reproducibility, forming a robust and reliable method for genotyping isolates of this mold.[85] As *MP1* is the only locus showing significant variation, this gene and its homologs are better than housekeeping genes for MLST typing in *T. marneffei*, due to the more rapid evolutionary rates of the latter.[86] Real-time PCR targeting the *MP1* gene encoding the Mplp cell wall protein, a key virulence factor specific from this mold, has been used for the diagnostic of *T. marneffei* in clinical isolates and plasma from patients.[56] This method showed a diagnostic sensitivity of 70.4% and a diagnostic specificity of 100%.[56,87]

Microsatellite typing. Microsatellite loci are widely used as genetic markers for genome mapping and population genetics because of their ubiquity, ease to score, co-dominance, reproducibility, assumed neutrality, and high level of polymorphism.[78] Several microsatellite loci were useful in strain typing and species recognition of cheese and meat starter strains of *Penicillium camemberti*, *Penicillium roqueforti*, *P. chrysogenum*, and *Penicillium nalgiovense*, as well as contaminating strains of *P. commune*.[88] However, fungal microsatellites are hard to isolate and exhibit lower polymorphism than those from other organisms, due to the scarcity and shortness of fungal microsatellite loci.[78] Therefore, these markers have rarely been used for *Penicillium* genetic studies. On the other hand, multilocus microsatellite typing (MLMT) based on 11 microsatellite loci was able to discriminate the genetic diversity of 169 clinical *T. marneffei* isolates, which makes it useful for epidemiological studies.[89]

19.2.1.5 Barcoding

The DNA barcoding system is among the most recently developed technologies for mold identification. The fungal barcoding site promotes the DNA barcoding of fungi, providing frameworks for advancing the use of DNA for species identifications (http://www.fungalbarcoding.org/). For this purpose, short-standardized segments from 500- to 800-bp of the fungal genome should be used. Unfortunately, no single gene is known that is invariant within species and variable between species as would be optimal for the barcode approach.[84]

The ITS region is considered to be the most widely sequenced marker for fungi. It was accepted as the official primary fungal barcode by the Consortium for the Barcode of Life, because of the ease of amplification, the very good level of sensitivity, the more clearly defined barcode gap, and the availability of universal primers.[90] ITS analysis combines the highest resolving power for reasonable discrimination of closely related species with a high PCR and sequencing success rate across a broad range of fungi.[90] Other rRNA genes might serve this purpose, such as the 28S nuclear ribosomal large subunit rRNA (LSU) and 18S nuclear ribosomal small subunit rRNA (SSU) genes[91] (Figure 19.1).

Some housekeeping genes have been proposed as secondary barcode markers or as a target for circumscribed taxonomic groups, such as the β-tubulin *Ben*A,[92,93] the translation elongation factor 1-α *TEF1α*,[94] the cytochrome c oxidase 1 *COX1*,[95] the calmodulin *CaM*,[96] the RNA-polymerase II largest subunit RPB1,[90] second-largest subunit *RPB2*,[94] the minichromosome maintenance complex MCM7, the pre-mRNA processing protein homolog Tsrl, and actin *ACT1*. However, the results obtained with these genes are not always accurate.

ITS. The ITS region has a high sensitivity for amplification, because of its large number of copies in the fungal genome. In addition, the high rate of evolution of ITS1 and ITS2 results in a higher degree of sequence variation that is typically species specific.[45] The ITS region is useful for the identification of the species of clinical importance from the genera *Aspergillus* and *Penicillium*.[97] Genetic information contained in ITS1 and ITS2 regions supports taxonomical, ecological, and physiological data for common foodborne *Penicillium* species, but the degree of ITS variability is too low to facilitate separation for all closely related taxa.[84,98] In fact, only 16 of 51 species of the *Penicillium* section *Aspergilloides* were identified using ITS sequences.[99] Therefore, studies using only rDNA sequences cannot solve all of the taxonomic and phylogenetic problems in *Penicillium* and its related teleomorphs.[96]

Other rRNA genes. LSU locus is considered the second preferred phylogenetic marker for molds after ITS because it has virtually no amplification, sequencing, alignment, or editing problems.[90] However, the results obtained are reliable for *Aspergillus* and *Talaromyces*, but not for the genus *Penicillium*.[91] SSU locus has very poor efficiency in determining the phylogeny of *Penicillium*, showing low species discrimination with almost no barcode gap; therefore, it can be eliminated as a candidate locus.[90,91] The intergenic spacer (IGS) region, located between the LSU and SSU (Figure 19.2), is the most variable region of rRNA genes and provides with a powerful tool for strain typing or molecular epidemiological investigation.[100] However, IGS sequence analysis is not appropriate for epidemiological investigations of *T. marneffei* due to the extremely unusual very high sequence similarity exhibited by clinical isolates of this species from patients of different countries.[100]

β-Tubulin gene. The β-tubulin *Ben*A is recommended as a secondary barcode for precise species identification in *Penicillium*.[92,93] Partial β-tubulin sequences were excellent markers at the species level in the subgenus *Penicillium*, providing species-specific sequences for 50 of 58 species,[93] although several species complexes have species sharing identical or highly similar sequences, and therefore cannot be distinguished.[72] In addition, the presence of numerous ambiguous regions in *Ben*A sequences make it difficult to align across different sections of *Penicillium*.[25]

Other housekeeping genes. TEF1α gene was difficult to amplify with standard primers, but according to the results obtained using a novel, high-fidelity primer pair, *TEF1α* has been proposed as one of the best candidates for non-rDNA barcodes because of its balanced 'trade-offs' and versatility among important fungal orders.[94] A fragment from the *COX1* gene has been used for barcoding animals, and it functions reasonably well as a barcode in the genus *Penicillium* with reliable primers. However, this gene is prone to having multiple introns, not all penicillia are genotypically different at the *COX1* locus, and it only allows to distinguish 38 of 58 species in the subgenus *Penicillium*.[95] *RPB1* is a ubiquitous and single-copy gene, with a slow rate of sequence divergence showing the highest levels of species discrimination of tested genes as the fungal barcode,[74] but it has PCR amplification problems, mainly related to primer failure.[90] Partial sequences of the *CaM* gene used for phylogenetic studies on *Eupenicillium*, *Talaromyces*, and *Penicillium* supported the classification of penicillia in the two subfamilies of the Trichocomaceae based on traditional methods.[96] However, the use of the sequence of *CaM* gene is not a good choice for the molecular identification of penicillia, including toxigenic strains, isolated from cheese.[44] In general, protein-coding genes, such as *CaM*, RPB1, RPB2, and MCM7, have a good species-resolving power in several groups of fungi, but PCR and sequencing failures are common because they occur as a single copy within the genome,[101] eliminating them as potential universal barcodes.[90]

Multiple DNA sequencing. Given the limited capacity to obtain an accurate classification using a single gene, multiple independent DNA sequencing could be necessary to obtain a robust identification of *Penicillium* species. The most effective two genes in the combined analysis for fungi identification were either ITS and RPB1 or LSU and RPB1.[90]

The obtained sequences should be compared with those registered in many available databases.[74,102] GenBank (http://www.ncbi.nlm.nih.gov) is the most inclusive database providing a quick and easy comparison of DNA sequences, but it often contains sequences from incorrectly or not well-characterized isolates, or listed under invalid species names or synonyms of others.[92,103] Other narrower databases containing well-characterized and ex-type cultures from *Penicillium* are available, such as MycoBank database (http://www.mycobank.org), Fungal Barcoding (http://www.fungalbarcoding.org/), the *Aspergillus* and *Penicillium* database (http://www.westerdijkinstitute.nl/aspergillus), the species and strains of subgenus *Penicillium* database (http://www.westerdijkinstitute.nl/penicillium), UNITE (https://unite.ut.ee/), or ARS Trichocomaceae database (http://199.133.98.43/Trichocomaceae). In addition, species identification in the last one is based on type isolates, genealogical concordance analysis of 3–7 genes (β-tubulin, calmodulin, ITS, ST, MCM7, RPB2, and TSR1), and phylogenetic species concepts.[103] These databases are not as inclusive as GenBank, but are highly reliable for the taxon range they cover. To guard against sequencing failures or unexpected sequence variation in any single locus, sequencing of three loci is recommended, and the best coverage of DNA reference sequences to identify *Penicillium* will be obtained using three databases.[103]

19.2.2 MALDI-TOF MS

Matrix-assisted laser desorption/ionization time-of-flight mass spectrometry (MALDI-TOF MS) has emerged as a rapid, accurate, sensitive, and reliable tool for the identification and classification of molds based on profiling of cell surface proteins. After proteome extraction, each mold is identified by specific biomarkers or by a typical spectrum (MS fingerprint) generated that are compared with the biomarkers or spectra contained in a reference database. This technique offers as main advantages a simple sample preparation, relatively high tolerance against sample impurities, short measurement times, the possibility of automation, and requirement of only a small amount of biological material.[104] Although the equipment cost is high, the cost per sample is negligible.[105]

MALDI-TOF MS shows enormous potential for the identification of molds at species or strain level.[104] However, it is not an alternative to DNA sequence-based species identification, but rather a complementary method.[105] With the intact cell or intact spore variant of this technique (IC/IS MALDI-TOF MS), the identification of molds can be carried out simply after an acidic extraction of proteins.[106] Even spores of *Penicillium* directly scratched from food surfaces contaminated by unknown fungi can be rapidly identified using MALDI-TOF MS analysis without any pretreatment.[107] However, due to the protective cell wall structures, a bead-beating procedure in conjunction with chemical extraction with ethanol, acetonitrile, formic acid, and so on improves the information content of the MALDI-TOF fingerprint mass spectra from *Penicillium* isolates.[108] IC/IS MALDI-TOF MS operates with a unique mass spectrometric profile directly acquired from the surface of unfractionated microorganisms or spores by the desorption of peptide or protein biomarkers.[109] IC/IS MALDI-TOF MS is used in the analysis of conidia and hyphae allowing the detection of specific biomarker ions in the *m/z* region of 1,000–20,000 to differentiate several *Penicillium* species. The mass spectral profiles differ among *Penicillium* species, but they are similar for different strains within the same species.[107,108] In addition, mass spectra from 12 *Penicillium* species allowed unambiguous discrimination between species, and a common biomarker to all *Penicillium* at *m/z* 13,900 was proposed as a potential *Penicillium* genus-specific biomarker.[108] On the other hand, given that the pattern of outer proteins of molds is related to specific physiological and morphological states, specific biomarkers could indicate the production of mycotoxins, as it has been shown for other secondary metabolites, such as penicillin in *P. chrysogenum*.[110]

However, the MALDI-TOF MS spectra are usually proprietary, only commercially available, heavily focused on medical applications, and do not recognize less commonly encountered fungi, newly discovered species, pathogens limited to certain geographical areas, or fungi commonly found in food samples.[105,111] This fact reveals that the technique is completely dependent on spectrum libraries built from type cultures. Moreover, the MALDI-TOF MS spectra depend on the fungal physiological state, the extraction procedure, and the settings of the equipment used for analysis.[105,108] Then, the fungal fingerprint mass spectrum of each *Penicillium* species is highly reproducible under the same cultural and methodological conditions, but may differ between laboratories.[107,108] The heterogeneous morphological phenotypes of molds cause heterogeneity in mass spectra between different strains of the same species as well as between subcultures of the same strain, which negatively impacts the reproducibility and could interfere with the identification process.[112]

Another problem with this methodology is the lack of a comprehensive public repository of MALDI-TOF MS reference spectra that allows fungal identification.[105] Commercial databases do not satisfy the needs for fungal identification.[113] For example, the commercial Bruker Filamentous Fungi Library 1.0 (Bruker Daltonik) contains 127 fungi, including only 24 identified *Penicillium* species. The free Spectra site hosted by the Public Health Agency of Sweden (http://spectra.folkhalsomyndigheten.se/spectra) offers spectra for several filamentous fungi, but only for two *Penicillium* species: *P. camemberti* and *P. chrysogenum*. MALDI-TOF MS failed to identify 60 *T. marneffei* strains using the Bruker original database combined with BDAL v4.0.0.1 and Filamentous Fungi Library 1.0, but when this combined database was expanded with the spectra from 21 *T. marneffei* strains, the remaining 39 *T. marneffei* strains were correctly identified to the species level.[111] However, strains of *P. brevicompactum*, *P. chrysogenum*, *Talaromyces aurantiacus*, and *Talaromyces stipitatus* could not be identified using the combined database with or without spectra from *T. marneffei*.[111] The use of MS spectra resulted in 44 (57.14%) correct identification of *Penicillium* isolates belonging to 16 human pathogen species.[114] However, only five species were present in the reference spectra library. Among the species present, MS-based identification resulted in 43 (70.49%) correct identifications of penicillia.[114] Then, as stated before, MS identification is possible only if the species is included in the reference spectra library.

19.2.3 Fatty Acid Profiles

Fatty acids are analyzed by gas-liquid chromatography after their extraction and methylation. The cellular fatty acid composition allows identification of species belonging to the genus *Penicillium*.[115] However, the lipid composition of a mold differs from mycelia to conidia, even when the ultrastructural morphology of the two stages is very similar.[115] On the other hand, the analysis of lipid composition by MALDI-MS has the potential for rapid species and strain differentiation of filamentous fungi, including *Penicillium* species.[113]

19.2.4 Profile of Secondary Metabolites

The profile of secondary metabolites or exometabolites is the set of small molecules, including mycotoxins, antibiotics, and other outward-directed compounds, produced during morphological and chemical differentiation of a mold on a substratum.[116] Given that exometabolites are taxonomically restricted to a limited number of species in a genus, an order, or even phylum, the profile of these compounds has a high differentiation power and it could be useful for identification at the species level.[1,43,93,116] Many species of genus *Penicillium* produce a large number of species-specific exometabolites with a high degree of chemoconsistency among isolates.[1] However, the number of exometabolites expressed by a mold isolate depends on the abiotic factors, especially temperature, water activity, pH, atmosphere, and culture medium. Besides, a mutation in a gene from an exometabolite gene-cluster will be sufficient for loss of phenotypic expression.[1,43] Then, fungal cultures for chemotaxonomic analysis must always be grown on the same medium, incubated at the same temperature, and extracted at the same time, to ensure the differences are due to fungal diversity and not to environmental conditions.[116] To standardize the determination, Czapek yeast autolysate agar and yeast extract sucrose agar have been proposed because they are very efficient for the production of a large number of exometabolites.[1,43] These compounds are commonly analyzed by HPLC and identified by their UV spectra and MS characteristics.[43,117]

19.2.5 Detection of Mycotoxins

Considering the potential detrimental effects of mycotoxins on human health, techniques and tools for detecting and quantifying these secondary metabolites in foods and drinks have been developed. The most common tools used for detecting mycotoxins in foods and drinks can be grouped in immunochemical and chromatographic methods.

Among all published immunological-based methods, different variants of the enzyme-linked immunosorbent assay (ELISA) are the most commonly used techniques for mycotoxin determination.[118] ELISA provides quick screening for most of the mycotoxins summarized in Table 19.1 and different methods have been validated for different food matrices.[119–124] However, commercial kits are only available for the most common mycotoxins. A large number of antibodies have been developed for mycotoxins analysis, including polyclonal, monoclonal, recombinant antibodies, and the promising nanobodies.[125] The competitive format, which is based on the competitive interactions between mycotoxins and toxin–enzyme conjugate for binding sites in the antibodies, is normally used due to the small molecular size of mycotoxins.[118] The main disadvantages of this technique are the single-mycotoxin detection provided by every kit, as well as the limited availability of commercial kits validated for the food matrix of interest. This means that positive ELISA results should be confirmed by a suitable chromatographic method if the matrix is different from those described by the manufacturer.[126]

The chromatographic techniques widely overcome this disadvantage. Among them, thin-layer chromatography (TLC), gas chromatography/electron capture detector (GC-ECD), gas chromatography/mass spectrometry (GC-MS), liquid chromatography/mass spectrometry (LC-MS), and ultra-high performance liquid chromatography (uHPLC) with a diode array detector (PDA) have been used for determination of different mycotoxins in foods.[127] Although TLC is no longer considered state of the art for chemical research, given the poor sensitivity and limitations for detecting several mycotoxins, it is still a useful technique for detecting colored or uncolored extrolites in laboratories with less sophisticated equipment.[92]

GC-ECD can be used for detecting mycotoxins produced by *Penicillium*, but requires a derivatization procedure before the injection into the chromatograph to detect the mycotoxin in the ECD.[128] This extra-step in the sample preparation means more time and cost for mycotoxin detection. Hence, GC has been largely superseded by modern developments in HPLC with UV and fluorescence detectors and, more recently, by MS detectors. The MS detectors seem to be the most suitable ones, given the high sensitivity, chemical structural information, and specificity based on *m/z*.[126] Additionally, the mycotoxin fragmentation through MS^n provides high selectivity to confirm the presence of a given mycotoxin. This methodology for maximizing the selectivity is not essential in high-resolution mass spectrometers (HRMS), which have the ability to discern between two very similar compounds with regard to *m/z* and facilitate detection and quantitation of multiple mycotoxins in a single run, including non-target analysis of fungal metabolites. Given the *m/z* measurement accuracy and sensitivity, these HRMS tools commonly use isotopes for further confirming the molecule identification. Additionally, in the case of even more precise identification requirements, an MS/MS fragmentation of the ion of interest is usually performed in combination with HRMS in hybrid mass spectrometers. These machines are endowed with a device that fragments the molecules for a subsequent *m/z* measurement. The combination of the accurate *m/z* measurement of the molecule together with the *m/z* measurements of the fragments allows the highest confidence in mycotoxin identification.

Low limits of detection (LOD) and quantification (LOQ), such as 0.1 and 0.3 $\mu g\ kg^{-1}$ respectively, have been achieved for OTA in dry-fermented meat products by using HPLC-MS.[129] Similarly, LOD and LOQ of 1 and 5 $\mu g\ kg^{-1}$ have been reached for patulin using this spectrometric tool.[28] These LOQs and LODs are lower than the maximum levels set in the European legislation for OTA and patulin, of 0.5 and 10 $\mu g\ kg^{-1}$ respectively, in products intended for infants and young children.[130]

19.3 FUTURE PERSPECTIVES

Penicillium and *Talaromyces* are related fungal genera that occur ubiquitously in the environments and represent emerging pathogenic and toxigenic agents for human populations, especially individuals with suppressed immune functions. Given the limitations associated with culture-based identification techniques, molecular approaches offer opportunities for dramatic improvement in *Penicillium* and *Talaromyces* detection and diagnosis.

However, as public sequence databases often contain misidentified and unspecified sequences due to incorrectly or not well-characterized *Penicillium* isolates, or listing under invalid species names or synonyms, there is an urgent need to curate and re-annotate ITS and other gene sequences in these databases to improve the accuracy and reliability of molecular identification of penicillia.

To enhance the capability and performance of MALDI-TOF MS for penicillia identification, standardized culture, sample preparation, and mass spectrometry conditions need to be established along with an expansion of MALDI-TOF MS databases.[108,131] The architecture of the libraries should be improved by incorporating an increased number of subcultures from each strain as well as by increasing the number of strains representing each species.[112]

REFERENCES

1. Frisvad JC. Taxonomy, chemodiversity and chemoconsistency of *Aspergillus, Penicillium* and *Talaromyces* species. *Front Microbiol* 2014;5:1–7.
2. Houbraken J, Kocsubé S, Visagie CM, et al. Classification of *Aspergillus, Penicillium, Talaromyces* and related genera (Eurotiales): An overview of families, genera, subgenera, sections, series and species. Stud Mycol. 2020;95:5–169.
3. NCBI. Taxonomy of *Penicillium* [Internet]. Available from: https://www.ncbi.nlm.nih.gov/Taxonomy/Browser/wwwtax.cgi?id=5073; accessed on Jan 11, 2020.
4. Samson RA, Yilmaz N, Houbraken J, et al. Phylogeny and nomenclature of the genus *Talaromyces* and taxa accommodated in *Penicillium* subgenus *Biverticillium. Stud Mycol* 2011;70:159–83.
5. Houbraken J, Samson RA. Phylogeny of *Penicillium* and the segregation of *Trichocomaceae* into three families. *Stud Mycol* 2011;70:1–51.
6. Samson RA, Houbraken J, Thrane U, Frisvad JC, Andersen B. *Food and Indoor Fungi*. Utrecht: CBS-KNAW Fungal Biodiversity Centre; 2010.
7. Perrone G, Susca A. Penicillium species and their associated mycotoxins. In: Moretti A, Susca A, editors. *Mycotoxigenic Fungi: Methods and Protocols*. New York: Humana Press; 2017. pp. 107–19.
8. Sweeney MJ, Dobson ADW. Mycotoxin production by *Aspergillus, Fusarium* and *Penicillium* species. *Int J Food Microbiol* 1998;43(3):141–58.
9. Ostry V, Malir F, Toman J, Grosse Y. Mycotoxins as human carcinogens-the IARC Monographs classification. *Mycotoxin Res* 2016;33:65–73.
10. Bezerra da Rocha ME, Oliveira Freire FDC, Feitosa Maia FE, Florindo Guedes MI, Rondina D. Mycotoxins and their effects on human and animal health. *Food Control* 2014;36(1):159–65.
11. Asefa DT, Gjerde RO, Sidhu MS, et al. Moulds contaminants on Norwegian dry-cured meat products. *Int J Food Microbiol* 2009;128:435–9.
12. Hayashi Y, Yoshizawa T. Analysis of cyclopiazonic acid in corn and rice by a newly developed method. *Food Chem* 2005;93:215–21.
13. El-Banna AA, Pitt JI, Leistner L. Production of mycotoxins by *Penicillium* species. *Syst Appl Microbiol* 1987;10(1):42–6.
14. Liu K, Yan J, Sachar M, et al. A metabolomic perspective of griseofulvin-induced liver injury in mice. *Biochem Pharmacol* 2015;98(3):493–501.

15. Del-Cid A, Gil-Durán C, Vaca I, et al. Identification and functional analysis of the mycophenolic acid gene cluster of *Penicillium roqueforti. PLoS One* 2016;11(1):1–15.
16. Regueira TB, Kildegaard KR, Hansen BG, Mortensen UH, Hertweck C, Nielsen J. Molecular basis for mycophenolic acid biosynthesis in *Penicillium brevicompactum. Appl Environ Microbiol* 2011;77(9):3035–43.
17. Hansen BG, Mnich E, Nielsen KF, et al. Involvement of a natural fusion of a cytochrome P450 and a hydrolase in mycophenolic acid biosynthesis. *Appl Environ Microbiol* 2012;78(14):4908–13.
18. Zhang W, Cao S, Qiu L, et al. Functional characterization of MpaG′, the O-methyltransferase involved in the biosynthesis of mycophenolic acid. *ChemBioChem* 2015;16(4):565–9.
19. Malekinejad H, Aghazadeh-Attari J, Rezabakhsh A, Sattari M, Ghasemsoltani-Momtaz B. Neurotoxicity of mycotoxins produced in vitro by *Penicillium roqueforti* isolated from maize and grass silage. *Hum Exp Toxicol* 2015;34(10):997–1005.
20. Frisvad JC, Smedsgaard J, Larsen TO, Samson RA. Mycotoxins, drugs and other extrolites produced by species in *Penicillium* subgenus *Penicillium. Stud Mycol* 2004;49:201–41.
21. Logrieco A, Bottalico A, Mulé G, Moretti A, Perrone G. Epidemiology of toxigenic fungi and their associated mycotoxins for some Mediterranean crops. *Eur J Plant Pathol* 2003;109(7):645–67.
22. Medina A, Schmidt-Heydt M, Rodríguez A, Parra R, Geisen R, Magan N. Impacts of environmental stress on growth, secondary metabolite biosynthetic gene clusters and metabolite production of xerotolerant/xerophilic fungi. *Curr Genet* 2014;61(3):325–34.
23. Hammami W, Al Thani R, Fiori S, et al. Patulin and patulin producing *Penicillium* spp. Occurrence in apples and apple-based products including baby food. *J Infect Dev Ctries* 2017;11(4):343–9.
24. Dombrink-Kurtzman MA. The sequence of the isoepoxydon dehydrogenase gene of the patulin biosynthetic pathway in *Penicillium* species. *Antonie Van Leeuwenhoek* 2007;91(2):179–89.
25. Eriksen GS, Moldes-Anaya A, Fæste CK. Penitrem A and analogues: Toxicokinetics, toxicodynamics including mechanism of action and clinical significance. *World Mycotoxin J* 2013;6(3):263–72.
26. Steyn PS. The isolation, structure and absolute configuration of secalonic acid D, the toxic metabolite of *Penicillium oxalicum. Tetrahedron* 1970;26(1):51–7.
27. Rodríguez A, Córdoba JJ, Gordillo R, Córdoba MG, Rodríguez M. Development of two quantitative real-time PCR methods based on SYBR Green and TaqMan to quantify sterigmatocystin-producing molds in foods. *Food Anal Methods* 2012;5(6):1514–25.
28. Núñez F, Díaz MC, Rodríguez M, Aranda E, Martín A, Asensio MA. Effects of substrate, water activity, and temperature on growth and verrucosidin production by *Penicillium polonicum* isolated from dry-cured ham. *J Food Prot* 2000;63(2):231–6.
29. DiSalvo A, Fickling A, Ajello L. Infection caused by *Penicillium marneffei*: Description of first natural infection in man. *Am J Clin Pathol* 1973;60(2):259–63.
30. Duong TA. Infection due to *Penicillium marneffei*, an emerging pathogen: Review of 155 reported cases. *Clin Infect Dis* 1996;23(1):125–30.
31. Hu Y, Zhang J, Li X, et al. *Penicillium marneffei* infection: An emerging disease in mainland China. *Mycopathologia* 2013;175(1–2):57–67.
32. Santos PE, Piontelli E, Shea YR, et al. *Penicillium piceum* infection: Diagnosis and successful treatment in chronic granulomatous disease. *Med Mycol* 2006;44(8):749–53.
33. Lyratzopoulos G, Ellis M, Nerringer R, Denning DW. Invasive infection due to *Penicillium* species other than *P. marneffei. J Infect* 2002;45(3): 184–207.
34. Barcus AL, Burdette SD, Herchline TE. Intestinal invasion and disseminated disease associated with *Penicillium chrysogenum. Ann Clin Microbiol Antimicrob* 2005;4:1–4.
35. Geltner C, Lass-Flörl C, Bonatti H, Müller L, Stelzmüller I. Invasive pulmonary mycosis due to *Penicillium chrysogenum*: A new invasive pathogen. *Transplantation* 2013;95(4):21–3.
36. Yike I. Fungal proteases and their pathophysiological effects. *Mycopathologia* 2011;171(5):299–323.
37. Zhang Z, Reponen T, Hershey GKK. Fungal exposure and asthma: IgE and non-IgE-mediated mechanisms. *Curr Allergy Asthma Rep* 2016;16:86.
38. Pongracic JA, O'Connor GT, Muilenberg ML, et al. Differential effects of outdoor versus indoor fungal spores on asthma morbidity in inner-city children. *J Allergy Clin Immunol* 2010;125(3):593–9.
39. González De León J, González Méndez R, Cadilla CL, Rivera-Mariani FE, Bolaños-Rosero B. Identification of immunoglobulin E-binding proteins of the xerophilic fungus *Aspergillus penicillioides* crude mycelial mat extract and serological reactivity assessment in subjects with different allergen reactivity profiles. *Int Arch Allergy Immunol* 2018;175(3):147–59.
40. Merget R, Sander I, Rozynek P, et al. Occupational immunoglobulin E-mediated asthma due to *Penicillium camemberti* in a dry-sausage packer. *Respiration* 2008;76(1):109–11.

41. Wantke F, Simon-Nobbe B, Pöll V, Götz M, Jarisch R, Hemmer W. Contact dermatitis caused by salami skin. *Contact Dermatitis* 2011;64(2):111–4.
42. Marvisi M, Balzarini L, Mancini C, Mouzakiti P. A new type of hypersensitivity pneumonitis: Salami brusher's disease. *Monaldi Arch Chest Dis - Pulm Ser* 2012;77(1):35–7.
43. Frisvad JC, Samson RA. Polyphasic taxonomy of *Penicillium* subgenus *Penicillium*: A guide to identification of food and air-borne terverticillate Penicillia and their mycotoxins. *Stud Mycol* 2004;49:1–173.
44. Decontardi S, Soares C, Lima N, Battilani P. Polyphasic identification of Penicillia and Aspergilli isolated from Italian grana cheese. *Food Microbiol* 2018;73:137–49.
45. Mahmoud MA, Abd-El-Aziz ARM, Al-Othman MR. Molecular and biochemical taxonomic tools for the identification and classification of plant-pathogenic *Penicillium* species. *Biotechnol Biotechnol Equip* 2016;30(6):1090–6.
46. Alanio A, Bretagne S. Difficulties with molecular diagnostic tests for mould and yeast infections: Where do we stand? *Clin Microbiol Infect* 2014;20(6):36–41.
47. Rodríguez A, Rodríguez M, Luque MI, Justesen AF, Córdoba JJ. Quantification of ochratoxin A-producing molds in food products by SYBR Green and TaqMan real-time PCR methods. *Int J Food Microbiol* 2011;149:226–35.
48. Rodríguez A, Córdoba JJ, Werning ML, Andrade MJ, Rodríguez M. Duplex real-time PCR method with internal amplification control for quantification of verrucosidin producing molds in dry-ripened foods. *Int J Food Microbiol* 2012;153(1–2):85–91.
49. Rodríguez A, Werning ML, Rodríguez M, Bermúdez E, Córdoba JJ. Quantitative real-time PCR method with internal amplification control to quantify cyclopiazonic acid producing molds in foods. *Food Microbiol* 2012;32(2):397–405.
50. Luque MI, Rodríguez A, Andrade MJ, Martín A, Córdoba JJ. Development of a PCR protocol to detect aflatoxigenic molds in food products. *J Food Prot* 2012;75(1):85–94.
51. Rodríguez A, Rodríguez M, Luque MI, Martín A, Córdoba JJ. Real-time PCR assays for detection and quantification of aflatoxin-producing molds in foods. *Food Microbiol* 2012;31(1):89–99.
52. Rodríguez A, Luque MI, Andrade MJ, Rodríguez M, Asensio MA, Córdoba JJ. Development of real-time PCR methods to quantify patulin-producing molds in food products. *Food Microbiol* 2011;28(6):1190–9.
53. Sanzani SM, Miazzi MM, Di Rienzo V, et al. A rapid assay to detect toxigenic *Penicillium* spp. contamination in wine and musts. *Toxins (Basel)* 2016;8:235.
54. Tannous J, Atoui A, El Khoury A, et al. Development of a real-time PCR assay for *Penicillium expansum* quantification and patulin estimation in apples. *Food Microbiol* 2015;50:28–37.
55. Okorski A, Polak-Śliwińska M, Karpiesiuk K, Pszczółkowska A, Kozera W. Real time PCR: A good tool to estimate mycotoxin contamination in pig diets. *World Mycotoxin J* 2017;10:219–28.
56. Hien HTA, Thanh TT, Thu NTM, et al. Development and evaluation of a real-time polymerase chain reaction assay for the rapid detection of *Talaromyces marneffei MP1* gene in human plasma. *Mycoses* 2016;59(12):773–80.
57. Pornprasert S, Praparattanapan J, Khamwan C, et al. Development of TaqMan real-time polymerase chain reaction for the detection and identification of *Penicillium marneffei*. *Mycoses* 2009;52(6):487–92.
58. Rodríguez A, Rodríguez M, Andrade MJ, Córdoba JJ. Development of a multiplex real-time PCR to quantify aflatoxin, ochratoxin A and patulin producing molds in foods. *Int J Food Microbiol* 2012;155:10–8.
59. Suanthie Y, Cousin MA, Woloshuk CP. Multiplex real-time PCR for detection and quantification of mycotoxigenic *Aspergillus, Penicillium* and *Fusarium*. *J Stored Prod Res* 2009;45(2):139–45.
60. Reddy KV, Priyanaka SR, Reddy IB. A multiplex PCR based method for the detection of agriculturally important aflatoxigenic and ochratoxigenic fungal species from cereals. *Int J Pharma Bio Sci* 2013;4(2):884–93.
61. Priyanka SR, Venkataramana M, Balakrishna K, Murali HS, Batra HV. Development and evaluation of a multiplex PCR assay for simultaneous detection of major mycotoxigenic fungi from cereals. *J Food Sci Technol* 2015;52(1):486–92.
62. Luque MI, Andrade MJ, Rodríguez A, Bermúdez E, Córdoba JJ. Development of a multiplex PCR method for the detection of patulin-, ochratoxin A- and aflatoxin-producing moulds in foods. *Food Anal Methods* 2012;6:1113–21.
63. Wong Y-P, Othman S, Lau Y-L, Son R, Chee H-Y. Loop mediated isothermal amplification (LAMP): A versatile technique for detection of microorganisms. *J Appl Microbiol* 2017;124:626–43.
64. Umesha S, Manukumar HM. Advanced molecular diagnostic techniques for detection of food-borne pathogens: Current applications and future challenges. *Crit Rev Food Sci Nutr* 2018;58(1):84–104.
65. Sun J, Li X, Zeng H, et al. Development and evaluation of loop-mediated isothermal amplification (LAMP) for the rapid diagnosis of *Penicillium marneffei* in archived tissue samples. *FEMS Immunol Med Microbiol* 2010;58(3):381–8.

66. Tone K, Fujisaki R, Yamazaki T, Makimura K. Enhancing melting curve analysis for the discrimination of loop-mediated isothermal amplification products from four pathogenic molds: Use of inorganic pyrophosphatase and its effect in reducing the variance in melting temperature values. *J Microbiol Methods* 2017;132:41–5.
67. Ferrara M, Perrone G, Gallo A, Epifani F, Visconti A, Susca A. Development of loop-mediated isothermal amplification (LAMP) assay for the rapid detection of *Penicillium nordicum* in dry-cured meat products. *Int J Food Microbiol* 2015;202: 42–7.
68. Vogt EI, Kupfer VM, Bechtner JD, Frisch LM, Vogel RF, Niessen L. Detection of *Penicillium oxalicum* in grapes with a loop-mediated isothermal amplification (LAMP) assay. *J Microbiol Biotechnol Food Sci* 2017;7(3):265–70.
69. Isshiki A, Takeharu H, Aoki S, et al. Development of a multiple detection technique for fungi by DNA microarray with the simultaneous use of internal transcribed spacer region of ribosomal RNA gene and -tubulin gene probes. *Biocontrol Sci* 2014;19(3):139–45.
70. Aoyama F, Miyamoto T. Development of a DNA Array for the simple identification of major filamentous fungi in the beverage manufacturing environment. *Biocontrol Sci* 2016;21(3):161–72.
71. Lezar S, Barros E. Oligonucleotide microarray for the identification of potential mycotoxigenic fungi. *BMC Microbiol* 2010;10:87.
72. Chen W, Seifert KA, Lévesque CA. A high density COX1 barcode oligonucleotide array for identification and detection of species of *Penicillium* subgenus *Penicillium*. *Mol Ecol Resour* 2009;9(S1):114–29.
73. Campa D, Tavanti A, Gemignani F, et al. DNA microarray based on arrayed-primer extension technique for identification of pathogenic fungi responsible for invasive and superficial mycoses. *J Clin Microbiol* 2008;46(3):909–15.
74. Russell R, Paterson M, Lima N. *Molecular Biology of Food and Water Borne Mycotoxigenic and Mycotic Fungi*. Boca Raton, FL: CRC Press; 2016.
75. Hsueh PR, Teng LJ, Hung CC, et al. Molecular evidence for strain dissemination of *Penicillium marneffei*: An emerging pathogen in Taiwan. *J Infect Dis* 2000;181(5):1706–12.
76. Lund F, Nielsen AB, Skouboe P. Distribution of *Penicillium commune* isolates in cheese dairies mapped using secondary metabolite profiles, morphotypes, RAPD and AFLP fingerprinting. *Food Microbiol* 2003;20(6):725–34.
77. Frisvad JC, Lund F, Elmholt S. Ochratoxin A producing *Penicillium verrucosum* isolates from cereals reveal large AFLP fingerprinting variability. *J Appl Microbiol* 2005;98(3):684–92.
78. Dutech C, Enjalbert J, Fournier E, et al. Challenges of microsatellite isolation in fungi. *Fungal Genet Biol* 2007;44(10):933–49.
79. Castella G, Ostenfeld Larsen T, Cabanes J, et al. Molecular characterization of ochratoxin A producing strains of the genus *Penicillium*. *Syst Appl Microbiol* 2002;25(1):74–83.
80. Rousseaux S, Guilloux-Bénatier M. PCR ITS-RFLP for *Penicillium* species and other genera. In: Morett A, Susca A, editors. *Mycotoxigenic Fungi: Methods and Protocols*. New York: Humana Press; 2017. pp. 321–33.
81. Dupont J, Dennetière B, Jacquet C, Roquebert MF. PCR-RFLP of ITS rDNA for the rapid identification of *Penicillium* subgenus *Biverticillium* species. *Rev Iberoam Micol* 2006;23(3):145–50.
82. Diguta CF, Vincent B, Guilloux-Benatier M, Alexandre H, Rousseaux S. PCR ITS-RFLP: A useful method for identifying filamentous fungi isolates on grapes. *Food Microbiol* 2011;28(6):1145–54.
83. Colombo F, Vallone L, Giaretti M, Dragoni I. Identification of *Penicillium aurantiogriseum* species with a method of polymerase chain reaction-restriction fragment length polymorphism. *Food Control* 2003;14(3):137–40.
84. Peterson SW. *Aspergillus* and *Penicillium* identification using DNA sequences: Barcode or MLST? *Appl Microbiol Biotechnol* 2012;95(2):339–44.
85. Lasker BA. Nucleotide sequence-based analysis for determining the molecular epidemiology of *Penicillium marneffei*. *J Clin Microbiol* 2006;44(9):3145–53.
86. Woo PCY, Lau CCY, Chong KTK, et al. MP1 homologue-based multilocus sequence system for typing the pathogenic fungus *Penicillium marneffei*: A novel approach using lineage-specific genes. *J Clin Microbiol* 2007;45(11):3647–54.
87. Woo PCY, Lau SKP, Lau CCY, et al. Mp1p Is a virulence factor in *Talaromyces* (*Penicillium*) *marneffei*. *PLoS Negl Trop Dis* 2016;10(8):1–16.
88. Giraud F, Giraud T, Aguileta G, et al. Microsatellite loci to recognize species for the cheese starter and contaminating strains associated with cheese manufacturing. *Int J Food Microbiol* 2010;137(2–3):204–13.
89. Li L, Hu F, Chen W, et al. Microsatellite analysis of clinical isolates of the opportunistic fungal pathogen *Penicillium marneffei* from AIDS patients in China. *Scand J Infect Dis* 2011;43(8):616–24.
90. Schoch CL, Seifert KA, Huhndorf S, et al. Nuclear ribosomal internal transcribed spacer (ITS) region as a universal DNA barcode marker for *Fungi*. *Proc Natl Acad Sci* 2012;109(16):6241–6.

91. Demirel R. Comparison of rDNA regions (ITS, LSU, and SSU) of some *Aspergillus, Penicillium*, and *Talaromyces* spp. *Turk J Botany* 2016;40(6):576–83.
92. Visagie CM, Houbraken J, Frisvad JC, et al. Identification and nomenclature of the genus *Penicillium. Stud Mycol* 2014;78: 343–71.
93. Samson RA, Seifert KA, Kuijpers AFA, Houbraken JAMP, Frisvad JC. Phylogenetic analysis of *Penicillium* subgenus *Penicillium* using partial β-tubulin sequences. *Stud Mycol* 2004;49:175–200.
94. Stielow JB, Lévesque CA, Seifert KA, et al. One fungus, which genes? Development and assessment of universal primers for potential secondary fungal DNA barcodes. *Persoonia* 2015;35:242–63.
95. Seifert KA, Samson RA, DeWaard JR, et al. Prospects for fungus identification using CO1 DNA barcodes, with *Penicillium* as a test case. *Proc Natl Acad Sci USA* 2007;104(10):3901–6.
96. Wang L, Zhuang WY. Phylogenetic analyses of penicillia based on partial calmodulin gene sequences. *BioSystems* 2007;88(1–2):113–26.
97. Plewa-Tutaj K, Lonc E. Molecular identification and biodiversity of potential allergenic molds (*Aspergillus* and *Penicillium*) in the poultry house: First report. *Aerobiologia (Bologna)* 2014;30(4):445–51.
98. Skouboe P, Frisvad JC, Taylor JW, Lauritsen D, Boysen M, Rossen L. Phylogenetic analysis of nucleotide sequences from the ITS region of terverticillate *Penicillium* species. *Mycol Res* 1999;103(7):873–81.
99. Houbraken J, Visagie CM, Meijer M, et al. A taxonomic and phylogenetic revision of *Penicillium* section *Aspergilloides. Stud Mycol* 2014;78:373–451.
100. Mekha N, Sugita T, Makimura K, et al. The intergenic spacer region of the ribosomal RNA gene of *Penicillium marneffei* shows almost no DNA sequence diversity. *Microbiol Immunol* 2010;54(11):714–6.
101. Raja HA, Miller AN, Pearce CJ, Oberlies NH. Fungal identification using molecular tools: A primer for the natural products research community. *J Nat Prod* 2017;80(3):756–70.
102. Yahr R, Schoch CL, Dentinger BTM. Scaling up discovery of hidden diversity in fungi: Impacts of barcoding approaches. *Philos Trans R Soc B Biol Sci* 2016;371(1702):20150336.
103. Peterson SW. Targeting conserved genes in *Penicillium* species. In: Moretti A, Susca A, editors. *Mycotoxigenic Fungi. Methods and Protocols.* New York: Humana Press; 2017. pp. 149–57.
104. Chalupová J, Raus M, Sedlářová M, Šebela M. Identification of fungal microorganisms by MALDI-TOF mass spectrometry. *Biotechnol Adv* 2014;32(1):230–41.
105. Drissner D, Freimoser FM. MALDI-TOF mass spectroscopy of yeasts and filamentous fungi for research and diagnostics in the agricultural value chain. *Chem Biol Technol Agric* 2017;4(1):1–12.
106. Welham K, Domin M, Johnson K, Jones L, Ashton D. Characterization of fungal spores by laser desorption/ionization time-of-flight mass spectrometry. *Rapid Commun Mass Spectrom* 2000;14:307–10.
107. Chen HY, Chen YC. Characterization of intact *Penicillium* spores by matrix-assisted laser desorption/ionization mass spectrometry. *Rapid Commun Mass Spectrom* 2005;19(23):3564–8.
108. Hettick JM, Green BJ, Buskirk AD, et al. Discrimination of *Penicillium* isolates by matrix-assisted laser desorption/ionization time-of-flight mass spectrometry fingerprinting. *Rapid Commun Mass Spectrom* 2008;22:2555–60.
109. Fenselau C, Demirev FA. Characterization of intact microorganisms by MALDI mass spectrometry. *Mass Spectrom Rev* 2001;20(4):157–71.
110. Posch AE, Koch C, Helmel M, et al. Combining light microscopy, dielectric spectroscopy, MALDI intact cell mass spectrometry, FTIR spectromicroscopy and multivariate data mining for morphological and physiological bioprocess characterization of filamentous organisms. *Fungal Genet Biol* 2013;51(1):1–11.
111. Lau SKP, Lam CSK, Ngan AHY, et al. Matrix-assisted laser desorption ionization time-of-flight mass spectrometry for rapid identification of mold and yeast cultures of *Penicillium marneffei. BMC Microbiol* 2016;16(1):36.
112. Normand AC, Cassagne C, Ranque S, et al. Assessment of various parameters to improve MALDI-TOF MS reference spectra libraries constructed for the routine identification of filamentous fungi. *BMC Microbiol* 2013;13(1):76.
113. Stübiger G, Wuczkowski M, Mancera L, Lopandic K, Sterflinger K, Belgacem O. Characterization of yeasts and filamentous fungi using MALDI lipid phenotyping. *J Microbiol Methods* 2016;130:27–37.
114. Ranque S, Normand A-C, Cassagne C, et al. MALDI-TOF mass spectrometry identification of filamentous fungi in the clinical laboratory. *Mycoses* 2014;57(3):135–40.
115. da Silva TL, de Sousa E, Pereira PT, Ferraìo AM, Roseiro JC. Cellular fatty acid profiles for the differentiation of *Penicillium* species. *FEMS Microbiol Lett* 1998;164: 303–10.
116. Frisvad JC, Andersen B, Thrane U. The use of secondary metabolite profiling in chemotaxonomy of filamentous fungi. *Mycol Res* 2008;112(2):231–40.
117. Núñez F, Westphal CD, Bermúdez E, Asensio MA. Production of secondary metabolites by some terverticillate penicillia on carbohydrate-rich and meat substrates. *J Food Prot* 2007;70(12):2829–36.

118. Alshannaq A, Yu JH. Occurrence, toxicity, and analysis of major mycotoxins in food. *Int J Environ Res Public Health* 2017;14(6):632.
119. Jin N, Ling S, Yang C, Wang S. Preparation and identification of monoclonal antibody against Citreoviridin and development of detection by Ic-ELISA. *Toxicon* 2014;90(1):226–36.
120. Li Y, Wang Y, Guo Y. Preparation of synthetic antigen and monoclonal antibody for indirect competitive ELISA of citrinin. *Food Agric Immunol* 2012;23(2):145–56.
121. Ostry V, Toman J, Grosse Y, Malir F. Cyclopiazonic acid: 50th anniversary of its discovery. *World Mycotoxin J* 2018;11:135–48.
122. Li S, Chen PY, Marquardt RR, Han Z, Clarke JR. Production of a sensitive monoclonal antibody to sterigmatocystin and its application to ELISA of wheat. *J Agric Food Chem* 1996;44:372–5.
123. Pleadin J, Malenica M, Vah N, Milone S, Safti L. Survey of aflatoxin B_1 and ochratoxin A occurrence in traditional meat products coming from Croatian households and markets. *Food Control* 2015;52: 71–7.
124. McElroy LJ, Weiss CM. The production of polyclonal antibodies against the mycotoxin derivative patulin hemiglutarate. *Can J Microbiol* 1993;39:861–3.
125. He T, Zhu J, Nie Y, et al. Nanobody technology for mycotoxin detection: Current status and prospects. *Toxin* 2018;10, 180.
126. Shephard GS. Current status of mycotoxin analysis: A critical review. *J AOAC Int* 2016;99(4):842–8.
127. Sun W, Han Z, Aerts J, et al. A reliable liquid chromatography-tandem mass spectrometry method for simultaneous determination of multiple mycotoxins in fresh fish and dried seafoods. *J Chromatogr A* 2015;1387:42–8.
128. Căpriţă A, Căpriţă R, Cozmiuc C, Maranescu B, Sărăndan H. Simultaneous determination of mycotoxins (ochratoxin a and deoxynivalenol) in biological samples. *J Agroaliment Process Technol* 2007;13(2):353–8.
129. Dall'Asta C, Galaverna G, Bertuzzi T, et al. Occurrence of ochratoxin A in raw ham muscle, salami and dry-cured ham from pigs fed with contaminated diet. *Food Chem* 2010;120(4):978–83.
130. Commission of the European Communities. Commision Regulation (EC) No 1881/2006 of 19 December 2006 setting maximum levels for certain contaminants in foodstuffs. *Off J Eur Union* 2006;L364:5–24.
131. Lima N, Santos C. MALDI-TOF MS for identification of food spoilage filamentous fungi. *Curr Opin Food Sci* 2017;13:26–30.

SUMMARY

Considered the "gold standard" methods for identification of *Penicillium* species for decades, culture-based techniques are noted by their laborious and time-consuming nature, requirement for specialized technical skills, and potential for result inconsistency. Molecular techniques, such as the sequencing of DNA barcoding markers and different PCR-based methods, overcome many shortcomings of culture-based techniques, allowing rapid and precise identification and detection of penicillia and improved assessment of their potential pathogenicity and toxigenicity. MALDI-TOF-MS based on profiling of cell surface proteins provides a complementary method to DNA sequence-based techniques for the identification of penicillia at the species or strain level. The profile of taxonomically restricted secondary metabolites, mainly mycotoxins, produced during mould differentiation, could be also useful for species-specific identification. Therefore, a polyphasic approach that encompasses morphological, chemical, and molecular analyses heralds a new trend in the identification and epidemiological investigation of penicillia.

20 Population Genetics of *Enterocytozoon bieneusi*

Yaoyu Feng and Lihua Xiao
South China Agricultural University

Dongyou Liu
Royal College of Pathologists of Australasia Quality Assurance Programs

Contents

20.1 INTRODUCTION

Enterocytozoon bieneusi is a human-pathogenic species of the genus *Enterocytozoon,* which also includes another species (*Enterocytozoon hepatopenaei*) affecting shrimps. Indeed, occurring in humans, other mammals, and birds, *E. bieneusi* accounts for nearly 90% of human microsporidiosis cases, while *Encephalitozoon intestinalis* causes about 10% of human microsporidiosis cases (see Chapter 17). Given that many human-pathogenic *E. bieneusi* genotypes also infect animals and that ingestion/drinking of contaminated food/water represents a key route of transmission, a thorough understanding of population genetics of this zoonotic parasite is crucial for its effective control and prevention.

Taxonomy. Classified taxonomically in the family Enterocytozoonidae, order Chytridiopsida, division Microsporidia, and kingdom Fungi, the genus *Enterocytozoon* comprises two unicellular, spore-forming species, i.e., *E. bieneusi* and *E. hepatopenaei,* in addition to several unassigned species.

Being obligate intracellular parasites, *E. bieneusi*, initially described in 1985 in enterocytes from an HIV-positive human, inhabits the small intestine of humans, other mammals, and birds and is responsible for about 90% of human microsporidiosis cases, while *E. hepatopenaei,* characterized in 2009, resides in the hepatopancreatic tissues of *Penaeus* shrimps, leading to retarded growth. Based on the analysis of an 848-bp fragment from its SSU rRNA gene, *E. hepatopenaei* shares 84% identity with *E. bieneusi* [1].

Considerable genetic variations exist among *E. bieneusi* isolates. Analysis of *E. bieneusi* ITS sequences differentiates this parasite into 11 phylogenetic groups (and >500 genotypes). Interestingly, while group 1 genotypes D, EbpC, and type IV are infective to both humans and animals, group 1 genotypes A, EbpB, and PigEBITS3 show host specificity. Apart from group 2 genotypes BEB4, BEB6, I, and J, most group 2 and virtually all groups 3–11 genotypes display strong host specificity [2,3].

Morphology. *E. bieneusi* is a unicellular, spore-forming organism that demonstrates morphological variations during its life cycle, from proliferative plasmodium (derived from sporoplasm of the spore; containing multiple nuclei, an electron-lucent inclusion, and electron-dense discs, the latter two of which are unique to the genus), early sporogonial plasmodium (multiple nuclei, an electron-lucent inclusion, and electron-dense discs that may form arcs), late sporogonial plasmodium (electron-dense arcs and polar vacuole), irregularly shaped sporoblast (two rows of polar tube coils), maturing sporoblast (one nucleus, a spore coat composed of three layers, polar tube coils, and lamellar polaroblast), to ovoid or ellipse mature spore (single nucleus, polar tube coils, and lamellar and vesicular polaroblast; 1.2–2 μm in size). Overall, ultrastructural examination of *E. bieneusi* reveals a lack of sporophorus vesicles or pansporoblastic membranes, direct contact of all stages with the host-cell cytoplasm, an elongated nuclei at proliferative and sporogonial stages, late thickening of the sporogonial plasmodium plasmalemma, electron-lucent inclusions throughout life cycle, precocious development of electron-dense discs before plasmodial division to sporoblasts, and presence of polar tube doublets in 5–7 coils within spores and sporoblasts [4].

Similarly, *E. hepatopenaei* sporogonal plasmodium is multinucleate and contains numerous small blebs at the surface. During early plasmodial development, plasmodial nuclei undergo binary fission and generate numerous pre-sporoblasts, which grow into early sporoblasts after the development of electron-dense disks and precursors of polar tubule in the cytoplasm of the plasmodium. Sporoblasts bud from the plasmodial surface and mature into oval spores (of 0.7 μm × 1.1 μm in size), which consist of a single nucleus, 5–6 coils of polar filament, a posterior vacuole, an anchoring disk attached to the polar filament, and a thick electron-dense wall. The spore wall is made up of a plasmalemma, an electron-lucent endospore (10 nm), and an electron-dense exospore (2 nm) [1].

Genome. *E. bieneusi* possesses a genome of ~6 Mb with 3,804 predicted protein-encoding genes, of which 1,702 encode proteins with assigned functions. Notably, the *E. bieneusi* genome contains a reduced number of genes for glycolysis, pentose phosphate, and trehalose metabolism; has no genes for spliceosomal intron processing; and lacks a fully functional pathway to generate ATP from glucose (and thus relies on transporters to import ATP from the host). Furthermore, the *E. bieneusi* genome harbors a partial 5.8S rRNA that fuses with the LSU (23S) rRNA, intergenic region, and part of the SSU (16S) rRNA genes [5,6].

Life cycle and epidemiology. *E. bieneusi* spore is infective and highly resistant to adverse environmental conditions. After ingestion, *E. bieneusi* spore extends its polar tubule and injects sporoplasm into the host cell through the polar tubule. Sporoplasm multiplies either by merogony (binary fission) or schizogony (multiple fission) and further develops by sporogony to mature spores with a resistant thick wall. Under pressure from an increasing number of spores, the host cell cytoplasm busts and releases spores in stool, which are ready to infect new cells.

With the capability to infect a wide range of hosts, including humans, monkeys, pigs, cattle, birds, and fish, *E. bieneusi* has been detected in the feces and intestinal tissue of 236 different animal species [2,3]. Transmission of *E. bieneusi* in humans mainly involves ingestion/drinking of contaminated food (e.g., watermelon)/water) or inhalation of aerosols, although vertical and horizontal transmissions are also possible. As pigs often carry human-pathogenic *E. bieneusi* genotypes and are in close contacts with humans, they appear to play a major role in human infection with this parasite. *E. bieneusi* infection in domestic animals and wildlife is likely transmitted through waterborne, foodborne, anthroponotic, and zoonotic routes [7].

Clinical features. *E. bieneusi* infection in individuals with HIV/AIDS, malnutritional children, and the recipients of immunosuppressive therapy often presents as acute or chronic diarrhea, abdominal pain, cholecystitis (inflammation of the gallbladder), weight loss, wasting syndrome, and occasional respiratory symptoms.

In immunocompetent individuals, *E. bieneusi* infection may be asymptomatic or associated with self-limiting diarrhea and malnutrition.

Pathogenesis. *E. bieneusi* is an obligate intracellular parasite armed with polar tube and other devices that facilitate its interaction with and entry (inject) into the host cell, utilization of host cell machineries for its own growth and replication, and evasion of host immune responses during infection processes.

Diagnosis. Observation of spores (1~5 μm) and other parasitic stages in stained clinical smears (e.g., fecal samples) by light microscopy and immunofluorescent microscopy provides a rapid and inexpensive diagnosis of microsporidia. Further identification of organelles by transmission electron microscopy enables species-specific determination of microsporidia. In addition, the use of nucleic acid amplification techniques (particularly PCR) permits precise, sensitive and speedy identification and genotyping of *E. bieneusi* [8,9].

Treatment and prevention. Anthelmintic albendazole (which inhibits tubulin) and antiprotozoal metronidazole (which inhibits nucleic acid synthesis by disrupting the DNA of microbial cells) may be used to treat *E. bieneusi* infection in humans with digestive symptoms. Antibiotic fumagillin (which inhibits methionine aminopeptidase type 2 or MetAP2) and immunomodulatory thalidomide (which reduces swelling and redness or inflammation) may also be considered in case other drugs are ineffective [10].

Prevention of human *E. bieneusi* infection focuses on avoidance of contaminated water/food, filtration of water supply, and maintenance of hygienic conditions. No vaccine for *E. bieneusi* is currently available.

20.2 POPULATION GENETICS OF *E. BIENEUSI*: RECENT DEVELOPMENTS

20.2.1 Overview of Population Genetics

As a part of evolutionary biology, population genetics seeks to understand how and why the frequencies of alleles and genes (genotypes) change over time within and between populations of organisms. The genetic diversity within a gene pool is influenced by factors such as population size, mutation, genetic drift, natural selection, environmental diversity, migration, and non-random mating patterns.

Population genetic analysis typically starts with an examination of allelic/genotypic changes using certain DNA or protein markers (genomics or proteomics) followed by an assessment of these changes through relevant models that help predict the occurrence of specific alleles in populations. To gain insights at the right evolutionary scale, appropriate DNA or protein markers that are reasonably polymorphic and reproducible are selected. While DNA markers with high mutation rates (e.g., microsatellites and minisatellites) provide clues into recent divergence, those with relatively low mutation rates (e.g., mitochondrial and nuclear loci) give inference about distant evolutionary history. Furthermore, the availability of increasingly affordable genome sequencing technology coupled to computer-based storage and comparison capacity allows construction of detailed evolutionary trees (or molecular clocks) that help determine the points where populations diverge and form new species.

20.2.2 Approaches for Population Genetic Analysis

DNA markers. The most widely applied DNA markers for population genetic analysis of *E. bieneusi* are ribosomal internal transcribed spacer (ITS) and microsatellites. Although *E. bieneusi* is classified as a fungus, its nuclear ribosomal RNA (rRNA) genes resemble their bacterial counterparts (i.e., presence of 16S, 23S, and 5S rRNA, but only partial presence of 5.8S rRNA), in comparison with fungal (rRNA) genes (i.e., presence of 18S, 26S, and 5S rRNA as well as 5.8S rRNA). Sequence comparison of *E. bieneusi* ITS region (measuring about 480 bp, of which a stretch of 243 bp is often targeted) has proven particularly useful for taxonomic determination and molecular phylogenetic analysis, due to its small size, multiple copy number, high degree of variation, and location between highly conserved flanking sequences [of small subunit (16S) and large subunit (23S) rRNA genes] for primer anchoring.

Furthermore, most eukaryotic genomes harbor microsatellites (tandem repeats of di-, tri-, or tetranucleotides flanked by non-repetitive regions) and minisatellites (tracts of GC-rich, non-protein-coding DNA sequences of 10–60 bp, typically repeated 5–50 times, with many repeated copies lying next to each other). Being codominant, highly polymorphic, relatively abundant, easily typed, and Mendelian inherited, microsatellites and minisatellites offer suitable markers for differentiation among closely related species and discrimination within the species. Notably, three microsatellite loci (MS1, MS5, and MS7) and one minisatellite locus (MS4) within the *E. bieneusi* genome have shown potential for population genetic study of *E. bieneusi*. PCR amplification of MS1, MS3, MS4, and MS7 yield products of 598 and 607 bp, 527 and 529 bp, 897 bp, 456 and 459 bp, and help delineate 12, 8, 7, and 11 genotypes of *E. bieneusi*, respectively [9].

Sequence analysis. Nucleotide sequences amplified from *E. bieneusi* isolates are first verified and then analyzed using relevant softwares to determine their genotypes and multilocus genotypes (MLG) and to reveal phylogenetic relationship and other population genetic details.

Specifically, *E. bieneusi* ITS, MS1, MS3, MS4, and MS7 sequences are verified through comparison with reference sequences deposited in the GenBank database using the Sequence Basic Local Alignment Search Tool (BLAST) and Clustal X 2.0.

E. bieneusi genotypes are determined based on distinct polymorphic sites found with the 243 bp fragment of the ITS region of the rRNA gene according to the established nomenclature system [2].

E. bieneusi MLG are ascertained through analysis of MS1, MS3, MS4, and MS7 sequences with consideration of both single-base nucleotide substitutions and short insertions and deletions (indels) polymorphisms [9].

Phylogenetic relationship among *E. bieneusi* genotypes is determined by concatenating the ITS sequence for ITS, MS1, MS3, MS4, and MS7 sequences using a neighbor-joining analysis in MEGA 7.01, as based on the Kimura 2-parameter model using 1,000 bootstrap replicates. Furthermore, median-joining phylogenies are generated using Network software version 5.0 under the default parameters. Networks are then arranged by hand and nodes colored using Network Publisher version 5.0.0.0. *E. bieneusi* isolates from different hosts and geographic origins may be included in the multilocus phylogeny and haplotype network analysis.

In population genetics, linkage disequilibrium (LD) refers to the non-random association (linkage) of alleles at different loci in a given population. When the frequency of association of independent and random alleles (loci) is higher or lower than expected, the loci are considered to be in LD. Specifically, intragenic LD for individual locus and concatenated multilocus data set is estimated from segregating sites without consideration of indels using DnaSP (applying exact tests with Markov chain parameters to allelic profile data) [11], and standardized index of association (I^S_A) values on five-loci haplotypes are calculated using LIAN 3.7, with I^S_A equaling to zero or a negative value indicating randomly mating populations and alleles in linkage equilibrium (LE), an I^S_A value greater than zero indicating that a non-panmictic population structure is exhibiting LD, a value less than L (95% critical value for the variance of pairwise differences V_D relative to the null hypothesis of panmixia) indicating that the population is panmictic and in LE, and with $V_D > L$ indicating that the population is non-panmictic and has some LD. Thus, when I^S_A is >0 and V_D is >L, the presence of LD and a clonal population structure for *E. bieneusi* are indicated.

E. bieneusi nucleotide diversity (Pi) and neutrality (Fu's neutrality tests or *Fs* and Tajima's D), genotype frequency, genetic diversity (Hd), and recombination rates [GENECONV, MaxChi, and SiScan in the software RDP (Recombination Detection Program) version 4.92 for confirming the occurrence of recombination] are calculated using DnaSP (based on segregating sites) and Arlequin version 3.5.2.2 (based on both segregating sites and indels) [11,12]. In addition, *E. bieneusi* population structure is assessed based on the intragenic and intergenic LD, I^S_A, neutrality, and recombination events (Rms). The degree of genetic differentiation between *E. bieneusi* populations is evaluated by Wright's fixation index (F_{ST}) using Arlequin and gene flow (*Nm*) using DnaSP.

Additionally, *E. bieneusi* prevalence (*ό*) between different geographical locations, ages, clinical signs, and other parameters is analyzed using the chi-squared test or other statistical tests. The impacts of the multiple variables are also evaluated by multivariable regression analysis using SPSS 22.0 version.

20.2.3 Current Data on Population Genetics

Genotype determination. Based on the hypermutation in the ITS sequence, at least 474 *E. bieneusi* ITS genotypes clustering to 11 phylogenetic groups (1–11) are identified from humans, non-human primates, porcines, ruminants, companion animals, equines, carnivores, rodents, lagomorphs, birds, water, and other sources [3]. Of these, ITS group 1 includes 314 genotypes originated from humans, other mammals, and birds; ITS group 2 contains 94 genotypes originated from humans (e.g., BEB4, BEB6, CHN3, I, and J), bovines, and ovines; and ITS groups 3–11 display strong host specificity and probably have limited public health importance (Figure 20.1) [13]. Nonetheless, among 106 genotypes identified in humans to date, 91 belong to group 1 (including A, EbpB, and PigEBITS3 with high host specificity; D, EbpC, and type IV with low host specificity, also affecting animals), six to group 2, two to group 5, five to group 6, one to group 10, and one as an outlier [3,13].

Phylogenetic and structural analysis. Although examination of the ITS sequence provides sufficient clarity on the genotype determination for most *E. bieneusi* isolates, this approach seems to be inadequate for a small number of isolates, leading to unequivocal results. Exploiting the intrinsic variability of microsatellites and minisatellites, MLST (multilocus sequence typing) analysis targeting three microsatellite loci (MS1, MS3, and MS7) and one minisatellite locus (MS4) generates MLG, which reveal a higher level of diversity than ITS-based genotypes and offer a practical genotyping solution to certain isolates for which ITS sequence analysis may fail.

These days, sequence analysis of five DNA markers [ITS, microsatellites (MS1, MS3, and MS7), and minisatellite (M4)] has formed the cornerstone of population genetic investigation of *E. bieneusi* parasites. Combined with relevant softwares, these DNA markers have helped clarify the population structures (combining intra-/intergenic LD, I^S_A, neutrality, and recombination events to assess *E. bieneusi* population structure as clonal or epidemic), substructures (using K-means partitional clustering and the admixture model in the Bayesian analysis tool STRUCTURE version 2.3.4 to analyze allelic profile data for determining subpopulations), and divergence [Arlequin for calculating Wright's fixation index (F_{ST}), a measure of population divergence, with $F_{ST}=0$ indicating similar polymorphisms across all markers and $F_{ST}=1$ indicating high levels of between-population divergence]; geographic segregation (a median-joining analysis implemented in the software Network version 5.0 for estimating the potential for geographic segregation); host adaptation; and transmission dynamics of *E. bieneusi* in humans and animal hosts from different geographic locations [14–33].

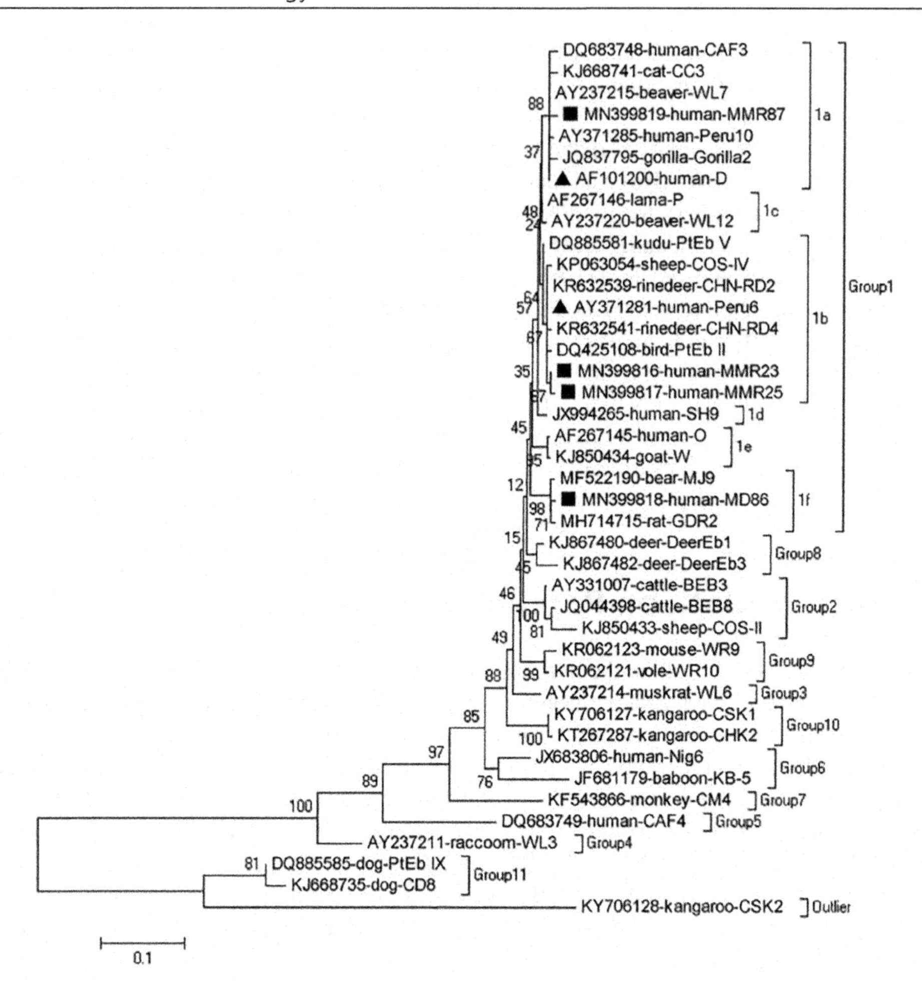

FIGURE 20.1 Phylogenetic relationship of *Enterocytozoon bieneusi* genotypes representing groups 1–11 based on neighbor-joining analysis of internal transcribed spacer (ITS) of nuclear ribosomal DNA sequence and calculation using the Kimura 2-parameter model. The numbers on the branches are percent bootstrap values from 1,000 replicates. Each sequence is identified by its accession number, host origin, and genotype designation. The triangles and squares filled in black indicate known genotypes and novel genotypes identified, respectively. (Photo credit: Shen Y, Gong B, Liu X, et al. *BMC Microbiol.* 2020;20(1):10.)

20.3 FUTURE PERSPECTIVES

Population genetic analysis of *E. bieneusi* using five DNA markers (ITS, M1, M3, M4, and M7) has helped uncover critical insights on the population structures, substructures, and divergence; geographic segregation; host adaptation; and transmission dynamics of this zoonotic parasite in both human and animal hosts worldwide, and facilitated the development and implementation of epidemiological control and prevention programs, leading to a significant reduction in the clinical cases of *E. bieneusi* infections in the human population.

The fact that *E. bieneusi* ITS groups 1 and 2 contain genotypes with the ability to infect humans or animals or both highlights the value of incorporating additional gene targets into the molecular epidemiological investigation for their differentiation. This will necessitate further input into the identification and characterization of novel virulence genes in *E. bieneusi* genotypes with specificity for humans or animals or both, which will yield crucial insights into the molecular mechanisms of *E. bieneusi* pathogenicity and contribute to improved management of *E. bieneusi* infection in humans as well as animals.

REFERENCES

1. Tourtip S, Wongtripop S, Stentiford GD, et al. *Enterocytozoon hepatopenaei* sp. nov. (Microsporida: Enterocytozoonidae), a parasite of the black tiger shrimp *Penaeus monodon* (Decapoda: Penaeidae): fine structure and phylogenetic relationships. *J Invertebr Pathol.* 2009;102(1):21–9.
2. Santín M, Fayer R. *Enterocytozoon bieneusi* genotype nomenclature based on the internal transcribed spacer sequence: a consensus. *J Eukaryot Microbiol.* 2009;56(1):34–8.
3. Li W, Feng Y, Santin M. Host specificity of *Enterocytozoon bieneusi* and public health implications. *Trends Parasitol.* 2019;35:436–51.
4. Chalifoux LV, MacKey J, Carville A, et al. Ultrastructural morphology of *Enterocytozoon bieneusi* in biliary epithelium of rhesus macaques (Macaca mulatta). *Vet Pathol.* 1998;35(4):292–6.
5. Akiyoshi DE, Morrison HG, Lei S, et al. Genomic survey of the non-cultivatable opportunistic human pathogen, *Enterocytozoon bieneusi. PLoS Pathog.* 2009;5(1):e1000261
6. Keeling PJ, Corradi N, Morrison HG, et al. The reduced genome of the parasitic microsporidian *Enterocytozoon bieneusi* lacks genes for core carbon metabolism. *Genome Biol Evol.* 2010;2:304–9.
7. Matos O, Lobo ML, Xiao L. Epidemiology of *Enterocytozoon bieneusi* infection in humans. *J Parasitol Res.* 2012;2012:981424.
8. Breton J, Bart-Delabesse E, Biligui S, et al. New highly divergent rRNA sequence among biodiverse genotypes of *Enterocytozoon bieneusi* strains isolated from humans in Gabon and Cameroon. *J Clin Microbiol.* 2007;45(8):2580–9.
9. Feng Y, Li N, Dearen T, et al. Development of a multilocus sequence typing tool for high-resolution genotyping of *Enterocytozoon bieneusi. Appl Environ Microbiol.* 2011;77(14):4822–8.
10. Han B, Weiss LM. Therapeutic targets for the treatment of microsporidiosis in humans. *Expert Opin Ther Targets.* 2018;22(11):903–915.
11. Librado P, Rozas J. DnaSP v5: a software for comprehensive analysis of DNA polymorphism data. *Bioinformatics.* 2009; 25(11):1451–2.
12. Excoffier L, Laval G, Schneider S. Arlequin (version 3.0): an integrated software package for population genetics data analysis. *Evol Bioinform Online.* 2007;1:47–50.
13. Shen Y, Gong B, Liu X. et al. First identification and genotyping of *Enterocytozoon bieneusi* in humans in Myanmar. *BMC Microbiol.* 2020;20:10.
14. Widmer G, Dilo J, Tumwine JK, Tzipori S, Akiyoshi DE. Frequent occurrence of mixed *Enterocytozoon bieneusi* infections in humans. *Appl Environ Microbiol.* 2013;79(17):5357–62.
15. da Silva Fiuza VR, Lopes CW, de Oliveira FC, Fayer R, Santin M. New findings of *Enterocytozoon bieneusi* in beef and dairy cattle in Brazil. *Vet Parasitol.* 2016;216:46–51.
16. Deng L, Li W, Yu X, et al. First report of the human-pathogenic *Enterocytozoon bieneusi* from red-bellied tree squirrels (*Callosciurus erythraeus*) in Sichuan, China. *PLoS One.* 2016;11(9):e0163605.
17. Deng L, Li W, Zhong Z, et al. Multi-locus genotypes of *Enterocytozoon bieneusi* in captive Asiatic black bears in southwestern China: high genetic diversity, broad host range, and zoonotic potential. *PLoS One.* 2017;12(2):e0171772.
18. Deng L, Li W, Zhong Z, et al. Molecular characterization and new genotypes of *Enterocytozoon bieneusi* in pet chipmunks (Eutamias asiaticus) in Sichuan province, China. *BMC Microbiol.* 2018;18(1):37.
19. Li W, Deng L, Yu X, et al. Multilocus genotypes and broad host-range of *Enterocytozoon bieneusi* in captive wildlife at zoological gardens in China. *Parasit Vectors.* 2016;9(1):395.
20. Li W, Song Y, Zhong Z, et al. Population genetics of *Enterocytozoon bieneusi* in captive giant pandas of China. *Parasit Vectors.* 2017;10(1):499.
21. Wan Q, Xiao L, Zhang X, et al. Clonal evolution of *Enterocytozoon bieneusi* populations in swine and genetic differentiation in subpopulations between isolates from swine and humans. *PLoS Negl Trop Dis.* 2016;10(8):e0004966.
22. Zhang XX, Cong W, Lou ZL, et al. Prevalence, risk factors and multilocus genotyping of *Enterocytozoon bieneusi* in farmed foxes (*Vulpes lagopus*), northern China. *Parasit Vectors.* 2016;9:72.

23. Yue DM, Ma JG, Li FC, et al. Occurrence of *Enterocytozoon bieneusi* in donkeys (*Equus asinus*) in China: a public health concern. *Front Microbiol.* 2017;8:565.
24. Zhong Z, Tian Y, Song Y, et al. Molecular characterization and multi-locus genotypes of *Enterocytozoon bieneusi* from captive red kangaroos (*Macropus rufus*) in Jiangsu province, China. *PLoS One.* 2017;12(8):e0183249.
25. Cong W, Qin SY, Meng QF. Molecular characterization and new genotypes of *Enterocytozoon bieneusi* in minks (*Neovison vison*) in China. *Parasite.* 2018;25: 34.
26. Greigert V, Pfaff AW, Abou-Bacar A, Candolfi E, Brunet J. Intestinal microsporidiosis in Strasbourg from 2014 to 2016: emergence of an *Enterocytozoon bieneusi* genotype of Asian origin. *Emerg Microbes Infect.* 2018;7(1):97.
27. Tang C, Cai M, Wang L, et al. Genetic diversity within dominant *Enterocytozoon bieneusi* genotypes in pre-weaned calves. *Parasit Vectors.* 2018;11(1):170.
28. Gong B, Yang Y, Liu X, et al. First survey of *Enterocytozoon bieneusi* and dominant genotype Peru6 among ethnic minority groups in southwestern China's Yunnan Province and assessment of risk factors. *PLoS Negl Trop Dis.* 2019;13(5):e0007356.
29. Li W, Xiao L. Multilocus sequence typing and population genetic analysis of *Enterocytozoon bieneusi*: host specificity and its impacts on public health. *Front Genet.* 2019;10:307.
30. Luo R, Xiang L, Liu H, et al. First report and multilocus genotyping of *Enterocytozoon bieneusi* from Tibetan pigs in southwestern China. *Parasite.* 2019;26:24.
31. Prado JBF, Ramos CADN, Fiuza VRDS, Terra VJB. Occurrence of zoonotic *Enterocytozoon bieneusi* in cats in Brazil. *Rev Bras Parasitol Vet.* 2019;28(1):80–90.
32. Wang HY, Qi M, Sun MF, et al. Prevalence and population genetics analysis of *Enterocytozoon bieneusi* in dairy cattle in China. *Front Microbiol.* 2019;10: 1399.
33. Zhang Y, Koehler AV, Wang T, Cunliffe D, Gasser RB. *Enterocytozoon bieneusi* genotypes in cats and dogs in Victoria, Australia. *BMC Microbiol.* 2019;19(1):183.

SUMMARY

Enterocytozoon bieneusi is a human-pathogenic species of the genus *Enterocytozoon,* which also includes another species (*Enterocytozoon hepatopenaei*) that affects shrimps. Parasitizing humans, other mammals, and birds, *E. bieneusi* is responsible for nearly 90% of human microsporidiosis cases (with the remaining 10% of human microsporidiosis cases attributed to *Encephalitozoon intestinalis*). Given that many human-pathogenic *E. bieneusi* genotypes also occur in animals and that ingestion/drinking of contaminated food/water represents a key route of transmission, a thorough understanding of population genetics of this zoonotic parasite is crucial for its control and prevention. Recent application of five DNA markers (ITS, M1, M3, M4, and M7) has helped uncover valuable insights on the population structures, substructures, and divergence; geographic segregation; host adaptation; and transmission dynamics of *E. bieneusi* in both human and animal hosts worldwide, and facilitated the implementation of appropriate control and prevention programs that lead to a significant reduction in the clinical cases of human *E. bieneusi* infections. Incorporation of additional gene targets into the molecular epidemiological investigation will enable differentiation of *E. bieneusi* ITS groups 1 and 2 that contain genotypes with the ability to infect humans or animals or both.

Molecular Mechanisms of Antifungal Resistance in *Candida*

21

Maurizio Sanguinetti and Brunella Posteraro
Fondazione Policlinico Universitario A. Gemelli IRCCS, Università Cattolica del Sacro Cuore

Patrizia Posteraro
GVM Ospedale San Carlo di Nancy

Contents

21.1 INTRODUCTION

Despite their diversity, abundance, and ubiquity, relatively few fungi are pathogenic to humans [1], and among the pathogenic fungal species, many are limited to causing opportunistic infections or diseases primarily in immunosuppressed hosts (e.g., *Candida*) [2,3]. Nonetheless, deaths due to fungal infections, ranging from superficial mucosal malaises to life-threatening diseases (e.g., disseminated hematogenous and invasive diseases), outnumber those caused by malaria or tuberculosis each year [3,4], highlighting that the host–fungus relationship is not always amicable.

Candida species are normal colonizers of the oral cavity, gastrointestinal tract, and vagina of healthy humans and, as a component of the resident microbiota [5–7], inflict little to no damage to the host [8]. However, if normal microbiota balance is perturbed or cell immunity is impaired, *Candida* overgrowth can lead to opportunistic infection [8], facilitated by its great adaptability to different host niches [9]. In addition to recurrent infections such as oral and vaginal candidiasis [10], *Candida* species cause invasive diseases (e.g., candidemia, endocarditis, intra-abdominal candidiasis) [11]. The fact that these infections are generally associated with high morbidity and mortality necessitates early implementation of appropriate antifungal therapy for successful patient outcome [12].

Taxonomically, *Candida* is a member of the hemiascomycetes, which form a subgroup of the fungal phylum ascomycetes (Figure 21.1). Composed of approximately 200 species, *Candida* represents the largest genus of medically important yeasts [2]. Although as many as 30 *Candida* species have been implicated in human infections (and the list continues to expand) [13], *C. albicans, C. glabrata, C. krusei, C. parapsilosis,* and *C. tropicalis* are responsible for >90% of invasive infections [14–16]. As the major pathogen in this group, *C. albicans* accounts for approximately 50% of bloodstream *Candida* isolates in the United States [14,15].

Primarily in *C. albicans*, various host-related phenotypes including morphogenesis, adhesion and biofilm formation, interactions with phagocytes, stress tolerance, and nutrient utilization correlate with virulence [17]. Indeed, the ability to reversibly switch between yeast and hyphae/pseudohyphae is a virulence trait that underscores the pathogenesis of *C. albicans* (Figure 21.2). The "CUG" clade – a group of species that use an alternative genetic code in which the codon specifies serine rather than leucine – encompasses all *Candida* species commonly isolated from patients other than *C. glabrata* and *C. krusei.* A study on the aforementioned phenotypes showed that two species deviated from expectations from clinical incidence and animal models [18]. One was *C. dubliniensis* (closely related to *C. albicans*, but notoriously a weaker pathogen) that appeared as robust as the most virulent species, *C. albicans* and *C. tropicalis.* Another was *C. parapsilosis* that performed in nearly all testing assays less well than might be expected from its clinical importance.

Among non-*albicans Candida* species, whose prevalence surpasses that of *C. albicans* in bloodstream infections, *C. parapsilosis* is the most heterogeneous [2,19,20]. Based on DNA sequences, the *C. parapsilosis* species complex is split into three morphologically and physiologically indistinguishable species, *C. parapsilosis, C. metapsilosis,* and *C. orthopsilosis* (termed "cryptic species"). *C. parapsilosis* is generally susceptible to antifungal agents, although minimum inhibitory concentration (MIC) values to echinocandins are elevated [21]. *C. metapsilosis* and *C. orthopsilosis* are at ~1% prevalent in most bloodstream isolate surveys [22] and respond well to antifungal agents [2]. Compared to *C. albicans,* which is a polymorphic organism showing a unicellular yeast form and one of two filamentous forms (pseudohyphae and hyphae), *C. dubliniensis* has a limited capability of undergoing yeast-to-hyphal transition [19], and thus invading deeper tissues [9]. As fluconazole resistance among *C. dubliniensis* remains very low (at <5%) [20], separating *C. dubliniensis* from *C. albicans* routinely may not be necessary for most sporadic isolates [2]. *C. glabrata* is one of the most difficult to treat because of its inherent

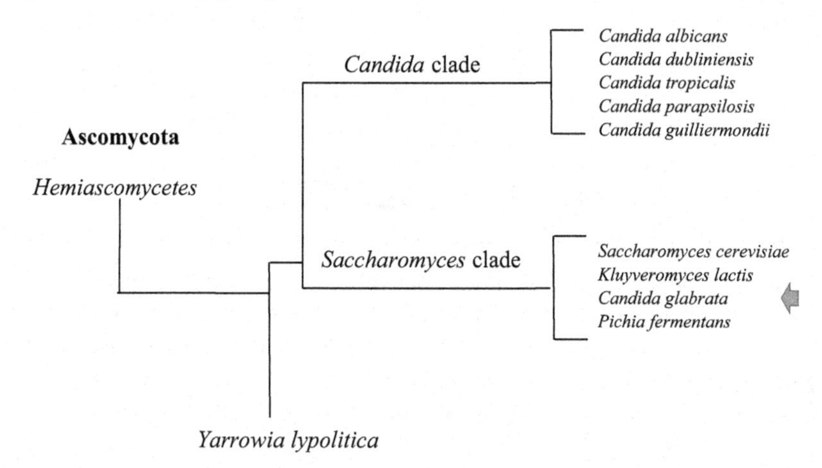

FIGURE 21.1 Tree-based grouping of the most prominently studied hemiascomycetes species. With exception of *C. glabrata* (indicated by arrow), all the pathogenic species belong to the *Candida* clade. (Adapted from Soll [16].)

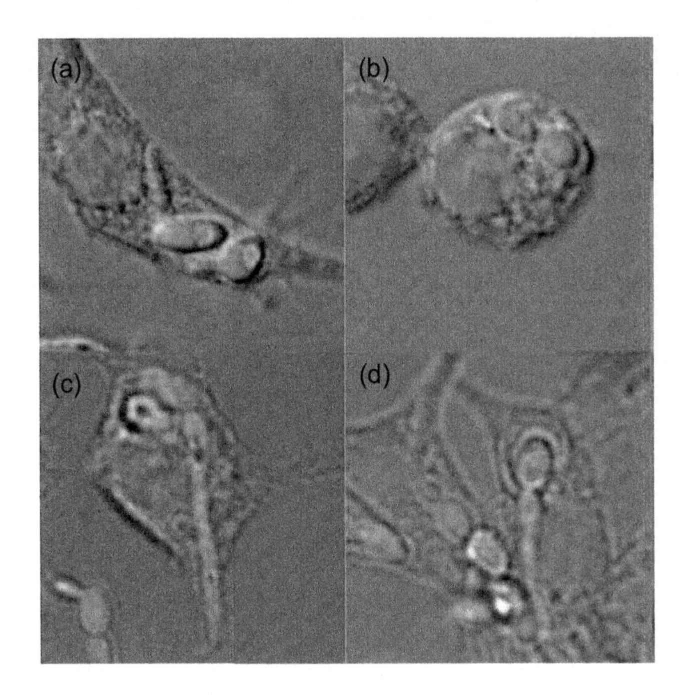

FIGURE 21.2 Morphogenesis of *Candida albicans. C. albicans* yeast cells germinate within macrophages after 1 h incubation (a and b); *C. albicans* forms hyphae/pseudohyphae that penetrate and disrupt macrophages after 6 h incubation (c and d). (Photo credit: Jiménez-López C, Lorenz MC. *PLoS Pathog.* 2013;9(11):e1003741.)

decreased susceptibility to azoles and its ability to acquire azole resistance rapidly [23]. As shown in Figure 21.1, this species is more closely related to *Saccharomyces cerevisiae* than to other *Candida* species. *C. bracarensis* and *C. nivariensis* are two *Candida* species genetically related to *C. glabrata.* The association between *Candida* species and antifungal susceptibility/resistance remains inadequately delineated for many of the cryptic species. While *C. bracarensis* shows a more variable susceptibility, *C. nivariensis* is uniformly resistant to multiple azole drugs [2]. Finally, *C. krusei* is the fifth most common cause of bloodstream infections in various surveys [11]. Together with other *non-albicans Candida* species, *C. krusei* is intrinsically resistant to fluconazole and less susceptible to amphotericin B. Genetically, this species is most closely related to *Pichia (Candida) norvegensis* and *Candida inconspicua*, which are both rare causes of human infections [2].

Although multidrug resistance (i.e., resistance to greater than one drug class) remains uncommon [24], it is notable that resistance to multiple antifungal drugs – azoles (especially fluconazole), echinocandins, and polyenes (namely amphotericin B) – has been increasingly reported, such as in *C. auris* [25]. Initially described from an external ear canal drainage in Japan in 2009 and later from a bloodstream infection in Korea in 2011 [26], multidrug-resistant *C. auris* has emerged worldwide as a healthcare-associated fungus causing invasive infections with high rates of clinical treatment failure [27]. Overall, 41% of *C. auris* isolates from 54 patients in Pakistan, India, South Africa, and Venezuela during 2012–2015 were resistant to two antifungal classes (multidrug-resistant) and 4% were resistant to three classes (azoles, echinocandins, and polyenes) [28].

As reviewed recently [24], resistance can be either intrinsic or acquired. Unlike bacteria, there is no exchange of resistance mechanisms between *Candida* species/strains. Therefore, acquired resistance occurs either in response to an antifungal selection pressure in the individual patient or, more rarely, in consequence to horizontal transmission of resistant strains between patients. *C. glabrata* can acquire resistance to azoles and echinocandins as single drug classes, as well as multidrug resistance involving all major drug classes. Conversely, intrinsic or primary resistance is inherent (not acquired) resistance, which is a trait featured by almost all strains belonging to a given species, and is predictive of clinic failure. *Candida* multidrug resistance appears to be evolved from species with intrinsic resistance, of which notable (and recent) examples are fluconazole-resistant *C. krusei* and the aforementioned *C. auris* strains. Although it may be rare, multidrug resistance can occur in normally susceptible species [24]. We are witnessing the changing epidemiology of fungal infections, which has resulted in a shift toward species with intrinsic resistance or less susceptibility to the antifungal agents commonly used in clinical practice [29].

Table 21.1 summarizes available data for some of the notorious *Candida* species that are intrinsically resistant or have elevated MICs to amphotericin B, echinocandins, and fluconazole [24]. As can be seen,

TABLE 21.1 Intrinsic susceptibility profiles for *Candida* species (Adapted from Arendrup and Patterson [24])

	ANTIFUNGAL AGENTS			
SPECIES	*AMPHOTERICIN B*	*ECHINOCANDINS*	*FLUCONAZOLE*	*NOTES*
Common Species				
C. albicans	S	S	S	
C. dubliniensis	S	S	S	Closely related to *C. albicans*; fluconazole resistance acquired in oral isolates [20]
C. glabrata	S	S	I	Efflux pumps often induced during azole therapy [23]
C. krusei	S	S	R	
C. parapsilosis	S	S/I	S	A naturally occurring *FKS* alteration Fks1p responsible for elevated echinocandins MICs
Uncommon Species				
C. auris	NC	NC	NC	93% resistant to fluconazole, 35% to amphotericin B, and 7% to echinocandins. 41% were resistant to two antifungal classes and 4% were resistant to three classes [28].
C. bracarensis			NC	Closely related to *C. glabrata*
C. metapsilosis		NC		Closely related to *C. parapsilosis*
C. nivariensis			NC	Closely related to *C. glabrata*
C. orthopsilosis		NC		Closely related to *C. parapsilosis*
C. incospicua			NC	
C. norvegensis				
C. robusta			NC	Known as *S. cerevisiae*; closely related to *C. glabrata*

Note: I, indeterminate; R, resistant; S, susceptible; NC, not classified.

uncommon *Candida* species are "not classified" (NC), although their MICs for the antifungal agent are higher than those for *C. albicans*. This is because CLSI (Clinical and Laboratory Standards Institute) and EUCAST (European Committee on Antimicrobial Susceptibility Testing) clinical breakpoints – the MIC thresholds used to classify isolates as susceptible (S), intermediate (I), or resistant (R) – have been established only for the common *Candida* species [30,31].

21.2 METHODOLOGICAL OUTLINES

The rise of *Candida* species that are resistant to the most common classes of antifungal agents (albeit relatively rare when compared to antibacterial agents) poses a serious threat to human health [32]. The recent introduction of nonculture fungal diagnostics provides substantial benefits to clinical outcome, antimicrobial stewardship, and control of antimicrobial resistance [33]. The most widely applied nonculture detection methods target nucleic acids [e.g., polymerase chain reaction (PCR), multiplex PCR, real-time PCR, nucleic acid sequence-based amplification (NASBA), and oligonucleotide DNA microarray]. Other nonculture detection methods include biosensor-based or immunological approaches.

To date, nucleic acid amplification methodologies have proven to be a promising alternative for the rapid and reliable detection and identification of fungal pathogens, including resistant organisms, although further standardization is clearly necessary regarding sample type, primer selection (panfungal, genus-specific, or species-specific), and methods (qualitative, quantitative, real time) [34,35]. In principle, nucleic acid amplification methodologies are highly sensitive (1–10 colony-forming units [CFU] ml^{-1}), especially those targeting high-copy ribosomal genes for genus- and species-level pathogen identification [36]. On the other hand, assays targeting resistance-associated genes (see below) may be problematic because many of these genes are found in single or low copy numbers in the genome.

As summarized in Table 21.2 (see Sanguinetti et al. 2015 [37] for a detailed review), various well-defined mechanisms can lead to azole resistance, including alteration or overexpression of the enzyme target lanosterol 14-α demethylase (encoded by the *ERG11* gene), induction of drug efflux pumps, and cellular changes due, in some cases, to nontarget effects induced by stress responses [38]. More than one resistance mechanism can occur in a single *Candida* isolate, and can produce additive effects or lead to the development of cross-resistance among azole drugs. While induction of multidrug efflux pumps, which decrease drug concentration at the enzyme target, is the most common mechanism of drug resistance,

TABLE 21.2 Resistance mechanisms reported in *Candida* species by antifungal drug class (Adapted from Perlin et al. [32])

RESISTANCE MECHANISM	*AZOLE*	*ECHINOCANDIN*	*POLYENE*	*GENE(S) INVOLVED*
Target site modification	+	+	–	Azole: *ERG11* Echinocandin: *FKS1* and *FKS2*
Target abundance	+	–	+	*ERG2*, *ERG3*, *ERG5*, *ERG6*, and *ERG11*
Target site upregulation	+	–	–	Azole: *ERG11* and *UPC2*
Drug efflux pump upregulation[a]	+	–	–	Azole: *CDR*, *MFS*, *CgSNQ2*, *PDH1* (*C. glabrata* specifically)
Nontarget effects	+	+	+	
Biofilm formation	+	+	+	

[a] The ATP-binding cassette (ABC) transporters, including *CDR1* and *CDR2*, and the major facilitator superfamily (MFS) transporters are regulated, respectively, by transcription factor regulators *TAC1* and *MMR1* in *C. albicans*. In *C. glabrata*, *CgCDR1*, *CgCDR2*, and *CgSNQ2* are regulated by transcription regulator *PDR1*.

overexpression of *ERG11*, caused by mutations in the transcription regulator gene *UPC2*, also confers azole resistance. Likewise, azole resistance can result from a loss of function of the sterol Δ5,6-desaturase gene (*ERG3*). Among more than 140 alterations described in the *ERG11* target gene, only some amino acid substitutions cause reduced target affinity and, thus, azole resistance.

Echinocandin resistance involves genetic acquisition of mutations in *FKS* genes, which encode the major catalytic subunit of the enzyme target glucan synthase. These mutations occur in two narrow hot spot regions of Fks1 for all *Candida* species (wild-type amino acid sequences for *C. albicans* hot spot 1: FLTLSLRDP and hot spot 2: DWIRRYTL) and Fks2 in *C. glabrata* (wild-type amino acid sequences hot spot 1: FLILSLRDP and hot spot 2: DWIRRYTL). A polymorphism at Pro-649 in hot spot 1 of the *C. parapsilosis* induces decreased *in-vitro* antifungal susceptibility, but the clinical importance of intrinsic, reduced drug susceptibility is unclear. The echinocandins are not substrates for multidrug transporters, and other mechanisms causing azole resistance are not cross-resistant with echinocandins. Despite this, reports of azole and echinocandin resistance in *C. glabrata* are beginning to emerge, and are due to the acquisition of mutations in *CgFKS2* and the overexpression of *CgCDR1* and *CgCDR2*.

The mechanism of resistance to amphotericin B, which belongs to the oldest antifungal drug class, involves a reduction in ergosterol content in the cell membrane. Ergosterol originates from a precursor, lanosterol, via several intermediate sterols, aided by a number of enzymes encoded by *ERG6*, *ERG11*, *ERG24*, *ERG25*, *ERG26*, *ERG27*, *ERG2*, *ERG3*, *ERG5*, and *ERG4*. Mutations in *ERG2*, *ERG3*, *ERG5*, *ERG6*, and *ERG11* – combined mutations in *ERG11* and in *ERG3* or *ERG5* and single mutations in *ERG6* or in *ERG2* – contribute to a range of resistance to polyenes in *Candida*, although acquired amphotericin resistance is to date a rare event.

Finally, fungal biofilms confer complete or partial resistance to most drug classes, and represent an important factor in the pathogenesis of *Candida* species [39], as well as a diagnostic challenge [40]. Biofilms are usually found in medical devices, such as prostheses, urinary and vascular catheters, and cardiac devices, where the extracellular matrix produced by *Candida* species functions as a barrier to drug diffusion [39]. Overexpression of drug pumps in *Candida* biofilm cells promotes the development of antifungal resistance in the early phase of the biofilm formation, while in the maturation process, changes in sterol composition seem to be more relevant [41]. At least two types of biofilms, one pathogenic and one sexual, have been described in *C. albicans*, depending upon the configuration of the mating type locus. Amazingly, in the first type, a biofilm forms that facilitates commensalism and pathogenesis, and in the second type, a biofilm forms that facilitates mating [16].

Molecular diagnostic techniques offer rapid and accurate assessment of both primary and secondary resistance to azole or echinocandin antifungal agents. Employing detection targets such as 18S or 28S ribosomal genes or ITS1 and ITS2 intervening noncoding regions, PCR and NASBA-based amplification platforms have the potential for rapid species determination. Real-time PCR using self-reporting fluorescent probes allows the kinetics of the amplification process to be analyzed yielding higher-quality diagnostic information [42]. Newer probing technologies enable simultaneous determinations of drug resistance by distinguishing single nucleotide changes (allele discrimination) and are suitable for multiplexed assays. They can be used with primary specimens because of their sensitivity to generate species-specific information that eliminates the need for microbial growth and guides treatment choices. For example, rapid identification of fungi that show a propensity for reduced azole susceptibility such as *C. glabrata* and *C. krusei* influences primary therapy, thereby allowing the initiation of appropriate therapy at an early disease stage to significantly improve the patient's outcome. Although acquired antifungal resistance mechanisms in *Candida* are multifaceted, their underlying genetic basis has provided the rationale for fine-tuning molecular probing technologies (Table 21.3) [43].

Mutations in the *ERG11* gene are easily assessed with high throughout DNA sequencing (e.g., pyrosequencing), allele-specific real-time molecular probes (e.g., TaqMan, Molecular Beacons), LightCycler™ melt curve analysis, or DNA microarray technology [43]. While these techniques are robust, they are technically demanding. Microarrays facilitate simultaneous expression profiling of many genes; however, this approach can be subjective when relating transcript levels to resistance phenotypes. The subjectivity

TABLE 21.3 Genetic mechanisms as a potential basis for molecular detection of antifungal resistance (Adapted from Perlin [43])

MECHANISMS	***CANDIDA*** *SPECIES*	*GENES*
Azoles		
Target site mutations	*C. albicans*	*ERG11*
Target site upregulation	*C. albicans*	*ERG11*
Transcription factor ABC	*C. albicans*	*TAC1*
	C. glabrata	*PDR1*
Transcription factor MFS	*C. albicans*	*MRR1*
Transcription factor ERG	*C. albicans*	*UPC2*
Echinocandins		
Target site mutations	*C. albicans*	*FKS1*
	C. glabrata	*FKS1, FKS2*

can be eliminated by evaluating gain-of-function mutations in transcription factors that promote expression of specific drug resistance genes (Table 21.3). These mutations can be directly targeted by microarray analysis, high throughout DNA sequencing, and multiplexed real-time PCR. In practice, due to the multifactorial nature of azole resistance, molecular diagnostics are not yet suitable for the clinical detection of resistance to this drug class, especially when targeting the differential expression of drug efflux pumps. While the recent elucidation of gain-of-function mutations in transcription factors paved the way for specific molecular targeting [43], clinical application will require a complete validation between mutations and relative azole resistance [44]. The exception may be *C. glabrata*, where azole resistance is closely linked to mutations in the transcription factor *PDR1* [45,46]. Otherwise, DNA sequencing may be the best choice for detecting resistance mutations that are broadly dispersed on the gene [44].

By contrast, the limited spectrum of mutations associated with echinocandin resistance in *Candida* is ideally suited for the development of robust diagnostic assays. Echinocandin resistance can be easily assessed by real-time PCR or DNA sequencing, whereas microarray analysis may be more useful to accommodate the large number of mutations in multiple *FKS* genes seen with *C. glabrata* [43]. Unlike the case of azoles, only a few mutations account for 65%–80% of observed echinocandin clinical resistance. In *C. albicans*, amino acid substitutions at residues Fks1-F641 and Fks1-S645 account for nearly 90% of echinocandin-resistant strains [47]. In *C. glabrata*, 88% of echinocandin-resistant strains showed amino acid substitutions at the Fks1-F625, Fks1-S629, Fks1-D632, Fks2-F659, and Fks2-S663 residues [48,49]. Furthermore, there is a large body of evidence that associates the presence of most frequent *FKS* mutations with poor echinocandin therapeutic response and clinical outcome [44]. These findings raise the possibility that molecular testing of clinical isolates may become a surrogate diagnostic method for resistance phenotype assessment [43].

21.3 RECENT ADVANCES

There are two representing examples of how a genetic target with a validated mechanism can confer (i) an elevated MIC in CLSI or EUCAST antifungal susceptibility testing, (ii) an altered drug target interaction or effector action (e.g., transcription factor), and (iii) documented clinical failure. One example is echinocandins with *Candida* species. Another is azoles with *Aspergillus fumigatus*, which is beyond of the topic of this chapter. Consistently, molecular assays are capable of resolving all known *FKS* mutations associated with drug resistance in *C. albicans* and *C. glabrata* within 3–4 h. As long as specific mutations are known and validated, the assays can be readily extended to other *Candida* species.

In 2006, a study first described an allele-specific real-time PCR molecular beacon assay that could distinguish changes in the codon for position Ser645 in *C. albicans* [50]. The assay was used to evaluate two large collections of spontaneous resistant mutants from different strain backgrounds. Mutations at this locus were reliably identified in both the heterozygous and homozygous states. In response, the Luminex MagPix technology using xMAP microspheres was proposed as a platform for the establishment of multiplexed assays, which permit the analysis of up to 100 different target molecules in a single test. In 2014, this technology was used along with asymmetric PCR to develop a microsphere-based assay for the rapid identification of prominent mutations in HS1 of both *FKS1* and *FKS2* in *C. glabrata*. The MagPix assay was initially validated using 102 isolates obtaining 100% of the results concordant with their DNA sequencing profiles; after that, the assay was further used to accurately screen 1032 *C. glabrata* surveillance isolates [51]. As a rapid and highly versatile format, *FKS1* HS1 and *FKS2* HS1 profiles of up to 95 isolates can be determined in as little as 5 h. Thus, the multiplex *FKS* MagPix might be used in place of DNA sequencing, particularly in laboratories that are already equipped with the Luminex technology.

Meanwhile, a simple and quick assay used classical PCR primer sets for wild-type and resistant (mutant) sequences to detect the ten most common mutations in *C. glabrata* within 4 h [48]. By this assay, 49 of 50 *C. glabrata* strains, including 16 *FKS1* and/or *FKS2* mutants, were correctly identified as echinocandin susceptible or resistant, and the PCR results were 98% concordant with those obtained from DNA sequencing. The one false result was due to an *FKS2* mutant, in which Fks2 carried a few-nucleotide deletion at the 659 residue (F659del), where the nucleotide sequence aligned with the primer did not change.

In 2016, a novel and highly accurate diagnostic platform for rapid *FKS* genotyping of *C. glabrata* isolates allowed to overcome the assay design and set-up complications of previously described assays [48,51]. Using asymmetric PCR in conjunction with allele-specific molecular beacon probes and melting curve analysis, a dual assay for *C. glabrata FKS1* and *FKS2* discriminated wild-type from mutated *FKS* genes within 3 h [52]. In this assay, signature melting profiles and corresponding values of the temperature at which the probe-target hybrids melt apart (T*m*) were generated from reference strains for 8 *FKS1* HS1 and 7 *FKS2* HS1 genotypes. Hence, the *FKS* genotype of the testing isolate could be easily identified by comparing the T*m* value with those of reference strains, representing all the mutations included in the dual assay. In the subsequent proof-of-concept clinical validation study, using a blinded panel of 188 *C. glabrata* isolates, both *FKS1* HS1 and *FKS2* HS1 assays were shown to have 100% concordance with DNA sequencing.

These assays hold the promise for rapidly identifying echinocandin resistance mechanisms in *Candida* species. As they have currently been evaluated only using cultures of clinical isolates, additional studies are required before their direct application to clinical samples to reduce the detection time for echinocandin resistance.

21.4 FUTURE PERSPECTIVES

The incidence of antifungal resistance in normally susceptible *Candida* species is moderate. However, the rise of fungi that are resistant to existing antifungal agents has become a cause for concern in the context of global increase in fungal infections. This has paradoxically been linked to progresses in medical cure, which lead to routine use of foreign bodies (central venous catheters, prosthetic joints, and permanent pacemakers, etc.) and successful management of immunosuppression in patients with cancer, inflammatory bowel disease, and other underlying diseases [53,54].

Thus, antifungal resistance has become a serious concern due to the limited number of available agents, which are losing effectiveness against fungi showing uncanny ability to acquire resistance. [55]. The recent emergence of multidrug-resistant *C. auris* [56,57] calls for novel antifungal agents and, in parallel, attention on the hurdles associated with antifungal drug development [58]. Despite the fact that fungi are eukaryotes and possess many potential target proteins that are also found in humans, the drug pipeline to treat resistant fungi may be expanded by taking advantage of functional genomics, which is a powerful tool in drug discovery [58].

Robust investment in research should be a priority to accomplish the goals of reducing antifungal drug resistance while improving outcomes in the hospital setting [59]. Therefore, it is important to understand the mechanisms of resistance to antifungal agents to help in the design of guidelines for choosing alternative therapies, and the development of molecular diagnostic tools for resistance detection. Nowadays, high-resolution melt analysis/PCR and DNA microarray have been utilized for addressing antifungal drug resistance [59]. However, ongoing whole-genome sequencing efforts [60] and microbiome research [61] will enhance our capability of recognizing antifungal drug resistance.

A recent study highlights the role of improving fungal diagnostic capabilities as a means of limiting antimicrobial (antibiotic and antifungal) usage [33]. The access to nonculture fungal diagnostics can substantially benefit clinical outcome, antimicrobial stewardship, and control of antimicrobial resistance [62]. Further refinement in diagnostic platforms will undoubtedly bring this goal one step closer.

REFERENCES

1. Stajich JE, Berbee ML, Blackwell M, et al. The fungi. *Curr Biol* 2009;19:R840–5.
2. Brandt ME, Lockhart SR. Recent taxonomic developments with *Candida* and other opportunistic yeasts. *Curr Fungal Infect Rep* 2012;6:170–7.
3. Brown GD, Denning DW, Gow NA, Levitz SM, Netea MG, White TC. Hidden killers: human fungal infections. *Sci Transl Med* 2012;4:165rv13.
4. Denning DW, Bromley MJ. Infectious disease. How to bolster the antifungal pipeline. *Science* 2015;347:1414–6.
5. Janus MM, Willems HM, Krom BP. *Candida albicans* in multispecies oral communities; a keystone commensal? *Adv Exp Med Biol* 2016;931:13–20.
6. Hallen-Adams HE, Suhr MJ. Fungi in the healthy human gastrointestinal tract. *Virulence* 2017;8:352–8.
7. Bradford LL, Ravel J. The vaginal mycobiome: a contemporary perspective on fungi in women's health and diseases. *Virulence* 2017;8:342–51.
8. Calderone RA. *Candida and Candidiasis*. Washington, DC: ASM Press, 2002.
9. Sardi JC, Scorzoni L, Bernardi T, Fusco-Almeida AM, Mendes Giannini MJ. *Candida* species: current epidemiology, pathogenicity, biofilm formation, natural antifungal products and new therapeutic options. *J Med Microbiol* 2013;62(1):10–24.
10. Smeekens SP, van de Veerdonk FL, Kullberg BJ, Netea MG. Genetic susceptibility to *Candida* infections. *EMBO Mol Med* 2013;5:805–13.
11. Kullberg BJ, Arendrup MC. Invasive candidiasis. *N Engl J Med* 2016;374:794–5.
12. Colombo AL, de Almeida Júnior JN, Slavin MA, Chen SC, Sorrell TC. *Candida* and invasive mould diseases in non-neutropenic critically ill patients and patients with haematological cancer. *Lancet Infect Dis* 2017;17:e344–56.
13. Miceli MH, Díaz JA, Lee SA. Emerging opportunistic yeast infections. *Lancet Infect Dis* 2011;11:142–51.
14. Pfaller MA, Diekema DJ. Epidemiology of invasive candidiasis: a persistent public health problem. *Clin Microbiol Rev* 2007;20:133–63.
15. McCarty TP, Pappas PG. Invasive candidiasis. *Infect Dis Clin North Am* 2016;30:103–24.
16. Soll DR. The evolution of alternative biofilms in an opportunistic fungal pathogen: an explanation for how new signal transduction pathways may evolve. *Infect Genet Evol* 2014;22:235–43.
17. Mayer FL, Wilson D, Hube B. *Candida albicans* pathogenicity mechanisms. *Virulence* 2013;4:119–28.
18. Priest SJ, Lorenz MC. Characterization of virulence-related phenotypes in *Candida* species of the CUG clade. *Eukaryot Cell* 2015;14:931–40.
19. Moran GP, Coleman DC, Sullivan DJ. *Candida albicans* versus *Candida dubliniensis*: why is *C. albicans* more pathogenic? *Int J Microbiol* 2012;2012:205921.
20. Moran GP, Sullivan DJ, Henman MC, et al. Antifungal drug susceptibilities of oral *Candida dubliniensis* isolates from human immunodeficiency virus (HIV)-infected and non-HIV-infected subjects and generation of stable fluconazole-resistant derivatives *in vitro*. *Antimicrob Agents Chemother* 1997;41:617–23.

21. Garcia-Effron G, Katiyar SK, Park S, Edlind TD, Perlin DS. A naturally occurring proline-to-alanine amino acid change in Fks1p in *Candida parapsilosis*, *Candida orthopsilosis*, and *Candida metapsilosis* accounts for reduced echinocandin susceptibility. *Antimicrob Agents Chemother* 2008;52:2305–12.
22. Falagas ME, Roussos N, Vardakas KZ. Relative frequency of albicans and the various non-albicans Candida spp among candidemia isolates from inpatients in various parts of the world: a systematic review. *Int J Infect Dis* 2010;14:e954–66.
23. Tumbarello M, Sanguinetti M, Trecarichi EM, et al. Fungaemia caused by *Candida glabrata* with reduced susceptibility to fluconazole due to altered gene expression: risk factors, antifungal treatment and outcome. *J Antimicrob Chemother* 2008;62:1379–85.
24. Arendrup MC, Patterson TF. Multidrug-resistant *Candida*: epidemiology, molecular mechanisms, and treatment. *J Infect Dis* 2017;216(S3):S445–51.
25. Jeffery-Smith A, Taori SK, Schelenz S, et al. *Candida auris*: a review of the literature. *Clin Microbiol Rev* 2018;31:e00029-17.
26. Chowdhary A, Voss A, Meis JF. Multidrug-resistant *Candida auris*: 'new kid on the block' in hospital-associated infections? *J Hosp Infect* 2016;94:209–12.
27. Vallabhaneni S, Kallen A, Tsay S, et al. Investigation of the first seven reported cases of *Candida auris*, a globally emerging invasive, multidrug-resistant fungus – United States, May 2013–August 2016. *Morb Mortal Wkly Rep* 2016;65:1234–7.
28. Lockhart SR, Etienne KA, Vallabhaneni S, et al. Simultaneous emergence of multidrug-resistant *Candida auris* on 3 continents confirmed by whole-genome sequencing and epidemiological analyses. *Clin Infect Dis* 2017;64:134–40.
29. Pappas PG, Kauffman CA, Andes DR, et al. Clinical practice guideline for the management of candidiasis: 2016 update by the Infectious Diseases Society of America. *Clin Infect Dis* 2016;62:e1–50.
30. Clinical and Laboratory Standards Institute. *Reference Method for Broth Dilution Antifungal Susceptibility Testing of Yeasts; Fourth Informational Supplement. Document M27-S4.* Wayne, PA: CLSI, 2012.
31. Arendrup MC, Cuenca-Estrella M, Lass-Florl C, Hope WW. European committee on antimicrobial susceptibility testing–subcommittee on antifungal susceptibility testing (EUCAST-AFST). EUCAST technical note on *Candida* and micafungin, anidulafungin and fluconazole. *Mycoses* 2014;57:377–9.
32. Perlin DS, Rautemaa-Richardson R, Alastruey-Izquierdo A. The global problem of antifungal resistance: prevalence, mechanisms, and management. *Lancet Infect Dis* 2017;17:e383–92.
33. Denning DW, Perlin DS, Muldoon EG, et al. Delivering on antimicrobial resistance agenda not possible without improving fungal diagnostic capabilities. *Emerg Infect Dis* 2017;23:177–83.
34. Clancy CJ, Nguyen MH. Diagnosing invasive candidiasis. *J Clin Microbiol* 2018;56:e01909–17.
35. McCarthy MW, Walsh TJ. PCR methodology and applications for the detection of human fungal pathogens. *Expert Rev Mol Diagn* 2016;16:1025–36.
36. Zhang SX, Wiederhold NP. Yeasts. *Microbiol Spectr* 2016;4:DMIH2-0030-2016.
37. Sanguinetti M, Posteraro B, Lass-Flörl C. Antifungal drug resistance among *Candida* species: mechanisms and clinical impact. *Mycoses* 2015;58(Suppl 2):2–13.
38. Shor E, Perlin DS. Coping with stress and the emergence of multidrug resistance in fungi. *PLoS Pathog.* 2015;11:e1004668.
39. Ramage G, Rajendran R, Sherry L, Williams C. Fungal biofilm resistance. *Int J Microbiol* 2012;2012:528521.
40. Sanguinetti M, Posteraro B. Diagnostic of fungal infections related to biofilms. *Adv Exp Med Biol* 2016; 931:63–82.
41. Cavalheiro M, Teixeira MC. *Candida* biofilms: threats, challenges, and promising strategies. *Front Med (Lausanne).* 2018;5:28.
42. Alanio A, Bretagne S. Difficulties with molecular diagnostic tests for mould and yeast infections: where do we stand? *Clin Microbiol Infect* 2014;20(Suppl 6):36–41.
43. Perlin DS. Antifungal drug resistance: do molecular methods provide a way forward? *Curr Opin Infect Dis* 2009;22:568–73.
44. Perlin DS, Wiederhold NP. Culture-independent molecular methods for detection of antifungal resistance mechanisms and fungal identification. *J Infect Dis* 2017;216(Suppl 3):S458–65.
45. Ferrari S, Ischer F, Calabrese D, et al. Gain of function mutations in *CgPDR1* of *Candida glabrata* not only mediate antifungal resistance but also enhance virulence. *PLoS Pathog* 2009;5:e1000268.
46. Caudle KE, Barker KS, Wiederhold NP, Xu L, Homayouni R, Rogers PD. Genomewide expression profile analysis of the *Candida glabrata* Pdr1 regulon. *Eukaryot Cell* 2011;10:373–83.
47. Perlin DS. Mechanisms of echinocandin antifungal drug resistance. *Ann NY Acad Sci* 2015;1354:1–11.
48. Dudiuk C, Gamarra S, Leonardeli F, et al. Set of classical PCRs for detection of mutations in *Candida glabrata FKS* genes linked with echinocandin resistance. *J Clin Microbiol* 2014;52:2609–14.

49. Castanheira M, Woosley LN, Messer SA, Diekema DJ, Jones RN, Pfaller MA. Frequency of fks mutations among *Candida glabrata* isolates from a 10-year global collection of bloodstream infection isolates. *Antimicrob Agents Chemother* 2014;58:577–80.
50. Balashov SV, Park S, Perlin DS. Assessing resistance to the echinocandin antifungal drug caspofungin in *Candida albicans* by profiling mutations in *FKS1*. *Antimicrob Agents Chemother* 2006;50:2058–63.
51. Pham CD, Bolden CB, Kuykendall RJ, Lockhart SR. Development of a Luminex-based multiplex assay for detection of mutations conferring resistance to echinocandins in *Candida glabrata*. *J Clin Microbiol* 2014;52:790–5.
52. Zhao Y, Nagasaki Y, Kordalewska M, et al. Rapid detection of *FKS*-associated echinocandin resistance in *Candida glabrata*. *Antimicrob Agents Chemother* 2016;60:6573–7.
53. Kontoyiannis DP, Perlin DS, Roilides E, Walsh TJ. What can we learn and what do we need to know amidst the iatrogenic outbreak of *Exserohilum rostratum* meningitis? *Clin Infect Dis* 2013;57:853–9.
54. Brandt ME, Park BJ. Think fungus–prevention and control of fungal infections. *Emerg Infect Dis* 2013;19:1688–9.
55. Perfect JR. The antifungal pipeline: a reality check. *Nat Rev Drug Discov* 2017;16:603–16.
56. Calvo B, Melo AS, Perozo-Mena A, et al. First report of *Candida auris* in America: clinical and microbiological aspects of 18 episodes of candidemia. *J Infect* 2016;73:369–74.
57. Schelenz S, Hagen F, Rhodes JL, et al. First hospital outbreak of the globally emerging *Candida auris* in a European hospital. *Antimicrob Resist Infect Control* 2016;5:35.
58. McCarthy MW, Kontoyiannis DP, Cornely OA, Perfect JR, Walsh TJ. Novel agents and drug targets to meet the challenges of resistant fungi. *J Infect Dis* 2017;216(Suppl n3):S474–83.
59. McCarthy MW, Denning DW, Walsh TJ. Future research priorities in fungal resistance. *J Infect Dis* 2017;216(Suppl 3):S484–92.
60. Zoll J, Snelders E, Verweij PE, Melchers WJ. Next-generation sequencing in the mycology lab. *Curr Fungal Infect Rep* 2016;10:37–42.
61. Enaud R, Vandenborght LE, Coron N, et al. The mycobiome: a neglected component in the microbiota-gut-brain axis. *Microorganisms* 2018;6(1):2.
62. Dubourg G, Raoult D. Emerging methodologies for pathogen identification in positive blood culture testing. *Expert Rev Mol Diagn* 2016;16:97–111.

SUMMARY

The hemiascomycetes *Candida* species are opportunistic pathogens primarily for immunosuppressed hosts. *C. albicans*, *C. glabrata*, *C. krusei*, *C. parapsilosis*, and *C. tropicalis* are responsible for >90% of invasive infections. The increasing number of *Candida* species that are resistant to the most common classes of antifungal agents, including the newly emerged *C. auris*, poses a substantial threat to human health. Various well-defined mechanisms can lead to azole resistance, including alteration or overexpression of the enzyme target lanosterol 14-α demethylase, induction of drug efflux pumps, and cellular changes. Conversely, echinocandin resistance involves genetic acquisition of mutations in *FKS* genes, which encode the major catalytic subunit of the enzyme target glucan synthase. Although antifungal resistance mechanisms in *Candida* are multifaceted, their underlying genetic basis provides the rationale for developing molecular detection assays. These assays hold the promise of rapidly and accurately identifying resistant fungal species, as well as resolving all known *FKS* mutations associated with echinocandin resistance in *C. albicans* and *C. glabrata* within 3–4 h. Additional efforts on the applicability of molecular assays in routine clinical practice are expected to extend their potentiality.

Genetic Manipulation and Genome Editing of *Saccharomyces cerevisiae*

22

Song Weining
Northwest A&F University

Dongyou Liu
Royal College of Pathologists of Australasia Quality Assurance Programs

Contents

22.1 INTRODUCTION

Saccharomyces cerevisiae (also known as budding yeast due to its tendency to reproduce by budding) is a small unicellular organism commonly used in baking, brewing, and winemaking. It is also an ingredient in cheese substitutes for vegans and vegetarians and a source of vitamins (especially B-complex vitamins), amino acids, and minerals for health-conscious community.

Noted by its rapid growth, ease of genetic manipulation, and high margin of biosafety, *S. cerevisiae* has been employed as a model organism in research laboratories and biotech industries for fundamental studies of eukaryotic biology and for production of proteins and biochemicals that constitute essential ingredients for pharmaceutical manufacturing.

Nonetheless, despite its broad utility in food and beverage production and biotech research and manufacturing, *S. cerevisiae* has been increasingly recognized as an opportunistic pathogen with the ability to take advantage of host's suppressed immune functions. Furthermore, therapeutic application of

S. cerevisiae var. *boulardii* for prevention or treatment of gastrointestinal diseases and allergic rhinitis is associated with possible fungemia in critically ill patients.

Taxonomy. Representing one of the 26 genera (i.e., *Ascobotryozyma*, *Ashbya*, *Candida*, *Citeromyces*, *Eremothecium*, *Hanseniaspora*, *Hansenula*, *Issatchenkia*, *Kazachstania*, *Kloeckera*, *Kluyveromyces*, *Komagataella*, *Kuraishia*, *Lachancea*, *Nakaseomyces*, *Nakazawaea*, *Naumovozyma*, *Pichia*, *Saccharomyces*, *Saturnispora*, *Tetrapisispora*, *Torulaspora*, *Vanderwaltozyma*, *Williopsis*, *Zygosaccharomyces*, and *Zygotorulaspora*) in the family Saccharomycetaceae, order Saccharomycetales, class Hemiascomycetes, phylum/division Ascomycota, and kingdom Fungi, the genus *Saccharomyces* currently includes at least 13 species (*S. bayanus*, *S. boulardii*, *S. cariocanus*, *S. castellii*, *S. douglasii*, *S. eubayanus*, *S. kudriavzevii*, *S. mikatae*, *S. paradoxus*, *S. pastorianus*, *S. rosinii*, *S. servazzii*, and *S. yakushimaensis*) of ascospore-forming yeasts. Among these, *S. cerevisiae* (often called baker's or brewer's yeast) is a key ingredient for the production of various food stuffs, wines, and beers; a model organism in fundamental research (e.g., mechanisms of pathogenicity and resistance to fungal drugs in light of its close genetic relationship to *Candida albicans*); and a generator of proteins and biochemicals for pharmaceutical industries. Furthermore, *S. boullardii* (also referred to as *S. cerevisiae* var. *boullardii*) is utilized as a therapy for intestinal disorders and allergic rhinitis.

Morphology. *Saccharomyces* vegetative cells are singular, globose, ellipsoid to elongated in shape, and approximately 5–10 μm in diameter (Figures 22.1a and 22.2a). Growing rapidly and maturing in 3 days on potato dextrose agar (PDA), *Saccharomyces* colonies are flat, smooth, moist, glistening or dull, and cream-to-tannish cream in color. Blastoconidia are present, but hyphae are absent (although rudimentary pseudohyphae are occasionally observed). On V-8 medium, acetate ascospore agar, or Gorodkowa medium (i.e., in the presence of a nonfermentable carbon source), *Saccharomyces* is starved for nitrogen and undergoes meiosis to form globose ascospores located in asci (each ascus containing one to four ascospores) (Figure 22.1b) [1,2]. Using Gram staining, *Saccharomyces* ascospores are Gram-negative while vegetative cells are Gram-positive.

Ultrastructurally, *Saccharomyces* vegetative cells possess cell wall, nucleus, mitochondria, endoplasmic reticulum (ER), Golgi apparatus, vacuoles, microbodies, secretory vesicles, and a complex extracellular and intracellular membrane network, similar to other higher eukaryotes (Figure 22.2b) [3].

Genome. *S. cerevisiae* genome comprises a 12,156,677 bp DNA organized in 16 chromosomes (A to P). Of the 6,275 predicted genes, about 5,800 genes have known functions, and about 31% of *S. cerevisiae* genes have human homologs [4,5]. Similarly, *S. cerevisiae* strain BAW-6 (formerly W-6), which is used for the production of barley shochu (a traditional Japanese distilled spirit), harbors a 11,872,199 bp DNA genome with the GC content of 38.3% and 5,425 predicted genes [6]. Further, *S. cerevisiae* strain Pf-1 from plum fruit (*Prunus mume*) consists of a 12,480,617 bp DNA genome with the G+C content of 38.5%, and 5,732 predicted genes [7].

Biology. *S. cerevisiae* is known to generate two forms of yeast cells, diploid (vegetative) cells and haploid cells (ascospores). Diploid cells (the preferential form of yeast) generally have a simple lifecycle

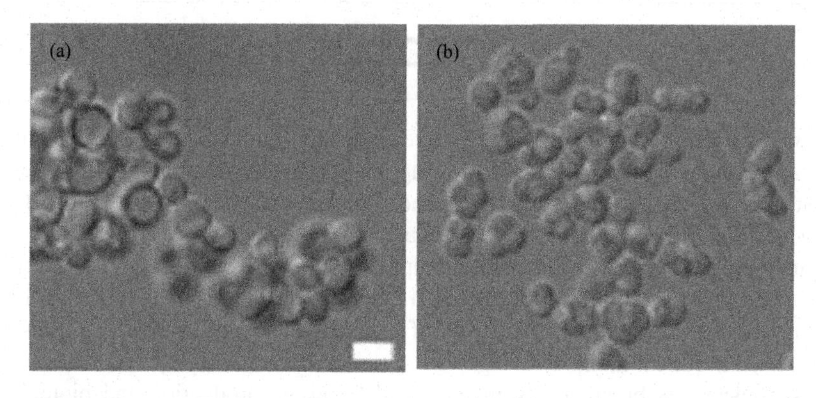

FIGURE 22.1 Differential interference contrast image of *Saccharomyces cerevisiae* strain AN390 vegetative cells (a) and asci containing ascospores (b); scale bar = 5 μm. (Photo credit: Coluccio AE, Rodriguez RK, Kernan MJ, Neiman AM. *PLoS One*. 2008;3(8):e2873.)

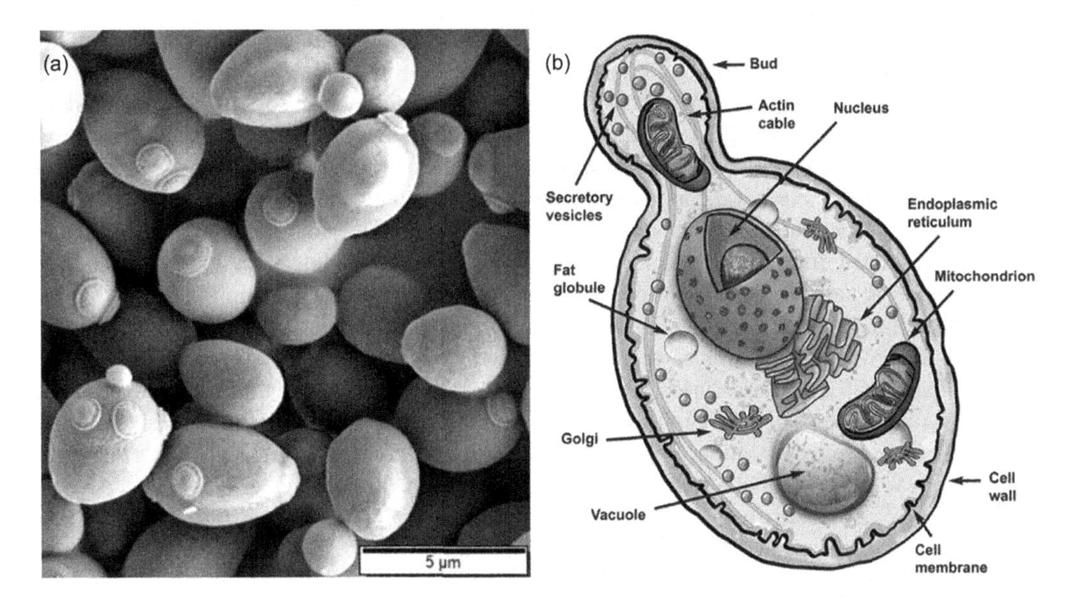

FIGURE 22.2 Electron micrograph (a) and structural representation (b) of *Saccharomyces cerevisiae* vegetative cells. (Photo credits: (a) Mogana Das Murtey and Patchamuthu Ramasamy at http://www.intechopen.com/books/modern-electron-microscopy-in-physical-and-life-sciences/sample-preparations-for-scanning-electron-microscopy-life-sciences and https://commons.wikimedia.org/w/index.php?curid=52254246; (b) Walker GM, Stewart GG. *Beverages*. 2016;2:30.)

of mitosis/budding and growth (asexual reproduction). However, under stressful conditions (e.g., in the presence of a nonfermentable carbon source), diploid cells starved for nitrogen will sporulate (2x meiosis) and form four haploid cells (ascospores) located inside ascus (which is a collapsed mother cell). Haploid cells (ascospores) may bud to generate additional haploid cells (asexual reproduction) or mate to produce diploid cells (sexual reproduction), which can either sporulate to form another generation of haploid cells or continue to exist as diploid cells [1]. Under optimal conditions, *S. cerevisiae* doubles every 100 min and may undertake average 26 cell divisions during its replicative lifespan.

Further studies indicate that *S. cerevisiae* grows aerobically on glucose, maltose, trehalose, galactose, and fructose, but not on lactose and cellobiose. This yeast thrives in warm [20°C–35°C (optimally 30°C–35°C)] and acidic (pH 4.5–6.5) environments in the presence of water (minimum a_w of around 0.65) and oxygen, which is a growth factor for membrane fatty acid (e.g., oleic acid) and sterol (e.g., ergosterol) biosynthesis. *S. cerevisiae* tends to overproduce glycerol or trehalose in response to water deficiency, leading to decreased ethanol yield in high gravity fermentation. It ferments various carbohydrates; utilizes ammonia and urea as the sole nitrogen source, as well as amino acids, small peptides, and nitrogen bases as nitrogen sources; but does not use nitrate (due to its inability to reduce nitrate to ammonium ions). It requires phosphorus (which is assimilated as a dihydrogen phosphate ion), sulfur (which is assimilated as a sulfate ion or as organic sulfur compounds such as amino acids methionine and cysteine), magnesium, iron, calcium, and zinc for good growth. In addition, it requires biotin and pantothenate for full growth.

Historically, *S. cerevisiae* had been linked to brewing and baking activity for at least 14,000 years [8–11], although its association with human society could have been well established before the domestication of plants and animals. Because its hydrophobic surface causes the flocs to adhere to CO_2 and rise to the top of the fermentation vessel during the fermentation process, *S. cerevisiae* is sometimes called top-fermenting or top-cropping yeast [3]. *S. cerevisiae* was later adopted in baking as CO_2 generated by fermentation acts as a leavening agent in bread and other baked goods. Nowadays, an estimated 600,000 tons of baker's yeast are being produced globally each year. In addition, the ability of *S. cerevisiae* to produce CO_2 (typically giving one bubble every 3–7 s when maintained in plastic bottles) is exploited by aquaculturists for providing CO_2 to underwater aquatic plants.

Epidemiology. *Saccharomyces* is essentially a saprotrophic yeast that occupies a range of environmental niches, including ripe fruits (e.g., grapes), trees, plants, olives, soil, wine, beer, insects (social wasps), birds, mammals and humans. *S. cerevisiae* is present in the skin, oral cavity, oropharynx, duodenal mucosa, digestive tract, and vagina of healthy humans. Indeed, *S. cerevisiae*'s close association with humans is reflected by its rare occurrence in environs removed from human habitation. Ingestion of food and beverage containing *S. cerevisiae* appears to contribute to its intestinal colonization and subsequent disease development as volunteer studies indicate that *S. cerevisiae* does not remain in the gut very long and is mostly eliminated after a few weeks. Generally considered nonpathogenic for immunocompetent hosts, *S. cerevisiae* is capable of causing opportunistic mycoses in immunocompromised individuals. Major risk factors for *S. cerevisiae* infection are severe immunosuppression, prolonged hospitalization, prior antibiotic therapy, and prosthetic cardiac valves.

Clinical features. *S. cerevisiae* infection is associated with pneumonia, endocarditis, liver abscess, vaginitis, fungemia, and sepsis mainly in immunocompromised individuals (HIV/AIDS, leukemia, other cancers, bone marrow transplantation, abdominal surgery). Additionally, overloaded *S. cerevisiae* var. *boullardii* therapy for diarrhea and allergic rhinitis may lead to fungemia in critically ill patients.

Pathogenesis. *S. cerevisiae* pathogenic strains demonstrate the ability to grow at 37°C–39°C (close to the body temperatures of humans and warm-blooded mammals), to adhere to mammalian cells, to generate proteases and phospholipases for host cell invasion, to survive in the presence of hydrogen peroxide released by macrophages, and to modulate/evade host immune responses. In contrast, *S. cerevisiae* environmental strains do not grow at above 35°C (optimal growth temperature for top-fermenting yeast is 21°C), and are incapable of coping with host immune attack.

Diagnosis. Diagnosis of *S. cerevisiae* infection relies on a combination of microscopic observation of vegetative cells and ascospores (using Kinyoun stain or ascospore stain) in clinical specimens, fermentation profiles/biochemical tests, serological assays, and molecular detection. Multipolar budding, production of ascospores, and fermentation profile aid in the identification of *Saccharomyces*, while molecular detection of specific gene targets provides additional confirmation.

Treatment and prevention. The current drug of choice for treatment of severe *S. cerevisiae* infection appears to be amphotericin B. Several new drugs (e.g., triazole, posaconazole, echinocandin, and anidulafungin) also hold promise.

As *S. cerevisiae* infection is likely caused by the consumption of food and beverage contaminated with yeast, individuals with suppressed immune function should be especially careful about proper food handling and cooking, personal hygiene, and consuming clean water and beverage.

22.2 MOLECULAR MANIPULATION AND GENOME EDITING OF *S. CEREVISIAE*

As a unicellular eukaryote, *S. cerevisiae* is remarkably fast growing (doubling time 1.25–2h at 30°C), amenable to genetic manipulation, and extremely safe to work with. While these features emulate those of prokaryotes (e.g., *Escherichia coli*), *S. cerevisiae* has added advantages of possessing eukaryotic properties that support the fundamental studies of eukaryotic systems' biology (e.g., genetics, genomics, transcriptomics, proteomics, metabolomics, phenomics), as well as of having enormous genetic versatilities that enable molecular manipulation (e.g., site-directed mutagenesis, gene editing, intrastrain cloning) for expression of humanized proteins, *de-novo* synthesis of complete chromosomes, and other biomedical applications [12–15]. Originated from a fig, *S. cerevisiae* strain S288c has been cultivated in laboratory conditions for over 100 years and has been used extensively in research laboratories worldwide [4,16].

Genetic manipulation and genome editing techniques. Methods that have been developed for genetic manipulation and genome editing of *S. cerevisiae* range from integrative plasmid, transformation-associated recombination (TAR), Cre-*loxP*-mediated recombination, delitto perfetto/meganucleases, to CRISPR/Cas9 [17].

Being the early generation technique for the transfer of target DNA, integrative plasmid (e.g., pRS or YIplac series) has played a significant role in the experimental introduction of targeted and specific modifications (e.g., gene deletion, insertion, or replacement) into the genome of *S. cerevisiae*. However, this technique has the risks of multiple integrations of the target DNA and the potential loss of integrated DNA during cell proliferation.

Developed in the late 1990s, TAR cloning exploits free DNA ends (one from yeast vector DNA and other from mammalian chromosomal fragment) as substrates for homologous recombination in yeast, and allows selective isolation of large genomic regions (up to 300 kb) from total mammalian genomic DNA as linear or circular yeast artificial chromosomes (YACs). Use of highly competent yeast spheroplasts and exclusion of a yeast origin of replication (*ARS* element) from a TAR vector further enhance the selectivity of gene isolation. In fact, TAR-generated YACs yield gene/genomic fragment-positive clones as high as 32% within 2 weeks, whereas genomic bacterial artificial chromosome (BAC) libraries based on DNA ligation has a gene-positive clone frequency of <0.003%. It is notable that propagation of TAR-generated YACs in yeast cells relies on *ARS*-like sequences from mammalian DNA (which contains one *ARS*-like sequence in every 20–30 kb) as an origin of replication in yeast. For chromosomal regions with multiple repetitive elements (e.g., the centromere and telomere) that are GC-rich and contain few *ARS*-like sequences, use of the TAR vector with an *ARS* and a counter-selectable marker is necessary to recover fragments of approximately 100 kb. TAR cloning may be coupled with human artificial chromosomes (HACs) for improved gene delivery and expression. TAR cloning is also valuable for the assembly of new microbe genomes and synthetic biology [18].

Cre-*loxP*-mediated recombination uses marker cassette that contains dominant heterologous antibiotic resistance marker to replace the gene of interest, and to facilitate the excision of the markers from the genome by Cre-mediated recombination between the two flanking *loxP* recognition sequences of 34 bp. Through the action of Cre recombinase, the selectable marker is rescued, and only one *loxP* scar is left in the genome. Cre-*loxP* mediated recombination provides a powerful tool to generate multiple gene deletions, inversions, or translocations of a chromosomal fragment, as well as *S. cerevisiae* synthetic chromosomes (SCRaMbLEing) [17].

Combining different synthetic oligonucleotides for targeting the gene of interest, *Delitto perfetto approach* utilizes various counter-selectable markers and reporter genes (CORE) for efficient genome editing via homologous recombination in *S. cerevisiae*, eliminating any marker sequences used for selection and leaving no foreign DNA in the genome. Incorporation of meganucleases (also known as homing endonucleases) to induce a single double strand break (DSB) at the locus to be modified and to stimulate oligonucleotide targeting further increases recombination of homology sequences and integration efficiency [17].

The CRISPR-Cas9 system is a relatively recent development for the modification of eukaryotic genomes including yeast. Mechanistically, it uses a Cas9 nuclease to generate a DNA DSB and a guide RNA (gRNA) to recruit the Cas9 nuclease to a target DNA sequence. This enables a marker-free integration involving a template DNA sequence that helps repair the DSB, and subsequent integration of the desired DNA sequence into the DSB site. Use of a set of CRISPR-Cas9 plasmids and corresponding repair plasmids (containing a linear target DNA sequence flanked by 50–500 bp sequences homologous to each side of the DSB) further improves the practicality, flexibility, and traceability of DNA sequence integration [17,19–24]. To help select appropriate gRNA target, various software may be utilized (e.g., ATUM, GenScript, ThermoFisher Scientific; ChopChop; E-CRISP; Breaking-Cas).

Use of *S. cerevisiae* for production of proteins and biochemicals. As many yeast protein-coding genes can be replaced with their human orthologs (e.g., those encoding cell cycle proteins, signaling proteins, and protein-processing enzymes), *S. cerevisiae* offers an important model for the study of human biology and disease [25]. Further, yeasts harboring foreign genes may be employed in two-hybrid system for studying protein interactions and tetrad analysis [26–28]. Additionally, genetic engineering of *S. cerevisiae* facilitates efficient production of various biochemicals [29,30].

Use of *S. cerevisiae* to assemble new microbe genomes. TAR cloning permits cloning and assembly of whole microbe genomes [e.g., human cytomegalovirus (HCMV) isolate Toledo (>230 kb), *M. pneumoniae* (0.8 Mb), *Acholeplasma laidlawii* (1.5 Mb), *Prochlorococcus marinus MED4* (1.6 Mb), *Phaeodactylum tricornutum* (27.4 Mb)] or individual chromosomes in *S. cerevisiae* [31,32]. This compares favorably to BAC, which has low efficiency, and may cause adaptive mutations and deletion of microbe genes to avoid oversized genomes.

Use of *S. cerevisiae* to study mitochondrial diseases. *S. cerevisiae* offers a useful system to model human mitochondrial disorders (>150 described to date, affecting 1 in 5,000 live births; including visual/hearing defects, encephalopathies, cardiomyopathies, myopathies, diabetes, liver and renal dysfunctions, diabetes, obesity, age-related neurodegenerative and cardiovascular diseases, cancer, and probably the aging process) and to identify new therapeutic compounds [33].

Use of *S. cerevisiae* to study cell death and aging. As *S. cerevisiae* possesses the basal machinery of apoptosis, necrosis, and autophagic cell death, it offers a critical model for the study of regulated cell death (RCD) and genes and cellular pathways involved in senescence (aging) [34].

22.3 FUTURE PERSPECTIVES

The budding yeast *Saccharomyces cerevisiae,* a unicellular eukaryote renowned for its utility in baking, brewing, and winemaking has attained a prominent role in the era of molecular biology, due largely to its easy culturability in defined media, ready amenability to genetic manipulation, and high safety margin. Besides offering a valuable model for the studies of eukaryotic biology, parasite–host interplay, gene–gene and gene–environment interactions, as well as mechanisms of mitochondrial diseases and aging *S. cerevisiae* functions as a reliable and effective producer of proteins and biochemicals that form essential ingredients for biotech and pharmaceutical industries. Relative to *Escherichia* coli, a well-characterized bacterium with extensive applications at both laboratory and industrial settings, *S. cerevisiae* is by far vastly understudied and under-utilized. It is envisaged that with continuing advances in molecular manipulation and gene editing techniques, *S. cerevisiae* will find its way to many previously unexplored biomedical applications, and will have an even bigger impact on the wellbeing of human society.

REFERENCES

1. Neiman AM. Ascospore formation in the yeast *Saccharomyces cerevisiae. Microbiol Mol Biol Rev.* 2005;69(4):565–84.
2. Coluccio AE, Rodriguez RK, Kernan MJ, Neiman AM. The yeast spore wall enables spores to survive passage through the digestive tract of *Drosophila. PLoS One*. 2008;3(8):e2873.
3. Walker, G.M.; Stewart, G.G. *Saccharomyces cerevisiae* in the production of fermented beverages. *Beverages.* 2016;2:30.
4. Engel SR, Dietrich FS, Fisk DG, et al. The reference genome sequence of *Saccharomyces cerevisiae*: Then and now. *G3 (Bethesda)*. 2014;4(3):389–98.
5. Peter J, De Chiara M, Friedrich A, et al. Genome evolution across 1,011 *Saccharomyces cerevisiae* isolates. *Nature*. 2018;556(7701):339–44.
6. Kajiwara Y, Mori K, Tashiro K, Higuchi Y, Takegawa K, Takashita H. Genomic sequence of *Saccharomyces cerevisiae* BAW-6, a yeast strain optimal for brewing barley shochu. *Genome Announc*. 2018;6(14):e00228-18.
7. Kanamasa S, Yamaguchi D, Machida C, et al. Draft genome sequence of *Saccharomyces cerevisiae* strain Pf-1, isolated from *Prunus mume. Microbiol Resour Announc*. 2019;8(46):e01169-19.
8. Pretorius IS, Curtin CD, Chambers PJ. The winemaker's bug: From ancient wisdom to opening new vistas with frontier yeast science. *Bioeng Bugs*. 2012;3(3):147–56.

9. Mulero-Cerezo J, Briz-Redón Á, Serrano-Aroca Á. *Saccharomyces cerevisiae* var. *boulardii*: Valuable probiotic starter for craft beer production. *Appl Sci.* 2019;9(16):3250.
10. Arranz-Otaegui A, Gonzalez Carretero L, Ramsey MN, Fuller DQ, Richter T. Archaeobotanical evidence reveals the origins of bread 14,400 years ago in northeastern Jordan. *Proc Natl Acad Sci USA.* 2018;115(31):7925–30.
11. Liu L, Wang J, Rosenberg D, Zhao H, Lengyel G, Nadel D. Fermented beverage and food storage in 13,000 y-old stone mortars at Raqefet Cave, Israel: Investigating Natufian ritual feasting. *J Archaeol Sci Rep.* 2018;21:783.
12. Díaz-Mejía JJ, Celaj A, Mellor JC, et al. Mapping DNA damage-dependent genetic interactions in yeast via party mating and barcode fusion genetics. *Mol Syst Biol.* 2018;14(5):e7985.
13. Rockenfeller P, Smolnig M, Diessl J, et al. Diacylglycerol triggers Rim101 pathway-dependent necrosis in yeast: A model for lipotoxicity. *Cell Death Differ.* 2018;25(4):767–83.
14. Wadhwa M, Srinivasan S, Bachhawat AK, Venkatesh KV. Role of phosphate limitation and pyruvate decarboxylase in rewiring of the metabolic network for increasing flux towards isoprenoid pathway in a TATA binding protein mutant of *Saccharomyces cerevisiae. Microb Cell Fact.* 2018;17(1):152.
15. Sarder HAM, Li X, Funaya C, Cordat E, Schmitt MJ, Becker B. *Saccharomyces cerevisiae*: First steps to a suitable model system to study the function and intracellular transport of human kidney anion exchanger 1. *mSphere.* 2020;5(1):e00802-19.
16. Espinosa MI, Williams TC, Pretorius IS, Paulsen IT. Benchmarking two *Saccharomyces cerevisiae* laboratory strains for growth and transcriptional response to methanol. *Synth Syst Biotechnol.* 2019;4(4):180–8.
17. Fraczek MG, Naseeb S, Delneri D. History of genome editing in yeast. *Yeast.* 2018;35(5):361–8.
18. Kouprina N, Larionov V. Transformation-associated recombination (TAR) cloning for genomics studies and synthetic biology. *Chromosoma.* 2016;125(4):621–32.
19. Roggenkamp E, Giersch RM, Wedeman E, et al. CRISPR-UnLOCK: Multipurpose Cas9-based strategies for conversion of yeast libraries and strains. *Front Microbiol.* 2017;8:1773.
20. Daniels PW, Mukherjee A, Goldman AS, Hu B. A set of novel CRISPR-based integrative vectors for *Saccharomyces cerevisiae. Wellcome Open Res.* 2018;3:72.
21. Fleiss A, O'Donnell S, Fournier T, et al. Reshuffling yeast chromosomes with CRISPR/Cas9. *PLoS Genet.* 2019;15(8):e1008332.
22. Kildegaard KR, Tramontin LRR, Chekina K, Li M, et al. CRISPR/Cas9-RNA interference system for combinatorial metabolic engineering of *Saccharomyces cerevisiae. Yeast.* 2019;36(5):237–47.
23. Ramachandran G, Bikard D. Editing the microbiome the CRISPR way. *Philos Trans R Soc Lond B Biol Sci.* 2019;374(1772):20180103.
24. Sanchez JC, Ollodart A, Large CRL, et al. Phenotypic and genotypic consequences of CRISPR/Cas9 editing of the replication origins in the rDNA of *Saccharomyces cerevisiae. Genetics.* 2019;213(1):229–49.
25. Laurent JM, Young JH, Kachroo AH, Marcotte EM. Efforts to make and apply humanized yeast. *Brief Funct Genomics.* 2016;15(2):155–63.
26. Gancedo C, Flores CL, Gancedo JM. The expanding landscape of moonlighting proteins in yeasts. *Microbiol Mol Biol Rev.* 2016;80(3):765–77.
27. Hammond TG, Allen PL, Gunter MA, et al. Physical forces modulate oxidative status and stress defense meditated metabolic adaptation of yeast colonies: Spaceflight and microgravity simulations. *Microgravity Sci Technol.* 2018;30(3):195–208.
28. Yadav A, Sinha H. Gene-gene and gene-environment interactions in complex traits in yeast. *Yeast.* 2018;35(6):403–16.
29. Lee YG, Seo JH. Production of 2,3-butanediol from glucose and cassava hydrolysates by metabolically engineered industrial polyploid *Saccharomyces cerevisiae. Biotechnol Biofuels.* 2019;12:204.
30. van der Hoek SA, Darbani B, Zugaj KE, et al. Engineering the yeast *Saccharomyces cerevisiae* for the production of L-(+)-ergothioneine. *Front Bioeng Biotechnol.* 2019;7:262.
31. Ishibashi K, Matsumoto-Yokoyama E, Ishikawa M. A tomato spotted wilt virus S RNA-based replicon system in yeast. *Sci Rep.* 2017;7(1):12647.
32. Vashee S, Stockwell TB, Alperovich N, et al. Cloning, assembly, and modification of the primary human cytomegalovirus isolate Toledo by yeast-based transformation-associated recombination. *mSphere.* 2017;2(5):e00331-17.
33. Lasserre JP, Dautant A, Aiyar RS, et al. Yeast as a system for modeling mitochondrial disease mechanisms and discovering therapies. *Dis Model Mech.* 2015;8(6):509–26.
34. Falcone C, Mazzoni C. External and internal triggers of cell death in yeast. *Cell Mol Life Sci.* 2016;73(11–12):2237–50.

SUMMARY

Saccharomyces cerevisiae (the budding yeast) is a unicellular eukaryote that is well known for its utility in baking, brewing, and winemaking, as well as for its value as an ingredient in cheese substitutes for vegans and vegetarians and a source of vitamins (especially B-complex vitamins), amino acids, and minerals for a health-conscious community. Being able to grow rapidly in defined media, amenable to genetic manipulation, and safe to work with, *S. cerevisiae* offers a valuable model for the study of eukaryotic biology, parasite–host interplay, gene–gene as well as gene–environment interactions, and mechanisms of mitochondrial diseases and aging. Furthermore, *S. cerevisiae* is a reliable and effective producer of proteins and biochemicals that make up important raw materials for biotech and pharmaceutical industries. With continuing advances in molecular manipulation and gene editing techniques, our understanding and appreciation of *S*. cerevisiae will one day catch up with those of *E. coli*. This will enable creative use of *S. cerevisiae* in many previously unexplored biomedical applications, and greatly enhance the wellbeing of human society.

SECTION IV

Molecular Analysis and Manipulation of Foodborne Parasites

Acanthamoeba Adhesion

23

Abdul Mannan Baig
Aga Khan University

Contents

23.1 INTRODUCTION

The genus *Acanthamoeba* contains nearly 25 free-living ameba (FLA), which are among the most communal protozoa in the natural habitat [1]. These amebae can persevere in diverse environments such as swimming pools, air conditioning systems, contact lenses, and operating rooms [2]. The fact that they cause infections in humans has remained a major reason for their research as they act as a pool for other microorganisms [3]. A well-known species of *Acanthamoeba*, that is, *A. castellanii*, gives rise to amebic keratitis (AK, an aching vision-jeopardizing inflammatory disease of the eyes), granulomatous amebic encephalitis (GAE, a deadly inflammation of the central nervous system), and cutaneous acanthamoebiasis [4].

In 1931, Volkonsky introduced the genus *Acanthamoeba* [5], which belongs to the family Acanthamoebidae. *Acanthamoeba* holds a multilayered microtubule-organizing center and can be pathogenic to humans [6]. On trophozoites, the presence of spiny surface projections, named acanthopodia, makes the finding of *Acanthamoeba* quite easy at the genus level. On the other hand, using morphological criteria to identify *Acanthamoeba* at the species level has been more challenging. Based on the cyst shape and size, *Acanthamoeba* spp. are classified into diverse groups (I, II, and III) [7,8]. Group I species have a large cyst compared to species of the other two groups. Group II species (including well-known human pathogenic *A. castellanii*) contain a crumpled ectocyst and an endocyst which could be of a polygonal, triangular, stellate, or oval shape. The Group III species characteristically possess a tinny, plane ectocyst and a round-appearing endocyst.

In 1958, *Acanthamoeba* pathogenicity to humans was noted by accident in the precautionary trial for the polio vaccine. During vaccine preparation, the cell cultures had a plaque-like appearance which was thought to be caused by a virus, but it was later discovered that the causative agents for these plaques were actually amebae. Immunization of monkeys and mice with the tissue culture fluid resulted in encephalitis-related death [9,10]. In the 1970s, *Acanthamoeba* was distinctly demonstrated to be human-pathogenic with a few reported cases of amebic encephalitis, AK, and cutaneous infections. Overtime, it has become progressively clear that FLA are pathogenic to humans. Upon further elucidation of the pathogenicity of FLA, human infections with other genera of amebae have also been described.

23.2 LIFECYCLE OF *ACANTHAMOEBA*

Acanthamoeba has a lifecycle that comprises two stages. The first is an actively feeding, separating trophozoite (Figure 23.1A), and the second is a dormant cyst. In its free-living stage, the trophozoite (of about 35 µm in size) consumes prokaryotes, yeast, and algae for nutrition. Furthermore, the trophozoite can survive abiotically on nutrients in fluid taken up by a process known as pinocytosis [11,12]. Food take-up by the trophozoite is preceded by projection of pseudopodia that phagocytoses or by creation of a food cup that consumes particulate material. Food cups on the surface of the ameba are impermanent structures that consume yeast, bacteria, or cells [13]. *Acanthamoeba* develops a hyaline pseudopodium for movement [14]. Investigation of ameboid motion, as exemplified by *A. castellanii*, has helped uncover valuable insights on the molecular mechanisms of actin polymerization [15,16].

Under electron microscopy, the cellular structures of *Acanthamoeba* trophozoites include free ribosome, Golgi complex, digestive vacuoles, mitochondria, smooth and rough endoplasmic reticula, and microtubules [17–20], which are all typically found in higher cells containing a true nucleus (eukaryotic). The cytoplasmic materials of the trophozoite are enclosed by a trilaminar cell membrane. The presence

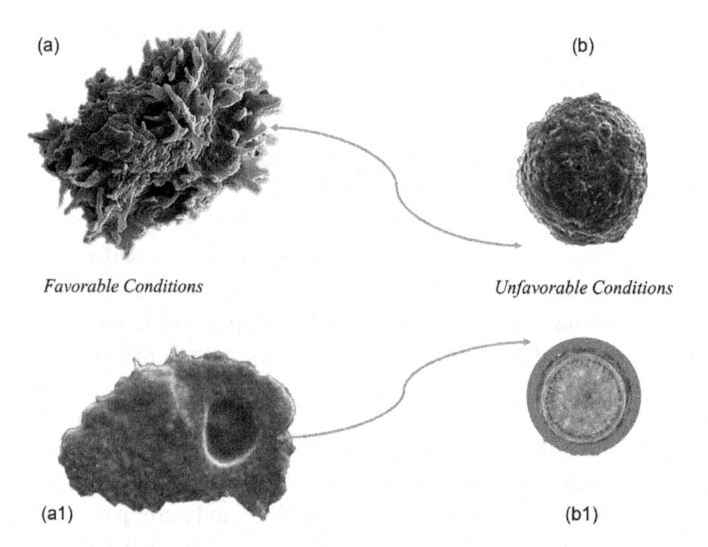

FIGURE 23.1 Shows the lifecycle of *Acanthamoeba* spp. Under favorable conditions of food, pH, and temperature, it lives in a trophozoite form (A–A1). When it falls short of food or senses an unfavorable temperature or pH, it assumes a cystic form (B–B1).

of thorny cell surface projections, named acanthopodia, is the characteristic feature of the trophozoite. In addition, within the cytoplasm, a striking contractile vacuole that regulates the water percentage within the intracellular space and a nucleus coupled with an oversized central nucleolus contribute to the distinguishing features of the trophozoite. Reproduction occurs via binary fission [7,21].

Acanthamoeba forms a double-walled crumpled cyst (of 13–22 μm in size), consisting of an ectocyst and an endocyst, under aggressive environmental circumstances such as food deficiency, dehydration, and variations in temperature and pH; the morphology of *Acanthamoeba* cyst may vary from one species to another [18,22,23]. *Acanthamoeba* cyst is resilient to antibiotics, biocides, and chlorination [24–27] and even persists in low temperatures between 0°C and 2°C [28]. However, it appears to be sensitive to methylene oxide, freon, and autoclaving [29]. Ameba can maintain viability within cysts for over two decades after retention at 4°C in the water, and remains immunogenic after being kept encysted for 24 years [30]. Thus, even though *Acanthamoeba* virulence is expected to decline overtime, its cyst form helps preserve its immunogenicity.

23.3 ADHESION OF *ACANTHAMOEBA* TO HUMAN TISSUES

Polymerization of a well-known protein called actin underscores various biological processes including cell adherence and adhesion to tissue components through the establishment of lamellipodia and filopodia [31]. As multiple membrane projections of *Acanthamoeba* trophozoites, acanthopodia are mainly formed from fibers and collections of actin (Figure 23.1A) [32]. Because these compositions allow for the organism to interact with the cell surface in the corneal epithelial cells, they play a significant role in disease processes [33]. A reduction in the number of binding sites on the host cells appears to lessen the host cell destruction [34]. The cell adhesion area of *A. castellanii* trophozoites relies on substrate rigidity [35]. Further, the number of acanthopodia of *Acanthamoeba* trophozoites corresponds with their pathogenicity, while nonpathogenic species display fewer acanthopodia [36]. Due to the ionic character of the lens, large amounts of water content support *Acanthamoeba* adhesions to contact lenses [37,38]. Populations who used silicone-containing hydrogel contact lenses had an increased incidence of contamination with *Acanthamoeba* spp. compared to populations who used hydrogel lenses containing no silicone [39–43]. The appearance of a bacterial biofilm to standard hydrogel contact lenses also influences *Acanthamoeba* adherence to the lens surface. The existence of a *Pseudomonas* biofilm on lens surface may increase the levels of *Acanthamoeba* adherence [44,45]. The methods by which contact lenses acquire sterility and are free of microbes are through the moist heating disinfection techniques [46,47] and with the use of hydrogen peroxide by the method of oxidation [48,49].

23.3.1 Cornea

The cytoskeleton has an active role in the adhesion of amebal trophozoites to the host cell [50]. The activity of cytochalasin and latrunculin is to reduce the rate of actin polymerization. This is facilitated by the emergence of a complex between the G-actin monomers and the drug molecules, inhibiting polymerization into F-actin [51,52]. The cytoskeleton–membrane interaction plays a key role in the development of diseases caused by protozoans. Many protozoan species adhere to the host tissue adhesion through the interaction of surface glycoprotein residues [53,54]. The adherence of *Acanthamoeba* spp. to corneal cells requires attachment of carbohydrate moieties expressed on the corneal cell surface [55]. In *A. castellanii*, the existence of a glycoprotein that contains the carbohydrate mannose residues on the cell surface of trophozoites termed mannose-binding-protein (MBP) has been described; MBP identifies glycoproteins in the mannosylated state in the cell membrane of the corneal epithelial cells to which *A. castellanii* adheres

[56–58]. Nevertheless, adhesion to the host cell is a complicated procedure which includes numerous proteins in the host–pathogen interactions. In addition to the MBP, use of monoclonal antibodies gives the hint of an adhesive molecule that also adheres to mannose moiety in *A. castellanii* [59]. The epithelial cell adherence by *Acanthamoeba* mannoproteins is only hindered by mannose and methyl-mannopyranoside, but not other sugars [55]. The epithelial cells of the cornea, a possible objective of *Acanthamoeba* infection, show proteins with GlcNAc and mannose residues on their cell surface [60], suggesting that the trophozoite forms of *Acanthamoeba* have an intricate configuration of corneal cell surface proteins and glycoproteins that likely serve as receptive adhesion molecules in the cornea (Figure 23.2).

The key step in establishing AK is the adhesion of ameba to the superficial corneal epithelium, which ultimately terminates the host cell [61]. The transmembrane proteins on the cell surface of *Acanthamoeba* interact with glycoproteins abundant in mannose remnants situated on corneal epithelial cells [62]. In fact, various adhesion-supporting proteins such as laminin and fibronectin contain a mannose component [57,58,63]. Contact emergence of *A. castellanii* commences between the sugar mannose on the cellular surface of host corneal cells and MBP expressed in the cell surface of *Acanthamoeba* spp. [62]c, which eventually enable the lytic elements to proceed to the host contact site and cause cell death (Figure 23.2) [63–65].

Mannose occluding and saturation studies produced slight adherence of *Acanthamoeba* to host corneal cells and decreased cytolysis, while no residue other than mannose has exhibited a comparable result [62,66,67]. In contrast to pathogenic species, the level of MBP in nonpathogenic *Acanthamoeba* is decreased, demonstrating a reliance of pathogenicity on the carbohydrate mannose [57,58].

Omaña-Molina and her team in 2010 found that *A. castellanii* trophozoites cocultivated with human corneal tissue created cytopathic modification on the corneal epithelium (Figure 23.2). On scanning

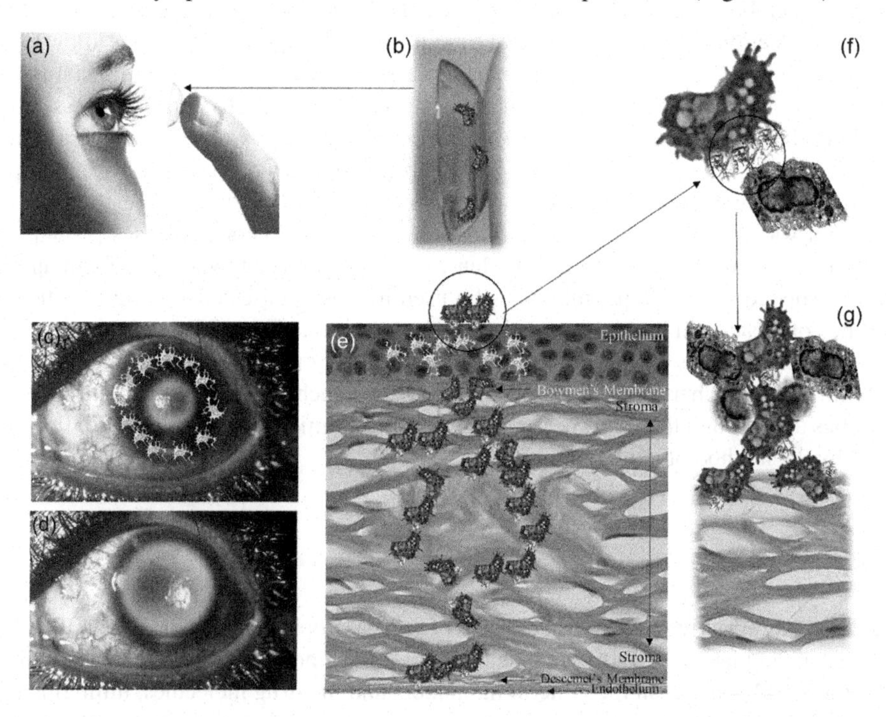

FIGURE 23.2 Use of a contact lens (a) can cause minor trauma to the cornea, and if the lens cleaning solution is contaminated with *Acanthamoeba* trophozoites (b), it enables them to infect the cornea. Corneal trauma incites growth of blood vessels from the limbus (sclerocorneal junction) that cause an inflammatory process to proceed. Progressive damage caused by *Acanthamoeba* (c) leads to corneal opacity (d) and blindness. Contact and adhesion of *Acanthamoeba* with corneal layers (e) involves mannose-binding protein and integrins that lead to epithelial cells damage (f) and progressive invasion, as well as destruction of deeper layers of cornea (f and g)

electron microscopy, initially, single or clustered trophozoites adhered to the corneal surface. Following the adherence, trophozoites shifted toward the borders of the cells, which caused the detachment of neighboring cells after 1 h of cocultivation. After 2–3 h, plenty of *Acanthamoeba* trophozoites appeared under the superficial corneal layers, forming a swelling on the corneal superficial layers in addition to edema under desquamating epithelial cells. Often, the trophozoites consumed shed corneal epithelial cells. Several ameba trophozoites persistently moved into deeper cell layers of the corneal epithelial cells and adhered to middle (wings) cells. Coculturing for 3 h led most of the superficial and middle cells of the corneal epithelial cells to be detached from the remaining corneal tissue. While the actual structure of corneal epithelial cells was missing, these cells did not show clear evidence of destruction, as they conserved their normal morphology. In other regions, few amebal trophozoites invaded the Bowman's membrane, where some *A. castellanii* trophozoites were seen clearly to engulf the basal corneal epithelial cells (Figure 23.2a–g). While trophozoites preserved their traditional shape as observed in axenic cultures, some morphological variations were apparent when the cellular projections of the trophozoites went through corneal epithelial cells intercellular junctions, where an elongated morphology was assumed to get through the tight intercellular spaces. This suggested that they may have gone through cell division. On light microscopy, trophozoites invaded the deeper layers of the human epithelial cells in cornea, subsequently extended to Bowman's membrane (Figure 23.2e), and ultimately phagocytosed currently separated corneal cells (Figure 23.2f). Throughout the corneal epithelium passage, only a few *A. castellanii* trophozoites demonstrated the typical morphology, in which the nucleus and nucleoli, in addition to the contractile vacuolar elements, were apparent, although other FLA altered the morphology to possess spaces between basal epithelium and proceed to the level of Bowman's membrane (Figure 23.2e) [68].

The surface of the eye is constantly exposed to the environs and pathogens, although it is defended by natural guarding procedures like the tears that remove pathogens. Immunoglobulin type A (IgA) is the chief antibody in tears where it forms a first line of defense in opposition to harmful pathogens, acting essentially as an immunologic blockade by hindering adhesion and ingestion of microorganisms [69]. The surface immune system of the mucosa provides a safeguard against *Acanthamoeba* keratitis [70]. Human tears from normal healthy individuals produce antiacanthamoebal IgA antibodies, which are decreased in patients with *Acanthamoeba* keratitis [71]. A 12 kDa secreted enzyme termed proteinase by *Acanthamoeba* is shown to destruct secretory immunoglobulin A (sIgA), IgM, and IgG, indicating that this proteinase is important in the process of cell injury [72]. The pathogenesis of *Acanthamoeba* keratitis appears to be a multistep process that includes adherence to the host cells, degradation of matrix, and invasion into deeper corneal tissues. *Acanthamoeba* trophozoites possibly adhere to an already damaged surface of corneal epithelium and enter into the stroma; therefore, a past injury to the corneal epithelium has to occur earlier than the keratitis [73]. Nevertheless, when trophozoites of *Acanthamoeba* are coincubated with unimpaired human corneal cells, the amebae can destroy the cornea without the need of former corneal trauma. In the initial stage of host–parasite interaction, individual trophozoites attach to the corneal surface, and then move to the cell borders. Subsequently, the intercellular junctions become detached and amebal trophozoites migrate to the intracellular spaces creating bulges and inducing cellular separation. As the invasion continues, the amebae wander into the deep layers of corneal epithelium, causing a commotion in the actual histological structure of the cornea. Moore et al. [74] showed that the adherence of amebal trophozoites is accompanied by surface epithelial destruction primarily at the outer peripheral zone of the cornea at the limbus, the sclerocorneal junction. The capability of *Acanthamoeba* to adhere to the corneal surface epithelium appears be a crucial stage in the pathological process that leads to *Acanthamoeba* keratitis. Therefore, the presence of MBP on the cellular surface of *Acanthamoeba* spp. indicates an early adherence to trophozoites in the corneal epithelial tissue and the formation of cytopathic elements that enhance the destruction [62,63,66,75,76]. *Acanthamoeba* adherence to corneal epithelial tissue initiates secondary processes such as phagocytosis, which involves an actin-dependent event. *Acanthamoeba* forms ameobastomes in the course of incubation with corneal cells that highlight the role of cellular events such as phagocytotic destruction in the progression of AK [33,77]. Phagocytosis as well as shedding of corneal epithelial cells is a frequently noticed occurrence when *Acanthamoeba* invades the deeper cellular layers of the corneal epithelial tissue, suggesting the capability of *A. castellanii* to utilize human corneal epithelial cells as a nutrient resource. The corneal epithelial cells serving as an abundant nutrient reserve allow the microorganisms to survive and grow rapidly for a prolonged duration. An equivalent series of events have been described in the course of the cocultivation of *A. castellanii* and *A. polyphaga* with hamster corneal epithelial tissue [33]. The proteases secreted by amebal trophozoites represent the principal element in pathogenesis as they are responsible for major exacerbating features in *Acanthamoeba* keratitis. This also appears to hold true for other protozoans [75,76,78–82]. *Acanthamoeba* proteinases are destructive to proteins in cornea, including major cellular structural proteins such as laminin collagen and fibronectin [80–82]. The secretory proteinases of *Acanthamoeba* demonstrate collagenolytic activity [83,84]. The pathogenic processes of *Acanthamoeba* contains contact-dependent and noncontact-dependent mechanisms. While mannose treatment causes *Acanthamoeba* to secrete cytolytic elements that rupture the epithelium of the cornea, the induced cytopathic activity is entirely hindered by the addition of a specific serine protease inhibitor to the medium [66]. It appears that serine proteases have a perilous role in the pathological process of AK as the introduction of a serine protease eliminates the cytotoxicity induced by *Acanthamoeba* conditioned medium on corneal epithelial cells [3].

23.3.2 Brain

GAE is a central nervous system (CNS) disease with fatal outcomes, although the key mechanism by which FLA interact to adhere with the blood–brain barrier (BBB) and disseminate to different parts of the CNS is unclear (Figure 23.3a and b).

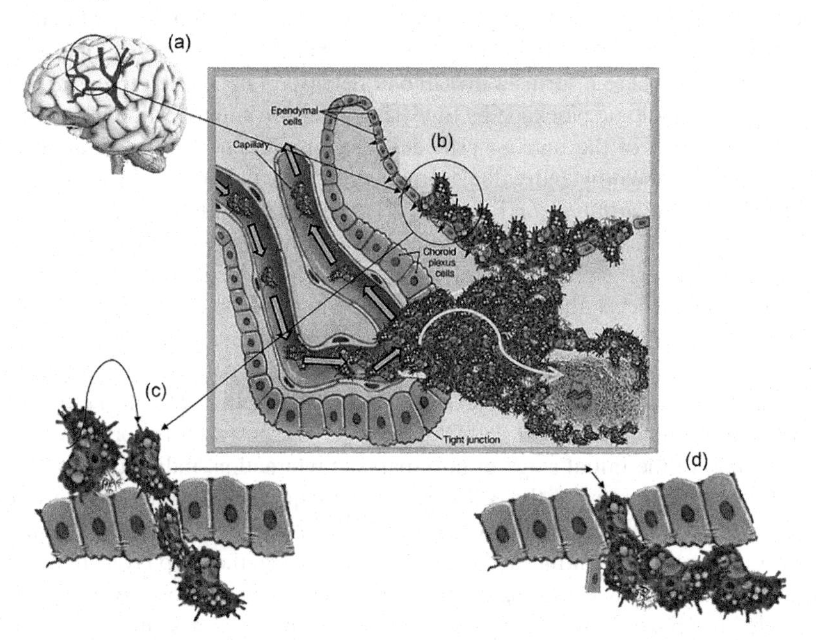

FIGURE 23.3 *Acanthamoeba* trophozoites induce encephalitis in brain (a) by accessing it via a hematogenous route. At the microvascular capillary level (b), the trophozoites reach the blood–brain barrier where the sluggish blood flow enables them to come in contact and adhere to the endothelium (b). The adhesion may be facilitated by expression of selectin, beta-alpha integrins and mannose-binding protein, which is followed by endothelial injury or transendothelial passage (b: circle, c, d). Prior exposure to *Acanthamoeba* antigens favors a granulomatous or granulomatoid immune reaction (b: white arrow).

When tested for adhesion to the human brain microvascular endothelial cells (HBMEC) to cause toxicity [85], *Acanthamoeba* genotype T3/T4 and T11 induced cytotoxicity on the HBMEC (Figure 23.3c), while T2 and T7 genotypes were weak or nonpathogenic. Pathogenic capabilities of *Acanthamoeba* isolates appear to correlate with MBP expression. Notably, the adherence of T7 isolate to HBMEC cells was not crucially hindered by alpha-mannose. The possible explanations for these observations include: (i) adherence of the T7 genotype is negligible to induce any changes regardless of the presence or absence of alpha-mannose, (ii) MBP is absent in this T7 isolate evaluated, or (iii) MBP has a confirmatively different structure in the T7 isolate tested. The variations in the adhesion of *Acanthamoeba* isolates to HBMEC can be the cause of conformational structural change in MBP. Therefore, an effective passage of *Acanthamoeba* across the BBB needs particular cross-talk with HBMEC, which forms an essential component of the BBB. The fact that the pathogenic genotypes of *Acanthamoeba* exert greater extracellular-protease actions than the nonpathogenic subtypes highlights the former's capability to develop cytotoxicity on corneal epithelial cells [86]. *Acanthamoeba* genotypes T3/T4 and T11 demonstrated greater MBP and developed greater cytotoxicity on HBMEC in contrast to the T2 and T7 isolates [85]. Therefore, MBP can also be designated as a possible marker for the disease-causing isolates of *Acanthamoeba*. The description of these injurious features can be helpful in understanding the pathogenesis of GAE.

Parasite adherence to the human host cells encourages the phosphorylation of tyrosine kinase of some HBMEC proteins. Acanthamoebal adhesion to the HBMEC leads to the stimulation of Rho-allied intracellular signaling sequence. Rho-linked pathways might interrupt the role of the tight junctions, thus causing augmented BBB permeability. Rho-A controls the phosphorylation of myosin light chain, subsequently leading to configurational alterations and redeployment of ZO-1 and the adhesion molecule occludin and causes augmented BBB permeability [87]. Persistent incubations lead to HBMEC apoptosis [88]. Use of a particular phosphatidylinositol 3-kinase (PI3K) inhibitor LY294002, along with HBMEC articulating mutant p85, that controls the subunit of PI3K (dominant-negative PI3K), has demonstrated that the PI3-K plays a critical part in *Acanthamoeba*-induced HBMEC death [88]. Further, *Acanthamoeba* by expressing genes like those encoding GADD45A and p130Rb induces a cell cycle arrest while inhibits the expression of other genes, similar to the ones that encode cyclin-dependent kinase-6 proteins, G1, and cyclins F, which are important for the progression of the cell cycle [89]. The general response of these events is due to the seizure of the cell cycle in HBMEC, which seems to be further assisted by the dephosphorylation of retinoblastoma protein, a strong inhibitor of G1/S cell cycle progression [89]. The capability of *Acanthamoeba* trophozoites to develop food cups (amebastomes) in course of coincubations with human cells indicates that cellular phagocytosis plays a crucial role in *Acanthamoeba* pathogenesis [13,77,90]. Favoring this, cytochalasin D that inhibits actin polymerization, inhibits *Acanthamoeba*-induced HBMEC death, affirming that actin-mediated cytoskeletal changes in *Acanthamoeba* are significant in engulfing HBMEC. Among noncontact-dependent processes, extracellular secretion of serine proteases helps amebae-mediated monolayer disruptions in the HBMEC [91]. Moreover, type I, III, and IV collagen, elastin, and fibronectin act as the substrate of serine proteases, which are the crucial elements of the human extracellular matrix, along with fibrinogen, IgG, IgA, albumin, plasminogen (required in proteolytic degradation of the extracellular matrix), and hemoglobin [81,92]. *Acanthamoeba* proteases are formed and are upregulated upon adhesion with the HBMEC (Figure 23.3c and d) [91]. *Acanthamoeba* proteases interfere with HBMEC monolayers, but this effect does not induce HBMEC cytotoxicity by the release of lactate dehydrogenase [91,92]. These results indicate that enzymes like proteases affect the BBB porosity, most probably by directing their actions toward cellular tight junctions (Figure 23.3c and d). Current researches have demonstrated that *Acanthamoeba* proteases target proteins like ZO-1 and occludin, which are the key constituents of the tight junctions [93].

23.4 ADHESION–INVASION CASCADE IN HUMAN TISSUE

Most of the destruction that is accomplished by *Acanthamoeba* trophozoites either in brain, skin, or corneal infections results from various pathogenic processes. Previously, it was assumed that *Acanthamoeba* did not have feeding cups known as ameobastomes, even though phagocytosis of the host cells was believed to be the cause of injury to host tissues. In 1996 [13], it was demonstrated by electron microscopy that there are a few *Acanthamoeba* species that produce ameobastomes while destroy the rat B103 neuroblastoma cells, and in due course pass the swallowed content into intracytoplasmic food vesicles. It is clear now that *Acanthamoeba* undertakes specific changes in its morphology when it is allowed to interact with hamster corneas, and forms food-cup surface patterns that consume separated corneal epithelial cells and therefore help in process of phagocytosis [33]. The serine protease secretory activity was discovered in the pathogenic isolates of *Acanthamoeba*, but only to a minute degree in nonpathogenic isolates [73]. Subsequent studies have demonstrated that *Acanthamoeba* secretes serine proteinases, cysteine proteinases, metalloproteinase, and induces chemotactic motion toward the corneal endothelial preparations [92].

23.4.1 Adhesion–Invasion of Cornea

Based on study of *Acanthamoeba* from ocular keratitis patients, the preliminary invasion process begins when a 136kDa MBP is expressed on the surface of the ameba which clings to mannosylated glycoproteins on the epithelial cells of cornea [58,94]. However, other proteinases are involved in the cytopathic effect on the corneal stroma (Figure 23.2). MBP is essential for the adhesion of the trophozoites to corneal epithelial cells and is primary for the pathogenicity of *Acanthamoeba* in AK. An anti-MBP IgA subtype created by oral immunization of hamsters was identified in the tear fluid, as well as from the animals suffering AK [57].

23.4.2 Adhesion–Invasion of Brain

The paths adopted by *Acanthamoeba* to enter into the brain and spinal cord were studied by Culbertson et al. [10] and Cerva [95]. It was found that intranasal, intrapulmonary, and intracardiac injections of *Acanthamoeba* into animals led to invasion of the brain and spinal cord, hence proving that *Acanthamoeba* may enter by various paths. Thus, the entrance of amebae in the brain is mostly by either the olfactory neuroepithelium path or through the blood. A previous study suggested that amebae enter the nasal mucosa and travel forward to the nerves, accompanied by invasion into the olfactory bulb. Nevertheless, recent analysis confirmed that *Acanthamoeba* gains access into the pulmonary tissues through the nasal path, then passes through the lungs into the bloodstream, via hematogenous route (Figure 23.3), and finally crosses the BBB (Figure 23.3b–d) into the brain and spinal cord, ultimately causing infections. Studies demonstrated that intranasally injected ameba into mice causes neurological signs as well as pulmonary disease [96].

The tissue invasion of *Acanthamoeba* into the brain and spinal cord occurs at BBB that is found at two areas: (i) the cerebral capillary endothelium and (ii) the choroid plexus (Figure 23.3b). Of these two sites, most of the invasion occurs at the cerebral capillary endothelium. At the cellular level, pathogens can invade the BBB either through a transcellular path or through a paracellular path (Figure 23.3b–d). The microorganisms may use the paracellular pathway either by damaging the cell or by crossing the cell while preserving the integrity of the cellular layers. Another mechanism by which *Acanthamoeba* may gain entrance into the brain and spinal cord is through leukocyte infiltration; this Trojan Horse mechanism is exercised by many intracellular microorganisms [97]. It is probable that *Acanthamoeba* invasion of the endothelium at the BBB takes place through the transcellular pathway by triggering endothelial cell injury (Figure 23.3d) or by the paracellular pathway by attacking the tight junctions.

23.4.3 Adhesion–Invasion of Skin

Skin lesions caused by *Acanthamoeba* spp. are very likely to occur in individuals that are HIV positive, and may or may not involve the brain and spinal cord [98–101]. Skin diseases have also been reported for HIV-negative individuals with amebic encephalitis for patients with immunological diseases [102–106], or for individuals receiving antirejection therapy for organ replacement [105,107]. Cutaneous acanthamoebiasis takes the form of a skin disease, and is categorized by the existence of hard reddish nodules (Figure 23.4) or cutaneous ulcers [4,108–115]. The initial appearance of the skin infection by *Acanthamoeba* comprises a firm pus-filled papulonodular lesion and subsequently develops into non-healing hardened ulcers [113,114] (Figure 23.4). Incidence of dispersed cutaneous wounds may be the immediate indication of an *Acanthamoeba* infection [101,115]. Whether cutaneous lesions signify an initial point of entry or are the consequences of hematogenous spread (Figure 23.4a–d) from different areas like the CNS, sinuses, or the respiratory tract is unknown [110]. Areas of necrosis enclosed by inflammatory cells, perivascular type of vasculitis, amebal trophozoite and cystic forms are usually seen on histologic examination of skin lesions [110]. Nevertheless, the microscopic manifestation of cutaneous lacerations may imitate that of mycobacteria, fungal skin lesions, viruses, or a foreign body-induced inflammation [102,109,116].

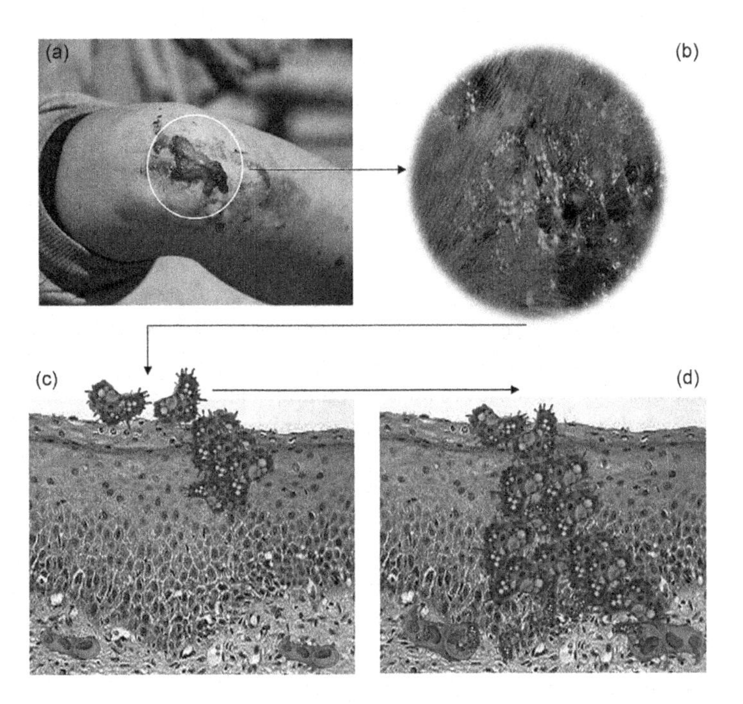

FIGURE 23.4 Wound infection due to contamination by *Acanthamoeba* spp. (a and b). An adhesion with the desquamated squamous epithelium is followed by invasion (c) into the deeper layers (d). After a variable period of skin infection, the trophozoite forms enter the blood vessels (c and d: lower left and right corners) to reach the circulation.

23.5 ADHESION–INJURY CASCADE IN HUMAN TISSUE

Acanthamoeba is a microbe that gets exposed to the host immune system during the lifetime of humans. The immunological response is a prime player in preservation against infection by this protist, but at the same time contributes to the BBB disruption and disease progression during CNS infection by *Acanthamoeba*. For instance, distinctive lesions in the CNS are attributed to the host immune response and are most probably formed by CD4 T-helper and cytotoxic T-killer CD8 T cells, B lymphocytes, plasma cells, macrophages, and parasites. The latter indicates the participation of inflammatory cytokines in immune defense along with pathophysiological processes. IFN-gamma is among the initial cytokines to participate and can play a key role in the stimulation of the immune cells. This and other cytokines and lymphokines can initiate the immune response to *Acanthamoeba* in the CNS [117]. Further, microglial cells secrete interleukin-IL-beta, interleukin-IL-alpha, and alpha-TNF in response to this protist pathogen [118]. Microglia equipped with IFN-gamma and TNF-alpha demonstrate amebicidal effects [119,120]. In contrast, microglia primed with IFN-gamma and IL-6 demonstrate antiproliferative effects [119,120]. The capability of these cells to show antimicrobial properties is most probably mediated by the production of free radicals, that is, nitric oxide (NO), and formation of inflammatory cytokines.

Parasite-mediated cytopathic effects have been observed *in vitro* by some researchers who showed that *Acanthamoeba* trophozoites can destroy a diverse sum of normal and cancerous mammalian cells [13,34,63,66,121,122]. Direct amebal trophozoite related cytolysis or necrotic destruction of normal and cancerous mammalian cells is calcium-dependent and is assisted by the calcium ionophore A23187 and hindered by a calcium channel blocker, Bepridil [34]. Cytoskeletal components are essential for *Acanthamoeba*-mediated cytolysis, as cytochalasin D treatment decreases cytopathic effects by ~98% [34]. Along with the elaborating soluble cytotoxic factors [121], *Acanthamoeba* spp. can swallow whole cells [13]. Mammalian cells uncovered to either live trophozoites or conditioned medium of trophozoites kill by a procedure that is similar to apoptosis [13,122]. Therefore, trophozoites cause epithelial cell disruption by three independent ways direct cytotoxicity, phagocytosis, and induction of apoptosis.

The immune defense mechanisms that work in opposition to *Acanthamoeba* have not been appropriately deciphered. *Acanthamoeba* infection appears to be frequent because the existence of immunoglobulins against *Acanthamoeba* has been observed in sera from the population who have no symptoms and are healthy [119,123–128]. However, even though the existence of *Acanthamoeba* cysts and trophozoites is at the peak in the environs, the prevalence of lethal disease seems low. *Acanthamoeba* has been isolated from the nose and pharynx of population who are evidently healthy [123,129–133]. Whether *Acanthamoeba* creates short-lived diseases in those populations and arouses host defense reactions which regulate infection and result in the eradication of the organism is unknown. Chappell et al. [124] proposed that even though severe eye infections and CNS diseases are rare, mucosal infections can result in countless numbers of unconfirmed sinus or pulmonary diseases. The occurrence of a greater incidence of *A. polyphaga*-specific immunoglobulin M (IgM) and reduced levels of IgG antibodies in serum from rheumatoid arthritis (RA) patients than from controls has also been described [134]. Defense against fatal diseases can include both acquired and innate immunity [125,126]. In experimental animal diseases, the age and strain of the animal, the host immune system, the number of trophozoites, temperature, and the genotype of the *Acanthamoeba* strain seems to be significant features in the infection outcomes of a murine infection [135]. Complement system gets activated by *Acanthamoeba* infection [136,137]. Complement activation can lead to creation of opsonic elements like C3b that functions in identification of amebae by human leukocytes [136]. On the other hand, complement can defend the host by rupturing amebae; however, *A. culbertsoni* are more resilient to complement-aided rupture than are nonpathogenic genotype of *Acanthamoeba* spp. [137]. Moreover, adherence of a complement product known as Clq by trophozoites inhibits adhesion sites for complement C1 that aids to hinder the classical pathway

[138]. Therefore, the development of resistance to complement-mediated rupture represents a type of immune evasion that helps in the progression of disease and spreading of *Acanthamoeba* inside the host [137,138]. Immunoglobulins can impede binding of *Acanthamoeba* to the host cells, hinder the movement of amebae, or counteract ameba cytolytic elements [125,126,139–141]. Although *Acanthamoeba* can destroy human IgG and IgA antibodies by serine proteases [81], in animal models, immunization with *A. culbertsoni* antigens using intraperitoneal, intranasal, oral, or intravenous route supposedly confers protection against a fatal trophozoite challenge [142]. However, immunization with other species of *Acanthamoeba* did not protect mice against challenge with *A. culbertsoni* [143–145]. An initial reaction of the host to amebae includes an infiltration of the granulocyte neutrophils in the area of the infection [139]. Although the neutrophils cannot eradicate *Acanthamoeba* except for those pretreated with alpha-TNF, *in-vitro* eradication of *Acanthamoeba* trophozoites by lymphokine-pretreated neutrophils needs the presence of both immunoglobulins and components of complement [139]. In mice exposed to *Acanthamoeba*, neutrophils and monocytes are the key inflammatory cells to infiltrate the site of infection. Monocytes/macrophages can play a more significant role than neutrophils in eradicating *Acanthamoeba*. These cells can damage *Acanthamoeba* and contain the cardinal cellular component for development of a granulomas often confronted in tissues containing *Acanthamoeba* cysts [140]. Masihi et al. [146] reviewed the consequence of the mycobacterium-derived immunopotentiation agents muramyl dipeptide and trehalose dimycolate against intranasal *Acanthamoeba* diseases in mice. *In-vitro* experiments with murine macrophages activated *in vivo* with TB vaccine Bacillus–Calmette–Guerin showed that the TB vaccine exposed activated macrophages were more effective in damaging trophozoites of *Acanthamoeba* than the unstimulated TB vaccine unexposed macrophages [140]. OH^- radicals, H_2O_2, and NO can be significant amebicidal elements because *Acanthamoeba* strains are easily harmed by hydrogen peroxide [147]. Further, rat macrophages, like the murine macrophages, display chemotaxis toward the amebae and kill trophozoites *in vitro* [148]. Further, macrophages activated with immunomodulators can phagocytose and damage amebae [140]. While pathogenic *Acanthamoeba* flee from the amebicidal activity of macrophages and monocyte/macrophage-like cells, weak or nonpathogenic species are attacked by these cells, and become ruptured, swallowed, and damaged [118,140,149]. Tanaka et al. [150] inspected T-lymphocyte reactions to *Acanthamoeba* antigens in normal individual. T-lymphocyte clones from these individuals were formed and examined for physiological reactions and activities. T-lymphocyte clones obtained from subjects who had no symptoms bred rapidly in reaction to exposure of *Acanthamoeba* antigens. T-cell clones were classified as being Thl cells involved in delayed-type hypersensitivity response in that they formed gamma interferon in response to amebic antigens. This is consistent with the finding of granulomatous response to *Acanthamoeba* in diseased individuals in an immunocompetent state. Immunocompromised individuals without sufficient Thl CD4+ve lymphocytes fail to mount a fully developed granulomatous response, with only perivascular cuffing called "granulomatoid" response [151].

23.6 RECENT ADVANCES IN MOLECULAR BASIS OF ADHESION OF *ACANTHAMOEBA* TO HUMAN TISSUES

As the understanding of the molecular basis of adhesion of *Acanthamoeba* to the human tissue plays a vital role in the disease pathogenesis such as AK, GAE, and wound infections, researchers have probed the adhesion molecules in vast details. Human leukocytic α integrins, CD1-a and CD11-b are known to help tight adherence prior to transendothelial migration and the ensuing entry into the blood vessel. Recently, with the help of bioinformatics computational tools molecules like a cell surface α integrin homolog on *Acanthamoeba* spp. termed integrin-a FG-GAP repeat-containing protein has been identified [152]. This amebal protein has been proposed for its interactions with the endothelium to initiate the wwadhesion in

TABLE 23.1 Novel hypothetical proteins that have structural homology with human cell adhesion proteins. The template-based models were developed by SWISS model database (data not shown) on submitting the *Acanthamoeba* proteins sequences, which were found to be resembling human cell adhesion (second column)

ACANTHAMOEBA PROTEIN MOLECULE	*HUMAN CELL ADHESION MOLECULE*	*FUNCTION IN HUMAN*	*SWISS MODEL TEMPLATE ID*
ACA1_369840-t26_1	E-selectin lectin	• Mitogenic with massive cytokine release	4c16.1
ACA1_188330	Integrin alpha II-Bb3	• Adhesion and invasion • Involved in cell–extracellular matrix adhesion and cell–cell adhesion	4gle.1
ACA1_080070-t26_1	Integrin beta-3	• Adhesion and invasion • Involved in cell–extracellular matrix adhesion and cell-cell adhesion	3fcs.1
ACA_1321210	Actinin	• α-Actinin binds actin filaments	1sjj.1
ACA1_036420 Calponin domain-containing protein	Dystrophin	• Is a vital part of a protein complex that connects the cytoskeleton of a muscle fiber to the surrounding extracellular matrix through the cell membrane	Idx.x.1
ACA 1_042800 FG-GAP Repeat-containing protein	Integrin – integrin alpha-V	• Interacts with extracellular matrix ligands	5ner.1

mediating interendothelial passage to cross the BBB. We investigated the ameba genome and proteome databases to explore the amebal counterparts of human adhesion molecules to understand the interaction of *Acanthamoeba* trophozoites with human tissues like endothelium, basement membranes, and connective tissue matrix.

Use of bioinformatics computational tools in providing evidence of novel adhesion molecules in *Acanthamoeba* spp. Of various methods used in elucidating the existence of adhesion molecules in *Acanthamoeba* spp., bioinformatics computational tools can be of immense help in exploring adhesion molecules. A list of known adhesion molecules expressed by human leukocytes and tissues can be searched for similar molecules in *Acanthamoeba* genome and proteome databases. Additionally, the level of expression of the molecules by *Acanthamoeba* trophozoites can also be examined by the use of the transcriptomics databases. Recently, the use of these bioinformatic computational tools and three-dimensional structural homology have helped discover novel voltage-gated ion channels [153,154], cell surface GPCRs [155], and hinted toward novel adhesion molecules [152].

A list of *Acanthamoeba* hypothetical and putative proteins (Table 23.1), the details of which are available from AmoebaDB.org NCBI and UniprotKB, are selected and explored at SWISS model database for structural homology with the human leukocyte adhesion proteins. The results show that some proteins previously designated as hypothetical and putative proteins turned out to have structural homology with human adhesion proteins selectins, integrins, and actins (Table 23.1). These adhesion proteins are well known in human leukocytes to promote adhesion with the endothelium and cell matrix. The proteins that were found to have homology with parallel proteins in *Acanthamoeba* include molecules like selectins and integrins is shown in (Figure 23.5). This method of discovery of adhesion molecules is expected to open the investigations into these previously reported proteins with unknown functions. This line of research can help establish the significance of the discovered adhesion molecules in *Acanthamoeba* adhesion to human cells like corneal epithelium, endothelium lining the BBB, cerebral neurons, and squamous epithelium of skin.

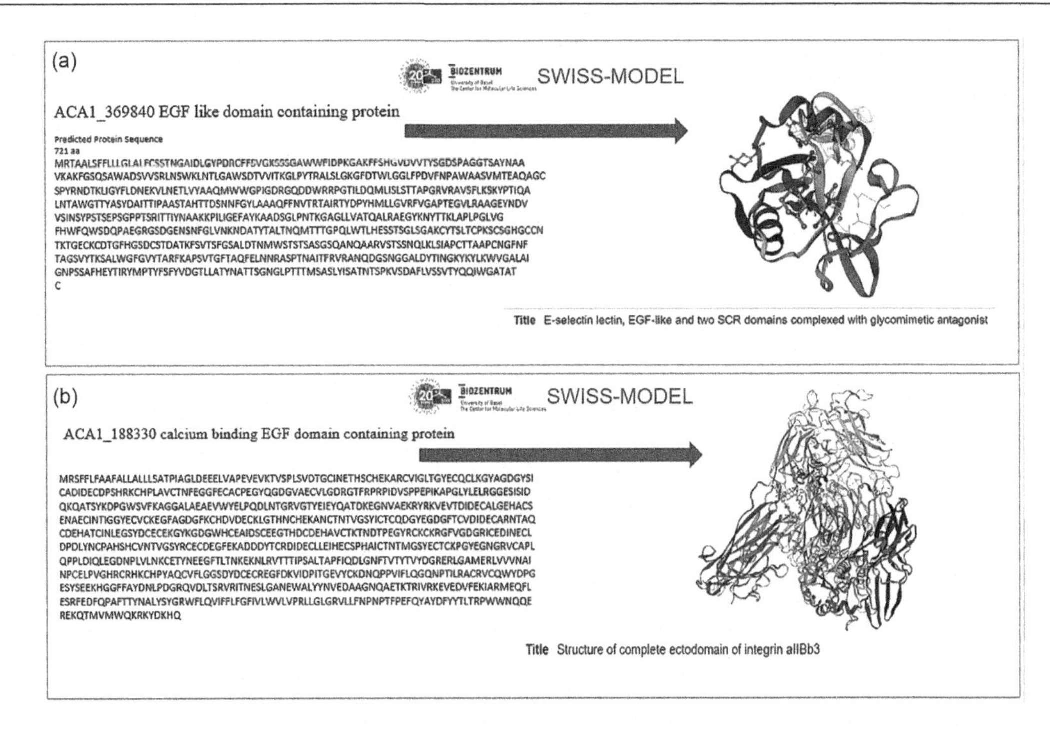

FIGURE 23.5 Methodology that uncovered selectins- and integrin-like proteins in *Acanthamoeba* spp. (a and b). The amino acid sequence of ACA1_369840 was submitted to SWISS model database that identified it to be an E-selectin (a). Similarly, ACA1_188330 was identified as IntegrinAlpha-2 Beta-3 (b). Both these proteins in humans are established adhesion molecules.

23.7 FUTURE PERSPECTIVES

Acanthamoeba is a genus of free-living ameba (FLA), which amount to nearly 25 recognized species. Of these, A. *castellanii* has the potential to cause amebic keratitis (AK), granulomatous amebic encephalitis (GAE), and cutaneous acanthamoebiasis in human populations, especially individuals with suppressed immune functions. Deciphering the role of a diverse list of *Acanthamoeba* adhesion molecules in detail through the use of contemporary imaging and computational tools is vital for the elucidation of its pathogenesis and subsequent development of novel approaches for its control and prevention. Direct visualization of *Acanthamoeba* trophozoites, with adhesion molecules tagged with fluorophore, in animal models with induced AK and GAE helps understand the factual role of the adhesion molecules in the disease process. Bioinformatics computational tools are promising in providing clues toward adhesion molecules of interests. The future use of miRNA and siRNA to knockout the genes discovered by computational methods will be instrumental for validating their significance in the biology of *Acanthamoeba* spp.

REFERENCES

1. Schuster, F. L., and G. S. Visvesvara. Free-living amoebae as opportunistic and non-opportunistic pathogens of humans and animals. *Int J Parasitol* 2004;34:1001–27.
2. Trabelsi, H., F. Dendana, A. Sellami, et al. Pathogenic free-living amoebae: epidemiology and clinical review. *Pathol Biol (Paris)* 2012;60:399–405.

3. Khan, N. A. Pathogenesis of *Acanthamoeba* infections. *Microb Pathog* 2003;34,277–85.
4. Torno, M. S., Jr., R. Babapour, A. Gurevitch, and M. D. Witt. Cutaneous acanthamoebiasis in AIDS. *J Am Acad Dermatol* 2000;42:351–4.
5. Volkonsky, M. *Hartmanella castellanii* Douglas, et classification des hartmannelles. *Arch Zool Exp Gen* 1931;72:317–39.
6. Patterson, D. J. The diversity of eukaryotes. *Am Nat* 1999;154:S96–124.
7. Page, F. C. Re-definition of the genus *Acanthamoeba* with descriptions of three species. *J Protozool* 1967;14:709–24.
8. Pussard, M., and R. Pons. Morphologies de la paroi kystique et taxonomie du genre *Acanthamoeba* (Protozoa, Amoebida). *Protistologica* 1977;13:557–610.
9. Culbertson, C. G., J. W. Smith, and H. Minner. *Acanthamoebae*: observations on animal pathogenicity. *Science* 1958;127:1506.
10. Culbertson, C. G., J. W. Smith, I. Cohen, and J. R. Minner. Experimental infection of mice and monkeys by *Acanthamoeba*. *Am J Pathol* 1959;35:185–97.
11. Bowers, B., and T. E. Olszewski. *Acanthamoeba* discriminates internally between digestible and indigestible particles. *J Cell Biol* 1983;97:317–22.
12. Bowers, B. Comparison of pinocytosis and phagocytosis in *Acanthamoeba castellanii*. *Exp Cell Res* 1977;110:409–17.
13. Pettit, D. A., J. Williamson, G. A. Cabral, and F. Marciano-Cabral. In vitro destruction of nerve cell cultures by *Acanthamoeba* spp.: a transmission and scanning electron microscopy study. *J Parasitol* 1996;82:769–77.
14. Preston, T. M., and C. A. King. Amoeboid locomotion of *Acanthamoeba castellanii* with special reference to cell-substratum interactions. *J Gen Microbiol* 1984;130:2317–23.
15. Kong, H. H., and T. D. Pollard. Intracellular localization and dynamics of myosin-II and myosin-IC in live *Acanthamoeba* by transient transfection of EGFP fusion proteins. *J Cell Sci* 2002;115:4993–5002.
16. Pollard, T. D., and E. M. Ostap. The chemical mechanism of myosin-I: implications for actin-based motility and the evolution of the myosin family of motor proteins. *Cell Struct Funct* 1996;21:351–6.
17. Bowers, B., and E. D. Korn. The fine structure of *Acanthamoeba castellanii*. I. The trophozoite. *J Cell Biol* 1968;39:95–111.
18. Bowers, B., and E. D. Korn. The fine structure of *Acanthamoeba castellanii* (Neff strain). II. Encystment. *J Cell Biol* 1969;41:786–805.
19. Gonzalez-Robles, A., A. Flores-Langarica, M. Omana-Molina, and M. Shibayama. *Acanthamoeba castellanii*: ultrastructure of trophozoites using fast freeze-fixation followed by freeze-substitution. *J. Electron Microsc. (Tokyo)* 2001; 50:423–7.
20. Rondanelli, E. G., G. Carosi, P. Lanzarini, and G. Filice. Ultrastructure of *Acanthamoeba*-Naegleria free-living amoebae, pp. 87–125. In E. G. Rondanelli (ed.), *Amphizoic Amoebae: Human Pathology*. Piccin Nuova Libraria, Padua, Italy, 1987.
21. Byers, T. J. Growth, reproduction, and differentiation in *Acanthamoeba*. *Int Rev Cytol* 1979;61:283–338.
22. Byers, T. J., R. A. Akins, B. J. Maynard, R. A. Lefken, and S. M. Martin. Rapid growth of *Acanthamoeba* in defined media; induction of encystment by glucose-acetate starvation. *J Protozool* 1980;27:216–9.
23. Chagla, A. H., and A. J. Griffiths. Growth and encystation of *Acanthamoeba castellanii*. *J Gen Microbiol.* 1974;85:139–45.
24. De Jonckheere, J., and H. Van de Voorde. Differences in destruction of cysts of pathogenic and nonpathogenic Naegleria and *Acanthamoeba* by chlorine. *Appl Environ Microbiol* 1976;31:294–7.
25. Khunkitti, W., D. Lloyd, J. R. Furr, and A. D. Russell. *Acanthamoeba castellanii*: growth, encystment, excystment and biocide susceptibility. *J Infect* 1998;36:43–8.
26. Lloyd, D., N. A. Turner, W. Khunkitti, A. C. Hann, J. R. Furr, and A. D. Russell. Encystation in *Acanthamoeba castellanii*: development of biocide resistance. *J Eukaryot Microbiol* 2001;48:11–6.
27. Turner, N. A., J. Harris, A. D. Russell, and D. Lloyd. Microbial differentiation and changes in susceptibility to antimicrobial agents. *J Appl Microbiol* 2000;89:751–9.
28. Brown, T. J., and R. T. Cursons. Pathogenic free-living amebae (PFLA) from frozen swimming areas in Oslo, Norway. *Scand J Infect Dis* 1977;9:237–40.
29. Meisler, D. M., I. Rutherford, F. E. Bican, et al. Susceptibility of *Acanthamoeba* to surgical instrument sterilization techniques. *Am J Ophthalmol.* 1985;99:724–5.
30. Mazur, T., E. Hadas, and I. Iwanicka. The duration of the cyst stage and the viability and virulence of *Acanthamoeba* isolates. *Trop Med Parasitol* 1995;46:106–8.
31. Pollard, T. D., and G. G. Borisy. Cellular motility driven by assembly and disassembly of actin filaments. *Cell* 2003;112:453–65.

32. González-Robles, A., G. Castañón, V. I. Hernández-Ramírez, et al. *Acanthamoeba castellanii*: identification and distribution of actin cytoskeleton. *Exp Parasitol* 2008;119:411–7.
33. Omána-Molina, M., F. Navarro-García, A. González-Robles, et al. Induction of morphological and electrophysiological changes in hamster cornea after in vitro interaction with trophozoites of *Acanthamoeba* spp. *Infect Immun* 2004;72:3245–51.
34. Taylor, W. M., M. S. Pidherney, H. Alizadeh, and J. Y. Niederkorn. In vitro characterization of *Acanthamoeba castellanii* cytopathic effect. *J Parasitol* 1995;81:603–9.
35. Gutekunst, S. B., C. Grabosch, A. Kovalev, S. N. Gorb, and C. Selhuber-Unkel. Influence of the PDMS substrate stiffness on the adhesion of *Acanthamoeba castellanii. Beilstein J Nanotechnol* 2014;5(1):1393–8.
36. Niederkorn, J. Y., H. Alizadeh, H. Leher, and J. P. McCulley. The pathogenesis of *Acanthamoeba* keratitis. *Microbes Infect* 1999;1(6):437–43.
37. Seal, D. V., E. S. Bennett, A. K. Mcfadyen, et al. Differential adherence of *Acanthamoeba* to contact lenses: effects of material characteristics. *Optom Vis Sci* 1995;72:23–8.
38. Gorlin, A. I., M. M. Gabriel, L. A. Wilson, et al. Binding of *Acanthamoeba* to hydrogel contact lenses. *Curr Eye Res* 1996;15:151–5.
39. Beattie, T. K., A. Tomlinson, A. K. McFadyen, et al. Enhanced attachment of *Acanthamoeba* to extended wear silicone hydrogel contact lenses – a new risk factor for infection. *Ophthalmology* 2003;110:765–71.
40. Beattie, T. K., A. Tomlinson, and A. K. McFadyen. Attachment of *Acanthamoeba* to first- and second-generation silicone hydrogel contact lenses. *Ophthalmology* 2006;113:117–25.
41. Teichroeb, J. H., J. A. Forrest, V. Ngai, et al. Imaging protein deposits on contact lens materials. *Optom Vis Sci* 2008;85:1151–64.
42. Kilvington, S., and J. Lonnen. A comparison of regimen methods for the removal and inactivation of bacteria, fungi and *Acanthamoeba* from two types of silicone hydrogel lenses. *Cont Lens Anterior Eye* 2009;32:73–7.
43. Giraldez, M. J., C. Serra, M. Lira, et al. Soft contact lens surface profile by atomic force microscopy. *Optom Vis Sci* 2010;87:475–81.
44. Gorlin, A. I., M. M. Gabriel, L. A. Wilson, and D. G. Ahearn. Effect of adhered bacteria on the binding of *Acanthamoeba* to hydrogel lenses. *Arch Ophthalmol* 1996;114:576–80.
45. Simmons, P. A., A. Tomlinson, and D. V. Seal. The role of *Pseudomonas aeruginosa* biofilm in the attachment of *Acanthamoeba* to four types of hydrogel contact lens materials. *Optom Vis Sci* 1998;75:860–6.
46. Kilvington, S. Moist heat disinfection of pathogenic *Acanthamoeba* cysts. *Lett Appl Microbiol* 1989;9:187–9.
47. Lindquist, T. D., D. J. Doughman, J. B. Rubinstein, et al. *Acanthamoeba* contaminated hydrogen contact lenses: susceptibility to disinfection. *Cornea* 1988:7;300–3.
48. Hughes, R., and S. Kilvington. Comparison of hydrogen peroxide contact lens disinfection systems and solutions against *Acanthamoeba polyphaga. Antimicrob Agents Chemother* 2001;45:2038–43.
49. Johnston, S. P., R. Sriram, Y. Qvarnstrom, et al. Resistance of *Acanthamoeba* cysts to disinfection in multiple contact lens solutions. *J Clin Microbiol* 2009;47:2040–5.
50. Soto-Arredondo, K. J., L. L. Flores-Villavicencio, J. J. Serrano-Luna, M. Shibayama, and M. Sabanero-Lopez. Biochemical and cellular mechanisms regulating *Acanthamoeba castellanii* adherence to host cells. *Parasitology* 2014;141(4):531–41.
51. Beckerle, M. C. Spatial control of actin filament review assembly: lessons from *Listeria. Cell* 1998;95:741–8.
52. Fürstner, A., D. Kirk, M. D. Fenster, C. Aïssa, D. De Souza, and O. Müller. Diverted total synthesis: preparation of a focused library of latrunculin analogues and evaluation of their actin-binding properties. *Proc Natl Acad Sci USA* 2005;102:8103–8.
53. Bailey, G. B., D. B. Day, and J. W. Gasque. Rapid polymerization of *Entamoeba histolytica* actin induced by interaction with target cells. *J ExpMed* 1985;162:546–58.
54. Pacheco-Yépez, J., R. Campos-Rodríguez, S. Rojas-Hernández, et al. Differential expression of surface glycoconjugates on *Entamoeba histolytica* and *Entamoeba dispar. Parasitol Int* 2009;58:171–7.
55. Morton, L. D., G. L. McLaughlin, and H. E. Whiteley. Effects of temperature, amebic strain, and carbohydrates on *Acanthamoeba* adherence to corneal epithelium in vitro. *Infect Immun* 1991;59:3819–22.
56. Garate, M., I. Cubillos, J. Marchant, and N. Panjwani. Biochemical characterization and functional studies of *Acanthamoeba* mannose-binding protein. *Infect Immun* 2005;73:5775–81.
57. Garate, M., H. Alizadeh, S. Neelam, J. Y. Niederkorn, and N. Panjwani. Oral immunization with *Acanthamoeba castellanii* mannose-binding protein ameliorates amoebic keratitis. *Infect Immun* 2006;74:7032–4.
58. Garate, M., J. Marchant, I. Cubillos, Z. Cao, N. A. Khan, and N. Panjwani. In vitro pathogenicity of *Acanthamoeba* is associated with the expression of the mannose-binding protein. *Invest Ophth Vis Sci* 2006;47(3):1056–62.
59. Kennett, M. J., R. R., Jr. Hook, C. L. Franklin, and L. K. Riley. *Acanthamoeba castellanii*: characterization of an adhesin molecule. *Exp Parasitol* 1999; 92, 161–9.

60. Panjwani, N., S. Ahmad, and M. B. Raizman. Cell surface glycoproteins of corneal epithelium. *Invest Ophthalmol Visual Sci* 1995;36:355–63.
61. Serrano-Luna, J. J., I. Cervantes-Sandoval, J. Calderón, F. Navarro-García, V. Tsutsumi, and M. Shibayama. Protease activities of *Acanthamoeba polyphaga* and *Acanthamoeba castellanii*. *Can J Microbiol* 2006;52:16–23.
62. Yang, Z., Z. Cao, and N. Panjwani. Pathogenesis of *Acanthamoeba* keratitis: carbohydrate-mediated host-parasite interactions. *Infect Immun* 1997;65:439–45.
63. Cao, Z., D. M. Jefferson, and N. Panjwani. Role of carbohydrate-mediated adherence in cytopathogenic mechanisms of *Acanthamoeba*. *J Biol Chem* 1998;273(25):15838–45.
64. Garate, M., Z. Cao, E. Bateman, and N. Panjwani. Cloning and characterization of a novel mannose-binding protein of *Acanthamoeba*. *J Biol Chem* 2004;279(28):29849–56.
65. Panjwani, N. Pathogenesis of *Acanthamoeba* keratitis. *Ocul Surf* 2010;8(2):70–9.
66. Leher, H., R. Silvany, H. Alizadeh, J. Huang, and J. Y. Niederkorn. Mannose induces the release of cytopathic factors from *Acanthamoeba castellanii*. *Infect Immun* 1998;66(1):5–10.
67. Kim, J. H., A. Matin, H. J. Shin, et al. Functional roles of mannose-binding protein in the adhesion, cytotoxicity and phagocytosis of *Acanthamoeba castellanii*. *Exp Parasitol* 2012;132(2):287–92.
68. Omaña-Molina, M, A. González-Robles, L. I. Salazar-Villatoro, et al. *Acanthamoeba castellanii*: morphological analysis of the interaction with human cornea. *Exp Parasitol* 2010;126:73–8.
69. Cao, Z., C. Saravanan, M. H. Goldstein, et al. Effect of human tears on *Acanthamoeba*-induced cytopathic effect. *Arch Ophthalmol* 2008;126:348–52.
70. Marciano-Cabral, F., and G. Cabral. *Acanthamoeba* spp. as agents of disease in humans. *Clin Microbiol Rev* 2003;16:273–307.
71. Niederkorn, J. Y., H. Alizadeh, H. Leher, et al. Role of tear anti-*Acanthamoeba* IgA in *Acanthamoeba* keratitis. *Adv Exp Med Biol* 2002;506:845–50.
72. Na, B. K., J. H. Cho, C. Y. Song, and T. S. Kim. Degradation of immunoglobulins, protease inhibitors and interleukin-1 by a secretory proteinase of *Acanthamoeba castellanii*. *Korean J Parasitol* 2002;2:93–9.
73. Khan, N. A. *Acanthamoeba*: biology and increasing importance in human health. *FEMS Microbiol Rev* 2006;30:564–95.
74. Moore, M. B., J. E. Ubelaker, J. H. Martin, et al. In vitro penetration of human corneal epithelium by *Acanthamoeba castellanii*: a scanning and transmission electron microscopy study. *Cornea* 1991;10:291–8.
75. Hurt, M., S. Neelam, J. Niederkorn, and H. Alizadeh. Pathogenic *Acanthamoeba* spp. secrete a mannose-induced cytolytic protein that correlates with the ability to cause disease. *Infect Immun* 2003;71:6243–55.
76. Hurt, M., J. Niederkorn, and H. Alizadeh. Effects of mannose on *Acanthamoeba castellanii* proliferation and cytolytic ability to corneal epithelial cells. *Invest Ophthalmol Visual Sci* 2003:44:3424–31.
77. Khan, N. A. Pathogenicity, morphology and differentiation of *Acanthamoeba*. *Curr Microbiol* 2001; 43:391–5.
78. Mckerrow, J. H., E. Sun, P. J. Rosenthal, and J. Bouvier. The proteases and pathogenicity of parasitic protozoa. *Annu Rev Microbiol* 1993;47:821–53.
79. Mitra, M. M., H. Alizadeh, R. D. Gerard, and J. Y. Niederkorn. Characterization of a plasminogen activator produced by *Acanthamoeba castellanii*. *Mol Biochem Parasitol* 1995;73:157–64.
80. Cho, J. H., B. K. Na, T. S. Kim, and C. Y. Song. Purification and characterization of an extracellular serine proteinase from *Acanthamoeba castellanii*. *IUBMB Life* 2000;50:209–14.
81. Kong, H. H., T. H. Kim, and D. I. Chung. Purification and characterization of a secretory proteinase of *Acanthamoeba* healyi isolates from GAE. *J Parasitol* 2000;86:12–17.
82. Na, B. K., J. C. Kim, and C. Y. Song. Characterization and pathogenetic role of proteinase from *Acanthamoeba castellanii*. *Microb Pathog* 2001;30:39–48.
83. He, Y. G., J. Y. Niederkorn, J. P. McCulley, et al. In vivo and in vitro collagenolytic activity of *Acanthamoeba castellanii*. *Invest Ophthalmol Visual Sci* 1990;11:2235–40.
84. Mitro, K., A. Bhagavathiammai, O. M. Zhou, et al. Partial characterization of the proteolytic secretions of *Acanthamoeba* polyphaga. *Exp Parasitol* 1994;4:377–85.
85. Alsam, S., K. S. Kim, M. Stins, A. Ortega Rivas, J. Sissons, and N. A. Khan. *Acanthamoeba* interactions with human brain microvascular endothelial cells. *Microb Pathog* 2013;35:235–41.
86. Khan, N. A., E. L. Jarroll, N. Panjwani, Z. Cao, and T. A. Paget. Proteases as markers of differentiation of pathogenic and non-pathogenic *Acanthamoeba*. *J Clin Microbiol* 2000;38:2858–61.
87. Shen, L., E. D. Black, E. D. Witkowski, et al. Myosin light chain phosphorylation regulates barrier function by remodeling tight junction structure. *J Cell Sci* 2006;119:2095–106.
88. Sissons, J., K. S. Kim, M. Stins, S. Jayasekera, S. Alsam, and N. A. Khan. *Acanthamoeba castellanii* induces host cell death via a phosphatidylinositol 3-kinase-dependent mechanism. *Infect Immun* 2005;73:2704–8.

89. Sissons, J., S. Alsam, S. Jayasekera, K. S. Kim, M. Stins, and N. A. Khan. *Acanthamoeba* induces cell-cycle arrest in the host cells. *J Med Microbiol* 2004;53:711–7.
90. Alsam, S., J. Sissons, R. Dudley, and N. A. Khan. Mechanisms associated with *Acanthamoeba castellanii* (T4) phagocytosis. *Parasitol Res* 2005;96:402–9.
91. Alsam, S., J. Sissons, S. Jayasekera, and N. A. Khan. Extracellular proteases of *Acanthamoeba castellanii* (encephalitis isolate belonging to T1 genotype) contribute to increased permeability in an in vitro model of the human blood–brain barrier. *J Infect* 2005;51:150–6.
92. Sissons, J., S. Alsam, G. Goldsworthy, M. Lightfoot, E. L. Jarroll, and N. A. Khan. Identification and properties of proteases from an *Acanthamoeba* isolate capable of producing granulomatous encephalitis. *BMC Microbiol* 2006;6:42.
93. Khan, N. A. *Acanthamoeba* invasion of the central nervous system. *Int J Parasitol* 2007;37:131–8.
94. Clarke, D. W., and J. Y. Niederkorn. The pathophysiology of *Acanthamoeba* keratitis. *Trends Parasitol* 2006;22:175–80.
95. Cerva, L. Intranasal, intrapulmonary and intracardial inoculation of experimental animals with *Hartmanella castellanii. Folia Parasitol (Prague)* 1967;14:207–15.
96. Martinez, A. J., S. M. Markowitz, and R. J. Duma. Experimental pneumonitis and encephalitis caused by *Acanthamoeba* in mice: pathogenesis and ultrastructural features. *J Infect Dis* 1975;131:692–99.
97. Drevets, D. A., and P. J. M. Leenen. Leukocyte-facilitated entry of intracellular pathogens into the central nervous system. *Microb Infect* 2000;2:1609–18.
98. Casper, T., D. Basset, C. Leclercq, J. Fabre, N. Peyron-Raison, and J. Reynes. Disseminated *Acanthamoeba* infection in a patient with AIDS: response to 5-fluorocytosine therapy. *Clin Infect Dis* 1999;29:944–5.
99. Duluol, A. M., M. F. Teilhac, and J. L. Poirot. Cutaneous lesions due to *Acanthamoeba* sp. in patients with AIDS. *J Eukaryot Microbiol* 1996;43:130–1.
100. Helton, J., M. Loveless, and C. R. White, Jr. Cutaneous *Acanthamoeba* infection associated with leukocytoclastic vasculitis in an AIDS patient. *Am J Dermatopathol* 1993;15:146–9.
101. Murakawa, G. J., T. McCalmont, J. Altman, et al. Disseminated acanthamebiasis in patients with AIDS. A report of five cases and a review of the literature. *Arch Dermatol* 1995;131:1291–6.
102. Gordon, S. M., J. P. Steinberg, M. H. DuPuis, P. E. Kozarsky, J. F. Nickerson, and G. S. Visvesvara. Culture isolation of *Acanthamoeba* species and leptomyxid amebas from patients with amebic meningoencephalitis, including two patients with AIDS. *Clin Infect Dis* 1992;15:1024–30.
103. Martinez, A. J. Acanthamoebiasis and immunosuppression. Case report. *J Neuropathol Exp Neurol* 1982; 41:548–57.
104. Oliva, S., M. Jantz, R. Tiernan, D. L. Cook, and M. A. Judson. Successful treatment of widely disseminated acanthamoebiasis. *South Med J* 1999;92:55–7.
105. Slater, C. A., J. Z. Sickel, G. S. Visvesvara, R. C. Pabico, and A. A. Gaspari. Brief report: successful treatment of disseminated *Acanthamoeba* infection in an immunocompromised patient. *N Engl J Med* 1994;331:85–7.
106. Visvesvara, G. S., S. S. Mirra, F. H. Brandt, D. M. Moss, H. M. Mathews, and A. J. Martinez. Isolation of two strains of *Acanthamoeba castellanii* from human tissue and their pathogenicity and isoenzyme profiles. *J Clin Microbiol* 1983;18:1405–12.
107. Van Hamme, C., M. Dumont, M. Delos, and J. M. Lachapelle. Acanthamibiase cutanee chez un transplante pulmonaire. *Ann Dermatol Venereol* 2001;128:1237–40.
108. Bonilla, H. F., A. Whitehurst, and C. A. Kauffman. *Acanthamoeba sinusitis* and disseminated infection in a patient with AIDS. *Infect Med* 1999;16:397–400.
109. Chandrasekar, P. H., P. S. Nandi, M. R. Fairfax, and L. R. Crane. Cutaneous infections due to *Acanthamoeba* in patients with acquired immunodeficiency syndrome. *Arch Intern Med* 1997; 157:569–72.
110. Friedland, L. R., S. A. Raphael, E. S. Deutsch, et al. Disseminated *Acanthamoeba* infection in a child with symptomatic human immunodeficiency virus infection. *Pediatr Infect Dis J* 1992;11:404–7.
111. Gonzalez, M. M., E. Gould, G. Dickinson, et al. Acquired immunodeficiency syndrome associated with *Acanthamoeba* infection and other opportunistic organisms. *Arch Pathol Lab Med* 1986;110:749–51.
112. Levine, S., A. E. Goldstein, M. Dahdouh, P. Blank, C. Hoffman, and C. A. Gropper. Cutaneous *Acanthamoeba* in a patient with AIDS: a case study with a review of new therapy. *Cutis* 2001;67:377–80.
113. May, L. P., G. S. Sidhu, and M. R. Buchness. Diagnosis of *Acanthamoeba* infection by cutaneous manifestations in a man seropositive to HIV. *J Am Acad Dermatol* 1992;26:352–5.
114. Rosenberg, A. S., and M. B. Morgan. Disseminated acanthamoebiasis presenting as lobular panniculitis with necrotizing vasculitis in a patient with AIDS. *J Cutan Pathol* 2001;28:307–13.

115. Tan, B., C. M. Weldon-Linne, D. P. Rhone, C. L. Penning, and G. S. Visvesvara. *Acanthamoeba* infection presenting as skin lesions in patients with the acquired immunodeficiency syndrome. *Arch Pathol Lab Med* 1993;117:1043–6.
116. Singhal, T., A. Bajpai, V. Kalra, et al. Successful treatment of *Acanthamoeba* meningitis with combination oral antimicrobials. *Pediatr Infect Dis J* 2001;20:623–7.
117. Benedetto, N., F. Rossano, F. Gorga, A. Folgore, M. Rao, and C. R. Carratelli. Defense mechanisms of IFN-γ and LPS-primed murine microglia against *Acanthamoeba castellanii* infection. *Int Immunopharmacol* 2003;3:825–34.
118. Marciano-Cabral, F., R. Puffenbarger, and G. A. Cabral. The increasing importance of *Acanthamoeba* infections. *J Eukaryot Microbiol* 2000;47:29–36.
119. Benedetto, N., and C. Auriault. Prolactin-cytokine network in defence against *Acanthamoeba castellanii* in murine microglia. *Eur Cytokine Netw* 2002;13:447–55.
120. Benedetto, N., and C. Auriault. Complex network of cytokines activating murine microglial cell activity against *Acanthamoeba castellanii. Eur Cytokine Netw* 2002;13:351–7.
121. Pidherney M. S., H. Alizadeh, G. L. Stewart, J. P. McCulley, and J. Y. Niederkorn. In vitro and in vivo tumoricidal properties of a pathogenic/free-living amoeba. *Cancer Lett* 1993;72:91–8.
122. Alizadeh H., M. S. Pidherney, J. P. McCulley, and J. Y. Niederkorn. Apoptosis as a mechanism of cytolysis of tumor cells by a pathogenic free-living amoeba. *Infect Immun* 1994;62:1298–303.
123. Cerva, L. *Acanthamoeba culbertsoni* and *Naegleria fowleri*: occurrence of antibodies in man. *J Hyg Epidemiol Microbiol Immunol* 1989;33:99–103.
124. Chappell, C. L., J. A. Wright, M. Coletta, and A. L. Newsome. Standardized method of measuring *Acanthamoeba* antibodies in sera from healthy human subjects. *Clin Diagn Lab Immunol* 2001;8:724–30.
125. Cursons, R. T., T. J. Brown, E. A. Keys, K. M. Moriarty, and D. Till. Immunity to pathogenic free-living amoebae: role of humoral antibody. *Infect Immun* 1980;29:401–7.
126. Cursons, R. T., T. J. Brown, E. A. Keys, K. M. Moriarty, and D. Till. Immunity to pathogenic free-living amoebae: role of cell-mediated immunity. *Infect Immun* 1980;29:408–10.
127. Newsome, A. L., F. T. Curtis, C. G. Culbertson, and S. D. Allen. Identification of *Acanthamoeba* in bronchoalveolar lavage specimens. *Diagn Cytopathol* 1992;8:231–4.
128. Powell, E. L., A. L. Newsome, S. D. Allen, and G. B. Knudson. Identification of antigens of pathogenic free-living amoebae by protein immunoblotting with rabbit immune and human sera. *Clin Diagn Lab Immunol* 1994;1:493–9.
129. Badenoch, P. R., T. R. Grimmond, J. Cadwgan, S. E. Deayton, and M. S. L. Essery. Nasal carriage of free-living amoebae. *Microb Ecol Health Dis* 1988;1:209–11.
130. Cerva, L., C. Serbus, and V. Skocil. Isolation of limax amoebae from the nasal mucosa of man. *Folia Parasitol (Prague)* 1973;20:97–103.
131. Michel, R., B. Hauroder-Philippczyk, K. Muller, and I. Weishaar. *Acanthamoeba* from human nasal mucosa infected with an obligate intra-cellular parasite. *Eur J Protistol* 1994;30:104–10.
132. Rivera, F., F. Lares, E. Ramirez, et al. Pathogenic *Acanthamoeba* isolated during an atmospheric survey in Mexico City. *Rev Infect Dis* 1991;13(Suppl. 5):S388–9.
133. Rivera, F., F. Medina, P. Ramirez, J. Alcocer, G. Vilaclara, and E. Robles. Pathogenic and free-living protozoa cultured from the nasopharyngeal and oral regions of dental patients. *Environ Res* 1984;33:428–40.
134. Jeansson, S., and T. K. Kvien. *Acanthamoeba polyphaga* in rheumatoid arthritis: possibility for a chronic infection. *Scand J Immunol* 2001;53:610–4.
135. Marciano-Cabral, F., T. Ferguson, S. G. Bradley, and G. Cabral. Delta-9-tetrahydrocannabinol (THC), the major psychoactive component of marijuana, exacerbates brain infection by *Acanthamoeba. J Eukaryot Microbiol* 2001;48:4S–5S.
136. Ferrante, A., and B. Rowan-Kelly. Activation of the alternative pathway of complement by *Acanthamoeba* culbertsoni. *Clin Exp Immunol* 1983;54:477–85.
137. Toney, D. M., and F. Marciano-Cabral. Resistance of *Acanthamoeba* species to complement lysis. *J Parasitol* 1998;84:338–44.
138. Walochnik, J., A. Obwaller, E. M. Haller-Schober, and H. Aspock. Differences in immunoreactivities and capacities to bind human complement subcomponent C1q between a pathogenic and a nonpathogenic *Acanthamoeba*, pp. 59–64. In S. Billot-Bonef, P. A. Cabanes, F. Marciano-Cabral, P. Pernin, and E. Pringuez (eds.), *IXth International Meeting on the Biology and Pathogenicity of Free-Living Amoebae Proceedings.* John Libbey Eurotext, Paris, France, 2001.
139. Ferrante, A., and T. J. Abell. Conditioned medium from stimulated mononuclear leukocytes augments human neutrophil-mediated killing of a virulent *Acanthamoeba* sp. *Infect Immun* 1986;51:607–17.

140. Marciano-Cabral, F., and D. M. Toney. The interaction of *Acanthamoeba* spp. with activated macrophages and with macrophage cell lines. *J Eukaryot Microbiol* 1998;45:452–8.
141. Stewart, G. L., K. Shupe, I. Kim, et al. Antibody-dependent neutrophil-mediated killing of *Acanthamoeba castellanii. Int J Parasitol* 1994;24:739–42.
142. Rowan-Kelly, B., and A. Ferrante. Immunization with killed *Acanthamoeba culbertsoni* antigen and amoeba culture supernatant antigen in experimental *Acanthamoeba meningoencephalitis. Trans R Soc Trop Med Hyg* 1984;78:179–82.
143. Bhaduri, C. R., K. Janitschke, and K. N. Masihi. Immunity to *Acanthamoeba culbertsoni*: experimental studies with *Acanthamoeba* and control antigen preparations. *Trans R Soc Trop Med Hyg* 1987;81:768–70.
144. Culbertson, C. G. The pathogenicity of soil amebas. *Annu Rev Microbiol* 1971;25:231–54.
145. Ferrante, A. Immunity to *Acanthamoeba. Rev Infect Dis* 1991;13(Suppl. 5):S403–9.
146. Masihi, K. N., C. R. Bhaduri, H. Werner, K. Janitschke, and W. Lange. Effects of muramyl dipeptide and trehalose dimycolate on resistance of mice to *Toxoplasma gondii* and *Acanthamoeba culbertsoni* infections. *Int Arch Allergy Appl Immunol* 1986;81:112–7.
147. Ferrante, A. Free-living amoebae: pathogenicity and immunity. *Parasite Immunol* 1991;13:31–47.
148. Stewart, G. L., I. Kim, K. Shupe, et al. Chemotactic response of macrophages to *Acanthamoeba castellanii* antigen and antibody-dependent macrophage-mediated killing of the parasite. *J Parasitol* 1992;78:849–55.
149. Shin, H. J., M. S. Cho, H. I. Kim, et al. Apoptosis of primary-culture rat microglial cells induced by pathogenic *Acanthamoeba* spp. *Clin Diagn Lab Immunol* 2000;7:510–4.
150. Baig, A. M. Granulomatous amoebic encephalitis: ghost response of an immunocompromised host. *J Med Microbiol* 2014;63(12):1763–6.
151. Tanaka, Y., S. Suguri, M. Harada, T. Hayabara, K. Suzumori, and N. Ohta. *Acanthamoeba*-specific human T-cell clones isolated from healthy individuals. *Parasitol Res* 1994;80:549–53.
152. Baig, A. M., and N. A. Khan. A proposed cascade of vascular events leading to granulomatous amoebic encephalitis. *Microb Pathog* 2015;88:48–51.
153. Baig, A. M., Z. Rana, M. Mohsin, T. Sumayya, and H. R. Ahmad. Antibiotic effects of loperamide: homology of human targets of loperamide with targets in *Acanthamoeba* spp. *Recent Pat Antiinfect Drug Discov* 2017;12(1):44–60.
154. Baig, A. M., Z. Rana, S. S. Tariq, and H. R. Ahmad. Bioinformatic insights on target receptors of amiodarone in human and *Acanthamoeba castellanii. Infect Disord Drug Targets* 2017;17(3):160–77.
155. Baig, A. M., and H. R. Ahmad. Evidence of a M1-muscarinic GPCR homolog in a unicellular eukaryote: featuring *Acanthamoeba* spp. bioinformatics 3D-modelling and experimentations. *J Recept Signal Transduct Res* 2017;37(3):267–75.

SUMMARY

As the etiological agents of ocular keratitis and encephalitis in humans, *Acanthamoeba* spp. are potentially deadly in patients with compromised immune systems such as AIDS and recipients of organ transplants. *Acanthamoeba* adhesion at the site of wounds like oral ulcers and eyes and the subsequent invasive infection underscore the pathogenesis of *Acanthamoeba* keratitis and fatal granulomatous encephalitis. The knowledge of *Acanthamoeba* adhesion molecules expressed at different stages of infection serves as next-generation drug targets to prevent and to treat complications of an ongoing *Acanthamoeba* infection. The *Acanthamoeba* genome and transcriptomics databases contribute to the exploration of the adhesion molecules expressed by *Acanthamoeba* spp. The comparison between human adhesion molecules like selectins and integrins and homologous molecules in *Acanthamoeba* lays the foundation for uncovering the adhesion molecules in protist pathogen. Use of siRNA and miRNA to knockout genes of the putative adhesion molecules further clarifies the role of the adhesion molecules. The development of multidrug resistance *Acanthamoeba* infection poses an increasing risk to patients with ADIS and those with organ transplant receiving corticosteroids. Newer drugs targeting *Acanthamoeba* adhesion molecules have the potential become next-generation antiamebic remedies.

Molecular Mechanisms of *Toxoplasma gondii* Invasion

24

Thiago Torres de Aguiar and Renato Augusto DaMatta
Universidade Estadual do Norte Fluminense Darcy Ribeiro

Andréa Rodrigues Ávila
Instituto Carlos Chagas (ICC)

Contents

24.1 INTRODUCTION

Toxoplasmosis is a widespread disease of homeothermic vertebrates including humans that is caused by an obligate intracellular protozoan belonging to the domain Eukarya, kingdom Alveolata, phylum Apicomplexan, class Coccidia, order Eucoccidiorida, family Sarcocystidae, genus *Toxoplasma*, and species *Toxoplasma gondii*.[1,2] It is estimated that one-third of the global population is infected by this protozoan,[1] and the risk factors for toxoplasmosis range from poverty to cultural habits.[3] In most cases, this parasite is transmitted through the ingestion of raw or undercooked meat containing infective forms within tissue cysts. However, oocysts shed by the definitive hosts may contaminate soil and food (vegetables and fruits) and cause outbreaks.[4,5]

The lifecycle of *T. gondii* is complex, with a sexual phase in the definite host of the Felidae family, including the domestic cat, and an asexual phase in the intermediate host that includes virtually all mammals and birds.[6] This parasite presents three basic infective forms: fast replicating tachyzoites in the acute phase of the disease, slow multiplying bradyzoites located in tissue cysts, and sporozoites within the sporulated oocytes that are shed in feces of species of the Felidae family. Most studies with this parasite are performed with tachyzoites, which can be easily maintained in cell culture, allowing genetic manipulation and selection.

One of the key aspects for the establishment of toxoplasmosis is the capacity of the infective forms of *T. gondii* to invade host cells. This invasion is an active process involving the apical pole of the parasite. The infective forms of this parasite are highly polarized, their length is approximately 7 µm, have an arc-shaped body (*toxon* in Greek means "arc"), and an apical pole composed of a conoid where the secretory organelles micronemes and rhoptries discharge their content.[7] The proteins secreted by micronemes (MICs) and rhoptries (rhoptry neck proteins – RONs) are involved, respectively, in parasite attachment to the host cells and formation of the moving junction (MJ) found at the interface of the parasite and the host cell during invasion.[8] After host cell invasion, the parasite is found inside a parasitophorous vacuole (PV) where it develops. Dense granule is another secretory organelle. Proteins from this organelle (GRAs) are secreted after host cell invasion and localize inside the PV space, in the PV membrane, and migrate to the host cell nucleus where gene expression is controlled.[9]

Although it is known today that active invasion of *T. gondii* into host cells is essential for its infectivity, it took a few decades to establish the importance of this active process for the success of the parasite. The first evidence that active penetration was being carried out by tachyzoites of *T. gondii* and that this parasite ends up in the PV came from the pioneer studies by Guimarães and Meyer in 1942.[10] Parasites freshly released can infect the neighboring cells at a fast rate[10] and are oriented with the apical pole pointing toward the host cell plasma membrane.[11] Later, it was shown that the fate of *T. gondii* can be determined by the way the parasite enters into professional phagocytes: survival, if by active penetration, or death, if by phagocytosis.[12]

Active penetration is driven by a gliding motility performed by all infective forms of distinct species from the Apicomplexans phylum. There is a molecular machinery that has been extensively studied using revolutionary molecular techniques that will be described below. This motility is powered by the glideosome, an actomyosin motor located at the pellicle of the parasite. The pellicle is formed by three units of biological membranes, including the plasma membrane, and two opposed inner membranes immediately below. These two opposed biological membranes are from flattened vesicles that originate at the endoplasmic reticulum. These vesicles form plates that are fused together, which can be seen by transmission electron microscopy as double membrane units. These opposed membranes form the inner membrane complex (IMC). Components of the actomyosin motor are located between the plasma membrane and the IMC.[12,13] Active penetration is characterized as a corkscrew-like movement in three-dimensional (3D) environments which powers penetration of infective forms into the host cell, where it replicates asexually within the PV. Eventually, parasites egress from the host cells using the same gliding motility, perpetuating its lifecycle.[14] The actomyosin motor is composed of myosins and a number of connectors, bridges, and enzyme proteins that are anchored to short filamentous actin internally and to the substrate/host cell externally by adhesin molecules that sustain motility (Figure 24.1). This complex system and how it works is further discussed in this chapter. To better understand this molecular machinery, genetic techniques were developed and are presented below.

24.2 METHODOLOGIES

24.2.1 Genetic Evidence for Active Invasion of *T. gondii* into Host Cells

The understanding of *T. gondii* biology has been greatly accelerated by the use of reverse genetic strategies that are based on the disruption or modification of a gene sequence and analysis of the phenotypic effects. In general, methods in reverse genetic aim to manipulate the genome to replace, mutate, or knockout endogenous genes, or to introduce exogenous sequences to a specific location in the genome.

Indeed, the manipulation of the pathogen genome is the most direct way to understand the function of specific genes and the proteins they express, mainly in the postgenomic era. Over recent decades, several protozoan genomes have been sequenced, providing datasets that drastically increased the list of novel candidates to drug targets, vaccines, or essential factors of biological pathways.[15–18] In addition, public databases currently have a high frequency of "hypothetical proteins," creating new challenges to validate their function and improve gene annotation. For this reason, several tools have been developed to identify and characterize the function of promising candidates for a variety of protozoan pathogens.[19–21]

Some functional genomic approaches are limited for some species of parasites[22] but a wide array of experimental tools is available for *T. gondii* experimentation, making it particularly suitable for genetic and biological studies.[23] This parasite is amenable to genetic manipulation mainly due to feasible transfection methods and isolation of considerable amounts of parasites for downstream molecular applications.[21,24]

The variety of experimental strategies available for *T. gondii* include positive and negative selectable markers, integrating vectors, homologous recombination to disrupt genes, nonhomologous random integration, and, more recently, systems based on genome editing by CRISPR/Cas9.[23] Here, we will discuss the principles of those strategies and how it supports the current invasion model proposed for *T. gondii*–host cell interaction (Figure 24.1).

24.2.2 Beginning of Genetic Manipulation in *T. gondii*

T. gondii was the first member of the Apicomplexa phylum to be transfected. Most of the commonly used selectable marker genes for eukaryotic cells are not suitable for selection of stable transformants in *T. gondii* because the parasites do not replicate outside host cells. However, the first systems for the transformation of *T. gondii* tachyzoites were developed based on the susceptibility of the parasite to chloramphenicol. The DNA encoding *Escherichia coli* chloramphenicol acetyltransferase (CAT) was introduced into *T. gondii* using cytomix buffer to electroporate plasmid vectors. CAT was transiently expressed and the addition of chloramphenicol in the culture resulted in the selection of resistant parasites, confirming CAT as an efficient selectable marker.[25] Further, a system for stable transformation was achieved by the introduction of linear plasmids containing CAT fused to *T. gondii* flanking sequences. It resulted in the integration of the *CAT* gene into the *T. gondii* genome, and after selection, stable parasites expressing CAT were obtained.[26]

Integration of foreign DNA in *T. gondii* is relatively efficient,[27,28] and efforts of different groups led to the establishment of stable transformation systems to select transfected parasites, favoring the following

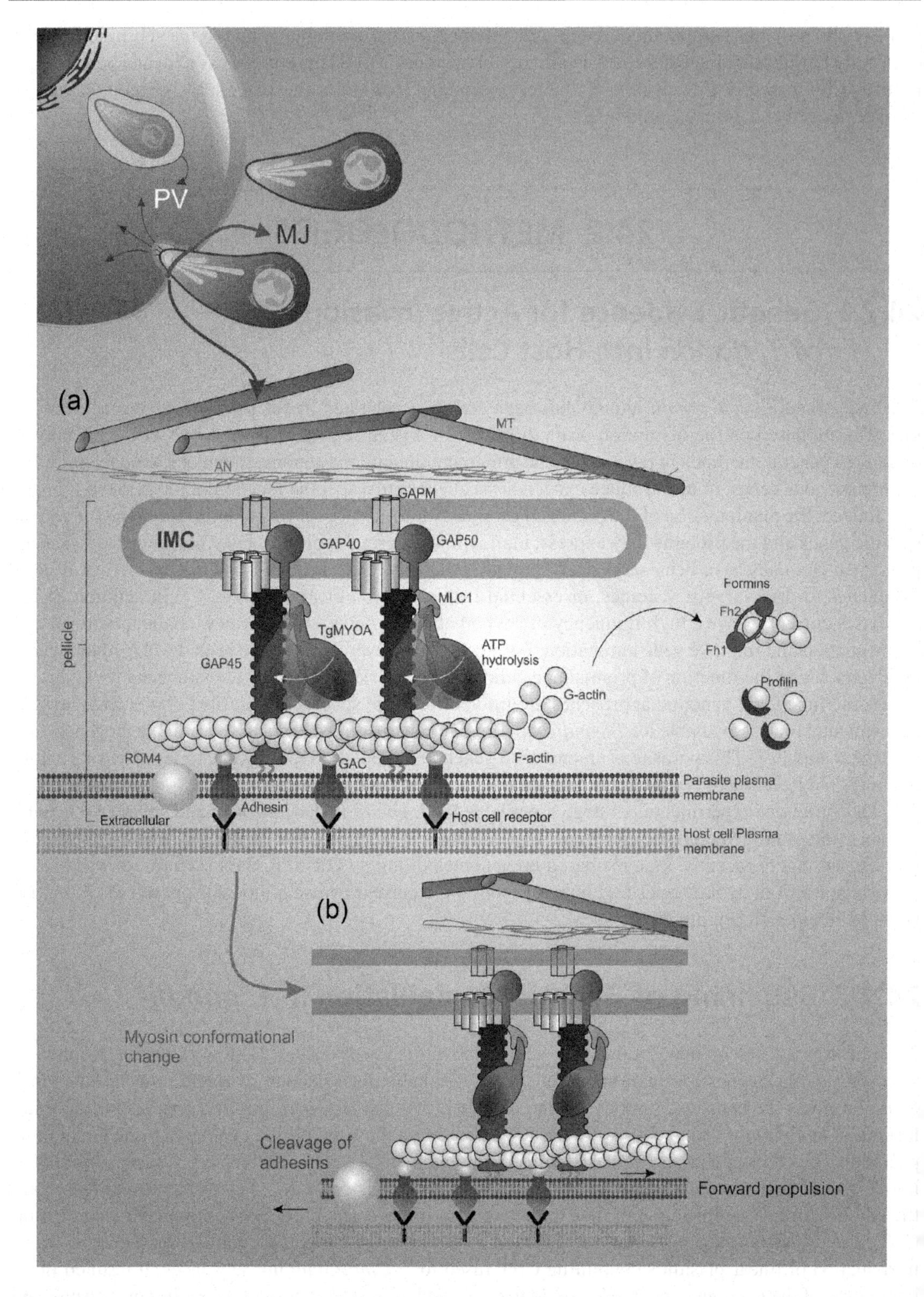

FIGURE 24.1 Molecular invasion model of *Toxoplasma gondii*. (a) Schematic overview of the glideosome machinery in *T. gondii* tachyzoites located in the pellicle, between the inner membrane complex (IMC) and the parasite plasma membrane. After attaching to the host cell and positioning the apical pole toward the host plasma membrane, tachyzoite forms the moving junction (MJ) that anchors both the cells together and

up of phenotypic effects. The establishment of CAT stable transformation rapidly led to the development of tools associated with DNA transformation, and was followed by the elaboration of diverse methods for the selection of stable transformants.[23,26,29–31]

One kind of selection system involves the use of the selectable marker gene that encodes the enzyme hypoxanthine-xanthine-guanine phosphoribosyltransferase (HXGPRT). The knockout of the *HXGPRT* gene resulted in parasites resistant to 6-thioxanthine because HXGPRT converts 6-thioxanthine into an inhibitor of guanosine monophosphate (GMP) synthase which kills the parasites, whereas parasites expressing *HXGPRT* are resistant to mycophenolic acid because this drug efficiently kills parasites lacking this enzyme. Thus, the *HXGPRT* gene can be used as both positive and negative selectable marker for the stable transformation of *T. gondii*.[30,32,33] The transfection and establishment of diverse selectable markers paved the way for molecular studies of different aspects of *T. gondii* biology and several groups started to dissect the function of proteins involved in both gliding motility system[14] and secretory pathway of effector proteins during host cell invasion.[34,35]

24.2.3 Protein Tagging: The First Step for Functional Characterization

T. gondii genome predicts a high proportion of genes unique to the phylum, which lack homologs in other model organisms [http://www.ToxoDB.org]. Finding the subcellular localization of these gene products is an important first step toward their functional characterization. Thus, protein tagging is an interesting approach to determine the fate of proteins when specific antibodies are not available, or to distinguish the protein of interest from endogenous ones. In general, protein tagging is performed by the introduction of a tag (which can be either an epitope sequence or a fluorescent report gene) at the C- or N-terminus of the protein. The protein is expressed conjugated with the tag, which can be recognized by antibodies or fluorescence detection systems (Figure 24.2a). Some peptide sequences were efficiently used as tags for epitope tagging of *T. gondii* genes, such as the Influenza hemagglutinin (HA) tag, the epitope of c-myc sequence,[36,37] or the Ty-1 epitope Tag.[38] For example, studies using gene tagging provided evidence that soluble proteins are escorted by transmembrane proteins to micronemes early in the secretory pathway, as well as to map some of the domains involved in protein–protein interactions and sorting.[38] Protein tagging has been successfully applied to understand the secretory pathway of MJ proteins. In this case, protein tagging distinguished deleted RON proteins from endogenous proteins in a study that showed the mislocalization of specific disrupted RON proteins, providing evidence of new RON proteins that form a high molecular complex required for proper sorting to the rhoptries.[39]

Alternatively, tools for fluorescent protein tagging have been optimized in Apicomplexan parasites. Efforts have been initially made to use the green fluorescence protein (GFP), from the Pacific jellyfish

supports the mechanical force generated by the parasite. A family of proteins, termed gliding-associated proteins (TgGAP40, TgGAP45, TgGAP50), help anchor the glideosome to the IMC and to the parasite plasma membrane, sustaining the integrity of the pellicle. Acting as a lever, myosin light chain 1 (MLC1) sustains myosin A (TgMyoA) and augments movement speed. To support the mechanical traction, actin filaments (F-actin) are formed from a pool of globular actin (G-actin), enabled by augmenters of F-actin turnover, such as formins (Fh1, Fh2), profilins, and coronin. Adhesin molecules secreted by the microneme attach the parasite to the host cell using glideosome-associated connectors (GAC). After sustaining traction, these adhesins are cleaved by rhomboid-like proteases (ROMs) and shed into the extracellular environment. (b) TgMyoA changes conformation after ATP hydrolysis, propelling the parasite forward. The synchrony of these processes enables the parasite to move forward to either glide or to invade the target host cell. After successful invasion, tachyzoites are found in a parasitophorous vacuole (PV), a compartment secluded from the endocytic pathways, which prevents the digestion and destruction of the parasite. After asexual multiplication, tachyzoites permeate the host cell with perforins and rupture the membrane mechanically to perform egress and infect neighbor cells. Other components of the cytoskeleton are depicted: microtubules (MT) and the alveolin network (AN). Figure based on Frénal et al.[14]

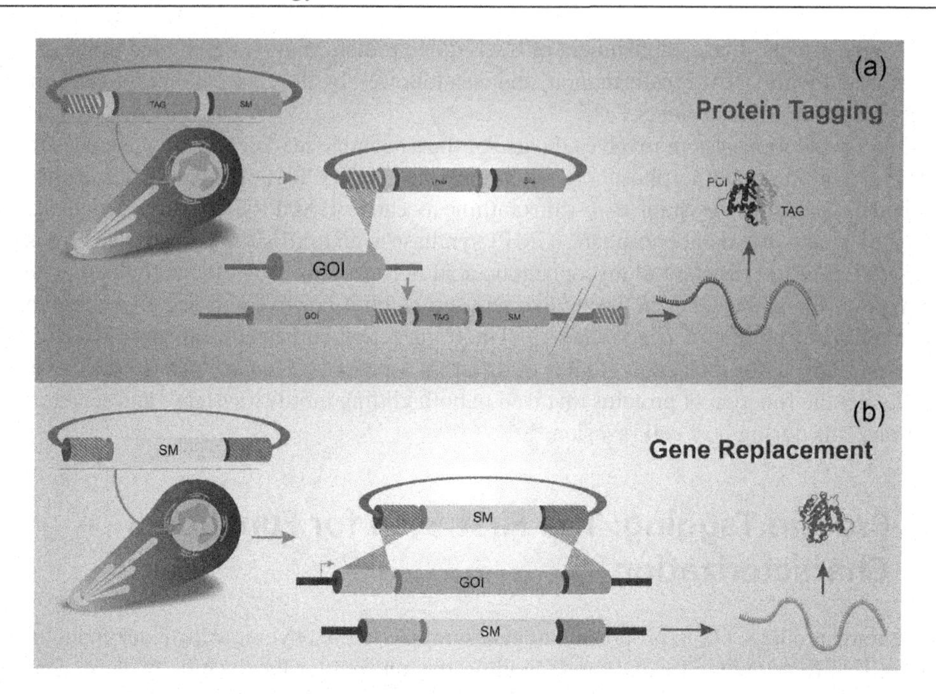

FIGURE 24.2 Genetic manipulation strategies on *Toxoplasma gondii.* (a) The protein tagging strategy is based on the introduction of a plasmid cassette containing the TAG sequence (TAG) and selectable marker gene (SM) in the extremity of the gene of interest (GOI). The selectable marker usually codifies a protein that confers resistance to a specific drug. The resistant parasites express the protein conjugated with the tag. (b) Gene replacement is a traditional strategy to knockout genes by a double homologous recombination leading to the replacement of the GOI by the selectable marker gene (SM) to select the mutant parasites.

Aequorea victoria, as reporter gene because GFP is a naturally fluorescent protein that has been exploited for studies on gene expression and protein tagging in many systems.[40–42] Localization of GFP-tagged proteins were revised by Joiner and Roos,[43] and the first studies in *T. gondii* using GFP-fused proteins identified distinct sets of secretory organelles involved in the invasion and formation of the PV. In addition, it has been demonstrated that color variants of GFP like yellow (YFP), cyan, and red fluorescent protein enable simultaneous detection of different report proteins within the same parasite, optimizing the localization analysis by microscopy.[44]

Importantly, protein tagging strategies are also feasible to genetic screen approaches using a library of plasmids to integrate the *YFP* gene in different regions of the parasite genome, allowing parasites screening by flow cytometry and fluorescence microscopy. Clones expressing tagged proteins in a wide variety of subcellular compartments were isolated and new proteins were identified.[44]

Epitope tagging at the endogenous loci is a suitable strategy to assess the native level of the expression of a gene and the subcellular localization of its product. However, the parasite cannot maintain episomes (nonintegrated DNA), possibly resulting in an inconvenient high rate of random integration of the plasmid into the parasite genome by nonhomologous recombination.[25] Unfortunately, homologous recombination is not favored over nonhomologous repair in wild-type parasite, which has hampered gene targeting approaches in *T. gondii*.[45] To overcome this limitation, the nonhomologous end-joining DNA repair pathway in *T. gondii* was disrupted by the deletion of the *KU80* gene. Indeed, knockout of the *KU80* gene (ΔKU80) in the RH strain (RHΔKU80 strain), a type I strain commonly used in laboratories globally, increased the efficiency of gene targeting via double-crossover homologous recombination at several genetic loci.[46] The RHΔKU80 strain was then used in a system to endogenously tag genes with YFP through a vector consisting of YFP preceded by a ligation-independent cloning (LIC) cassette, which allows the insertion of PCR products containing complementary LIC sequences by a single crossover event.[47] The findings demonstrated that the combination of the ΔKU80 strain and LIC vectors is a time-consuming strategy and reduces the cost required to determine localization of a new gene of interest (GOI), increasing the chances to perform larger-scale studies of novel *T. gondii* genes. The nonhomologous recombination pathway was also abolished in type II strains, which exhibit normal growth rates and behavior, both *in vivo* and *in vitro*, and essentially 100% frequency of homologous recombination.[48]

It is worth mentioning that KU80 is critical to many cellular activities, including chromosome stability. For this reason, ΔKU80 strain is more sensible to double-strand DNA breaks and may not be suitable for protocols that expose the parasite to stresses that causes such breaks, such as chemical or UV mutagenesis,[22] or selection methods that also induces double-strand DNA breaks, such as the use of phleomycin.[49] Nevertheless, both type I and type II ΔKU80 strains are widely used in the field, increasing the success of genetic manipulation, as well as the variety of applications involving gene tagging or gene disrupting.[50]

24.2.4 Disrupting Genes: From Gene Replacement to CRISPR/Cas9 System

Gene disruption at a specific locus can be achieved by a double homologous recombination leading to the replacement of the targeted gene by the selectable marker gene (Figure 24.2b). For example, the GRA2 locus was disrupted by replacement with the selectable marker for phleomycin resistance and the parasites became less virulent in mice. Although Δgra2 mutants grew normally *in vitro*, lower virulence in mice indicates that GRA2 plays an important role during *in vivo* infection.[51] Further studies using Δgra2 strains identified functional domains of the protein responsible for PV formation.[52] Gene replacement was also the method of choice to disrupt the expression of MICs, leading to genetic evidence of the critical role of these soluble proteins in parasite invasion of host cells.[53] However, the classical knockout by gene replacement cannot be used to characterize genes that are essential in a haploid organism such as *T. gondii*. To understand the function of essential genes it is necessary to employ conditional systems that allow a tight and reliable regulation of their activity or the conditional removal of their coding sequences. Currently, inducible systems are available in *T. gondii* to control gene expression at both transcriptional and post-transcriptional levels.

At the posttranscriptional level, gene expression can be controlled by specifically affecting protein stability.[54,55] In one system, the stability of a protein can be regulated by fusing it with a ligand-controlled destabilization domain known as ddFKBP. The addition of the ligand (Shield-1) rapidly stabilizes the protein through binding to ddFKBP, which prevents the degradation by the proteasome (Figure 24.3a). The control of this degradation is fast and suitable to generate dominant-negative mutants, in which the overexpression of genes results in phenotypic effects.[55–60]

More recently, an auxin-inducible degron (AID) system in *T. gondii* was established.[61,62] The auxin system basically depends on two components: the plant auxin receptor 1 and a protein of interest tagged with AID. Auxin activates an ubiquitin ligase complex that targets the AID-tagged protein for ubiquitin-dependent proteasomal degradation (Figure 24.3b). An RH strain expressing plant auxin receptor 1 was generated and essential GOIs were tagged with AID, allowing their expression regulation with auxin. The method was successful to allow conditional knockdown in *T. gondii*, and the protocol to establish the system in other Apicomplexan parasites has been published.[61]

At the transcriptional level, the development of a Tetracycline-transactivator system (Tet-system) was efficient to allow tight regulation of the expression of myosin A (myoA), and the conditional removal of this myosin provided proof of pathogenic function of the actomyosin motor during *in vivo* infection.[63] In this system, a parasite strain expressing a functional transactivator (TATi-1) was generated establishing an inducible system. TATi-1, if not bound to tetracycline, or one of its derivatives, binds to a transactivator responsive promoter (consisting of TetO sequences placed upstream of a minimal promoter), inducing the expression of the GOI. Thus, in the absence of tetracycline, the promoter is active because the transactivator is bound to it, while addition of tetracycline abolishes DNA binding. The random integration of a second copy of myoA (myoAi) under control of the TATi-1 allowed the expression of myoAi in the

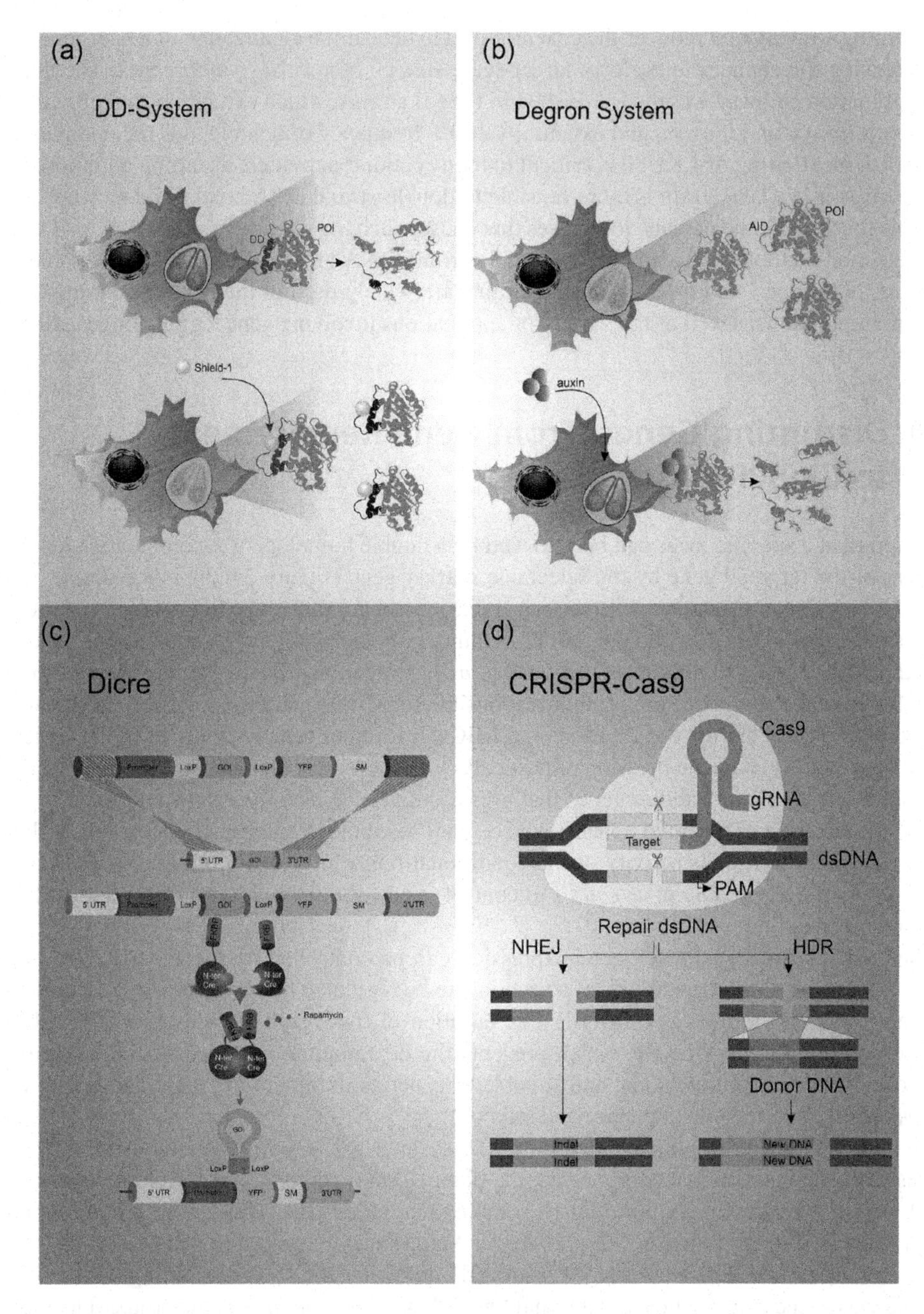

FIGURE 24.3 Examples of inducible systems established in *Toxoplasma gondii*. (a) DD-system allows the control of the expression of proteins of interest (POI) that are tagged with destabilization domain (DD). The presence of DD targets the protein to degradation by the proteasome, while the presence of the ligand Shield-1 prevents the degradation. (b) The degron system is based on the activation of an ubiquitin ligase complex that targets the protein tagged with auxin-inducible degron (AID) for ubiquitin-dependent proteasomal degradation. (c) The dimerizable Cre recombinase (DiCre) system is performed in an RH strain expressing a DiCre. Initially, a gene-swap strategy is applied to replace the gene of interest (GOI) by a cassette containing a promoter, the cDNA of GOI flanked by LoxP recombination sites (LoxP), the yellow fluorescent protein gene (YFP), and selectable marker gene (SM). The drug-resistant parasites can also be isolated by the detection of YFP expression. Further, the selected parasites are used to knockout the GOI by the addition of rapamycin that induces the reconstitution of Cre activity and, consequently, the excision of GOI flanked by LoxP. (d) The

absence of anhydrotetracycline (ATc), and, consequently, the endogenous myoA could be removed by double homologous recombination. Subsequently, the inducible copy was switched off by the addition of anhydrotetracycline, leading to an efficient depletion of myoA and the assessment of its importance in cell infection.

The availability of a parasite strain expressing this transactivator (RHΔKU80 TATi1) allowed the Tet-system to be used in different approaches to generate conditional mutants in *T. gondii*.[64,65] The Tet-system has contributed to identify some of the major proteins involved in host cell invasion machinery. For example, conditional suppression of MIC2 provided the first opportunity to directly determine the role of this protein in gliding motility.[66] However, Tet-system may not achieve the complete downregulation of gene expression, and residual expression might mask the phenotypes associated with the GOI.[22,23,67]

Although the systems that control gene expression have been useful to generate relevant data regarding the mechanisms of *T. gondii* invasion, the residual expression of the gene or phenotypes caused by dominant-negative mutants may lead to different interpretations, and care is necessary when analyzing the phenotypes caused by inducible systems.[67] Thereby, only null mutants (expression of a gene is completely abolished) can unequivocally demonstrate the function of a protein. Nevertheless, the traditional knockout using gene replacement tools is not feasible to study essential genes in *T. gondii*, and the development of conditional knockout has brought light to these issues.

In mammals, null mutants can be generated by a conditional system of site-specific recombination based on Cre-LoxP technology that allows temporal excision of DNA sequences flanked by LoxP sites by the recombinase Cre.[68–70] Excision of DNA sequence between two LoxP sites by Cre recombinase was first established in *T. gondii* without any kind of control.[71] Further, knockout mutants of invasion factors were generated based on the regulation of Cre recombinase,[72] which was based on a system previously established.[73] In this system, Cre is split into two inactive fragments and each fragment is fused to a rapamycin-binding domain (FKBP12 and FRB), allowing the enzyme to be efficiently heterodimerized by rapamycin. Addition of rapamycin forms a dimerizable Cre recombinase (DiCre), leading to the reconstitution of Cre activity and, consequently, the excision of genes flanked by LoxP. The DiCre system was further applied to generate a gene-swap strategy to remove cDNA of GOI flanked by LoxP sites.[72] First, the endogenous copy of GOI is replaced by a cassette containing the cDNA of GOI flanked by LoxP site, the YFP gene, and the selectable marker gene by homologous recombination in a strain expressing DiCre (Figure 24.3c). Subsequently, the addition of rapamycin in the culture activates the removal of the cDNA by Cre-recombinase-mediated recombination. This also results in YFP-expressing parasites, allowing the selection of null mutant parasites by the expression of YFP. This new strategy generated conditional knockout mutants for the invasion factors: myoA (gliding motility), MIC2 (microneme), and act1 (motor).[72]

Intriguingly, the data obtained by the conditional knockout of myoA and MIC2 generated by DiCre indicates that the parasite is capable of employing alternative invasion mechanism(s), as those proteins are shown to be dispensable for host cell invasion, contradicting previous results of knockdown generated by the Tet-system that described both proteins as essential for host cell invasion.[72] Similarly, the mutants obtained by the inactivation of *ama1* gene mediated by DiCre system are still capable of penetrating the host cell,[74] although previous work using the Tet-system proposed that apical membrane antigen 1 (TgAMA1) is essential for host cell invasion.[75,76] In addition, studies using tools to generate knockout mutants for different rhomboid-like proteases (ROMs) indicate that the processing of these proteins is important but not essential, emphasizing that essential processes such as invasion typically relay on redundant pathways to ensure survival.[77] Some authors highlighted the occurrence of residual expression using the DiCre inducible method that can lead to misinterpretation on phenotypes,[78] but the isolation of

disruption of genes based on CRISPR-Cas9 system is triggered by the annealing of single-guide RNA (gRNA) that targets the GOI (target) in the double-strand DNA (dsDNA). The complex gRNA+dsDNA directs Cas9 to the specific genomic location. Cas9 induces double-strand breaks and, consequently, two different recombination pathways can mediate dsDNA repair. The nonhomologous end-joining DNA repair pathway (NHEJ) can introduce insertions or deletions (Indel), causing disruption of target gene. The homology direct repair pathway (HDR) inserts a "donor DNA," a new DNA sequence that leads to gene disruption.

both viable null mutants and invasive null mutants supports the idea that host cell invasion is more complex than previously thought and may employ alternative mechanisms.

Recent applications of CRISPR/Cas9 system to manipulate the genome of several organisms, including protozoan parasites, are creating exciting new opportunities to understand gene function and reveal important biological insights.[19,79] The chronology of CRISPR discovery and its adaptation for genome editing was revised elsewhere.[80] The first study for genome editing based on CRISPR/Cas9 in an eukaryotic pathogen was reported in *T. gondii* in 2014,[81,82] and excellent reviews that discuss the application of CRISPR/Cas9 as well as illustrate the various strategies in Apicomplexans have been published recently.[21,83]

CRISPR-Cas9 is an adaptive immune system used by several bacteria to protect them from viruses, and it involves different stages that allow bacteria to get rid of foreign DNA elements. Basically, two components are responsible for triggering the system: CRISPR that consists of a family of DNA repeats (Clustered Regularly Interspaced Short Palindromic Repeats) and Cas proteins (CRISPR-associated proteins) that carry typical domains of nucleotide-binding proteins, and their genes (cas genes) are adjacent to CRISPR.[84,85] Thus, when foreign DNA sequences are inserted into the CRISPR locus, the cas genes are expressed and the CRISPR array is transcribed and processed into mature CRISPR RNA duplex (crRNA). The target DNA is recognized by the complex crRNA and Cas and foreign DNA is cleaved.[86,87] Consequently, two different recombination pathways can mediate DNA repair that will generate either mutations by insertions or deletions (Indel) or the insertion of a new sequence provided by a DNA donor (new DNA), leading to gene disruption (Figure 24.3d).

Different CRISPR/Cas systems have been described,[88–90] and the type II system is the most commonly used for genome editing in many organisms.[91–93] The tool based on type II system requires an endonuclease (Cas9) and a crRNA that targets DNA, and have been adapted in Apicomplexan parasites.[83] The first study involved an engineered vector to express the RNA duplex called single-guide RNA (sgRNA) and Cas9 fused to a nuclear localization signal and GFP (Cas-NLS-GFP), which facilitated site-specific insertion in the absence of homology.[81] Using CRISPR/Cas9, the group generated gene rop18 knockouts in the type I GT1 strain, a natural isolate, revealing that ROP18 contributed greatly to acute pathogenesis in GT1 strain than in the laboratory-type I RH strain. One month later, another group described the generation of knockouts and epitope tags into genome using RNA-guided Cas9 nuclease in *T. gondii*.[94] In this work, the authors used strategies to transfect parasites of diverse strains with or without a DNA donor to induce different mediated DNA repairs, allowing homologous and nonhomologous recombination, respectively. Thus, CRISPR/Cas9 is a versatile methodology that opened the possibility of performing genetic manipulation in any *T. gondii* strain[83] and protocols are available to the scientific community.[82]

Different works have been describing functional studies of proteins as the first description of CRISPR/Cas9 tool in *T. gondii*,[62,82,95–97] and this technology led to the first genome-wide genetic screen of an Apicomplexans.[98] In this work, the CRISPR/Cas9 tool was adapted to assess the contribution of each gene of *T. gondii* during infection of human cells, leading to identification of essential genes for invasion, including the claudin-like Apicomplexa microneme protein (CLAMP), using sgRNA libraries to perform genome-wide knockouts in a parasite population. Such approaches allowed measuring the fitness contribution of every gene in the tachyzoite genome and this contribution is now available (www.toxodb.org).

One aspect of Cas9 is the impact of genomic off-targeting due to breaks in genomic sites other than the specific sgRNA. However, two highly specific Cas9, eSpCA9 and Cas9-HF, were engineered to significantly reduce the possibility of producing off-targeting, and most of the CRISPR/Cas9 systems are now being adapted to include those high-fidelity versions.[99–101] In Apicomplexans, CRISPR assays were performed using the original version of Cas9, which might be a major concern in these parasites, considering their small genomes. However, there is no evidence of off-target mutations introduced by Cas9 in *Plasmodium* parasites,[102,103] probably due to the absence of the nonhomologous pathway end-joining, which makes the system very specific in these parasites. In addition, the off-target effects are not explored in *T. gondii* that have the nonhomologous end-joining pathway as a major repair mechanism,[83] and we cannot still affirm the real effects in *T. gondii*.

Overall, most of the cutting-edge technologies and strategies for genome manipulation that are developed in other organisms have been successfully adapted for *T. gondii*, confirming that this parasite is very amenable to genome editing. The fast publication of new studies using the high-fidelity version of Cas9 to improve functional genetics in relevant human pathogens would not be a surprise.

24.3 RECENT ADVANCES

Motility in *T. gondii* tachyzoites is performed by the glideosome. This complex set of proteins includes adhesive proteins, connectors, and myosins (Figure 24.1). Adhesive proteins adhere the parasite to the substrate/host cell, while the connectors anchor the glideosome machinery to the parasite cytoskeleton. This allows the machinery to function correctly and the conformational change of the myosin head after ATP hydrolysis provides the mechanical force for invasion. Another set of molecules regulate motility by modulating the turnover of globular (G-actin) to filamentous actin (F-actin) (Figure 24.1a). Additionally, microneme secretion, crucial for motility, is regulated through different pathways. The invasion model is discussed below.

24.3.1 Early Evidence of Invasion Models

In Apicomplexan parasites, such as *T. gondii*, gliding motility is intrinsically related to host cell invasion.[14,104] This has been demonstrated by knowledge accumulated through recent decades. The glideosome, responsible for Apicomplexan motility, plays a pivotal role in cell invasion in these parasites.[105] This complex structure is responsible for three distinct types of locomotion: circular gliding, twirling, and helical rotation,[106] which are observed in two-dimensional environments (*in vitro*). In 3D substrates, such as Matrigel and live tissues, *T. gondii* tachyzoites perform a corkscrew-like movement, with the characteristic crescent shape of the parasite playing a major role in this type of motion.[107] The gliding movement is substrate-dependent, and is different from the motility observed in cells that possess flagella.[108]

Host cell invasion was previously described in *T. gondii*[10] and in *P. berghei yoeli* and *P. gallinaceum*[109] in early 1960s and further investigated in *P. knowlesi* in 1978.[110] An early capping model has been described as responsible for the motility of Eimeria sporozoites.[111] In this model, surface ligands are secreted by the parasite, and through a submembranous microfilament system, these ligands are capped from the anterior to the posterior pole of the parasite, which results in locomotion. This model was extrapolated across the Apicomplexan phylum. These parasites direct their apical pole toward the host cell plasma membrane before invasion. The region of contact between the parasite and the host cell forms a depression in the host cell plasma membrane, which results in the formation of the MJ (Figure 24.1). During the invasion process, the host plasma membrane does not disrupt, allowing the parasite to penetrate the host cell and to be kept inside a forming PV.[110] The process of cell penetration was also observed in the Apicomplexan *Isospora canis*.[112] Similarly, the host cell membrane is not disrupted, and the content of rhoptries and micronemes are discharged during host cell invasion, as noted by the presence of partially empty organelles in parasites that recently invaded host cells.[112] In fact, Luo and colleagues found a wide array of proteins that are secreted by *T. gondii* to enable motility and invasion. Most of these proteins are MICs, which support the importance of micronemes for Apicomplexans in general. Additionally, the endoplasmic reticulum is the organelle responsible for the emergence and folding of most of these proteins.[113]

Initially, it was verified that invasion was dependent on the secretion of proteins (MICs and unidentified rhoptry proteins), the presence of the MJ, and the parasite's apical pole directed toward the host cell plasma membrane, indicating an active process. Further research allowed the dissection of the components leading to a description of a molecular machinery that will be briefly reviewed in subsequent sections.

24.3.2 Cytoskeleton and Micronemal Proteins

First evidence supporting the role of cell motility in parasite invasion arose when myosin was blocked with butanedione monoxime, a myosin ATPase inhibitor, which impaired *T. gondii* cell invasion.[114] Dobrowolski and Sibley demonstrated that the parasite actin is necessary for active penetration into the host cell.[115] Cyt-1 host cells, which are resistant to cytochalasin (a drug that disrupts actin microfilaments), were infected by *T. gondii* in the presence of this drug. In these infection assays, the rate of infection decreased, indicating the importance of the parasite actin in the infection of the host cell. Furthermore, a cytochalasin-resistant *T. gondii* mutant strain was obtained through chemical mutagenesis induced by ethyl nitrosourea and selective culture. This mutant strain was used to infect host cells in the presence of cytochalasin, and the infection ratio was higher than that of wild-type (WT) parasites, indicating that an actin-based contractile system present in the parasite is important for host cell invasion. Cytochalasin impairs host cell invasion and gliding motility of WT *T. gondii*, but not of cytochalasin-resistant parasites.[115] In fact, cytochalasin also prevents conoid extrusion, a key step to motility.[116] Using DiCre system to knockout MIC2, myoA and actin genes in *T. gondii* (RH strain) have impaired invasion of host cells. Some parasites can still invade cells, but the lack of expression of actin prevents the replication of apicoplasts,[117] an essential relict organelle for Apicomplexans' survival.[118,119]

Actin is a protein that exists in globular and filamentous forms. G-actin polymerizes into F-actin that can change the cell shape, enabling endo and exocytosis and cell motility.[120] The actin network creates a structure that provides mechanical stability for the cell to migrate on a substrate. In Apicomplexans, actin polymerization is regulated differently than most organisms.[14] This protein is kept mainly in its globular form,[121] scattered around the cytosol, forming pools in both poles of the parasite. Apicomplexans lack an important regulator of actin polymerization, the actin-related protein 2/3 (Arp2/3) complex,[122] which plays a crucial role in actin nucleation/polymerization in most eukaryotic cells.[123] Instead, Apicomplexans have a set of proteins to regulate actin turnover. In *T. gondii*, formins (TgFRM) 1 and 2 nucleate actin monomers, promoting the growth of actin filaments. TgFRM1 and 2 are expressed in the pellicle.[124] TgFRM1 is located at the apical pole of the parasite.[125] Later, a third formin, TgFRM3, was also characterized. TgFRM3 has a lower molecular weight than TgFRM1 and 2, is expressed more diffusely, and is not necessary for *T. gondii* survival.[126] FRMs possess common regulatory domains known as "formin homology domains FH1 and FH2." The knockout of TgFRM1 impairs invasion and cell motility,[124] which supports the important role of actin polymerization in both scenarios. Profilin is another important protein in actin homeostasis. Together with formin, both can increase the rate of elongation of actin filaments.[120] The disruption of profilin expression in *T. gondii* (TgPRF) using the Tet-system impairs its capacity to infect host cells, and makes this parasite nonlethal to mice, unlike its nonmutant counterpart.[127] Another actin-binding protein that regulates actin polymerization is coronin (COR), a protein that regulates actin dynamics in different cell types and organisms.[128] In *T. gondii*, TgCOR has a WD40 conserved domain. This protein binds to actin and increases the rate of polymerization. Although TgCOR is not essential for invasion and egress of host cell, it contributes to these processes.[129]

Actin polymerization needs to be precisely regulated during *T. gondii* motility and host cell invasion (Figure 24.1a). Treating tachyzoites with jasplakinolide, a drug that stabilizes and promotes actin polymerization,[130] impairs invasion and cell motility.[131] Actin depolymerizing factors (ADF) are ubiquitous eukaryotic proteins that regulate actin dynamics.[132] In *T. gondii*, TgADF knockdown with Tet-system renders tachyzoites with erratic motility and impaired invasion and egress. Furthermore, more F-actin was detected in knockdown parasites,[133] which can be explained by the fact that TgADF regulates actin turnover from F-actin to G-actin by sequestering actin monomers in actin-rich environments.[134] This bundling of actin monomers generates pools of readily polymerizable actin, necessary for the rapid turnover that enables fast Apicomplexans motility.[135] G-actin polymerization into F-actin is a tightly regulated process in *T. gondii*, necessary for gliding and host cell invasion, and even cell survival.

Another key player for host cell invasion by *T. gondii* is the secretory organelle microneme. Thrombospondin-related anonymous protein (TRAP) is an adhesin protein stored in micronemes and secreted at the apical pole of species of the *Plasmodium* genus. Upon its secretion, it is translocated to the parasite surface membrane. Through its binding capacities, TRAP attaches firmly to ligands, and through interaction with the parasite molecular motor, it is translocated backward, propelling the parasite forward.[136] The extracellular domain of TRAP contains two matrix and/or cell-binding portions.[137] The amount of thrombospondin family of proteins present at the surface of Apicomplexan members is usually low, but increases when the parasite needs tight attachment to the host cell.[136] *P. berghei* mutants that do not express TRAP have impaired capacity to invade mosquito salivary gland cells and rat hepatocytes.[137] Additionally, mutant sporozoites have impaired gliding motility.[137] A TRAP homolog, MIC2, is found in *T. gondii*.[138] As described above, MIC2-deficient *T. gondii*, obtained using a transcriptional Tet-transactivator, is less efficient in attaching and invading host cells, and are less lethal to mice.[139] To enable further motion, thrombospondin family proteins must be released after being translocated backward from the apical pole.[136] The C-terminal domain of the complex MIC2-M2AP (MIC2-associated protein, discussed below) is cleaved by the microneme protein protease (MPP)-1. Afterwards, the N-terminal domain is cleaved by MPP2, and the protein complex is released into the extracellular environment.[140]

Secretion of MIC2 by micronemes in *T. gondii* is regulated by intracellular Ca^{2+} ions, but not by extracellular Ca^{2+} ions.[141] It is a temperature-dependent process that occurs between 25°C and 40°C.[8] When tachyzoites are treated with ionomycin to raise intracellular Ca^{2+}, MIC2 is detected on culture supernatant through Western Blot. The discharge might be prevented when parasites are pretreated with an ion chelator. However, Ca^{2+} does not induce rhoptry discharge. Afterwards, MIC2 released by the microneme on the anterior pole of the parasite is translocated backward to the posterior pole.[141] An accessory protein, M2AP, is necessary for the proper translation, assembly, and functionality of MIC2. Although the knockout of M2AP impairs the parasite's capacity to invade host cells, the invasion is not completely disrupted. This suggests that other MIC complexes play a role in *T. gondii*–host cell invasion. This also suggests that the invasion mechanism is redundant, with alternate mechanisms that compensate when necessary. MIC proteins might have evolved to recognize and attach to different host cell receptors and molecular motifs, enabling *T. gondii* to invade a greater selection of cells.[53] One of these motifs might be sialic acid, a glycan ubiquitously found in the glycocalyx of a wide array of vertebrate cells. This motif is used by Apicomplexans as a binding site for cell attachment and invasion.[142]

24.3.3 Myosins and Adhesins

T. gondii myosin A (TgMyoA) has properties similar to fast muscle myosins, probably limited to the rate of ATP cleavage and ADP release. Myo undergoes conformational changes at the head portion of the molecule after ATP hydrolysis[143] (Figure 24.1a and b). It associates with *T. gondii* myosin light chain 1 (TgMLC1) and a myristoylated protein (TgMADP) to form the TgMyoA complex.[144] TgMyoA requires a species-specific chaperone, TgUNC, to fold the protein correctly. Furthermore, an essential light chain protein (TgELC1) is part of the TgMyoA complex and together with TgMLC1 enables fast actin movement. These two components locate at the C-terminus of the heavy chain of TgMyoA lever portion, increasing actin movement speed by extending the length of the lever arm.[145] It is speculated that TgELC1 shares homology with calmodulin, which suggests that TgMyoA motor complex responds to Ca^{2+} signaling.[146] Additionally, the modification of TgMLC1 impairs parasite motility, demonstrating its importance in the TgMyoA complex.[147] Besides TgMyoA, a family of small myosins found in Apicomplexans, known as myosins XIV,[105] includes TgMyoB and TgMyoC, along with two types found exclusively in bradyzoites, TgMyoD and TgMyoE.[148] TgMyoA is permanently anchored to the plasma membrane through TgMADP105,144 and is distributed throughout the entire cytoplasm, including both poles.[114]

Other components of the TgMyoA complex include the gliding-associated proteins (GAP) (Figure 24.1a). TgGAP50 anchors the glideosome to the IMC.[149] TgGAP45 is anchored to the plasma membrane and to the IMC. The C-terminal region of TgGAP45 recruits MLC1-myoA to the IMC, while the N-terminal acylation and coiled coil domain preserve the pellicle during invasion. TgGAP45 might help maintain a predetermined distance between the plasma membrane and the IMC. Furthermore,

TgGAP45 knockout parasites using the Tet-system have impaired gliding, invasion, and egress capabilities, although it does not seem to affect parasite replication. The same authors describe a fifth component of the glideosome, the TgGAP40, which might also act as an anchor due to its numerous transmembrane domains. Therefore, the glideosome complex might be composed of five elements: TgMyoA, TgMLC1, TgGAP45, TgGAP50, TgGAP40.[150] However, another set of proteins, that seem to have similar anchoring roles, have been described; this family of proteins is termed glideosome-associated proteins with multiple-membrane spans (TgGAPM) and seems to interact with TgGAPs, the glideosome, the IMC, and the cytoskeleton.[151] While GAP45 is found across the Apicomplexan phylum, GAP70 and GAP80 are restricted to coccidians. Both proteins share similarities with TgGAP45. While TgGAP70 is found in the apical pole of *T. gondii*, TgCAP80 is found at the basal pole. At the basal pole, TgGAP80 is a member of the MyoC-glideosome complex, while TgGAP70 is a component of the apical MyoA-glideosome complex. IMC-associated protein 1 (IAP1) is responsible for the localization of the MyoC-glideosome complex to the posterior pole.[152] TgMyoA depleted tachyzoites are less capable of invading host cells and have impaired gliding motility and egress capacity. However, this depletion does not affect parasite growth.[153] TgMyoH is a novel class of myosin associated to the conoid, and is likely the initiator of the process of motility, acting together with TgMyoA. TgMyoH was later acknowledged as a fundamental player in motility, cell invasion, and egress.[154]

The invasion process of the host cell requires interaction of adhesive ligands from the parasite with host cell receptors. The adhesive complexes form clusters to increase host cell–parasite affinity, and may function to hide potential antigenic epitopes from immunity responders.[155] MIC proteins (TRAP family homologs) are secreted by micronemes. MICs have adhesive motifs that mediate protein–protein and protein–carbohydrate interactions, and have a crucial role during the invasion process in *T. gondii*, including host cell attachment. Additionally, recombinant TgMIC2 binds to laminin and collagen. Other key MIC proteins are the TgMIC3 and TgMIC4, which have been shown to be secreted and relocalized during host cell invasion. TgMIC6 and TgMIC8 appear to function as escorters for TgMIC1, TgMIC3, and TgMIC4. MICs are capped to the posterior end of the parasite after forming connections between the host cell and the actomyosin motor.[156] Surface antigen 1 (SAG1) is another protein involved in adhesion. It is expressed at the surface of *T. gondii* and has been considered one of the most promising candidates for vaccines, as it elicits efficient immunological response. Indeed, in some mice models, SAG1-based vaccines, conjugated with other MICs, were capable of reducing brain tissue burden and mortality. The blockage of SAG1 with antibodies in tachyzoites impairs adhesion.[157–159]

Another adhesin conserved across Apicomplexans is AMA1. TgAMA1-depleted tachyzoites, using the conditional Tet-system, are able to move, extrude the conoid, and secrete MICs, but fail to establish intimate contact with the membrane of target cells. This suggests that TgAMA1 functions as a secondary adhesin, presumably together with other TgMICs. The critical role of TgAMA1 in host cell invasion is evident when knockout tachyzoites are less able to invade.[160] The residual invasion observed in TgAMA1-depleted parasites might be possible due to redundant mechanisms present in the parasite, such as the AMA1-like proteins AMA2 and AMA4.[161] However, another article demonstrates that *AMA1* deletion, by DiCre-LoxP recombination, in *T. gondii* tachyzoites and by direct homologous recombination in *P. berghei* merozoites and sporozoites, impairs host cell invasion and binding, although these parasites are still able to invade. Further, the depletion of AMA1 does not seem to affect tachyzoite motility. AMA1 seems to be important in aligning the parasite in relation to the target cell.[74] To sustain the MJ, when AMA1 is complexed with RON2, it becomes less susceptible to rhomboid processing. Furthermore, AMA1 may act as an outside-in signal transducer as the cytosolic tail is dephosphorylated when it binds to RON2.[162]

Adhesins, such as TgMIC2 and AMA1, need to connect to the glideosome to be able to transduce the mechanical effort into movement. Initially, it was suggested that aldolase played this role. Later, it was demonstrated that this protein is conjugated to TgMIC2. Additionally, aldolase binds to F-actin, but it also plays a key role in glycolysis.[14,163] After demonstrating that *T. gondii* tachyzoites depleted of aldolase had impaired motility and invasion capacity, further investigations showed that this behavior could be due to incomplete glycolysis. When tachyzoites were grown in glucose-depleted media, no such behavior was observed, which raised doubts about the role of aldolase in glideosome association.[14] Later, a glideosome-associated connector (TgGAC) was unveiled as the key player connecting adhesin (in this case, TgMIC2) to the actomyosin motor (binding and stabilizing F-actin). TgGAC relocalizes from the apical to the basal pole of the parasite in a similar manner to myosins and adhesins.[125]

24.3.4 Cleavage of Adhesins, Signaling, and Aftermath

A gradient of adhesins seems to be responsible for the correct positioning of the parasite in relation to the host cell when facing the host cell membrane with its apical pole.[14] After performing its role and being translocated rearward from the apical pole, adhesins need to be cleaved for the parasite to move forward (Figure 24.1b). ROM4 has been implicated to play a role in this event. TgROM4-depleted tachyzoites, engineered with conditional Tet-system, adhere to the target cell but fail to form the MJ, a crucial structure in the invasion process. It also exhibits impaired motility and accumulates adhesins in the membrane surface, specifically MIC2.[164] The role of TgROM4 in shedding MICs is evident when TgROM4-depleted tachyzoites are found adhered to the host cell, but not performing invasion. *T. gondii* tachyzoites possess a set of ROMs, including ROM4, ROM1, and ROM5 that are important, but not essential, for host cell invasion. This set of proteases is not required for penetration but are necessary for efficient adhesion. ROM4 is the most prolific protease among this set and appears to be responsible for the cleavage of MIC2 and AMA1, although ROM5 is responsible for the cleavage of AMA1 in ROM4-depleted parasites.[77,165]

MPP is a family of rhomboid proteases that contribute to the processing and shedding of the MIC2-M2AP complex, although their identity is only speculated.[166] Following a proposed two-step process, MPP2 and MPP3 proteolytically act over MIC2-M2AP complex. Afterwards, the integral membrane protease MPP1 sheds MIC2-M2AP into the extracellular environment.[167] It is speculated that MPP1 might in fact be TgROM4.[166] Another set of proteases that trim adhesive complexes from *T. gondii* surface include subtilisin-like serine protease (TgSUB1). Knockdown of this protease decreases motility efficiency and invasion rates.[168]

During host cell invasion, *T. gondii* actively excludes host transmembrane proteins from the forming PV membrane. Hereafter, *T. gondii* resides in the PV that is a nonfusogenic vacuole secluded from the cellular endocytic pathway.[169] This prevents the destruction of the parasite by lisosomal digestion.[170] This exclusion process occurs in the MJ and is likely responsible for not allowing the fusion of the forming PV to the endocytic and exocytic pathways.[171] Distinct vacuoles are formed after host cell invasion. Live tachyzoites reside in a tight vacuole, while opsonized or effete tachyzoites are phagocytosed into loose vacuoles. In some cases, phagocytosed tachyzoites are able to escape from loose vacuoles that are targeted for digestion and reside in a new vacuole that is not fused with other endosomes/lysosomes.[12]

The glideosome might also play a role in parasite egress from the host cell,[172] although the use of mechanical force might not be required for egress.[173] Through cell division, tachyzoites may create an internal pressure that possibly ruptures the host cell and allows egress.[173] This is contradictory to an earlier suggestion that the glideosome plays a role in egress as cytochalasin-treated tachyzoites were unable to egress from host cell.[172] Together with the membrane pore-forming molecule perforin-like protein 1[174] and host cell calpain-1,[175] the parasite motility ruptures the PV and the host cell membrane through mechanical force.[14] The rupture of the PV, that precedes the dismantlement of the host cell membrane, releases Ca^{2+} into the host cell cytosol, which might be the activator of calpain.[176]

During invasion, *T. gondii* is intimately associated with the host cell plasma membrane by the MJ where tachyzoites secrete toxofilin, an actin-binding protein that promotes host cell actin disassembly, disrupting the host cell cytoskeleton, which facilitates invasion.[177] The MJ is characterized by a distinct host cell plasma membrane reorganization, forming a ring-like structure, plus the clustering of RON-2. This protein is secreted from the rhoptry into the extruded conoid, binds to the host cell plasma membrane, and anchors the parasite, possibly by binding to AMA1 at the parasite end.[178,179] Together, AMA1-RON2, RON4, RON5, and RON8 delineate the MJ, a site at which the cytoskeleton of the parasite and the host cell are connected, and enable the application of traction to help invasion.[14] Additionally, target cell Arp2/3 complex and cortactin, a nucleation promotion factor that interacts with Arp2/3, are recruited to the MJ to promote actin polymerization, thus enabling the parasite anchoring to the host cell. The MJ might function as a bridge connecting the scaffold of actin in both the parasite and the target cell.[180] This becomes more evident when it is shown that RON proteins recruit a series of connector proteins in the host cell (ALIX, TSG101, CIN85, and CD2AP) that are known to interact with actin and tubulin.[181] Indeed, *P. falciparum*, after the formation of the MJ, secretes rhoptry proteins into the attached erythrocyte, which could act to promote changes necessary for successful invasion.[179]

After invasion, the PV involving *T. gondii* buds from and is primarily composed by the host cell plasma membrane. This process maintains the plasma membrane sealed and intact.[182] After constricting through the MJ and invading the host cell, tachyzoites twist causing the new formed PV to be severed from the cell membrane.[183]

Parasites that reside in the PV and have replicated respond to the rupture of host cell membrane a few minutes prior to egress: it leads to a decrease in host cell K^+ levels, which triggers an increase of Ca^{2+} in the parasite cytoplasm through phospholipase C activity. Two pathways play a role in this signaling: (a) the calcium-dependent, but calmodulin-independent, activation of protein kinase (TgCDPK1) and (b) calmodulin-dependent protein phosphatase calcineurin activation.[172] It has been demonstrated that microneme secretion depends on Ca^{2+} signaling. This was further demonstrated using a conditional Tet-system to abrogate expression of TgCDPK1, which depends on Ca^+, resulting in impairment of the gliding motility and invasion, as expected.[74] Besides egress, gliding motility, conoid extrusion, and host cell invasion is influenced by Ca^{2+} signaling. Upon contact with the host cell, the levels of intracellular Ca^{2+} usually increase, and the use of a Ca^{2+} chelator inhibits invasion,[184] corroborating evidence that invasion depends on Ca^{2+} signaling. Additionally, culture of tachyzoites in medium containing Ca^{2+} promotes conoid extrusion.[116]

Apicomplexans have a wide variety of genes related to Ca^{2+} signaling. As stated before, micronemal secretion of adhesins is regulated by Ca^{2+}. Apicomplexans have Ca^{2+} storages, such as the endoplasmic reticulum, Golgi apparatus, mitochondria, acidocalcisomes, food vacuoles in *Plasmodium* spp., and a plant-like vacuole in *T. gondii*. Acidocalcisomes in *T. gondii* possess a Ca^{2+}-ATPase for Ca^{2+} uptake that has been termed TgA1. The importance of calcium signaling in *T. gondii* is evident as tachyzoites with *TgA1* disrupted with a plasmid have impaired invasion capacity, growth rate, micronemal secretion, and *in vivo* virulence. These compartments are maintained acidified by proton pumps, which helps maintain the Ca^{2+} levels, as the use of alkalinizing agents trigger Ca^{2+} release into the cytoplasm.[185,186] In addition, motility might be triggered by acidification of the environment that triggers micronemal secretion dependent of Ca^{2+}.[187] Another calcium-dependent pathway involves DOC2.1 protein, possibly downstream of CDPKs pathway. This protein mediates microneme exocytosis by regulating membrane fusion in a Ca^{2+}-dependent manner.[188]

Apart from calcium signaling, a broader variety of proteins and molecules can also regulate Apicomplexans motility in a Ca^{2+}-independent manner. For example, a cGMP-dependent protein kinase, diacylglycerol, phosphatidic acid, diacylglycerol-kinase-1, acylated pleckstrin-homology PH domain-containing protein, and RNG2 are among the constituents of Ca^{2+}-independent signaling pathways that regulate microneme secretion[189–191] (reviewed in[14]). Additionally, TgGAC has a pleckstrin-homology domain that binds to phosphatidic acid, and this supports the role of this lipid in signaling pathways.[125]

24.4 FUTURE PERSPECTIVE

Recent progresses in our understanding of the motility and invasion of *T. gondii* tachyzoites have come in leaps and bounds. The use of knockout and knockdown parasites for specific target genes has made a big impact in expanding the current knowledge. A wide variety of proteins, connectors, and enzymes have been implied in Apicomplexan gliding motility as well as host cell invasion. Some of these proteins are key to motility, while the abrogation of others had a negligible impact. In some cases, the deletion of target proteins did not completely impair the capacity of parasites to infect host cells, and some residual infection was still observed. This indicates redundancy in the molecular mechanism regarding parasite motility; although this might be a limitation of the on-off switch-like conditional expression systems, such as the Tet-system used to disable GOIs, as these protocols might permit residual expression of GOIs. Therefore, further studies are needed to enrich the current knowledge about *T. gondii* motility and invasion. This is not only important for understanding the invasion biology of this parasite but to open new avenues for the development of efficient compounds that may target the diverse aspects regarding each step of parasite host cell invasion and egress. Key players in these events that are exclusively found in Apicomplexans might serve as promising candidates as compounds/vaccine targets.

ACKNOWLEDGMENTS

The authors thank Wagner Nagib de Souza Birbeire for the graphic design of the figures in this chapter and Sheila Nardelli e Markus Meissner for critical suggestions regarding the text. In addition, the authors acknowledge the funding support from Universidade Estadual do Norte Fluminense Darcy Ribeiro, FAPERJ, CNPq, CAPES, Fundação Oswaldo Cruz Paraná – Instituto Carlos Chagas, and Fundação Araucária.

REFERENCES

1. Tenter AM, Heckeroth AR, Weiss LM. *Toxoplasma gondii*: from animals to humans. *Int J Parasitol* 2000;30:1217–58.
2. Gould SB, Tham WH, Cowman AF, McFadden GI, Waller RF. Alveolins, a new family of cortical proteins that define the protist infrakingdom Alveolata. *Mol Biol Evol* 2008;25:1219–30.
3. Pappas G, Roussos N, Falagas ME. Toxoplasmosis snapshots: global status of *Toxoplasma gondii* seroprevalence and implications for pregnancy and congenital toxoplasmosis. *Int J Parasitol* 2009;39:1385–94.
4. Bahia-Oliveira LM, Jones JL, Azevedo-Silva J, Alves CC, Orefice F, Addiss DG. Highly endemic, waterborne toxoplasmosis in north Rio de Janeiro state, Brazil. *Emerg Infect Dis* 2003;9:55–62.
5. Vaudaux JD, Muccioli C, James ER, et al. Identification of an atypical strain of *Toxoplasma gondii* as the cause of a waterborne outbreak of toxoplasmosis in Santa Isabel do Ivai, Brazil. *J Infect Dis* 2010;202:1226–33.
6. Dubey JP. Toxoplasmosis – a waterborne zoonosis. *Vet Parasitol* 2004;126:57–72.
7. Hill DE, Chirukandoth S, Dubey JP. Biology and epidemiology of *Toxoplasma gondii* in man and animals. *Anim Health Res Rev* 2005;6:41–61.
8. Carruthers VB, Giddings OK, Sibley LD. Secretion of micronemal proteins is associated with *Toxoplasma* invasion of host cells. *Cell Microbiol* 1999;1:225–35.
9. Hakimi MA, Olias P, Sibley LD. *Toxoplasma* effectors targeting host signaling and transcription. *Clin Microbiol Rev* 2017;30:615–45.
10. Lund E, Lycke E, Sourander P. A cinematographic study of *Toxoplasma gondii* in cell cultures. *Br J Exp Pathol* 1961;42:357–62.
11. Werk R. How does *Toxoplasma gondii* enter host cells? *Rev Infect Dis* 1985;7:449–57.
12. Morisaki JH, Heuser JE, Sibley LD. Invasion of *Toxoplasma gondii* occurs by active penetration of the host cell. *J Cell Sci* 1995;108(Pt 6):2457–64.
13. Dubremetz JF, Ferguson DJ. The role played by electron microscopy in advancing our understanding of *Toxoplasma gondii* and other Apicomplexans. *Int J Parasitol* 2009;39:883–93.
14. Frénal K, Dubremetz J-F, Lebrun M, Soldati-Favre D. Gliding motility powers invasion and egress in Apicomplexa. *Nat Rev Microbiol* 2017;15:645–60.
15. Carlton JM, Angiuoli SV, Suh BB, et al. Genome sequence and comparative analysis of the model rodent malaria parasite *Plasmodium yoelii yoelii*. *Nature* 2002;419:512–9.

16. El-Sayed NM. The genome sequence of *Trypanosoma cruzi*, etiologic agent of Chagas disease. *Science* 2005;309:409–15.
17. Gardner MJ, Shallom SJ, Carlton JM, et al. Sequence of *Plasmodium falciparum* chromosomes 2, 10, 11 and 14. *Nature* 2002;419:531–4.
18. Ivens AC. The genome of the Kinetoplastid parasite, *Leishmania major. Science* 2005;309:436–42.
19. Lander N, Chiurillo MA, Docampo R. Genome editing by CRISPR/Cas9: a game change in the genetic manipulation of protists. *J Eukaryot Microbiol* 2016;63:679–90.
20. Meissner M, Breinich MS, Gilson PR, Crabb BS. Molecular genetic tools in *Toxoplasma* and *Plasmodium*: achievements and future needs. *Curr Opin Microbiol* 2007;10:349–56.
21. Suarez CE, Bishop RP, Alzan HF, Poole WA, Cooke BM. Advances in the application of genetic manipulation methods to Apicomplexan parasites. *Int J Parasitol* 2017;47:701–10.
22. Limenitakis J, Soldati-Favre D. Functional genetics in Apicomplexa: potentials and limits. *FEBS Lett* 2011;585:1579–88.
23. Wang J-L, Huang S-Y, Behnke MS, Chen K, Shen B, Zhu X-Q. The past, present, and future of genetic manipulation in *Toxoplasma gondii. Trends Parasitol* 2016;32:542–53.
24. Seeber F, Steinfelder S. Recent advances in understanding Apicomplexan parasites. *F1000Research* 2016;5:1369.
25. Soldati D, Boothroyd JC. Transient transfection and expression in the obligate intracellular parasite *Toxoplasma gondii. Science* 1993;260:349–52.
26. Kim K, Soldati D, Boothroyd JC. Gene replacement in *Toxoplasma gondii* with chloramphenicol acetyltransferase as selectable marker. *Science* 1993;262:911–4.
27. Black M, Seeber F, Soldati D, Kim K, Boothroyd JC. Restriction enzyme-mediated integration elevates transformation frequency and enables co-transfection of *Toxoplasma gondii. Mol Biochem Parasitol* 1995;74:55–63.
28. Roos DS, Sullivan WJ, Striepen B, Bohne W, Donald RG. Tagging genes and trapping promoters in *Toxoplasma gondii* by insertional mutagenesis. *Methods* 1997;13:112–22.
29. Donald RG, Roos DS. Stable molecular transformation of *Toxoplasma gondii*: a selectable dihydrofolate reductase-thymidylate synthase marker based on drug-resistance mutations in malaria. *Proc Natl Acad Sci USA* 1993;90:11703–7.
30. Donald RG, Carter D, Ullman B, Roos DS. Insertional tagging, cloning, and expression of the *Toxoplasma gondii* hypoxanthine-xanthine-guanine phosphoribosyltransferase gene. Use as a selectable marker for stable transformation. *J Biol Chem* 1996;271:14010–9.
31. Sibley LD, Messina M, Niesman IR. Stable DNA transformation in the obligate intracellular parasite *Toxoplasma gondii* by complementation of tryptophan auxotrophy. *Proc Natl Acad Sci USA* 1994;91:5508–12.
32. Chaudhary K, Darling JA, Fohl LM, et al. Purine salvage pathways in the Apicomplexan parasite *Toxoplasma gondii. J Biol Chem* 2004;279:31221–7.
33. Donald RGK, Roos DS. Gene knock-outs and allelic replacements in *Toxoplasma gondii*: HXGPRT as a selectable marker for hit-and-run mutagenesis. *Mol Biochem Parasitol* 1998;91:295–305.
34. Blader IJ, Koshy AA. *Toxoplasma gondii* development of its replicative niche: in its host cell and beyond. *Eukaryot Cell* 2014;13:965–76.
35. English ED, Adomako-Ankomah Y, Boyle JP. Secreted effectors in *Toxoplasma gondii* and related species: determinants of host range and pathogenesis? *Parasite Immunol* 2015;37:127–40.
36. Karsten V, Qi H, Beckers CJ, Joiner KA. Targeting the secretory pathway of *Toxoplasma gondii. Methods* 1997;13:103–11.
37. Karsten V, Qi H, Beckers CJ, et al. The protozoan parasite *Toxoplasma gondii* targets proteins to dense granules and the vacuolar space using both conserved and unusual mechanisms. *J Cell Biol* 1998;141:1323–33.
38. Reiss M, Viebig N, Brecht S, et al. Identification and characterization of an escorter for two secretory adhesins in *Toxoplasma gondii. J Cell Biol* 2001;152:563–78.
39. Lamarque MH, Papoin J, Finizio A-L, et al. Identification of a new rhoptry neck complex RON9/RON10 in the Apicomplexa parasite *Toxoplasma gondii. PLoS One* 2012;7:e32457.
40. Chalfie M, Tu Y, Euskirchen G, Ward WW, Prasher DC. Green fluorescent protein as a marker for gene expression. *Science* 1994;263:802–5.
41. Cubitt AB, Heim R, Adams SR, Boyd AE, Gross LA, Tsien RY. Understanding, improving and using green fluorescent proteins. *Trends Biochem Sci* 1995;20:448–55.
42. Heim R, Cubitt AB, Tsien RY. Improved green fluorescence. *Nature* 1995;373:663–4.
43. Joiner KA, Roos DS. Secretory traffic in the eukaryotic parasite *Toxoplasma gondii*: less is more. *J Cell Biol* 2002;157:557–63.
44. Gubbels M-J, Striepen B. Studying the cell biology of Apicomplexan parasites using fluorescent proteins. *Microsc Microanal* 2004;10:568–79.

45. Donald RG, Roos DS. Homologous recombination and gene replacement at the dihydrofolate reductase-thymidylate synthase locus in *Toxoplasma gondii. Mol Biochem Parasitol* 1994;63:243–53.
46. Fox BA, Ristuccia JG, Gigley JP, Bzik DJ. Efficient gene replacements in *Toxoplasma gondii* strains deficient for nonhomologous end joining. *Eukaryot Cell* 2009;8:520–9.
47. Huynh MH, Carruthers VB. Tagging of endogenous genes in a *Toxoplasma gondii* strain lacking Ku80. *Eukaryot Cell* 2009;8:530–9.
48. Fox BA, Falla A, Rommereim LM, et al. Type II *Toxoplasma gondii* KU80 knockout strains enable functional analysis of genes required for cyst development and latent infection. *Eukaryot Cell* 2011;10:1193–206.
49. Messina M, Niesman I, Mercier C, Sibley LD. Stable DNA transformation of *Toxoplasma gondii* using phleomycin selection. *Gene* 1995;165:213–7.
50. Rommereim LM, Hortua Triana MA, Falla A, et al. Genetic manipulation in Δku80 strains for functional genomic analysis of *Toxoplasma gondii. J Vis Exp* 2013;77:e50598.
51. Mercier C, Dubremetz J-F, Rauscher B, Lecordier L, Sibley LD, Cesbron-Delauw M-F. Biogenesis of nanotubular network in *Toxoplasma* parasitophorous vacuole induced by parasite proteins. *Mol Biol Cell* 2002;13:2397–409.
52. Travier L, Mondragon R, Dubremetz J-F, et al. Functional domains of the *Toxoplasma* GRA2 protein in the formation of the membranous nanotubular network of the parasitophorous vacuole. *Int J Parasitol* 2008;38:757–73.
53. Huynh MH, Rabenau KE, Harper JM, Beatty WL, Sibley LD, Carruthers VB. Rapid invasion of host cells by *Toxoplasma* requires secretion of the MIC2-M2AP adhesive protein complex. *EMBO J* 2003;22:2082–90.
54. Armstrong CM, Goldberg DE. An FKBP destabilization domain modulates protein levels in *Plasmodium falciparum. Nat Methods* 2007;4:1007–9.
55. Herm-Götz A, Agop-Nersesian C, Münter S, et al. Rapid control of protein level in the Apicomplexan *Toxoplasma gondii. Nat Methods* 2007;4:1003–5.
56. Agop-Nersesian C, Naissant B, Ben Rached F, et al. Rab11A-controlled assembly of the inner membrane complex is required for completion of Apicomplexan cytokinesis. *PLoS Pathog* 2009;5:e1000270.
57. Agop-Nersesian C, Egarter S, Langsley G, Foth BJ, Ferguson DJP, Meissner M. Biogenesis of the inner membrane complex is dependent on vesicular transport by the alveolate specific GTPase Rab11B. *PLoS Pathog* 2010;6:e1001029.
58. Breinich MS, Ferguson DJP, Foth BJ, et al. A dynamin is required for the biogenesis of secretory organelles in *Toxoplasma gondii. Curr Biol* 2009;19:277–86.
59. van Dooren GG, Reiff SB, Tomova C, Meissner M, Humbel BM, Striepen B. A novel dynamin-related protein has been recruited for apicoplast fission in *Toxoplasma gondii. Curr Biol* 2009;19:267–76.
60. Kremer K, Kamin D, Rittweger E, et al. An overexpression screen of *Toxoplasma gondii* Rab-GTPases reveals distinct transport routes to the micronemes. *PLoS Pathog* 2013;9:e1003213.
61. Brown K, Long S, Sibley L. Conditional knockdown of proteins using auxin-inducible degron (AID) fusions in *Toxoplasma gondii. Bio Protoc* 2018;8(4):e2728.
62. Long S, Brown KM, Drewry LL, Anthony B, Phan IQH, Sibley LD. Calmodulin-like proteins localized to the conoid regulate motility and cell invasion by *Toxoplasma gondii. PLoS Pathog* 2017;13:e1006379.
63. Meissner M, Schlüter D, Soldati D. Role of *Toxoplasma gondii* myosin A in powering parasite gliding and host cell invasion. *Science* 2002;298:837–40.
64. van Poppel NFJ, Welagen J, Duisters RFJJ, Vermeulen AN, Schaap D. Tight control of transcription in *Toxoplasma gondii* using an alternative tet repressor. *Int J Parasitol* 2006;36:443–52.
65. Sheiner L, Demerly JL, Poulsen N, et al. A systematic screen to discover and analyze apicoplast proteins identifies a conserved and essential protein import factor. *PLoS Pathog* 2011;7:e1002392.
66. Huynh M-H, Carruthers VB. *Toxoplasma* MIC2 is a major determinant of invasion and virulence. *PLoS Pathog* 2006;2:e84.
67. JiméNez-Ruiz E, Wong EH, Pall GS, Meissner M. Advantages and disadvantages of conditional systems for characterization of essential genes in *Toxoplasma gondii. Parasitology* 2014;141:1390–8.
68. Deng C-X. Conditional knockout mouse models of cancer. *Cold Spring Harb Protoc* 2014;2014:1217–33.
69. García-Otín AL, Guillou F. Mammalian genome targeting using site-specific recombinases. *Front Biosci* 2006;11:1108–36.
70. Zhang J, Zhao J, Jiang W-J, Shan X-W, Yang X-M, Gao J-G. Conditional gene manipulation: creating a new biological era. *J Zhejiang Univ Sci B* 2012;13:511–24.
71. Brecht S, Erdhart H, Soete M, Soldati D. Genome engineering of *Toxoplasma gondii* using the site-specific recombinase Cre. *Gene* 1999;234:239–47.
72. Andenmatten N, Egarter S, Jackson AJ, Jullien N, Herman J-P, Meissner M. Conditional genome engineering in *Toxoplasma gondii* uncovers alternative invasion mechanisms. *Nat Methods* 2013;10:125–7.

73. Jullien N. Regulation of Cre recombinase by ligand-induced complementation of inactive fragments. *Nucleic Acids Res* 2003;31:e131.
74. Bargieri DY, Andenmatten N, Lagal V, et al. Apical membrane antigen 1 mediates Apicomplexan parasite attachment but is dispensable for host cell invasion. *Nat Commun* 2013;4:2552.
75. Besteiro S, Dubremetz J-F, Lebrun M. The moving junction of Apicomplexan parasites: a key structure for invasion. *Cell Microbiol* 2011;13:797–805.
76. Parussini F, Tang Q, Moin SM, Mital J, Urban S, Ward GE. Intramembrane proteolysis of *Toxoplasma* apical membrane antigen 1 facilitates host-cell invasion but is dispensable for replication. *Proc Natl Acad Sci USA* 2012;109:7463–8.
77. Shen B, Buguliskis JS, Lee TD, Sibley LD. Functional analysis of rhomboid proteases during *Toxoplasma* invasion. *mBio* 2014;5:e01795-14.
78. Drewry LL, Sibley LD. *Toxoplasma* actin is required for efficient host cell invasion. *mBio* 2015;6:e00557–15.
79. Grzybek M, Golonko A, Górska A, et al. The CRISPR/Cas9 system sheds new lights on the biology of protozoan parasites. *Appl Microbiol Biotechnol* 2018;102(11):4629–40.
80. Lander Eric S. The heroes of CRISPR. *Cell* 2016;164:18–28.
81. Shen B, Brown KM, Lee TD, Sibley LD. Efficient gene disruption in diverse strains of *Toxoplasma gondii* using CRISPR/CAS9. *mBio* 2014;5:e01114–14.
82. Shen B, Brown K, Long S, Sibley LD. Development of CRISPR/Cas9 for efficient genome editing in *Toxoplasma gondii*. *Methods Mol Biol* 2017;1498:79–103.
83. Di Cristina M, Carruthers VB. New and emerging uses of CRISPR/Cas9 to genetically manipulate Apicomplexan parasites. *Parasitology* 2018;145(9):1119–26.
84. Grissa I, Vergnaud G, Pourcel C. CRISPRFinder: a web tool to identify clustered regularly interspaced short palindromic repeats. *Nucleic Acids Res* 2007;35:W52–7.
85. Horvath P, Barrangou R. CRISPR/Cas, the immune system of bacteria and archaea. *Science* 2010;327:167–70.
86. Garneau JE, Dupuis M-È, Villion M, et al. The CRISPR/Cas bacterial immune system cleaves bacteriophage and plasmid DNA. *Nature* 2010;468:67–71.
87. Rath D, Amlinger L, Hoekzema M, Devulapally PR, Lundgren M. Efficient programmable gene silencing by Cascade. *Nucleic Acids Res* 2015;43:237–46.
88. Charpentier E, Doudna JA. Biotechnology: rewriting a genome. *Nature* 2013;495:50–1.
89. Makarova KS, Haft DH, Barrangou R, et al. Evolution and classification of the CRISPR-Cas systems. *Nat Rev Microbiol* 2011;9:467–77.
90. Makarova KS, Aravind L, Wolf YI, Koonin EV. Unification of Cas protein families and a simple scenario for the origin and evolution of CRISPR-Cas systems. *Biol Direct* 2011;6:38.
91. Doudna JA, Charpentier E. Genome editing. The new frontier of genome engineering with CRISPR-Cas9. *Science* 2014;346:1258096.
92. Hsu PD, Lander ES, Zhang F. Development and applications of CRISPR-Cas9 for genome engineering. *Cell* 2014;157:1262–78.
93. Yang L, Mali P, Kim-Kiselak C, Church G. CRISPR-Cas-mediated targeted genome editing in human cells. *Methods Mol Biol* 2014;1114:245–67.
94. Sidik SM, Hackett CG, Tran F, Westwood NJ, Lourido S. Efficient genome engineering of *Toxoplasma gondii* using CRISPR/Cas9. *PLoS One* 2014;9:e100450.
95. Long S, Wang Q, Sibley LD. Analysis of noncanonical calcium-dependent protein kinases in *Toxoplasma gondii* by targeted gene deletion using CRISPR/Cas9. *Infect Immun* 2016;84:1262–73.
96. Wang J-L, Huang S-Y, Li T-T, Chen K, Ning H-R, Zhu X-Q. Evaluation of the basic functions of six calcium-dependent protein kinases in *Toxoplasma gondii* using CRISPR-Cas9 system. *Parasitol Res* 2016;115:697–702.
97. Wang M, Cao S, Du N, et al. The moving junction protein RON4, although not critical, facilitates host cell invasion and stabilizes MJ members. *Parasitology* 2017;144:1490–7.
98. Sidik SM, Huet D, Ganesan SM, et al. A genome-wide CRISPR screen in *Toxoplasma* identifies essential Apicomplexan genes. *Cell* 2016;166:1423–35.
99. Kleinstiver BP, Prew MS, Tsai SQ, et al. Engineered CRISPR-Cas9 nucleases with altered PAM specificities. *Nature* 2015;523:481–5.
100. Kleinstiver BP, Pattanayak V, Prew MS, et al. High-fidelity CRISPR-Cas9 nucleases with no detectable genome-wide off-target effects. *Nature* 2016;529:490–5.
101. Slaymaker IM, Gao L, Zetsche B, Scott DA, Yan WX, Zhang F. Rationally engineered Cas9 nucleases with improved specificity. *Science* 2016;351:84–8.
102. Ghorbal M, Gorman M, Macpherson CR, Martins RM, Scherf A, Lopez-Rubio J-J. Genome editing in the human malaria parasite *Plasmodium falciparum* using the CRISPR-Cas9 system. *Nat Biotechnol* 2014;32:819–21.

103. Wagner JC, Platt RJ, Goldfless SJ, Zhang F, Niles JC. Efficient CRISPR-Cas9-mediated genome editing in *Plasmodium falciparum*. *Nat Methods* 2014;11:915–8.
104. Walker DM, Oghumu S, Gupta G, McGwire BS, Drew ME, Satoskar AR. Mechanisms of cellular invasion by intracellular parasites. *Cell Mol Life Sci* 2014;71:1245–63.
105. Opitz C, Soldati D. 'The glideosome': a dynamic complex powering gliding motion and host cell invasion by *Toxoplasma gondii*. *Mol Microbiol* 2002;45:597–604.
106. Hakansson S, Morisaki H, Heuser J, Sibley LD. Time-lapse video microscopy of gliding motility in *Toxoplasma gondii* reveals a novel, biphasic mechanism of cell locomotion. *Mol Biol Cell* 1999;10:3539–47.
107. Leung JM, Rould MA, Konradt C, Hunter CA, Ward GE. Disruption of TgPHIL1 alters specific parameters of *Toxoplasma gondii* motility measured in a quantitative, three-dimensional live motility assay. *PLoS One* 2014;9:e85763.
108. King CA. Cell motility of sporozoan protozoa. *Parasitol Today* 1988;4:315–9.
109. Ladda R, Aikawa M, Sprinz H. Penetration of erythrocytes by merozoites of mammalian and avian malarial parasites. *J Parasitol* 1969;55:633–44.
110. Aikawa M, Miller LH, Johnson J, Rabbege J. Erythrocyte entry by malarial parasites. A moving junction between erythrocyte and parasite. *J Cell Biol* 1978;77:72–82.
111. Russell DG, Sinden RE. The role of the cytoskeleton in the motility of coccidian sporozoites. *J Cell Sci* 1981;50:345–59.
112. Jensen JB, Edgar SA. Fine structure of penetration of cultured cells by *Isospora canis* sporozoites. *J Protozool* 1978;25:169–73.
113. Zhou XW, Kafsack BF, Cole RN, Beckett P, Shen RF, Carruthers VB. The opportunistic pathogen *Toxoplasma gondii* deploys a diverse legion of invasion and survival proteins. *J Biol Chem* 2005;280:34233–44.
114. Dobrowolski JM, Carruthers VB, Sibley LD. Participation of myosin in gliding motility and host cell invasion by *Toxoplasma gondii*. *Mol Microbiol* 1997;26:163–73.
115. Dobrowolski JM, Sibley LD. Toxoplasma invasion of mammalian cells is powered by the actin cytoskeleton of the parasite. *Cell* 1996;84:933–9.
116. Mondragon R, Frixione E. Ca(2+)-dependence of conoid extrusion in *Toxoplasma gondii* tachyzoites. *J Eukaryot Microbiol* 1996;43:120–7.
117. Andenmatten N, Egarter S, Jackson AJ, Jullien N, Herman JP, Meissner M. Conditional genome engineering in *Toxoplasma gondii* uncovers alternative invasion mechanisms. *Nat Methods* 2013;10:125–7.
118. Ralph SA, D'Ombrain MC, McFadden GI. The apicoplast as an antimalarial drug target. *Drug Resist Updat* 2001;4:145–51.
119. He CY, Shaw MK, Pletcher CH, Striepen B, Tilney LG, Roos DS. A plastid segregation defect in the protozoan parasite *Toxoplasma gondii*. *EMBO J* 2001;20:330–9.
120. Blanchoin L, Boujemaa-Paterski R, Sykes C, Plastino J. Actin dynamics, architecture, and mechanics in cell motility. *Physiol Rev* 2014;94:235–63.
121. Dobrowolski JM, Niesman IR, Sibley LD. Actin in the parasite *Toxoplasma gondii* is encoded by a single copy gene, *ACT1* and exists primarily in a globular form. *Cell Motil Cytoskeleton* 1997;37:253–62.
122. Gordon JL, Sibley LD. Comparative genome analysis reveals a conserved family of actin-like proteins in Apicomplexan parasites. *BMC Genom* 2005;6:179.
123. Goley ED, Welch MD. The ARP2/3 complex: an actin nucleator comes of age. *Nat Rev Mol Cell Biol* 2006;7:713.
124. Daher W, Plattner F, Carlier MF, Soldati-Favre D. Concerted action of two formins in gliding motility and host cell invasion by *Toxoplasma gondii*. *PLoS Pathog* 2010;6:e1001132.
125. Jacot D, Tosetti N, Pires I, et al. An Apicomplexan actin-binding protein serves as a connector and lipid sensor to coordinate motility and invasion. *Cell Host Microbe* 2016;20:731–43.
126. Daher W, Klages N, Carlier MF, Soldati-Favre D. Molecular characterization of *Toxoplasma gondii* formin 3, an actin nucleator dispensable for tachyzoite growth and motility. *Eukaryot Cell* 2012;11:343–52.
127. Plattner F, Yarovinsky F, Romero S, et al. *Toxoplasma* profilin is essential for host cell invasion and TLR11-dependent induction of an interleukin-12 response. *Cell Host Microbe* 2008;3:77–87.
128. Uetrecht AC, Bear JE. Coronins: the return of the crown. *Trends Cell Biol* 2006;16:421–6.
129. Salamun J, Kallio JP, Daher W, Soldati-Favre D, Kursula I. Structure of *Toxoplasma gondii* coronin, an actin-binding protein that relocalizes to the posterior pole of invasive parasites and contributes to invasion and egress. *FASEB J* 2014;28:4729–47.
130. Holzinger A. Jasplakinolide: an actin-specific reagent that promotes actin polymerization. *Methods Mol Biol* 2009;586:71–87.
131. Wetzel D, Håkansson S, Hu K, Roos D, Sibley L. Actin filament polymerization regulates gliding motility by Apicomplexan parasites. *Mol Biol Cell* 2003;14:396–406.

132. Bamburg JR. Proteins of the ADF/cofilin family: essential regulators of actin dynamics. *Annu Rev Cell Dev Biol* 1999;15:185–230.
133. Mehta S, Sibley LD. Actin depolymerizing factor controls actin turnover and gliding motility in *Toxoplasma gondii. Mol Biol Cell* 2011;22:1290–9.
134. Mehta S, Sibley LD. *Toxoplasma gondii* actin depolymerizing factor acts primarily to sequester G-actin. *J Biol Chem* 2010;285:6835–47.
135. Sattler JM, Ganter M, Hliscs M, Matuschewski K, Schuler H. Actin regulation in the malaria parasite. *Eur J Cell Biol* 2011;90:966–71.
136. Naitza S, Spano F, Robson KJ, Crisanti A. The thrombospondin-related protein family of Apicomplexan parasites: the gears of the cell invasion machinery. *Parasitol Today* 1998;14:479–84.
137. Sultan AA, Thathy V, Frevert U, et al. TRAP is necessary for gliding motility and infectivity of *Plasmodium* sporozoites. *Cell* 1997;90:511–22.
138. Wan KL, Carruthers VB, Sibley LD, Ajioka JW. Molecular characterisation of an expressed sequence tag locus of *Toxoplasma gondii* encoding the micronemal protein MIC2. *Mol Biochem Parasitol* 1997;84:203–14.
139. Huynh MH, Carruthers VB. *Toxoplasma* MIC2 is a major determinant of invasion and virulence. *PLoS Pathog* 2006;2:e84.
140. Carruthers VB, Sherman GD, Sibley LD. The *Toxoplasma* adhesive protein MIC2 is proteolytically processed at multiple sites by two parasite-derived proteases. *J Biol Chem* 2000;275:14346–53.
141. Carruthers VB, Sibley LD. Mobilization of intracellular calcium stimulates microneme discharge in *Toxoplasma gondii. Mol Microbiol* 1999;31:421–8.
142. Friedrich N, Matthews S, Soldati-Favre D. Sialic acids: key determinants for invasion by the Apicomplexa. *Int J Parasitol* 2010;40:1145–54.
143. Sugimoto Y, Tokunaga M, Takezawa Y, Ikebe M, Wakabayashi K. Conformational changes of the myosin heads during hydrolysis of ATP as analyzed by x-ray solution scattering. *Biophys J* 1995;68:S29–33; discussion S33–4.
144. Herm-Gotz A, Weiss S, Stratmann R, et al. *Toxoplasma gondii* myosin A and its light chain: a fast, single-headed, plus-end-directed motor. *EMBO J* 2002;21:2149–58.
145. Bookwalter CS, Kelsen A, Leung JM, Ward GE, Trybus KM. A *Toxoplasma gondii* class XIV myosin, expressed in Sf9 cells with a parasite co-chaperone, requires two light chains for fast motility. *J Biol Chem* 2014;289:30832–41.
146. Nebl T, Prieto JH, Kapp E, et al. Quantitative in vivo analyses reveal calcium-dependent phosphorylation sites and identifies a novel component of the *Toxoplasma* invasion motor complex. *PLoS Pathog* 2011;7:e1002222.
147. Heaslip AT, Leung JM, Carey KL, et al. A small-molecule inhibitor of *T. gondii* motility induces the posttranslational modification of myosin light chain-1 and inhibits myosin motor activity. *PLoS Pathog* 2010;6:e1000720.
148. Heintzelman MB, Schwartzman JD. A novel class of unconventional myosins from *Toxoplasma gondii. J Mol Biol* 1997;271:139–46.
149. Gaskins E, Gilk S, DeVore N, Mann T, Ward G, Beckers C. Identification of the membrane receptor of a class XIV myosin in *Toxoplasma gondii. J Cell Biol* 2004;165:383.
150. Frenal K, Polonais V, Marq JB, Stratmann R, Limenitakis J, Soldati-Favre D. Functional dissection of the Apicomplexan glideosome molecular architecture. *Cell Host Microbe* 2010;8:343–57.
151. Bullen HE, Tonkin CJ, O'Donnell RA, et al. A novel family of Apicomplexan glideosome-associated proteins with an inner membrane-anchoring role. *J Biol Chem* 2009;284:25353–63.
152. Frenal K, Marq JB, Jacot D, Polonais V, Soldati-Favre D. Plasticity between MyoC- and MyoA-glideosomes: an example of functional compensation in *Toxoplasma gondii* invasion. *PLoS Pathog* 2014;10:e1004504.
153. Meissner M, Schluter D, Soldati D. Role of *Toxoplasma gondii* myosin A in powering parasite gliding and host cell invasion. *Science* 2002;298:837–40.
154. Graindorge A, Frenal K, Jacot D, Salamun J, Marq JB, Soldati-Favre D. The conoid associated motor MyoH is indispensable for *Toxoplasma gondii* entry and exit from host cells. *PLoS Pathog* 2016;12:e1005388.
155. Paing MM, Tolia NH. Multimeric assembly of host-pathogen adhesion complexes involved in Apicomplexan invasion. *PLoS Pathog* 2014;10:e1004120.
156. Soldati D, Dubremetz JF, Lebrun M. Microneme proteins: structural and functional requirements to promote adhesion and invasion by the Apicomplexan parasite *Toxoplasma gondii. Int J Parasitol* 2001;31:1293–302.
157. Mineo JR, Kasper LH. Attachment of *Toxoplasma gondii* to host cells involves major surface protein, SAG-1 (P30). *Exp Parasitol* 1994;79:11–20.
158. Wang Y, Yin H. Research progress on surface antigen 1 (SAG1) of *Toxoplasma gondii. Parasit Vectors* 2014;7:180.
159. Jongert E, Roberts CW, Gargano N, Forster-Waldl E, Petersen E. Vaccines against *Toxoplasma gondii*: challenges and opportunities. *Mem Inst Oswaldo Cruz* 2009;104:252–66.

160. Mital J, Meissner M, Soldati D, Ward GE. Conditional expression of *Toxoplasma gondii* apical membrane antigen-1 (TgAMA1) demonstrates that TgAMA1 plays a critical role in host cell invasion. *Mol Biol Cell* 2005;16:4341–9.
161. Lamarque MH, Roques M, Kong-Hap M, et al. Plasticity and redundancy among AMA-RON pairs ensure host cell entry of *Toxoplasma* parasites. *Nat Commun* 2014;5:4098.
162. Krishnamurthy S, Deng B, Del Rio R, et al. Not a simple tether: binding of *Toxoplasma gondii* AMA1 to RON2 during invasion protects AMA1 from rhomboid-mediated cleavage and leads to dephosphorylation of its cytosolic tail. *mBio* 2016;7(5):e00754-16.
163. Jewett TJ, Sibley LD. Aldolase forms a bridge between cell surface adhesins and the actin cytoskeleton in Apicomplexan parasites. *Mol Cell* 2003;11:885–94.
164. Buguliskis JS, Brossier F, Shuman J, Sibley LD. Rhomboid 4 (ROM4) affects the processing of surface adhesins and facilitates host cell invasion by *Toxoplasma gondii*. *PLoS Pathog* 2010;6:e1000858.
165. Rugarabamu G, Marq J-B, Guérin A, Lebrun M, Soldati-Favre D. Distinct contribution of *Toxoplasma gondii* rhomboid proteases 4 and 5 to micronemal protein protease 1 activity during invasion: ROM4 and ROM5 contribute to MPP1 activity. *Mol Microbiol* 2015;97:244–62.
166. Dowse TJ, Pascall JC, Brown KD, Soldati D. Apicomplexan rhomboids have a potential role in microneme protein cleavage during host cell invasion. *Int J Parasitol* 2005;35:747–56.
167. Zhou XW, Blackman MJ, Howell SA, Carruthers VB. Proteomic analysis of cleavage events reveals a dynamic two-step mechanism for proteolysis of a key parasite adhesive complex. *Mol Cell Proteom* 2004;3:565–76.
168. Lagal V, Binder EM, Huynh MH, et al. *Toxoplasma gondii* protease TgSUB1 is required for cell surface processing of micronemal adhesive complexes and efficient adhesion of tachyzoites. *Cell Microbiol* 2010;12:1792–808.
169. Mordue DG, Hakansson S, Niesman I, Sibley LD. *Toxoplasma gondii* resides in a vacuole that avoids fusion with host cell endocytic and exocytic vesicular trafficking pathways. *Exp Parasitol* 1999;92:87–99.
170. Jones TC, Hirsch JG. The interaction between *Toxoplasma gondii* and mammalian cells. II. The absence of lysosomal fusion with phagocytic vacuoles containing living parasites. *J Exp Med* 1972;136:1173–94.
171. Mordue DG, Desai N, Dustin M, Sibley LD. Invasion by *Toxoplasma gondii* establishes a moving junction that selectively excludes host cell plasma membrane proteins on the basis of their membrane anchoring. *J Exp Med* 1999;190:1783–92.
172. Moudy R, Manning TJ, Beckers CJ. The loss of cytoplasmic potassium upon host cell breakdown triggers egress of *Toxoplasma gondii*. *J Biol Chem* 2001;276:41492–501.
173. Lavine MD, Arrizabalaga G. Exit from host cells by the pathogenic parasite *Toxoplasma gondii* does not require motility. *Eukaryot Cell* 2008;7:131–40.
174. Kafsack BF, Pena JD, Coppens I, Ravindran S, Boothroyd JC, Carruthers VB. Rapid membrane disruption by a perforin-like protein facilitates parasite exit from host cells. *Science* 2009;323:530–3.
175. Chandramohanadas R, Davis PH, Beiting DP, et al. Apicomplexan parasites co-opt host calpains to facilitate their escape from infected cells. *Science* 2009;324:794–7.
176. Blackman MJ, Carruthers VB. Recent insights into Apicomplexan parasite egress provide new views to a kill. *Curr Opin Microbiol* 2013;16:459–64.
177. Delorme-Walker V, Abrivard M, Lagal V, et al. Toxofilin upregulates the host cortical actin cytoskeleton dynamics, facilitating *Toxoplasma* invasion. *J Cell Sci* 2012;125:4333–42.
178. Bichet M, Joly C, Hadj Henni A, et al. The *Toxoplasma*-host cell junction is anchored to the cell cortex to sustain parasite invasive force. *BMC Biol* 2014;12:773.
179. Riglar DT, Richard D, Wilson DW, et al. Super-resolution dissection of coordinated events during malaria parasite invasion of the human erythrocyte. *Cell Host Microbe* 2011;9:9–20.
180. Gonzalez V, Combe A, David V, et al. Host cell entry by Apicomplexa parasites requires actin polymerization in the host cell. *Cell Host Microbe* 2009;5:259–72.
181. Guérin A, Corrales RM, Parker ML, et al. Efficient invasion by *Toxoplasma* depends on the subversion of host protein networks. *Nat Microbiol* 2017;2:1358–66.
182. Suss-Toby E, Zimmerberg J, Ward GE. *Toxoplasma* invasion: the parasitophorous vacuole is formed from host cell plasma membrane and pinches off via a fission pore. *Proc Natl Acad Sci USA* 1996;93:8413–8.
183. Pavlou G, Biesaga M, Touquet B, et al. *Toxoplasma* parasite twisting motion mechanically induces host cell membrane fission to complete invasion within a protective vacuole. *Cell Host Microbe* 2018;24:81–96.
184. Vieira MC, Moreno SN. Mobilization of intracellular calcium upon attachment of *Toxoplasma gondii* tachyzoites to human fibroblasts is required for invasion. *Mol Biochem Parasitol* 2000;106:157–62.
185. Lourido S, Moreno SN. The calcium signaling toolkit of the Apicomplexan parasites *Toxoplasma gondii* and *Plasmodium* spp. *Cell Calcium* 2015;57:186–93.
186. Luo S, Ruiz FA, Moreno SN. The acidocalcisome Ca^{2+}-ATPase (*TgA1*) of *Toxoplasma gondii* is required for polyphosphate storage, intracellular calcium homeostasis and virulence. *Mol Microbiol* 2005;55:1034–45.

187. Roiko MS, Svezhova N, Carruthers VB. Acidification activates *Toxoplasma gondii* motility and egress by enhancing protein secretion and cytolytic activity. *PLoS Pathog* 2014;10:e1004488.
188. Farrell A, Thirugnanam S, Lorestani A, et al. A DOC2 protein identified by mutational profiling is essential for Apicomplexan parasite exocytosis. *Science* 2012;335:218–21.
189. Lourido S, Tang K, Sibley LD. Distinct signalling pathways control *Toxoplasma* egress and host-cell invasion. *EMBO J* 2012;31:4524–34.
190. Bullen HE, Jia Y, Yamaryo-Botte Y, et al. Phosphatidic acid-mediated signaling regulates microneme secretion in *Toxoplasma. Cell Host Microbe* 2016;19:349–60.
191. Katris NJ, van Dooren GG, McMillan PJ, Hanssen E, Tilley L, Waller RF. The apical complex provides a regulated gateway for secretion of invasion factors in *Toxoplasma. PLoS Pathog* 2014;10:e1004074.

SUMMARY

Toxoplasma gondii is a widespread Apicomplexan parasite that may cause opportunistic infection in immunocompetent individuals for whom no treatment is necessary, but severe damage to the developing fetuses and/or prolonged illnesses in immunocompromised individuals. Members of the Apicomplexan phylum are known to invade and replicate in host cells and share a common form of motility, termed gliding, that is related to cell invasion and egress. Currently, the most accepted model states that gliding motility involves an actomyosin motor located at the periphery of *T. gondii* tachyzoites that acts in synergy with microneme secretion, rhoptry discharge, and conoid extrusion. The structural support is provided by actin filaments in a tightly regulated process of globular actin polymerization into filamentous actin. Tachyzoites also induce conformational changes in host cell cytoskeleton to facilitate penetration and formation of a parasitophorous vacuole. These changes occur in the moving junction, an interface of intimate contact between the parasite and the host cell. The parasitophorous vacuole remains excluded from the endocytic pathway, preventing the killing of the parasite by possible digestion. After replicating, tachyzoites combine the secretion of membrane pore-forming molecules with motility to escape the cell and perpetuate its lifecycle. This chapter reviews a broad range of structural and signaling proteins and enzymes, and their contribution to *T. gondii* motility, invasion, and egress.

Molecular Pathogenesis of *Sarcocystis* Infection

25

Fen Li and Guohua Liu
Hunan Agricultural University

Dongyou Liu
Royal College of Pathologists of Australasia Quality Assurance Programs

Contents

25.1 INTRODUCTION

The genus *Sarcocystis* consists of >200 protozoan species that utilize carnivores/omnivores as definitive hosts and herbivores as intermediate hosts. Although the type species of this genus was first described by Meischer in 1843 as long, thin, white cysts (initially called Meischer's tubules, and renamed *Sarcocystis meischeriana* by Labbe in 1899) in muscles of deer mouse in Switzerland, the heterogenous lifecycle of *Sarcocystis* involving carnivore/omnivore (definitive host) and herbivore (intermediate host) was only established in the 1970s after *in vitro* and *in vivo* experimentations. In definitive carnivore/omnivore host (e.g., dog and cat), *Sarcocystis* produces intestinal sarcocystosis, whereas in intermediate herbivore host (e.g., cattle, pigs, and sheep), *Sarcocystis* causes muscular sarcocystosis, with clinical presentations ranging from abortion, reduced meat yield, increased meat contamination, to frequent death. Given the role of *Sarcocystis* in the causation of both intestinal and muscular sarcocystosis in human population, and the current lack of effective therapy against this zoonotic parasite, further elucidation on the molecular mechanisms of *Sarcocystis* pathogenicity is vital for the development of innovative strategies toward its control and prevention [1].

Taxonomy. Classified in the phylum *Apicomplexa*, class *Conoidasida*, order *Eucoccidiorida*, suborder *Eimeriorina*, family *Sarcocystidae*, which also includes genera *Besnoitia*, *Cystoisospora*, *Frenkelia*, *Hyaloklossia*, *Nephroisospora*, *Neospora*, and *Toxoplasma*, the genus *Sarcocystis* (Greek *sarx*, flesh;

kystis, bladder) encompasses over 200 protozoan species that occur in a wide range of animal hosts, such as herbivores (intermediate host) and carnivores or omnivores (definitive host) [2].

Of the three human-infecting *Sarcocystis* species (*S. hominis*, *S. suihominis*, and *S. nesbitti*), *S. hominis* and *S. suihominis* are responsible for producing muscular sarcocystosis in cattle and pigs (intermediate host) and intestinal sarcocystosis in humans (definitive host) after ingestion of mature sarcocysts in undercooked beef and pork, respectively. First described in 1969 in a rhesus macaque (*Macaca mulatta*) in Northern India, *S. nesbitti* is associated with intestinal sarcocystosis in snakes (definitive host) and muscular sarcocystosis in nonhuman primates (intermediate host), and occasionally in humans (aberrant intermediate host). There is a possibility that *S. nesbitti* may have a snake-rodent lifecycle, with primates and humans acting as aberrant intermediate hosts [3].

Morphology. *S. hominis* oocysts measure approximately 15 μm × 17 μm, each containing two sporocysts of 9.3 μm × 14.7 μm in size, which, in turn, harbor four sporozoites each (Figure 25.1) [4]. Similarly, *S. suihominis* oocysts measure 12.3–14.6 μm × 18.5–20.0 μm, each containing two sporocysts of 10.5 μm × 13.5 μm in size, which, in turn, comprise four sporozoites each. The dimensions and structures of *S. nesbitti* oocysts and sporocysts resemble those of *S. hominis* and *S. suihominis* oocysts and sporocysts (Figure 25.1) [3].

S. hominis sarcocysts from the muscles of infected cattle are of 1,220–4,460 μm × 80–384 μm in size with a 3–6 μm thick wall (Figure 25.2a). The sarcocyst wall appears radially striated (due to the presence of palisade-like villar protrusions on the surface), and the villar protrusions have dimensions of 3.1–4.3 μm × 0.7–1.1 μm and contain many microtubules in the core [5].

S. suihominis sarcocysts from the heart and diaphragm of infected pigs measure 1,080–2,040 μm × 106–170 μm in size (Figure 25.2b), with a striated wall (or parasitophorous vacuole membrane) of 4–6 μm in thickness and many palisade-like villar protrusions of 6–6 μm × 0.3–0.5 μm in size, which contain a small number of microtubules in the core. Bradysoites are average 15 μm × 4 μm in size.

S. nesbitti sarcocysts from the muscle of infected humans measure approximately 190 μm × 60 μm. The sarcocyst wall is 0.5 μm thick, and appears smooth without any protrusions (Figure 25.2c) [6].

Electron microscopy reveals numerous structural units in *Sarcocystis* sporozoites, including apical polar ring, rhoptries, micronemes, conoid, apicoplast, Golgi apparatus, dense granules, mitochondrion, nucleus, alveoli, and posterior ring. Collectively, rhoptries, micronemes, apical polar ring, and conoid are referred to as the apical complex, which plays a key role in *Sarcocystis* entry into host cells [7].

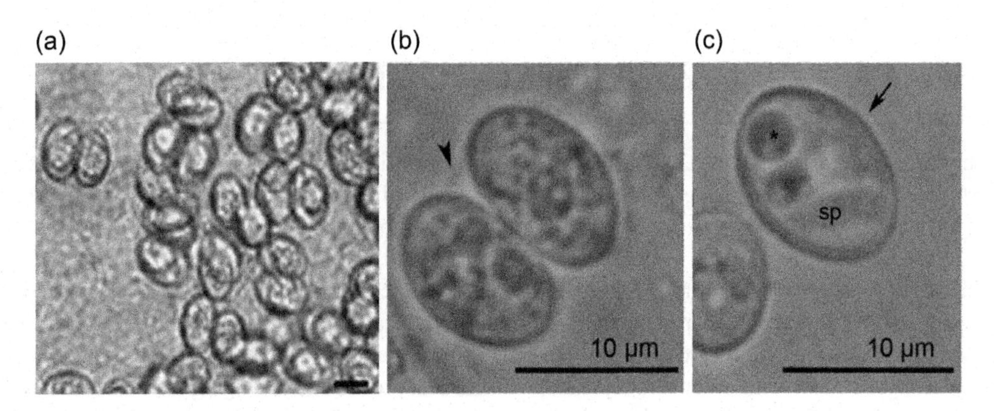

FIGURE 25.1 (a) *Sarcocystis hominis* oocysts (average 17 μm × 15 μm; arrows) each containing two sporocysts (average 9 μm × 14 μm) in fecal specimen from a 19-year-old male with intestinal sarcocystosis (wet mount; scale bar, 12 μm). (b) *Sarcocystis nesbitti* oocyst (thin wall indicated by arrowhead) containing two sporocysts from an Australian scrub python (scale bar, 10 μm). (c) *Sarcocystis nesbitti* sporocyst (arrow) consisting of sporozoites (sp) (scale bar, 10 μm). (Photo credits: (a) Nimri L. *JMM Case Rep*. 2014;1(4):e004069; (b and c) Wassermann M, Raisch L, Lyons JA, et al. *PLoS One*. 2017;12(11):e0187984.)

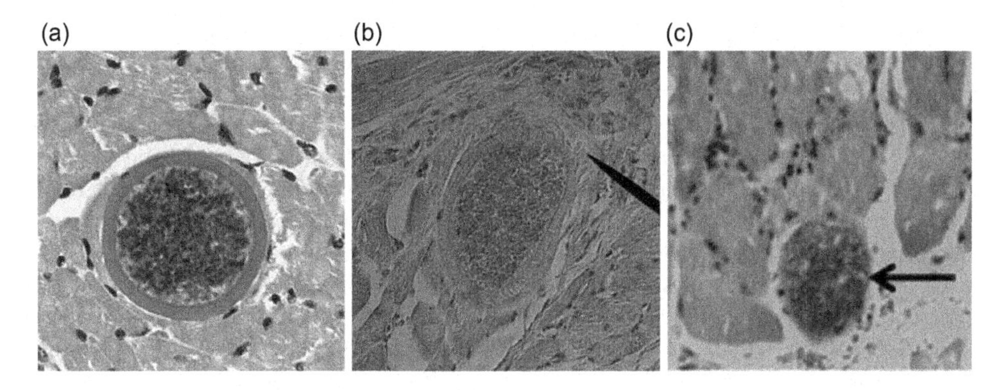

FIGURE 25.2 (a) *Sarcocystis hominis* sarcocyst from bovine tongue. (b) *Sarcocystis suihominis* sarcocyst from porcine skeletal muscle (note the clearly defined striated border around the sarcocyst that differentiates from Toxoplasm). (c) *Sarcocystis nesbitti* sarcocyst (arrow) in the muscle fibers of an affected human patient. (Photo credits: (a) https://commons.wikimedia.org/wiki/File:Sarcocystis_hominis_tissue_cyst_in_beef_tongue.jpg. (b) https://commons.wikimedia.org/wiki/File:Sarcocystis_in_pig_muscle.jpg. (c) Italiano CM, Wong KT, AbuBakar S, et al. *PLoS Negl Trop Dis*. 2014;8:e2876.)

Lifecycle and epidemiology. During its lifecycle, *Sarcocystis* requires both intermediate (prey) and definitive (predator) hosts and undergoes both asexual and sexual development. *Sarcocystis* infection begins when an intermediate host (usually herbivore) ingests *Sarcocystis* oocysts contained in the feces of a definitive host (e.g., dog or cat), which release motile sporozoites upon bile exposure. Sporozoites then enter endothelial cells of the small arteries, undergo asexual development (merogony/schizogony), and gives rise to motile, crescent-shaped merozoites, which mature into sarcocysts containing infective bradyzoites in muscles (of the heart, tongue, esophagus, diaphragm, skeletal muscle, and, rarely, of the central nervous system and gut). Sarcocysts in an intermediate host may rupture and release bradyzoites which often die without initiating new infection [1].

After being taken up by a definitive host, *Sarcocystis* sarcocysts contained in the muscle of an intermediate host release bradyzoites, which reach intracellularly in villi of the small intestine, undergo gametogony (sexual development) in lamina propria or goblet cells, and produce microgametes (males) and macrogametes (females). Microgamete then fuses with macrogamete, developing into an oocyst, which undergoes sporogony, forming two sporocysts each containing four sporozoites, and excretes in feces [1].

Human intestinal sarcocystosis occurs in many parts of the world apart from Africa and the Middle East. The prevalence of human intestinal sarcocystosis associated with ingestion of beef containing *S. hominis* sarcocysts is estimated at 6.2%–97.4% in Europe, 53.5% in Asia, and 94% in South America [8,9]. However, the prevalence of human intestinal sarcocystosis caused by *S. suihominis* is unclear. Further, human muscular sarcocystosis resulting from the consumption of water or food containing *S. nesbitti* oocysts/sporocysts has an estimated prevalence of 0%–3.6% in western countries and 21% in Malaysia [10,11].

Clinical features. Most individuals with intestinal sarcocystosis are asymptomatic and only show mild signs, although 10% of affected individuals develop nausea, loss of appetite, vomiting, abdominal pain, and diarrhea that last as long as 48 h, as well as blood eosinophilia, after ingestion of uncooked beef or pork containing *S. hominis* or *S. suihominis* sarcocysts [1].

In one experiment, consumption of *S. hominis*-infected beef induced symptoms (e.g., anorexia, nausea, abdominal pain, distension, diarrhea, vomiting, dyspnea, and tachycardia) within 3–6 h, which lasted about 36 h. In a second experiment, symptoms (e.g., abdominal pain, distension, watery diarrhea, and eosinophilia) emerged at 1 week and resolved after 3 weeks. Based on experimental studies in Germany, two volunteers who consumed 500 g of ground raw beef diaphragm harboring *S. hominis* sarcocysts began excreting sporocysts 9 days later, and continued for 40 days or longer; whereas three of four volunteers ingested ground raw pork diaphragm harboring *S. suihominis* sarcocysts began excreting sporocysts 9, 13, and 17 days later, and continued for at least 30 days [1].

Individuals with muscular sarcocystosis may experience acute fever, myalgia, myositis (muscle tenderness upon physical examination), vasculitis, bronchospasm, and pruritic rashes. Some patients may develop rheumatic fever/nonspecific rheumatic disease, glomerulonephritis, bilateral pulmonary thrombosis with infarction, chronic cardiopathology, and possible malignancies. Other nonspecific features include slightly elevated levels of hepatic enzymes (aspartate aminotransferase [AST] and alanine aminotransferase [ALT]), inflammatory markers (C-reactive protein and erythrocyte sedimentation rate [ESR]), markers of general cell damage (lactic dehydrogenase [LDH]), serum creatinine phosphokinase (CPK), and blood eosinophilia [12].

In a survey involving 89 Malaysian patients with *S. nesbitti* muscular sarcocystosis, frequencies of clinical presentations ranged from fever (94%), relapsing fever (57%), myalgia (any muscle group 91%; lower limbs 70%; back 60%; upper limbs 56%; neck 42%, face/jaw 6%), headache (86%), cough (40%), joint pain (39%), nausea (28%), vomiting (18%), diarrhea (18%), to rash (4%). While a vast majority of patients recovered spontaneously, a few received corticosteroids and other medications (e.g., sulfadoxine/pyrimethamine) [6].

S. nesbitti-affected travelers to Malaysia showed a biphasic course of muscular sarcocystosis, with fever, myalgia, fatigue, headache, arthralgia, elevated eosinophil counts (32%), moderately elevated liver enzymes (60%–70%), and elevated CPK (15%) emerging in the first phase about 2 weeks (range 9–26 days) after return; myalgia, fever, fatigue, arthralgia, headache, myositis (frequently accompanied by pain and swelling in the thighs or the facial/cervical region), transient rash, eosinophilia (67%), elevated CPK (90%), and moderately elevated liver enzymes (60%–70%) in the second phase around 6 weeks after return [13].

Diagnosis. Individuals with a history of gastroenteritis and habit of eating raw or undercooked beef or pork may be suspected of intestinal sarcocystosis. As *S. hominis* and *S. suihominis* have prepatent periods of 14–18 days and 11–13 days, respectively, *Sarcocystis* oocysts/sporocysts may be absent in feces during the acute/early stage of intestinal infection. Individuals with compatible symptoms (e.g., fever, myalgia, headache, cough, episodic weakness, fatigue, arthralgia and muscle pain), history of travel or residence in places with native wildlife (e.g., Southeast Asia), and negative test results for *Toxoplasma* and *Trichinella* may be suspected of muscular sarcocystosis [1].

Microscopic detection of *Sarcocystis* oocysts (of 15 μm × 17 μm in size)/sporocysts (of 10 μm × 15 μm in size) in fecal sample provides a useful means of diagnosing intestinal sarcocystosis (Figure 25.1). The oocysts of *Sarcocystis* spp. differ notably from the oocysts of *Cryptosporidium*, *Cyclospora*, and *Cystoisospora* in shape and size [14]. Use of flotation technique (mixing feces with concentrated zinc sulfate, sucrose, sodium or cesium chloride solutions, or Percoll, followed by centrifugation at 500 × *g* to sediment fecal debris and concentrate oocysts/sporocysts at the surface), modified Kato thick smear, formalin-ether technique, direct smear, and periodic acid-Schiff stain helps increase the sensitivity of microscopic detection. Because *Sarcocystis* oocysts/sporocysts do not take up acid-fast stain well, they may be easily missed by Ziehl-Neelsen stain that is commonly used for the detection of apicomplexan parasites in stool [1].

Similarly, microscopic observation of sarcocysts in muscles is of diagnostic value for muscular sarcocystosis. Indeed, most human muscular sarcocystosis cases reported to date are confirmed with a positive periodic acid-Schiff reaction of sarcocyst wall from incidental muscle biopsies (or biopsies from regions of maximal tenderness, swelling, or increased warmth), or at autopsies (tongue muscle). Isolation and molecular detection of *Sarcocystis* DNA from muscle biopsy offers another means of diagnostic confirmation [15]. In addition, enzyme-linked immunosorbent assays (ELISA) and immunofluorescence assays (IFA) using bradyzoite antigens may be utilized for the serological diagnosis of muscular sarcocystosis [16].

Conventional technique for *Sarcocystis* speciation is dependent on variations in sarcocyst wall structure, which is subject to change with age. Recent analysis of small-subunit (SSU or 18S) ribosomal RNA (rRNA), internal transcribed spacer (ITS), and cytochrome C oxidase subunit I [COI, also known as CO1, COX1, or mitochondrially encoded cytochrome c oxidase I (MT-CO1)] sequences offers a more accurate, less cumbersome approach for *Sarcocystis* determination. Considering the fact that COI DNA sequences are easier than SSU rRNA sequences to align, maximum-parsimony analysis of COI data represents a method of choice to delineate closely related *Sarcocystis* species [17–19]. In fact, precise discrimination of *S. hominis* and *S. sinensis* sarcocysts in beef is critical in helping to avoid unnecessary meat contamination [20].

Treatment and prevention. In general, human intestinal sarcocystosis is self-limiting as it does not involve multiplication. However, some cases may persist due to absence of protective immunity or reinfection. No treatment is currently available for human intestinal sarcocystosis. Similarly, most human cases of muscular sarcocystosis resolve without medical intervention, although albendazole alone or in combination with prednisone may be prescribed to improve the clinical outcome of some cases. In addition, antiprotozoal drug co-trimoxazole (trimethoprim and sulfamethoxazole) may be administered to relieve muscular pain (myositis) [6].

Prevention of human intestinal sarcocystosis should focus on thorough cooking (e.g., at 60°C for 20 min, at 70°C for 15 min, or at 100°C for 5 min) or freezing (at −4°C for 48 h or at −20°C for 24 h) of meat to kill bradyzoites and destroy the toxins of *S. suihominis* and *S. suihominis* sarcocysts. Prevention of human muscular sarcocystosis involves filtration, proper cleaning/cooking, and chemical disinfection (with chlorine) to reduce exposure to *Sarcocystis* oocysts/sporocysts in contaminated drinking water, fresh produce, recreational water, and soil [1].

25.2 MOLECULAR PATHOGENESIS OF *SARCOCYSTIS* INFECTION: CURRENT STATUS

The genus *Sarcocystis* comprises over 200 zoonotic protozoan species for which mammals (74%), birds (14%), reptiles (10%), and fish (0.5%) function as intermediate hosts; while mammals (27%), reptiles (11%), birds (6%), and unknown (56%) act as definitive hosts. Although animals infected with *Sarcocystis* spp. rarely present any symptoms, humans affected by sarcocystosis may develop a range of clinical diseases, including myositis and vasculitis (muscular sarcocystosis) and nausea, abdominal pain, and diarrhea (intestinal sarcocystosis). The fact that *Sarcocystis* spp. are infective to a diverse range of animal hosts and induce a broad spectrum of clinical diseases in humans suggests their ability to invades host cells, evade host immune responses, and take advantage of host cellular machineries in their transition from free-living, photosynthetic organisms to obligate intracellular parasites [1].

Invasion. As obligate intracellular parasites, apicomplexans including *Sarcocystis* spp. utilize several functionally distinct secretory organelles (i.e., micronemes, rhoptries, and dense granules) for entry into the host cells and employ their gliding motility to move inside and between host cells. Located at the apical end, micronemes secrete several proteins for binding to host cell receptors after adhesion to the host. Through the combined actions of myosin A-motor complex (or glideosome), parasite actin, and micronemal transmembrane proteins of the thrombospondin-related anonymous protein (TRAP) family, along with motor-independent invasion mechanism, apicomplexans glide circularly, upright-twirlingly, and helically into the host cells. Further, micronemes produce apical membrane antigen 1 (AMA-1) protein for rhoptry release and pore-forming protein for parasite egress. Attached to the very apical end, rhoptries are pear- or club-shaped organelles that generate the rhoptry neck (RON) proteins to form the tight junction between the parasite and the host cell, rhoptry bulb proteins to modify the parasitophorous vacuolar membrane (PVM, which assists nutrient transport from the host cell, and blocks lysosomal fusion that harms the parasites) and the host cell, and rhoptry kinases to modulate virulence. Dense granules found throughout apicomplexans secrete proteins to modify the host cell as well as the vacuolar compartment, helping to transform the parasitophorous vacuoles into functionally active organelles (to allow the parasites to grow and replicate) [7,21].

Immune modulation. *Sarcocystis* spp. appear to induce minimal immune responses and transient immune memory in both definitive and intermediate hosts, and thus facilitate their evasion

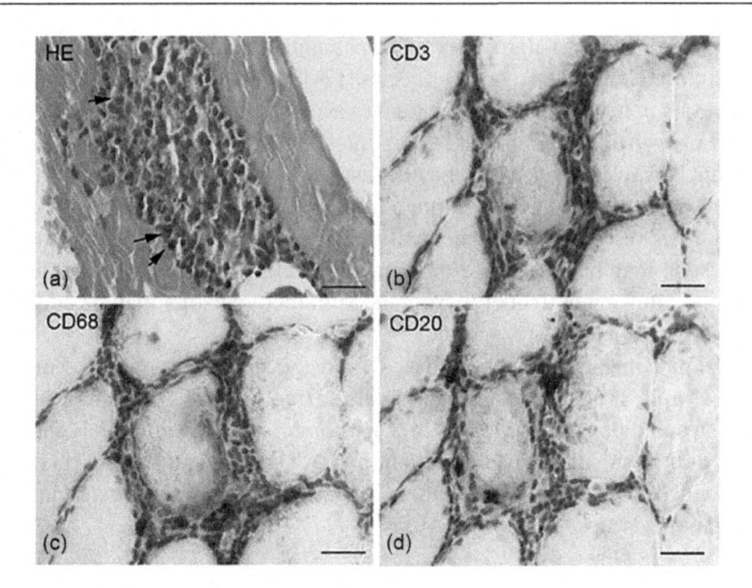

FIGURE 25.3 *Sarcocystis nesbitti*-related muscular sarcocystosis in a 46-year-old male showing endomysial inflammatory cell infiltrate with eosinophils (arrows) (a, H and E stain); invasion of muscle fibers by (CD3) T lymphocytes (b), (CD68) macrophages (c), and (CD20) B lymphocytes (scale bar, 40 μM). (Photo credit: Harris VC, van Vugt M, Aronica E, et al. *Curr Infect Dis Rep.* 2015;17(8):495.)

from host immune surveillance. This is exemplified by the marginal clinical symptoms associated with intestinal sarcocystosis (nausea, abdominal pain, loss of appetite, vomiting, and diarrhea) and muscular sarcocystosis (muscle pain) by the accumulation of immune cells (e.g., eosinophils, lymphocytes, and macrophages) that are insufficient to outright eliminate *S. nesbitti* sarcocysts in muscle (Figures 25.3 and 25.4) [16] and to stimulate a long-lasting immune memory against *S. nesbitti* reinfection; and by altered antigen presentation and/or lack of antigen recognition (leading

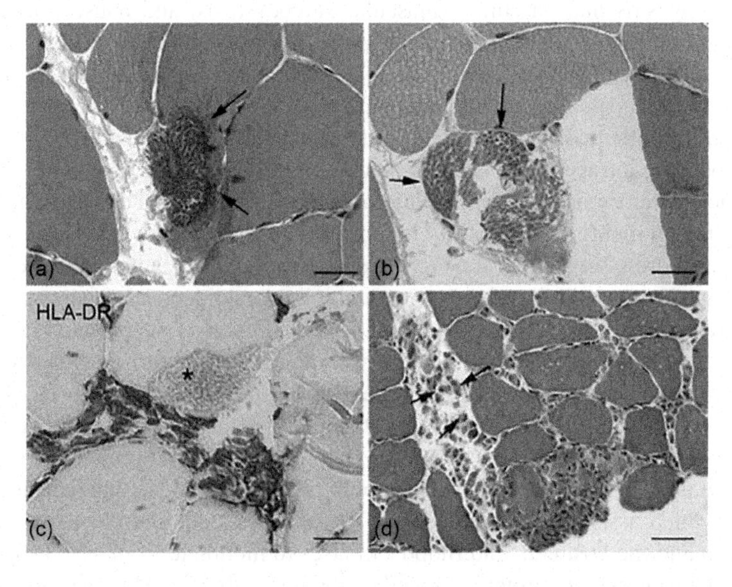

FIGURE 25.4 *Sarcocystis nesbitti*-related muscular sarcocystosis showing sarcocysts (arrows) within muscle fibers from a 46-year-old male (a) and a 51-year-old-female (b); major histocompatibility complex class II (MHC-II) inflammatory cell infiltrate around sarcocyst (asterisk) (46-year-old male, (c); and endomysial inflammatory cell infiltrate with eosinophils (arrows) (51-year-old-female, d) (H and E stain; scale bar, 40 μM). (Photo credit: Harris VC, van Vugt M, Aronica E, et al. *Curr Infect Dis Rep.* 2015;17(8):495.)

to decreased cytokine production) at the initial stage of infection, suggesting an early-phase immunosuppression, which may aid in the establishment of infection and promote possible development of *Sarcocystis*-related malignancy [22,23]. Indeed, it has been shown that *S. calchasi* (causal agent of pigeon protozoal encephalitis) adopts an immune evasion strategy (through down-regulation of interleukin [IL]-12/IL-18/interferon [IFN]-γ) during the early/schizogonic phase, and a delayed-type hypersensitivity reaction (as evidenced by upregulation of IFN-γ and tumor necrosis factor [TNF] α-related cytokines, in correlation with prominent MHC-II protein expression in areas of mononuclear cell infiltration and necrosis) during the late phase [24].

One type of molecules that assist apicomplexans in host cell invasion, cell cycle progression, and host immune evasion are protein kinases (PK). Indeed, within the genome of *S. neurona*, which causes equine protozoal myeloencephalitis (EPM) – a degenerative neurological disease of horses, 97 putative PK belonging to the classes of PKs A, G, and C (AGC), CMGC (including cyclin-dependent kinases [CDKs], mitogen-activated protein kinases [MAP kinases], glycogen synthase kinases [GSK], calmodulin/calcium-dependent PK [CAMK], casein kinase [CK], tyrosine kinase-like kinases [TKL], atypical PK [aPK], and other PK sub family [OPK]) are present. These molecules modify host cell structure and function and contribute to *Sarcocystis* invasion, proliferation, and differentiation [25].

Other mechanisms. The sequenced genome (130 Mb) of *S. neurona* SO SN1 strain includes numerous metabolic and molecular innovations that facilitate effective exploitation of alternative sources of energy for growth and survival and promote a heterogenous lifecycle involving a wide range of definitive and intermediate hosts [26].

25.3 FUTURE PERSPECTIVES

Members of the genus *Sarcocystis* are obligate intracellular parasites that infect a broad range of animals including humans. Specifically, human intestinal sarcocystosis is attributed to the consumption of raw/undercooked beef and pork harboring *S. hominis* and *S. suihominis* sarcocysts, whereas human muscular sarcocystosis is caused by the ingestion/drinking of food and water contaminated with *S. nesbitti* oocysts/sporocysts. Relative to other zoonotic apicomplexan pathogens (e.g., *Toxoplasma*), *Sarcocystis* spp. are inadequately studied. Unsurprisingly, many unanswered questions remain about *Sarcocystis* and sarcocystosis, such as the modes of transmission, immunobiological characteristics of the disease, and intervention strategies.

One of the important questions relates to the intermediate hosts of *S. nesbitti*. In Southeast Asia, snakes are regarded as a definitive host and nonhuman primates as an intermediate host for *S. nesbitti*. However, in Australia where nonhuman primates are absent, the identification of *S. nesbitti* in Australian snakes suggests the existence of alternative intermediate host(s) for this parasite [3,27]. While rodents are thought to assume such a role, there is an obvious need for experimental evidence to support this claim.

Another key question concerns the immune evasion strategy used by *Sarcocystis* spp., and its impact on the survival and establishment of parasite and its possible link to malignancy that often accompanies human sarcocystosis [1]. While elucidation of its impact on parasite survival and establishment is beneficial for designing innovative measures to disrupt *Sarcocystis* infection, clarification of its possible link to malignancy helps rule out or implement sarcocystosis control as part of malignancy management.

A third question hangs over the inadequacy/absence of intervention strategies (e.g., vaccines) for sarcocystosis in human and animal hosts [1]. Obviously, this conundrum stems from the poor understanding of molecular mechanisms of *Sarcocystis* pathogenicity as well as insufficient appreciation of the intricacies between host–parasite interplays. In view of the current availability of various sophisticated molecular, immunological, and biochemical tools at our disposal, it is only a matter of time (and money) that this question will be satisfactorily resolved in the not-too-distant future.

REFERENCES

1. Fayer R, Esposito DH, Dubey JP. Human infections with *Sarcocystis* species. *Clin Microbiol Rev.* 2015;28(2): 295–311.
2. Abe N, Matsuo K, Moribe J, et al. Morphological and molecular characteristics of seven *Sarcocystis* species from sika deer (*Cervus nippon centralis*) in Japan, including three new species. *Int J Parasitol Parasites Wildl.* 2019;10:252–62.
3. Wassermann M, Raisch L, Lyons JA, et al. Examination of *Sarcocystis* spp. of giant snakes from Australia and Southeast Asia confirms presence of a known pathogen – *Sarcocystis nesbitti. PLoS One.* 2017;12(11):e0187984.
4. Nimri L. Unusual case presentation of intestinal *Sarcocystis hominis* infection in a healthy adult. *JMM Case Rep.* 2014;1(4):e004069.
5. Ahmadi MM, Hajimohammadi B, Eslami G, et al. First identification of *Sarcocystis hominis* in Iranian traditional hamburger. *J Parasit Dis.* 2015;39(4):770–2.
6. Italiano CM, Wong KT, AbuBakar S, et al. *Sarcocystis nesbitti* causes acute, relapsing febrile myositis with a high attack rate: Description of a large outbreak of muscular sarcocystosis in Pangkor Island, Malaysia, 2012. *PLoS Negl Trop Dis.* 2014;8:e2876.
7. Morrissette NS, Sibley D. Cytoskeleton of apicomplexan parasites. *Microbiol Mol Biol Rev.* 2002;66(1):21–38.
8. Chhabra MB, Samantaray S. *Sarcocystis* and sarcocystosis in India: Status and emerging perspectives. *J Parasit Dis.* 2013;37(1):1–10.
9. Poulsen CS, Stensvold CR. Current status of epidemiology and diagnosis of human sarcocystosis. *J Clin Microbiol.* 2014;52(10):3524–30.
10. Abubakar S, Teoh BT, Sam SS, et al. Outbreak of human infection with *Sarcocystis nesbitti*, Malaysia, 2012. *Emerg Infect Dis.* 2013;19(12):1989–91.
11. Shahari S, Tengku-Idris TI, Fong MY, Lau YL. Molecular evidence of *Sarcocystis nesbitti* in water samples of Tioman Island, Malaysia. *Parasit Vectors.* 2016;9(1):598.
12. Tappe D, Abdullah S, Heo CC, Kannan Kutty M, Latif B. Human and animal invasive muscular sarcocystosis in Malaysia – Recent cases, review and hypotheses. *Trop Biomed.* 2013;30(3):355–66.
13. Esposito DH, Stich A, Epelboin L, et al. Acute muscular sarcocystosis: An international investigation among ill returned travelers returning from Tioman Island, Malaysia, 2011–2012. *Clin Infect Dis.* 2014;59:1401–10.
14. Meistro S, Peletto S, Pezzolato M, et al. *Sarcocystis* spp. prevalence in bovine minced meat: A histological and molecular study. *Ital J Food Saf.* 2015;4(2):4626.
15. Kwok CY, Ting Y. Atypical presentation of human acute muscular sarcocystosis: *Sarcocystis nesbitti* confirmed on molecular testing. *Am J Case Rep.* 2019;20:499–502.
16. Harris VC, van Vugt M, Aronica E, et al. Human extraintestinal sarcocystosis: What we know, and what we don't know. *Curr Infect Dis Rep.* 2015;17(8):495.
17. Hooshyar H, Abbaszadeh Z, Sharafati-Chaleshtori R, Arbabi M. Molecular identification of *Sarcocystis* species in raw hamburgers using PCR-RFLP method in Kashan, central Iran. *J Parasit Dis.* 2017;41(4):1001–5.
18. Hoeve-Bakker BJA, van der Giessen JWB, Franssen FFJ. Molecular identification targeting cox1 and 18S genes confirms the high prevalence of *Sarcocystis* spp. in cattle in the Netherlands. *Int J Parasitol.* 2019;49(11):859–66.
19. Rubiola S, Chiesa F, Zanet S, Civera T. Molecular identification of *Sarcocystis* spp. in cattle: Partial sequencing of cytochrome C oxidase subunit 1 (COI). *Ital J Food Saf.* 2019;7(4):7725.
20. Robertson LJ, Clark CG, Debenham JJ, et al. Are molecular tools clarifying or confusing our understanding of the public health threat from zoonotic enteric protozoa in wildlife? *Int J Parasitol Parasites Wildl.* 2019;9:323–41.
21. Gubbels MJ, Duraisingh MT. Evolution of apicomplexan secretory organelles. *Int J Parasitol.* 2012;42(12): 1071–81
22. Lewis SR, Ellison SP, Dascanio JJ, et al. Effects of experimental *Sarcocystis neurona*-induced infection on immunity in an equine model. *J Vet Med.* 2014;2014:239495.
23. Agholi M, Shahabadi SN, Motazedian MH, Hatam GR. Prevalence of enteric protozoan oocysts with special reference to *Sarcocystis cruzi* among fecal samples of diarrheic immunodeficient patients in Iran. *Korean J Parasitol.* 2016;54(3):339–44.
24. Olias P, Meyer A, Klopfleisch R, Lierz M, Kaspers B, Gruber AD. Modulation of the host Th1 immune response in pigeon protozoal encephalitis caused by *Sarcocystis calchasi. Vet Res.* 2013;44:10.

25. Murungi EK, Kariithi HM. Genome-wide identification and evolutionary analysis of *Sarcocystis neurona* protein kinases. *Pathogens*. 2017;6(1):12.
26. Blazejewski T, Nursimulu N, Pszenny V. et al. Systems-based analysis of the *Sarcocystis neurona* genome identifies pathways that contribute to a heteroxenous life cycle. *mBio*. 2015;6(1):e02445-14.
27. Lau YL, Chang PY, Subramaniam V, et al. Genetic assemblage of *Sarcocystis* spp. in Malaysian snakes. *Parasit Vectors*. 2013;6:257.

SUMMARY

Members of the genus *Sarcocystis* are obligate intracellular parasites that circulate between carnivores/omnivores (as definitive host) and herbivores (as intermediate host). In definitive carnivore/omnivore host (e.g., dog and cat), *Sarcocystis* is associated with intestinal sarcocystosis, whereas in intermediate herbivore host (e.g., cattle, pigs, and sheep), *Sarcocystis* causes muscular sarcocystosis, leading to abortion, reduced meat yield, increase meat contamination, and frequent death. In humans, intestinal sarcocystosis results from the consumption of raw/undercooked beef and pork harboring *S. hominis* and *S. suihominis* sarcocysts, whereas muscular sarcocystosis is due to the ingestion/drinking of food and water contaminated with *S. nesbitti* oocysts/sporocysts. Considering the current poor understanding of *Sarcocystis* transmission, immune evasion strategy, and molecular pathogenicity, further studies are essential to enable development of novel intervention strategies against this emerging zoonotic parasite.

Molecular Pathogenesis of *Trichinella spiralis* Infection

26

Dongyou Liu
RCPAQAP

Contents

26.1 INTRODUCTION

Trichinella spiralis is a member of the nematode genus *Trichinella* characterized by a relatively simple lifecycle that can be completed in a single host, by the production of various serine proteases and other molecules that aid its invasion of host cells, and by the formation of a collagenous capsule that shields from host immune attack. Since its first description in 1881 in pork in Xiamen City, China, *T. spiralis* has been shown to occur in >150 species of mammals, birds, and reptiles. Typically acquired through consumption of raw or undercooked port harboring *Trichinella* muscle larvae, humans trichinellosis has been documented in 55 countries around the world and is responsible for approximately 10,000 new cases each year. While the early stage of human trichinellosis may be asymptomatic or linked to mild illnesses (e.g., transient diarrhea and nausea, abdominal pain, vomiting, malaise, and low-grade fever), the late stage is associated with clinical diseases of differing severity (from fever, headache, cough, myositis, diffuse myalgia, periorbital/facial edema, conjunctivitis/subconjunctival hemorrhage, skin rash, dyspnea, dysphagia, hepatomegaly, myocarditis, encephalitis, to renal failure). Given the current absence of effective treatment and control measures for trichinellosis, an improved understanding of molecular mechanisms of *T. spiralis* pathogenicity is vital for the design of innovative therapeutics and vaccines to eradicate this zoonotic parasite.

Taxonomy. The genus *Trichinella* covers 12 species or genotypes, which can be separated into two distinct clades: *Trichinella* clade I is encapsulated (the invaded host muscle cells have a collagen capsule) and comprises six species and three genotypes [i.e., *T. spiralis* (T1), *T. nativa* (T2), *T. britovi* (T3), *T. murrelli* (T5), *T. nelsoni* (T7), *T. patagoniensis* (T12), and genotypes *Trichinella* T6, T8, and T9] and parasitize only mammals (with *T. spiralis* being basal to all encapsulated species, and *T. nativa* and *Trichinella* T6 being freeze-tolerant), whereas *Trichinella* clade II is nonencapsulated and consists of three species [i.e., *T. pseudospiralis* (T4), *T. papuae* (T10), and *T. zimbabwensis* (T11)], which infect mammals and birds (T4s) or mammals and reptiles (T10 and T11) [1]. It is noteworthy that *T. pseudospiralis* (T4) exhibits varying degrees of intraspecific genetic variability, and that *T. papuae* (T10) and *T. zimbabwensis* (T11) are capable of completing their lifecycle regardless of whether the host is warm-blooded or cold-blooded. There is evidence that the nonencapsulated and encapsulated clades diverged about 15–20 million years ago (MYA), and species likely emerged from the bases of the encapsulated clade, that is, *T. spiralis* and *T. nelsoni*, less than 10 MYA.

Although *T. spiralis* is the most common cause of human trichinellosis, other species (i.e., *T. nativa*, *T. nelson*, *T. britovi*, *T. pseudospiralis*, *T. murelli*, *T. papuae*) are occasionally involved in human diseases. Interestingly, a recent phylogenetic and genotyping study based on seven polymorphic microsatellite loci (TS103, TS128, TS130, TS1122, TS1131, TS1380, and TS1444), two mitochondrial (*cox*1, *cyt*b), and four nuclear (5S ISR, ESV, ITS1, 18S rRNA) genetic markers separated *T. spiralis* isolates from China into two clusters (i.e., GX-td and Tibet-lz) [2].

Morphology. *T. spiralis* is a relatively small nematode with adult males measuring 0.62–1.58 mm in length and 25–33 µm in width; and adult females measuring 0.952–3.35 mm in length and 26–43 µm in width. Dwelling in the epithelial tissues in the small intestine, adult females produce newborn larvae (NBL; average 110 µm in length and 7 µm in width), which move via the bloodstream to skeletal muscles, form collagen capsule around muscle larvae (male larvae of 0.641–1.07 mm in length and 26–38 µm in width; female larvae of 0.71–1.09 mm in length and 25–40 µm in width), which can remain viable for a long time (Figure 26.1) [3,4].

Similar to other *Trichinella* spp., *T. spiralis* possesses a stichocyte esophagus composed of stacks of cylindrical cells (as opposed to a muscular esophagus in *Caenorhabditis elegans*), has longitudinal rows of bacillary band cells that run along the length of the body, and shows anterior positioning of the endodermal precursor (as opposed to posterior positioning in *C. elegans*).

Genomics. The genomes of 12 currently recognized *Trichinella* taxa range from 46.1 to 51.5 Mb (mean 49.0 Mb) in size, with median GC% of 33, which on average contain 2.4% (range: 1.1%–3.7%) retrotransposons, 2.9% (0.7%–5.5%) DNA transposons, 7.7% (0.04%–10.6%) unclassified dispersed elements, and 4.8% (4.1%–5.8%) simple repeats; and which possess 11,006–16,067 (mean: 13,912) protein-encoding genes, with 7,691–12,000 (mean: 9,730) linked to biological (KEGG) pathways and 314–414 (mean: 363) to excretory/secretory (ES) proteins. Comparative genomic analyses of 12 *Trichinella* taxa indicate that in the encapsulated taxa, a lineage with *T. spiralis* + *T. nelsoni* is the sister to other taxa [including *T. patagoniensis*, and pairs of sister species (*T. nativa* + T6, *T. murrelli* + T9 and *T. britovi* + T8)], and in nonencapsulated taxa, the pair *T. papuae* + *T. zimbabwensis* is basal to *T. pseudospiralis* [1,5].

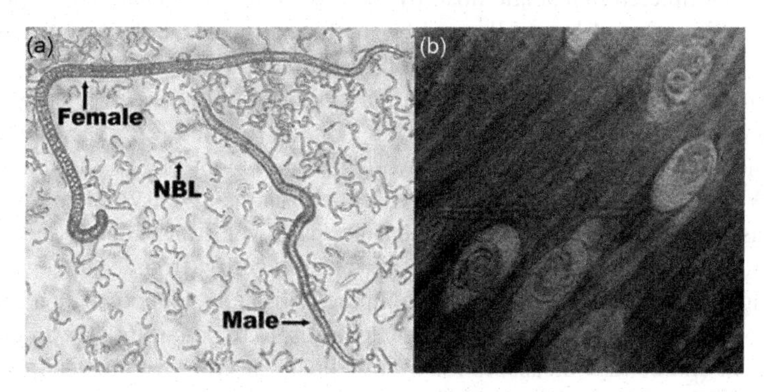

FIGURE 26.1 *Trichinella spiralis* adult worms and new born larvae (NBL) (a) and infective muscle larvae (b). (Photo credit: Gao F, Wang R, Liu M. *Front Physiol.* 2014;4:410.)

Trichinella mitochondrial DNA (mtDNA) is approximately 16,656–16,706 in length (due largely to a size polymorphism of a region downstream from *nad1* and *nad2*), with 37 or 36 genes. Specifically, *T. spiralis* mtDNA consists of 37 genes, including 13 protein-coding genes [i.e., subunits 6 and 8 of the F_0 ATPase (*atp6*, 276 bp and *atp8*, 41 bp), cytochrome *c* oxidase subunits 1–3 (*cox1–cox3*, 514, 225, and 257 bp, respectively), apocytochrome *b* (*cob*, 371 bp), and NADH dehydrogenase subunits 1–6 (*nad1–6*, 299, 295, 116, 411, 518, and 156 bp, respectively) and 4L (*nad4L*, 81 bp)], 2 ribosomal RNAs [small and large subunit rRNAs (*rrnS*, 688 bp and *rrnL*, 947 bp)], and 22 transfer RNAs (tRNAs) [with the two leucine and two serine tRNAs differing in their anticodon sequences (uag/uaa and ucu/uga, respectively), namely, *trnL(uag)*, *trnL(uaa)*, *trnS(ucu)*, and *trnS(uga)*]. In contrast, all non-*spiralis Trichinella* mtDNA harbor 36 genes, with the notable absence of the protein-encoding gene *atp8*. It is noteworthy that *T. spiralis* mtDNA contains two strands (α and β) with significantly different nucleotide composition: the α-strand is AC rich and comprises the sense sequence of nine mRNAs, both ribosomal RNAs, and 12 tRNAs; while the β-strand is GT rich and includes all other genes [6].

Lifecycle. *T. spiralis* is an intracellular parasite with a direct lifecycle that is completed in a single host (i.e., adult worms in the small intestine and muscle larvae in the muscles of the same host). The lifecycle begins when meat containing first-stage muscle larvae is ingested by humans or animals, in which the capsule-like cyst is dissolved by digestive juices (pepsin and hydrochloric acid) from the stomach to release the larvae (0.9 h postinfection). After passage to the small intestine, the larvae penetrate into the columnar epithelium, where they molt four times (10–28 h postinfection) and mature into adults (30–34 h postinfection). Upon mating, females produce at 5 days postinfection NBL (about 1,500 during its life span of 6 weeks), which invade the tissue, then enter the lymphatics and blood circulation at the thoracic duct, and, finally, reach the skeletal muscle fibers via the capillaries (tiny blood vessels). Inside the muscle fibers, the larvae form capsule, become infective within 15 days, and remain viable for 4–6 month in humans [7].

Epidemiology. *Trichinella* spp. have a global distribution, from the arctic to the tropics, with encapsulated species showing better adaptation to cold climates than nonencaspulated species. Over >150 domestic animal (e.g., swine, horses, and rodents) and wildlife (e.g., bear, wild boar, moose, seals, walrus, birds, and reptiles) species are affected [7].

Human infection (trichinellosis) results mainly from consumption of raw or inadequately cooked pork harboring muscle larvae. To date, >10 million people in 55 countries have been infected, and about 10,000 new cases are reported each year. In countries (e.g., China) where pork is commonly consumed, human trichinellosis is particularly common. An increasing incidence is noted in Europe due to recent preference for antibiotic-free meat. On the other hand, annual human trichinellosis cases in the United States have dropped from around 400 in the 1940s to around 20 between 2008 and 2010, with hunters and others who eat meat from wildlife being at a greater risk.

Clinical features. Patients with the early stage (enteric or gastrointestinal phase) of trichinellosis (2–7 days after consumption of raw or undercooked meat harboring *Trichinella* muscle larvae), during which larvae are released in the stomach, invade the small intestine, and grow into adult worms, may be asymptomatic or associated with mild transient diarrhea and nausea (due to larval and adult worm invasion of the intestinal mucosa), abdominal pain, vomiting, malaise, and low-grade fever [7].

Patients with the late stage (systemic or parenteral phase) of trichinellosis (2–6 weeks postinfection), during which NBL enter the lymphatic and blood circulation and migrate to the skeletal muscles, myocardium, and brain often present with fever, headache, cough, myositis (chiefly in the mid-abdomen, face/masseter, and chest/intercostal muscles), diffuse myalgia, periorbital/facial edema, conjunctivitis/subconjunctival hemorrhage (50% of cases), skin rash (urticaria, splinter hemorrhage on nailbeds, retinal hemorrhage, and petechiae), dyspnea, dysphagia, hepatomegaly, myocarditis, encephalitis, and renal failure [7].

Diagnosis. Initial diagnosis of trichinellosis is based on observation of clinical signs and symptoms, and further confirmation is obtained with biochemical tests, serological assays, imaging studies, muscle biopsy/histopathology, and molecular identification.

Complete blood count may reveal leukocytosis and eosinophilia, and biochemical tests may show elevated creatine kinase, lactate dehydrogenase, aldolase, and aminotransferases (due to parasites causing muscle destruction), as well as hypokalemia, hypoalbuminemia, and increased serum IgE levels.

Serological assays such as enzyme-linked immunosorbent assay (ELISA), indirect immunofluorescence, latex agglutination test, and western blot provide additional evidence for late trichinellosis, but not during the first 3 weeks of the disease. Indeed, ELISA using ES antigens permits accurate diagnosis of human trichinellosis and effective epidemiological investigation in swine and human populations [8].

X-ray imaging may show calcific densities in the muscles, CT helps rule out other causes of neurological dysfunction, and ECG may uncover features of pericarditis, ischemia, or myocarditis.

Muscle biopsy (performed at 4 weeks or more after infection) facilitates microscopic detection of muscle larvae (directly or after digestion in HCL-pepsin or histological staining) and gives a definitive diagnosis of trichinellosis.

Molecular identification enables species-specific determination and epidemiological tracking of *Trichinella* [9,10]. Specifically, PCR amplification and sequencing analysis of the internal transcribed spacers ITS1, ITS2, and expansion segment ESV of the rRNA repeat, as well as CO I mitochondrial gene facilitate unequivocal identification of *Trichinella* spp. [4].

For diagnosis of *T. spiralis* in domestic swine, the preferred method involves digestion of meat samples (e.g., diaphragm, masseters, tongue, or other muscles) with HCL-pepsin followed by microscopic inspection for larvae. Alternatively, ELISA using ES antigen or tyvelose-based epitope permits rapid detection of late *T. spiralis* infection in serum samples.

Treatment. Human trichinellosis is mostly self-limiting and patients often recover within 2–6 months of infection with minimal treatment. Individuals with mild infection may be managed with antipyretics and anti-inflammatory agents, while those with systemic complications are treated with antihelminthic agents (albendazole 500 mg twice daily given orally for 10–14 days or mebendazole 200–400 mg thrice daily for 3 days, and 400–500 mg three times daily for 10 days) and corticosteroids (prednisone 30–60 mg daily for 10–14 days). Pregnant women with trichinellosis may be prescribed mebendazole, albendazole, or levamisole after the first trimester [7].

Prevention. Prevention of trichinellosis involves avoidance of raw or undercooked pork or wildlife harboring *Trichinella* muscle larvae, proper cooking and freezing of pork, and control of *Trichinella* infection in pigs and other animals [11,12].

26.2 MOLECULAR PATHOGENESIS OF *T. SPIRALIS* INFECTION: RECENT FINDINGS

The pathogenesis of *T. spiralis* infection is built essentially upon the ability of this parasite: (i) to produce large numbers of NBL that ensure survival during tissue migration; (ii) to secrete various functional enzymes and proteins that remove host barrier and facilitate invasion of host cells; (iii) to form collagen capsule that shields muscle larvae from host-acquired immune responses; and (iv) to manipulate host immune system for immune evasion.

Fecundity. Despite a relatively short life span (about 6 weeks), *T. spiralis* adult females compensate by having a high fecundity and producing about 1,500 NBL each during their life time. This ensures that sufficient numbers of NBL remain during the tissue migration phase before they settle in the skeletal muscles and other sites [4].

Functional enzymes and proteins. *T. spiralis* genome consists of 50 Mb DNA and 14,745 putative coding genes. Among 185 *T. spiralis* proteins characterized to date, 60 display hydrolase activity (including serine protease, cytosol aminopeptidase, chymotrypsin-like protease, chymotrypsin-like protease, and enteropeptidase). Serine proteases with chymotrypsin-like, elastase-like, or trypsin-like activities are found abundantly in the ES products and crude extract *T. spiralis* larvae and adult worms, and play critical roles in parasite development and nutrition, host tissue and cell invasion, anticoagulation, and immune evasion (Table 26.1) [13–22]. For instance, Ts31 is a 31.3 kDa, 285 aa ES protein localized at the

TABLE 26.1 Biological and molecular characteristics of *Trichinella spiralis* enzymes and proteins involved in host cell invasion and immune evasion

PROTEIN	*MOLECULAR MASS (KDA)/ AMINO ACIDS*	*ORIGIN*	*FUNCTION*
TS15-1	71.6/667	TS15-1 (TsSerp) is localized outside as well as on the inner layer of the cuticle and stichocytes of muscle larvae (i.e., during nurse cell formation 1–4 weeks postinfection)	TS15-1 displays serine protease activity and acts as a key host collagen-inducing factor, leading to increased expression of type I collagen by host muscle tissue, as well as upregulation of TGF-βI pathway-related signal proteins (Smad2/3)
Ts31	31.3/285	An excretory/secretory (ES) protein localized principally at the stichosome (secretory organ) and cuticle of *T. spiralis* muscle larvae, immature larvae, and adult worm	Ts31 has a domain of trypsin-like serine protease with an active site carrying a classic catalytic triad (serine–histidine–aspartate) for proteolysis and plays a role in *T. spiralis* penetration of intestinal epithelium
TspSP-1	44.9/454	TspSP-1 (Ts32-2) is an ES product of *T. spiralis* muscle larvae and adult worms	TspSP-1 is a serine protease involved in degradation of intercellular or cytoplasmic proteins, and facilitates intestinal larval invasion and nurse cell formation
TspSP1.2	35.5/313	*T. spiralis* serine protease 1.2 (TsSP1.2) is an ES as well as a surface protein of *T. spiralis* muscle larvae and intestinal infectious larvae	TspSP1.2 2 has a domain of trypsin-like serine protease involved in larval invasion of enterocytes, as well as in initiation of a concurrent Th1/Th2 immune response and intestinal mucosal IgA response
TsSP	45.2/411	TsSP is a protein component from somatic and ES products of *T. spiralis* muscle larvae, which is detectable in the cuticles and stichosomes of muscle larva, intestinal infectious larva, adult worm, and the embryos within uterus of female adult at 3 days postinfection	TsSP contains a domain of trypsin-like serine protease and is involved in larval invasion and growth, as well as in stimulation of Th2 predominant immune response
TsSP-ZH68	47.5/429	TsSP-ZH68 is one of ES proteins of *T. spiralis* adult worms, and recognized by sera of early patients with trichinellosis at 19 and 35 days postinfection	TsSP-ZH68 belongs to peptidase S1A, chymotrypsin family, with the ability to break down proteins and polypeptide, and facilitates larval invasion of intestinal epithelial cells
TsE	47.3/449	*T. spiralis* elastase-1 (TsE or TsEla) is highly expressed by muscle larvae and intestinal infectious larvae	Being a trypsin-like serine protease, TsE participates in parasite penetration, and stimulates a mixed Th1/Th2 response, leading to elevated levels of Th1 (IFN-γ, IL-2) and Th2 (IL-4, IL-10) cytokines
ES L1		ES L1 is a mix of ES antigens of *T. spiralis* muscle larvae released during the chronic phase of infection	ES L1 is involved in suppression of host inflammatory immune response against *T. spiralis*, and enhancement of Th2-type response that mitigates unwanted immune responses to autoantigens and allergens

(*Continued*)

TABLE 26.1 (*Continued*) Biological and molecular characteristics of *Trichinella spiralis* enzymes and proteins involved in host cell invasion and immune evasion

PROTEIN	*MOLECULAR MASS (KDA)/ AMINO ACIDS*	*ORIGIN*	*FUNCTION*
TsSPI	39.6/349	*T. spiralis* serine protease inhibitor (TsSPI) is an ES protein of adult worms highly expressed at the enteral stage (adult worm and newborn larvae) and distributed mainly in the cuticle and stichosome	TsSPI displays inhibitory activity against porcine trypsin and may assist larval invasion of intestinal endothelial cells
TsASP2	43/406	*T. spiralis* aspartic protease-2 (TsASP2) is expressed in all developmental stages and located in the hindgut, midgut, and muscle cells of muscle larvae and intestinal infectious larvae, as well as in areas surrounding embryos within the female uterus	TsASP2 shows aspartic protease activity and plays a role in hemoglobin, collagen, and IgM cleavage, facilitating larval invasion of intestinal epithelial cells
TsCB	40.2/356	*T. spiralis* cathepsin B (TsCB) is expressed in all lifecycle stages (adult worm, newborn larvae, muscle larvae, intestinal infective L1 larvae), and is primarily located in the cuticle and stichosome of the parasite	TsCB is a cysteine protease that facilitates intestinal infective L1 intrusion of host enteral epithelium during infection
TsCP	57/508	TsCP is a muscle larval crude protein located mainly in the cuticle, stichosome, and reproductive organs	TsCP is a cysteine protease involved in larval invasion of intestinal epithelial cells and orientation of Th2-predominant immune response
TsCstN	13/117	*T. spiralis* cystatin (TsCstN) is a component of ES L1 mixture produced by muscle larvae	TsCstN is a cysteine protease inhibitor with the ability to suppress proinflammatory cytokines and to interfere with the antigen presentation process through depletion of MHC class II expression
TsENO	52/473	*T. spiralis* enolase (TsENO) is distributed on the surface of parasite throughout the entire lifecycle	TsENO is a 2-phospho-D-glycerate hydrolyase that binds and activates host plasminogen (PLG) and participates in invasion of host intestinal wall
Tsgal	29.1/284	*T. spiralis* galectin (Tsgal) is located at the cuticles and stichosomes of larvae	Tsgal promotes larval invasion of intestinal epithelial cells
*Ts*Pmy	102/885	*T. spiralis* paramyosin (*Ts*Pmy) is expressed on the surface of *larvae* and adult worms	*Ts*Pmy interacts with complements C1q and C8/C9 and compromises their activation and function, thus forming part of an immune evasion strategy

stichosome (secretory organ) and cuticle of *T. spiralis* muscle larvae, immature larvae, and adult worm; harboring a domain of trypsin-like serine protease with an active site for proteolysis, Ts31 has been shown to participate in *T. spiralis* invasion of intestinal epithelial cells (Table 26.1) [23]. Another serine protease identified from somatic and ES products of *T. spiralis* muscle larvae, that is, TsSP, appears to occur in all developmental stages; besides potential involvement in larval invasion and growth, TsSP may elicit a Th2 predominant immune response in the host (Figure 26.2) [18,19]. Further, *T. spiralis* synthesizes a serine protease inhibitor (TsSPI), which demonstrates inhibitory effects on porcine trypsin and plays a possible role in larval invasion of intestinal endothelial cells [24].

In addition to serine proteases, *T. spiralis* also generates aspartic protease (e.g., TsASP2), cysteine protease (e.g., TsCB and TsCP), and enolase (e.g., TsENO) that facilitate larval invasion of intestinal epithelial cells, and orientation of Th2-predominant immune response in the host (Table 26.1) [25–27]. Moreover, *T. spiralis* produces galectin (Tsgal) and paramyosin (*Ts*Pmy) that promote larval invasion of intestinal epithelial cells and immune evasion, respectively (Table 26.1) [28,29].

Capsule formation. Once reaching the skeletal muscles, *T. spiralis* NBL infect muscle fibers by disrupting myofibrils, enlarging and centralizing host muscle nuclei, inducing host expression of acid phosphatase and syndecan-1, and forming capsule (or collagen capsule, which is composed of host-derived nurse cell and collagenous wall) that encloses the larvae (or muscle larvae) within 10–15 days. While the collagenous wall protects the parasite, the nurse cell provides metabolic support for (or nurses) the muscle larva.

Mechanistically, nurse cell formation shows analogy to muscle cell regeneration after injury or trauma. Specifically, following injury-induced muscle degeneration (and macrophage infiltration and cellular debris clean-up), muscle cell regeneration initiates satellite cell (a distinct subpopulation of myoblasts that fails to differentiate into mononucleated myocytes, which later fuse to each other to form a multinucleated syncytium and then muscle fibers, but that remains associated with the surface of the developing myofiber) activation, proliferation, differentiation/fusion, and self-renewal. In a similar manner, *Trichinella* infection of muscle cells induces satellite cell proliferation, differentiation, and fusion to each other or with the infected muscle cell, which dedifferentiates, reenters cell cycle, and arrests at G2/M, leading to nurse cell formation. The mature nurse cell contains increased amount of sarcoplasmic matrix; increased size and number of nuclei that move from the periphery to the center of muscle fiber; increased size of affected myofibers, increased number of mitochondria, DNA and RNA content, and free ribosomes; and increased proliferation of rough endoplasmic reticulum and smooth sarcoplasmic reticulum. It is notable that the nurse cell undergoes a transition from basophilic to eosinophilic cytoplasm. While the former results from transformation of the infected muscle cell after NBL invasion (basophilic transformation), the latter originates from satellite cells and expands as the former recedes [30,31].

At the molecular level, nurse cell formation requires mobilization of multiple genes (related to cell differentiation, proliferation, cell cycle control, and apoptosis) and signaling pathways (e.g., mitochondrial pathway comprising BAX, Apaf-1, and caspase 9; death receptor pathway consisting of TNF-α, TNF receptor I, TRADD, caspase 8, and caspase 3; TGF-β signaling pathway involving TGF-β, Smad2, and Smad4) [30].

The capsule wall (collagenous wall) is a noncellular structure not shared by normal muscle cell. Made up of a simple thickening of the basal lamina, which is a single noncellular sheet covering muscle cells and associated myoblasts (satellite cell) in normal muscles, the capsule wall comprises two layers: the inner layer is produced by the nurse cell and the outer layer is produced by fibroblasts around the capsule [30].

Overall, capsule formation (or cystogenesis) involves complex steps and takes up to >20 days from larval invasion to nurse cell formation.

Interaction with host immune system. *T. spiralis* infection stimulates T-helper 2 (Th2)-based responses (involving CD4+ T lymphocytes, mast cells, Th2 cytokine IL-13, and Th2 cellular signaling pathways) from the host, which (especially host mastocytosis) act on host cells lining mucosal surfaces (e.g., through IL-17-related intestinal muscle hypercontractility) and drive adult worm expulsion [31–35]. However, the level of Th2 immune responses is tightly regulated by the host as well as the parasite (which, e.g., secretes a 13 kDa cystatin or TsCstN with ability to inhibit cysteine proteases, suppress

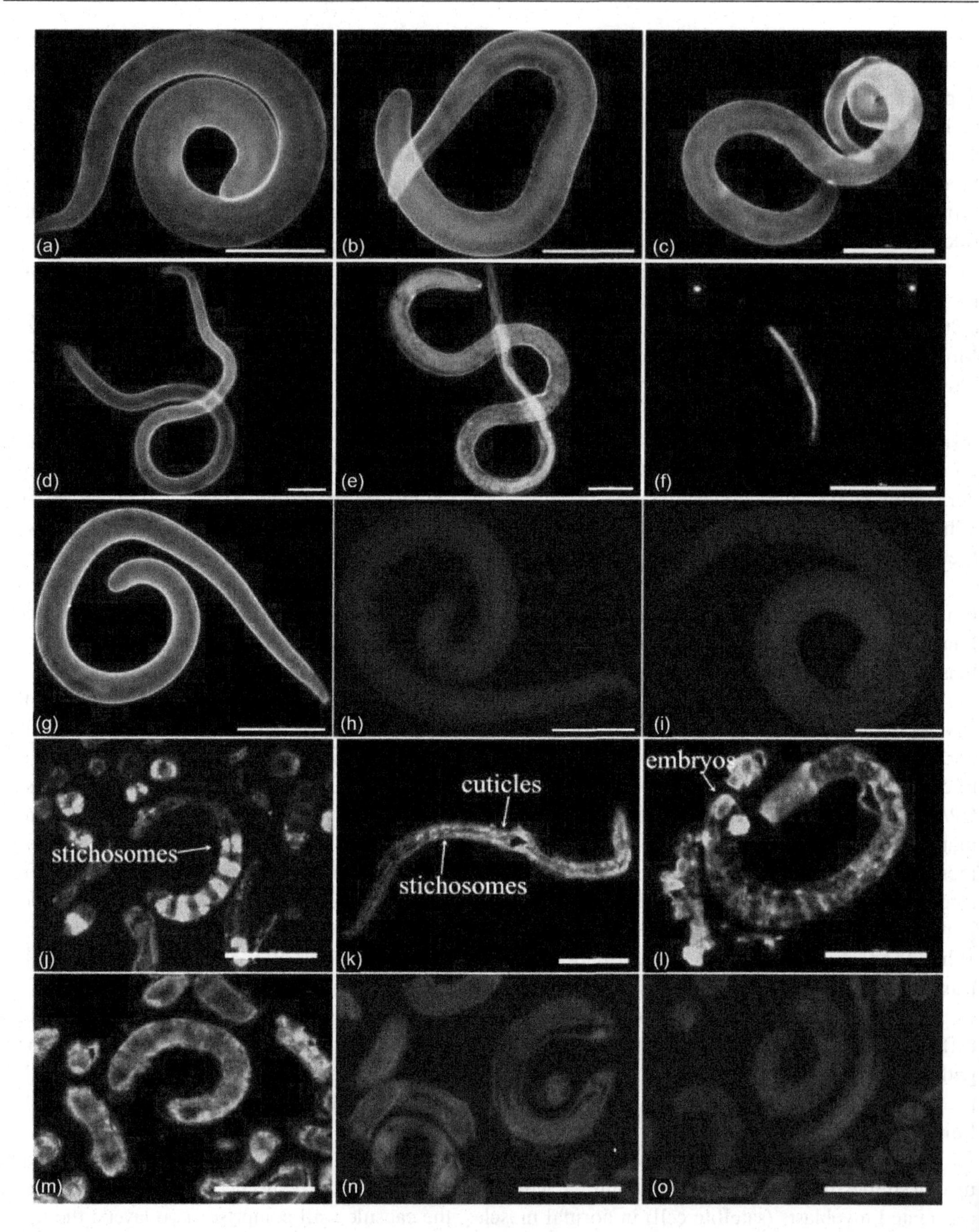

FIGURE 26.2 Immunolocalization of TsSP at *Trichinella spiralis* cuticles of muscle larva (a), intestinal infectious larva, 6h postinfection (b), intestinal infectious larva, 24h postinfection (c), female adult, 3days postinfection (d), female adult, 6days postinfection (e), and newborn larva (f), using anti-rTsSP serum in immunofluorescent assay, with muscle larva probed by infection serum as positive serum control (g), and muscle larvae incubated with normal mouse serum (h) and PBS (i) as negative controls. Similarly, positive staining is observed in the cuticles and stichosomes of muscle larvae (j), intestinal infectious larva (k), and female adult at 3days postinfection (l), using anti-rTsSP serum, with muscle larva probed by infection serum (m) as a positive serum control, and muscle larvae incubated with normal mouse serum (n) and PBS (o) as a negative control. Scale bars: 100µm. (Photo credit: Sun GG, Song YY, Jiang P, et al. *PLoS Negl Trop Dis*. 2018;12(5):e0006485.)

proinflammatory cytokines, and interfere with the antigen presentation process through depletion of MHC class II expression) to prevent unnecessary immunopathology to the host and to retain a suitable environment for long-term survival of the parasite [36,37]. In murine intestinal infection, *T. spiralis* inhibits Th1 immune response from 6h to 6days postinfection (dpi) and significantly increases Th2 immune response at 6 dpi in the spleen, while induces a Th1/Th2 mixed immune response from 3 dpi to 6 dpi and suppresses Th1 immune response at 6 dpi in the mesenteric lymph node [24,38–43].

26.3 PERSPECTIVES

Relative to other foodborne pathogens that cause significant morbidity and mortality, *T. spiralis* infection is mild and seldom fatal. This may partly explain the insufficient attention paid to *T. spiralis* in the past. With the increasing realization regarding its capability to modulate host immune responses for its own advantage, *T. spiralis* has become a model organism through which further molecular insights into the parasite–host relationship can be gained. The availability of such information will not only lead to innovative countermeasures against trichinellosis but also open an avenue for exploiting *T. spiralis* for potential remediation of other immune disorders (e.g., intestinal bowel disease, allergic and autoimmune diseases) and possibly tumors [44–50]. Therefore, additional experimental studies in these areas are important and necessary to transform this vista into reality.

REFERENCES

1. Korhonen PK, Pozio E, La Rosa G, et al. Phylogenomic and biogeographic reconstruction of the *Trichinella* complex. *Nat Commun.* 2016;7:10513.
2. Zhang X, Han LL, Hong X, et al. Genotyping and phylogenetic position of *Trichinella spiralis* isolates from different geographical locations in China. *Front Genet.* 2019;10:1093.
3. Gao F, Wang R, Liu M. *Trichinella spiralis*, potential model nematode for epigenetics and its implication in metazoan parasitism. *Front Physiol.* 2014;4:410.
4. Pozio E, Zarlenga D. International Commission on Trichinellosis: Recommendations for genotyping *Trichinella* muscle stage larvae. *Food Waterborne Parasitol.* 2019;15:e00033.
5. Mitreva M, Jasmer DP, Zarlenga DS, et al. The draft genome of the parasitic nematode *Trichinella spiralis. Nat Genet.* 2011;43(3):228–35.
6. Lavrov DV, Brown WM. *Trichinella spiralis* mtDNA: A nematode mitochondrial genome that encodes a putative ATP8 and normally structured tRNAS and has a gene arrangement relatable to those of coelomate metazoans. Genetics. 2001;157(2):621–37.
7. Rawla P, Sharma S. *Trichinella spiralis (trichnellosis). StatPearls [Internet].* Treasure Island, FL: StatPearls Publishing; 2019 Jan–2019 Nov 13.
8. Bruschi F, Gómez-Morales MA, Hill DE. International Commission on Trichinellosis: Recommendations on the use of serological tests for the detection of *Trichinella* infection in animals and humans. *Food Waterborne Parasitol.* 2019;14:e00032.
9. Almeida M, Bishop H, Nascimento FS, Mathison B, Bradbury RS, Silva AD. Multiplex TaqMan qPCR assay for specific identification of encapsulated *Trichinella* species prevalent in North America. *Mem Inst Oswaldo Cruz.* 2018;113(11):e180305.
10. Gómez-Morales MA, Ludovisi A, Amati M, Cherchi S, Tonanzi D, Pozio E. Differentiation of *Trichinella* species (*Trichinella spiralis*/*Trichinella britovi* versus *Trichinella pseudospiralis*) using western blot. *Parasit Vectors.* 2018;11(1):631.
11. Gamble HR, Alban L, Hill D, Pyburn D, Scandrett B. International Commission on Trichinellosis: Recommendations on pre-harvest control of *Trichinella* in food animals. *Food Waterborne Parasitol.* 2019;15:e00039.

12. Noeckler K, Pozio E, van der Giessen J, Hill DE, Gamble HR. International Commission on Trichinellosis: Recommendations on post-harvest control of *Trichinella* in food animals. *Food Waterborne Parasitol.* 2019;14:e00041.
13. Yang Y, Wen Yj, Cai YN, et al. Serine proteases of parasitic helminths. *Korean J Parasitol.* 2015;53(1):1–11.
14. Li JF, Guo KX, Qi X, et al. Protective immunity against *Trichinella spiralis* in mice elicited by oral vaccination with attenuated *Salmonella*-delivered TsSP1.2 DNA. *Vet Res.* 2018;49(1):87.
15. Park MK, Kim HJ, Cho MK, et al. Identification of a host collagen inducing factor from the excretory secretory proteins of *Trichinella spiralis. PLoS Negl Trop Dis.* 2018;12(11):e0006516.
16. Song YY, Zhang Y, Ren HN, et al. Characterization of a serine protease inhibitor from *Trichinella spiralis* and its participation in larval invasion of host's intestinal epithelial cells. *Parasit Vectors.* 2018;11(1):499.
17. Song YY, Wang LA, Na Ren H, et al. Cloning, expression and characterisation of a cysteine protease from Trichinella spiralis. *Folia Parasitol (Praha).* 2018;65:1–11. pii: 2018.007.
18. Sun GG, Ren HN, Liu RD, et al. Molecular characterization of a putative serine protease from *Trichinella spiralis* and its elicited immune protection. *Vet Res.* 2018;49(1):59.
19. Sun GG, Song YY, Jiang P, et al. Characterization of a *Trichinella spiralis* putative serine protease. Study of its potential as sero-diagnostic tool. *PLoS Negl Trop Dis.* 2018;12(5):e0006485.
20. Steel N, Faniyi AA, Rahman S, et al. TGFβ-activation by dendritic cells drives Th17 induction and intestinal contractility and augments the expulsion of the parasite *Trichinella spiralis* in mice. *PLoS Pathog.* 2019;15(4):e1007657.
21. Hu CX, Jiang P, Yue X, et al. Molecular characterization of a *Trichinella spiralis* elastase-1 and its potential as a diagnostic antigen for trichinellosis. *Parasit Vectors.* 2020;13(1):97.
22. Zhang XZ, Sun XY, Bai Y, et al. Protective immunity in mice vaccinated with a novel elastase-1 significantly decreases *Trichinella spiralis* fecundity and infection. *Vet Res.* 2020;51(1):43.
23. Ren HN, Guo KX, Zhang Y, et al. Molecular characterization of a 31 kDa protein from *Trichinella spiralis* and its induced immune protection in BALB/c mice. *Parasit Vectors.* 2018;11(1):625.
24. Xu J, Yu P, Wu L, Liu M, Lu Y. Regulatory effect of two Trichinella spiralis serine protease inhibitors on the host's immune system. *Sci Rep.* 2019;9(1):17045.
25. Cui J, Han Y, Yue X, et al. Vaccination of mice with a recombinant novel cathepsin B inhibits *Trichinella spiralis* development, reduces the fecundity and worm burden. Parasit Vectors. 2019;12(1):581.
26. Jiang P, Zao YJ, Yan SW, et al. Molecular characterization of a *Trichinella spiralis* enolase and its interaction with the host's plasminogen. *Vet Res.* 2019;50(1):106.
27. Xu J, Liu RD, Bai SJ, et al. Molecular characterization of a *Trichinella spiralis* aspartic protease and its facilitation role in larval invasion of host intestinal epithelial cells. *PLoS Negl Trop Dis.* 2020; 14(4): e0008269.
28. Wang Z, Hao C, Huang J, Zhuang Q, Zhan B, Zhu X. Mapping of the complement C1q binding site on *Trichinella spiralis* paramyosin. *Parasit Vectors.* 2018;11(1):666.
29. Xu J, Yang F, Yang DQ, et al. Molecular characterization of *Trichinella spiralis* galectin and its participation in larval invasion of host's intestinal epithelial cells. *Vet Res.* 2018;49(1):79.
30. Wu Z, Sofronic-Milosavljevic L, Nagano I, et al. *Trichinella spiralis*: Nurse cell formation with emphasis on analogy to muscle cell repair. *Parasit Vectors.* 2008;1: 27.
31. Ock MS, Cha HJ, Choi YH. Verifiable hypotheses for thymosin β4-dependent and -independent angiogenic induction of *Trichinella spiralis*-triggered nurse cell formation. *Int J Mol Sci.* 2013;14(12):23492–8.
32. Fabre MV, Beiting DP, Bliss SK, Appleton JA. Immunity to *Trichinella spiralis* muscle infection. *Vet Parasitol.* 2009;159(3–4):245–8.
33. Angkasekwinai P, Srimanote P, Wang YH, et al. Interleukin-25 (IL-25) promotes efficient protective immunity against *Trichinella spiralis* infection by enhancing the antigen-specific IL-9 response. *Infect Immun.* 2013;81(10):3731–41.
34. Luo XC, Chen ZH, Xue JB, et al. Infection by the parasitic helminth *Trichinella spiralis* activates a Tas2r-mediated signaling pathway in intestinal tuft cells. *Proc Natl Acad Sci USA.* 2019;116(12):5564–9.
35. Wang N, Bai X, Tang B, et al. Primary characterization of the immune response in pigs infected with Trichinella spiralis. *Vet Res.* 2020;51(1):17.
36. Cho MK, Park MK, Kang SA, Choi SH, Ahn SC, Yu HS. *Trichinella spiralis* infection suppressed gut inflammation with CD4(+)CD25(+)Foxp3(+) T cell recruitment. *Korean J Parasitol.* 2012;50(4):385–90.
37. Kobpornchai P, Flynn RJ, Reamtong O, et al. A novel cystatin derived from *Trichinella spiralis* suppresses macrophage-mediated inflammatory responses. *PLoS Negl Trop Dis.* 2020;14(4):e0008192.
38. Blum LK, Mohanan S, Fabre MV, Yafawi RE, Appleton JA. Intestinal infection with *Trichinella spiralis* induces distinct, regional immune responses. *Vet Parasitol.* 2013;194(2–4):101–5.
39. Ilic N, Gruden-Movsesijan A, Cvetkovic J, et al. *Trichinella spiralis* excretory-secretory products induce tolerogenic properties in human dendritic cells *via* Toll-like receptors 2 and 4. *Front Immunol.* 2018;9:11.

40. Farid AS, Fath EM, Mido S, Nonaka N, Horii Y. Hepatoprotective immune response during *Trichinella spiralis* infection in mice. *J Vet Med Sci.* 2019;81(2):169–76.
41. Zhang R, Sun Q, Chen Y, et al. Ts-Hsp70 induces protective immunity against *Trichinella spiralis* infection in mouse by activating dendritic cells through TLR2 and TLR4. *PLoS Negl Trop Dis.* 2018;12(5):e0006502.
42. Song Y, Xu J, Wang X, et al. Regulation of host immune cells and cytokine production induced by *Trichinella spiralis* infection. *Parasite.* 2019;26:74.
43. Sun XM, Guo K, Hao CY, Zhan B, Huang JJ, Zhu X. Trichinella spiralis excretory-secretory products stimulate host regulatory T cell differentiation through activating dendritic cells. *Cells.* 2019;8(11):1404.
44. Sofronic-Milosavljevic L, Ilic N, Pinelli E, Gruden-Movsesijan A. Secretory products of *Trichinella spiralis* muscle larvae and immunomodulation: Implication for autoimmune diseases, allergies, and malignancies. *J Immunol Res.* 2015;2015:523875.
45. Bai X, Hu X, Liu X, Tang B, Liu M. Current research of trichinellosis in China. *Front Microbiol.* 2017;8:1472.
46. Cheng Y, Zhu X, Wang X, et al. *Trichinella spiralis* infection mitigates collagen-induced arthritis *via* programmed death 1-mediated immunomodulation. *Front Immunol.* 2018;9:1566.
47. Liao C, Cheng X, Liu M, Wang X, Boireau P. *Trichinella spiralis* and tumors: Cause, coincidence or treatment? *Anticancer Agents Med Chem.* 2018;18(8):1091–9.
48. Patel N, Kreider T, Urban JF Jr, Gause WC. Characterisation of effector mechanisms at the host:parasite interface during the immune response to tissue-dwelling intestinal nematode parasites. *Int J Parasitol.* 2009;39(1):13–21.
49. Liu S, Pan J, Meng X, Zhu J, Zhou J, Zhu X. *Trichinella spiralis* infection decreases the diversity of the intestinal flora in the infected mouse. *J Microbiol Immunol Infect.* 2019. pii: S1684-1182(19)30165-3.
50. Sun S, Li H, Yuan Y, et al. Preventive and therapeutic effects of *Trichinella spiralis* adult extracts on allergic inflammation in an experimental asthma mouse model. *Parasit Vectors.* 2019;12(1):326.

SUMMARY

The genus *Trichinella* covers 12 nematode species and genotypes that parasitize over 150 different domestic animals and wildlife. Human infection with *Trichinella* spp., particularly *T. spiralis*, results largely from consumption of raw or inadequately cooked pork harboring muscle larvae. With the capacity to complete its lifecycle in a single host, to produce various serine proteases and other molecules for intestinal and tissue invasion, and to form collagenous capsule for immune evasion, *T. spiralis* is responsible for causing a diverse spectrum of clinical symptoms in affected patients. While histological, serological, and molecular analyses provide a useful means of confirming the diagnosis of human trichinellosis, antihelminthic drugs may be applied for its treatment. Further investigation on the molecular basis of *T. spiralis* pathogenicity is crucial for developing novel control measures against this overlooked zoonotic parasite.

Molecular Identification of *Cyclospora cayetanensis*

27

Dongyou Liu
Royal College of Pathologists of Australasia Quality Assurance Programs

Contents

27.1 INTRODUCTION

Cyclospora is a genus of apicomplexan parasites that affect various insects, reptiles, and mammals. From its initial finding in mole in 1870, and subsequent observation in millipede in 1881, when the genus was formally proposed, *Cyclospora* has attracted scant attention from the scientific community because of its apparent medical irrelevance. However, a series of events that took place since the 1970s have changed our perceptions about *Cyclospora*. These included the reports of diarrheal cases in Papua New Guinea in 1979; traveler's diarrhea in Haiti, Nepal, and Peru in the 1980s; and chronic diarrhea in acquired immunodeficiency syndrome (AIDS) patients in Canada and United States in the 1990s, which were all attributed to *C. cayetanensis*, a species formally recognized in 1994 after successful *in vitro* excystation and sporulation of its oocysts. Consequently, strenuous efforts have been directed to ward the studies of *Cyclospora* spp., including development and application of molecular techniques for their detection, typing, and phylogenetic analysis, leading to the identification of a total of 21 species to date, of which *C. cayetanensis* is the only species implicated in human cyclosporiasis (Table 27.1) [1].

TABLE 27.1 Morphological and biological features of *Cyclospora* spp.

SPECIES	*OOCYST SIZE (MM)*	*PRINCIPAL HOST*	*YEAR OF INITIAL DESCRIPTION*
C. glomericola	25–36×9–10	Insect	1881
C. caryolytica	18×12.5	Small mammal	1902
C. viperae	16.8×10.5	Reptile	1923
C. babaulti	17×10	Reptile	1924
C. scinci	10×7	Reptile	1924
C. tropidonoti	17×10	Reptile	1924
C. zamenis	17×10	Reptile	1924
C. niniae	14.6×13.3	Reptile	1965
C. talpae	15-18×10–12	Small mammal	1968
C. megacephali	18.5×15.7	Small mammal	1988
C. ashtabulensis	18×14.3	Small mammal	1989
C. parascalopi	16.5×13.6	Small mammal	1989
C. angimurinensis	19–24×16–22	Small mammal	1990
C. cayetanensis	8.6×8.6	Human	1994
C. colobi	8–9×8–9	Primate	1999
C. cercopitheci	8–10×8–10	Primate	1999
C. papionis	8–10×8–10	Primate	1999
C. schneideri	19.8×16.6	Reptile	2005
C. macacae	8–9×8–9	Primate	2015
C. duszynskii	11.4×10	Small mammal	2018
C. yatesi	17×15.2	Small mammal	2018

Taxonomy. Classified taxonomically in the family Eimeriidae, order Eucoccidiorida, subclass Coccidiasina, phylum Apicomplexa, and kingdom Protista, the genus *Cyclospora* currently comprises 21 valid species, of which seven occur in reptiles, one (*C. glomericola*) in insects (millipede), eight in small mammals (moles and rodents), four (*C. colobi, C. cercopitheci, C. papionis* and *C. macacae*) in nonhuman primates, and one (*C. cayetanensis*) in humans (causing diarrhea and flu-like illness known as cyclosporiasis) (Table 27.1) [2–4].

Morphology. *Cyclospora* demonstrates considerable morphological similarities to other coccidian parasites such as *Cryptosporidium*, *Isospora*, and *Sarcocystis*. However, *Cyclospora* (whose oocysts possess two sporocysts each containing two sporozoites) (Figure 27.1) differs from *Cryptosporidium* (whose oocysts contain no sporocyst but four sporozoites), *Isospora*, and *Sarcocystis* (whose oocysts harbor two sporocysts each containing four sporozoites). In addition, *Cyclospora* oocysts (8–36 µm×7–22 µm in size) (Table 27.1) are larger than *Cryptosporidium* oocysts (5.2 µm×4.9 µm), but smaller than *I. belli* oocysts (30 µm×12 µm) and *Sarcocystis* oocysts (30 µm ×15 µm) [5].

Specifically, *C. cayetanensis* oocysts measure about 8–10 µm in diameter, appear spheroidal/spherical, and show a characteristic blue or green outer ring fluorescence under UV light (Figure 27.1) [6], in contrast to the larger, differently shaped oocysts of other *Cyclospora* spp. Further, *C. cayetanensis* oocysts possess a thin (<1 µm), bilayered, colorless wall, polar, body and oocyst residuum. A sporulated oocyst consists of two sporocysts, each harboring two sporozoites. Sporocysts (4×6 µm in size) are ovoidal and contain both Stieda and substieda bodies and a large residuum, whereas sporozoites (1×9 µm in size) are elongated and lack crystalloid body or refractile bodies [7].

Genome. The genome of *C. cayetanensis* is about 44.36 Mb in length, including 6,043 predicted coding sequences (and 5,793 identifiable proteins), 118 tRNA genes, 9 rRNA genes, 20 other RNA genes, and 103 pseudogenes, with an overall G+C content of 51.9%. Among the 5,793 identifiable proteins, some are involved in essential metabolism (e.g., glycolysis, tricarboxylic acid cycle, pentose phosphate pathway)

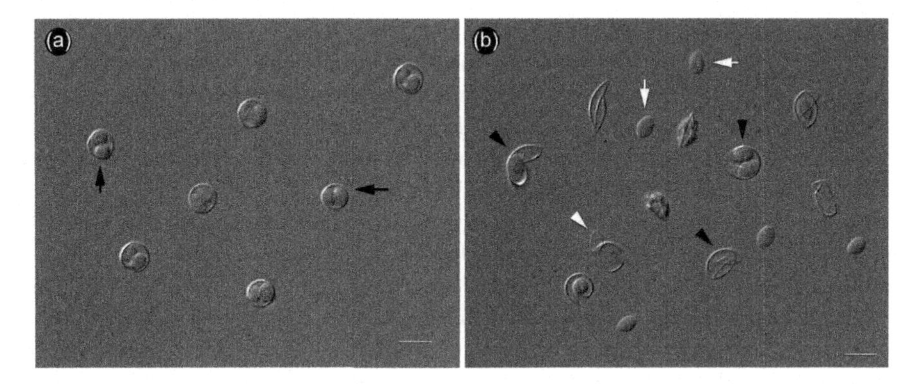

FIGURE 27.1 *Cyclospora cayetanensis* oocysts (8–10 μm in size) before (a) and after (b) 15 cycles of freeze and thaw, with intact oocysts indicated by black arrows; partially or completely empty oocysts by black arrowheads; sporocysts by white arrows; and empty sporocysts by white arrowhead (scale bar=10 μm). (Photo credit: Qvarnstrom Y, Wei-Pridgeon Y, Van Roey E, et al. *Gut Pathog.* 2018;10:45.)

and invasion mechanism (although a smaller number of rhoptry protein kinases, phosphatases and serine protease inhibitors are identified in *C. cayetanensis* than other coccidians) [8–10].

The mitochondrial genome of *C. cayetanensis* is linear concatemer or circular mapping topology and consists of 6,273 bp, including 3 protein-coding genes [i.e., *cytb* (cytochrome), 1,080 bp (nt 128–1,207); *cox1* (cytochrome C oxidase subunit 1), 1,443 bp (nt 1,248–2,690); *cox3* (cytochrome C oxidase subunit 3), 780 bp (nt 4,226–5,005)], 14 large subunit (28S), and nine small-subunit (18S) rRNA genes, with an overall GC content of 33% [i.e., A (30%), T (36%), C (16%), and G (17%)] (Figure 27.2) [11–13].

The apicoplast genome of *C. cayetanensis* is circular and consists of 34,155 bp, encoding the complete machinery for protein biosynthesis, and contains two inverted repeats that differ slightly in 18S rRNA gene sequences [14,15].

Lifecycle and biology. *C. cayetanensis* is an obligate intracellular parasite with the ability to complete its lifecycle in the epithelial cells of human gastrointestinal tract. Following ingestion of contaminated food or water, sporulated oocysts (8–10 μm) excyst in the gut with the help of bile, trypsin, and proteolytic enzymes, and release crescent-shaped sporozoites (1.2×9.0 μm in size), which invade the epithelial cells of the jejunum as well as duodenum. Sporozoites undergo sexual reproduction (merogony) in the parasitophorous vacuoles of the epithelial cells, yielding 8–12 merozoites (0.5×3–4 μm), which are capable of infecting new host cells or initiating another cycle of merogony, yielding four merozoites (0.8×12–15 μm). The four merozoites from the second round of merogony infect new cells and undergo sexual reproduction (gametogony) in the parasitophorous vacuoles to generate both microgamonts and macrogamonts, which mature into flagellated sperm-like microgametes and macrogametes. The fusion between microgametes and macrogametes yields zygotes, which then grow into oocysts (i.e., unsporulated oocysts consisting of two sporocysts or small oocysts). By disrupting the host cell, oocysts exit along

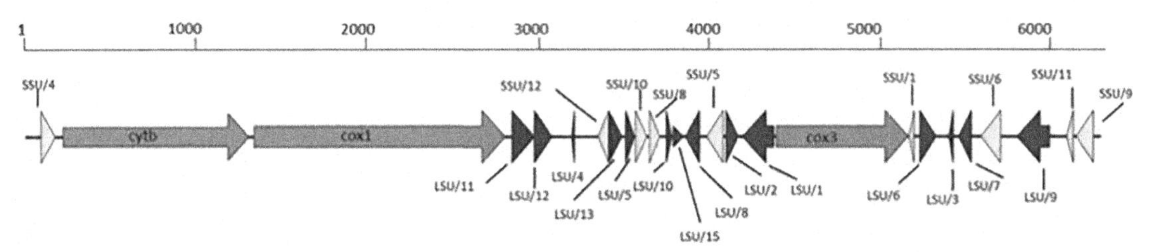

FIGURE 27.2 Organization of *C. cayetanensis* mitochondrial genome. Gray arrow indicates coding genes, black arrow head indicates fragments of large-subunit rRNA, and clear arrow head indicates fragments of small-subunit rRNA. Transcriptional direction is indicated by the arrowed end. (Photo credit: Cinar HN, Gopinath G, Jarvis K, Murphy HR. *PLoS One.* 2015;10(6):e0128645.)

with stool. Being unsporulated, noninfectious, able to survive in water at 4°C for 2 months and at 37°C for 7 days, and resistant to chlorine and most other chemical disinfectants, unsporulated oocysts sporulate (with two sporocysts each developing into two banana-shaped sporozoites) and become infectious in about 7–15 days at temperatures of 22°C–32°C. Interestingly, sporulation of *Cyclospora* oocyst may be induced with 2.5% potassium dichromate, or assisted by the physical rupture of *Cyclospora* oocyst wall and exposure to 0.5% trypsin and 1.5% sodium taurocholate in PBS [1,5].

Epidemiology. Being the only host for *C. cayetanensis*, humans typically acquire the infection through consumption of human stool-contaminated food (e.g., raspberries, blackberries, fresh basil, fresh baby lettuce leaves, potato salad) or water. Therefore, untreated or poorly treated water used for irrigation of fruits and vegetables contributes to transmission of human cyclosporiasis, while animals are unlikely to play a direct role in its transmission. Human cyclosporiasis has been diagnosed in Americas, Asia, Africa, Europe, and Australia, with an overall prevalence of 3.6% [16,17]. It is notable that 97.8% of human cyclosporiasis cases are reported during spring and summer months (April through July) in the northern hemisphere. Major risk factors consist of immunocompromised status, young age (children of 1.5–9 years old are five times more likely to be positive for *C. cayetanensis* in stool samples than adults), and a history of travel to and consumption of raw fresh fruits and produce from endemic countries (e.g., the Dominican Republic, Mexico, Guatemala, Haiti, Peru, or Nepal, among others) [18,19].

Clinical features. With an incubation period of 7 days (range 2–11 days), human cyclosporiasis typically presents with gastrointestinal symptoms [e.g., loose or watery diarrhea (three or more stools per day), nausea, vomiting, abdominal cramps, bloating, flatulence, constipation, loss of appetite, unexplained weight loss (3.5 kg for non-AIDS patients, and 7.2 kg for AIDS patients)] and other non-specific, flu-like symptoms (e.g., fever, chills, headache, muscle aches, joint aches, and fatigue) [20]. Patients with immune dysfunction (e.g., AIDS) may develop severe, protracted, or chronic watery diarrhea that lasts for 199 days in comparison to 57 days in immunocompetent patients. Left untreated, symptoms may persist for 1 month to 6 weeks, with potential complications ranging from Guillain–Barré syndrome, reactive arthritis, to acalculous cholecystitis. However, some patients may be asymptomatic or show no watery diarrhea [1,21].

Pathogenesis. The *C. cayetanensis* genome harbors genes encoding myosin, gliding-associated protein, and other proteins that form the motor complex involved in gliding and invasion, as well as genes encoding anchor molecules for attaching to host cell cytoskeleton. In addition, *C. cayetanensis* produces three groups of proteins in the apical complex: (i) adhesins to bind and interact with host cells during initial invasion; (ii) peptidases and proteases to process parasite rhoptry and micronemal proteins and degrade host proteins after crossing the plasma membrane; and (iii) signaling proteins (e.g., protein phosphatases and kinases) to modulate host cell signaling pathways or immune responses and enhance parasite survival. Nonetheless, as a monoxenous organism, *C. cayetanensis* appears to possess very limited capabilities to modulate host nuclear activities and signaling pathways during invasion in comparison with heterogenous organisms such as *Toxoplamsa gondii* [1,5].

Diagnosis. As human cyclosporiasis presents symptoms that are indistinguishable to those caused by other coccidian parasites, bacterial, and viral pathogens, its diagnosis requires: (i) microscopic observation of *C. cayetanensis* oocysts (of spherical shape, and 8–10 µm in size) in stool using modified acid-fast stain (with oocysts appearing light pink to deep red, some containing granules or having a bubbly look) (Figure 27.3) or other stains, and autofluorescence with UV light (*C. cayetanensis* oocysts appear dark blue under a 365-nm dichromatic filter, and mint green under a 450- to 490-nm dichromatic filter; while *C. parvum* oocysts failing to autofluoresce under UV light); premounting centrifugation of fecal sample in sucrose solution increases the sensitivity to 84% [22]; (ii) recovery of oocysts in intestinal fluid or biopsy specimens; (iii) microscopic observation of two sporozoites each containing two sporocysts after oocyst sporulation *in vitro* (involving a 2-week incubation in 2.5% potassium dichromate solution); and (iv) molecular detection of *C. cayetanensis* DNA (see section 27.2) [1,5].

Treatment and prevention. The treatment of choice for human cyclosporiasis is trimethoprim-sulfamethoxazole (TMP-SMZ; also known as cotrimoxazole), which is given orally at 160/800 mg twice

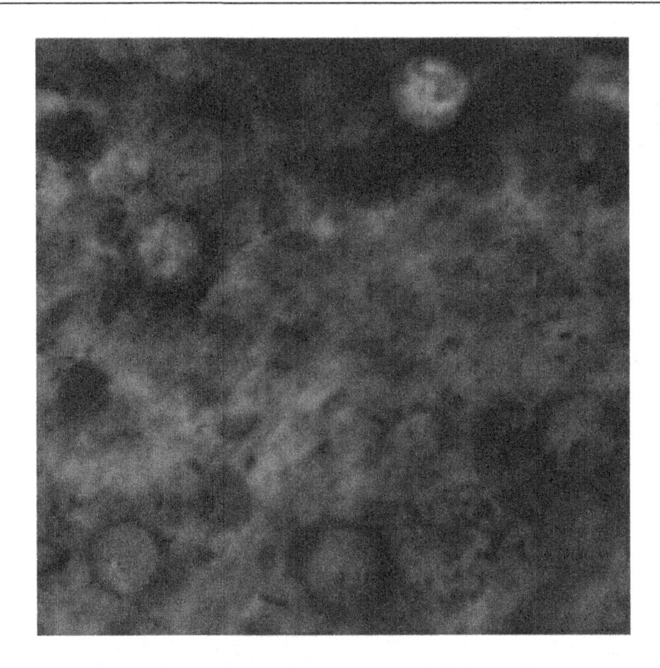

FIGURE 27.3 *Cyclospora cayetanensis* oocysts (8–10 μm in size) from fecal smear stained by the acid-fast modified carbolfuchsin method. (Photo credit: Kaminsky RG, Lagos J, Raudales Santos G, Urrutia S. *BMC Infect Dis.* 2016;16:66.)

a day for 7 days or 160/800 mg four times a day for 10 days in AIDS patients. Patients with sulfonamide allergies may be treated with ciprofloxacin (500 mg twice daily for 7 days) or nitazoxanide (100 mg twice daily for 3 days). Oral or intravenous rehydration may be considered if necessary [1,5].

Prevention of human cyclosporiasis should center on personal hygiene and sanitary conditions, including careful preparation of food and fresh produce, drinking filtered water, and proper treatment of human sewage.

27.2 MOLECULAR IDENTIFICATION OF *C. CAYETANENSIS*

Notwithstanding their relative simplicity and low cost, conventional phenotypic techniques (e.g., microscopy involving filtration, centrifugation, or flotation followed by staining) for diagnosis of human cyclosporiasis are encumbered by considerable labor input, insufficient sensitivity, and variable outcomes. Genotypic techniques (particularly PCR) overcome these drawbacks and provide a sensitive, specific, and rapid approach for species-specific identification, typing, and phylogenetic analysis of *C. cayetanensis*.

27.2.1 Species-Specific Identification

Molecular identification of *C. cayetanensis* is based on the premise that this apicomplexan protozoan possesses certain unique genes or nucleotide variations in these genes, detection of which leads to the confirmation of its species identity. Numerous *C. cayetanensis* genes have been targeted for identification, including small- and large-subunit ribosomal RNA (18S and 28S rRNA), internal transcribed spacers 1

(ITS1) and 2 (ITS2), and other genes. Through *in vitro* amplification of these genes or gene regions, rapid, sensitive, and specific identification of *C. cayetanensis* is possible. Among various nucleic acid amplification procedures, PCR and its derivatives (including standard, RFLP, multiplex, nested, quantitative, real time, reverse transcription) have been widely applied in clinical and research laboratories for improved identification of *C. cayetanensis* [6,23–33]. By incorporating an FTA filter-based protocol that eliminates inhibitive substances (e.g., heme, hemoglobulin, lactoferrin, immunoglobulin G, leukocyte DNA, and polysaccharides in blood; phenolics, glycogen, calcium ions, fat, and other organic substances in foods; phenolics, humic acids, and heavy metals in environmental samples) and increases the concentration of *C. cayetanensis*, it allows detection of 3–10 *C. cayetanensis* oocysts directly applied to filters, and 10–30 oocysts in complex mixed samples (e.g., 100 g of fresh raspberries) (equivalent to 0.3 oocysts per g of raspberries) by PCR. A commercial diagnostic kit (BioFire FilmArray gastrointestinal panel, BioMerieux) covering 12 enteric bacterial pathogens, five groups of viruses, and protozoan parasites *C. cayetanensis*, *Cryptosporidium*, *Giardia*, and *Entamoeba histolytica* may be utilized for molecular detection of human cyclosporiasis and other gastrointestinal infections [1,5].

27.2.2 Typing

In addition to specific identification of *C. cayetanensis*, it is extremely valuable to be able to track the culprit organism involved in the outbreaks and implement appropriate control and prevention measures accordingly. Fortunately, through PCR amplification and sequencing analysis of certain conserved genes (e.g., 18S and 28S rRNA, ITS, heat shock protein genes, mitochondrial and apicoplast genomes), molecular approaches offer a useful means of typing and tracking *C. cayetanensis* isolates [34].

Specifically, *C. cayetanensis* ITS1, located between the 16S and 28S rRNA genes, shows sequence variability, and isolates from different geographic origins may contain varied number of ITS1 copies, suggesting their distinct clonalities [35].

C. cayetanensis mitochondrial genome demonstrates strain-level diversity that can be utilized to link outbreak cases to the source [36]. In particular, the polymorphic region of *C. cayetanensis* mitochondrial genome provides extremely informative genotyping markers for epidemiological traceback, case-linkage, source-tracking, and distinct case cluster investigations on this emerging foodborne pathogen [37–39].

C. cayetanensis apicoplast genome encompasses 29 peptides, several high-quality single nucleotide polymorphisms, and small insertion/deletions, which can be also targeted for its identification and differentiation. The observation of a 30-bp sequence repeat at the terminal spacer region of the apicoplast genome in a *C. cayetanensis* isolate from Nepal highlights the potential of the apicoplast genome for outbreak investigations.

In addition, using five microsatellite loci (CYC3, CYC13, CYC15, CYC21, and CYC22) in multilocus sequence typing (MLST) procedure, *C. cayetanensis* isolates can be separated into three distinct subpopulations [40]. Furthermore, the application of microsatellite loci CYC21 and CYC22 in MLST allows identification of 17 different concatenated sequence types among *C. cayetanensis* isolates [41].

27.2.3 Phylogenetic Analysis

Through PCR amplification and sequencing examination of 18S rRNA, apicoplast and mitochondrial genomes as well as other genes, phylogenetic analysis helps reveal insights on the genetic relationship between *Cyclospora* and other apicomplexans, among *Cyclospora* species, and confirms the genetic identity of newly isolated *Cyclospora* species [42].

Based on the alignment of the 1677 bp 18S rRNA sequence, *C. cayetanensis* and four *Cyclospora* species (i.e., *C. macacae*, *C. colobi*, *C. cercopitheci*, and *C. papionis*) from nonhuman primates form a distinct monophyletic group related to avian *Eimeria* species [3]. In addition, comparison genomic analysis shows that *C. cayetanensis* and *E. tenella* share 85.6% and 90.4% nucleotide sequence similarities in the apicoplast and mitochondrial genomes, respectively [11,14].

27.3 FUTURE PERSPECTIVES

Resulting from infection with apicomplexan protozoan *C. cayetanensis*, human cyclosporiasis is an emerging foodborne disease that typically presents with gastrointestinal as well as flu-like symptoms. Given the close morphological, biological, molecular, and clinical similarities between *C. cayetanensis* and other apicomplexan protozoa (e.g., *Cryptosporidium*, *Isospora*, and *Sarcocystis*), it is critical to develop the technical capability for their rapid, sensitive, and specific identification and differentiation. Fortunately, the recent development and application of molecular techniques have overcome many obstacles in the species identification, typing, and phylogenetic analysis of *C. cayetanensis*. However, because clinical, food, and environmental water samples often harbor many inhibitory substances and relatively few *C. cayetanensis* oocysts, which pose a challenge for molecular assays to amplify and detect, there is ample room for continuing improvement in sample preparation procedures to help concentrate the oocysts and eliminate inhibitors at the same time. In addition, there is a need to further understand the molecular mechanisms of *C. cayetanensis* pathogenicity to develop innovative control measures for human cyclosporiasis.

REFERENCES

1. Almeria S, Cinar HN, Dubey JP. *Cyclospora cayetanensis* and cyclosporiasis: an update. *Microorganisms.* 2019;7(9):E317.
2. Lainson R. The genus *Cyclospora* (Apicomplexa: Eimeriidae), with a description of *Cyclospora schneideri* n. sp. in the snake *Anilius scytale* (Aniliidae) from Amazonian Brazil-a review. *Mem Inst Oswaldo Cruz.* 2005;100(2):103–10.
3. Li N, Ye J, Arrowood MJ, et al. Identification and morphologic and molecular characterization of *Cyclospora macacae* n. sp. from rhesus monkeys in China. *Parasitol Res.* 2015;114(5):1811–6.
4. McAllister CT, Motriuk-Smith D, Kerr CM. Three new coccidians (Cyclospora, Eimeria) from eastern moles, *Scalopus aquaticus* (Linnaeus) (Mammalia: Soricomorpha: Talpidae) from Arkansas, USA. *Syst Parasitol.* 2018;95(2–3):271–9.
5. Li J, Cui Z, Qi M, Zhang L. Advances in cyclosporiasis diagnosis and therapeutic intervention. *Front Cell Infect Microbiol.* 2020;10:43.
6. Qvarnstrom Y, Wei-Pridgeon Y, Van Roey E, et al. Purification of *Cyclospora cayetanensis* oocysts obtained from human stool specimens for whole genome sequencing. *Gut Pathog.* 2018;10:45.
7. Ortega YR, Sanchez R. Update on *Cyclospora cayetanensis*, a food-borne and waterborne parasite. *Clin Microbiol Rev.* 2010;23(1):218–34.
8. Qvarnstrom Y, Wei-Pridgeon Y, Li W, et al. Draft genome sequences from *Cyclospora cayetanensis* oocysts purified from a human stool sample. *Genome Announc.* 2015;3(6):e01324–15.
9. Qvarnstrom Y, Benedict T, Marcet PL, Wiegand RE, Herwaldt BL, da Silva AJ. Molecular detection of *Cyclospora cayetanensis* in human stool specimens using UNEX-based DNA extraction and real-time PCR. *Parasitology.* 2018;145(7):865–70.
10. Liu S, Wang L, Zheng H, et al. Comparative genomics reveals *Cyclospora cayetanensis* possesses coccidia-like metabolism and invasion components but unique surface antigens. *BMC Genomics.* 2016;17:316.
11. Cinar HN, Gopinath G, Jarvis K, Murphy HR. The complete mitochondrial genome of the foodborne parasitic pathogen *Cyclospora cayetanensis. PLoS One.* 2015;10(6):e0128645.

12. Ogedengbe ME, Qvarnstrom Y, da Silva AJ, Arrowood MJ, Barta JR. A linear mitochondrial genome of *Cyclospora cayetanensis* (Eimeriidae, Eucoccidiorida, Coccidiasina, Apicomplexa) suggests the ancestral start position within mitochondrial genomes of eimeriid coccidia. *Int J Parasitol.* 2015;45(6):361–5.
13. Gopinath GR, Cinar HN, Murphy HR, et al. A hybrid reference-guided de novo assembly approach for generating *Cyclospora* mitochondrion genomes. *Gut Pathog.* 2018;10: 15.
14. Tang K, Guo Y, Zhang L, et al. Genetic similarities between *Cyclospora cayetanensis* and cecum-infecting avian *Eimeria* spp. in apicoplast and mitochondrial genomes. *Parasit Vectors.* 2015;8:358.
15. Cinar HN, Qvarnstrom Y, Wei-Pridgeon Y, et al. Comparative sequence analysis of *Cyclospora cayetanensis* apicoplast genomes originating from diverse geographical regions. *Parasit Vectors.* 2016;9(1):611.
16. Yazar S, Yalcln S, Sahin I. Human cyclosporiosis in Turkey. *World J Gastroenterol.* 2004;10(12):1844–7.
17. Karanja RM, Gatei W, Wamae N. Cyclosporiasis: an emerging public health concern around the world and in Africa. *Afr Health Sci.* 2007;7(2):62–7.
18. Marques DFP, Alexander CL, Chalmers RM, et al. Cyclosporiasis in travellers returning to the United Kingdom from Mexico in summer 2017: lessons from the recent past to inform the future. *Euro Surveill.* 2017;22(32):30592.
19. Siddiqui ZA. An overview of parasitic infections of the gastro-intestinal tract in developed countries affecting immunocompromised individuals. *J Parasit Dis.* 2017;41(3):621–6.
20. Keaton AA, Hall NB, Chancey RJ, et al. Notes from the field: cyclosporiasis cases associated with dining at a Mediterranean-style restaurant chain - Texas, 2017. *Morb Mortal Wkly Rep.* 2018;67(21):609–10.
21. Alfano-Sobsey EM, Eberhard ML, Seed JR, et al. Human challenge pilot study with *Cyclospora cayetanensis. Emerg Infect Dis.* 2004;10(4):726–8.
22. Kaminsky RG, Lagos J, Raudales Santos G, Urrutia S. Marked seasonality of *Cyclospora cayetanensis* infections: ten-year observation of hospital cases, Honduras. *BMC Infect Dis.* 2016;16:66.
23. Almeria S, da Silva AJ, Blessington T, et al. Evaluation of the U.S. Food and Drug Administration validated method for detection of *Cyclospora cayetanensis* in high-risk fresh produce matrices and a method modification for a prepared dish. *Food Microbiol.* 2018;76:497–503.
24. Bhattachan B, Sherchand JB, Tandukar S, Dhoubhadel BG, Gauchan L, Rai G. Detection of *Cryptosporidium parvum* and *Cyclospora cayetanensis* infections among people living in a slum area in Kathmandu valley, Nepal. *BMC Res Notes.* 2017;10(1):464.
25. Bilung LM, Tahar AS, Yunos NE, et al. Detection of *Cryptosporidium* and *Cyclospora* oocysts from environmental water for drinking and recreational activities in Sarawak, Malaysia. *Biomed Res Int.* 2017;2017:4636420.
26. Sim S, Won J, Kim JW, Kim K, Park WY, Yu JR. Simultaneous molecular detection of *Cryptosporidium* and *Cyclospora* from raw vegetables in Korea. *Korean J Parasitol.* 2017;55(2):137–42.
27. Murphy HR, Cinar HN, Gopinath G, et al. Interlaboratory validation of an improved method for detection of *Cyclospora cayetanensis* in produce using a real-time PCR assay. *Food Microbiol.* 2018;69:170–8.
28. Shin JH, Lee SE, Kim TS, et al. Development of molecular diagnosis using multiplex real-time PCR and T4 phage internal control to simultaneously detect *Cryptosporidium parvum*, *Giardia lamblia*, and *Cyclospora cayetanensis* from human stool samples. *Korean J Parasitol.* 2018;56(5):419–27.
29. Casillas SM, Hall RL, Herwaldt BL. Cyclosporiasis surveillance - United States, 2011-2015. *MMWR Surveill Summ.* 2019;68(3):1–16.
30. Temesgen TT, Tysnes KR, Robertson LJ. A new protocol for molecular detection of *Cyclospora cayetanensis* as contaminants of berry fruits. *Front Microbiol.* 2019;10:1939.
31. Assurian A, Murphy H, Ewing L, Cinar HN, da Silva A, Almeria S. Evaluation of the U.S. Food and Drug Administration validated molecular method for detection of *Cyclospora cayetanensis* oocysts on fresh and frozen berries. *Food Microbiol.* 2020;87:103397.
32. Puebla LEJ, Núñez Fernández FA, Nodarse JF, et al. Sporadic cyclosporiasis in symptomatic Cuban patients: confirmation of positive results from conventional diagnostic methods by molecular assay. *Diagn Microbiol Infect Dis.* 2020: (3): 115048.
33. Resendiz-Nava CN, Orozco-Mosqueda GE, Mercado-Silva EM, Flores-Robles S, Silva-Rojas HV, Nava GM. A molecular tool for rapid detection and traceability of *Cyclospora cayetanensis* in fresh berries and berry farm soils. *Foods.* 2020;9(3):261.
34. Houghton KA, Lomsadze A, Park S, et al. Development of a workflow for identification of nuclear genotyping markers for *Cyclospora cayetanensis. Parasite.* 2020;27:24.
35. Adam RD, Ortega YR, Gilman RH, Sterling CR. Intervening transcribed spacer region 1 variability in *Cyclospora cayetanensis. J Clin Microbiol.* 2000;38(6):2339–43.
36. Cinar HN, Gopinath G, Murphy HR, et al. Molecular typing of *Cyclospora cayetanensis* in produce and clinical samples using targeted enrichment of complete mitochondrial genomes and next-generation sequencing. *Parasit Vectors.* 2020;13(1):122.

37. Barratt JLN, Park S, Nascimento FS, et al. Genotyping genetically heterogeneous *Cyclospora cayetanensis* infections to complement epidemiological case linkage. *Parasitology.* 2019;146(10):1275–83.
38. Guo Y, Wang Y, Wang X, Zhang L, Ortega Y, Feng Y. Mitochondrial genome sequence variation as a useful marker for assessing genetic heterogeneity among *Cyclospora cayetanensis* isolates and source-tracking. *Parasit Vectors.* 2019;12(1):47.
39. Nascimento FS, Barta JR, Whale J, et al. Mitochondrial junction region as genotyping marker for *Cyclospora cayetanensis. Emerg Infect Dis.* 2019;25(7):1314–9.
40. Guo Y, Roellig DM, Li N, et al. Multilocus sequence typing tool for *Cyclospora cayetanensis. Emerg Infect Dis.* 2016;22(8):1464–7.
41. Hofstetter JN, Nascimento FS, Park S, et al. Evaluation of multilocus sequence typing of *Cyclospora cayetanensis* based on microsatellite markers. *Parasite.* 2019;26:3.
42. Zhao GH, Cong MM, Bian QQ, et al. Molecular characterization of *Cyclospora*-like organisms from golden snub-nosed monkeys in Qinling Mountain in Shaanxi province, northwestern China. *PLoS One.* 2013;8(2):e58216.

SUMMARY

Cyclospora is a genus of apicomplexan parasites that affect various insects, reptiles, and mammals. Although members of this genus were identified in mole and millipede in late 19th century, *Cyclospora* has attracted little attention from the scientific community due to its apparent medical irrelevance. However, since the 1970s, a series of clinical cases linking to a previously unknown parasite have led to the identification of *C. cayetanensis* in 1994 as the causal agent for diarrheal diseases in several parts of the world. Subsequent development and application of molecular techniques have facilitated rapid and accurate detection, typing, and phylogenetic analysis of this emerging foodborne protozoan. As clinical, food, and environmental water samples often contain many inhibitory substances and relatively few *C. cayetanensis* oocysts, which pose challenge for molecular assays to amplify and detect, there is a need for continuing improvement in sample preparation procedures that help concentrate the oocysts and eliminate inhibitors at the same time. Additionally, there is an urgent need to further understand the molecular mechanisms of *C. cayetanensis* pathogenicity with the goal to develop innovative control measures for human cyclosporiasis.

Anisakis From Molecular Identification to Omic Studies

28

S. Cavallero, and S. D'Amelio
Sapienza University of Rome

Contents

28.1 INTRODUCTION

Anisakis is a gastrointestinal parasitic nematode genus in the family Anisakidae, superfamily Ascaridoidea. Members of the Anisakidae family demonstrate a cosmopolitan distribution and utilize diverse aquatic hosts for the successful completion of their lifecycle.[1] *Anisakis* spp. have an indirect and complex life-cycle based on prey–predator relation, with marine mammals as definitive hosts, and crustaceans, fishes, and squids as intermediate/paratenic hosts. *Anisakis* adults live in the gastrointestinal portion of several marine mammal species and eggs are released via feces into the sea. After molts, embryonated eggs are ingested by crustaceans, which, in turn, are taken up by fish or squids. Boring through the digestive tract wall and passing into the visceral body cavity of fish or squids where the host-induced encapsulation of *Anisakis* larvae take place.[2] Even though experimental evidence support the presence of third-stage larva inside eggs,[3,4] molting cannot be excluded in crustaceans. Therefore, crustaceans are considered intermediate hosts, while fish and squids are regarded as paratenic hosts. While the majority of the larvae are found in the visceral body cavity of infected fish, some larvae may migrate to the flesh, even before the death of the intermediate hosts.[5,6] To complete the lifecycle, infected paratenic hosts are then ingested by marine mammals, in the stomach or intestine of which *Anisakis* spp. undergo two final molts and develop into a sexually mature adult nematode.

The occurrence of larval nematodes in fish fillets is of medical and economic concern: in addition to the effects on the marketability of fish meat, *Anisakis* larvae are the causative agent of fish-borne anisakiasis when infected raw or undercooked fish or squids are ingested by humans. Parasitic larvae are unable to become adults in humans, but they still cause mild-to-severe diseases, including gastric anisakiasis, intestinal anisakiasis (IA), or ectopic/extragastrointestinal disease (Figure 28.1), depending on the location and behavior of the larvae.[7] Another form of anisakiasis is known as gastroallergic anisakiasis (GAA) due mostly to an IgE-mediated allergic reaction.

Anisakiasis is a fish-borne zoonosis and its transmission often reflects local/regional food tradition that involves consumption of raw and undercooked fish. Around 20,000 cases have been reported by now mostly from Japan, followed by the Netherlands, Germany, France, Spain, Croatia, and Italy.[8] Fish dishes that are considered to be of high risk for human anisakiasis include Japanese sushi and sashimi, Scandinavian gravlax, Dutch salted and marinated herring, Spanish boquerones, and Italian marinated anchovies. Owing to the popularity of Japanese dishes as well as the consumption of small pickled fish and the overall increased consumption of fish worldwide, an increasing number of anisakiasis cases have been reported in recent years.[9,10] However, due to the nonmandatory requirement for its notification and complexity in its diagnosis, human anisakiasis remains an underestimated zoonosis.

Among the genera and species of the Anisakidae family, *A. simplex* sensu lato (s.l.) and *Pseudoterranova decipiens* s.l are mainly responsible for anisakidosis in humans.[11,12] Rarely, *Contracaecum* spp. and *Hysterothylacium* spp. (Raphidascaridae family) are associated with accidental larval recovery in gastric/intestinal tract,[13,14] in particular, the latter is commonly considered nonpathogenic to humans.

Because anisakids can cause a human illness, their precise identification to the species level is important for epidemiological and controlling purposes.

Traditionally, morphological characteristics have played a crucial role in the identification of nematodes including anisakids. However, morphological characters of taxonomic significance are scarce and relevant to anisakid adult specimens only, such as the excretory system, the alimentary canal, the number and distribution of male caudal papillae, the position of the vulva, and the length of the spicules.[15–17] Furthermore, the causative agent of anisakidosis is the infective larval stage L3 and informative morphological parameters are even less marked in these developing larval forms.

Anisakis larvae are determined to the genus level mainly on the presence of larval tooth, on the morphology and length of the glandular part of the esophagus (i.e., the "ventriculus"), and the presence/absence of a caudal spine ("mucron"), which differs between the Type I and Type II larvae (sensu Berland[18]).

The limited taxonomic significance of morphological characters and the speciation processes virtually without morphological differentiation advocate the use of molecular approaches for inferring the

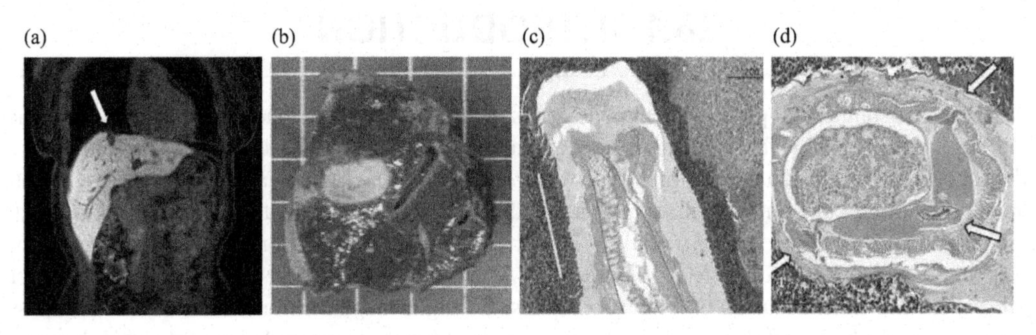

FIGURE 28.1 *Anisakis simplex* s.s. infection in the liver of a 44 year-old female. (a) Coronal slice on enhanced MRI in the hepatobiliary phase showing a mass (arrow). (b) Macroscopic appearance of the resected liver showing a white node with a regular border of about 2 cm in diameter. (c) Sagittal slice of the larva showing a wave-shaped (cross-striated) border (white line). (d) Axial slice of the larva showing a Y-shaped lateral cord (arrows 1) and a renette cell (arrow 2), characteristics of *Anisakis*. (Photo credit: Nogami Y, Fujii-Nishimura Y, Banno K, et al. *BMC Med Imaging*. 2016;16:31.)

systematic and evolutionary relationships among anisakid nematodes and for epidemiologic tracking of parasitic anisakid species.

Combining classical morphological identification with modern molecular approach (integrative taxonomy) allows classification of at least nine species into three phylogenetic clades within the *Anisakis* genus.

Clade I comprises members of the *A. simplex* sensu lato complex and the two closely related taxa *A. ziphidarum*[16] and *A. nascettii*[19]. The *Anisakis simplex* s.l. complex includes species mainly involved in human disease, such as *Anisakis simplex* sensu stricto and *A. pegreffii* (formerly *A. simplex* A of Nascetti et al.[20]). Putative hybrids between these two sibling species have been firstly described in sympatric areas by the nuclear ribosomal ITS marker,[21] which depicted a recombinant genotype between fixed diagnostic polymorphisms of the two species. Hybridization of the two species was then confirmed by the use of an alternative nuclear constitutive gene, encoding the elongation factor.[22,23]

In contrast to the increasing records of heterozygote larval individuals in fish paratenic hosts, such evidences in adult nematodes are rare,[24,25] suggesting some reduction in the fitness of hybrids. The third cryptic species of the simplex complex is *A. berlandi*[17] (formerly *A. simplex* C of Mattiucci et al.[26]).

Clade II contains three closely related species *A. physeteris*, *A. brevispiculata*, and *A. paggiae*. Clade III is represented by *A. typica* which forms the basal sister branch in the genus,[27] probably including still undetected taxa, as recently suggested by Palm et al.,[28] who reported the presence of *A. typica* var. *indonesiensis* from Indonesian waters. Finally, *A. schupakovi*, a parasite of the relict phocid species of the Caspian Sea, *Phoca caspica*, was considered as *inquirendae* by Davey in his revision of the genus *Anisakis*, but retained as a valid species by Kurochkin.[29]

It is important to remember that human disease is caused by the larval stage, which cannot be identified by morphological traits only. The limited value of morphological analyses makes the use of genetic/molecular methods mandatory for the identification of *Anisakis* species. Moreover, a precise identification of human pathogens is fundamental for the comprehension of pathogenic mechanisms and infectivity potential, as well as the management of control and prevention procedures.

28.2 MOLECULAR IDENTIFICATION OF *ANISAKIS* SPECIES

Identification of anisakid nematodes have traditionally relied on microscopic observation of informative features,[30,31] such as those involved in important biological mechanisms, including feeding (esophagus and ventricular bulb) and mating (caudal papillae, spicules).

However, usable morphological diagnostic traits are present only in adult males, while females and immature specimens can be identified only at the genus level. Additionally, the existence of cryptic species seems to be highly frequent in marine ascaridoid. In fact, with the advent of molecular techniques based initially on multilocus allozymes and later on polymerase chain reaction (PCR) amplification, the systematics of the family and genera have been dramatically revised as a great occurrence of morphological convergence was discovered.[26,27,32,33] Multilocus allozyme electrophoresis (MAE), first used at the beginning of the 1980s, revealed a high genetic heterogeneity within anisakid morphospecies, such as *A. simplex*, *P. decipiens* and *C. osculatum*.[20,34–36] MAE allowed the discovery of new biological species with specific ecological traits,[16,19] addressing questions on population genetics and on evolutionary biology.

Notwithstanding the bulk of data obtained so far from the application of MAE, the development of molecular markers for the accurate identification of related species using PCR-based approaches is preferable considering latter's superior sensitivity (thus requiring small amount of fresh or ethanol-fixed parasite material for analysis), speed, reliability, relatively low cost, and, more importantly, interlaboratory reproducibility, which is one of the main pitfalls of MAE approach.

A method widely used by the scientific community and by official accredited laboratories (see, e.g., in Italy, http://www.izssicilia.it/centri-di-referenza/c-re-n-a) is PCR-based restriction fragment length polymorphism analysis (PCR-RFLP) of the ribosomal RNA (rRNA) internal transcribed spacers (ITS-1 and ITS-2). This approach proved fruitful for the specific identification of all the nine species belonging to the genus *Anisakis*, irrespective of developmental stages or sex, with the use of three different endonucleases.[21,27,37,38]

Direct DNA sequencing of nuclear and mitochondrial genes represents another useful tool for the identification of different morphospecies and sibling species of the genus. Sequence data are now available for almost all of the recognized species, with the only exception of *A. schupakovi*.

Nuclear ribosomal DNA spanning the final part of the 18S subunit and the very beginning of the 28S subunit named ITS (ITS1-5.8S-ITS2) exhibits a significant degree of variation between closely related species and it is therefore useful for species discrimination,[39] as a transcribed noncoding region is. Intraspecific variability is commonly less than the interspecific variability because concerted evolution tends to minimize intraspecific variation allowing an unambiguous attribution of one specified sequence to one corresponding species.[40,41] However, as ribosomal DNA consists of tandem-repeated copies that are inherited as a unit, the finding of heterogeneity in ITS markers may be attributable to the retention of ancestral polymorphisms, incomplete gene conversion, or possible hybridization events. To confirm hybridization, the simultaneous use of a single-copy gene is needed, as D'Amelio et al.[22] and Mattiucci et al.[23] carried out within *A. simplex* complex adding molecular data on elongation factor gene (ef1-alpha), thus confirming the gene flow between *A. simplex* s.s. and *A. pegreffii*.

Similar to nuclear genetic markers, two regions of the mitochondrial DNA have been sequenced for all *Anisakis* spp.: the subunit 2 of the cytochrome oxidase gene (*cox*2)[17,19,27,42] and the small subunit of the ribosomal RNA in the mitochondrial genome *rrn*S, named 12S.[17,43,44] These studies confirmed the taxonomic status of all the nine species, helping clarify the phylogenetic relationships among morphospecies. The entire mitochondrial genome of *A. simplex* is also available.[45]

Modifications of the standard PCR, such as multiplex PCR and real-time PCR, have been developed and used. Umehara et al.[46] proposed a method based on multiplex PCR which recognizes six different species of anisakids, including *A. simplex* s.s. and *A. pegreffii* as well as Abe[47] for the rapid discrimination of the two main human pathogenic species.

Another PCR derivative used in the past for the identification of anisakid nematodes is SSCP (single-strand conformation polymorphism). This is a direct mutation detection method, in which amplicons are separated as single-strand filaments by electrophoretic run in polyacrylamide gel under nondenaturing conditions. A single base difference is sufficient to discriminate the filament through a change in the shape and, consequently, in mobility. SSCP-based markers have been developed for anisakids belonging to the genera *Pseudoterranova* and *Contracaecum*,[48,49] but have not been applied to members of the genus *Anisakis*.

Mladineo et al.[50] provided data on sequence repeats or microsatellites in noncoding regions, analyzing a panel of highly polymorphic microsatellite loci, both in *A. simplex* s.s. and *A. pegreffii*. Genetic characterization of individuals showed a low level of intergroup heterogeneity, suggesting that the existing mosaicism is likely a retention of an ancestral polymorphism rather than a recent recombination event. However, this study has been conducted in allopatric areas and therefore this evidence requires confirmation in geographical areas where the two species co-exist (e.g., around Gibraltar Strait).

Anisakids are considered among the most important biological hazards present in "seafood" products, and several methods for their detection and identification in fish and fish products, even of small amount of parasitic material, have been developed so far. Most detection methods are based on visual inspections of the coelomic cavity and fillets, such as candling and UV visualization;[51] recently, a modified protocol of chloropeptic digestion routinely used for *Trichinella* spp. was proposed.[52,53] The Baermann technique, based on active migration of larvae from the host under positive hydrotropism, also proved to be effective.[54]

The use of molecular approaches, besides the potential improvement in sensitivity, has the advantage of discriminating between species and genera, according to the specificity of the probes and primers selected. This is particularly relevant in the evaluation of risk related to anisakidosis, given that pathogenic and nonpathogenic anisakid species may be present in fishes and fish products, as well as in the investigation of alimentary frauds as these nematodes may be used as valuable biological tags for fish stock discrimination.[55,56]

Screening and survey methods to detect and quantify anisakid presence based on real-time PCR assays targeting nuclear and mitochondrial DNA have been recently developed, with particular regard to food products.[57–63] However, such methods did not undergo thorough validation processes, which are highly desirable or mandatory in the case of testing laboratories accredited for the ISO IEC 17025:2017.[61] A recent genotyping approach based on high-resolution melting curve analysis of DNA has been used to confirm the taxonomic profiles of the two species *A. simplex* s.s. and *A. pegreffii*, and of the hybrid individuals revealing the great potentiality of real-time PCR as a screening technique.[25]

The *A. simplex* genome has become available recently and the advent of the next-generation sequencing (NGS) and of the so-called third-generation sequencing technologies permit further discovery of new markers for molecular systematics, diagnosis and control, and exploration of key genes related to specific biological issues, such as host–parasite interaction or pathogenicity as proposed and shown in studies on other parasitic nematodes such as *Ascaris*.[64,65]

28.3 ANISAKIS PROTEOMICS AND TRANSCRIPTOMICS: RECENT DATA

With the use of MAE and PCR techniques, several aspects of molecular systematics, epidemiology, and ecology of anisakids have been analyzed in depth.[66,67] Further investigations on the immunology and host–parasite interaction provide opportunity to fill the knowledge gaps in the understanding of complex biological processes related to this zoonotic parasite and to improve its diagnosis.

The advent of NGS methodologies allows the scientific community for the first time to collect a large bulk of data on biological mechanisms for various organisms. However, predicting biological outcomes on the basis of genomic data remains a major challenge because of the complexity in the interactions of specific genetic profiles with external factors (i.e., host, other organisms, environment), as well as of the variability in the analytical and bioinformatics modeling levels.[68]

Parasitic nematodes undergo a great spectrum of interactions with the host.[69] The absence of reliable, appropriate *in-vivo* and *in-vitro* systems including suitable animal models has impeded research progresses on anthelmintic discovery and immunological screening test, as well as on effective control and diagnosis of infections. There is a growing interest in elucidating the rationale on the possible interactions between the microbiota, immune response, inflammatory processes, and intestinal parasites. The use of the parasitic whipworm nematode *Trichuris suis* as treatment for inflammatory bowel diseases (IBDs) such as Crohn's disease and ulcerative colitis and autoimmune disorders is also being explored[70–72].

High-throughput molecular and computer technologies are instrumental for biological explorations of parasites. Transcriptomes of different sex or developmental stages of parasitic nematodes may yield clues on the gene expression, regulation, and function in a parasite, which underscore its biology, physiology, and interactions with the host.[73]

The free-living model organism *Caenorhabditis elegans* is a valuable resource for comparative studies of parasitic nematodes, as well as for the development of new biotechnological tools (e.g., RNA interference) for their control. For example, among phylogenetically related parasitic nematodes, genes associated with spermatogenesis and gonad development in *A. suum* have been identified[74] and a large portion showed gonadal genes-related RNAi phenotypes in *C. elegans*, providing potential RNAi targets for parasite control.

A draft sequence of genome for *A. simplex* (Wormbase BioProject PRJEB496; http://parasite.wormbase.org/Anisakis_simplex_prjeb496/Info/Index/) is shown to contain around 20,000 putative genes.[75]

The first study using an "omic" approach on *Anisakis* spp. was an investigation of proteomic profiling and characterization of allergens in the two sibling species *A. simplex* s.s., *A. pegreffii*, and in their hybrid forms.[76] Differentially expressed proteins for the three taxonomic units have been identified: 28 different allergenic proteins are classified and described as potentially new allergens in *Anisakis*. A second investigation by Baird et al.[77] explored the transcriptomic repertoire of third-stage infective larvae of the abovementioned species to refine current diagnostics for anisakiasis. Serological tests used for diagnostic purposes are not reliable and often nonspecific, but are still used as indicative of exposure. They rely on antigenic molecules such as Ani-s1-14,[78] which are known to cross-react with similar molecules from other organisms, such as arthropods, as they belong to "ecdysozoa," organisms united by the tendency to molt or cast off the outer covering (a process called ecdysis). The development of a reliable, sensitive, and specific serological tool is critical for studies on the exposure level of a population, leading to improved diagnostic and epidemiology outcomes. The identification of around 30 molecules with antigenic properties by Baird and colleagues provided a resource for fundamental and applied studies on allergy-causing parasites. Further clarification on the mechanisms that underlie infectivity, with particular regard to nematodes as group of parasites capable of maintaining a prolonged infection in complex hosts is important. A wide-scale pathogenic nematode RNA-seq datasets could be relevant to depict general and specific mechanisms related to pathogenic behavior, and in a broad-spectrum may favor the development of new diagnostic and screening tools based on antigen–antibody study and may favor drugs target discovery. Moreover, information of this kind will aid in the understanding of both conserved and divergent aspects of nematode biology.

A deep exploration of transcriptomic and proteomic datasets of *Anisakis* spp. may have implications in understanding their interactions with different natural and accidental (humans) hosts. This approach may allow unraveling patterns and processes involved in differential schemes of host adaptations, tissue migration, pathogenicity, and immunobiology, with respect to zoonotic diseases in humans.

Tissue-specific transcriptomic analyses of *A. simplex* s.s. and *A. pegreffii* have shed light onto the pathological processes during infection of human accidental hosts.[79] In particular, high-throughput RNA-seq and bioinformatics analyses of sequence data have been applied to the characterization of the whole sets of transcripts expressed by infective larvae, as well as their pharyngeal tissues, in a bid to identify transcripts potentially involved in tissue invasion and host–pathogen interplay. This work provided the scientific community with a list of key transcripts for future functional studies of biological pathways specifically related to anisakids.

28.4 FUTURE PERSPECTIVES

Members of the genus *Anisakis* are responsible for anisakiasis, a fish-borne zoonosis that presents with gastrointestinal disturbances and/or allergic reactions after ingestion of raw or undercooked fish containing *Anisakis* third-stage larvae. Prevalent in countries where large quantities of raw fish are consumed (e.g., Japan), anisakiasis is now recognized as an emerging parasitic disease of considerable economic relevance worldwide. Given their limited morphological variations, molecular approaches are increasingly relied upon for inferring the systematic and evolutionary relationships among anisakid nematodes. Recent transcriptomic and proteomic studies on anisakids have helped uncover new insights on the host–parasite interactions and pathogenic mechanisms, further empowering allergen-specific characterization of *Anisakis* spp. Given the unavailability of effective vaccines against *Anisakis* parasites, additional studies to identify protective agents are justified.

REFERENCES

1. Anderson RC. The superfamily Ascaridoidea. In: Anderson, RC (ed.), *Nematodes Parasites of Vertebrates: Their Development and Transmission*. CAB International, London; 1992; 253–6.
2. Levsen A, Berland B. *Anisakis* species. In: Woo PTK, Buchmann K (eds.), *Fish Parasites: Pathobiology and Protection*. CAB International, London; 2012; 298–309.
3. Køie M, Berland B, Burt MDB. Development to third-stage larva occurs in eggs of *Anisakis simplex* and *Pseudoterranova decipiens* (Nematoda, Ascaridoidea, Anisakidae). *Can J Fish Aquat Sci*. 1995;52(1):134–9.
4. Højgaard D. Impact of temperature, salinity and light on hatching of eggs of *Anisakis simplex* (Nematoda: Anisakidae), isolated by a new method, and some remark on survival of larvae. *Sarsia*. 1998;83:21–8.
5. Karl H, Baumann F, Ostermeyer U, Kuhn T, Klimpel S. *Anisakis simplex* (s.s.) larvae in wild Alaska salmons: no indication of post-mortem migration from viscera into flesh. *Dis Aquat Org*. 2011;94:201–9.
6. Cipriani P, Acerra V, Bellisario B, et al. Larval migration of the zoonotic parasite *Anisakis pegreffii* (Nematoda: Anisakidae) in European anchovy, *Engraulis encrasicolus*: Implications to seafood safety. *Food Control*. 2015;59:148e157.
7. Audicana MT, Ansotegui IJ, de Corres LF, Kennedy MW. *Anisakis simplex*: dangerous-dead and alive? *Trends Parasitol*. 2002;18(1):20.
8. Mattiucci S, D'Amelio S. Anisakiasis. In: Bruschi F, editor. *Helminth Infection and Their Impact on Global Public Health*. Springer, Vienna; 2014.
9. Audicana MT, Kennedy MW. *Anisakis simplex*: from obscure infectious worm to inducer of immune hypersensitivity. *Clin Microbiol Rev*. 2011; 21:360–79.
10. Fayer R. Introduction and public health importance of foodborne parasites. In: Xiao L, Ryan U and Feng Y (eds.), *Biology of Foodborne Parasites*. CRC Press, Boca Raton, FL; 2014; 3–19.
11. Arizono N, Miura T, Yamada M, Tegoshi T, Onishi K. Human infection with *Pseudoterranova azarasi* roundworm. *Emerg Infect Dis*. 2011;17:555–6.
12. Mattiucci, S., Fazii, P., De Rosa, A., et al. Anisakiasis and gastroallergic reactions associated with *Anisakis pegreffii* infection, Italy. *Emerg Infect Dis*. 2013;19(3):496–9.
13. Shamsi S, Butcher AR. First report of human anisakidosis in Australia. *Med J Aust*. 2011; 194(4):199–200.
14. Yagi K, Nagasawa K, Ishikura H, et al. Female worm *Hysterothylacium aduncum* excreted from human: a case report. *Jpn J Parasitol*. 1996;45:12–23.
15. Fagerholm HP. Intra-specific variability of the morphology in a single population of the seal parasite *Contracaecum osculatum* (Rudolphi) (Nematoda, Ascaridoidea), with a redescription of the species. *Zool Scripta*. 1989;18:33–41.
16. Paggi L, Nascetti G, Webb SC, Mattiucci S, Cianchi R, Bullini L. A new species of *Anisakis* Dujardin, 1845 (Nematoda: Anisakidae) from beaked whale (Ziphiidae): allozyme and morphological evidence. *Syst Parasitol*. 1998;40:161–74.
17. Mattiucci S., Cipriani P, Webb SC, et al. Genetic and morphological approaches distinguishing the three sibling species of the *Anisakis simplex* species complex, with a species designation as *Anisakis berlandi* n. sp. for *A. simplex* sp. C (Nematoda: Anisakidae). *J Parasitol*. 2014; 100(2):199–214.
18. Berland B. Nematodes from some Norwegian marine fishes. *Sarsia*. 1961;2:1–50.
19. Mattiucci S, Paoletti M, Webb SC. *Anisakis nascettii* n. sp. (Nematoda: Anisakidae) from beaked whales of the southern hemisphere: morphological description, genetic relationships between congeners and ecological data. *Syst Parasitol*. 2009;74(3):199–217.
20. Nascetti G, Paggi L, Orecchia P, Smith JW, Mattiucci S, Bullini L. Electrophoretic studies on the *Anisakis simplex* complex (Ascaridida: Anisakidae) from the Mediterranean and North-East Atlantic. *Int J Parasitol*. 1986;16:633–40.
21. Abollo E, Paggi L, Pascual S, D'Amelio S. Occurrence of recombinant genotypes of *Anisakis simplex* s.s. and *Anisakis pegreffii* (Nematoda: Anisakidae) in an area of sympatry. *Infect Genet Evol* 2003:3:175–81.
22. D'Amelio S, Busi M, Farjallah S, Ingrosso S, Cavallero S, Paggi L. On the significance of heterozygote genotypes in sibling species of the *Anisakis simplex* complex. XXVth Congress of the Italian Society of Parasitology (SOIPA). Pisa, Italy, June 21–24, 2008. *Parassitologia*. 2008;50(suppl.1):18.
23. Mattiucci S, Acerra V, Paoletti M, et al. No more time to stay 'single' in the detection of *Anisakis pegreffii, A. simplex* (s. s.) and hybridization events between them: a multi-marker nuclear genotyping approach. *Parasitology*. 2016;143:998–1011.

24. Umehara A, Kawakami Y, Araki J, Uchida A. Molecular identification of the etiological agent of the human anisakiasis in Japan. *Parasitol Int.* 2007;56(3):211–5.
25. Cavallero S, Costa A, Caracappa S, Gambetta B, D'Amelio S. Putative hybrids between two *Anisakis* cryptic species: molecular genotyping using high resolution melting. *Exp Parasitol.* 2014; 146:87–93.
26. Mattiucci S, Nascetti G, Cianchi R, et al. Genetic and ecological data on the *Anisakis simplex* complex, with evidence for a new species (Nematoda, Ascaridoidea, Anisakidae). *J Parasitol.* 1997;86:401–16.
27. Cavallero S, Nadler SA, Paggi L, Barros NB, D'Amelio S. Molecular characterization and phylogeny of anisakid nematodes from cetaceans from southeastern Atlantic coasts of USA, Gulf of Mexico, and Caribbean Sea. *Parasitol Res.* 2011 108(4):781–92.
28. Palm HW, Theisen S, Damriyasa IM, Kusmintarsih ES, Oka IB, Setyowati EA, Suratma NA, Wibowo S, Kleinertz S. *Anisakis* (Nematoda: Ascaridoidea) from Indonesia. *Dis Aquat Organ.* 2017;123(2):141–57.
29. Kurochkin, YV. Parasites of the Caspian seal *Pusa caspica.* Rapport et proces-verbaux des reunion, *Cons Int Expl Mer.* 1975;169;363.
30. Hartwich G. Keys to genera Ascaridoidea. In: Anderson R.C., Chabaud A.G., Wilmott S. (eds.), *CIH Keys to Nematode Parasites of Vertebrates.* Farnham Royal, Commonwealth Agricolture Bureau, Richmond; 1974; 153.
31. Fagerholm HP. Systematic implications of male caudal morphology in ascaridoid nematode parasites. *Syst Parasitol.* 1991;19:215–28.
32. Mattiucci S, Nascetti G, Dailey M, et al. Evidence for a new species of *Anisakis* Dujardin, 1845: morphological description and genetic relationships between congeners (Nematoda: Anisakidae). *Syst Parasitol.* 2005;61:157–71.
33. D'Amelio S, Barros NB, Ingrosso S, Fauquier DA, Russo R, Paggi L. Genetic characterization of members of the genus Contracaecum (Nematoda: Anisakidae) from fish-eating birds from west-central Florida, USA, with evidence of new species. *Parasitology.* 2007;134(7):1041–51.
34. Paggi L., Nascetti G., Cianchi R., et al. Genetic evidence for three sealworm species within *Pseudoterranova decipiens* (Nematoda, Ascaridida, Ascaridoidea) in the North Atlantic and Norwegian and Barents Sea. *Int J Parasitol.* 1991;21(2):195–212.
35. Nascetti G, Cianchi R, Mattiucci S, et al. Three sibling species within *Contracaecum osculatum* (Nematoda, Ascaridida, Ascaridoidea) from the Atlantic Arctic-Boreal region: reproductive isolation and host preferences. *Int J Parasitol.* 1993;23(1):105–20.
36. Orecchia P, Mattiucci S, D'Amelio S, et al. Two members in the *Contracaecum osculatum* complex (Nematoda, Ascaridida, Ascaridoidea) from the Atlantic Ocean. *Int J Parasitol.* 1994;24:367–77.
37. D'Amelio S, Mathiopoulos K, Santos CP, et al. Genetic markers in ribosomal DNA for the identification of members of the genus *Anisakis* (Nematoda: Ascaridoidea) defined by polymerase chain reaction-based restriction fragment length polymorphism. *Int J Parasitol.* 2000;30:223–6.
38. Pontes T, D'Amelio S, Costa G, Paggi L. Molecular characterization of larval anisakid nematodes from marine fishes of Madeira by a PCR-based approach, with evidence for a new species. *J Parasitol.* 2005;91(6):1430–4.
39. Nadler and Hudspeth. Ribosomal DNA and phylogeny of the Ascaridoidea (Nemata: Secernentea): Implications for morphological evolution and classification. *Mol Phylogenet Evol.* 1998;10:221–36.
40. Gerbi SA. The evolution of eukariotic ribosomal DNA. *Biosystems.* 1986;19:247–58.
41. Gasser RB, Monti RJ. Identification of parasitic nematodes by PCR-SSCP of ITS-2 rDNA. *Mol Cell Probes.* 1997;11:201–9.
42. Valentini A, Mattiucci S, Bondanelli P, et al. Genetic relationships among *Anisakis* species (Nematoda: Anisakidae) inferred from mitochondrial cox2 sequences, and comparison with allozyme data. *J Parasitol.* 2006;92(1):156–66.
43. Nadler SA, D'Amelio S, Dailey MD, Paggi L, Siu S, Sakanari JA. Molecular phylogenetics and diagnosis of *Anisakis, Pseudoterranova* and *Contracaecum* from northern Pacific marine mammals. *J Parasitol.* 2005;91(6):1413–29. Erratum in: J Parasitol. 2007;93(2):444.
44. D'Amelio S., Cavallero S., Busi M., et al. *Anisakis* (Chapter 45). In: Liu D (ed.), M*olecular Detection of Human Parasitic Pathogens.* CRC Press, Boca Raton, FL; 2012.
45. Kim KH, Eom KS, Park JK. The complete mitochondrial genome of *Anisakis simplex* (Ascaridida: Nematoda) and phylogenetic implications. *Int J Parasitol.* 2006;36:319–28.
46. Umehara A, Kawakami Y, Araki J, Uchida A. Multiplex PCR for the identification of *Anisakis simplex* sensu stricto, *Anisakis pegreffii* and the other anisakid nematodes. *Parasitol Int.* 2008;57(1):49–53
47. Abe N. Application of the PCR-sequence-specific primers for the discrimination among larval *Anisakis simplex* complex. *Parasitol Res.* 2008;102(5):1073–5.

48. Hu M, D'Amelio S, Zhu XQ, Paggi L, Gasser RB. Mutation scanning for sequence variation in three mitochondrial DNA regions for members of the *Contracaecum osculatum* (Nematoda: Ascaridoidea) complex. *Electrophoresis*. 2001;22:1069–75.
49. Zhu XQ, D'Amelio S, Palm HW, Paggi L, George-Nascimento M, Gasser RB. SSCP-based identification of members within the *Pseudoterranova decipiens* complex (Nematoda: Ascaridoidea: Anisakidae) using genetic markers in the internal transcribed spacers of ribosomal DNA. *Parasitology*. 2002;124:615–23.
50. Mladineo I, Trumbić Ž, Radonić I, Vrbatović A, Hrabar J, Bušelić I. *Anisakis simplex* complex: ecological significance of recombinant genotypes in an allopatric area of the Adriatic Sea inferred by genome-derived simple sequence repeats. *Int J Parasitol*. 2017;47(4):215–23.
51. Yang X, Nian R, Lin H, Duan C, Sui J, Cao L. Detection of anisakid larvae in cod fillets by UV fluorescent imaging based on principal component analysis and gray value analysis. *J Food Prot*. 2013;76(7):1288–92.
52. Fraulo P, Morena C, Costa A. Recovery of anisakid larvae by means of chloro-peptic digestion and proposal of the method for the official control. *Acta Parasitol*. 2014;59(4):629–34.
53. Cammilleri G, Chetta M, Costa A, et al. Validation of the TrichinEasy® digestion system for the detection of Anisakidae larvae in fish products. *Acta Parasitol*. 2016;61(2):369–75.
54. Cavallero S, Magnabosco C, Civettini M, et al. Survey of *Anisakis* sp. and *Hysterothylacium* sp. in sardines and anchovies from the North Adriatic Sea. *Int J Food Microbiol*. 2015;200:18–21.
55. Mattiucci S, Garcia A, Cipriani P, Santos MN, Nascetti G, Cimmaruta R. Metazoan parasite infection in the swordfish, *Xiphias gladius*, from the Mediterranean Sea and comparison with Atlantic populations: implications for its stock characterization. *Parasite*. 2014;21:35.
56. Cavallero S, Scribano D, D'Amelio S. First case report of invasive pseudoterranoviasis in Italy. *Parasitol Int*. 2016; 65(5):488–90.
57. Fang W, Liu F, Zhang S, Lin J, Xu S, Luo D. *Anisakis pegreffii*: a quantitative fluorescence PCR assay for detection in situ. *Exp Parasitol*. 2011;127(2):587–92.
58. Herrero B, Vieites JM, Espiñeira M. Detection of anisakids in fish and seafood products by real-time PCR. *Food Control*. 2011;22:933–9.
59. Lopez I, Pardo MA. Evaluation of a real-time polymerase chain reaction (PCR) assay for detection of *Anisakis simplex* parasite as a food-borne allergen source in seafood products. *J Agric Food Chem*. 2010;58(3):1469–77.
60. Mossali C, Palermo S, Capra E, et al. Sensitive detection and quantification of anisakid parasite residues in food products. *Foodborne Pathog Dis*. 2010;7(4):391–7.
61. Cavallero S, Bruno A, Arletti E, et al. Validation of a commercial kit aimed to the detection of pathogenic anisakid nematodes in fish products. *Int J Food Microbiol*. 2017;257:75–9.
62. Godínez-González C, Roca-Geronès X, Cancino-Faure B, Montoliu I, Fisa R. Quantitative SYBR Green qPCR technique for the detection of the nematode parasite *Anisakis* in commercial fish-derived food. *Int J Food Microbiol*. 2017;261:89–94.
63. Paoletti M, Mattiucci S, Colantoni A, Levsen A, Gay M, Nascetti G. Species-specific Real Time-PCR primers/probe systems to identify fish parasites of the genera *Anisakis*, *Pseudoterranova* and *Hysterothylacium* (Nematoda: Ascaridoidea). *Fish Res*. 2017;202:38–48.
64. Jex AR, Liu S, Li B, et al. *Ascaris suum* draft genome. *Nature*. 2011;479:529–33.
65. Wang T, Van Steendam K, Dhaenens M, Vlaminck J, Deforce D, Jex AR. Proteomic analysis of the excretory-secretory products from larval stages of *Ascaris suum* reveals high abundance of glycosyl hydrolases. *PLoS Negl Trop Dis*. 2013;7(10):e2467.
66. Mattiucci S, Nascetti G. Advances and trends in the molecular systematics of anisakid nematodes, with implications for their evolutionary ecology and host-parasite co-evolutionary processes. *Adv Parasitol*. 2008;66:47–148.
67. Mattiucci S, Cipriani P, Levsen A, Paoletti M, Nascetti G. Molecular epidemiology of *Anisakis* and anisakiasis: An ecological and evolutionary road map. *Adv Parasitol*, 2018; 99:93–263.
68. Nicholson JK, Holmes E, Lindon JC, Wilson ID. The challenges of modeling mammalian biocomplexity. *Nat Biotechnol*. 2004;22(10):1268–74.
69. Viney M. The genomic basis of nematode parasitism. *Brief Funct Genomics*, 2018; 17(1):8–14.
70. Berrilli F, Di Cave D, Cavallero S, D'Amelio S. Interactions between parasites and microbial communities in the human gut. *Front Cell Infect Microbiol*. 2012;2:68–73.
71. Summers RW, Elliott DE, Urban J.F Jr, Thompson RA, Weinstock, JV. *Trichuris suis* therapy for active ulcerative colitis: a randomized controlled trial. *Gastroenterology*. 2005;128:825–32.
72. Summers RW, Elliott DE, Urban JF Jr, Thompson RA, Weinstock, JV. *Trichuris suis* therapy in Crohn's disease. *Gut*. 2005;54:87–90.

73. Cantacessi C, Hofmann A, Campbell B, Gasser RB. Impact of next-generation technologies on exploring socio-economically important parasites and developing new interventions. In: Cunha, MV, and Inácio J (Eds.), *Veterinary Infection Biology: Molecular Diagnostics and High-Throughput Strategies, Methods in Molecular Biology*. Springer, New York, 2015; pp. 437–74.
74. Ma X, Zhu Y, Li C, Shang Y, Meng F, Chen S, Miao L. Comparative transcriptome sequencing of germline and somatic tissues of the *Ascaris suum* gonad. *BMC Genomics*. 2011;12:481.
75. Howe KL, Bolt BJ, Shafie M, Kersey P, Berriman M. WormBase ParaSite - a comprehensive resource for helminth genomics. *Mol Biochem Parasitol*. 2017;215:2–10.
76. Arcos SC, Ciordia S, Roberston L, et al. Proteomic profiling and characterization of differential allergens in the nematodes *Anisakis simplex* sensu stricto and *A. pegreffii*. *Proteomics* 2014;14:1547–68.
77. Baird FJ, Su X, Aibinu I, et al. The *Anisakis* transcriptome provides a resource for fundamental and applied studies on allergy-causing parasites. *PLoS Negl Trop Dis*. 2016;10(7):e0004845.
78. Daschner A, Cuellar C, Rodero M. The *Anisakis* allergy debate: does an evolutionary approach help? *Trends Parasitol*. 2012;28:9–15.
79. Cavallero S, Lombardo F, Su X, Salvemini M, Cantacessi C, D'Amelio S. Tissue-specific transcriptomes of *Anisakis simplex* (sensu stricto) and *Anisakis pegreffii* reveal potential molecular mechanisms involved in pathogenicity. *Parasit Vectors*. 2018;11(1):31.

SUMMARY

Parasitic nematodes of the genus *Anisakis* are responsible for a fish-borne zoonosis known as anisakiasis, which is associated with gastrointestinal disturbances and/or allergic reactions in humans after ingestion of raw or undercooked fish containing *Anisakis* third-stage larvae. Prevalent in countries where large quantities of raw fish are consumed (e.g., Japan), anisakiasis is now recognized as an emerging parasitic disease of considerable economic relevance worldwide. Given their limited morphological variations, molecular approaches are increasingly relied upon for inferring the systematic and evolutionary relationships among anisakid nematodes. Recent transcriptomic and proteomic studies on anisakids have helped uncover new insights on the host–parasite interactions and pathogenic mechanisms, further empowering allergen-specific characterization of *Anisakis* spp.

Molecular Epidemiology of *Dientamoeba fragilis*

29

Dongyou Liu
Royal College of Pathologists of Australasia Quality Assurance Programs

Contents

29.1 INTRODUCTION

Initially observed by Wenyon in 1909 and fully documented by Jepps and Dobell in 1918, *Dientamoeba fragilis* [a name reflecting its morphological (an ameba with a binucleate structure) as well as biological (fragility, with the tendency to disintegrate quickly outside the human body) characteristics] is a unflagellated flagellate (i.e., a flagellate that has lost its flagella) that is a grossly neglected foodborne pathogen found in 0.4% of patients with gastrointestinal discomfort and up to 82.9% of children with gastrointestinal protozoan diseases [1,2]. The application of molecular techniques has helped clarify its taxonomic status, enable its sensitive and specific diagnosis as well as epidemiological tracking, and facilitate further investigation on its pathogenic mechanisms, along with development of innovative measure for its ultimate control and elimination.

Taxonomy. Making up one of the four genera (*Dientamoeba*, *Histomonas*, *Parahistomonas*, and *Protrichomonas*) in the family Dientamoebidae, order Trichomonadida, class Tritrichomonadidae, and phylum Parabasalia, the genus *Dientamoeba* currently comprises one recognized species, that is, *Dientamoeba fragilis*. Based on the differences in the 18S rRNA (by 2%–4%), actin, and elongation factor 1α genes (by ~3%), *D. fragilis* is separated into two genotypes, with genotype 1 being more common than genotype 2 (also known as the Bi/PA strain) [2].

Dientamoeba differs from *Entamoeba*, a more common protozoan cause of human diarrhea, by its predominately binucleated form, its apparent lack of cyst stage, especially in clinical specimens, and its similarities to other flagellates (both morphologically and antigenically).

Morphology. *Dientamoeba* undergoes three developmental stages during its lifecycle: trophozoite, cyst, and precyst. *D. fragilis* trophozoite is pleomorphic and about 4–20 µm in size (range 5–15 µm; as large as 40 µm in culture). In stained smear, *D. fragilis* trophozoite displays a finely granular cytoplasm consisting of vacuoles and food inclusions and ingested debris, two nuclei (80%), and, occasionally, one nucleus (20%), with delicate nuclear membrane and fragmented karyosome containing 4–8 chromatin granules (Figure 29.1) [3]. In the absence of external flagella, *D. fragilis* trophozoite moves by crawling action associated with cytoplasmic streaming of angular, serrated, or broad-lobed and almost transparent pseudopodia [2,4].

D. fragilis cyst appears oval or round (amoeboid-shaped), and measures about 4–6 µm in diameter. Having a distinct cyst wall with a clear zone, *D. fragilis* cyst often shows two nuclei, each containing a large central karyosome with a delicate nuclear membrane and chromatin granules, an axostyle, flagellar axonemes, pelta, and a costa. Being nonmotile and infective, *D. fragilis* cyst is rarely seen in clinical samples (<5%) [2,4].

D. fragilis precyst (about 3.5–5 µm in diameter) is a small spherical organism showing finely granular and uniform cytoplasm without food particles, and has one nucleus or two nuclei. Although *D. fragilis* precyst is observed in up to 5% in clinical samples, its infectivity remains unknown [4,5].

Genomics. *D. fragilis* genome is yet to be sequenced. Transcriptomic analysis of *D. fragilis* genotype 1 trophozoite uncovers >6,000 novel nucleotide sequences, demonstrating about 30% similarity to those of *Trichomonas vaginalis*, such as the expansion in BspA-like leucine-rich repeats and actin family genes, as well as the presence of cytotoxic cysteine proteases [6].

Interestingly, *D. fragilis* 18S rRNA gene has a lower G+C content and ~100 more nucleotides than other trichomonads, permitting accurate differentiation between *D. fragilis* and other trichomonads [7]. Further, unusual among protozoa, *D. fragilis* harbors multicopy internal transcribed spacers (ITS), which form the basis of C-profiling technique for distinguishing isolates [8].

Biology and epidemiology. As the preferred host, humans acquire *D. fragilis* infection after ingestion of food or water containing trophozoites. Once inside, trophozoites migrate to the large intestine and multiply via binary fission to generate precysts, cysts, and trophozoites, which are subsequently discharged in the feces [9]. In addition to contaminated food or water, *D. fragilis* may be plausibly transmitted in the ova of the human pinworm, *Enterobius vermicularis*, as a strong association between the incidence of *E. vermicularis* and *D. fragilis* infection is noted [10,11]. *D. fragilis* trophozoites are infectious to rodents, gorillas, and pigs, but not other mammals, which may possibly play a role in *D. fragilis* epidemiology [12].

D. fragilis has been reported in all continents, with incidence ranging from 0.4% in the general population to 82% in children [13–21]. Based on an experimental study, oral doses of 4×10^6, 10^5 and 10^3 *D. fragilis* trophozoites are needed to infect 100%, 37.5%, and 0% of mice, respectively [22].

Clinical features. Most (85%) of patients with *D. fragilis* infection develop diarrhea (watery/loose stools; 64%) often ≥2 weeks (50%), abdominal pain and cramps (55%), flatus, abdominal swelling and discomfort (33%), nausea (13%), weight loss (12%), vomiting (9%), fatigue (9%), fever (9%), constipation (7%),

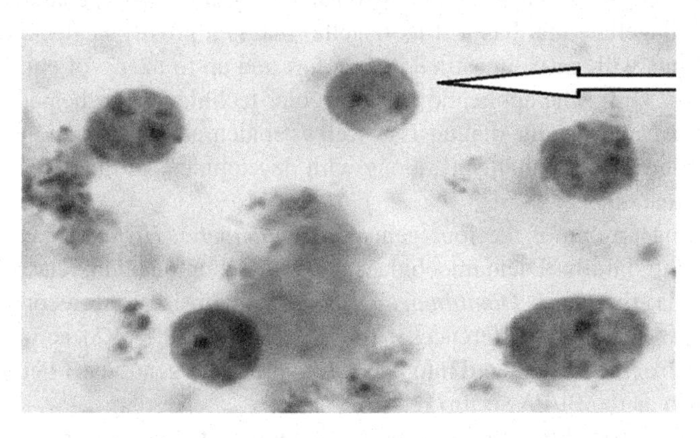

FIGURE 29.1 *Dientamoeba fragilis* trophozoites in stool exhibit an ameba-like morphology and are often binucleated (arrow). (Photo credit: Ali S, Khetpal N, Khan MT, Rasheed M, Asad-Ur-Rahman F, Echeverria-Beltran K. *Cureus*. 2017;9(12):e1992.)

anal pruritus (7%), fecal urgency (4%), bloody stool (3%), abnormally smelly stools (2%), and heartburn (1%). Some patients (15%) may be asymptomatic [2,23,24].

Pathogenesis. There have been some debates about the pathogenicity of *D. fragilis*. The fact that 85% of people infected with *D. fragilis* display gastrointestinal symptoms suggests its pathogenic potential. Recent development of a rodent model, which helps confirm *D. fragilis*' ability to produce unformed stools, stimulate inflammatory response, and induce statistically significant weight loss, lends further support to its pathogenicity [22]. Moreover, transcriptomic identification of genes encoding cathepsin L-like cysteine proteases, lectins, calmodulin, amoebapore-like proteins, and putative immunomodulatory proteins in *D. fragilis* genotype 1 trophzoites yields additional insights on the pathogenic mechanisms of this protozoan.

Diagnosis. Diagnosis of human dientamoebiasis involves microscopy, culture, biochemical tests, serology, and molecular techniques. Approximately half of the patients have eosinophilia.

Light microscopy of saline or iodine preparations from stool reveals *D. fragilis* trophozoites as nonspecific rounded masses. Use of a fixative [e.g., mercury-based compound, sodium acetate-acetic acid-formalin (SAF), phenol alcohol-formalin, modified Schaudinn's fixative, thimerosal (merthiolate)-iodine-formalinon, polyvinyl alcohol (PVA)] followed by a permanent stain (e.g., iron-hematoxylin, trichrome, Chlorazol black stain, modified Field's stain, Mayer's hemalum and Lawless' stain) is critical for observing internal structures. A combination of SAF and modified iron-hematoxylin or Chlorazol black stain appears highly effective, relatively easy, and fast to perform. Collection of three specimens on consecutive days helps increase positive rate for *D. fragilis* detection in permanently stained smears [4,25].

Dientamoeba can be cultured in Boeck and Drbohlav's medium, Robinson medium, Dobell and Laidlaw's medium, Cleveland–Collier medium, Balamuth's medium, and TYGM-9. It appears that Loeffler's slope medium overlaid with Earle's balanced salt solution supplemented with cholesterol and ferric ammonium citrate in a microaerophilic atmosphere at 42°C allows optimal growth of *D. fragilis* trophozoites. Furthermore, modified Robinson medium or modified Boeck and Drbohlav's medium is helpful for increased detection of *D. fragilis* in comparison with microscopy. Prompt inoculation of stool specimens into the medium is critical for the success of *D. fragilis* culture [2,4].

Serological tests [e.g., immunofluorescence microscopy, enzyme immunoassays (EIA), and immunochromatographic tests (ICT)] are valuable for sensitive and specific detection of *D. fragilis* [2,4]. Further, MALDI-TOF-MS may also be used for specific identification of *D. fragilis* [26].

Molecular techniques (especially PCR) targeting 18S rRNA, 5.8S rRNA, ITS, and other genes enable rapid identification of *D. fragilis* directly from clinical, food, and environmental samples [27–33].

Treatment and prevention. Treatment options for symptomatic dientamoebiasis include clioquinol (40 mg kg^{-1} of body weight/day for 10–21 days), iodoquinol (650 mg per oral, three times daily for 20 days), metronidazole (30 mg $(kg\ day)^{-1}$ per oral for 10 days), ornidazole (30 mg kg^{-1} for child, 2 g for adult), paromomycin (25–35 mg kg^{-1} daily for 4–5 days), secnidazole (30 mg kg^{-1} for child, 2 g for adult), and tinidazole [34,35].

Prevention of *D. fragilis* infection should focus on its fecal-oral transmission, including proper cooking and preparation of food, drinking filtered water, and reduction of fecal contamination through appropriate hygiene and sanitary measures.

29.2 MOLECULAR EPIDEMIOLOGY OF *D. FRAGILIS*

D. fragilis demonstrates considerable genetic homology, and early study targeting the 18S rRNA gene has resulted in the identification of two genotypes, with genotype 1 predominating clinical isolations from various geographic areas [7]. A combination of PCR amplification of 18S rRNA gene and restriction fragment length polymorphism (RFLP) failed to reveal additional genetic resolution among *D. fragilis* isolates. While examination of three *D. fragilis* housekeeping genes (actin, EF-1α, and 18S rRNA genes) confirms the validity of the two known genotypes, application of these genes in multilocus sequencing analysis does not seem to enhance genetic discrimination [36]. Other potential targets for phylogenetic analysis and epidemiological tracking of *D. fragilis* isolates include the internal transcribed spacer 1 (ITS-1)-5.8S rRNA gene-ITS-2 region, and RNA polymerase II largest-subunit protein gene [2,8].

Comparative examination of the ITS-1-5.8S gene-ITS-2 region reveals that *D. fragilis* and *Histomonas meleagridis* possess short, highly homogenous 5.8S rRNA genes (111 bp) in comparison with other trichomonads (~150 bp); and that *D. fragilis* ITS-1 region contains nucleotide variations that allow separation into two different groups, consistent with previous 18S rRNA gene results [8]. Additionally, comparison of RNA polymerase II largest-subunit protein sequences from *D. fragilis* and other trichomonads helps confirm the close genetic relationship between *D. fragilis* and *Histomonas meleagridis* [2]. Moreover, PCR amplification of 18S rRNA gene followed by high-resolution melting curve permits detection of genetic variability among *D. fragilis* isolates and leads to the identification of four profiles (subtypes), with profile 1 accounting for 50% of isolates, profile 2 for 20%, profiles 3 and 4 for 16.7% and 13.4%, respectively. Interestingly, 73.4% of profile 1 and 75% of profile 4 patients have chronic intermittent diarrhea; all profile 2 patients show acute diarrhea and all profile 3 patients suffer diarrhea alternating with constipation [37].

29.3 FUTURE PERSPECTIVES

Dientamoeba fragilis is a single-celled trichomonad parasite with the ability to induce gastrointestinal diseases in human population, especially children [38–40]. Compared to other diarrhea-causing protozoa (e.g., *Entamoeba*), *D. fragilis* is understudied and overtly neglected. Even today, the genome of *D. fragilis* is yet to be sequenced. Obviously, this negligence prevents targeted analyses on the molecular mechanisms of *D. fragilis* pathogenicity and delays the development of innovative measures against human dientamoebiasis. In addition, there is much to learn about *D. fragilis* lifecycle and mode of transmission, the complete elucidation of which is critical for effective monitoring and control of this emerging foodborne pathogen [41,42].

REFERENCES

1. Johnson EH, Windsor JJ, Clark CG. Emerging from obscurity: biological, clinical, and diagnostic aspects of *Dientamoeba fragilis*. *Clin Microbiol Rev*. 2004;17(3):553–70.
2. Stark D, Barratt J, Chan D, Ellis JT. *Dientamoeba fragilis*, the neglected trichomonad of the human bowel. *Clin Microbiol Rev*. 2016;29(3):553–80.
3. Ali S, Khetpal N, Khan MT, Rasheed M, Asad-Ur-Rahman F, Echeverria-Beltran K. A Mexican honeymoon marred by gastrointestinal upset: a case of *Dientamoeba fragilis* causing post-infectious irritable bowel syndrome. *Cureus*. 2017;9(12):e1992.
4. Garcia LS. *Dientamoeba fragilis*, one of the neglected intestinal protozoa. *J Clin Microbiol*. 2016;54(9):2243–50.
5. Stark D, Garcia LS, Barratt JL, et al. Description of *Dientamoeba fragilis* cyst and precystic forms from human samples. *J Clin Microbiol*. 2014;52(7):2680–3.
6. Barratt JL, Cao M, Stark DJ, Ellis JT. The transcriptome sequence of *Dientamoeba fragilis* offers new biological insights on its metabolism, kinome, degradome and potential mechanisms of pathogenicity. *Protist*. 2015;166(4):389–408.
7. Windsor JJ, Macfarlane L, Clark CG. Internal transcribed spacer dimorphism and diversity in *Dientamoeba fragilis*. *J Eukaryot Microbiol*. 2006;53(3):188–92.
8. Bart A, van der Heijden HM, Greve S, Speijer D, Landman WJ, van Gool T. Intragenomic variation in the internal transcribed spacer 1 region of *Dientamoeba fragilis* as a molecular epidemiological marker. *J Clin Microbiol*. 2008;46(10):3270–5.

9. Miguel L, Salvador F, Sulleiro E, et al. Clinical and epidemiological characteristics of patients with *Dientamoeba fragilis* infection. *Am J Trop Med Hyg.* 2018;99(5):1170–3.
10. Ögren J, Dienus O, Löfgren S, Iveroth P, Matussek A. *Dientamoeba fragilis* DNA detection in *Enterobius vermicularis* eggs. *Pathog Dis.* 2013;69(2):157–8.
11. Clark CG, Röser D, Stensvold CR. Transmission of *Dientamoeba fragilis*: pinworm or cysts? *Trends Parasitol.* 2014;30(3):136–40.
12. Cacciò SM, Sannella AR, Manuali E, et al. Pigs as natural hosts of *Dientamoeba fragilis* genotypes found in humans. *Emerg Infect Dis.* 2012;18(5):838–41.
13. Stark D, Barratt J, Ellis J, Harkness J, Marriott D. Repeated *Dientamoeba fragilis* infections: a case report of two families from Sydney, Australia. *Infect Dis Rep.* 2009;1(1):e4.
14. Al-Hindi AI, Shammala BM. *Dientamoeba fragilis* in Gaza Strip: a neglected protozoan parasite. *Iran J Parasitol.* 2013;8(2):249–55.
15. Ögren J, Dienus O, Löfgren S, Einemo IM, Iveroth P, Matussek A. *Dientamoeba fragilis* prevalence coincides with gastrointestinal symptoms in children less than 11 years old in Sweden. *Eur J Clin Microbiol Infect Dis.* 2015;34(10):1995–8.
16. Osman M, El Safadi D, Cian A, et al. Prevalence and risk factors for intestinal protozoan infections with *Cryptosporidium, Giardia, Blastocystis* and *Dientamoeba* among school children in Tripoli, Lebanon. *PLoS Negl Trop Dis.* 2016;10(3):e0004496.
17. Boughattas S, Behnke JM, Al-Ansari K, et al. Molecular analysis of the enteric protozoa associated with acute diarrhea in hospitalized children. *Front Cell Infect Microbiol.* 2017;7:343.
18. Jokelainen P, Hebbelstrup Jensen B, et al. *Dientamoeba fragilis*, a commensal in children in Danish day care centers. *J Clin Microbiol.* 2017;55(6):1707–13.
19. Ibrahim AN, Al-Ashkar AM, Nazeer JT. Additional glance on the role of *Dientamoeba fragilis* & *Blastocystis hominis* in patients with irritable bowel syndrome. *Iran J Parasitol.* 2018;13(1):100–7.
20. Brands MR, Van de Vijver E, Haisma SM, Heida A, van Rheenen PF. No association between abdominal pain and *Dientamoeba* in Dutch and Belgian children. *Arch Dis Child.* 2019;104(7):686–9.
21. Pietilä JP, Meri T, Siikamäki H, et al. *Dientamoeba fragilis* - the most common intestinal protozoan in the Helsinki Metropolitan Area, Finland, 2007 to 2017. *Euro Surveill.* 2019;24(29): 1800546.
22. El-Gayar EK, Mokhtar AB, Hassan WA. Study of the pathogenic potential of *Dientamoeba fragilis* in experimentally infected mice. *Parasite Epidemiol Control.* 2016;1(2):136–43.
23. Stark D, Barratt J, Roberts T, Marriott D, Harkness J, Ellis J. A review of the clinical presentation of dientamoebiasis. *Am J Trop Med Hyg.* 2010;82(4):614–9.
24. Vassalou E, Vassalos CM, Spanakos G, et al. First report of *Dientamoeba fragilis* infection explaining acute non-specific abdominal pain. *Indian J Med Microbiol.* 2016;34(1):106–8.
25. Garcia JA, Cimerman S. Detection of *Dientamoeba fragilis* in patients with HIV/AIDS by using a simplified iron hematoxylin technique. *Rev Soc Bras Med Trop.* 2012;45(2):156–8.
26. Calderaro A, Buttrini M, Montecchini S, et al. MALDI-TOF MS as a new tool for the identification of *Dientamoeba fragilis. Parasit Vectors.* 2018;11(1):11.
27. Stark D, Al-Qassab SE, Barratt JL, et al. Evaluation of multiplex tandem real-time PCR for detection of *Cryptosporidium* spp., *Dientamoeba fragilis*, *Entamoeba histolytica*, and *Giardia intestinalis* in clinical stool samples. *J Clin Microbiol.* 2011;49(1):257–62.
28. Stark D, Roberts T, Marriott D, Harkness J, Ellis JT. Detection and transmission of *Dientamoeba fragilis* from environmental and household samples. *Am J Trop Med Hyg.* 2012; 86(2):233–6.
29. Dunwell D. ME/CFS and *Blastocystis* spp or *Dientamoeba fragilis*, an in-house comparison. *Br J Gen Pract.* 2013;63(607):73–4.
30. Sarafraz S, Farajnia S, Jamali J, Khodabakhsh F, Khanipour F. Detection of *Dientamoeba fragilis* among diarrheal patients referred to Tabriz health care centers by nested PCR. *Trop Biomed.* 2013;30(1):113–8.
31. Caradonna T, Marangi M, Del Chierico F, et al. Detection and prevalence of protozoan parasites in ready-to-eat packaged salads on sale in Italy. *Food Microbiol.* 2017;67:67–75.
32. Hamidi N, Meamar AR, Akhlaghi L, Rampisheh Z, Razmjou E. *Dientamoeba fragilis* diagnosis by fecal screening: relative effectiveness of traditional techniques and molecular methods. *J Infect Dev Ctries.* 2018;12(1):52–9.
33. Gough R, Ellis J, Stark D. Comparison and recommendations for use of *Dientamoeba fragilis* real-time PCR assays. *J Clin Microbiol.* 2019;57(5):e01466–18.
34. Nagata N, Marriott D, Harkness J, Ellis JT, Stark D. Current treatment options for *Dientamoeba fragilis* infections. *Int J Parasitol Drugs Drug Resist.* 2012;2:204–15.
35. Nagata N, Marriott D, Harkness J, Ellis JT, Stark D. In vitro susceptibility testing of *Dientamoeba fragilis. Antimicrob Agents Chemother.* 2012;56(1):487–94.

36. Stensvold CR, Clark CG, Röser D. Limited intra-genetic diversity in *Dientamoeba fragilis* housekeeping genes. *Infect Genet Evol.* 2013;18:284–6.
37. Hussein EM, Al-Mohammed HI, Hussein AM. Genetic diversity of *Dientamoeba fragilis* isolates of irritable bowel syndrome patients by high-resolution melting-curve (HRM) analysis. *Parasitol Res.* 2009;105(4):1053–60.
38. Wong ZW, Faulder K, Robinson JL. Does *Dientamoeba fragilis* cause diarrhea? A systematic review. *Parasitol Res.* 2018;117(4):971–80.
39. Aykur M, Calıskan Kurt C, Dirim Erdogan D, et al. Investigation of *Dientamoeba fragilis* prevalence and evaluation of sociodemographic and clinical features in patients with gastrointestinal symptoms. *Acta Parasitol.* 2019;64(1):162–70.
40. van Gestel RS, Kusters JG, Monkelbaan JF. A clinical guideline on *Dientamoeba fragilis* infections. *Parasitology.* 2019;146(9):1131–9.
41. Cacciò SM, Sannella AR, Bruno A, et al. Multilocus sequence typing of *Dientamoeba fragilis* identified a major clone with widespread geographical distribution. *Int J Parasitol.* 2016;46(12):793–8.
42. Cacciò SM. Molecular epidemiology of *Dientamoeba fragilis. Acta Trop.* 2018;184:73–7.

SUMMARY

Dientamoeba fragilis is a single-celled trichomonad parasite that occurs in 0.4% of patients with gastrointestinal discomfort and up to 82.9% of children with gastrointestinal protozoan diseases. In comparison with other diarrhea-causing protozoa (e.g., *Entamoeba*), *D. fragilis* is grossly neglected. This negligence has contributed to an inadequate understanding of the lifecycle, mode of transmission, and molecular pathogenesis of *D. fragilis* and hampered the efforts in the treatment and control of human dientamoebiasis. There is much to be learned about this parasite, and sequencing analysis of *D. fragilis* genome should be a top priority. Further application of state-of-the-art omic techniques will help uncover genetic, transcriptomic, proteomic, and metabolomic insights on the parasite–host interplay during *D. fragilis* infection, and lay a solid foundation for future development of innovative countermeasures against this emerging foodborne pathogen.

30 Molecular Epidemiology of *Echinococcus* spp.

Wenbao Zhang and Jun Li
Prevention and Treatment of High Incidence Diseases in Central Asia and WHO Collaborating Centre for Prevention and Care Management of Echinococcosis
The First Affiliated Hospital of Xinjiang Medical University

Dongyou Liu
Royal College of Pathologists of Australasia Quality Assurance Programs

Contents

30.1 INTRODUCTION

Echinococcus (the name referring to the small, round, spiny protoscolices located in the cysts or metacestodes) is a genus of zoonotic tapeworms that involve carnivores as definitive hosts and herbivores and rodents as intermediate hosts. Besides affecting 2–3 million people globally and producing >200,000 new cases each year, *Echinococcus* has a devastating impact on animal industry through condemnation of infected viscera, reduced meat/milk/wool production, delayed fecundity/growth, and decreased performance.

Taxonomy. Since the description of its first species from a sheep in Germany by Batsch in 1786, the genus *Echinococcus*, which constitutes one of four genera (i.e., *Echinococcus*, *Hydatigera, Taenia*, and *Versteria*) in the family *Taeniidae*, order *Cyclophyllidea*, class *Cestoda*, phylum *Platyhelminthes* [1,2], has expanded to 11 species: *E. granulosus* sensu stricto, *E. equinus*, *E. ortleppi*, *E. intermedius*, *E. borealis*, *E. canadensis*, *E. felidis*, *E. multilocularis*, *E. oligarthrus*, *E. shiquicus*, and *E. vogeli* based on differences in morphology, development, host specificity, mitochondrial genome, and nuclear DNA sequences.

Notably, seven of the 11 *Echinococcus* species (i.e., *E. granulosus* sensu stricto, *E. equinus*, *E. ortleppi*, *E. intermedius*, *E. borealis*, *E. canadensis*, and *E. felidis*) are carved up from *E. granulosus* sensu lato (the small dog tapeworm), an umberalla species that covers ten strains/genotypes (G1–10) [i.e., *E. granulosus* sensu stricto (G1, G3), *E. equinus* (G4), *E. ortleppi* (G5), *E. intermedius* (G6, G7), *E. borealis* (G8), *E. canadensis* (G10), and *E. felidis*, with G2 (Tasmanian sheep strain) being a variant of G3 (buffalo strain) and G9 (variant pig strain) being a microvariant of G7 (pig strain)] and *E. felidis* (Table 30.1) [3].

TABLE 30.1 Biological and clinical features of *Echinococcus* spp.

SPECIES	*GENOTYPES AND STRAINS*	*HOSTS*	*GEOGRAPHIC DISTRIBUTION*	*HUMAN DISEASE*
E. granulosus sensu stricto (small dog tapeworm)	G1 (sheep strain), G3 (buffalo strain)	Adult worm: dog Metacestode: sheep, cattle, human	Mediterranean areas, Eastern Europe, parts of South America, parts of Africa, Central Asia/Western China	Cystic echinococcosis (CE) or hydatidosis (G1 accounts for ~200,000 or 88% of annual CE cases)
E. equinus	G4 (horse strain)	Adult worm: dog, jackal, lion, captive lemur Metacestode: horse, donkey, zebra	UK, Ireland, Germany, Italy, Spain, Tunisia, Egypt, Namibia	No human infection documented
E. ortleppi	G5 (cattle strain)	Adult worm: dog, jackal Metacestode: cattle, buffalo, goat, sheep, pig, monkey, captive reindeer	South Africa, Switzerland, Germany, the Netherlands, Sudan, Kenya, South Africa, Brazil, Italy, France, India, Vietnam	Cystic echinococcosis (seven cases reported)
E. intermedius	G6 (camel strain), G7 (pig strain)	Adult worm: dog, wolf Metacestode:camel, pig, goat, reindeer	Middle East/Africa/ Russia (camel strain), Eastern Europe/ Mexico (pig strain)	Cystic echinococcosis (G6 and G7 are responsible for ~14,000 or 7.3% and ~6,000 or 4.7% of annual CE cases, respectively)
E. borealis	G8 (American cervid strain)	Adult worm: wolf Metacestode: moose, wapiti	Sub-Arctic region of North America	Cystic echinococcosis (two cases reported)
E. canadensis	G10 (Fennoscandian cervid strain)	Adult worm: wolf Metacestode: moose, reindeer	Arctic region of North America	Suspected to be human-infective, but no clinical CE cases documented
E. felidis	Lion strain	Adult worm: lion, spotted hyena Metacestode: warthog	Uganda, Kenya, South Africa	No human infection documented

(Continued)

TABLE 30.1 (*Continued*) Biological and clinical features of *Echinococcus* spp.

SPECIES	*GENOTYPES AND STRAINS*	*HOSTS*	*GEOGRAPHIC DISTRIBUTION*	*HUMAN DISEASE*
E. multilocularis (small fox tapeworm)	M1 (Europe), M2 (China, Alaska, North America)	Adult worm: red fox, cat, dog Metacestode: field mouse, human	Central Asia, northwestern China, northern Russia, northern Japan, north-central United States, northwestern Alaska, northwestern Canada, France, Germany, Austria, Switzerland	Alveolar echinococcosis (AE) resulting in ~18,000 new cases each year
E. oligarthrus		Adult worm: wildcat Metacestode: agoutis, spiny rat, paca, human	Central and South America (e.g., Brazil, Colombia, Ecuador)	Polycystic/unicystic echinococcosis (four cases reported)
E. shiquicus		Adult worm: Tibetan fox Metacestode: plateau pika	Qinghai-Tibet Plateau	No human infection documented
E. vogeli	Monophyletic	Adult worm: bush dog Metacestode: paca, agoutis, armadillo, human	Central and South America (e.g., Brazil, Colombia, Ecuador)	Polycystic echinococcosis (~220 cases reported)

Note: E. granulosus s.s. G2 (Tasmanian sheep strain) is now regarded as a variant of E. granulosus s.s. G3 (buffalo strain), while E. canadensis G9 (variant pig strain) represents a variant of E. intermedius G7 (pig strain)

Out of the 11 *Echinococcus* species, *E. granulosus* s.s. and to a lesser extent *E. intermedius* (G6/G7) [3] are principal agents of cystic echinococcosis (CE), *E. multilocularis* (the small fox tapeworm) causes alveolar echinococcosis (AE), and *E. vogeli* and *E. oligarthrus* are implicated in polycystic/unicystic echinococcosis (Table 30.1) [1,4].

Morphology. *Echinococcus* undergoes four developmental stages: adult worm, egg, metacestode (larva), and protoscolex. *Echinococcus* adult worms are about 1.2–12 mm in length (2–11 mm for *E. granulosus*, 1.2–4.5 mm for *E. multilocularis*, 2.9 mm for *E. oligarthrus*, 1.3–1.7 mm for *E. shiquicus*, average 5.6 mm and maybe up to 12 mm for *E. vogeli*) and possess a scolex (with four suckers and a rostellum decorated by a double ring of 20–50 hooks), a neck, and a strobilus (consisting of 2–6 segments/proglottids for *E. granulosus*; 2–6 segments for *E. multilocularis*; 2–6 segments for *E. oligarthrus*, 2–3 segments for *E. shiquicus*, 2–6 segments for *E. vogeli*) (Figure 30.1) [5]. *Echinococcus* eggs are spherical and of 35–50 µm in diameter, and have a thick striated wall that encloses an embryo (oncosphere) (Figure 30.2). *Echinococcus* metacestodes (larvae) are cystic, alveolar, or polycystic. Specifically, *E. granulosus* metacestodes are fluid-filled unilocular cysts (also called hydatid cysts) of 1–10 cm (up to 20 cm) in diameter consisting of a three-layered wall (adventitious layer, laminated layer, and germinal layer) and cyst fluid (which contains brood capsules, protoscoleces, and daughter cysts generated endogenously by the germinal layer) (Figure 30.2) [6]. *E. multilocularis* metacestodes are alveolar and contain numerous fluidless small cysts (a few mm to 2 cm in diameter) with the germinal layer budding exogenously to form multiple cysts, which are surrounded by a mix of immune cells, fibrosis, and necrosis. *E. vogeli* and *E. oligarthrus* metacestodes are polycystic and composed of multichambered fluid-filled cysts (a few cm in diameter) with the germinal layer budding exogenously to form new cysts and endogenously to form septae. *E. oligarthrus* metacestodes may be unicystic at times [7].

Genome. *E. granulosus* s.s. G1 genome measures 151.6 Mb in size and contains 11,325 genes. Among the expressed genes are the EgAgB family genes for interaction with and redirection of host immune

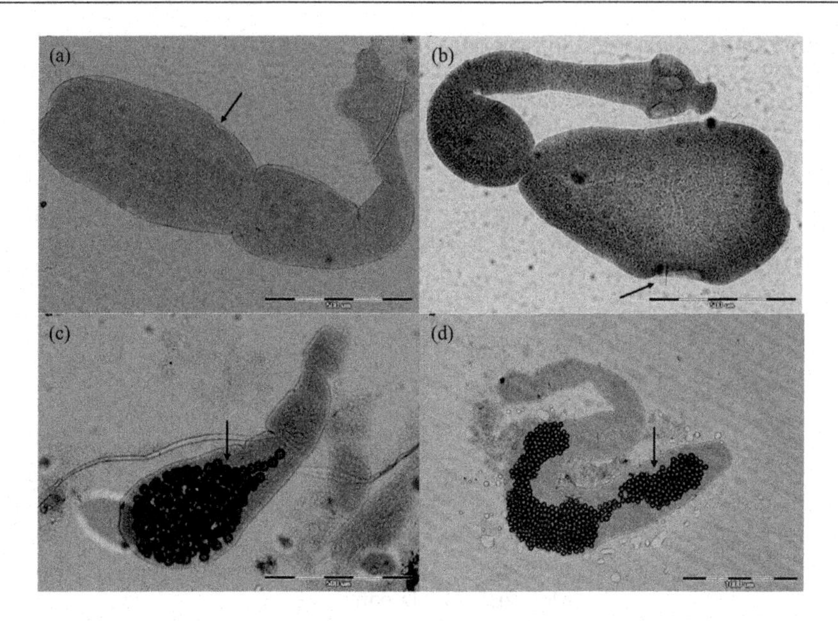

FIGURE 30.1 *Echinococcus multilocularis* (a) and *Echinococcus granulosus* (b) genital pore (arrow) in gravid proglottid (anterior to mid-length in a; posterior to mid-length in b); scale bars: a and b, 500 μm. *E. multilocularis* (c) and *E. granulosus* (d) uterus (arrow) with eggs in gravid proglottid (sac-like uterus in c; laterally branching uterus in d); scale bars: c, 500 μm; d, 1,000 μm. (Photo credit: Heidari Z, Sharbatkhori M, Mobedi I, et al. *Parasit Vectors*. 2019;12(1):606.)

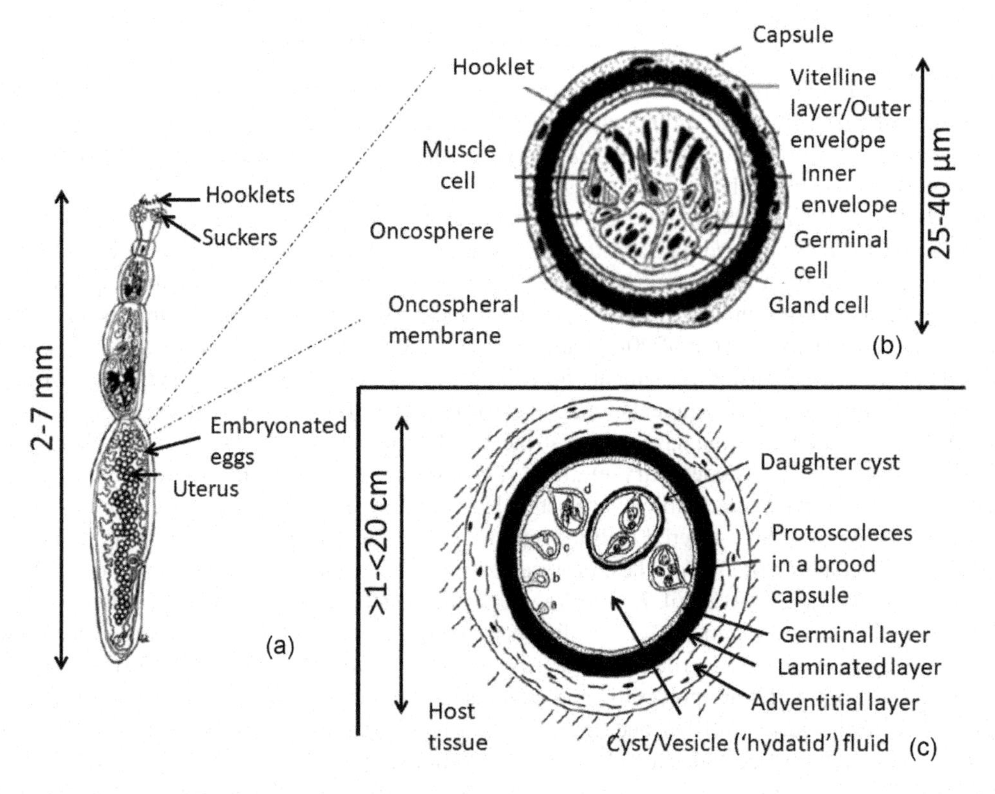

FIGURE 30.2 Key morphological and structural features of *Echinococcus granulosus*. (a) adult worm, (b) egg, (c) metacestode (larva). (Photo credit: Vuitton D, Zhang W, Giraudoux P. http://www.waterpathogens.org/book/echinococcus 2017.)

responses, proteases for digestion of host proteins, solute carrier family proteins for transporting amino acids from the mammalian host, proteins in bile salt pathways for controlling the bidirectional development, and calcium channel subunit EgCavβ1 related to praziquantel sensitivity [8–10].

E. multilocularis genome is 115 Mb in size with 10,345 genes, and has a highly reduced gene repertoire (similar to that of *E. granulosus*) for metabolic functions (including loss of the ability to synthesize most amino acids), but displays increased capacity to absorb nutrients from the host [8]. Further, *E. multilocularis* mitochondrial DNA (13,738 bp) consists of two major noncoding regions and 36 genes (12 for proteins involved in oxidative phosphorylation, two for rRNAs and 22 for tRNAs) without ATPase subunit 8 gene [11].

E. intermedius (*E. canadensis*) G7 genome is 115 Mb in length with 11,449 genes and demonstrates a high nucleotide sequence divergence from *E. granulosus* s.s. G1 genome. The syntenic genes between *E. intermedius* (*E. canadensis*) G7 and *E. granulosus* s.s. G1 average about 98.3%, while those between *E. intermedius* (*E. canadensis*) G7 and *E. multilocularis* average about 94.6% [12].

E. oligarthrus genome is 86.22 Mb in size and shows ~90% identity and 76.3% coverage with *E. multilocularis*. As one of the basal species of the genus *Echinococcus*, *E. oligarthrus* is phylogenetically closer to *E. multilocularis* and displays a higher genetic distance from *E. granulosus* s.l. species [13].

Lifecycle. *Echinococcus* spp. have a complex lifecycle involving two mammalian hosts. *Echinococcus* adult worms reside in the small intestines of definitive hosts (carnivores) and discharge the gravid proglottids or eggs via feces into soil, pasture land, livestock farms, and water. After ingestion by intermediate hosts (herbivores, rodents, and humans), *Echinococcus* eggs hatch and release oncospheres (embryos) in the duodenum, which penetrate the lamina propria of the intestine; move via the blood circulation or lymphatics into the liver (65%), lungs (25%), and other internal organs; and develop into cystic, alveolar, or polycystic metacestodes. When raw organs (e.g., the liver and lungs) from infected intermediate hosts are consumed by definitive hosts, the cysts are digested, then protoscoleces evaginate and attach to the mucosa of the small intestine and grow into adult worms [1].

Epidemiology. *Echinococcus* spp. are infective to >40 companion animal, livestock, and wildlife species in many parts of the world [14–17]. Human CE caused by *E. granulosus* s.l. occurs in pastoral areas of Eurasia, Africa, America, and Oceania, with >200,000 new cases each year. Human AE due to *E. multilocularis* is seen in the rural areas of the northern hemisphere, with 18,000 new cases each year. Human polycystic/unicystic echinococcosis resulting from *E. vogeli* and *E. oligarthrus* infections is rare and limited to Central and South America [1,6].

Clinical features. CE may not show any symptoms until the growing cysts create pressure effects on surrounding tissues or organs [e.g., the liver (65%), lungs (25%), muscles, (5%), bones (3%), kidneys, spleen, brain, and eyes], with clinical manifestations ranging from fever, urticaria, eosinophilia, to anaphylactic shock (largely as a consequence of spontaneous or traumatic cyst rupture that elicits an IgE-mediated hypersensitivity reaction). Specifically, hepatic hydatidosis (65% of cases) is associated with abdominal pain, decreased appetite, hepatomegaly, a palpable mass, and abdominal distention, with rupture of liver cysts into the biliary tree causing biliary obstruction, superinfection of the cyst, and secondary peritonitis, as well as formation of hepatic fibrosis inducing jaundice, cholecystitis, and portal hypertension. On the other hand, pulmonary hydatidosis (25% of cases) is linked to chronic cough, chest pain, and shortness of breath, with rupture of lung cysts into the bronchial tree resulting in pneumonitis, pneumothorax, pleural effusion, and secondary pleuritis [18,19]. Interestingly, *E. intermedius* G7 produces smaller liver cysts than *E. granulosus* s.s. G1, while *E. intermedius* G6 tends to occur in the brain.

AE often manifests as a slow growing, destructive tumor in the liver that causes abdominal pain, biliary obstruction, liver failure, and hepatic encephalopathy.

Polycystic echinococcosis may present with a slow growing tumor in the liver.

Pathogenesis. *Echinococcus* adult worms utilize their suckers and hooklets to hold onto the intestinal mucosa, while *Echinococcus* metacestodes (larvae) generate the EgAgB family for interacting with and redirecting host immune responses, proteases for digesting host proteins, and solute carrier family proteins for transporting amino acids from the mammalian host. Further, *Echinococcus* metacestodes (larvae) settling in the liver and lungs activate host cells and cause imbalanced deposition and degradation of the extracellular matrix (ECM), leading to the formation of peripheral fiberboards (collagen fibers) around the metacestodes and fibrosis which increase resistance to antiparasitic drugs [20,21].

Diagnosis. As patients infected with *Echinococcus* spp. often present with nonspecific clinical symptoms, accurate diagnosis of echinococcosis requires detailed history and physical exam, as well as input from various laboratory procedures (e.g., imaging, microscopy, biochemical tests, immunological assays, molecular detection).

Detailed history and physical exam evaluates past exposure or immigration from endemic areas (e.g., pastoral, semi-agricultural, and semi-pastoral areas) and assesses physical abnormalities suggestive of echinococcosis.

Imaging (e.g., ultrasonography, radiography, CT, MRI, ultrasound-guided fine needle aspiration, endoscopic retrograde cholangiopancreatography) delineates the lesions associated with pulmonary or hepatic echinococcosis. Radiography reveals the location of hydatid cysts (with pulmonary hydatid cysts more frequently located in the middle and lower lobes of the right lung) and detects ring-like calcifications in up to 30% of cases. CT shows water-like cystic fluid and moderate wall thickness, and uncovers cyst rupture, underlying infection, and biliary or vascular involvement. Whereas chest CT is essential for diagnosing pulmonary echinococcosis, and portable ultrasound is sufficient for screening hepatic echinococcosis [1].

Microscopic examination reveals three distinct features of hydatid cysts: (i) thick, acellular, laminated layer with acidophilic staining; (ii) cellular germinal layer; (iii) brood capsules or protoscolices.

Biochemical tests (e.g., complete blood count, liver function test) may show eosinophilia and elevated alkaline phosphatase.

Immunological assays (e.g., Casoni test, ELISA, immunoelectrophoresis, immunoblot) are valuable for screening and confirming echinococcosis. Casoni test involves intradermal injection of 0.1–0.3 ml of filtrated/disinfected sheep hydatid cyst fluid and subsequent observation of skin redness and swelling (larger than 1/2 cm in diameter about 5–10 min later), and gives a positive predictive value of 80%–95% [22].

Molecular detection of *Echinococcus* species-specific genes using standard, multiplex, real-time, or real-time multiplex-nested PCR allows rapid, sensitive, and specific confirmation and genotyping of echinococcosis [23–29].

Differential diagnoses for echinococcosis include pulmonary/hepatic abscess, pulmonary/hepatic cysts, encapsulated pleural effusion, Budd–Chiari syndrome, biliary colic, biliary cirrhosis, tuberculosis, and primary pulmonary/hepatic carcinoma (e.g., teratoma, type B2 thymoma).

Treatment and prevention. Treatment of echinococcosis involves surgery (e.g., enucleation for small pulmonary hydatid cysts with little risk of rupture; pericystectomy for hydatid cyst along with the pericyst allows complete removal of the parasite; cystotomy with capitonnage; cystostomy with closure of the bronchial openings and capitonnage; cystostomy with the closure of the bronchial openings alone; puncture, aspiration, injection, and reaspiration or PAIR; segmental resection for ruptured hydatid cyst; lobectomy for cyst involving over 50% of the lobe), radiotherapy (e.g., radiosurgery with a gamma knife, stereotactic body radiotherapy), and chemotherapy (e.g., albendazole at a dose of 10–15 mg (kg day)$^{-1}$ twice daily for at least 3–6 months, with 30% cure rate and 30%–50% improvement rate) [1].

Prevention of *Echinococcus* infections should focus on adoption of general hygienic measures (e.g., washing hands before eating, avoiding contact between mouth and unwashed hands, thorough washing of raw vegetables, avoiding direct contact with possibly infected dogs and foxes, use of a safe water source), public education to not feed raw livestock viscera to dogs, and regular administration of anthelmintics (e.g., praziquantel at a frequency of 8–12 times per year) to dogs [30]. A recently developed subunit vaccine EG95 has been promoted as a complementary intervention to eliminate CE transmission and the approach has been trialed in China and South America [31]. Dog vaccination would be an effective intervention for controlling echinococcosis transmission although progress has been slow [32].

30.2 MOLECULAR EPIDEMIOLOGY OF *ECHINOCOCCUS* SPP.

PCR amplification and sequence analysis of nuclear (e.g., 18S ribosomal RNA, ITS) and mitochondrial (e.g., *cob*, *nad2*, *cox1*) genes as well as microsatellite markers offer a useful way to ascertain the phylogenetic relationship and epidemiological tracking of *Echinococcus* spp. [33].

For example, examination of nuclear 18S ribosomal DNA and concatenated exon regions of protein-coding genes (phosphoenolpyruvate carboxykinase and DNA polymerase delta) along with 12 mitochondrial genes enabled resurrection of *Hydatigera* for *T. parva*, *T. krepkogorski*, and *T. taeniaeformis* and establishment of a new genus, *Versteria*, for *T. mustelae* and *T. brachyacantha* in the family *Taeniidae* [2].

Further, analysis of nuclear and mitochondrial sequences as well as SNPs permits accurate discrimination of *E. granulosus* s.l. genotypes (G1–10 and the "lion strain") (Figure 30.3) [34,35]. Indeed, examination of the complete mitochondrial genomes identified *E. granulosus* s.s. (G1–3), *E. equinus* (G4), *E. ortleppi* (G5), *E. intermedius* (G6–7), *E. borealis* (G8), *E. canadensis* (G10), and *E. felidis* (lion strain)

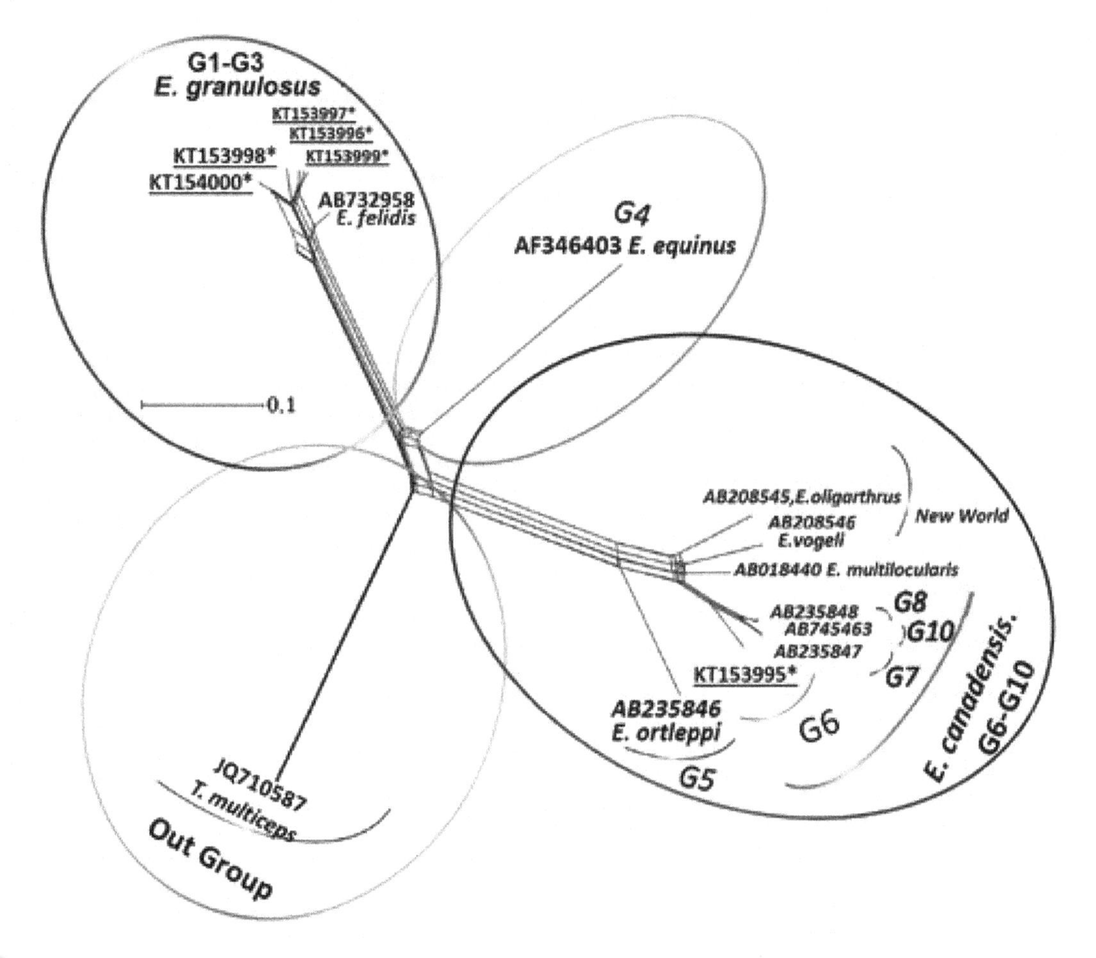

FIGURE 30.3 NeighborNet graph highlighting phylogenetic relationship among *Echinococcus granulosus* s.l. genotypes/species based on *cox1* sequences (mitogenome) and the Kimura-2 parameter model. (Photo credit: Shariatzadeh SA, Spotin A, Gholami S, et al. *Parasit Vectors*. 2015; 8:409.)

(Table 30.1) [36–38]. In addition, examination of the 1,609 bp *cox1* gene identified 24 *cox1* haplotypes among *E. granulosus* s.s. isolates [39].

Analysis of mitochondrial cytochrome *c* oxidase subunit 1 (*cox1*, 366 nt) and NADH dehydrogenase subunit 1 (*nad1*, 471 nt) as well as nuclear 18S rRNA and homeobox gene sequences allowed differentiation of *E. multilocularis* isolates into two geographic genotypes, M1 (Europe) or M2 (Japan, China, Alaska, and North America) [40,41]. Subsequent use of multiple mitochondrial loci and microsatellites identified additional diversity among *E. multilocularis* isolates [42]. In fact, based on sequencing of three protein-coding gene in mitochondrial DNA, 17 regional haplotypes were identified from *E. multilocularis* 76 isolates sampled in Europe (five haplotypes), Asia (ten haplotypes), and North America (two haplotypes). Further, sequencing of a 370 nt region in the *nad1* gene generated 17 new haplotypes (A–Q) among *E. multilocularis* isolates in Canada [43]. In addition, using the microsatellite marker EmsB, which is tandemly repeated microsatellite with a (CA)n (GA)n pattern, 32 EmsB profiles (G1 to G32) were recognized among *E. multilocularis* isolates from Europe [44].

Other phylogenetic findings include the placement of *E. shiquicus* with *E. multilocularis* in a common clade of ancestor with bootstrap value >70%, as well as relevant genetic similarities between *E. multilocuaris* and *E. oligarthrus*.

30.3 FUTURE PERSPECTIVES

Despite having significant structural, biological, and molecular similarities, *Echinococcus* spp. show differences in their host preference, pathogenicity, and clinical presentations (cystic, alveolar, and polycystic echinococcosis). For effective control of human echinococcosis, it is critical to be able to track *Echinococcus* isolates implicated in epidemics. Currently, various nuclear (e.g., 18S ribosomal RNA, ITS) and mitochondrial (e.g., *cob, nad2, cox1*) genes as well as microsatellite markers have been applied for phylogenetic analysis of *Echinococcus* spp., leading to the recognition of five species (*E. granulosus* s.l., *E. multilocularis, E. oligarthrus, E. shiquicus*, and *E. vogeli*) within the genus, and further separation of *E. granulosus* s.l. into 10 genotypes (G1–10), along with *E. felidis*. Nonetheless, as the target genes that have been used for interrogating *Echinococcus* phylogenetic relationship may not be ideal representatives for these tapeworms, the results are predictably variable. Considering that the recent availability of complete genome sequences of *E. granulosus* s.s. G1, *E. intermedius* (*E. canadensis*) G7, *E. multilocularis*, and *E. oligarthrus* has yielded exciting insights into their phylogenetic relationships, sequencing studies on other *Echinococcus* species and genotypes are invaluable and keenly anticipated. Moreover, additional genomic and transcriptomic comparisons between *E. granulosus* and *E. multilocularis* are necessary to uncover distinct gene expression profiles that underpin their pathological differences, and open another way for designing tailor-made therapeutics for these important zoonotic pathogens.

REFERENCES

1. Wen H, Vuitton L, Tuxun T, et al. Echinococcosis: Advances in the 21st century. *Clin Microbiol Rev.* 2019;32(2):e00075–18.
2. Nakao M, Lavikainen A, Iwaki T, et al. Molecular phylogeny of the genus *Taenia* (Cestoda: Taeniidae): proposals for the resurrection of *Hydatigera* Lamarck, 1816 and the creation of a new genus *Versteria. Int J Parasitol.* 2013;43(6):427–37.
3. Romig T, Ebi D, Wassermann M. Taxonomy and molecular epidemiology of *Echinococcus granulosus* sensu lato. *Vet Parasitol.* 2015;213(3–4):76–84.

4. Piccoli L, Bazzocchi C, Brunetti E, et al. Molecular characterization of *Echinococcus granulosus* in south-eastern Romania: evidence of G1-G3 and G6-G10 complexes in humans. *Clin Microbiol Infect.* 2013;19(6):578–82.
5. Heidari Z, Sharbatkhori M, Mobedi I, et al. *Echinococcus multilocularis* and *Echinococcus granulosus* in canines in North-Khorasan Province, northeastern Iran, identified using morphology and genetic characterization of mitochondrial DNA. *Parasit Vectors.* 2019;12(1):606.
6. Vuitton D, Zhang W, Giraudoux P. Echinococcus spp. In: J.B. Rose and B. Jiménez-Cisneros, (eds) *Global Water Pathogen Project.* http://www.waterpathogens.org (Robertson, L (eds) Part 4 Helminths) http://www.waterpathogens.org/book/echinococcus Michigan State University, E. Lansing, MI, 2017.
7. Gottstein B, Wang J, Blagosklonov O, et al. Echinococcus metacestode: in search of viability markers. *Parasite.* 2014;21:63.
8. Tsai IJ, Zarowiecki M, Holroyd N, et al. The genomes of four tapeworm species reveal adaptations to parasitism. *Nature.* 2013;496(7443):57–63.
9. Zheng H, Zhang W, Zhang L, et al. The genome of the hydatid tapeworm *Echinococcus granulosus. Nat Genet.* 2013;45(10):1168–75.
10. Zhang W, Wang S, McManus DP. *Echinococcus granulosus* genomics: a new dawn for improved diagnosis, treatment, and control of echinococcosis. *Parasite.* 2014;21:66.
11. Nakao M, Yokoyama N, Sako Y, Fukunaga M, Ito A. The complete mitochondrial DNA sequence of the cestode *Echinococcus multilocularis* (Cyclophyllidea: Taeniidae). *Mitochondrion.* 2002;1(6):497–509.
12. Maldonado LL, Assis J, Araújo FM, et al. The *Echinococcus canadensis* (G7) genome: a key knowledge of parasitic platyhelminth human diseases. *BMC Genomics.* 2017;18(1):204.
13. Maldonado LL, Arrabal JP, Rosenzvit MC, Oliveira GC, Kamenetzky L. Revisiting the phylogenetic history of helminths through genomics, the case of the new *Echinococcus oligarthrus* genome. *Front Genet.* 2019;10:708.
14. das Neves LB, Teixeira PE, Silva S, et al. First molecular identification of *Echinococcus vogeli* and *Echinococcus granulosus* (sensu stricto) G1 revealed in feces of domestic dogs (Canis familiaris) from Acre, Brazil. *Parasit Vectors.* 2017;10(1):28.
15. Grocholski S, Agabawi S, Kadkhoda K, Hammond G. *Echinococcus granulosus* hydatid cyst in rural Manitoba, Canada: Case report and review of the literature. *IDCases.* 2019;18:e00632.
16. Guo B, Zhang Z, Zheng X, et al. Prevalence and molecular characterization of *Echinococcus granulosus* sensu stricto in Northern Xinjiang, China. *Korean J Parasitol.* 2019;57(2):153–9.
17. Khan A, Ahmed H, Simsek S, et al. Molecular characterization of human *Echinococcus* isolates and the first report of *E. canadensis* (G6/G7) and *E. multilocularis* from the Punjab Province of Pakistan using sequence analysis. *BMC Infect Dis.* 2020;20(1):262.
18. Mao T, Chungda D, Phuntsok L, et al. Pulmonary echinococcosis in China. *J Thorac Dis.* 2019;11(7):3146–55.
19. Rawat S, Kumar R, Raja J, Singh RS, Thingnam SKS. Pulmonary hydatid cyst: Review of literature. *J Family Med Prim Care.* 2019;8(9):2774–8.
20. Niu F, Chong S, Qin M, Li S, Wei R, Zhao Y. Mechanism of fibrosis induced by *Echinococcus* spp. *Diseases.* 2019;7(3):51.
21. He Z, Yan T, Yuan Y, Yang D, Yang G. miRNAs and lncRNAs in *Echinococcus* and echinococcosis. *Int J Mol Sci.* 2020;21(3):730.
22. Han X, Kim JG, Wang H, et al. Survey of echinococcoses in southeastern Qinghai Province, China, and serodiagnostic insights of recombinant *Echinococcus granulosus* antigen B isoforms. *Parasit Vectors.* 2019;12(1):323
23. Boubaker G, Macchiaroli N, Prada L, et al. A multiplex PCR for the simultaneous detection and genotyping of the *Echinococcus granulosus* complex. *PLoS Negl Trop Dis.* 2013;7(1):e2017.
24. Ito A, Dorjsuren T, Davaasuren A, et al. Cystic echinococcoses in Mongolia: molecular identification, serology and risk factors. *PLoS Negl Trop Dis.* 2014;8(6):e2937.
25. Hu D, Song X, Wang N, et al. Molecular identification of *Echinococcus granulosus* on the Tibetan Plateau using mitochondrial DNA markers. *Genet Mol Res.* 2015;14(4):13915–23.
26. Liu CN, Lou ZZ, Li L, et al. Discrimination between *E. granulosus* sensu stricto, *E. multilocularis* and *E. shiquicus* using a multiplex PCR assay. *PLoS Negl Trop Dis.* 2015;9(9):e0004084.
27. Safa AH, Harandi MF, Tajaddini M, Rostami-Nejad M, Mohtashami-Pour M, Pestehchian N. Rapid identification of *Echinococcus granulosus* and *E. canadensis* using high-resolution melting (HRM) analysis by focusing on a single nucleotide polymorphism. *Jpn J Infect Dis.* 2016;69(4):300–5.
28. Corrêa F, Stoore C, Horlacher P, et al. First description of *Echinococcus ortleppi* and cystic echinococcosis infection status in Chile. *PLoS One.* 2018;13(5):e0197620.
29. Shang J, Zhang G, Yu W, et al. Molecular characterization of human echinococcosis in Sichuan, Western China. *Acta Trop.* 2019;190:45–51.

30. Saelens G, Gabriël S. Currently available monitoring and surveillance systems for *Taenia* spp., *Echinococcus* spp., *Schistosoma* spp., and soil-transmitted helminths at the control/elimination stage: A systematic review. *Pathogens.* 2020;9(1):47.
31. Larrieu E and Zanini F. Critical analysis of cystic echinococcosis control programs and praziquantel use in South America, 1974–2010. *Rev Panam Salud Publica.* 2012;31(1):81–7.
32. Zhang W and McManus DP. Vaccination of dogs against *Echinococcus granulosus*: a means to control hydatid disease? *Trends Parasitol.*2008;24(9):419–24.
33. Kinkar L, Korhonen PK, Cai H, et al. Long-read sequencing reveals a 4.4 kb tandem repeat region in the mitogenome of *Echinococcus granulosus* (sensu stricto) genotype G1. *Parasit Vectors.* 2019;12(1):238.
34. Shariatzadeh SA, Spotin A, Gholami S, et al. The first morphometric and phylogenetic perspective on molecular epidemiology of *Echinococcus granulosus* sensu lato in stray dogs in a hyperendemic Middle East focus, northwestern Iran. *Parasit Vectors.* 2015;8:409.
35. Alvarez Rojas CA, Kronenberg PA, Aitbaev S, et al. Genetic diversity of *Echinococcus multilocularis* and *Echinococcus granulosus* sensu lato in Kyrgyzstan: The A2 haplotype of *E. multilocularis* is the predominant variant infecting humans. *PLoS Negl Trop Dis.* 2020;14(5):e0008242.
36. Nakao M, McManus DP, Schantz PM, Craig PS, Ito A. A molecular phylogeny of the genus *Echinococcus* inferred from complete mitochondrial genomes. *Parasitology.* 2007;134(5):713–22.
37. Sharma M, Sehgal R, Fomda BA, Malhotra A, Malla N. Molecular characterization of *Echinococcus granulosus* cysts in north Indian patients: identification of G1, G3, G5 and G6 genotypes. *PLoS Negl Trop Dis.* 2013;7(6):e2262.
38. Sgroi G, Varcasia A, Dessi G, et al. Cystic echinococcosis in wild boars (*Sus scrofa*) from southern Italy: Epidemiological survey and molecular characterization. *Int J Parasitol Parasites Wildl.* 2019;9:305–11.
39. Bonelli P, Dei Giudici S, Peruzzu A, et al. Genetic diversity of *Echinococcus granulosus* sensu stricto in Sardinia (Italy). *Parasitol Int.* 2020;77:102120.
40. Knapp J, Gottstein B, Saarma U, Millon L. Taxonomy, phylogeny and molecular epidemiology of *Echinococcus multilocularis*: From fundamental knowledge to health ecology. *Vet Parasitol.* 2015;213(3–4):85–91.
41. Wu C, Zhang W, Ran B. et al. Genetic variation of mitochondrial genes among *Echinococcus multilocularis* isolates collected in western China. *Parasit Vectors.* 2017;10:265.
42. Cerda JR, Buttke DE, Ballweber LR. *Echinococcus* spp. tapeworms in North America. *Emerg Infect Dis.* 2018;24(2):230–5.
43. Gesy KM, Schurer JM, Massolo A, et al. Unexpected diversity of the cestode *Echinococcus multilocularis* in wildlife in Canada. *Int J Parasitol Parasites Wildl.* 2014;3(2):81–7.
44. Knapp J, Gottstein B, Bretagne S, et al. Genotyping *Echinococcusmultilocularis* in human alveolar echinococcosis patients: An EmsB microsatellite analysis. *Pathogens.* 2020;9(4):282.

SUMMARY

Echinococcus is a genus of zoonotic tapeworms that involve carnivores as definitive hosts and herbivores and rodents as intermediate hosts. Since the description of its first species from a sheep in Germany by Batsch in 1786, the genus *Echinococcus* has expanded to 11 species. While *E. granulosus* s.s. and to a lesser extent *E. intermedius* (G6/G7) are principal agents of cystic echinococcosis (CE), *E. multilocularis* causes alveolar echinococcosis (AE), and *E. vogeli* and *E. oligarthrus* are implicated in polycystic/unicystic echinococcosis. In addition to affecting 2–3 million people globally and producing >200,000 new cases each year, *Echinococcus* has a devastating impact on animal industry through condemnation of infected viscera, reduced meat/milk/wool production, delayed fecundity/growth, and decreased performance. The availability of molecular techniques to track *Echinococcus* outbreak isolates is critical for its effective control. As the target genes currently used for interrogating *Echinococcus* phylogenetic relationship may not be ideally representatives of these tapeworms, further examination of complete genome sequences of *Echinococcus* spp. will be necessary to identify innovative markers for improved epidemiological tracking and therapeutical management of these important zoonotic pathogens.

Proteomic Analysis of *Giardia* Infection

31

Xun Suo
China Agricultural University

Xinming Tang
Institute of Animal Science

Dongyou Liu
Royal College of Pathologists of Australasia Quality Assurance Programs

Contents

31.1 INTRODUCTION

First described by Antonie van Leeuwenhoek in 1681 and named after Alfred Mathieu Giard, the genus *Giardia* encompasses a large number of unicellular, flagellated protozoan species that colonize the upper portion of the small intestine in humans and animals, leading to giardiasis (watery diarrhea, epigastric pain, nausea, vomiting, and weight loss). With a relatively simple lifecycle involving a cyst stage and a trophozoite stage, *Giardia*, specifically *G. lamblia* (synonyms *G. duodenalis* and *G. intestinalis*), affects approximately 280 million people each year worldwide. Having an estimated prevalence of 1%–7.6% in high-income countries and 20%–30% in low- and middle-income countries, giardiasis exerts an enormous burden on global healthcare systems and compromises the quality of life among millions of sufferers. Apart from timely detection and tracking of *Giardia* outbreaks and implementation of appropriate treatment and prevention measures, a detailed understanding of molecular mechanisms of *Giardia* infection is crucial for the development of intervention strategies for its ultimate elimination. As *Giardia* genomes harbor ~5,000 protein-coding genes, of which up to 50% have unknown functions, proteomic analysis offers a valuable tool for the characterization of *Giardia* proteins and contributes toward the full elucidation of *Giardia* pathogenesis.

Taxonomy. Classified in the family Giardiidae, order Giardida, class Metamonadea, phylum Metamonada, kingdom Protozoa, the genus *Giardia* comprises about 40 described species. While many of these described species are likely synonymous, at least eight are morphologically distinct and infective to humans and animals, including *G. lamblia* (synonyms: *G. duodenalis* and *G. intestinalis*) in humans and other mammals; *G. cricetidarum*, *G. microti*, *G. muris*, and *G. peramelids* in rodents; *G. ardeae* and *G. psittaci* in birds; *G. agilis* in amphibians and marine mammals. Furthermore, use of genetic and biochemical techniques helps reveal additional heterogeneity within the species *G. lamblia*, leading to the identification of at least eight assemblages (lineages or cryptic species), that is, A (*G. lamblia*; mainly affecting humans, other primates, cats, and dogs; further distinguished into A1 and A2), B (once referred to as *G. enterica*; humans, other primates, dogs, cats, and some wild animals), C/D (once referred to as *G. canis*; dogs and other canids), E (once referred to as *G. bovis*; cattle and other hoofed animals), F (once referred to as *G. cati*; cats), G (once referred to as *G. simondi*; rats), and H (now established as *G. peramelids*; southern brown bandicoot and hamsters) [1–3].

Morphology. *Giardia* is a flagellated protozoan with a lifecycle that alternates between an ovoid or ellipsoid cyst (of 8–14 um in size, which is larger than Cryptosporidium oocyst of 4–6 in size) and a crescent- or pear-shaped trophozoite (of 12–15 μm in length and 5–9 μm in width in the case of *G. lamblia*) (Figure 31.1) [4,5].

Giardia cysts have two nuclei (binucleate) in the immature stage and four nuclei (tetranucleate) in the mature stage (in which flagella and sucking disks may be seen in the cytoplasm), and are infective and environmentally resistant.

Giardia trophozoites possess a binuclear structure (each consisting of a complete set of genome), four pairs of flagella, a ventral disk (composed of microtubules arranged in a coiled array around a bare area, microribbons protruding into the cytoplasm, and cross-bridges connecting adjacent microtubules; a structure notably absent in other closely related metamonads), a median body (a microtubular element of 0.2–1.8 μm in thickness located dorsally to the ventral disk), and funis (two sheets of microtubules that accompany the axonemes of the caudal flagella) (Figure 31.2) [6,7].

Genome. From the sequencing data available, *Giardia* genomes vary from 10.7 to 12.8 Mb in size, including 12.8 Mb in A1 assemblage (with 6,027 genes), 10.7 Mb in A2 assemblage (with 5,147 genes), 11.0–12.0 Mb in B assemblage (with 4,471–6,094 genes), 11.2–12.0 Mb in C/D assemblages (with 2,885–3,917 genes), and 11.5 Mb in E assemblage (with 5,068 genes) [8,9]. It is notable that up to 50% of *Giardia* protein-coding genes (ORFs) have unassigned functions. Genome sequence comparisons further suggest that the E assemblage is more closely related to the A assemblage than to the B assemblage, supporting the establishment of *G. enterica* [8].

Lifecycle and epidemiology. The lifecycle of *G. lamblia* shows two major stages of development, that is, an infectious and environmentally resistant cyst stage, and a motile trophozoite stage. Shed in feces, *G. lamblia* infective cysts enter the host via contaminated food/water, pass through the stomach, and differentiate into flagellated excyzoites. After two rounds of cell division (binary fission), *G. lamblia* flagellated excyzoite turns into four fully developed (flagellated pear-shaped binucleate) trophozoites, which attach to the epithelium of the small intestine and stimulate host responses, resulting in a spectrum of clinical manifestations (giardiasis). Under the influence of bile, available lipids, or pH, trophozoites undergo further differentiation and transform into oval tetranucleate, environmentally resistant cysts. The complete lifecycle of *G. lamblia*, including excystation and encystation (cyst formation), has been reproduced *in vitro*.

G. lamblia (synonyms *G. duodenalis*, *G. intestinalis*) is a unicellular parasite that is transmitted via oral-fecal route. Ingestion/consumption of water/food contaminated with *Giardia* assemblages A2 and B, but rarely A1 infectious cysts result in the development of trophozoites in the small intestine (and giardiasis), and subsequent discharge of environmentally resistant cysts in feces facilitates further spread of infection. As a main protozoan causative agent of diarrhea, *G. lamblia* is estimated to affect 280 million people worldwide, with prevalence ranging from 1% to 7.6% in high-income countries (due mainly to waterborne outbreaks or overseas travel) to 30% in low- and middle-income countries (due to poor sanitary facilities, contaminated food/water) [5].

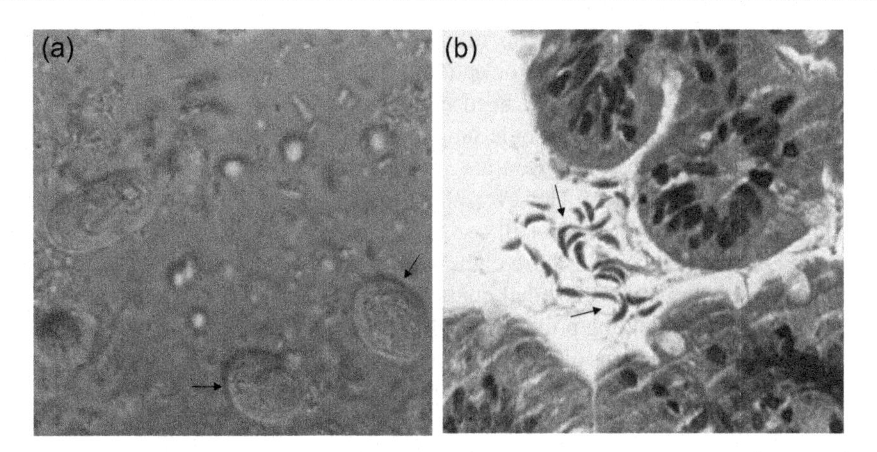

FIGURE 31.1 (a) *Giardia duodenalis* (synonym: *G. lamblia*) cysts (arrows) from infected human fecal specimen (wet mount, fixation with merthiolate-iodine-formaldehyde solution). (b) *G. lamblia* trophozoites (arrows) from infected human tissue biopsy (noting the presence of scattered crescent-shaped randomly oriented trophozoites on the luminal surface of the duodenal wall (H and E stain). (Photo credits: (a) Wang Y, Gonzalez-Moreno O, Roellig DM, et al. *Parasit Vectors*. 2019;12(1):432. (b) Corleto VD, Di Marino VP, Galli G, et al. *BMC Gastroenterol*. 2018;18(1):162.)

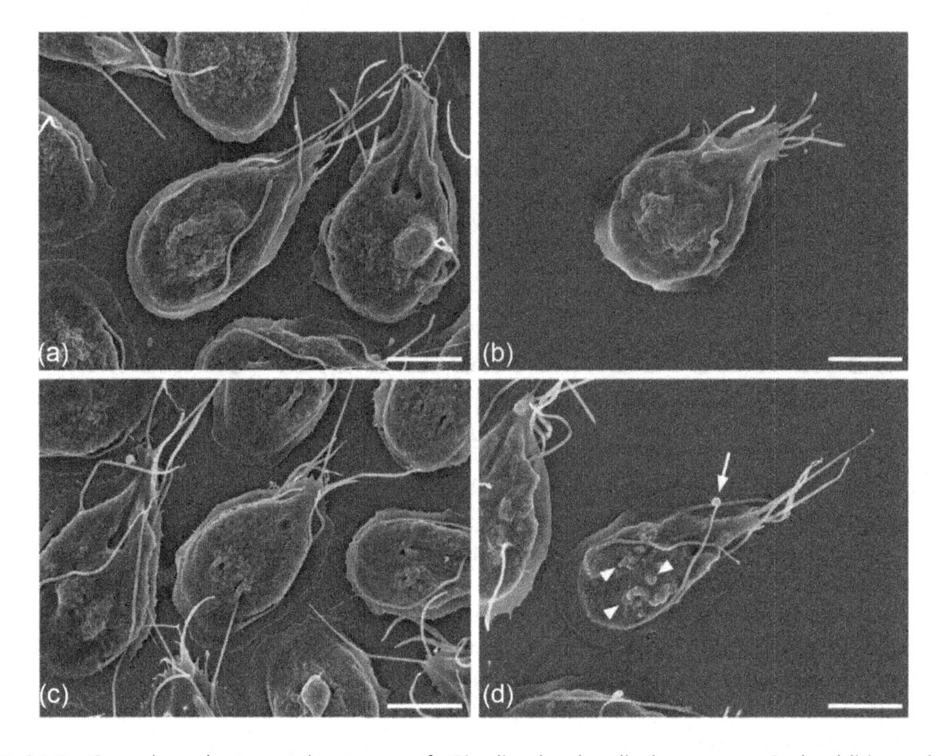

FIGURE 31.2 Scanning electron microscopy of *Giardia duodenalis* (synonym: *G. lamblia*) trophozoites. (a) solvent control (DMSO, 0.1% in medium; showing typical pear-shaped cell body and characteristic arrangement of eight flagella); (b) metronidazole (MTZ) treatment (11 μM; showing shorter flagella than solvent control a); (c) solvent control (ethanol, 0.12% in medium; similar appearance to a); (d) tetrahydrolipstatin (orlistat) treatment [5 μM; appearing shrunken in comparison to solvent control c and showing blebs on the dorsal surface (arrowheads) and at the tip of the flagella (arrow)]. (scale bar, 5 μm). (Photo credit: Hahn J, Seeber F, Kolodziej H, et al. *PLoS One*. 2013;8(8):e71597.)

Clinical features. The clinical symptoms of giardiasis often appear 2 days after infection, ranging from violent diarrhea, excess gas, stomach or abdominal cramps, upset stomach, and nausea, leading to dehydration and nutritional loss that may need immediate treatment. Possible postinfectious sequela may include irritable bowel syndrome, chronic fatigue, and lactose intolerance (which can persist after the eradication of *Giardia* from the digestive tract). In many cases, the infection may last between 2 and 6 weeks and resolve without treatment. However, some infected individuals may become asymptomatic carriers and transmit the parasite to others [10].

Pathogenesis. *Giardia* trophozoites have eight flagella and a ventral disk (made of tubulin and giardins) for motility and attachment to intestinal epithelial cells. *Giardia* produces at least ten basal body proteins (including universal signaling proteins) involved in the regulation of mitosis and interaction with host, the introduction of host reactions and pathological changes, and the causation of atrophy and flattening in the villi of the small intestine and subsequent malabsorption in the intestine [11,12].

Diagnosis. Diagnosis of giardiasis relies on microscopy, immunofluorescent antibody tests, ELISA, and, more recently, nucleic acid detection. The latter permits not only species-specific identification as well as discrimination between viable and nonviable cysts/oocysts but also genotype/subtype determination [4]. Use of *Giardia* microsatellites allows for species- and genetic assemblage-specific detection (monomorphic markers) and for high-resolution studies (polymorphic markers) [13]. The availability of rapid and increasingly affordable whole genome sequencing platform facilitates further improvement in molecular epidemiological tracking of *Giardia* outbreaks [14,15].

Treatment and prevention. Giardiasis due to *G. lamblia* (synonyms *G. duodenalis*, *G. intestinalis*) infection may be treated with nitroheterocyclics, especially metronidazole (MTZ), which is activated by *Giardia* enzymes and proteins [e.g., pyruvate-flavodoxin oxidoreductases (PFOR-1 and 2), thiol-cycling-associated enzyme thioredoxin reductase (TrxR), nitroreductase (NR)-1, iron-sulfur cluster binding proteins, and ferredoxins]. This leads to protein malfunction, multiplication disruption, DNA damage, oxidative stress, and death in trophozoites attached to the epithelial lining of the proximal small intestine, with efficacy of 73%–100% (Figure 31.2) [7,10].

Prevention of *Giardia* infection involves avoiding ingestion of contaminated water and food, provision of sewage treatment facility, and education.

31.2 PROTEOMIC ANALYSIS OF *GIARDIA* INFECTION: RECENT DEVELOPMENTS

Giardia spp. possess 10.7–12.8 Mb genomes that harbor ~5,000 protein-coding genes. Of these, about 50% of the encoded proteins participate in external survival, host cell invasion, evasion of host immune surveillance, initiation of pathological changes in host, and other functions, while up to 50% encoded proteins have unknown functions. Considering that *Giardia* genomes have undergone a long period of evolution, any genes (and their encoded proteins) that remain have exquisite functions that enhance the competitiveness of this parasite, deciphering the precise roles of each of these proteins in *Giardia* lifecycle is clearly necessary for improved understanding of its pathogenesis. In the sections below, a brief overview of key proteomic techniques is presented. This is followed by several exemplary studies that highlight the utility of proteomics approach for the characterization of *Giardia* proteins involved in the formation of ventral disk, excystation/encystation, virulence, mitosome, and resistance to nitro drugs.

Key proteomic techniques. Given their important functions in biological and pathological processes, proteins and their characterization have long attracted the attention of biomedical science community. Various procedures (e.g., sodium sulfate precipitation, spectrophotometry, Edman degradation, SDS-PAGE, ELISA, and western blot) have been developed over the years to separate, detect, quantify, and characterize proteins related to specific diseases or conditions.

As a scientific discipline that focuses on proteins and their altered levels of expression as well as posttranslational modification during a disease state relative to a healthy state, proteomics (deriving from proteome, which was coined in 1994 from *prote*in and gen*ome*) detects, quantifies, and characterizes the complete range of proteins in a cell or tissue in association with a disease state or effect of therapy. In this sense, the first proteomics study could be traced to the introduction of two-dimensional gel separation and visualization of proteins from *Escherichia coli* in 1975. Subsequent development and adoption of other techniques (particularly mass spectrophotometry or MS) make large-scale study of proteins possible and further strengthen the capability of proteomic determination of the overall level of protein composition, structure, and activity in a biological system. Indeed, as protein purification and mass spectrometry play a dominant role in protein detection and characterization these days, their combined application is sometimes considered synonymous to proteomics.

The development of MS following the discovery of soft ionization [e.g., matrix-assisted laser desorption/ionization (MALDI) and electrospray ionization (ESI)] in the 1980s heralded the beginning of a new era in proteomics and enabled rapid, sensitive, high-throughput identification and quantification of proteins in systematic manner. Procedurally, MS-based protein analysis involves five steps: (i) protein extraction, purification, and quantification from matrices/samples, (ii) direct analysis (top-down approach) or enzymatic digestion (bottom-up approach), (iii) optional protein/peptide separation based on liquid chromatography, (iv) mass-to-charge and intensity detection of protein/peptide and induced fragments by MS, and (v) protein identification and quantification by de novo or database-driven data analysis [16].

Of the two current MS-based methods for protein profiling, the first relies on high-resolution, two-dimensional electrophoresis (2-DE) to separate proteins from different samples in parallel, and selection and staining of differentially expressed proteins for MS identification; the second employs stable isotope tags to differentially label proteins from two different mixtures, and after digestion and combination of the labeled mixtures, the peptides are separated by multidimensional liquid chromatography and analyzed by tandem MS.

Other variations of MS-based methods include monodimensional gel electrophoresis liquid chromatography–electrospray ionization–tandem mass spectrometry (1-DE LC-ESI-MS/MS), bidimensional matrix-assisted laser desorption/ionization time-of-flight mass spectrometry (2-DE MALDI TOF)], multidimensional protein identification technology (MudPit), gas chromatography-mass spectrometry (GC-MS), mass spectrometric immunoassay (MSIA), and stable isotope standard capture with antipeptide antibodies (SISCAPA). Furthermore, flow cytometry, OFFGEL fractionation by isoelectric focusing (OGE), capillary electrophoresis (CE), and proton nuclear magnetic resonance spectroscopy (H NMR) can be also utilized along with MS [17].

A suite of software may be utilized to infer and/or determine protein structures and functions. For example, comparison of a test sequence with those of functionally characterized proteins, hidden Markov models (HMM) and Basic Local Alignment Search Tool (BLAST; heuristic local alignment) give some ideas about the likely function of a protein in question [18]. Furthermore,; a combination of artificial neural networks (TargetP), hidden Markov models (SignalP), and PsortII to predict the subcellular localizations of the proteins; a program based on hidden Markov models (TMHMM), and Memsat3 to predict the secondary structures and topologies of alpha-helical integral membrane proteins.

***Giardia* cytoskeleton proteins**. *Giardia* possesses a complex cytoskeleton that comprises four pairs of flagella (anterior, posterior-lateral, ventral, and caudal; each flagellar axoneme/core is templated in a basal body located between and slightly anterior to the two nuclei), a ventral disk (composed of microtubules, microribbons, and cross-bridges), a median body (an microtubular element of 0.2–1.8 µm located dorsally to the ventral disk), and funis (two sheets of microtubules accompanying the axonemes of the caudal flagella) [6].

Formed in trophozoites in the small intestine after transition from cyst (excystation) under influence of gastric acid, *Giardia* cytoskeleton is involved in its mobility (swimming, steering), attachment to small intestinal enterocytes, and subsequent differentiation into cyst (encystation).

Using proteomic approaches, a number of proteins have been identified in *Giardia* cytoskeleton, including those linked to the flagella (e.g., α giardins and kinesin 2 in flagella; γ-tubulin and centrin in the basal bodies), ventral disk (α and β tubulins; β, γ and δ giardins; SALP-1; Mplp; TTHERM), median body (e.g., tubulins; β giardin; actin, α actinin; centrin; kinesin 13, EB1; MBP), and funis (α tubulin; actin) (Figure 31.3) [19,20].

***Giardia* excystation and encystation proteins.** Upon ingestion by vertebrate host, *Giardia* cysts transform into trophozoites in the stomach, which then move to the small intestine and attach to the enterocytes through their ventral suck and flagella. Utilizing their specialized endocytic organelles (termed peripheral vesicles or PV; which are ~150 nm compartments located under the plasma membrane on the exposed dorsal side as well as in a specialized region of the ventral disk), trophozoites obtain fluids and nutrients (e.g., exogenous ferritin) from host small intestine for proliferation. Trophozoites may further reproduce by binary division or encyst depending on specific environmental cues. During encystation, trophozoites generate encystation-specific vesicles (ESV; which are Golgi-like organelles due to their reliance on active endoplasmic reticulum (ER) exit sites, their link to known Golgi-specific protein trafficking components, their sensitivity to brefeldin A, and their capacity to interact with cyst wall cargo during secretory transport) for subsequent secretion of cyst wall material.

In light of their involvement in *Giardia* proliferation and transmission, identification of protein components associated with PV and ESV is fundamental to the understanding of *Giardia* biology and pathogenicity. Through flow cytometry-based organelle sorting of a mixed microsome fraction containing green fluorescent ESV organelles in condensed cores and red-labeled PV along with MS analysis and subtraction of overlapping hits to increase identification of organelle-specific candidates, Lourenço et al. [21] revealed enrichment of motor proteins (e.g., dyneins and kinesins) prior to ESV neogenesis, as well as several protein targets regulated during encystation. Furthermore, Faso et al. [22] employed label-free shotgun proteomics to quantify >1,000 proteins in nonencysting trophozoites and cells induced to encyst, including decrease of endoplasmic reticulum-targeted variant-specific surface proteins and increase of cytoskeleton regulatory components, NEK protein kinases, and proteins involved in protein folding and glycolysis during early stages of encystation; and similar proteome composition to nonencysting trophozoites during the later stages of encystation.

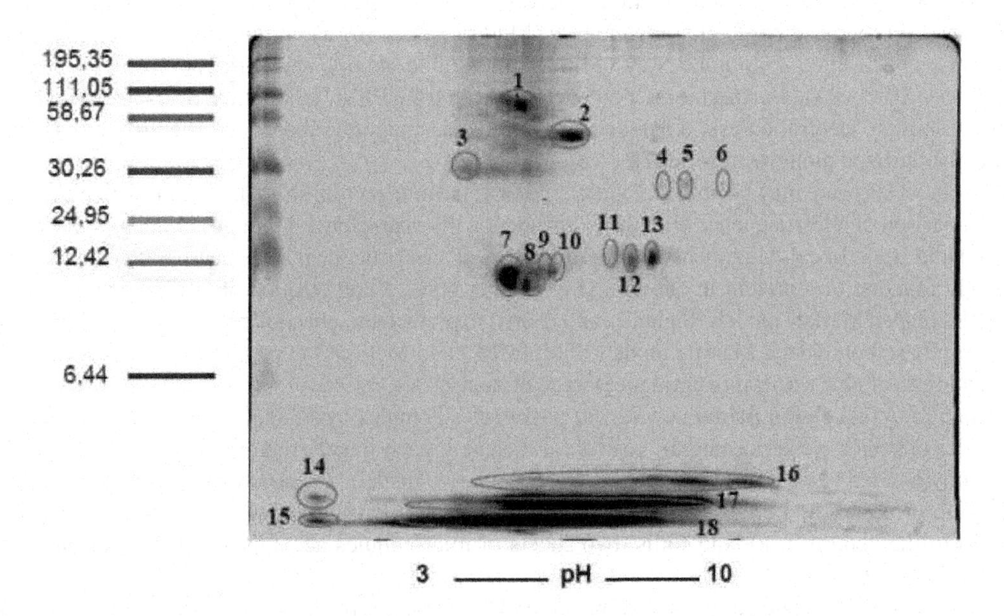

FIGURE 31.3 Visualization of *Giardia lamblia* ventral disk proteins by two-dimensional electrophoresis gel, which are further identified by LC-MS/MS after excision and trypsin in gel digestion of spots 1–18. (Photo credit: Lourenço D, Andrade Ida S, Terra LL, Guimarães PR, Zingali RB, de Souza W. *BMC Res Notes*. 2012;5:41.)

***Giardia* tenascins as key components of pathogenesis**. Tenascins, first found in vertebrates, are a family of large hexameric extracellular matrix proteins involved in regulating numerous developmental processes, such as morphogenetic cell migration and organogenesis [23]. The tenascin family consists of repeated structural modules including heptad repeats, epidermal growth factor (EGF)-like repeats, fibronectin type III repeats, and a globular domain shared with the fibrinogens [24]. Examination of secretome of *Giardia* assemblages A and B revealed a new class of virulence factors, the *Giardia* tenascins, along with known virulence factors such as cathepsin B cysteine proteases and members of a *Giardia* superfamily of cysteine-rich proteins comprising variant surface proteins and high-cysteine membrane proteins [25]. It is proposed that tenascins work as a virulence factor through interference with the host epidermal growth factor [26].

***G. lamblia* mitosome proteins**. *Giardia* possesses highly reduced mitochondria (so-called mitosome). Bounded by double membranes, *Giardia* mitosomes are not involved in core ATP synthesis, although they appear to function in the assembly of iron–sulfur clusters [27,28]. Found in many "amitochondrial" eukaryotic organisms, mitosomes offer ideal models for the study of mitochondrial evolution and for the establishment of the minimal set of proteins required for the biogenesis of an endosymbiosis-derived organelle [29]. By proteomic analysis of a mitosome-rich cellular fraction and immunofluorescence microscopy, Rada et al. identified a novel mitosomal protein homologous to monothiol glutaredoxins containing a CGFS motif at the active site, adding new components to the mitosomal FeS cluster biosynthetic pathway [28]. In another quantitative proteomic assessment of *Giardia* mitosomes, 20 proteins were confirmed as mitosome proteins, including nine components of the FeS cluster assembly machinery, a novel diflavo-protein with NADPH reductase, a novel VAMP-associated protein, and a key component of the outer membrane protein translocase. These works lend support that the *Giardia* mitosomes are of mitochondrial origin and their function is limited to the FeS cluster assembly pathway [30].

***Giardia* proteins implicated in resistance to nitroheterocyclics**. Nitroheterocyclics [e.g., metronidazole (MET), nitazoxanide (NTZ)] and non-nitro drug albendazole (ALB) are currently used to treat giardiasis [7]. However, treatment failure may occur due to *Giardia* resistance. MS shotgun analysis of proteome from trophozoites of nitro drug-resistant *G. lamblia* assemblage AI strains (i.e., C4, 1062ID10, and 713M3) in comparison with corresponding wild-types strains (i.e., WBC6, 106, and 713) identified 200–500 differentially expressed proteins, including enzymes involved in reduction and detoxification of nitro radicals, NO or O_2, highlighting the complex of drug resistance mechanism in *Giardia* [31–34].

31.3 FUTURE PERSPECTIVES

G. lamblia (synonyms: *G. duodenalis* and *G. intestinalis*) is a flagellated, binucleated protozoan that is responsible for causing persistent diarrhea (giardiasis) worldwide. While the lifecycle of *Giardia* parasites has been elucidated, much remains unknown about their interactions with hosts and their pathogenic mechanisms underlying the development of clinical diseases.

Proteomic approach offers an exciting option to unravel the molecular intricacies between *Giardia* parasites and their hosts. Preliminary application of proteomic techniques has yielded interesting data about the composition and diversity of proteins associated with *Giardia* cytoskeleton, encystation/excystation, virulence, mitosome, and nitro compound resistance, as well as other biological processes [30,36–40]. Although many proteins do not show obvious changes in concentrations, they often modulate their functions through posttranslational modifications. Therefore, future development of proteomic methodologies that enable detection and monitoring of posttranslational modifications of *Giardia* proteins in healthy and diseased cells is invaluable. In addition, the ability to quantify proteins originating from *Giardia* parasites and their hosts during various stages of infection is also critical to the understanding of *Giardia* pathogenesis [41]. Further advanced proteomic technologies will also help realize *Giardia*'s long proposed potential as a model parasite and eukaryote [42].

REFERENCES

1. Coelho CH, Costa AO, Silva AC, et al. Genotyping and descriptive proteomics of a potential zoonotic canine strain of *Giardia duodenalis*, infective to mice. *PLoS One.* 2016;11(10):e0164946.
2. Heyworth MF. *Giardia duodenalis* genetic assemblages and hosts. *Parasite.* 2016;23:13.
3. Thompson RCA, Ash A. Molecular epidemiology of *Giardia* and *Cryptosporidium* infections – what's new? *Infect Genet Evol.* 2019;75:103951.
4. Corleto VD, Di Marino VP, Galli G, et al. Improving basic skills in celiac-like disease diagnosis: a case report. *BMC Gastroenterol.* 2018;18(1):162.
5. Wang Y, Gonzalez-Moreno O, Roellig DM, et al. Epidemiological distribution of genotypes of *Giardia duodenalis* in humans in Spain. *Parasit Vectors.* 2019;12(1):432.
6. Dawson SC, House SA. Life with eight flagella: flagellar assembly and division in *Giardia. Curr Opin Microbiol.* 2010;13(4):480–90.
7. Hahn J, Seeber F, Kolodziej H, et al. High sensitivity of *Giardia duodenalis* to tetrahydrolipstatin (orlistat) in vitro. *PLoS One.* 2013;8(8):e71597.
8. Adam RD, Dahlstrom EW, Martens CA. et al. Genome sequencing of *Giardia lamblia* genotypes A2 and B isolates (DH and GS) and comparative analysis with the genomes of genotypes A1 and E (WB and Pig). *Genome Biol Evol.* 2013;5(12):2498–511.
9. Kooyman FNJ, Wagenaar JA, Zomer A. Whole-genome sequencing of dog-specific assemblages C and D of *Giardia duodenalis* from single and pooled cysts indicates host-associated genes. *Microb Genom.* 2019;5(12):e000302.
10. Dunn N, Juergens AL. *Giardiasis. StatPearls* [Internet]. Treasure Island (FL): StatPearls Publishing; 2019.
11. Emery SJ, Mirzaei M, Vuong D, et al. Induction of virulence factors in *Giardia duodenalis* independent of host attachment. *Sci Rep.* 2016;6:20765.
12. Dubourg A, Xia D, Winpenny JP, et al. *Giardia* secretome highlights secreted tenascins as a key component of pathogenesis. *Gigascience.* 2018;7(3):1–13.
13. Durigan M, Cardoso-Silva CB, Ciampi-Guillardi M, et al. Molecular genotyping, diversity studies and high-resolution molecular markers unveiled by microsatellites in *Giardia duodenalis. PLoS Negl Trop Dis.* 2018;12(11):e0006928.
14. Hanevik K, Bakken R, Brattbakk HR, Saghaug CS, Langeland N. Whole genome sequencing of clinical isolates of *Giardia lamblia. Clin Microbiol Infect.* 2015;21(2):192.e1–3.
15. Hooshyar H, Rostamkhani P, Arbabi M, Delavari M. *Giardia lamblia* infection: review of current diagnostic strategies. *Gastroenterol Hepatol Bed Bench.* 2019;12(1):3–12.
16. Aebersold R, Mann M. Mass spectrometry-based proteomics. *Nature.* 2003;422(6928):198–207.
17. Bond A, Vernon A, Reade S, et al. Investigation of volatile organic compounds emitted from faeces for the diagnosis of giardiasis. *J Gastrointestin Liver Dis.* 2015;24(3):281–6.
18. Ansell BRE, Pope BJ, Georgeson P, Emery-Corbin SJ, Jex AR. Annotation of the *Giardia* proteome through structure-based homology and machine learning. *Gigascience.* 2019;8(1):giy150.
19. Lauwaet T, Smith AJ, Reiner DS, et al. Mining the *Giardia* genome and proteome for conserved and unique basal body proteins. *Int J Parasitol.* 2011;41(10):1079–92.
20. Wampfler PB, Tosevski V, Nanni P, Spycher C, Hehl AB. Proteomics of secretory and endocytic organelles in *Giardia lamblia. PLoS One.* 2014;9(4):e94089.
21. Lourenço D, Andrade Ida S, Terra LL, Guimarães PR, Zingali RB, de Souza W. Proteomic analysis of the ventral disc of *Giardia lamblia. BMC Res Notes.* 2012;5:41.
22. Faso C, Bischof S, Hehl AB. The proteome landscape of *Giardia lamblia* encystation. *PLoS One.* 2013;8(12):e83207.
23. Chiquet-Ehrismann R, Hagios C, Matsumoto K. The tenascin gene family. *Perspect Dev Neurobiol.* 1994;2(1):3–7.
24. Chiquet-Ehrismann R. Tenascins, a growing family of extracellular matrix proteins. *Experientia.* 1995;51(9–10):853–62.
25. Dubourg A, Xia D, Winpenny JP, et al. *Giardia* secretome highlights secreted tenascins as a key component of pathogenesis. *Gigascience.* 2018;7(3):1–13.
26. Ortega-Pierres MG, Argüello-García R. *Giardia duodenalis*: role of secreted molecules as virulent factors in the cytotoxic effect on epithelial cells. *Adv Parasitol.* 2019;106:129–69.

27. Tovar J, León-Avila G, Sánchez LB, et al. Mitochondrial remnant organelles of *Giardia* function in iron-sulphur protein maturation. *Nature.* 2003;426(6963):172–6.
28. Rada P, Smíd O, Sutak R, et al. The monothiol single-domain glutaredoxin is conserved in the highly reduced mitochondria of *Giardia intestinalis. Eukaryot Cell.* 2009;8(10):1584–91.
29. Regoes A, Zourmpanou D, León-Avila G, van der Giezen M, Tovar J, Hehl AB. Protein import, replication, and inheritance of a vestigial mitochondrion. *J Biol Chem.* 2005;280(34):30557–63.
30. Jedelský PL, Doležal P, Rada P, et al. The minimal proteome in the reduced mitochondrion of the parasitic protist *Giardia intestinalis. PLoS One.* 2011;6(2):e17285.
31. Emery SJ, Baker L, Ansell BRE, Differential protein expression and post-translational modifications in metronidazole-resistant *Giardia duodenalis. Gigascience.* 2018;7(4).
32. Müller J, Hemphill A, Müller N. Physiological aspects of nitro drug resistance in *Giardia lamblia. Int J Parasitol Drugs Drug Resist.* 2018;8(2):271–7.
33. Müller J, Braga S, Heller M, Müller N. Resistance formation to nitro drugs in *Giardia lamblia*: no common markers identified by comparative proteomics. *Int J Parasitol Drugs Drug Resist.* 2019;9:112–9.
34. Saghaug CS, Klotz C, Kallio JP, et al. Genetic variation in metronidazole metabolism and oxidative stress pathways in clinical *Giardia lamblia* assemblage A and B isolates. *Infect Drug Resist.* 2019;12:1221–35.
35. Jerlström-Hultqvist J, Stadelmann B, Birkestedt S, Hellman U, Svärd SG. Plasmid vectors for proteomic analyses in *Giardia*: purification of virulence factors and analysis of the proteasome. *Eukaryot Cell.* 2012;11(7):864–73.
36. Emery SJ, Lacey E, Haynes PA. Data from a proteomic baseline study of assemblage A in *Giardia duodenalis. Data Brief.* 2015;5:23–7.
37. Camerini S, Bocedi A, Cecchetti S, et al. Proteomic and functional analyses reveal pleiotropic action of the anti-tumoral compound NBDHEX in *Giardia duodenalis. Int J Parasitol Drugs Drug Resist.* 2017;7(2):147–58.
38. Frontera LS, Moyano S, Quassollo G, Lanfredi-Rangel A, Rópolo AS, Touz MC. Lactoferrin and lactoferricin endocytosis halt *Giardia* cell growth and prevent infective cyst production. *Sci Rep.* 2018;8(1):18020.
39. Touz MC, Feliziani C, Rópolo AS. Membrane-associated proteins in *Giardia lamblia. Genes.* 2018;9(8):404.
40. Duarte TT, Ellis CC, Grajeda BI, De Chatterjee A, Almeida IC, Das S. A targeted mass spectrometric analysis reveals the presence of a reduced but dynamic sphingolipid metabolic pathway in an ancient protozoan, *Giardia lamblia. Front Cell Infect Microbiol.* 2019;9:245.
41. Marzano V, Mancinelli L, Bracaglia G, et al. "Omic" investigations of protozoa and worms for a deeper understanding of the human gut "parasitome". *PLoS Negl Trop Dis.* 2017;11(11):e0005916.
42. Jex AR, Svärd S, Hagen KD, et al. Recent advances in functional research in *Giardia intestinalis. Adv Parasitol.* 2020;107:97–137.

SUMMARY

The genus *Giardia* encompasses a large number of unicellular, flagellated protozoa species that colonize the upper portion of the small intestine in humans and animals, leading to giardiasis (watery diarrhea, epigastic pain, nausea, vomiting, and weight loss). With a relatively simple lifecycle involving a cyst stage and a trophozoite stage, *Giardia*, specifically *G. lamblia* (synonyms *G. duodenalis* and *G. intestinalis*) affects approximately 280 million people each year worldwide. Showing an estimated prevalence of 1%–7.6% in high-income countries and 20%–30% in low- and middle-income countries, giardiasis exerts an enormous burden on global healthcare systems, and compromises the quality of life among millions of sufferers. Apart from timely detection and tracking of *Giardia* outbreaks and implementation of appropriate treatment and prevention measures, a detailed understanding of molecular mechanisms of *Giardia* infection is crucial for the development of novel intervention strategies for ultimate eradication of giardiasis in human population. As *Giardia* genomes contain ~5,000 protein-coding genes, of which up to 50% have unknown functions, proteomic analysis plays a fundamental role in the characterization of *Giardia* proteins and contributes toward the full elucidation of *Giardia* pathogenesis.

Metabolomic Analysis of *Fasciola* Infection

32

Dongyou Liu
Royal College of Pathologists of Australasia Quality Assurance Programs

Contents

32.1 INTRODUCTION

The genus *Fasciola* consists of two zoonotic trematode species (i.e., *F. hepatica* and *F. gigantica*) that disseminate between ruminants (definitive hosts) and snails (intermediate hosts). *Fasciola* infection (fascioliasis) affects >700 million domestic ruminants (cattle, sheep, pig, donkey, buffalo, and goats) worldwide and causes significant economic losses through liver condemnation, high mortality, and reduced production of meat, milk, and wool. Additionally, as accidental host, humans may acquire fascioliasis through consumption of vegetables (e.g., watercress) and water containing *Fasciola* larvae, leading to various clinical diseases and significant health costs. Although *Fasciola* has a cosmopolitan distribution, fascioliasis tends to target people in developing countries and is recognized by the WHO as one of the neglected tropical diseases (NTDs). Given our limited understanding about the molecular mechanisms of *Fasciola* pathogenesis, metabolomic analysis has a critical role in helping decipher the intricacies between the parasite and the host and identify new biomarkers for improved diagnosis and control [1,2].

Taxonomy. Constituting one of the five genera (i.e., *Fasciola*, *Fascioloides*, *Fasciolopsis*, *Parafasciolopsis*, and *Protofasciola*) in the family Fasciolidae, order Echinostomida, subclass Digenea, class Trematoda, phylum Platyhelminthes, the genus *Fasciola* comprises two valid species (*F. hepatica* and *F. gigantica*) together with a large number of unassigned species. A former *Fasciola* species (i.e., *F. jacksoni*, which parasitizes elephants) is now reclassified in the genus *Fascioloides* as *Fascioloides jacksoni* after examination of both mitochondrial (ND1) and nuclear (28S rRNA, ITS1, and ITS2) gene sequences [3].

F. hepatica (also known as the common liver fluke or the temperate fluke) and *F. gigantica* (also known as the giant liver fluke or the tropical fluke) are parasites of various ruminants as well as humans, and responsible for significant economic losses worldwide.

Morphology. *F. hepatica* adult worm (average 28 mm×11 mm) is leaf-shaped, broadly flattened, and gray-brown. The anterior end is broader than the posterior end and has a cone (consisting of an oral sucker at the end and a ventral sucker at the base), which marks off from the body (Figure 32.1a) [4]. With backwardly projecting spines covering the tegument, the hermaphroditic adult worm displays highly branched ovaries and testes for egg production. The eggs (about 140 μm×70 μm) are oval, operculate, and yellow (Figure 32.1b) [5]. The juvenile evolving from the egg is lancet-like and 1–2 mm long (at the time of entry into the liver).

F. gigantica adult worm (average 41 mm×8 mm) is leaf-like and notably longer than *F. hepatica* adult worm (Figure 32.1c) [4]; *F. gigantica* eggs (about 135 μm×80 μm) are slightly broader than *F. hepatica* eggs.

Genome. The complete genomes of two *F. hepatica* strains are shown to be 1.13 and 1.27 Gb, with 14,642 and 22,676 predicted genes, respectively. Interestingly, 855 (5.8%) proteins (many showing cysteine-type endopeptidase activity) contain predicted signal peptides without transmembrane domains, suggesting the possibility of their being secreted or excreted. Enjoying the richer milieu in the host, this hepatobiliary trematode is metabolically less constrained than schistosomes and cestodes. Its expanded protease (e.g., cathepsins L and B) and tubulin families facilitate migration from the gut across the peritoneum and through the liver for growth and maturation in the bile ducts within a broad range of definitive mammalian and intermediate molluscan hosts. Furthermore, its expansion of cathepsins, fatty acid-binding proteins, protein disulfide-isomerases, and molecular chaperones highlights the significance of excretory-secretory proteins in this liver-dwelling fluke, while the presence of G-protein-coupled receptors may play a critical key role in adaptation of physiology and behavior in mammalian hosts. In addition, its possession of antioxidant and detoxification pathways underlies its resistance to anthelmintic drugs (e.g., praziquantel) [6,7]. The finding of an 859,205 bp genome belonging to a *Neorickettsia* endobacterium, which closely relates to the etiological agents of human Sennetsu and Potomac horse fevers, suggests the possible involvement of this bacterium in the evolution of *F. hepatica* [8].

Similarly, the complete genomes of two *F. gigantica* strains are available and range from 1.04 (with 20,858 predicted genes) to 1.13 Gb (with 13,940 predicted genes). This parasite contains the genes of

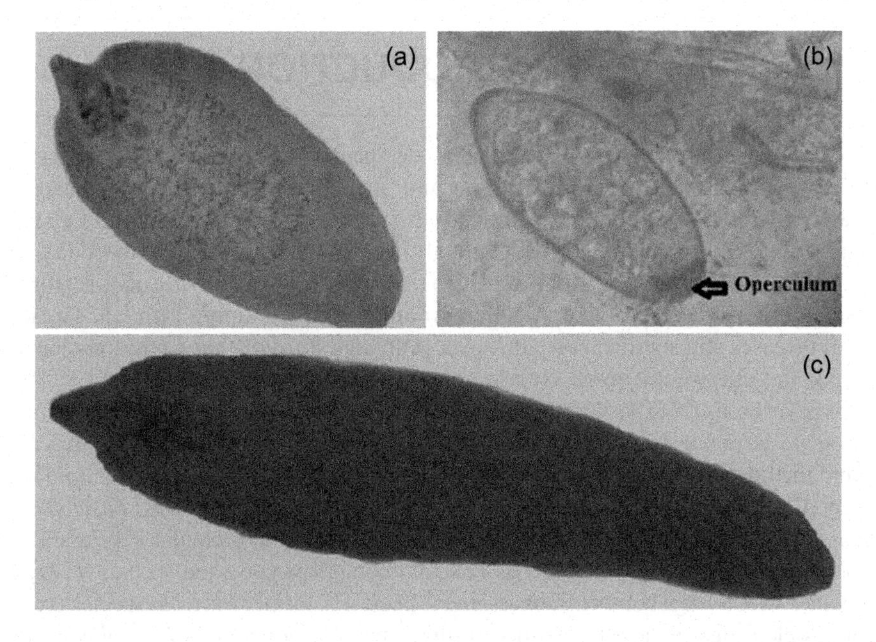

FIGURE 32.1 *Fasciola hepatica* adult worm (28 mm×11 mm) from sheep (a) and egg (140 μm×70 μm) from human stool (b); *F. gigantica* adult worm (41 mm×8 mm) from cattle (c). (Photo credits: (a, c) Shafiei R, Sarkari B, Sadjjadi SM, Mowlavi GR, Moshfe A. *Vet Med Int.* 2014;2014:405740. (b) Sah R, Khadka S, Khadka M, et al. *BMC Res Notes.* 2017;10(1):439.)

glycolysis, Krebs cycle, and fatty acid metabolism pathways, but lacks the key genes of the fatty acid biosynthesis pathway, suggesting its reliance on the host bile for fatty acid requirement [9].

The complete mitochondrial genomes of *F. hepatic* and *F. gigantica* are 14,462 and 14,478 bp in size, respectively, each containing 12 protein-encoding (*cox* 1-3, *nad* 1-6, *nad* 4L, *cyt* b and *atp* 6, which are organized in the order of *nad* 5 > *cox* 1 > *nad* 4 > *cyt* b > *nad* 1 > *nad* 2 > *cox* 3 > *cox* 2 > *atp* 6 > *nad* 6 > *nad* 3 > *nad* 4L), two ribosomal (*rrn* S and *rrn* L), and 22 transfer RNA genes, but lacking an *atp* 8 gene [10].

Lifecycle. *Fasciola* has a complex lifecycle involving seven stages (i.e., egg, miracidium, sporocyst, redia, cercaria, metacercaria, and adult worm) in definitive mammal hosts and intermediate snail hosts.

Residing in the bile ducts of mammalian hosts, *F. hepatica* hermaphroditic adult worms reproduce sexually (through self- or cross-fertilization) and lay up to 25,000 eggs per day per fluke, which move along the bile ducts to the gastrointestinal tract and exit in feces, contaminating pastures and grazing areas near water bodies. Eggs that reach fresh water embryonate in 2 weeks and release ciliated miracidiae (singular, miracidium), which actively seek for and infect suitable *Lymanaea* snail hosts (e.g., *Galba truncatula* for *F. hepatica* and African *Radix natalensis* and Eurasian *R. auricularia* for *F. gigantica*) within 24 h of hatching. Inside snail, miracidium loses cilia and grows into a sac of germinating cells (called a sporocyst), with each cell developing into a redia (plural, rediae). Bursting through the sporocyst, rediae migrate in the hepato-pancreas of the snail and grow into cercariae (singular, cercaria). During this process, a single miracidium turns into 600–4,000 cercariae. Moving out of the snail host, cercariae attach to aquatic vegetation (e.g., watercress), and form encysted metacercariae (singular, metacercaria) that remain viable for months. After encysted metacercariae attaching to vegetation (e.g., watercress) or in water are consumed by a mammalian host, they excyst in the small intestine, traverse the intestinal wall, migrate to the peritoneal cavity, penetrate the Glisson's capsule of the liver (provoking reactions associated with acute infection, leading to fever, nausea and abdominal pain), and finally settle in the bile ducts, in which young flukes mature into adult worms within a month in ruminants and 3–4 months in humans (inducing anemia, inflammation, fibrosis, cholangitis, and biliary stasis) [11].

F. gigantica differs from *F. hepatica* in its prepatent period (10–14 weeks versus 8–12 weeks for *F. hepatica*) and its snail host (deeper water snail versus and amphibious snail close to the edge of slow moving or stagnant water for *F. hepatica*).

Epidemiology. The natural hosts for *Fasciola* spp. are ruminants (definitive host) and Lymnaeidae snails (intermediate host). *F. hepatica* (the temperate fluke) shows a cosmopolitan distribution and is prevalent in temperate areas with extensive sheep and cattle raising, affecting >700 million livestock worldwide. *F. gigantica* (the tropical fluke) mainly occurs in tropical regions of Africa, the Middle East, and Asia (especially India, Pakistan, Indonesia, Indochina, and the Philippines), as well as Hawaii, where 25%–100% of total cattle population are infected. As accidental hosts, about 2.4–17 million people in >60 countries across Africa, Europe, the Middle East, Southeast Asia, and Latin America are infected with *Fasciola* spp., and >180 million people are at risk of infection [12]. It is notable that the incidence of animal fascioliasis does not correspond directly to that of human fascioliasis, which may be influenced by culture and customs, dietary and hygiene practices, and general sanitary conditions [13].

Clinical features. About 50% of patients with *Fasciola* infections are asymptomatic while others may develop acute, chronic, halzoun, and ectopic diseases, with clinical manifestations ranging from abdominal pain (72% cases), fever (57%), constitutional symptoms (44%), urticaria (13%), itching (10%), respiratory symptoms (8%), headache (5%), and cardiac symptoms (1%), in addition to blood [eosinophilia (>500 mm^{-3}) in 96% cases, leukocytosis (>10,000 mm^{-3}) in 64% cases], microscopic (fecal eggs in 40% cases), and imaging (positive CT in 70% cases) abnormalities [11].

Acute fascioliasis may become evident about 4–7 days after ingestion of a large number of metacercariae (1,000–5,000) at once, which migrate throughout the liver parenchyma for 6–8 weeks (causing necrosis, hemorrhage, and inflammation) before settling and maturing in the bile ducts. Clinically, patients with acute fascioliasis may have fever (40°C–42°C), abdominal pain, right upper-quadrant tenderness, hepatosplenomegaly, vomiting, loss of appetite, flatulence, diarrhea, urticaria (hives), cough, jaundice, ascites, and anemia [11].

Chronic fascioliasis results from ingestion of a smaller number of metacercariae (200–1,000) or evolves from acute fascioliasis. Due to the presence of large adult worms, fragments, debris, and metabolic wastes in the bile ducts (causing biliary obstruction and inflammation), patients with chronic fascioliasis may experience biliary cholic, right upper-quadrant abdominal tenderness, hepatomegaly, fatty food intolerance, nausea, pruritus, jaundice, cardiac disorder, pancreatitis, calculi, bacterobilia, cholangitis, cholecystitis, severe anemia in children, portal cirrhosis, liver fibrosis (pipestem liver), and death [11].

Halzoun is linked to the consumption of *Fasciola*-infected raw liver. Young adult worms attaching to the pharyngeal mucosa may induce pain, edema, bleeding, and respiration difficulty, while adult worms living in the biliary ducts may cause symptoms for up to 10 years.

Ectopic fascioliasis is caused by *Fasciola* that accidentally migrate to and settle in the lungs, diaphragm, intestinal wall, kidneys, subcutaneous tissue, lymph nodes, eye, and other locations, instead of the liver and biliary ducts.

Pathogenesis. *Fasciola* infection (fascioliasis) in definitive hosts consists of two phases: parenchymal (migratory) and biliary.

In the parenchymal (migratory) phase, *Fasciola* juvenile flukes pierce through the intestinal wall and move to the liver or other organs, causing mechanical damages; at the same time, they secrete various proteases and other enzymes to assist their invasion of host cells and migrate within tissues, provoking host immune responses [14–20]. In the biliary phase, *Fasciola* adult worms utilize their oral sucker at the end and ventral sucker at the base of the cone as well as spines on the tegument for attaching to the biliary ducts and feed on blood and nutrients, resulting in hypertrophy of biliary ducts and anemia; furthermore, accumulation of adult worms, fragments, debris, and metabolic products obstruct the lumen and induce other pathological changes.

Among immunomodulatory molecules identified from *F. hepatica* are helminth defense molecule-1 (molecular mimicry of antimicrobial peptide CAP18/LL-37 through binding to LPS and reducing its activity; preventing acidification of the endolysosomal compartments and antigen processing; inhibiting NLRP3 inflammasome activation), TGF-like molecule (ligation of mammalian TGF-β receptor and induction of IL-10 and arginase in macrophages), fatty acid-binding protein (suppression of LPS-induced activation via binding and blocking of CD14; inducing alternatively activated macrophages), mucin-like polypeptide (promoting TLR4 activation of DCs and Th1 cell induction), cathepsin L peptidases (degradation of fibrinogen and fibrin), peroxiredoxin (inactivation of ROS and induction of AA-MF in mouse models), and thioredoxin peroxidase (induction of AA-MF with increased IL-10 and PGE2 responses). These molecules contribute to *F. hepatica* evasion of host immune responses and survival within host environment [21,22].

Diagnosis. Diagnosis of human fascioliasis involves medical history review and physical examination, imaging (e.g., X-ray, ultrasound, CT, and MRI to visualize adult flukes in the bile ducts, and to reveal the burrow tracts and dilation of the bile ducts produced by the worms), microscopy (to identify *Fasciola* eggs in stool, duodenal aspirate, or biliary aspirate), biochemical tests [to detect anemia, hypoalbuminemia, eosinophilia, and elevation of liver enzymes such as glutamate dehydrogenase (GLDH), gamma-glutamyl transferase, and lactate dehydrogenase (LDH)], serology (e.g., ELISA and Western blot), and molecular detection [targeting mitochondrial (e.g. NDI, COI, rRNA, ITS1, and ITS2) and nuclear genes] [11,23,24].

Treatment and prevention. *Fasciola* appears to be resistant to praziquantel, and the drug of choice for treating human fascioliasis human is triclabendazole (TCBZ, at one to two oral doses of 10 mg kg^{-1} body weight in 24 h), which is effective against both juveniles and adult liver flukes. Alternative drugs include nitazoxanide and bithionol (30 mg kg^{-1} body weight).

Prevention of human fascioliasis requires attention to water source contamination by human and nonhuman hosts (ruminants and snails) and dietary practices (e.g., consumption of raw, untreated aquatic vegetation). *Fasciola* encysted metacercariae on vegetables may be destroyed by either 6% vinegar or potassium permanganate for 5 to 10 min. Treatment of ruminants with halogenated phenols (e.g., bithionol, hexachlorophene, nitroxynil), salicylanilides (e.g., closantel, rafoxanide), benzimidazoles (e.g., triclabendazole, albendazole, mebendazol, luxabendazole), sulphonamides (e.g., clorsulon) or phenoxyalkanes (e.g., diamphenetide) helps decrease the incidence of animal fascioliasis. Application of molluscicides contributes to reduction of the population of *Lymnaea* snails [11,13].

32.2 METABOLOMIC ANALYSIS OF *FASCIOLA* INFECTION

Overview. Metabolomics is a newly developed technique that focuses on metabolites in a biosystem. As low-molecular-weight small molecules secreted into body fluids or tissues by host and microbial cells in response to environmental, nutritional, and immunological stimuli, the types and levels of metabolites can be determined by high-throughput mass spectrometry nuclear magnetic resonance spectroscopy, and aligned against libraries of known biochemicals. This systematic, nonbiased analysis of metabolites (or metabolome analysis) helps reveal molecular details on the host–parasite interplay and mechanisms of health and disease, and also permits identification of new biomarkers for improved disease diagnosis and control.

Currently two major technologies (i.e., mass spectrometry and nuclear magnetic resonance spectroscopy) are commonly applied for metabolome analyses. While mass spectrometry coupled to gas chromatography (GC-MS) detects volatile, thermally stable metabolites, mass spectrometry in association with liquid chromatography (LC-MS) identifies nonvolatile polar and nonpolar compounds with nanomolar resolution. Because both GC-MS and LC-MS require prior preparation of compounds, their performance may be influenced by the bias and inevitably metabolite losses introduced by sample preparation methods (e.g., protein precipitation, liquid–liquid or solid-phase extraction/microextraction, use of molecularly imprinted polymers and restricted-access materials). On the other hand, nuclear magnetic resonance (NMR) spectroscopy involves minimal sample preparation and allows detection of a full spectrum of metabolites, but at a reduced level of resolution (at or above a millimolar level).

Following mass spectrometry or NMR spectroscopy separation and detection, metabolites are aligned to libraries of known biochemicals, identified, and categorized by multivariate statistical approaches.

Multiplatform-based metabotyping for *F. hepatica* metabolome. Upon evaluation of five different analytical methods in conjunction with extraction protocol optimization, Savic et al. [2] showed that ultra-performance liquid chromatography and capillary electrophoresis coupled with mass spectroscopy (UPLC-MS and CE-MS) enable identification of 142 metabolites out of 14,724 features from *F. hepatica*-infected cow liver extracts, and that the chloroform:methanol:water proportion of 15:59:26 represents the best composition for metabolite extraction for UPLC-MS and CE-MS platforms, accommodating different columns and ionization modes (Figure 32.2).

Metabolomic analysis of *F. hepatica* extracellular vesicles. Extracellular vesicles (EV or exosomes) are nano-sized, membrane-contained vesicles (40–150 nm in diameter) containing different biomolecules (e.g., proteins, lipids, nucleic acids, and sugars) that are released into the extracellular microenvironment (e.g., blood, urine, semen, saliva, cerebrospinal fluid, breast milk, and tissue) by all types of eukaryotic cells (as well as a few prokaryotic cells) for tissue repair, neural communication, immunological response, and transfer of pathogenic proteins [25–29]. Metabolomic analysis of *F. hepatica* EV identified 79 different parasite proteins, including proteases (e.g., cathepsins and leucine aminopeptidase, LAP), detoxifying enzymes (e.g., heat-shock proteins and fatty acid-binding proteins), which account for 52% of the

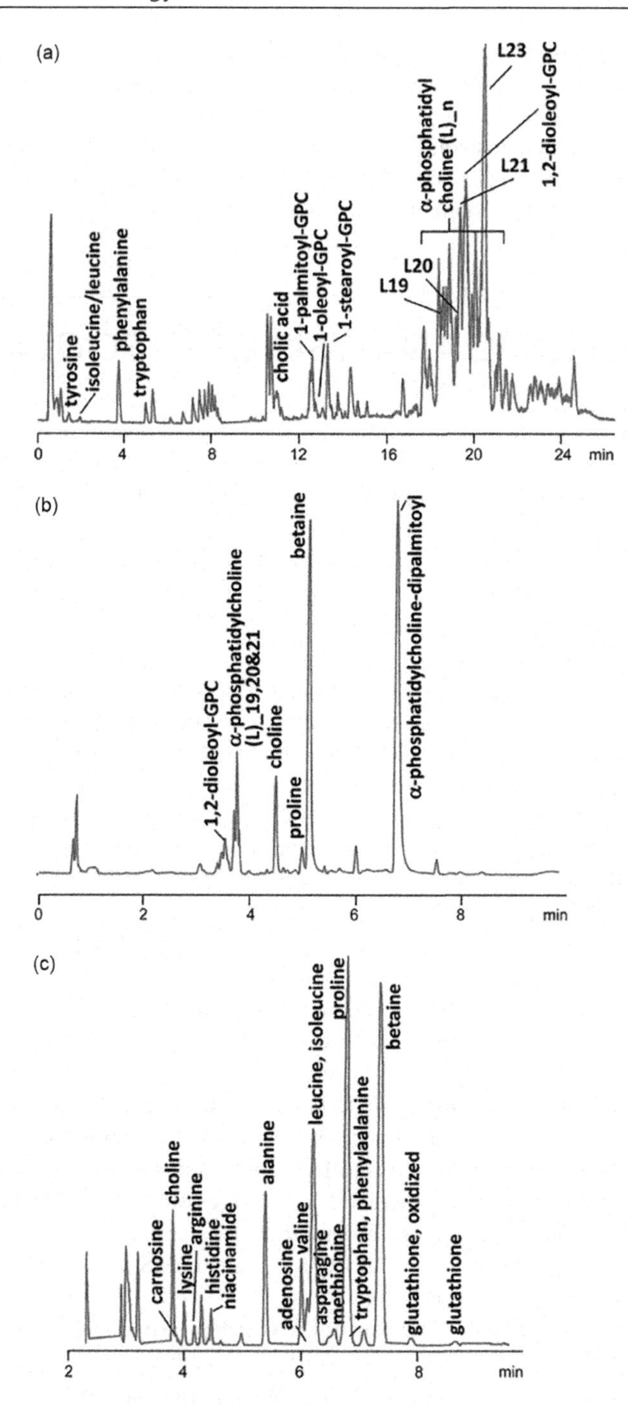

FIGURE 32.2 Typical mass chromatograms and electropherograms of 80% methanol extracts of *Fasciola hepatica* flukes acquired at ESI+ mode in (a) C18 column, (b) HILIC column, and (c) CE capillary. Key: GPC, glycerophosphocholine; the remaining terms have the nomenclature α-phosphatidylcholine (L)_*n* (which represents a mixture of α-phosphatidylcholines (L)_*n* numbered 1 to 23 by order of elution in the RPLC-MS method; tentative identification based on *m/z* searches at http://www.lipidmaps.org/data/structure/LMSDSearch.php): L19, 1-hexadecanoyl-2-octadecadienoyl-sn-glycero-3-phosphocholine; L20, 1-nonanoyl-2-tricosanoyl-sn-glycero-3-phosphocholine; L21, 1-hexadecanoyl-2-octadecenoyl-sn-glycero-3-phosphocholine; and L23, 1-octadecanoyl-2-octadecenoyl-sn-glycero-3-phosphocholine. (Photo credit: Saric J, Want EJ, Duthaler U, et al. *Anal Chem.* 2012;84(16):6963–72.)

secretome. In addition, 19 host proteins (including immunoglobulins, metabolic enzymes, exosomal molecules such as CD19) were also detected in *F. hepatica* vesicles. The presence of actin, enolase, and LAP in purified *F. hepatica* vesicles was subsequently confirmed by immunohistochemistry and transelectron microscopy (TEM).

32.3 FUTURE PERSPECTIVES

Although the first description of sheep liver fluke was made as early as in 1379, *Fasciola* infections in animals and humans have remained unabated due partly to negligence and also to inadequate understanding of the pathogenic mechanisms of fascioliasis. Rapid advances in molecular techniques in the past decade or so have laid a solid foundation for omic (including genomic, metagenomic, transcriptomic, proteomic, metabolomic, glycomic) analysis of parasite–host interplay as well as targeted research on potential biomarkers for accurate diagnosis and effective control of parasitic infections [30,31]. Nonetheless, application of omic techniques to the study of fascioliasis is still in its infancy, and relatively few publications on *Fasciola* researches (e.g., mechanisms of protective immunity, pathogenesis associated with parasite migration through the liver, and evolution of antihelminthic resistance) are available in the public domain [32–42]. Unquestionably, this glaring gap in our knowledge about *Fasciola* and fascioliasis is hindering the progress in the development of effective countermeasures against this neglected tropical disease [43]. The onus is on us all to step up our efforts toward the complete elucidation of *Fasciola* pathogenesis and design of innovative measures for ultimate eradication of fascioliasis in human and ruminant hosts.

REFERENCES

1. Saric J, Li JV, Utzinger J, et al. Systems parasitology: Effects of *Fasciola hepatica* on the neurochemical profile in the rat brain. *Mol Syst Biol*. 2010;6:396.
2. Saric J, Want EJ, Duthaler U, et al. Systematic evaluation of extraction methods for multiplatform-based metabotyping: Application to the *Fasciola hepatica* metabolome. *Anal Chem*. 2012;84(16):6963–72.
3. Heneberg P. Phylogenetic data suggest the reclassification of *Fasciola jacksoni* (Digenea: Fasciolidae) as *Fascioloides jacksoni* comb. nov. *Parasitol Res*. 2013;112:1679–89.
4. Sah R, Khadka S, Khadka M, et al. Human fascioliasis by *Fasciola hepatica*: The first case report in Nepal. *BMC Res Notes*. 2017;10(1):439.
5. Shafiei R, Sarkari B, Sadjjadi SM, Mowlavi GR, Moshfe A. Molecular and morphological characterization of *Fasciola* spp. isolated from different host species in a newly emerging focus of human fascioliasis in Iran. *Vet Med Int*. 2014;2014:405740.
6. Cwiklinski K, Dalton JP, Dufresne PJ, et al. The *Fasciola hepatica* genome: Gene duplication and polymorphism reveals adaptation to the host environment and the capacity for rapid evolution. *Genome Biol*. 2015;16(1):71.
7. Choi YJ, Fontenla S, Fischer PU, et al. Adaptive radiation of the flukes of the family Fasciolidae inferred from genome-wide comparisons of key species. *Mol Biol Evol*. 2020;37(1):84–99.
8. McNulty SN, Tort JF, Rinaldi G, et al. Genomes of *Fasciola hepatica* from the Americas reveal colonization with *Neorickettsia* endobacteria related to the agents of potomac horse and human sennetsu fevers. *PLoS Genet*. 2017;13(1):e1006537.
9. Pandey T, Ghosh A, Todur VN, et al. Draft genome of the liver fluke *Fasciola gigantica*. *ACS Omega*. 2020;5(19):11084–91.
10. Liu GH, Gasser RB, Young ND, Song HQ, Ai L, Zhu XQ. Complete mitochondrial genomes of the 'intermediate form' of *Fasciola* and *Fasciola gigantica*, and their comparison with *F. hepatica*. *Parasit Vectors*. 2014;7:150.

11. Tolan RW Jr. Fascioliasis due to *Fasciola hepatica* and *Fasciola gigantica* infection: An update on this 'neglected' neglected tropical disease. *Lab Med.* 2011;42(2):107–16.
12. Goral V, Senturk S, Mete O, Cicek M, Ebik B, Kaya B. A case of biliary dascioliasis by *Fasciola gigantica* in Turkey. *Korean J Parasitol.* 2011;49(1):65–8.
13. Cwiklinski K, O'Neill SM, Donnelly S, Dalton JP. A prospective view of animal and human fasciolosis. *Parasite Immunol.* 2016;38(9):558–68.
14. Morphew RM, Wright HA, LaCourse EJ, Woods DJ, Brophy PM. Comparative proteomics of excretory-secretory proteins released by the liver fluke *Fasciola hepatica* in sheep host bile and during in vitro culture ex host. *Mol Cell Proteomics.* 2007;6(6):963–72.
15. Morphew RM, Wright HA, Lacourse EJ, et al. Towards delineating functions within the fasciola secreted cathepsin l protease family by integrating in vivo based sub-proteomics and phylogenetics. *PLoS Negl Trop Dis.* 2011;5(1):e937.
16. Morphew RM, Hamilton CM, Wright HA, Dowling DJ, O'Neill SM, Brophy PM. Identification of the major proteins of an immune modulating fraction from adult *Fasciola hepatica* released by Nonidet P40. *Vet Parasitol.* 2013;191(3–4):379–85.
17. Marcilla A, De la Rubia JE, Sotillo J, et al. Leucine aminopeptidase is an immunodominant antigen of *Fasciola hepatica* excretory and secretory products in human infections. *Clin Vaccine Immunol.* 2008;15(1):95–100.
18. LaCourse EJ, Perally S, Morphew RM, et al. The Sigma class glutathione transferase from the liver fluke *Fasciola hepatica. PLoS Negl Trop Dis.* 2012;6(5):e1666.
19. Di Maggio LS, Tirloni L, Pinto AF, et al. Across intra-mammalian stages of the liver fluke *Fasciola hepatica*: A proteomic study. *Sci Rep.* 2016;6:32796.
20. Garcia-Campos A, Ravidà A, Nguyen DL, et al. Tegument glycoproteins and cathepsins of newly excysted juvenile *Fasciola hepatica* carry mannosidic and paucimannosidic N-glycans. *PLoS Negl Trop Dis.* 2016;10(5):e0004688.
21. Ruiz-Campillo MT, Molina Hernandez V, Escamilla A, et al. Immune signatures of pathogenesis in the peritoneal compartment during early infection of sheep with *Fasciola hepatica. Sci Rep.* 2017;7(1):2782.
22. Cwiklinski K, Jewhurst H, McVeigh P, et al. Infection by the helminth parasite *Fasciola hepatica* requires rapid regulation of metabolic, virulence, and invasive factors to adjust to its mammalian host. *Mol Cell Proteomics.* 2018;17(4):792–809.
23. Kang BK, Jung BK, Lee YS, et al. A case of *Fasciola hepatica* infection mimicking cholangiocarcinoma and ITS-1 sequencing of the worm. *Korean J Parasitol.* 2014;52(2):193–6.
24. Rokni MB, Bozorgomid A, Heydarian P, Aryaeipour M. Molecular evidence of human fasciolosis due to *Fasciola gigantica* in Iran: A case report. *Iran J Public Health.* 2018;47(5):750–4.
25. Marcilla A, Trelis M, Cortés A, et al. Extracellular vesicles from parasitic helminths contain specific excretory/secretory proteins and are internalized in intestinal host cells. *PLoS One.* 2012;7(9):e45974.
26. Cwiklinski K, de la Torre-Escudero E, Trelis M, et al. The extracellular vesicles of the helminth pathogen, *Fasciola hepatica*: Biogenesis pathways and cargo molecules involved in parasite pathogenesis. *Mol Cell Proteomics.* 2015;14(12):3258–73.
27. Yáñez-Mó M, Siljander PR, Andreu Z, et al. Biological properties of extracellular vesicles and their physiological functions. *J Extracell Vesicles.* 2015;4:27066.
28. Davis CN, Phillips H, Tomes JJ, et al. The importance of extracellular vesicle purification for downstream analysis: A comparison of differential centrifugation and size exclusion chromatography for helminth pathogens. *PLoS Negl Trop Dis.* 2019;13(2):e0007191.
29. de la Torre-Escudero E, Gerlach JQ, Bennett APS, et al. Surface molecules of extracellular vesicles secreted by the helminth pathogen *Fasciola hepatica* direct their internalisation by host cells. *PLoS Negl Trop Dis.* 2019;13(1):e0007087.
30. Preidis GA, Hotez PJ. The newest "omics" – Metagenomics and metabolomics – Enter the battle against the neglected tropical diseases. *PLoS Negl Trop Dis.* 2015;9(2):e0003382.
31. Marzano V, Mancinelli L, Bracaglia G, et al. "Omic" investigations of protozoa and worms for a deeper understanding of the human gut "parasitome". *PLoS Negl Trop Dis.* 2017;11(11):e0005916.
32. Robinson MW, Menon R, Donnelly SM, Dalton JP, Ranganathan S. An integrated transcriptomics and proteomics analysis of the secretome of the helminth pathogen *Fasciola hepatica*: Proteins associated with invasion and infection of the mammalian host. *Mol Cell Proteomics.* 2009;8(8):1891–907.
33. Cancela M, Ruétalo N, Dell'Oca N, et al. Survey of transcripts expressed by the invasive juvenile stage of the liver fluke *Fasciola hepatica. BMC Genomics.* 2010;11:227.
34. Young ND, Jex AR, Cantacessi C, et al. A portrait of the transcriptome of the neglected trematode, *Fasciola gigantica* – Biological and biotechnological implications. *PLoS Negl Trop Dis.* 2011;5(2):e1004.

35. Faridi A, Farahnak A, Golmohammadi T, Eshraghian M, Sharifi Y, Molaei Rad M. Triclabendazole (anthelmintic drug) effects on the excretory-secretory proteome of *Fasciola hepatica* in two dimension electrophoresis gel. *Iran J Parasitol.* 2014;9(2):202–8.
36. Figueroa-Santiago O, Espino AM. *Fasciola hepatica* fatty acid binding protein induces the alternative activation of human macrophages. *Infect Immun.* 2014;82(12):5005–12.
37. Khan YA, Khan MA, Abidi SM. 2D-PAGE analysis of the soluble proteins of the tropical liver fluke, *Fasciola gigantica* and biliary amphistome, *Gigantocotyle explanatum*, concurrently infecting *Bubalus bubalis. J Parasit Dis.* 2016;40(3):910–13.
38. Khoontawad J, Intuyod K, Rucksaken R, et al. Discovering proteins for chemoprevention and chemotherapy by curcumin in liver fluke infection-induced bile duct cancer. *PLoS One.* 2018;13(11):e0207405.
39. Radio S, Fontenla S, Solana V, et al. Pleiotropic alterations in gene expression in Latin American *Fasciola hepatica* isolates with different susceptibility to drugs. *Parasit Vectors.* 2018;11(1):56.
40. Di Maggio LS, Tirloni L, Pinto AFM, et al. Dataset supporting the proteomic differences found between excretion/secretion products from two isolates of *Fasciola hepatica* newly excysted juveniles (NEJ) derived from different snail hosts. *Data Brief.* 2019;25:104272.
41. Rasouli A, Farahnak A, Zali H, Rezaeian M, Golestani A, Molaei Rad MB. Protein detection of excretory-secretory products and somatic extracts from *Fasciola hepatica* and *F. gigantica* using two-dimensional electrophoresis. *Iran J Parasitol.* 2019;14(3):379–86.
42. Zhang FK, Hu RS, Elsheikha HM, et al. Global serum proteomic changes in water buffaloes infected with *Fasciola gigantica. Parasit Vectors.* 2019;12(1):281.
43. Stutzer C, Richards SA, Ferreira M, Baron S, Maritz-Olivier C. Metazoan parasite vaccines: Present status and future prospects. *Front Cell Infect Microbiol.* 2018;8:67.

SUMMARY

The genus *Fasciola* consists of two zoonotic trematode species (i.e., *F. hepatica* and *F. gigantica*) that circulate between ruminants (definitive hosts) and snails (intermediate hosts). *Fasciola* infection (fasciolia-sis) affects >700 million domestic ruminants (cattle, sheep, pig, donkey, buffalo, and goats) worldwide and causes significant economic losses through liver condemnation, high mortality, and reduced production of meat, milk, and wool. Additionally, as accidental host, humans may acquire fascioliasis through consumption of vegetables (e.g., watercress) and water containing *Fasciola* larvae, leading to various clinical diseases and significant health costs. Although *Fasciola* has a cosmopolitan distribution, fascioliasis tends to target people in developing countries and is recognized by the WHO as one of the NTDs. Given our limited understanding about the molecular mechanisms of *Fasciola* pathogenesis, metabolomic analysis plays a critical role in helping decipher the intricacies between the parasite and the host and identify new biomarkers for improved diagnosis and control.

Molecular Mechanisms of Host Immune Responses to *Entamoeba*

33

Moisés Martínez-Castillo, Jesús Serrano-Luna, and Daniel Coronado-Velázquez
Center for Research and Advanced Studies of the National Polytechnic Institute (Cinvestav-IPN)

Judith Pacheco-Yépez
Escuela Superior de Medicina, IPN

Mireya de la Garza and Mineko Shibayama
Center for Research and Advanced Studies of the National Polytechnic Institute (Cinvestav-IPN)

Contents

33.1 INTRODUCTION

Entamoeba histolytica is an extracellular parasitic protozoan that causes amebiasis, a human disease transmitted by the fecal-oral route involving food and water contaminated with cysts. This disease is the third leading cause of death by parasites worldwide, after malaria and schistosomiasis, and mainly occurs in developing countries. According to the World Health Organization (WHO), 500 million people are infected with *Entamoeba*,[1] although many more are colonized by *E. dispar*, a nonpathogenic ameba. *E. dispar* and *E. moshkovskii*, another nonpathogenic ameba, have been isolated from patients; they are morphologically indistinguishable from *E. histolytica*.[2] In contrast, as a potent human pathogen, *E. histolytica* is responsible for 50 million cases of disease and 40,000–100,000 deaths per year worldwide.[3]

E. histolytica trophozoites live in the lumen of the large intestine, where they survive by feeding on bacteria and cell debris. However, they are able to invade the tissue, leading to the destruction of the intestinal mucosa through the secretion of diverse virulence factors. In this process, both parasite and host contribute to the damage. With both innate and specific immune responses working to limit the infection, a deep understanding of the immune mechanisms related to *E. histolytica* infection is vital for the development of vaccines against amebiasis.[4–6] It is well known that reactive oxygen species (ROS) and nitric oxide (NO) generated by neutrophils and macrophages underscore innate responses against *E. histolytica* trophozoites' invasion of the intestinal mucosa.[7] The expression of antimicrobial peptides, the activities of alternative and classical pathways of the complement system, and cytokine regulation of the production of inflammatory mediators also contribute to the elimination of *E. histolytica*. Use of animal models of amebic colitis and amebic liver abscess (ALA) has also shown that innate and specific immunity play important roles in protecting tissues against *E. histolytica* invasion.

In this chapter, we present an overview regarding host immune responses against amebiasis, and highlight the potential of secondary metabolites from traditional plants as a promising and natural strategy to combat *E. histolytica* with limited side effects. Furthermore, we examine specific molecules with immunogenic properties that can be applied as vaccine candidates by different routes of administration (e.g., oral, intraperitoneal, and intrahepatic) to prevent infections in the bowel and the liver of susceptible animals before they are introduced into humans.

33.2 GENERAL FEATURES

33.2.1 Lifecycle and Morphology

The lifecycle of *E. histolytica* consist of two stages, the cyst (infective form) and the trophozoite (invasive stage). The cyst enters the host through contaminated food and water, survives several host stress conditions such as the acidic pH of the stomach (pH 3.0), and then travels to the digestive tract. After excystation in the ileum-cecal portion, the trophozoite migrates and multiplies in the large intestine (colon), where *E. histolytica* may produce amebic colitis or a more severe disease called amebic dysentery. In addition, trophozoite can differentiate into cyst, which is discharged through feces for reinitiating a new lifecycle.[3] Unfortunately, the complete lifecycle and the molecular mechanisms that trigger encystation and excystation have not been reproduced in an *in vitro* system.

E. histolytica trophozoites are typically pleomorphic and produce cytoplasmic projections forming pseudopods, which participate in the directional motility.[5] The morphology and size of the nucleus provide valuable parameters for diagnosis in clinical laboratories.[8] Usually, the cytoplasm contains several vacuoles and vesicles that can be observed by light microscopy (Figure 33.1A).[9] *E. histolytica* trophozoites have the ability to phagocytize bacteria, erythrocytes, and epithelial cells, on which the trophozoites survive and proliferate.[10,11]

E. histolytica cysts are round or oval and usually appear hyaline with a refractive wall. Each cyst contains four nuclei in the cytoplasm, glycogen, and chromatoid bodies.[3] Ultrastructural analysis shows that the cyst wall is 125–150 nm thick and includes five to seven layers of fibers of β-1,4-linked N-acetylglucosamine (GlcNAc) (chitin).[12]

33.2.2 Pathogenic Mechanisms

During the invasion process, *E. histolytica* interacts with different components of the immune system and utilizes diverse strategies to survive, colonize, and multiply in host tissues. Some of the most important pathogenic mechanisms include cap formation, adherence, protease activity, and phagocytosis. Cap formation was observed in an *in vivo* susceptible model (hamster), and the authors demonstrated that *E. histolytica* trophozoites are surrounded by inflammatory cells, at which point the ameba is capable of reorganizing its cytoplasmic membrane to eliminate antibodies and immune cells (Figure 33.1B).[13] In contrast, *in vitro* assays have shown that nonpathogenic *E. dispar* trophozoites do not form well-defined caps under concanavalin A treatment compared with pathogenic *E. histolytica*.[14]

Adhesion is a crucial step of colonization and invasion by pathogens. In *E. histolytica*, galactose/*N*-acetyl-galactosamine (Gal/GalNAc) lectin is the main glycoprotein involved in the adhesion to target cells (erythrocytes, epithelial cells, and others).[15] This protein is a 260-kDa heterodimer of disulfide-linked surface proteins that contains heavy and light subunits.[16] The role of this lectin was demonstrated using GalNAc and galactose sugars to inhibit the adherence and cytolysis of host cells. Moreover, incubation with a monoclonal antibody against the 170-kDa lectin subunit reduces adherence to mucin oligosaccharides in human and hamster hepatocytes.[17–19] In addition, the ameba presents a complex membrane protein called EhADHCP112, which is formed by two polypeptides. One is the 75-kDa EhADH112 peptide; this adhesin is involved in adherence to target cells and phagocytosis. The other, a 49-kDa polypeptide,

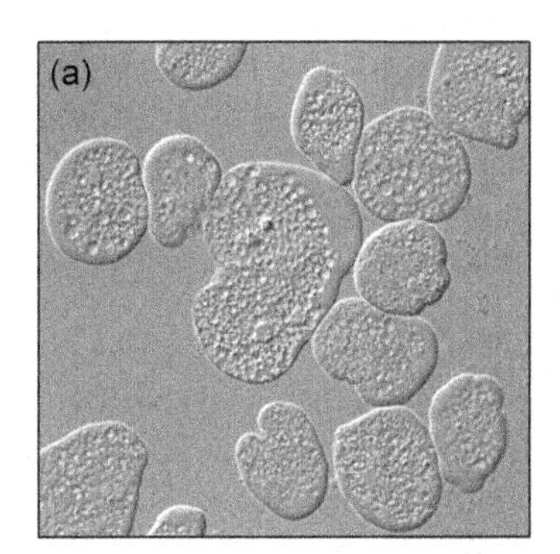

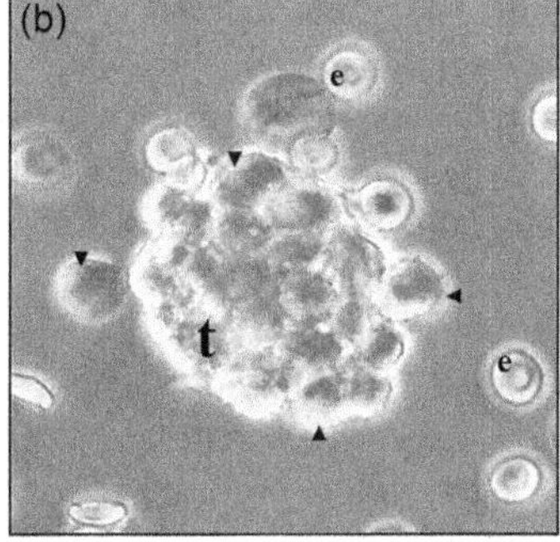

FIGURE 33.1 (a) Axenic culture of *E. histolytica* trophozoites. Normasky Optic. 40×. (b) Trophozoite (t) surrounded by inflammatory cells (arrow-heads), erythrocytes (e). Phase contrast microscopy. 40×.

is a cysteine protease (EhCP112) that participates in the alteration of cell junctions and the destruction of epithelial cells.[20,21] *E. histolytica* shows cysteine protease (CP) expression levels approximately 1,000 times greater than *E. dispar.*[22] In addition to involvement in the cleavage of extracellular matrix proteins such as collagen, elastin, fibronectin, and laminin,[23,24] CPs participate in the degradation of IgG and IgA, complement proteins (C3a and C5a), and cytokines.[25,26] In intestinal models of amebiasis, the most important CPs are EhCP1, EhCP4, and EhCP5. The inhibition of EhCP1 causes an important reduction of the number of trophozoites in human colon xenografts (approximately 90%).[27] Furthermore, in the mouse cecal model (strain C3H/HeJ), the inhibition of EhCP4 significantly reduces parasite burden (80%) and inflammation of the cecum.[28] In an *ex vivo* human intestinal model, the silencing of *ehcp5* prevented the ameba from causing intestinal damage.[29] *E. histolytica* is also capable of inducing apoptosis in cecal epithelial cells, hepatocytes, inflammatory cells, and erythrocytes.[30–32] These cells, which express apoptotic markers, are preferentially ingested by *E. histolytica* trophozoites.[33] Recently, trogocytosis has been described in *E. histolytica*, and trophozoites can kill human living cells by ingesting Caco-2, Jurkat, and red blood cells.[34]

Thus, *E. histolytica* has the ability to evade the immune system and produces diverse virulence factors that cause tissue damage.

33.3 METHODOLOGIES

33.3.1 Experimental Amebiasis (Intestine and Liver): Immunological Techniques

The classical methods for the diagnosis of *E. histolytica* (cysts and trophozoites) from fresh stool samples include concentration (Ritchie's formaldehyde ethyl acetate method), flotation (zinc sulfate method), and staining with Lugol's solution.[35,36] However, it is important to mention that, while these methods provide morphological identification, they do not differentiate between pathogenic and nonpathogenic amebas (e.g., *E. dispar* and *E. moshkovskii*) in human feces.[37,38] In recent years, immunological techniques have been used for accurate identification of *E. histolytica* in both clinical and experimental samples.[35,39,40] Based on antibodies against specific parasite molecules, immunological techniques enable highly sensitive and precise detection of *E. histolytica*. The most widely used antibody-based techniques are enzyme-linked immunosorbent assay (ELISA), western blot (Wb), immunofluorescence assays (IFA), and immunohistochemistry (IHC).[40–44]

In clinical diagnosis, the most extensively validated protein for ELISA detection of *E. histolytica*-specific antibodies is the Gal/GalNAc lectin (170 kDa subunit).[16] This lectin can be found in stool samples of asymptomatic and symptomatic patients using the ELISA method, with a sensitivity of 80%–94%. In ALA patients, the sensitivity is 96% in serum and liver abscess aspirates.[45,46] Experimental models provide important information regarding host responses during intestinal and liver amebiasis. Histopathological techniques such as hematoxylin and eosin (H&E) staining and IHC use specific antibodies against the ameba or purified molecules.[17,47]

IHC assays have been used to detect the Gal/GalNAc lectin in ALA in hamsters. This lectin is released by *E. histolytica* during invasion and can be localized in the liver parenchyma, sinusoidal space, and in the inflammatory cells, where cellular damage is evident.[44] In recent years, direct and indirect IFA using antibodies coupled to fluorochromes has been employed to identify *E. histolytica* Gal/GalNAc lectin, lipophosphoglycans (LPGs), a cysteine-rich 29 kDa amebic protein, a serine-rich *E. histolytica* protein (SHREP), amebapores, and CPs (EhCP2 and EhCP5).[16,48–53] Additionally, Wb technique based on horseradish peroxidase (HRP)-coupled antibodies for detection of specific amebic proteins helps to confirm the results of IFA assays, thus enhancing the diagnosis of *E. histolytica*.[16,48,54,55]

PCR-based techniques (e.g., conventional PCR, quantitative PCR, or qRT-PCR) offer superior sensitivity and specificity for the diagnosis of *E. histolytica* infection.[56] Several genes including the 18S rRNA (small subunit rRNA) from *E. histolytica*, cysteine proteinases (*ehcp2*, *ehcp5*), serine-rich *E. histolytica* gene (*srehp*), and *gal/galnac* lectin have been well studied.[57–59] These virulence factors have been evaluated in experimental ALA by qRT-PCR. It was noted that the genes were overexpressed after *in vivo* interactions in a time-dependent manner. However, lectin showed greater overexpression in this context than in the trophozoites from axenic culture.[59]

Thus, the antibody–antigen-based techniques and molecular methodologies provide excellent detection of amebic proteins.

33.4 RECENT ADVANCES IN MOLECULAR MECHANISMS

Several *in vivo* and *in vitro* models that mimic *E. histolytica* infection in the intestine and liver have allowed us to determine the molecular mechanisms of the complex interplay between the ameba and its host.[4,60] During the first stages of infection, the innate response against *E. histolytica* is orchestrated mainly by several barriers, including physical and biochemical factors (enterocytes, complement proteins, antimicrobial peptides, lactoferrin, among others). Inflammatory cells such as neutrophils and macrophages are also involved.[4,61,62] Once the infection is established, it is controlled and eliminated. However, if the innate responses are insufficient, specific immunity is responsible for the elimination of this pathogen through the production of cytokines and immunoglobulins, as well as by Th1 and Th2 responses.[4,63] Through evolution, *E. histolytica* has developed strategies to evade the immune responses of the host. The study of these immune molecular mechanisms has provided valuable information that could be used to enhance the immune response to eliminate *E. histolytica.*

33.4.1 Host Response during Intestinal Invasion

The intestinal mucosa is responsible for absorption; the gastrointestinal mucosa must tightly regulate immune responses to allow commensal flora to flourish while fighting potential pathogens. It is known that *E. histolytica* can exist as a commensal (90% of asymptomatic patients).[5] However, through mechanisms that are not fully understood, the ameba can change to a pathogenic phenotype. It has been reported that the host microbiota shows a direct correlation with the overexpression of virulence factors such as Gal/GalNAc lectin, amebapore protein, and proteases.[64,65] Moreover, the microbiota composition has been associated with higher frequency of diarrhea caused by *E. histolytica*.[4,66,67] In the 1960s, it was reported that germ-free guinea pigs infected intracecally with trophozoites did not develop intestinal infections; however, the infection was induced when a single bacteria was present.[68] Interestingly, some studies have reported the ability of the enteric microbiota to suppress the inflammatory responses; it is possible that this immune suppression allows the transition from nonpathogenic to pathogenic *E. histolytica*.[69] Nonetheless, it is necessary to perform more studies to elucidate the complex intestinal microenvironment during the interaction of the ameba microbiota and the host immune responses. Another important barrier in the human intestine is mucus. This natural barrier is a polysaccharide network that contains a broad spectrum of antimicrobial proteins such as mucins (mainly MUC5AC and MUC2), complement proteins, defensins, lactoferrin, and immunoglobulins (sIgA, IgG and IgM); all of these components of the human immune response provide protection against pathogen invasion in the intestine.[70]

However, the ameba can display different strategies that allow it to establish an intestinal infection. In *in vitro* and *ex vivo* studies, it has been reported that *E. histolytica* induces the production of mucus (Figure 33.2); moreover, trophozoites can induce the rearrangement of mucus by the production of β-N-acetyl-D-glucosaminidase, β-N-acetyl-D-galactosaminidase, β-D-galactosidase, and cysteine proteases

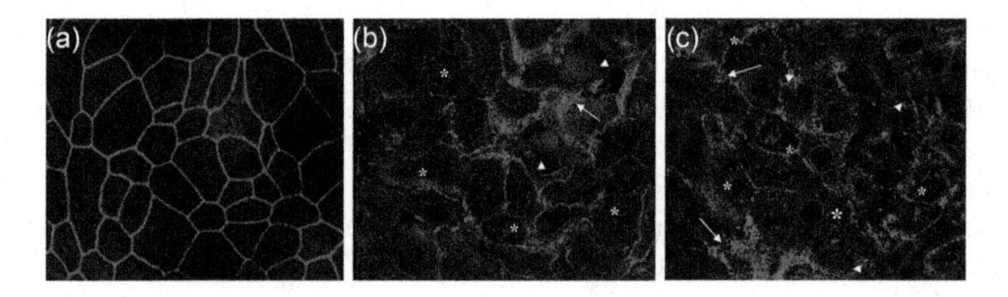

FIGURE 33.2 Production of MUC5AC by *E. histolytica*. Confocal microscopy. (a) Mucoepithelial cells without *E. histolytica* trophozoites (control), actin cytoskeleton was stained with phalloidin rhodamine. (b and c) Amebas (arrow-heads) induced the production of mucin MUC5AC (asterisks) in the mucoepithelial cells at 5 and 15 min of coincubation. Cytoskeleton alterations (stress fibers) were observed at 15 min (arrows). 63×.

(EhCP2 and EhCP5). These proteins cause mucin degradation, and, consequently, disrupt the rheological functions of the mucus.[71,72]

Additionally, it was reported that complement proteins induce the lysis of *E. dispar*; however, *E. histolytica* showed resistance to lysis by the complement system. The authors suggest that resistance to these proteins is due to Gal/GalNAc lectin in the pathogenic ameba, which has antigenic cross-reactivity with CD59 (a natural membrane inhibitor of C5b-9 present in human blood cells).[73,74] Additionally, the resistance to complement and anaphylatoxins (C3a and C5a) has been associated with the secretion of N-acetylglucosaminidase and proteases.[25,75]

In vitro studies revealed that pathogen-associated molecular patterns such as the amebic Gal/GalNAc lectin can induce the production of human β-defensin 2 (HBD2) in human colonic Caco-2 cells via TLR-2 and TLR-4. Moreover, the antimicrobial peptides produced by the enterocytes show the ability to damage amebas (50%–60%); the authors suggest that HDB2 has a possible role in the elimination of *E. histolytica* trophozoites.[76] Another antimicrobial protein is apo-lactoferrin, which is a natural iron chelator. Interestingly, *in vitro* studies have shown that apo-lactoferrin (1 mg ml^{-1}) has significant amebicidal activity (40%) at 3 h postincubation.[77] Furthermore, oral administration of apo-lactoferrin (20 mg $(kg\ week)^{-1}$) could eliminate infection in 63% of C3H/HeJ mice.[78] Nevertheless, more studies are needed to demonstrate the potential effects of these antimicrobial proteins.

In *E. histolytica*, it has been reported that proteases play a crucial role during the invasion and degradation of host tissues. In particular, amebic EhCP5 can block IgG and cleave human sIgA.[79,80] When trophozoites escape the immune system, they can attach to the intestinal epithelium. The Gal/GalNAc lectin mediates adherence to the colon surface. Additionally, the recognition of this lectin is mediated by Toll-like receptors (TLR-2 and TLR-4), which promote the production of proinflammatory cytokines (TNF-α, IL-8 and IL-1β) in human colonic cells (Caco-2).[29,81] Furthermore, vaccination with Gal/GalNAc lectin prevents cecal infection in a C3H/HeJ mouse model.[82] In a similar manner, the L220 kDa are involved in extracellular matrix adherence. Recently, it was demonstrated that EhCP112 and prostaglandin E2 from *E. histolytica* participate in the degradation of tight junction proteins, which increases the permeability of the epithelium and facilitates invasion. [23,24,53,54,83] In response to the tissue damage, the host initiates a robust inflammatory response; in a SCID mouse–human intestinal xenograft model, the presence of cytokines (IL-8 and IL-1β) was observed, and neutrophils, macrophages, plasma cells, lymphocytes, and eosinophils were present in the lamina propria.[63] However, it is important to mention that this exacerbated host response causes necrosis of the lamina propria and induces deep flask-shaped ulcers in the colonic mucosa; the ulcers also damage the muscularis and serosa layers (Figure 33.3).[63,84] In amebic dysentery, the adaptive immune response plays a crucial role in resolving or maintaining invasive intestinal infections. Human clinical cases and *in vitro* and *in vivo* studies have shown that the Th1 response participates in the elimination of trophozoites from the intestine; in fact, it was reported that INF-γ production reduces susceptibility to secondary episodes of dysentery and diarrhea in children.[85] Moreover, in *in vitro* studies, the activation of macrophages and neutrophils by INF-γ resulted in an effective oxidative burst and

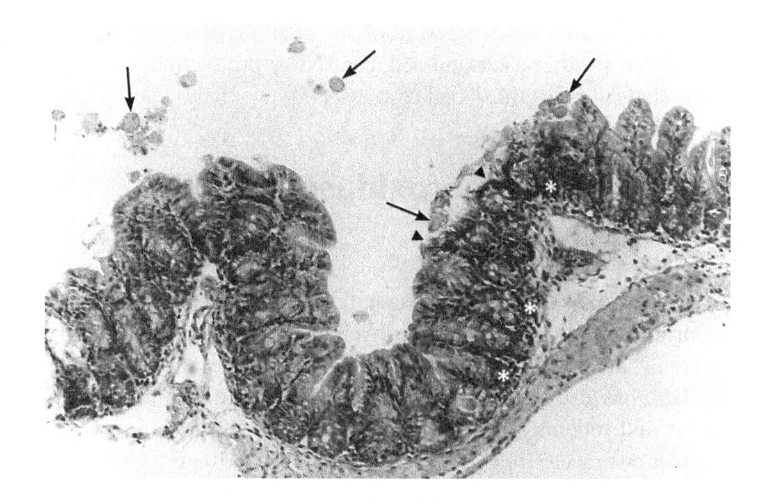

FIGURE 33.3 Intestinal lesions in the mouse model (H&E stain). C57BL/6 mouse was inoculated by intracecal route with *E. histolytica* trophozoites. At 3 h postinoculation, the cecum presents areas of ulceration (arrow heads), inflammatory reaction (asterisks), and numerous trophozoites (arrows). 20×.

production of nitric oxide (NO), which is critical for killing the ameba.[86,87] In most cases, intestinal invasive amebiasis is resolved, but in some cases, the trophozoites can migrate by portal circulation, causing ALA; the conditions that permit this migration are not well understood.

33.4.2 Host Immune Response in Amebic Liver Abscess (ALA)

During migration to the liver, the trophozoites are not recognized by the host immune response. This ability of the ameba has been associated with the presence of membrane molecules such as Gal/GalNAc, lipophosphoglycan, lipopeptidophosphoglycan (LPPG), and β1-integrin-like molecule (β1EhFNR). These molecules confer resistance to complement attack and blood cell recognition.[73,88,89]

In the liver, the cytokines and immune resident cells, which include NKT cells, Kupffer cells, and T cells, confer immunological tolerance and discrimination between harmless and pathogenic microorganisms.[90] During the early stages of liver invasion in C57BL/6 mice, it was observed that several trophozoites are recognized and eliminated by iNKT cells. The importance of iNKT cells in the clearance of *E. histolytica* was demonstrated using Jα18 chain knockout mice (which are unable to produce functional iNKT cells). These mice produce larger abscesses than control animals. Moreover, it was reported that LPPG from *E. histolytica* is recognized directly by iNKT cells through TLR-2, TLR-4, and CD1d, promoting the production of INF-γ.[91,92] Furthermore, it has been reported in hamsters that the establishment and survival of the ameba depends on CP activity.[93] The surviving trophozoites are surrounded by neutrophils and macrophages to eliminate the amebas; however, these cells exacerbate the immune response, and the secretory components of *E. histolytica* promote liver damage.[94] Recently, it was demonstrated that the evolution of ALA can be modulated by neutrophils in a murine model. Neutrophils from BALB/c mice are more efficient to reduce ALA and diminish the number of *E. histolytica* trophozoites compared with neutrophils from hamsters, which cannot efficiently eliminate the ameba. These differences are related to the level of myeloperoxidase activity. Clearly, more studies are needed to evaluate the mechanisms involved in the activation and modulation of the neutrophil response.[95]

Interestingly, PGE_2 from *E. histolytica* can regulate the immune response by suppressing macrophages and T cells. In BALB/c mice, it was reported that PGE_2 inhibits the correct expression of MHC II in macrophages; additionally, PGE_2 causes downregulation of inducible nitric oxide synthase (iNOS), nitric oxide (NO), and TNF-α production.[63,96] In human ALA, the total number of T cells was normal, but there were more $CD8^+$ cells than $CD4^+$ cells.[97] In addition, a coculture of T cells with sera from recently cured ALA patients inhibits cell proliferation and reduces INF-γ production; these results suggest the presence of amebic proteins that suppress the T cell response.[97]

33.4.3 Cytokines and Myeloperoxidase System (Intestine and Liver)

Cytokines play a pivotal role in both the innate and specific immune responses. Research on amebiasis has demonstrated the importance of cytokines and the myeloperoxidase system.

In the bowel, the epithelial cells produce the cytokine IL-25, which is an important inducer of antimicrobial peptides. This mediator has also been reported to participate in the regulation of TNF-α. In a study of amebic colitis in humans and rodents, IL-25 was suppressed during infection with *E. histolytica* in both humans and mice. When mice were treated by the intraperitoneal route with recombinant IL-25, the animals were protected against amebic colitis. Therefore, IL-25 negatively regulates TNF-α, inhibiting inflammation and limiting the severity of the disease.[98] A study of human fecal samples showed the role of cytokines in symptomatic and asymptomatic patients with amebiasis; the results revealed higher expression of IL-10 and TGF-β in the symptomatic group than in the asymptomatic and normal healthy subjects. Thus, symptomatic amebiasis presented a suppressive immune response caused by *E. histolytica*.[99]

Regarding ALA, several studies have shown that neutrophils and macrophages activated with IFN-γ and TNF-α can kill amebas. *In vivo* studies using C.B-17 SCID mice (which have altered IFN-γ receptor) inoculated with *E. histolytica* trophozoites showed an increase in ALA size, suggesting that IFN-γ plays an important role in preventing liver tissue damage.[100] In protection studies using gerbils that were previously immunized intramuscularly with Gal/GalNAc lectin and posteriorly challenged with virulent amebas, the authors found increased IL-12 and IFN-γ production, which protected the animals against developing abscesses.[101] However, in an *in situ* study using a susceptible ALA model, the authors demonstrated the presence and expression of proinflammatory and regulatory cytokines (TNF-α, IFN-γ, IL-1β, IL-8, and IL-10); the proinflammatory mediators were overexpressed during ALA evolution in the hamster. In contrast, IL-10 was downregulated during ALA development. Thus, regulatory cytokines do not exert their function, and therefore, proinflammatory mediators contribute to liver damage.[102] The important role of regulatory cytokines has been recently studied in mouse models. Noll and colleagues (2016) reported that the IL-23/IL-17 axis plays an important role in the pathogenesis of ALA. Additionally, they evaluated the role of IL-13 in disease outcome; the results suggested that the IL-23/IL-17 axis plays a critical role in the immunopathology of ALA. Interestingly, IL-13, which is secreted by monocytes, participated in the resolution of liver tissue damage.[103]

Promising advances have been made in understanding the complex intercellular signals mediated by cytokines that regulate the switch of the Th1 and Th2 responses. However, it is necessary to explore in more detail their participation in the elimination of *E. histolytica* trophozoites.

It is known that the inflammatory reaction mounted by the host against the ameba is mainly mediated by neutrophils and macrophages. Neutrophils, which are leucocytes of the innate immune responses, are the first cells to contact *E. histolytica* at the lesion site. The neutrophils present a cationic enzyme, myeloperoxidase (MPO), which releases hypochlorous acid (HOCl), a highly cytotoxic molecule.[104] HOCl is the most efficient microbicide system in neutrophils.[105] It has been reported that MPO isolated from peritoneal exudates of hamster shows significant amebicidal activity. This MPO binds to the surface of *E. histolytica* trophozoites to produce morphological and ultrastructural changes in the ameba, eventually leading to parasite death.[106] More recently, *in vivo* studies have shown the important role of MPO in the resolution of ALA in mice. The presence, expression, and activity of MPO were analyzed in the livers of BALB/c mice inoculated with *E. histolytica*. The presence of MPO was observed during ALA evolution. The expression of *mpo* increased significantly at different times in hepatic lesions. MPO enzymatic activity also increased in a time-dependent manner. Finally, mice treated with MPO inhibitor developed larger abscesses than untreated mice. These results show that MPO can participate in the elimination of *E. histolytica*.[95] Further studies are needed to explore the detailed mechanisms of the activation and inhibition of MPO in both models (susceptible and resistant animals). This knowledge will allow exploring novel strategies to stimulate neutrophils to eliminate the ameba.

33.5 FUTURE PERSPECTIVES

33.5.1 Novel Perspectives of Alternative Treatments in Amebiasis

33.5.1.1 Antiamebic Activity of Medicinal Plants

The scientific community has taken strong interest in the search for alternative treatments for infectious diseases. Medicinal plants present compounds that could be used as drug precursors and drug prototypes to avoid or cure diseases.[107,108] Amebic infection is treated with metronidazole (MTZ), but this drug has undesirable side effects; therefore, new safe drugs are needed. Natural products have been shown to be effective in the development of new antiparasitic drugs. In the case of amebiasis, several plants have been tested. *Codiaeum variegatum*, a traditional plant used for the treatment of bloody diarrhea, was used as antiamebic treatment. Mfotie et al. examined aqueous extract from leaves and various methanolic fractions against *E. histolytica* from axenic culture. The authors reported significant antiamebic activity, accompanied by morphological changes such as cell membrane disorganization and lysis of trophozoites. Using differential gene expression analysis, the authors characterized the mechanisms of action of *C. variegatum*; they reported a disturbance in ceramide biosynthesis, which is involved in biochemical processes in the cell membrane including differentiation, cell growth arrest, and apoptosis.[109]

Velázquez-Domínguez and colleagues reported that the herbaceous medicinal plant *Decachaeta incompta* expresses the sesquiterpenlactone incomptine A. Proteomic and mass spectrometry experiments showed that this molecule has an effect on amebic metabolism. Three glycolytic enzymes were downregulated after *E. histolytica* treatment with incomptine A: enolase, pyruvate:ferredoxin oxidoreductase, and fructose 1,6-biphosphate aldolase. The treatment of trophozoites with incomptine A also induced important ultrastructural changes: an increase in the number of glycogen deposits, a reduction in the number of vacuoles, and the rearrangement of chromatin. Thus, incomptine A affects ameba growth by altering glucose metabolism.[110]

The effect of an ethanolic extract from the stem of *Bursera fagoroides* (a flowering plant) was evaluated. *B. fagoroides* extract was added to ameba cultures; the results showed that the ethanolic extracts inhibited the growth of *E. histolytica*, affected amebic size, formed aggregates, and decreased amebic mobility. The activity of *B. fagoroides* extract is exerted through the inhibition of the activity of ornithine decarboxylase, a key enzyme in the regulation of polyamine biosynthesis, which is related to cell growth and proliferation.[111]

The antiamebic effect of *Thymus vulgaris* (thyme; a small, shrubby plant) was investigated. Coincubation of *E. histolytica* trophozoites with *T. vulgaris* extracts and their essential oils caused growth inhibition. The minimal inhibitory concentration for the *T. vulgaris* hydroalcoholic fraction and the essential oil after 24 h of coculture was 4 mg ml^{-1} and 0.7 mg ml^{-1}, respectively. *T. vulgaris* essential oil was the most active chemical compound against *E. histolytica*. The activity of thyme oil is related to its content of thymol, carvacrol, tannins, triterpenic acids, and flavonoids, which are the most abundant compounds.[112]

Herrera-Martínez and colleagues reported the antiamebic activity of an ethyl acetate extract from the root of *Adenophyllum aurantium*. The ethyl acetate extract inhibited adhesion to fibronectin and erythrophagocytosis, and it also affected the actin cytoskeleton structure and prevented the formation of contractile rings. Ultrastructural analysis of trophozoites treated with ethyl acetate extract showed a decrease in the number of vacuoles. In *in vivo* studies using a hamster model, intraperitoneal administration of ethyl acetate extract (2.5 or 5 mg) and challenge with *E. histolytica* trophozoites inhibited the development of ALA in 48.5% and 89.0% of cases, respectively. Thus, *A. aurantium* ethyl acetate extract could be effective for the treatment of amebiasis.[113]

Recently, Bashyal et al. analyzed the antiamebic activity of *Larrea tridentate* (creosote bush), using an ATP bioluminescence-based assay. The results demonstrated that secondary metabolites from *L. tridentata* display dose-response antiparasitic activity. The secondary metabolites from the creosote bush represent natural products that can used as an effective therapy against *E. histolytica*. It appears that natural secondary metabolites such as lignan have broad, clinically relevant biological activities, including antioxidant, antiviral, antibacterial, immunosuppressive, anti-inflammatory, and anticancer properties.[114] Therefore, lignan from the creosote bush represents a class of natural products that can be optimized through medicinal chemistry to produce more effective therapeutic options for amebiasis.

33.5.1.2 Lactoferrin as Antiamebic Agent

Lactoferrin (Lf) is an iron-binding glycoprotein of the mammalian innate defense system. This protein is secreted by exocrine glands in the digestive, reproductive, and respiratory tracts. It is also a component of saliva and tears.[115,116] Lf is also secreted by neutrophils at sites of acute inflammation caused by pathogenic microorganisms. The secreted Lf is free of iron (apo-Lf).[116,117] In the inflammation area, Lf has two functions: it removes all of the free iron from the site of inflammation, so that microorganisms do not grow, and it eliminates microorganisms via microbicide domains that interact with components of the surfaces of microorganisms and damage them.[118,119] Bovine apo-Lf and its derivative peptide lactoferricin (bLfcin) were able to kill *E. histolytica* trophozoites in *in vitro* studies. When Lf or bLfcin was combined with MTZ, a synergistic effect against amebas was observed.[77] There was also an increased amebicidal effect when bovine apo-Lf was combined with human milk, sIgA, or lysozyme.[77] Another study tested the microbicidal effects of peptides derived from Lf such as Lfcin 17-30 aa, Lactoferrampin 265-284 aa, and a Lactoferrin chimera (combined Lfcin and Lactoferrampin) on *E. histolytica*. The Lactoferrin chimera resulted in 75% cell death.[120] The effect of b-Lf administered orally in a model of amebic infection in the cecum using C3H/HeJ mice was also analyzed. The results showed that 20 mg of Lf per kg, administered orally for a week, can eliminate infection in 63% of animals; in the remaining 37% of the animals, epithelial tissue damage was reduced.[78] Finally, b-Lf was used alone or in combination with MTZ in hamsters with ALA, and both treatments decreased the ALA.[121] Therefore, Apo-Lf or its derivatives could be used alone or in combination with other antiamebic agents as therapies against amebic colitis and ALA. These proteins of the innate immune response could act as immune regulators in amebiasis; they may work to prevent the inflammatory process in the intestine during ALA development.

33.5.2 Antigen Candidates for Vaccine Development

Parasitic diseases are among the main health problems in developing countries. *E. histolytica* is the protozoan parasite of the human being capable of causing amebic colitis and ALA. In developed countries, it has been possible to prevent the presence of this disease in large part due to the implementation of hygiene and health measures. However, in developing countries, where a large percentage of the population is living in poverty with food deficiency and poor hygiene and health conditions, the disease is still present. In some countries, such as Mexico, amebiasis is endemic; in a study using antibodies against *E. histolytica*, it was shown that amebiasis is present in 8.4% of the population.[122] In a seroepidemiological study conducted in the state of Fortaleza in Brazil, it was observed that approximately 25% of the people studied had antiamebic antibodies.[123] Thus, although there are effective drugs against *E. histolytica*, it has not been possible to reduce the incidence of amebiasis in the marginalized areas of these countries. Several researchers from different parts of the world have proposed vaccination against this parasite because epidemiological studies suggest the presence of protective acquired immunity in humans.[124,125] Unlike other parasites, *E. histolytica* has a simple lifecycle, which makes it a good candidate for vaccine-based approaches. In studies using animal models such as hamster, guinea pig, or mouse, different amebic proteins with protective immunogenic properties have been found, the most relevant of which are the Gal/GalNAc, amebapore, 25-kDa serine-rich protein, and 29 kDa *E. histolytica* antigens.

By far, the antigen of *E. histolytica* that has been most studied as a vaccine candidate is the Gal/GalNac lectin. The heavy chain of the lectin contains an immunogenic region composed of a cysteine-rich carbohydrate recognition domain. Monoclonal antibodies directed against this region can inhibit amebic adherence to target cells.[19] The prevention of intestinal amebiasis by vaccination with purified lectin and with the lectin carbohydrate recognition domain has also been demonstrated in a C3H/HeJ mouse model.[84] Gal/GalNac lectin has been used with the adjuvant cytosine guanine oligodeoxynucleotide (CpG oligodeoxynucleotide) to increase Th1 cytokine profile in macrophages and spleen *in vitro*, where it was able to completely protect against the development of ALA in gerbils.[101]

Some researchers have worked with G3, a genetically modified virulence-attenuated *E. histolytica* HM1-IMSS strain that lacks the amebapore proteins. Gene silencing is a stable and persistent event, and this ameba was found to be nonvirulent. Intraperitoneal immunization of hamsters with G3 trophozoites resulted in IgG anti-HM1-IMSS antibodies.[126]

The 25-kDa serine-rich *E. histolytica* protein (SREHP) is another antigen that localizes to the ameba surface by a mechanism that is not well defined, and its exact function is also not known. However, it acts as a chemoattractant for *E. histolytica* trophozoites,[127] and mediates the binding of amebas to the cell target.[128] A recombinant SREHP/MBP fusion protein was an effective vaccine in preventing ALA in gerbils.[128,129]

Several immunogenic molecules from *E. histolytica* can be administered by different routes, such as oral, intraperitoneal, and intrahepatic. They are able to protect against amebiasis in the bowel and liver in susceptible animal models. Although much effort has been devoted to developing an antiamebic vaccine, there is still no strong scientific evidence that supports its effectiveness. This must be properly studied and validated before a vaccine is introduced in human beings.

33.6 CONCLUSIONS

Clinical and experimental investigations of the host–parasite interaction in amebiasis have improved our understanding regarding the molecular mechanisms of host immune responses against *E. histolytica*. *In vitro* and *in vivo* models have helped uncover the roles of both innate and adaptive responses in the pathogenesis of amebiasis. Nonetheless, further research is necessary for inexpensive diagnostic tests to become accessible to all global communities, and for effective therapies without undesirable side effects to be designed. Elucidation of host immune mechanisms and identification of novel amebic antigens should form part of this research effort.

ACKNOWLEDGMENTS

We thank MSc. Angélica Silva-Olivares for her valuable assistance in the amebic culture. This work was supported by grant number 237523 (Shibayama M) from CONACyT, Mexico.

REFERENCES

1. World Health Organization, (WHO). Amoebiasis. *Wkly Epidemiol Rec* 1997;72:97–9.
2. Ali IK, Clark CG, Petri WA, Jr. Molecular epidemiology of amebiasis. *Infect Genet Evol* 2008;8:698–707.
3. Stanley SL. Amoebiasis. *Lancet* 2003;361:1025–34.
4. Nakada-Tsukui K, Nozaki T. Immune response of amebiasis and immune evasion by *Entamoeba histolytica*. *Front Immunol* 2016;7:175.
5. Haque R, Huston CD, Hughes M, Houpt E, Petri WA, Jr. Amebiasis. *N Engl J Med* 2003;348:1565–73.
6. Stanley SL, Jr. Vaccines for amoebiasis: barriers and opportunities. *Parasitol* 2006;133 Suppl:S81–6.
7. Seydel KB, Li E, Swanson PE, Stanley SL, Jr. Human intestinal epithelial cells produce proinflammatory cytokines in response to infection in a SCID mouse-human intestinal xenograft model of amebiasis. *Infect Immun* 1997;65:1631–9.
8. Ravdin JI. Amebiasis now. *Am J Trop Med Hyg* 1989;41:40–8.
9. Martínez-Palomo A. The pathogenesis of amoebiasis. *Parasitol Today* 1987;3:111–8.
10. Talamás-Lara D, Chávez-Munguía B, González-Robles A, et al. Erythrophagocytosis in *Entamoeba histolytica* and *Entamoeba dispar*: a comparative study. *Biomed Res Int* 2014;2014:626259.
11. Boettner DR, Huston CD, Linford AS, et al. *Entamoeba histolytica* phagocytosis of human erythrocytes involves PATMK, a member of the transmembrane kinase family. *PLoS Pathog* 2008;4:e8.
12. Chávez-Munguía B, Martínez-Palomo A. High-resolution electron microscopical study of cyst walls of *Entamoeba* spp. *J Eukaryot Microbiol* 2011;58:480–6.
13. Shibayama M, Campos-Rodríguez R, Ramírez-Rosales A, et al. *Entamoeba histolytica*: liver invasion and abscess production by intraperitoneal inoculation of trophozoites in hamsters, *Mesocricetus auratus*. *Exp Parasitol* 1998;88:20–7.
14. Chávez-Munguía B, Talamás-Rohana P, Castanon G, Salazar-Villatoro L, Hernández-Ramírez V, Martínez-Palomo A. Differences in cap formation between invasive *Entamoeba histolytica* and non-invasive *Entamoeba dispar*. *Parasitol Res* 2012;111:215–21.
15. Ravdin JI, Guerrant RL. Role of adherence in cytopathogenic mechanisms of *Entamoeba histolytica*. Study with mammalian tissue culture cells and human erythrocytes. *J Clin Invest* 1981;68:1305–13.
16. Petri WA, Jr., Joyce MP, Broman J, Smith RD, Murphy CF, Ravdin JI. Recognition of the galactose- or N-acetylgalactosamine-binding lectin of *Entamoeba histolytica* by human immune sera. Infect Immun 1987;55:2327–31.
17. Pacheco J, Shibayama M, Campos R, et al. *In vitro* and *in vivo* interaction of *Entamoeba histolytica* Gal/GalNAc lectin with various target cells: an immunocytochemical analysis. *Parasitol Int* 2004;53:35–47.
18. Petri WA, Jr., Snodgrass TL, Jackson TF, et al. Monoclonal antibodies directed against the galactose-binding lectin of *Entamoeba histolytica* enhance adherence. *J Immunol* 1990;144:4803–9.
19. Chadee K, Petri WA, Jr., Innes DJ, Ravdin JI. Rat and human colonic mucins bind to and inhibit adherence lectin of *Entamoeba histolytica*. *J Clin Invest* 1987;80:1245–54.
20. Ocadíz R, Orozco E, Carrillo E, et al. EhCP112 is an *Entamoeba histolytica* secreted cysteine protease that may be involved in the parasite-virulence. *Cell Microbiol* 2005;7:221–32.
21. Rodríguez MA, Orozco E. Isolation and characterization of phagocytosis- and virulence-deficient mutants of *Entamoeba histolytica*. *J Infect Dis* 1986;154:27–32.
22. Reed SL, Keene WE, McKerrow JH. Thiol proteinase expression and pathogenicity of *Entamoeba histolytica*. *J Clin Microbiol* 1989;27:2772–7.
23. Que X, Reed SL. Cysteine proteinases and the pathogenesis of amebiasis. *Clin Microbiol Rev* 2000;13:196–206.
24. Serrano-Luna J, Piña-Vázquez C, Reyes-López M, Ortíz-Estrada G, de la Garza M. Proteases from *Entamoeba spp*. and pathogenic free-living amoebae as virulence factors. *J Trop Med* 2013;2013:890603.
25. Reed SL, Ember JA, Herdman DS, DiScipio RG, Hugli TE, Gigli I. The extracellular neutral cysteine proteinase of *Entamoeba histolytica* degrades anaphylatoxins C3a and C5a. *J Immunol* 1995;155:266–74.
26. Que X, Kim SH, Sajid M, et al. A surface amebic cysteine proteinase inactivates interleukin-18. *Infect Immun* 2003;71:1274–80.
27. Meléndez-López SG, Herdman S, Hirata K, et al. Use of recombinant *Entamoeba histolytica* cysteine proteinase I to identify a potent inhibitor of amebic invasion in a human colonic model. *Eukaryotic Cell* 2007;6:1130–6.

28. He C, Nora GP, Schneider EL, et al. A novel *Entamoeba histolytica* cysteine proteinase, EhCP4, is key for invasive amebiasis and a therapeutic target. *J Biol Chem* 2010;285:18516–27.
29. Bansal D, Ave P, Kerneis S, et al. An *ex vivo* human intestinal model to study *Entamoeba histolytica* pathogenesis. *PLoS Negl Trop Dis* 2009;3:e551.
30. Seydel KB, Li E, Zhang Z, Stanley SL, Jr. Epithelial cell-initiated inflammation plays a crucial role in early tissue damage in amebic infection of human intestine. *Gastroenterol* 1998;115:1446–53.
31. Ralston KS, Petri WA, Jr. Tissue destruction and invasion by *Entamoeba histolytica. Trends Parasitol* 2011;27:254–63.
32. Becker SM, Cho KN, Guo X, et al. Epithelial cell apoptosis facilitates *Entamoeba histolytica* infection in the gut. *Am J Pathol* 2010;176:1316–22.
33. Huston CD, Boettner DR, Miller-Sims V, Petri WA, Jr. Apoptotic killing and phagocytosis of host cells by the parasite *Entamoeba histolytica. Infect Immun* 2003;71:964–72.
34. Ralston KS. Taking a bite: Amoebic trogocytosis in *Entamoeba histolytica* and beyond. *Curr Opin Microbiol* 2015;28:26–35.
35. Parija SC, Mandal J, Ponnambath DK. Laboratory methods of identification of *Entamoeba histolytica* and its differentiation from look-alike *Entamoeba* spp. *Trop Parasitol* 2014;4:90–5.
36. Uslu H, Aktas O, Uyanik MH. Comparison of various methods in the diagnosis of *Entamoeba histolytica* in stool and serum specimens. *Eurasian J Med.* 2016 Jun;48(2):124–9.
37. Fotedar R, Stark D, Beebe N, Marriott D, Ellis J, Harkness J. PCR detection of *Entamoeba histolytica, Entamoeba dispar,* and *Entamoeba moshkovskii* in stool samples from Sydney, Australia. *J Clin Microbiol* 2007;45:1035–7.
38. Tanyuksel M, Yilmaz H, Ulukanligil M, et al. Comparison of two methods (microscopy and enzyme-linked immunosorbent assay) for the diagnosis of amebiasis. *Exp Parasitol* 2005;110:322–6.
39. Li E, Stanley SL, Jr. Protozoa. Amebiasis. *Gastroenterol Clin North Am* 1996;25:471–92.
40. Tanyuksel M, Petri WA, Jr. Laboratory diagnosis of amebiasis. *Clin Microbiol Rev* 2003;16:713–29.
41. Fotedar R, Stark D, Beebe N, Marriott D, Ellis J, Harkness J. Laboratory diagnostic techniques for *Entamoeba* species. *Clin Microbiol Rev* 2007;20:511–32.
42. Zhang YY, Li E, Jackson TFHG, Zhang TH, Gathiram V, Stanley SL. Use of a recombinant 170-Kilodalton surface-antigen of *Entamoeba-histolytica* for serodiagnosis of amebiasis and identification of immunodominant domains of the native molecule. *J Clin Microbiol* 1992;30:2788–92.
43. Ximénez C, Sosa O, Leyva O, Morán P, Melendro EI, Ramiro M. Western blot of *Entamoeba histolytica* antigenic fractions: reactivity analysis with sera from intestinal amoebiasis patients. *Ann Trop Med Parasitol* 1992;86:121–7.
44. Pacheco-Yépez J, Campos-Rodríguez R, Serrano-Luna J, et al. *Entamoeba histolytica*: localization of the Gal/GalNAc adherence lectin in experimental amebic liver abscess. *Arch Med Res* 2000;31:S242–4.
45. Haque R, Neville LM, Hahn P, Petri WA. Rapid diagnosis of *Entamoeba* infection by using *Entamoeba* and *Entamoeba histolytica* stool antigen-detection kits. *J Clin Microbiol* 1995;33:2558–61.
46. Haque R, Mollah NU, Ali IK, et al. Diagnosis of amebic liver abscess and intestinal infection with the TechLab *Entamoeba histolytica* II antigen detection and antibody tests. *J Clin Microbiol* 2000;38:3235–9.
47. Campos-Rodríguez R, Jarillo-Luna A, Ventura-Juarez J, et al. Interaction of antibodies with *Entamoeba histolytica* trophozoites from experimental amebic liver abscess: an immunocytochemical study. *Parasitol Res* 2000;86:603–7.
48. Chaudhry OA, Petri WA, Jr. Vaccine prospects for amebiasis. *Expert Review of Vaccines* 2005;4:657–68.
49. Zhang T, Stanley SL, Jr. Expression of the serine rich *Entamoeba histolytica* protein (SREHP) in the avirulent vaccine strain *Salmonella typhi* TY2 chi 4297 (Δcya Δcrp Δasd): safety and immunogenicity in mice. *Vaccine* 1997;15:1319–22.
50. Reed SL, Flores BM, Batzer MA, et al. Molecular and cellular characterization of the 29-kilodalton peripheral membrane protein of *Entamoeba histolytica*: differentiation between pathogenic and nonpathogenic isolates. *Infect Immun* 1992;60:542–9.
51. Zhang Z, Duchene M, Stanley SL. A monoclonal antibody to the amebic lipophosphoglycan-proteophosphoglycan antigens can prevent disease in human intestinal xenografts infected with *Entamoeba histolytica. Infect Immun* 2002;70:5873–6.
52. Andra J, Berninghausen O, Leippe M. Membrane lipid composition protects *Entamoeba histolytica* from self-destruction by its pore-forming toxins. *FEBS Lett* 2004;564:109–15.
53. Bruchhaus I, Loftus BJ, Hall N, Tannich E. The intestinal protozoan parasite *Entamoeba histolytica* contains 20 cysteine protease genes, of which only a small subset is expressed during in vitro cultivation. *Eukaryot Cell* 2003;2:501–9.

54. Cuellar P, Hernández-Nava E, García-Rivera G, et al. *Entamoeba histolytica* EhCP112 dislocates and degrades claudin-1 and claudin-2 at tight junctions of the intestinal epithelium. *Front Cell Infect Microbiol* 2017;7:372.
55. López-Vancell R, Arreguín Espinosa R, González-Canto A, et al. *Entamoeba histolytica*: expression and localization of Gal/GalNAc lectin in virulent and non-virulent variants from HM1:IMSS strain. *Exp Parasitol* 2010;125:244–50.
56. Sanuki J, Asai T, Okuzawa E, Kobayashi S, Takeuchi T. Identification of *Entamoeba histolytica* and *E. dispar* cysts in stool by polymerase chain reaction. *Parasitol Res* 1997;83: 96–8.
57. Santos HL, Bandea R, Martins LA, et al. Differential identification of *Entamoeba* spp. based on the analysis of 18S rRNA. *Parasitol Res* 2010;106:883–8.
58. ElBakri A, Samie A, Ezzedine S, Odeh R. Genetic variability of the serine-rich *Entamoeba histolytica* protein gene in clinical isolates from the United Arab Emirates. *Trop Biomed* 2014;31:370–7.
59. Sánchez V, Serrano-Luna J, Ramírez-Moreno E, Tsutsumi V, Shibayama M. *Entamoeba histolytica*: Overexpression of the *gal/galnac* lectin, *ehcp2* and *ehcp5* genes in an *in vivo* model of amebiasis. *Parasitol Int* 2016;65:665–7.
60. Faust DM, Guillen N. Virulence and virulence factors in *Entamoeba histolytica*, the agent of human amoebiasis. *Microbes Infect* 2012;14:1428–41.
61. Campos-Rodríguez R, Gutiérrez-Meza M, Jarillo-Luna RA, et al. A review of the proposed role of neutrophils in rodent amebic liver abscess models. *Parasite* 2016;23:6.
62. Marie C, Petri WA, Jr. Regulation of virulence of *Entamoeba histolytica*. *Annu Rev Microbiol* 2014;68: 493–520.
63. Mortimer L, Chadee K. The immunopathogenesis of *Entamoeba histolytica*. *Exp Parasitol* 2010;126:366–80.
64. Galván-Moroyoqui JM, Domínguez-Robles MD, Franco E, Meza I. The interplay between *Entamoeba* and enteropathogenic bacterial modulates epithelial cell damage. *PLoS Neglec Trop Dis* 2008;2:e266.
65. Burgess SL, Petri WA, Jr. The intestinal bacterial microbiome and *E. histolytica* infection. *Curr Trop Med Rep* 2016;3:71–4.
66. Gilchrist CA, Petri SE, Schneider BN, et al. Role of the gut microbiota of children in diarrhea due to the protozoan parasite *Entamoeba histolytica*. *J Infect Dis* 2016;213:1579–85.
67. Morton ER, Lynch J, Froment A, et al. Variation in rural african gut microbiota is strongly correlated with colonization by *Entamoeba* and subsistence. *PLoS Genet* 2015;11:e1005658.
68. Phillips BP. Studies on the ameba-bacteria relationship in amebiasis. III. Induced amebic lesions in the germ-free guinea pig. *Am J Trop Med Hyg* 1964;13:391–5.
69. Belkaid Y, Hand TW. Role of the microbiota in immunity and inflammation. *Cell* 2014;157:121–41.
70. Thornton DJ, Sheehan JK. From mucins to mucus: toward a more coherent understanding of this essential barrier. *Proc Am Thorac Soc* 2004;1:54–61.
71. Moncada D, Keller K, Ankri S, Mirelman D, Chadee K. Antisense inhibition of *Entamoeba histolytica* cysteine proteases inhibits colonic mucus degradation. *Gastroenterology* 2006;130:721–30.
72. Moncada D, Keller K, Chadee K. *Entamoeba histolytica* secreted products degrade colonic mucin oligosaccharides. *Infect Immun* 2005;73:3790–3.
73. Ventura-Juárez J, Campos-Rodríguez R, Jarillo-Luna RA, et al. Trophozoites of *Entamoeba histolytica* express a CD59-like molecule in human colon. *Parasitol Res* 2009;104:821–6.
74. Reed SL, Curd JG, Gigli I, Gillin FD, Braude AI. Activation of complement by pathogenic and nonpathogenic *Entamoeba histolytica*. *J Immunol* 1986;136:2265–70.
75. Arias-Negrete S, Sabanero-López M, Villagómez-Castro JC. Biochemical analysis of *Entamoeba histolytica* HM1 strain resistant to complement lysis. *Arch Med Res* 1992;23:135–7.
76. Ayala-Sumuano JT, Téllez-López VM, Domínguez-Robles Mdel C, Shibayama-Salas M, Meza I. Toll-like receptor signaling activation by *Entamoeba histolytica* induces beta defensin 2 in human colonic epithelial cells: its possible role as an element of the innate immune response. *PLoS Negl Trop Dis* 2013;7:e2083.
77. León-Sicairos N, López-Soto F, Reyes-López M, Godínez-Vargas D, Ordaz-Pichardo C, de la Garza M. Amoebicidal activity of milk, apo-lactoferrin, sIgA and lysozyme. *Clin Med Res* 2006;4:106–13.
78. León-Sicairos N, Martínez-Pardo L, Sánchez-Hernández B, de la Garza M, Carrero JC. Oral lactoferrin treatment resolves amoebic intracecal infection in C3H/HeJ mice. *Biochem Cell Biol* 2012;90:435–41.
79. Tran VQ, Herdman DS, Torian BE, Reed SL. The neutral cysteine proteinase of *Entamoeba histolytica* degrades IgG and prevents its binding. *J Infect Dis* 1998;177:508–11.
80. García-Nieto RM, Rico-Mata R, Arias-Negrete S, Ávila EE. Degradation of human secretory IgA1 and IgA2 by *Entamoeba histolytica* surface-associated proteolytic activity. *Parasitol Int* 2008;57:417–23.

81. Galván-Moroyoqui JM, Del Carmen Domínguez-Robles M, Meza I. Pathogenic bacteria prime the induction of Toll-like receptor signalling in human colonic cells by the Gal/GalNAc lectin carbohydrate recognition domain of *Entamoeba histolytica*. *Int J Parasitol* 2011;41:1101–12.
82. Houpt E, Barroso L, Lockhart L, et al. Prevention of intestinal amebiasis by vaccination with the *Entamoeba histolytica* Gal/GalNac lectin. *Vaccine* 2004;22:611–7.
83. Lejeune M, Moreau F, Chadee K. Prostaglandin E2 produced by *Entamoeba histolytica* signals via EP4 receptor and alters claudin-4 to increase ion permeability of tight junctions. *Am J Pathol* 2011;179:807–18.
84. Ghosh PK, Mancilla R, Ortiz-Ortiz L. Intestinal amebiasis: histopathologic features in experimentally infected mice. *Arch Med Res* 1994;25:297–302.
85. Haque R, Mondal D, Shu J, et al. Correlation of interferon-gamma production by peripheral blood mononuclear cells with childhood malnutrition and susceptibility to amebiasis. *Am J Trop Med Hyg* 2007;76: 340–4.
86. Siman-Tov R, Ankri S. Nitric oxide inhibits cysteine proteinases and alcohol dehydrogenase 2 of *Entamoeba histolytica*. *Parasitol Res* 2003;89:146–9.
87. Lin JY, Seguin R, Keller K, Chadee K. Tumor necrosis factor alpha augments nitric oxide-dependent macrophage cytotoxicity against *Entamoeba histolytica* by enhanced expression of the nitric oxide synthase gene. *Infect Immun* 1994;62:1534–41.
88. Talamás-Rohana P, Hernández-Ramírez VI, Pérez-García JN, Ventura-Juárez J. *Entamoeba histolytica* contains a beta 1 integrin-like molecule similar to fibronectin receptors from eukaryotic cells. *J Eukaryot Microbiol* 1998;45:356–60.
89. Braga LL, Ninomiya H, McCoy JJ, et al. Inhibition of the complement membrane attack complex by the galactose-specific adhesion of *Entamoeba histolytica*. *J Clin Invest* 1992;90:1131–7.
90. Racanelli V, Rehermann B. The liver as an immunological organ. *Hepatology* 2006;43:54–62.
91. Lotter H, Jacobs T, Gaworski I, Tannich E. Sexual dimorphism in the control of amebic liver abscess in a mouse model of disease. *Infect Immun* 2006;74:118–24.
92. Lotter H, González-Roldan N, Lindner B, et al. Natural killer T cells activated by a lipopeptidophosphoglycan from *Entamoeba histolytica* are critically important to control amebic liver abscess. *PLoS Pathog* 2009;5:e1000434.
93. Olivos-García A, Tello E, Nequiz-Avendano M, et al. Cysteine proteinase activity is required for survival of the parasite in experimental acute amoebic liver abscesses in hamsters. *Parasitology* 2004;129:19–25.
94. Tsutsumi V, Martínez-Palomo A. Inflammatory reaction in experimental hepatic amebiasis. An ultrastructural study. *Am J Pathol* 1988;130:112–9.
95. Cruz-Baquero A, Cárdenas Jaramillo LM, Gutíerrez-Meza M, et al. Different behavior of myeloperoxidase in two rodent amoebic liver abscess models. *PLoS One* 2017;12:e0182480.
96. Denis M, Chadee K. *In vitro* and *in vivo* studies of macrophage functions in amebiasis. *Infect Immun* 1988;56:3126–31.
97. Salata RA, Martínez-Palomo A, Murray HW, et al. Patients treated for amebic liver abscess develop cell-mediated immune responses effective in vitro against *Entamoeba histolytica*. *J Immunol* 1986;136:2633–9.
98. Noor Z, Watanabe K, Abhyankar MM, et al. Role of eosinophils and tumor necrosis factor alpha in interleukin-25-mediated protection from amebic colitis. *MBio* 2017;8:1–10.
99. Bansal D, Sehgal R, Chawla Y, Malla N, Mahajan RC. Cytokine mRNA expressions in symptomatic vs. asymptomatic amoebiasis patients. *Parasite Immunol* 2005;27:37–43.
100. Seydel KB, Smith SJ, Stanley SL. Innate immunity to amebic liver abscess is dependent on gamma interferon and nitric oxide in a murine model of disease. *Infect Immun* 2000;68:400–2.
101. Ivory CP, Keller K, Chadee K. CpG-oligodeoxynucleotide is a potent adjuvant with an *Entamoeba histolytica* Gal-inhibitable lectin vaccine against amoebic liver abscess in gerbils. *Infect Immun* 2006;74:528–36.
102. Pacheco-Yépez J, Galván-Moroyoqui JM, Meza I, Tsutsumi V, Shibayama M. Expression of cytokines and their regulation during amoebic liver abscess development. *Parasite Immunol* 2011;33:56–64.
103. Noll J, Helk E, Fehling H, et al. IL-23 prevents IL-13-dependent tissue repair associated with Ly6C(lo) monocytes in *Entamoeba histolytica*-induced liver damage. *J Hepatol* 2016;64:1147–57.
104. Klebanoff SJ. Myeloperoxidase: friend and foe. *J Leukoc Biol* 2005;77:598–625.
105. Hampton MB, Kettle AJ, Winterbourn CC. Inside the neutrophil phagosome: oxidants, myeloperoxidase, and bacterial killing. *Blood* 1998;92:3007–17.
106. Pacheco-Yépez J, Rivera-Aguilar V, Barbosa-Cabrera E, Rojas Hernández S, Jarillo-Luna RA, Campos-Rodríguez R. Myeloperoxidase binds to and kills *Entamoeba histolytica* trophozoites. *Parasite Immunol* 2011;33:255–64.

107. Al-Jaber HI, Mosleh IM, Mallouh A, Abu Salim OM, Abu Zarga MH. Chemical constituents of *Osyris alba* and their antiparasitic activities. *J Asian Nat Prod Res* 2010;12:814–20.
108. Calixto JB. Efficacy, safety, quality control, marketing and regulatory guidelines for herbal medicines (phytotherapeutic agents). *Braz J Med Biol Res* 2000;33:179–89.
109. Mfotie Njoya E, Weber C, Hernandez-Cuevas NA, et al. Bioassay-guided fractionation of extracts from *Codiaeum variegatum* against *Entamoeba histolytica* discovers compounds that modify expression of ceramide biosynthesis related genes. *PLoS Negl Trop Dis* 2014;8:e2607.
110. Velázquez-Domínguez J, Marchat LA, López-Camarillo C, et al. Effect of the sesquiterpene lactone incomptine A in the energy metabolism of *Entamoeba histolytica. Exp Parasitol* 2013;135:503–10.
111. Rosas-Arreguín P, Arteaga-Nieto P, Reynoso-Orozco R, et al. *Bursera fagaroides*, effect of an ethanolic extract on ornithine decarboxylase (ODC) activity in vitro and on the growth of *Entamoeba histolytica. Exp Parasitol* 2008;119:398–402.
112. Behnia M, Haghighi A, Komeylizadeh H, Tabaei SJ, Abadi A. Inhibitory effects of Iranian *Thymus vulgaris* extracts on in vitro growth of *Entamoeba histolytica. Korean J Parasitol* 2008;46:153–6.
113. Herrera-Martínez M, Hernández-Ramírez VI, Hernández-Carlos B, Chávez-Munguía B, Calderón-Oropeza MA, Talamás-Rohana P. Antiamoebic activity of *Adenophyllum aurantium* (L.) strother and its effect on the actin cytoskeleton of *Entamoeba histolytica. Front Pharmacol* 2016;7:169.
114. Bashyal B, Li L, Bains T, Debnath A, LaBarbera DV. *Larrea tridentata*: A novel source for anti-parasitic agents active against *Entamoeba histolytica, Giardia lamblia* and *Naegleria fowleri. PLoS Negl Trop Dis* 2017;11:e0005832.
115. Brock JH. The physiology of lactoferrin. *Biochem Cell Biol* 2002;80:1–6.
116. Arnold RR, Russell JE, Champion WJ, Brewer M, Gauthier JJ. Bactericidal activity of human lactoferrin: differentiation from the stasis of iron deprivation. *Infect Immun* 1982;35:792–9.
117. Lonnerdal B, Iyer S. Lactoferrin: molecular structure and biological function. *Annu Rev Nutr* 1995;15:93–110.
118. Bolscher JGM, Adao R, Nazmi K, et al. Bactericidal activity of Lfchimera is stronger and less sensitive to ionic strength than its constituent lactoferricin and lactoferrampin peptides. *Biochimie* 2009;91:123–32.
119. Bellamy W, Takase M, Wakabayashi H, Kawase K, Tomita M. Antibacterial spectrum of lactoferricin B, a potent bactericidal peptide derived from the N-terminal region of bovine lactoferrin. *J Appl Bacteriol* 1992;73:472–9.
120. López-Soto F, León-Sicairos N, Nazmi K, Bolscher JG, de la Garza M. Microbicidal effect of the lactoferrin peptides lactoferricin17-30, lactoferrampin265-284, and lactoferrin chimera on the parasite *Entamoeba histolytica. Biometals* 2010;23:563–8.
121. Ordaz-Pichardo C, León-Sicairos N, Hernández-Ramírez VI, Talamás-Rohana P, de la Garza M. Effect of bovine lactoferrin in a therapeutic hamster model of hepatic amoebiasis. *Biochem Cell Biol* 2012;90: 425–34.
122. Caballero-Salcedo A, Viveros Rogel M, Salva Tierra B, et al. Seroepidemiology of amebiasis in Mexico. *Am J Trop Med Hyg* 1994;50:412–9.
123. Braga LL, Lima AA, Sears CL, et al. Seroepidemiology of *Entamoeba histolytica* in a slum in northeastern Brazil. *Am J Trop Med Hyg* 1996;55:693–7.
124. Haque R, Duggal P, Ali IM, et al. Innate and acquired resistance to amebiasis in bangladeshi children. *J Infect Dis* 2002;186:547–52.
125. Haque R, Ali IM, Sack RB, Farr BM, Ramakrishnan G, Petri WA, Jr. Amebiasis and mucosal IgA antibody against the *Entamoeba histolytica* adherence lectin in Bangladeshi children. *J Infect Dis* 2001;183:1787–93.
126. Bujanover S, Katz U, Bracha R, Mirelman D. A virulence attenuated amoebapore-less mutant of *Entamoeba histolytica* and its interaction with host cells. *Int J Parasitol* 2003;33:1655–63.
127. Stanley SL, Jr., Zhang T, Rubin D, Li E. Role of the *Entamoeba histolytica* cysteine proteinase in amebic liver abscess formation in severe combined immunodeficient mice. *Infect Immun* 1995;63:1587–90.
128. Zhang T, Cieslak PR, Stanley SL, Jr. Protection of gerbils from amebic liver abscess by immunization with a recombinant *Entamoeba histolytica* antigen. *Infect Immun* 1994;62:1166–70.
129. Zhang T, Cieslak PR, Foster L, Kunz-Jenkins C, Stanley SL, Jr. Antibodies to the serine rich *Entamoeba histolytica* protein (SREHP) prevent amoebic liver abscess in severe combined immunodeficient (SCID) mice. *Parasite Immunol* 1994;16:225–30.

SUMMARY

E. histolytica is the etiological agent of amebiasis. As a parasite associated with high morbidity and mortality rates, mainly in developing countries, *E. histolytica* often forms part of the intestinal microbiota in humans; however, under certain conditions, it is capable of inducing amebic colitis and amebic liver abscess (ALA). During infection, both innate and adaptive immune responses occur; complement proteins, NK, iNKT, neutrophils, macrophages, and cytokines, among others, are activated to control and eliminate invading *E. histolytica*. However, this ameba employs several pathogenic mechanisms that aid its survival (e.g., capping, adhesion, proteases and phagocytosis). Considering the crucial role of the host immune system in the amelioration or exacerbation of tissue damage, a better understanding of the host–parasite interaction could lead to novel therapeutic strategies against *E. histolytica*. Owing to their low side effects, medicinal plants with potential amebicidal effects represent alternative or complementary therapies against amebiasis. Further identification of protective proteins from *E. histolytica* is also important for future vaccine development.

Molecular Control of *Taenia solium* Cysticercosis

34

Wei Liu
Hunan Agricultural University

Dongyou Liu
Royal College of Pathologists of Australasia Quality Assurance Programs

Contents

34.1 INTRODUCTION

Taenia solium (pork tapeworm) is a zoonotic cestode that typically involves humans as definitive host (after consumption of raw or poorly cooked pork containing *T. solium* larva/cysticercus, leading to intestinal taeniasis) and pigs as intermediate host (due to ingestion of *T. solium* eggs discharged by infected humans, leading to cysticercosis). However, accidental ingestion of *T. solium* eggs via contaminated food/water or fecal-oral autoinfection by humans may result in muscular, ocular, and neurocysticercosis, the latter of which accounts for 60%–90% of cases and contributes to late-onset epilepsy and possible death.

T. solium-related taeniasis and cysticercosis in humans and pigs have been known since ancient times, with the domestication of pigs over 11,000 years ago, early mention of the disease about 2000 BC, and naming of the culprit parasite adult worm as taenia (meaning ribbon, bandage, lace, or stripe in Greek) in the 1st century BC. In fact, the involvement of pork in the causation of human diseases might have underscored the formulation of Jewish and Islamic dietary laws that prohibit pork consumption. Suspected as the cause of epilepsy in 1536, the relationship between *T. solium* cyst (cysticercus) and adult worm was noted in 1784, and the term "cellulosae" was introduced in 1818 to describe *T. solium* cyst (larva). By 1854, the connection between *T. solium* taeniasis and cysticercosis was clearly established after several experimental studies.

Ranked as the top foodborne parasite by the World Health Organization (WHO) in 2014, *T. solium* is estimated to affect 50 million people globally, leading to 50,000 deaths per year. In addition, this parasite is responsible for significant reduction of the market value of pork worldwide. While human taeniasis due to *T. solium* adult worm is treatable with niclosamide or praziquantel, a combination of TSOL18 vaccine and oxfendazole treatment is crucial for eradication of cysticercosis in pigs, along with improvement in pig rearing and meat inspection practices.

Taxonomy. Belonging to the family *Taeniidae*, order *Cyclophyllidea* (terrestrial cycles, scolex with suckers), subclass *Eucestoda* (segmented, hermaphroditic), class Cestoda (tapeworms), phylum *Platyhelminthes* (flatworms), and kingdom *Animalia*, the genus *Taenia* consists of 45 recognized tapeworm species that affect a diverse range of hosts. Human-infecting *Taenia* species include *T. solium* (commonly referred to as pork tapeworm), *T. saginata* (beef tapeworm), and *T. asiatica* (Asian tapeworm), whose adult worms inhabit the small intestine of humans, and larvae (*C. cellulosae*, *C. bovis* and *C. viscerotropica*) reside in the muscles and other tissues of pigs/humans, cattle, and pigs, respectively. Additionally, larvae of *T. multiceps*, *T. serialis*, *T. crassiceps*, and *T. taeniaeformis* are occasionally found in human internal tissues. Application of molecular procedures allows further differentiation between two endemic *T. solium* genotypes from Asia and America/Africa, which cause neurocysticercosis and subcutaneous cysticercosis and neurocysticercosis without subcutaneous cysticercosis, respectively.

Morphology. Residing in the small intestine of humans, *T. solium* adult worm (2–8 m in length) possesses a spheroidal/rectangular scolex (1.2 mm in width, which contains four suckers/acetabula, and a rostellum of 1 mm in diameter surrounded by 22–32 spiny hooks arranged in two rows), a short neck, and a ribbon-like, flattened, segmented body (strobila) (Figure 34.1) [1]. The strobila consists of about 1,000 segments/proglottids, ranging from immature (near the neck), mature (in the middle), to gravid/ripe proglottids (at the posterior end, which have no posterior protuberances). Inside each segment exists a set of male and female reproductive organs (hermaphroditic), including uteri with 5–10 branches on each side (filled with eggs in gravid proglottids), ovary, testes, genital pore, and vitelline gland, with testes and ovary opening into a common genital pore located on the side. Without body cavity or digestive system, *Taenia* adult worm depends on its tegument for nutrient adsorption.

T. solium proglottids (segments) each produce 50,000 eggs of 33–48 μm in diameter, which are of spherical shape with a thick striated wall and a six-hooked larva (hexacanth, also known as oncosphere) inside. Being highly resistant to desiccation and sewage treatment, *T. solium* eggs are discharged in feces along with

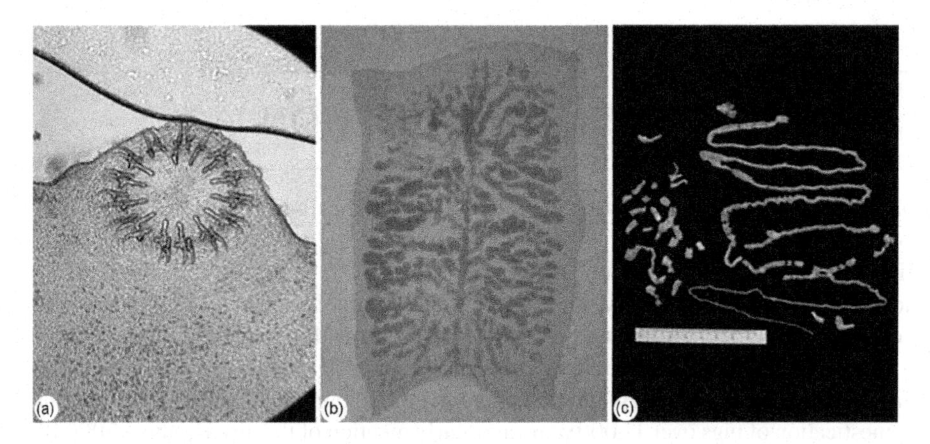

FIGURE 34.1 *Taenia solium* adult worm recovered from a 15-year-old male (case code no. 1614) in Mbulu, Tanzania; (a) unstained scolex showing rostellum with two rows of spiny hooks; (b) gravid proglottid (acetocarmine-stained); (c) strobila. (Photo credit: Eom KS, Chai JY, Yong TS, et al. Korean J Parasitol. 2011;49(4):399–403.)

detached gravid proglottids. After ingestion by intermediate host, *T. solium* eggs hatch and release oncospheres in the intestine, which penetrate the intestinal wall and reach the muscles of pigs, visceral organs (e.g., central nervous system, liver, or lung) of humans and dogs, where they transform into cysticerci [2].

T. solium larvae (metacestodes) are pearly-white, fluid-filled cysts (also known as cysticerci cellulosae or bladder-worms) of 5–8 mm×3–6 mm in size, which contain a single invaginated protoscolex with hooks but no external wart-like formations (infective stage) (Figure 34.2) [3].

Genome. *T. solium* genomes is about 131 Mb in length (with 11,902 coding genes) compared to 169 Mb (13,161 coding genes) and 168 Mb (13,323 coding genes) in of *T. saginata* and *T. asiatica* genomes, respectively. It is notable that *T. solium* genome includes 200 open reading frames encoding proteases (e.g., calpains, cytosolic, mitochondrial signal peptidases, ubiquitylation-related proteins) [4].

Lifecycle and epidemiology. *Taenia* lifecycle involves a carnivore/omnivore (definitive host) for its adult stage, and a herbivore/omnivore (intermediate host) for its larval stage (metacestode).

Specifically, *T. solium* eggs (or gravid proglottids) excreted in human feces are ingested by pigs, in which embryonated eggs hatch and release motile oncospheres in the duodenum. After piercing through the intestinal wall, oncospheres migrate via the bloodstream to the muscles, liver, lung, brain, eyes, and skin, where they transform into metacestodes (cysticerci; porcine cysticercosis) within 70 days [5]. Humans who accidentally ingest *T. solium* eggs (through consumption of contaminated food/water or autoinfection) may also develop cysticerci in the central nervous system and muscles (neurocysticercosis), which display three morphologically distinct forms: (i) the ordinary "cellulose" cysticercus, with a fluid-filled cyst and an invaginated scolex; (ii) the intermediate form with a scolex; and (iii) the "racemose" form (20 cm in length and 60 ml of fluid) with no evident scolex.

When raw or undercooked pork containing cysticerci is consumed by a human, the protoscolex within the cyst evaginates in the presence of bile and intestinal enzymes and attaches to the mucosa of the small intestine, in which it matures into hermaphroditic adult worm, which is expected to live up to 5 years (intestinal taeniasis). After fertilization, the proglottids containing mature eggs are expelled in feces to the environment to initiate another round of infection.

T. solium (taeniasis/cysticercosis) has a cosmopolitan distribution and shows high prevalence in places where humans live in close proximity with pigs and dogs (e.g., Mexico, Guatemala, Bolivia, Peru, West Africa, South East Asia, China, Nepal, India, and Russia). Poor sanitation conditions and inadequate supply of clean water increase human exposure to *T. solium* eggs, and thus human cases of neurocysticercosis. Travel to and migration from endemic countries may also contribute to the development of human neurocysticercosis as shown by recent cases in the United States, Canada, and Europe. Indeed, 10%–25% of people in *T. solium* endemic regions display specific anti *T. solium* serum antibody responses. It is estimated that *T. solium* cysticercosis affects about 50 million people, leading to 50,000 deaths per year.

Clinical features. *T. solium* adult worm infection in humans (intestinal taeniasis) is often asymptomatic or displays mild-to-moderate symptoms, which range from passage of 6–8 proglottids per 24 h,

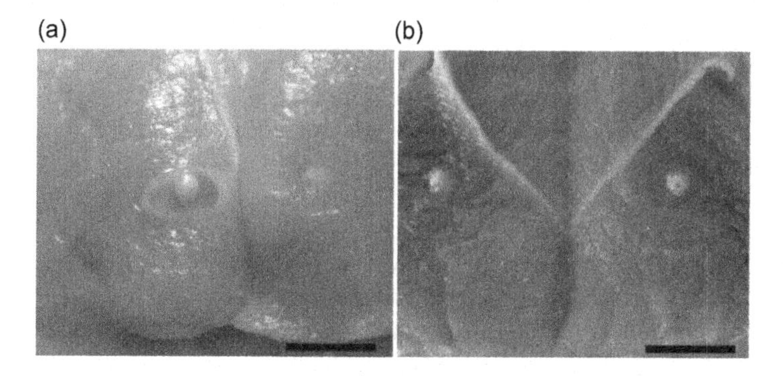

FIGURE 34.2 *Taenia solium* cysticercosis in pigs; (a) viable *T. solium* cyst in muscle (scale bar: 1 cm); (b) non-viable *T. solium* cyst in muscle (scale bar: 1 cm). (Photo credit: Gauci CG, Ayebazibwe C, Nsadha Z, et al. PLoS Negl Trop Dis. 2019;13(6):e0007408.)

colicky abdominal pain, flatulence, nausea, decreased or increased appetite, vomiting, intermittent diarrhea or constipation, fever, headache, weight loss, anemia, emaciation, dizziness, weakness, anal pruritis, irritability/hyperexcitability, damage or bleeding in stomach, to eosinophilia. Possible complications include appendicitis, cholecystitis, cholangitis, obstruction of bile ducts, pancreatic duct, and tapeworm growth in ectopic locations (e.g., middle ear, adenoid tissue, uterine cavity).

Resulting from accidental ingestion of embryonated eggs or proglottids (via contaminated food/water or autoinfection), *T. solium* larval infection (cysticercosis) in humans may appear as muscular and dermatologic cysticercosis, ocular cysticercosis, and neurocysticercosis (60%–90% of cases). Specifically, muscular and dermatologic cysticercosis is associated with muscle mass (acute myositis, muscular pseudohypertrophy) and subcutaneous nodules (1–2 cm in size) near the trunk and extremities, which become painful and swollen and then resolve; ocular (ophthalmic) cysticercosis affects the subretinal space, vitreous, and conjunctiva, with severe pain and blurred or lost vision; and neurocysticercosis is linked to acute-onset, recurrent, and unprovoked seizures or epilepsy (30%–50% of cases) (due to localization of cysticerci in brain parenchyma), intracranial hypertension (obstruction of CSF by intraventricular brain cysts), headache, nausea, vomiting, vertigo, papilledema, psychiatric disturbances (changes in personality and mental status, changes in behavior, and learning disabilities), and death.

Pathogenesis. *T. solium* adult worms utilize their suckers and hooks to anchor the intestinal mucosa, and their strobila to adsorb nutrients, leading to local damage, disturbance, pain, and inflammatory response (with the recruitment of mast cells, goblet cells, plasma cells, lymphocytes, neutrophils and eosinophils). *T. solium* cysticerci employ their hooks to settle in the tissues, use their cyst wall to block host complement system, and mask themselves with host immunoglobulins to evade immune surveillance [6]. As the cysticerci grow in number and size, they obstruct cerebrospinal fluid flow (hydrocephalus) and increase intracranial pressure, triggering mass inflammation, arachnoiditis, and edema.

Diagnosis. Diagnosis of *T. solium* adult worm infection (taeniasis) in humans is dependent on observation of *T. solium* proglottids or eggs. Use of serological and molecular techniques allows detection of genus-specific coproantigens, species-specific serum antibodies, and taeniid nucleic acids [7].

Diagnosis of *T. solium* larval infection (cysticercosis) in humans is facilitated by the application of computer tomography (CT) and magnetic resonance imaging (MRI), which reveal *T. solium* live cysts as small, round areas easily distinguishable from the brain parenchyma, and *T. solium* scolex as a nodule of high density within the cyst of the brain; and show *T. solium* calcified cysticerci to be surrounded by an area of perilesional edema in the subcutaneous and muscle tissues. Further, detection of *T. solium* antigens, specific antibodies, and nucleic acids is possible through serological and molecular approaches [8].

Diagnosis of porcine cysticercosis is based on *antemortem* inspection of tongue and muscle, serological detection of antibodies and antigen, and *postmortem* inspection of carcass inspection and molecular identification of *T. solium* nucleic acids [3].

Treatment and prevention. Treatment options for *T. solium* adult worm infection (intestinal taeniasis) in humans include praziquantel, niclosamide, and albendazole, while those for *T. solium* larval infection in humans (neurocysticercosis) consist of corticosteroids (e.g., prednisone or dexamethasone), anticonvulsants (for patients with multiple degenerating lesions and high risk of seizures), antiepileptics (e.g., phenobarbitone and carbamazepine), anthelmintics (e.g., albendazole and/or praziquantel), and surgery (e.g., direct excision of ventricular cysts, shunting procedure for fluid drainage and resolution of hydrocephalus, and endoscopic removal of cysts).

Prevention of *T. solium* infection in humans requires stringent meat inspection, condemnation of infected carcasses, proper cooking (80°C) or freezing (–10°C for 9 days) of meat, and careful washing of salad vegetables, accompanied by strict personal hygiene (e.g., careful handwashing for food handlers), improved access to clean water, and identification and treatment of carriers of *T. solium* adult tapeworms.

Furthermore, prevention of *T. solium* infection in pigs involves proper disposal of human waste, prohibition of using untreated sewage effluent in pasture and food crop irrigation intended for pigs, avoidance of the exposure of pigs to environments contaminated with human waste, and vaccination of pigs against *T. solium* with TSOL18 vaccine consisting of recombinant 45W-4B antigens) or S3PVAC composed of three peptides of 8 (KETc12), 12 (KETc1), and 18 amino acids (GK1) [9].

34.2 MOLECULAR CONTROL OF *T. SOLIUM* CYSTICERCOSIS: CURRENT STATUS

34.2.1 Protective Antigens

Prior to the 1970s, it was widely believed that immunity to metazoan parasites invariably requires direct exposure to living parasites. However, the paradigm shifted following the landmark reports in 1971, in which larvae of *T. ovis* and *T. taeniaformis* confined in diffusion chambers in the peritoneal cavity of their respective intermediate hosts (i.e., sheep and rats) induced protective immunity to challenge with infective eggs [10,11]. This allowed the identification of the proteins associated with such immunity, facilitated the genetic engineering of a vaccine against *T. ovis* [12], and thus opened ways for subsequent development of recombinant vaccines against other taeniid parasites such as *T. solium, T. saginata and Echinococcus granulosus* [13].

***T. solium* TSOL18, TSOL45–1A, and TSOL16.** *T. solium* oncospheres and cysts contain various antigens that are capable of stimulating protective immune responses in the hosts [14–18]. Among protective antigens identified from *T. solium* oncospheres, three (i.e., TSOL18, TSOL45-1A, and TSOL16) are prominent. Cloned from mRNA of *T. solium* hatched and activated oncospheres, TSOL18 shows close homology to host-protective oncosphere antigens from *T. ovis* To18 and *T. saginata* TSA-18 [13,19,20]. Similarly, TSOL45-1A is a protein produced by *T. solium* oncospheres that possesses putative N-linked glycosylation sites, an amino terminal secretory signal, a hydrophobic carboxy terminal sequence characteristic of GPI-anchored proteins and fibronectin type III motifs and displays homology to the host-protective antigens of *T. ovis* 45W and *T. saginata* TSA9. Furthermore, TSOL16 is a conserved protein (secretory signal and fibronectin type III domain) specifically expressed in the larval oncosphere stage of *T. solium* that infects pigs. Showing homology to *T. ovis* TO16 and *T. saginata* Tsa18, TSOL16 is associated with the penetration gland cells of *T. solium* and possibly involved in host infection and parasite survival.

Interestingly, both anti-TSOL18 and TSOL45-1A antibodies strongly recognize the surface of *T. solium* oncospheres while anti-TSOL21 antibodies react with *T. solium* scolex and immature proglottids (Figures 34.3 and 34.4) [18]. This suggests that TSOL18 is present on the surfaces of *T. solium* oncosphere, scolex, and immature proglottid, and TSOL45-1A locates on the surface of *T. solium* oncosphere only, but not the surfaces of scolex and immature proglottid.

Screening of an expressed sequence tag (EST) library produced from *T. solium* cysticerci of the pigs uncovered 25 most highly expressed genes, including four (T24, Tsol18/HP6, Tso31d, and Tsol15) that encode proteins related to those found on the surface of *T. solium* oncospheres. Of these, Tsol15 displays homology to *T. ovis* 45W antigen [22]. This indicates the common occurrence of T24, Tsol18/HP6, Tso31d, and Tsol15 in *T. solium* oncosphere and larva.

***T. solium* KE7 (TsKE7).** *T. solium* KE7 (TsKE7) belongs to a ubiquitously distributed family of proteins associated with membrane and possibly involved in several vital cell pathways. *T. solium* KE7 (TsKE7) is a 31-kDa protein of 264 amino acids, including a 18-amino-acid-long peptide (GK-1) at the carboxyl-terminus. GK-1 harbors a B-cell epitope in the N terminus (YYYPS) and a class II T-cell epitope in the C terminus, the latter of which has the capacity to activate cytotoxic T lymphocytes and helper T lymphocytes. Occurring also in *T. crassiceps*, *E. granulosus*, and *E. multilocularis*, GK-1 shows the potential to be a vaccine candidate against the different cestodes [23–25].

34.2.2 Vaccine Trials

Although TSOL18, TSOL45-1A, and TSOL16 as well as TsKE7 from *T. solium* oncospheres are all capable of inducing near complete protection in pigs against *T. solium* challenge under experimental conditions, TSOL18 is exceptional, achieving 100% and 99.5% protection in trials conducted in Mexico, 99.9%

in Peru, 99.3% in Honduras, and 100% in Cameroon [26–34]. This may be partly explained by the findings that TSOL18 is present on the surface of *T. solium* oncosphere, scolex, and immature proglottid, whereas TSOL45-1A is only found on the surface of *T. solium* oncosphere, but not scolex and immature proglottid (Figures 34.3 and 34.4) [21].

By linking a 579-bp fragment of *T. solium*-activated oncosphere gene (*TSOL18*) with the gene encoding the signal secretion protein SPUSP45 and subsequently inserting *SP-TSOL18* into the multiple cloning sites (MCS) of plasmid pMG36e, Zhou et al [35] generated recombinant plasmid pMG36e-SP-TSOL18 for transformation into *L. lactics* strain. Mice immunized orally with recombinant pMG36e-SP-TSOL18/*L. lactis* strain developed strong cellular, humoral, and mucosal immune responses to *T. solium* parasite. This offers an option of using *L. lactis* strain for delivery of *T. solium* protective antigens [35].

34.2.3 Elimination Strategies

Optimal strategies for eliminating *T. solium* infection in humans and pigs cover several aspects, including (i) treatment of human taeniasis, (ii) treatment of porcine cysticercosis, (iii) vaccination of pigs, (iv) improved meat inspection, hygiene and sanitation, and (v) integrated cross-sectoral control [36–38].

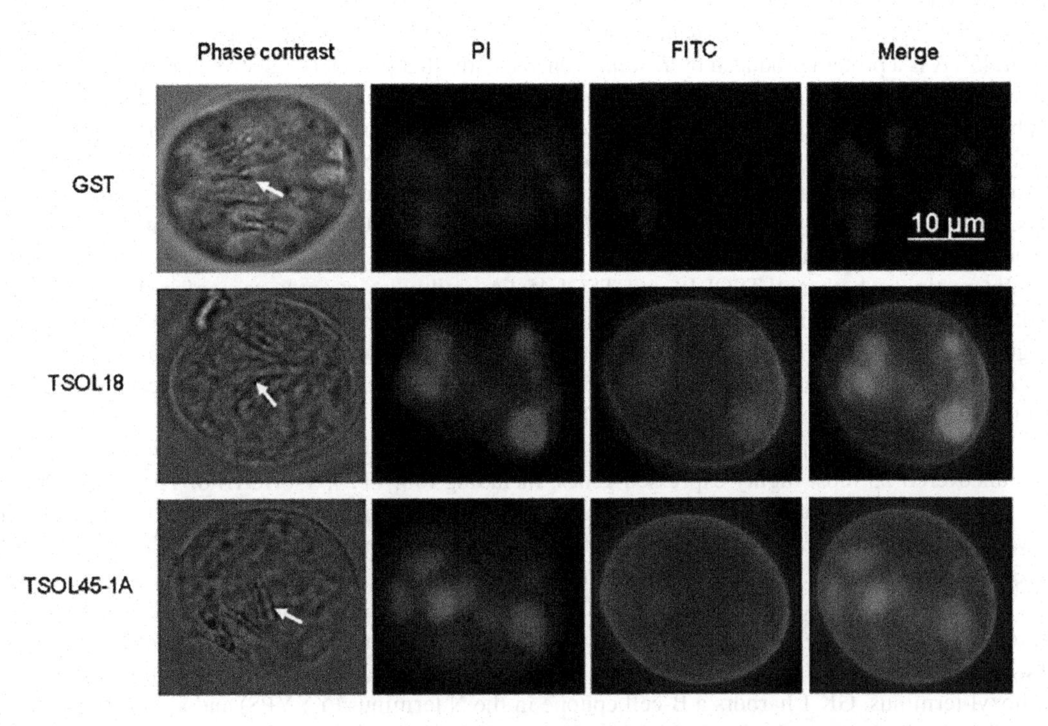

FIGURE 34.3 *Taenia solium* oncospheres from a tapeworm carrier are examined under contrast phase microscopy, propidium iodide (PI), fluorescein isothiocyanate (FITC), and a merged image of both stains. The top row images are oncospheres incubated with antiglutathione S-transferase (GST) IgG as a negative control; the middle row images are oncospheres incubated with anti-TSOL18 antibodies; and the bottom row images are oncospheres incubated with anti-TSOL45-1A antibodies. *T. solium* hooks of the oncospheres are indicated by arrows in the left column. Positive reactions with anti-TSOL18 and anti-TSOL45-1A antibodies are only on the surface of the oncospheres as seen in the merged image. Nonspecific background fluorescence on the same internal structures is observed in propidium iodide (PI) and fluorescein isothiocyanate (FITC)-stained images. (Photo credit: Martinez-Ocaña J, Romero-Valdovinos M, de Kaminsky RG, Maravilla P, Flisser A. Parasit Vectors. 2011;4:3.)

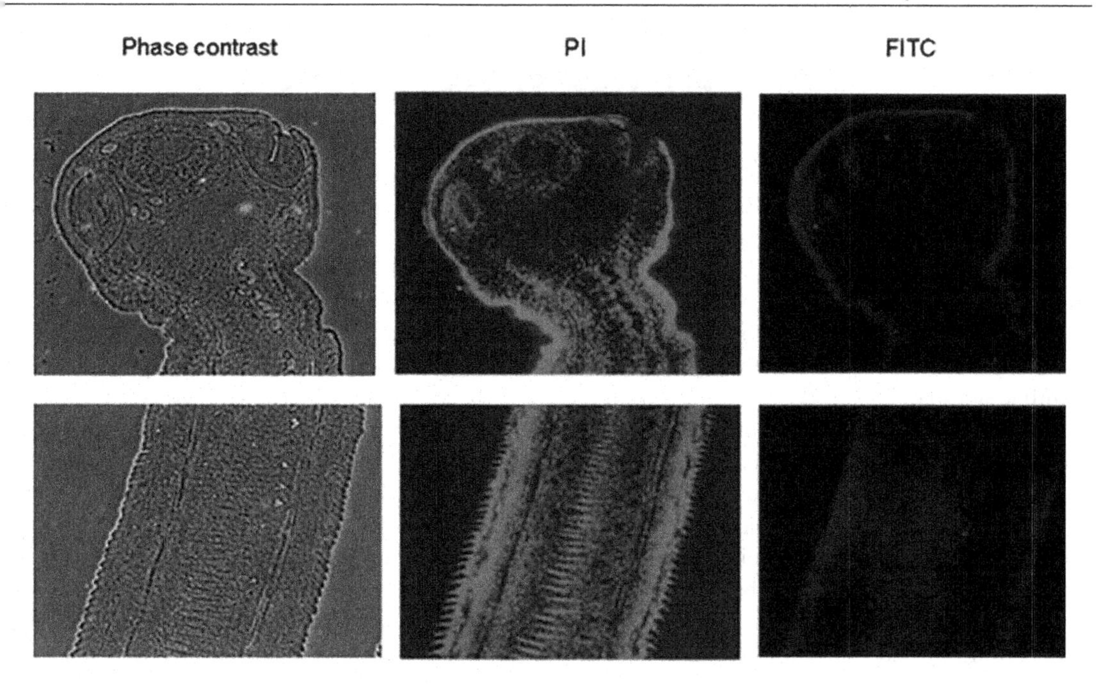

FIGURE 34.4 *Taenia solium* scolex-neck and immature proglottid recovered from an experimentally infected hamster are examined under contrast phase microscopy or with filters for propidium iodide (PI) and fluorescein isothiocyanate (FITC) after incubation with anti-TSOL18 antibodies (as incubation with anti-TSOL45-1A or GST produces negative results). The top row images show scolex-neck and the bottom row images show immature proglottid. (Photo credit: Martinez-Ocaña J, Romero-Valdovinos M, de Kaminsky RG, Maravilla P, Flisser A. Parasit Vectors. 2011;4:3.)

Treatment of human taeniasis. Preventive chemotherapy (mass drug administration) is one of the five main intervention strategies (the others being intensified disease management, vector control, improve water quality and sanitation, and zoonotic disease management) outlined by the WHO for control of the neglected tropical diseases such as *T. solium* infection. Out of the two currently registered drugs for taeniasis (i.e., praziquantel and niclosamide), annual, or biennial administration of praziquantel (40 mg kg^{-1}) is suggested for school-aged children in areas where both taeniasis and schistosomiasis are endemic, and may reduce human taeniasis prevalence from 2% to 1% in the entire population.

Treatment of porcine cysticercosis. Among drugs tested for porcine cysticercosis, anthelmintic oxfendazole is remarkably effective against muscle cysts (but not brain cysts) with minimal side-effects. Similar to human intestinal taeniasis, annual treatment of pigs with oxfendazole may decrease the prevalence of porcine cysticercosis from nearly 20% to 15%.

Vaccination of pigs. Two vaccines [i.e., TSOL18 vaccine (consisting of recombinant 45W-4B antigens) and S3PVAC (consisting of KETc12, KETc1 and GK1) are currently available for vaccination of pigs against *T. solium*. Half yearly administration of TSOL18 vaccine and oxfendazole to pigs of 1–9-month age may help reduce taeniasis to <1% and porcine cysticercosis to <5%.

Improved meat inspection, hygiene, and sanitation. These measures reduce potential human infections and are expected to deliver long-term benefits, although in the short term, their effects are difficult to quantify.

Integrated cross-sectoral control. A combination of annual mass drug administration to school-aged children (90% coverage and 100% efficacy) and half yearly vaccination and treatment of pigs aged 1–9 months of age (100% efficacious vaccine, 90% efficacious treatment) together with annual education at the community level over a 4-year period will enable simultaneous reduction of *T. solium* infection in humans and pigs, not complete elimination in neither humans nor pigs.

34.3 FUTURE PERSPECTIVES

T. solium is a zoonotic tapeworm that typically affects humans (definitive host as well as accidental intermediate host) and pigs (intermediate host). Although human intestinal taeniasis is largely asymptomatic and clinically insignificant, human cysticercosis (especially neurocysticercosis) is associated with acute onset, recurrent, and unprovoked seizures or epilepsy, intracranial hypertension, headache, nausea, vomiting, vertigo, papilledema, psychiatric disturbances, and death. It is estimated that *T. solium* is responsible for about 50 million cases and 50,000 deaths per year worldwide. Furthermore, porcine cysticercosis contributes to a reduction in the market value of pork and exerts an enormous economic burden on pig farming industry.

Undoubtedly, there is an obvious need as well as economic incentive to control and eradicate *T. solium* taeniasis/cysticercosis. We have gone a long way from the elucidation of the immunobiology of *T. solium*, the identification of protective antigens in *T. solium* oncospheres, to the development of recombinant TSOL18 and S3PVAC vaccines. Although these vaccines are highly effective for controlling *T. solium* cysticercosis in pigs, their value for reducing *T. solium* taeniasis and cysticercosis in humans remains to be determined. Further study is also necessary to acquire knowledge about *T. solium* local epidemiology and transmission, which is fundamental to implementation of tailor-made eradication measures against this zoonotic parasite.

REFERENCES

1. Eom KS, Chai JY, Yong TS, et al. Morphologic and genetic identification of *Taenia* tapeworms in Tanzania and DNA genotyping of *Taenia solium*. *Korean J Parasitol.* 2011;49(4):399–403.
2. Chile N, Clark T, Arana Y, et al. In vitro study of *Taenia solium* postoncospheral form. *PLoS Negl Trop Dis.* 2016;10(2):e0004396.
3. Gauci CG, Ayebazibwe C, Nsadha Z, et al. Accurate diagnosis of lesions suspected of being caused by *Taenia solium* in body organs of pigs with naturally acquired porcine cysticercosis. *PLoS Negl Trop Dis.* 2019;13(6):e0007408.
4. Yan HB, Lou ZZ, Li L, Brindley PJ, et al. Genome-wide analysis of regulatory proteases sequences identified through bioinformatics data mining in *Taenia solium*. *BMC Genomics.* 2014;15:428.
5. Palma S, Chile N, Carmen-Orozco RP, et al. In vitro model of postoncosphere development, and in vivo infection abilities of *Taenia solium* and *Taenia saginata*. *PLoS Negl Trop Dis.* 2019;13(3):e0007261.
6. Garcia HH, Rodriguez S, Friedland JS; Cysticercosis Working Group in Peru. Immunology of *Taenia solium* taeniasis and human cysticercosis. *Parasite Immunol.* 2014;36(8):388–96.
7. Bustos JA, Rodriguez S, Jimenez JA, et al. Detection of *Taenia solium* taeniasis coproantigen is an early indicator of treatment failure for taeniasis. *Clin Vaccine Immunol.* 2012; 19(4):570–3.
8. Handali S, Klarman M, Gaspard AN, et al. Multiantigen print immunoassay for comparison of diagnostic antigens for 5 *solium* cysticercosis and taeniasis. *Clin Vaccine Immunol.* 2010;17(1):68–72.
9. Lightowlers MW. Fact or hypothesis: *Taenia crassiceps* as a model for *Taenia solium*, and the S3Pvac vaccine. *Parasite Immunol.* 2010;32(11–12):701–9.
10. Rickard MD, Bell KJ. Induction of immunity of lambs to a larval cestode by diffusible antigens. *Nature.* 1971;232(5306):120.
11. Rickard MD, Bell KJ. Immunity produced against *Taenia ovis and T. taeniaeformis* infection in lambs and rats following in vivo growth of their larvae in filtration membrane diffusion chambers. *J Parasitol.* 1971;57(3):571–5.
12. Johnson KS, Harrison GB, Lightowlers MW, et al. Vaccination against ovine cysticercosis using a defined recombinant antigen. *Nature.* 1989;338(6216):585–7.
13. Gauci C, Lightowlers MW. Genes encoding homologous antigens in taeniid cestode parasites: Implications for development of recombinant vaccines produced in *Escherichia coli*. *Bioengineered.* 2013;4(3):168–71.

14. Mayta H, Hancock K, Levine MZ, et al. Characterization of a novel *Taenia solium* oncosphere antigen. *Mol Biochem Parasitol.* 2007;156(2):154–61.
15. Esquivel-Velázquez M, Larralde C, Morales J, Ostoa-Saloma P. Protein and antigen diversity in the vesicular fluid of *Taenia solium* cysticerci dissected from naturally infected pigs. *Int J Biol Sci.* 2011;7(9):1287–97.
16. Zimic M, Pajuelo M, Gilman RH, et al. The highly antigenic 53/25 kDa *Taenia solium* protein fraction with cathepsin-L like activity is present in the oncosphere/cysticercus and induces non-protective IgG antibodies in pigs. *Vet Immunol Immunopathol.* 2012;145(1–2):171–8.
17. Gauci C, Jayashi C, Lightowlers MW. Vaccine development against the *Taenia solium* parasite: the role of recombinant protein expression in *Escherichia coli. Bioengineered.* 2013;4(5):343–7.
18. Gauci C, Lightowlers MW. Genes encoding homologous antigens in taeniid cestode parasites: Implications for development of recombinant vaccines produced in *Escherichia coli. Bioengineered.* 2013;4(3):168–71.
19. Cai X, Yuan G, Zheng Y, et al. Effective production and purification of the glycosylated TSOL18 antigen, which is protective against pig cysticercosis. *Infect Immun.* 2008;76(2):767–70.
20. Zimic M, Gutiérrez AH, Gilman RH, et al. Immunoinformatics prediction of linear epitopes from *Taenia solium* TSOL18. *Bioinformation.* 2011;6(7):271–4.
21. Martinez-Ocaña J, Romero-Valdovinos M, de Kaminsky RG, Maravilla P, Flisser A. Immunolocalization of TSOL18 and TSOL45-1A, the successful protective peptides against porcine cysticercosis, in *Taenia solium* oncospheres. *Parasit Vectors.* 2011;4:3.
22. Lundström J, Salazar-Anton F, Sherwood E, Andersson B, Lindh J. Analyses of an expressed sequence tag library from *Taenia solium*, cysticerca. *PLoS Negl Trop Dis.* 2010; 4(12):e919.
23. Rassy D, Bobes RJ, Rosas G, et al. Characterization of S3Pvac anti-cysticercosis vaccine components: implications for the development of an anti-cestodiasis vaccine. *PLoS One.* 2010;5(6):e11287.
24. Capelli-Peixoto J, Chávez-Olórtegui C, Chaves-Moreira D, et al. Evaluation of the protective potential of a *Taenia solium* cysticercus mimotope on murine cysticercosis. *Vaccine.* 2011;29(51):9473–9.
25. Bobes RJ, Navarrete-Perea J, Ochoa-Leyva A, et al. Experimental and theoretical approaches to investigate the immunogenicity of *Taenia solium*-derived KE7 antigen. *Infect Immun.* 2017;85(12):e00395–17.
26. Flisser A, Gauci CG, Zoli A, et al. Induction of protection against porcine cysticercosis by vaccination with recombinant oncosphere antigens. *Infect Immun.* 2004;72(9):5292–7.
27. Kyngdon CT, Gauci CG, Gonzalez AE, et al. Antibody responses and epitope specificities to the *Taenia solium* cysticercosis vaccines TSOL18 and TSOL45-1A. *Parasite Immunol.* 2006;28(5):191–9.
28. Guo YJ, Wu D, Wang KY, Sun SH. Adjuvant effects of bacillus Calmette-Guerin DNA or CpG-oligonucleotide in the immune response to *Taenia solium* cysticercosis vaccine in porcine. *Scand J Immunol.* 2007;66(6):619–27.
29. Assana E, Gauci CG, Kyngdon CT, et al. Antibody responses to the host-protective T*aenia solium* oncosphere protein TSOL18 in pigs are directed against conformational epitopes. *Parasite Immunol.* 2010;32(6):399–405.
30. Assana E, Kyngdon CT, Gauci CG, et al. Elimination of *Taenia solium* transmission to pigs in a field trial of the TSOL18 vaccine in Cameroon. *Int J Parasitol.* 2010;40(5):515–9.
31. Gauci CG, Jayashi CM, Gonzalez AE, Lackenby J, Lightowlers MW. Protection of pigs against *Taenia solium* cysticercosis by immunization with novel recombinant antigens. Vaccine. 2012;30(26):3824–8.
32. Jayashi CM, Kyngdon CT, Gauci CG, Gonzalez AE, Lightowlers MW. Successful immunization of naturally reared pigs against porcine cysticercosis with a recombinant oncosphere antigen vaccine. *Vet Parasitol.* 2012;188(3–4):261–7.
33. Garcia HH, Gonzalez AE, Tsang VC, et al. Elimination of *Taenia solium* transmission in Northern Peru. *N Engl J Med.* 2016;374(24):2335–44.
34. Poudel I, Sah K, Subedi S, et al. Implementation of a practical and effective pilot intervention against transmission of *Taenia solium* by pigs in the Banke district of Nepal. *PLoS Negl Trop Dis.* 2019;13(2):e0006838.
35. Zhou BY, Sun JC, Li X, et al. Analysis of immune responses in mice orally immunized with recombinant pMG36e-SP-TSOL18/*Lactococcus lactis* and pMG36e-TSOL18/ *Lactococcus lactis* vaccines of *Taenia solium. J Immunol Res.* 2018;2018:9262631.
36. Johansen MV, Trevisan C, Gabriël S, Magnussen P, Braae UC. Are we ready for *Taenia solium* cysticercosis elimination in sub-Saharan Africa? *Parasitology.* 2017;144(1):59–64.
37. Samorek-Pieróg M, Karamon J, Cencek T. Identification and control of sources of *Taenia solium* infection - the attempts to eradicate the parasite. *J Vet Res.* 2018;62(1):27–34.
38. Sánchez-Torres NY, Bobadilla JR, Laclette JP, José MV. How to eliminate taeniasis/ cysticercosis: porcine vaccination and human chemotherapy (Part 2). *Theor Biol Med Model.* 2019;16(1):4.

SUMMARY

T. solium (pork tapeworm) is a zoonotic tapeworm that typically affects humans (definitive host as well as accidental intermediate host) and pigs (intermediate host). Resulting from consumption of raw or poorly cooked pork containing *T. solium* larva/cysticercus, human intestinal taeniasis is largely asymptomatic and clinically insignificant. On the other hand, due to accidental ingestion of *T. solium* eggs via contaminated food/water or fecal-oral autoinfection, human cysticercosis may present with a range of clinical symptoms depending on places where the parasite settles (muscular, ocular, and neurocysticercosis). Ranked as the top foodborne parasite by the WHO in 2014, *T. solium* is estimated to affect 50 million people globally, leading to 50,000 deaths per year. Furthermore, porcine cysticercosis is responsible for significant reduction of the market value of pork worldwide. While human taeniasis due to *T. solium* adult worm is treatable with niclosamide or praziquantel, a combination of TSOL18 vaccine and oxfendazole treatment is crucial for eradication of cysticercosis in pigs, along with improvement in pig rearing and meat inspection practices.

35 Genetic Manipulation of *Cryptosporidium*

Wanyi Huang, Yaoyu Feng, and Lihua Xiao
South China Agricultural University

Dongyou Liu
Royal College of Pathologists of Australasia Quality Assurance Programs

Contents

35.1 INTRODUCTION

Cryptosporidium is a genus of unicellular organisms that parasitize a wide range of animal hosts including humans. Early description of *Cryptosporidium* was made in 1907 by Tyzzer, who identified *Cryptosporidium muris*, the type species of the genus, in the gastric epithelium of laboratory mice. Shortly afterwards, he described another species, *Cryptosporidium parvum*, in the intestine of mice. However, its potential for causing human disease was not realized until 1976, when a child and an adult presenting with acute, nonbloody watery diarrhea (cryptosporidiosis) were found to harbor *Cryptosporidium* parasites. The emergence of HIV/AIDS in the 1980s further highlighted the disease-causing capability of this parasite in individuals with compromised immune functions. As one of the most prevalent waterborne parasites in the world, *Cryptosporidium*-induced diarrhea is responsible for significant morbidity and mortality, especially in severely immunocompromised people and in young children at low-resource settings. Given its resistance to commonly used disinfectants in water treatment plants, *Cryptosporidium* poses a major public health risk worldwide.

Taxonomy. Classified in the family *Cryptosporidiidae*, suborder *Eimeriorina*, order *Eucoccidiorida*, subclass *Coccidia*, class *Sporozoasida*, phylum *Apicomplexa*, domain *Eukarya*, the genus *Cryptosporidium* (Tyzzer, 1907) currently includes approximately 40 recognized species. Although anthroponotic *C. hominis* and zoonotic *C. parvum* infect 8–10 million people globally each year, accounting for >90% of human

TABLE 35.1 *Cryptosporidium* species with potential for human infections

SPECIES	*PRINCIPAL HOST*	*SITE OF INFECTION*
C. hominis	Human	Small intestine
C. parvum	Ruminants	Small intestine
C. meleagridis	Turkey, birds, human	Small intestine
C. canis	Dogs	Small intestine
C. felis	Cats	Small intestine
C. ubiquitum	Cattle, rodents, primates	Intestine
C. cuniculus	European rabbits	Intestine
C. viatorum	Human	Small intestine
C. muris	Rodents	Stomach
C. andersoni	Cattle	Abomasum
C. bovis	Cattle	Small intestine
C. suis	Pigs	Small and large intestine
C. scrofarum	Pigs	Intestine
C. xiaoi	Sheep, goats	
C. tyzzeri	Rodents	Small intestine
C. erinacei	European hedgehog, horses	
C. fayeri	Kangaroo, marsupials	Small intestine

cryptosporidiosis cases, at least 15 other species are also implicated in human clinical diseases to some lesser extents (Table 35.1) [1–3].

Cryptosporidium species were characterized historically according to their phenotypic properties (e.g., host specificity, infection site, and oocyst morphology), which have led to the unfortunate creation of multiple names for some identical species. Application of molecular techniques has helped rectify this overdescription and demonstrated a close genetic relationship among major human-pathogenic species (i.e., *C. hominis*, *C. parvum*, *C. cuniculus*, and *C. meleagridis*) [4].

Morphology. *Cryptosporidium* oocysts are round to oval in shape and measure 4–6 μm in size. Specifically, *C. hominis* oocysts have average length of 4.91 μm and width of 4.28 μm, whereas *C. parvum* oocysts are of 4.85 μm in length and 4.39 μm in width. Having a thick wall composed of a surface glycocalyx layer, carbohydrates, fatty acids, aliphatic hydrocarbons, hydrophobic proteins, and an inner glycoprotein layer, *Cryptosporidium* oocysts are often refractile on wet smear, and display black dots or small vacuoles upon modified acid-fast staining. *Cryptosporidium* sporozoites (of 2 μm×0.8 μm in size) and merozoites show apical complex (microneme, roptry, conoid, preconoidal ring) at the anterior part, a feature that helps define the genus and family (Figures 35.1 and 35.2) [5]. Further, *Cryptosporidium* possesses only a rudimentary mitochondrion (mitosome), and no secondary plastid called apicoplast. Although both *Cryptosporidium* and related apicomplexan *Toxoplasma gondii* have four sporozoites in oocysts, the former differs from the latter by its monogenous lifecycle – completing its entire cycle within a single host instead of two hosts [5,6].

Genome. *C. hominis* genome is organized in eight chromosomes totaling 9.05 Mb DNA with 3,819 protein-coding genes. Similarly, *C. parvum* genome possesses eight chromosomes, 9.10 Mb DNA and 3,941 protein-coding genes. Relative to other apicomplexan parasites, *Cryptosporidium* harbors a reduced genome, which lacks all *de-novo* nutrient synthesis genes and relies on protein recycling and its own reserves of amylopectin to survive shifts in environmental conditions. Because *C. hominis* and *C. parvum* genomes contain nearly identical sets of protein- and RNA-coding genes, their phenotypic differences (e.g., host range) are possibly attributable to copy number variations in several subtelomeric gene families, genetic polymorphisms (e.g., SNPs and indels), and differential gene regulation [7,8].

Life cycle and epidemiology. *Cryptosporidium* is an intracellular parasite, with its monogenous lifecycle of six developmental stages taking place in a single host, and involving both sexual and asexual reproduction. This includes: (i) orally ingested oocysts excyst in the host small intestine (in response to

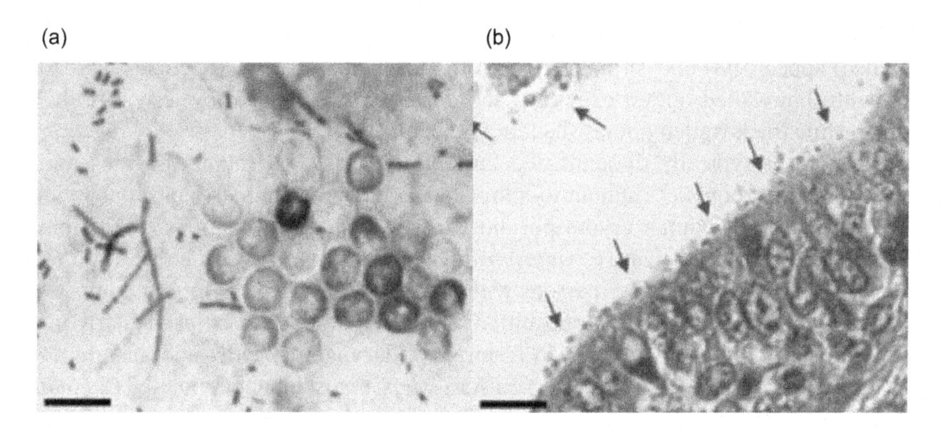

FIGURE 35.1 (a) *Cryptosporidium parvum* oocysts (circular bodies) in stool sample (modified Ziehl–Neelsen stain; scale bar, 7.5 μm). (b) *C. parvum*-infected intestinal tissue showing parasitophorous vacuoles as round bodies on the surface of the epithelial layer (arrows; H and E stain; scale bar, 260 μm). (Photo credit: Hemphill A, Müller N, Müller J. *Pathogens.* 2019;8(3):116.)

acid, bile salts, and an increase in temperature), releasing four motile sporozoites each; (ii) sporozoite uses its apical end to attach to the surface of the intestinal mucosa and invade intestinal epithelial cells; located in a parasitophorous vacuole formed of both host and parasite membranes, the sporozoite transforms into replicative trophozoite (characterized by rounded shape and enlarged nucleus; after 24 h of infection) and initiates several rounds of asexual multiplication (merogony) to become meronts (each containing six to eight merozoites; after 48 h of infection); (iii) merozoites can either develop into trophozoites and become meronts containing merozoites, or differentiate into microgametes and macrogametes (gametogony after

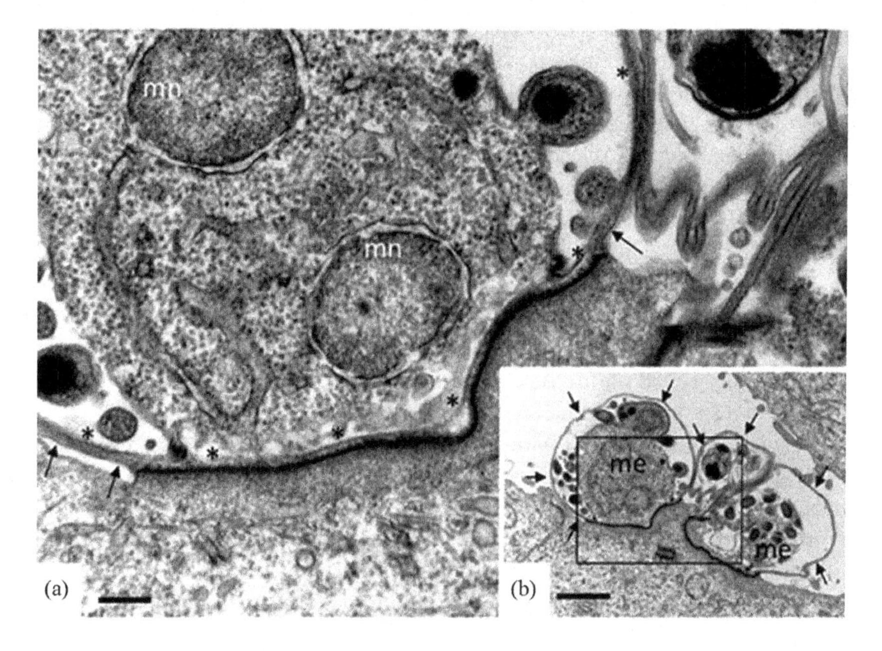

FIGURE 35.2 (a) *Cryptosporidium parvum* sporozoites in Madin Darbey canine kidney (MDCK) cells (72 h post-infection) showing parasitophorous vacuoles (PV) [asterisks (*) indicating the membrane of PV; arrows pointing toward host cell surface membrane; mn indicating nuclei of developing merozoites; scale bar, 2.5 μm). (b) A low magnification view of three PV showing two developing meronts (me) and outer host cell surface membrane (arrows; scale bar, 12 μm). (Photo credit: Hemphill A, Müller N, Müller J. *Pathogens.* 2019;8(3):116.)

96 h post-infection); (iv) gametes fertilize and undergo sexual replication; (v) oocysts (both thin-walled and thick-walled) appear; (vi) oocyst matures (with the formation of new, infectious sporozoites inside) (sporogony), with thin-walled oocyst excysting and continuing replication cycle within the same host (autoinfection), while thick-walled oocyst shedding in feces [9,10].

Cryptosporidium is typically disseminated through the fecal-oral route, involving animal-to-animal, animal-to-human (zoonotic), human-to-animal, and human-to-human (anthroponotic) transmissions [11–13]. Indeed, most human cryptosporidiosis cases in industrial nations result from ingestion/drinking of water contaminated with *Cryptosporidium* oocysts [the 50% infective dose (ID50) of *C. parvum* is only 132 oocysts for healthy persons with no previous serological immunity to cryptosporidiosis]. Occasionally, *Cryptosporidium* is transmitted via inhalation of oocysts, particularly in children or immunocompromised individuals, leading to respiratory (laryngotracheitis) and gastrointestinal (mild diarrhea) abnormalities. While the main reservoir for *C. parvum* is zoonotic, that for *C. hominis* may be asymptomatic carriage in young children [14].

Clinical features. *Cryptosporidium* infection in immunocompetent individuals may manifest as an acute, yet self-limiting diarrheal illness (1–2 week duration) and other symptoms (frequent watery diarrhea, nausea, vomiting, and abdominal cramps).

On the other hand, *Cryptosporidium* infection in immunocompromised persons (CD4 counts of <200 cell mm^{-3}) can lead to debilitating, cholera-like diarrhea (up to 20 l day^{-1}), severe abdominal cramps, malaise, low-grade fever, weight loss, and anorexia.

Interestingly, *C. hominis* tends to cause diarrhea, nausea, vomiting, and malaise in HIV-infected persons and in children, while *C. parvum*, *C. meleagridis*, *C. canis*, and *C. felis* are responsible for mostly diarrhea. Moreover, *C. hominis*, but not *C. parvum*, is linked to nonintestinal sequelae (joint pain, eye pain, recurrent headache, and fatigue) [3].

Pathogenesis. Under influence of serine protease and aminopeptidase, *Cryptosporidium* oocyst releases four sporozoites (excystation), which generate lectin adherence factor and other adhesion molecules (e.g., CSL, gp900, gp60/40/15, P23, P30, TRAP-C1, Cp47, CPS-500, and CpMIC1) that aid the binding of the apical end to the surface of the intestinal mucosa. Further, *Cryptosporidium* secretes invasion proteins (e.g., Cp2, Cpa135, secretory phospholipase, CpSUB, and CpMuc), along with adhesion molecules, stimulate production of cytokines/chemokines from host epithelial mucosa cells, which activate phagocytes, enhance phagocytosis, increase intestinal secretion of water and chloride (causing diarrhea), and inhibit absorption (causing nutrient malabsorption), leading to T cell-mediated inflammation, villus atrophy, and crypt hyperplasia (causing apoptosis). *Cryptosporidium* also produces other proteins (e.g., CpABC, CpATPase2, CpATPase3, HSP70, HSP90, cysteine protease, and acetyl-co-synthase) that enhance its intracellular multiplication and survival in host. Additionally, *Cryptosporidium*-induced recruitment of inflammatory monocytes to the infected mucosa may compromise epithelial integrity and suppress intestinal dendritic cells that normally function as immunological sentinels and active effectors against invading parasites [15,16].

Diagnosis. As *Cryptosporidium* infection in humans is associated with nonspecific symptoms, its diagnosis has traditionally relied on microscopic observation of oocysts in stool samples, and parasitic components in intestinal tissues, with the help of tinctorial/acid-fast stain (e.g., modified Ziehl–Neelsen stain), fluorescent stain (e.g., auramine O), or immunofluorescent stain (Figures 35.1 and 35.2). Further, immunoassays [e.g., enzyme-linked immunoabsorbent assay (ELISA), immunofluorescence assay (IFA), immunochromatographic lateral flow (ICLF) assay] targeting parasite antigens offer an alternative approach for *Cryptosporidium* diagnosis. Increasingly, molecular tests (e.g., PCR) are employed for specific amplification and detection of *Cryptosporidium* 18S rRNA gene and other genes [e.g., *gp60* (60 kDa glycoprotein gene), *hsp70* (70 kDa heat shock protein gene), *cowp* (oocyst wall protein gene), and actin gene] [17–19].

Treatment and prevention. Cryptosporidiosis is a self-limiting illness in immunocompetent individuals, for whom general and supportive care (e.g., oral or intravenous rehydration and replacement of electrolytes) is adequate, although nitazoxanide may be used to ameliorate/resolve diarrhea. However, nitazoxanide shows little benefit for malnourished children and immunocompromised patients in this regard [20].

Prevention of human cryptosporidiosis centers on avoidance of contaminated food/water, provision of boiled/filtrated/purified water (use of 1 µm filter effectively removes *Cryptosporidium* oocysts of 4 µm in size), and disinfection of contaminated ground and facility. *Cryptosporidium* oocysts are extremely hardy and resistant to many common disinfectants. However, oocysts are sensitive to UV radiation (90% inactivation at 3 mJ cm^{-2} UV) and can be inactivated by heating to 55°C or boiling for 5 min.

35.2 GENETIC MANIPULATION OF *CRYPTOSPORIDIUM*: RECENT DEVELOPMENTS

35.2.1 Overview of Apicomplexan Parasites

Apicomplexans are obligate intracellular protozoan parasites characterized by the presence of an apical complex structure, and the ability to invade multiple hosts and cells. Of the >6,000 apicomplexan species recognized to date, many (particularly of the genera *Toxoplasma*, *Eimeria*, *Babesia*, *Theileria*, *Cryptosporidium*, and *Neospora*) are known to cause serious, sometimes life-threatening diseases in domestic, food, and wild animals as well as humans.

Past research has uncovered some critical insights on apicomplexan parasites, including their reliance on myosin motors for motility, their use of apical complex (consisting of secretory organelles such as micronemes/microspheres, rhoptries, dense granules/spherical bodies) for specific interaction with and invasion into host cells, and their possession of haploid genomes that facilitate both asexual production (mitosis such as merogony and sporogony) and sexual production (fusion of gametes generated by gametogony, derived from merozoites; concomitant with meiosis occurring in zygotes).

Nonetheless, much remains unknown about apicomplexan parasites in relation to the genetic regulation during life stage transition, the molecular mechanisms of host cell invasion, and the ability to evade host immune surveillance. Consequently, no effective treatment and immune prophylaxis are currently available for many apicomplexan infections in humans and livestock. Recent developments in gene manipulation methods have opened an important avenue to identify critical pathways in apicomplexan parasites and to enhance rational design of novel control measures (e.g., vaccines and drugs) against apicomplexan infections [21].

35.2.2 Techniques for Genetic Manipulation of Apicomplexan Parasites

Transfection-based genetic manipulation. Transfection is a classic method for introduction of exogenous nucleic acid (DNA or RNA) via a plasmid into eukaryotic cells using physical (e.g., electroporation, nucleofection, biolistic delivery/gene gun, or microinjection) or chemical (e.g., liposomes) means [22,23].

When transfected nucleic acid (usually DNA) is located in extrachromosomal replicating episomes within the nucleus of host cells, its effect tends to be transient (so-called transient transfection). Specifically, in transient transfection, exogenous DNA (e.g., a reporter gene) contained in a circular or linearized plasmid is not integrated into the genome of the target cells and not replicated in subsequent cell generations, and it thus has only transient effect. Transient transfection is particularly useful in in the investigation of the short-term (24–96 h) expression of altered gene or gene product, as the altered gene is lost through cell division or other factors.

When transferred nucleic acid is stably inserted via homologous recombination mechanism into chromosomes, its effect is often long term (so-called stable transfection). Specifically, in stable transfection, exogenous DNA in the form of plasmid is integrated into chromosomes and replicated in subsequent cell generations, it thus has long-term effect. Stable transfection allows selection of cells with additional trait(s) (e.g., resistance to a particular antibiotic).

Transfection typically requires: (i) selection of a genetic marker; (ii) preparation of a transfection plasmid that has a suitable promoter for control of the genetic marker and 5′ and 3′ regions for integration into the genome; (iii) use of a transfection protocol (e.g., liposome, electroporation, nucleofection) with minimal impact on the viability of target cells; (iv) use of a selection protocol for transfected parasites [22].

Programmable nuclease-based genetic manipulation. As transfection-based on homologous recombination has low efficiency, alternative methods utilizing programmable nucleases have been developed to enhance gene editing outcome. Programmable nucleases demonstrate the capacity to induce blunt-ended double-strand breaks (DSB) in the target DNA. Subsequent repair of the DSB, either through nonhomologous end-joining (NHEJ) pathway in the presence of homologous donor DNA (in the form of a plasmid), or through homology-directed repair (HDR) pathway in the presence of homologous double or single-strand donor DNA (in the form of a plasmid), results in insertion of exogenous nucleic acid or deletion (indel)/knockout of the target gene. In addition, microhomology-mediated end-joining (MMEJ, also known as alternative nonhomologous end-joining or Alt-NHEJ) is involved in 10% of DSB repair, particularly in cases where DSB is resected but a sister chromatid is not available for homologous recombination. MMEJ uses microhomologous sequences during the alignment of broken ends before joining, thereby resulting in deletion of the DNA sequence between the microhomologies. Increased MMEJ is often observed in cells that are deficient in either NHEJ or HDR [23,24].

Specifically, the generation of a DSB with programmable nuclease followed by the NHEJ-based repair causes frameshift in open reading frame and functional knockout (indel/deletion) of the gene. On the other hand, the creation of a DSB with programmable nuclease followed by the HDR-based repair in the presence of a donor plasmid DNA containing the gene to be inserted together with homologous flanking regions corresponding to sequences upstream and downstream the Cas9 cut site leads to the insertion of gene. As *Cryptosporidium* lacks the NHEJ pathway, gene editing in this organism is essentially mediated by the HDR pathway [25].

Programming nucleases that are commonly used for gene editing include zinc-finger nucleases (ZFN), transcription activator-like effector nucleases (TALEN), and RNA-guided engineered nucleases [RGEN; represented by CRISPR (clustered regularly interspaced short palindromic repeats)/Cas9 (CRISPR-associated gene 9)]. Because ZFN and TALEN require two cutting domains to cleave double-stranded DNA, they are inferior to CRISPR/Cas9, for which a single effector Cas9 protein is sufficient (Figure 35.3) [26].

As part of bacterial adaptive immune system for controlling bacteriophage, the CRISPR/Cas9 has the ability to copy and specifically cleave exogenous genetic materials through a two-stage recognition process. Specifically, a CRISPR RNA (crRNA, which is encoded by CRISPR) and a trans-activating crRNA (tracrRNA) are combined into a single synthetic guide RNA (sgRNA), which interacts with Cas 9 to form Cas9-sgRNA complex, thus redirecting the Cas9 enzyme to a target sequence, if a specific 20 nucleotide sequence (protospacer) matching the target gene of interest is included in the sgRNA [26].

The widely applied CRISPR/Cas9 system is based on the type II CRISPR system from *Streptococcus pyogenes*, whose Cas9 endonuclease incorporates nuclear localization signals (NLS) and requires an NGG (N, any nucleotide base; GG, two guanine bases) sequence named protospacer-adjacent motif (PAM, whose upstream contains the protospacer) for recognition (although NAG PAM and NGA PAM are sometimes recognized). Then, the 20-bp sequence (protospacer) in the guide RNA recognizes and binds to the homologous DNA target sequence, resulting in allosteric activation of two nuclease domains (RuvC and HNH) of Cas9, which create a DSB in the target sequence.

To help design guide RNA (gRNA) containing appropriate 20 nucleotide sequence (protospacer), an online tool known as the "Eukaryotic Pathogen gRNA Design Tool" (EuPaGDT; available online at http://grna.ctegd.uga.edu) may be utilized. The sequences encoding the gRNA, Cas9 (including NLS), and donor DNA are incorporated into a plasmid (or multiple plasmids) and delivered to the target cells for expression [23]

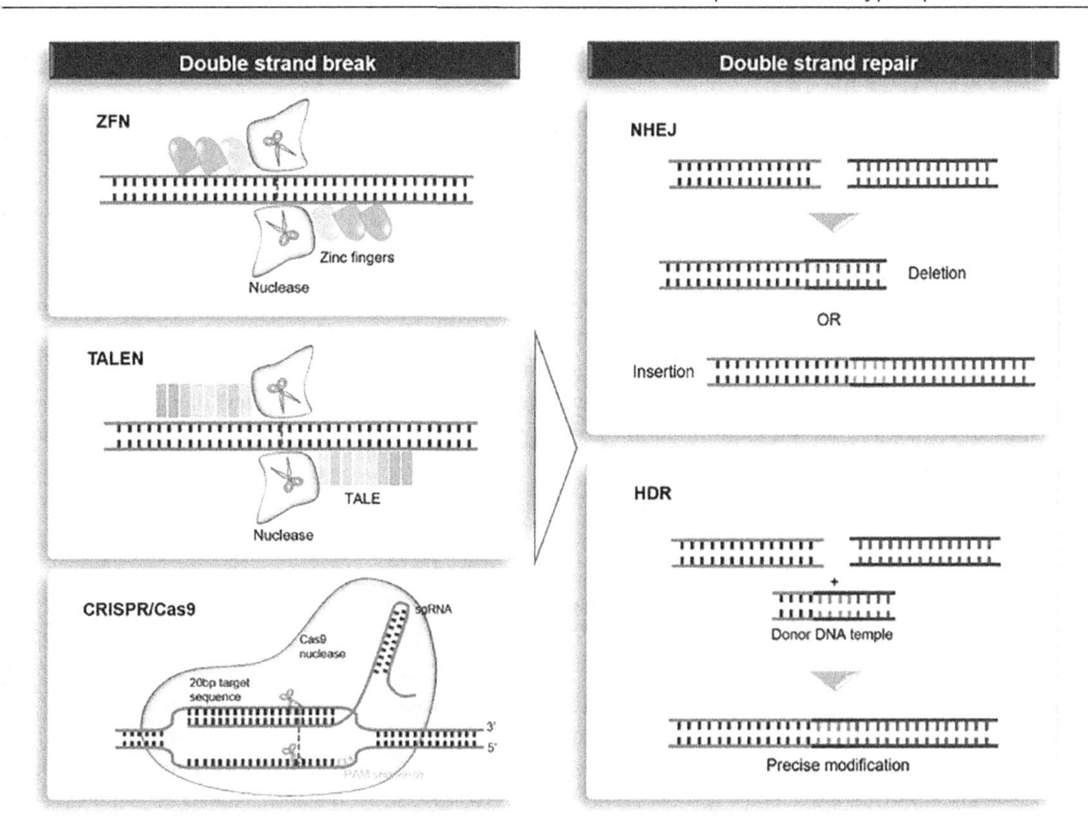

FIGURE 35.3 Double-strand break (DBS) repair and genome editing. Programmable nucleases [zinc-finger nuclease (ZFN), TALEN transcription activator-like effector nuclease (TALEN), and clustered regularly interspaced short palindromic repeat associated nine nuclease (CRISPR/Cas9)] induce DSB, which can be repaired by nonhomologous end-joining (NHEJ; leading to indel/deletion) or by homology-directed repair (HDR; in the presence of donor template; leading to insertion). (Photo credit: Li H, Yang Y, Hong W, Huang M, Wu M, Zhao X. *Signal Transduct Target Ther.* 2020;5:1.)

35.2.3 Recent Developments in the Genetic Manipulation of *Cryptosporidium*

Compared to other apicomplexan parasites (e.g., *Toxoplasma*), *Cryptosporidium* is inadequately studied using genetic manipulation techniques [27]. The first transient transfection system for *C. parvum* was described in 2009, in which a dsRNA virus (CPV) that naturally occurs in *C. parvum* was used to insert a green fluorescent protein (GFP) gene into the parasite, resulting in transient expression of GFP in oocysts and excysted sporozoites [28].

After a few years of relative inactivity, a CRISPR/Cas9 gene editing system was developed in 2015 for repairing a cotransfected mutated nanoluciferase (nLuc) gene as well as for knocking out the thymidine kinase (TK) gene in *C. parvum* [29]. Namely, a plasmid vector encoding a guide RNA with specificity for TK and *S. pyogenes* Cas9 endonuclease, driven by *C. parvum* U6 RNA promoter and parasite regulatory sequences, is transfected along with a donor DNA (encoding a fusion protein of the Nluc luciferase reporter gene and the Neo^R resistance gene conferring resistance against paromomycin, which is driven by *C. parvum* enolase promoter; together with flanking sequences complementary to regions upstream and downstream of the endogenous TK gene for homologous recombination) into *C. parvum*. After transfection, the *S. pyogenes* Cas9 endonuclease introduces a DSB in the TK locus, and the donor

DNA serves as a template to repair the double-stranded break introduced by Cas9, leading to a stable *C. parvum* TK knockout line [29].

A more recent report further highlighted the usefulness of the CRISPR/Cas9 gene editing system for functional analysis of *C. parvum* genes. The results indicated that without dihydrofolate reductase-thymidylate synthase (DHFR-TS) and inosine monophosphate dehydrogenase (IMPDH), *C. parvum* compensates through TK, which provides an alternative route to thymidine monophosphate in the absence of DHFR-TS [30].

Overall, these new genetic manipulation tools have provided unprecedented opportunities for detailed studies of *Cryptosporidium* basic biology and pathogenic mechanisms [31,32]. Indeed, genetically engineered strains have been used fruitfully in the experimental investigations on the lifecycle progression and sexual development of *C. parvum* and host protective immunity against *Cryptosporidium* spp. [33,34]. Transgenic parasite lines created using CRISPR/Cas9 have enabled the genetic cross of parasites in vitro [35] and the identification of a *Cryptosporidium* PI(4)K inhibitor as a drug candidate for cryptosporidiosis [36].

35.3 FUTURE PERSPECTIVES

As the second most important diarrheal pathogen after rotavirus in young children worldwide, *Cryptosporidium* is responsible for significant morbidity and mortality in affected patients. Due to the lack of efficient genetic manipulation techniques, studies on the biology, epidemiology, immunology, pathogenesis, and treatment of *Cryptosporidium* spp. have lagged well behind those of other apicomplexan parasites [37,38].

The successful development of a CRISPR/Cas9 system for genome editing in *C. parvum* opens endless possibilities for the molecular manipulation and investigation on this and other related parasites, including in-depth understanding of specific gene functions in *Cryptosporidium* spp., streamlined identification of novel drug targets for the treatment and management of cryptosporidiosis, and accelerated design of attenuated parasites as vaccines for the prevention of *Cryptosporidium* infection [29–36].

REFERENCES

1. Xiao L, Fayer R, Ryan U, Upton SJ. Cryptosporidium taxonomy: recent advances and implications for public health. *Clin Microbiol Rev.* 2004;17(1):72–97.
2. Fayer R. Taxonomy and species delimitation in *Cryptosporidium*. *Exp Parasitol.* 2010;124(1):90–7.
3. Chalmers RM, Davies AP, Tyler K. *Cryptosporidium*. *Microbiology.* 2019;165(5):500–2.
4. Xu Z, Guo Y, Roellig DM, Feng Y, Xiao L. Comparative analysis reveals conservation in genome organization among intestinal *Cryptosporidium* species and sequence divergence in potential secreted pathogenesis determinants among major human-infecting species. *BMC Genomics.* 2019;20(1):406.
5. Hemphill A, Müller N, Müller J. Comparative pathobiology of the intestinal protozoan parasites *Giardia lamblia*, *Entamoeba histolytica*, and *Cryptosporidium parvum*. *Pathogens.* 2019;8(3):116.
6. Seeber F, Steinfelder S. Recent advances in understanding apicomplexan parasites. *F1000Res.* 2016;5:F1000 Faculty Rev-1369.
7. Rider SD Jr, Zhu G. *Cryptosporidium*: genomic and biochemical features. *Exp Parasitol.* 2010;124(1):2–9.
8. Widmer G, Sullivan S. Genomics and population biology of *Cryptosporidium* species. *Parasite Immunol.* 2012;34(2–3):61–71.
9. O'Hara SP, Chen XM. The cell biology of *Cryptosporidium* infection. *Microbes Infect.* 2011;13(8–9):721–30.

10. Cui Z, Song D, Qi M, et al. Revisiting the infectivity and pathogenicity of *Cryptosporidium avium* provides new information on parasitic sites within the host. *Parasit Vectors.* 2018;11(1):514.
11. Cacciò SM, Chalmers RM. Human cryptosporidiosis in Europe. *Clin Microbiol Infect.* 2016;22(6):471–80.
12. Ryan U, Zahedi A, Paparini A. *Cryptosporidium* in humans and animals – a one health approach to prophylaxis. *Parasite Immunol.* 2016;38(9):535–47.
13. Chen L, Hu S, Jiang W, et al. *Cryptosporidium parvum* and *Cryptosporidium hominis* subtypes in crab-eating macaques. *Parasit Vectors.* 2019;12(1):350.
14. Feng Y, Xiao L. Molecular epidemiology of cryptosporidiosis in China. *Front Microbiol.* 2017;8:1701.
15. Bouzid M, Hunter PR, Chalmers RM, Tyler KM. *Cryptosporidium* pathogenicity and virulence. *Clin Microbiol Rev.* 2013;26(1):115–34.
16. Laurent F, Lacroix-Lamandé S. Innate immune responses play a key role in controlling infection of the intestinal epithelium by *Cryptosporidium. Int J Parasitol.* 2017;47(12):711–21.
17. Laude A, Valot S, Desoubeaux G, et al. Is real-time PCR-based diagnosis similar in performance to routine parasitological examination for the identification of *Giardia intestinalis, Cryptosporidium parvum/Cryptosporidium hominis* and *Entamoeba histolytica* from stool samples? Evaluation of a new commercial multiplex PCR assay and literature review. *Clin Microbiol Infect.* 2016;22(2):190.e1–8.
18. Cunha FS, Peralta RHS, Peralta JM. New insights into the detection and molecular characterization of *Cryptosporidium* with emphasis in Brazilian studies: a review. *Rev Inst Med Trop Sao Paulo.* 2019;61:e28.
19. Morris A, Robinson G, Swain MT, Chalmers RM. Direct sequencing of *Cryptosporidium* in stool samples for public health. *Front Public Health.* 2019;7:360.
20. Checkley W, White AC Jr, Jaganath D, et al. A review of the global burden, novel diagnostics, therapeutics, and vaccine targets for *Cryptosporidium. Lancet Infect Dis.* 2015;15(1):85–94.
21. Bhalchandra S, Cardenas D, Ward HD. Recent breakthroughs and ongoing limitations in *Cryptosporidium* research. *F1000Res.* 2018;7:F1000 Faculty Rev1-1380.
22. Di Cristina M, Carruthers VB. New and emerging uses of CRISPR/Cas9 to genetically manipulate apicomplexan parasites. *Parasitology.* 2018;145(9):1119–26.
23. Suarez CE, Bishop RP, Alzan HF, Poole WA, Cooke BM. Advances in the application of genetic manipulation methods to apicomplexan parasites. *Int J Parasitol.* 2017;47(12):701–10.
24. Kang Y, Chu C, Wang F, Niu Y. CRISPR/Cas9-mediated genome editing in nonhuman primates. *Dis Model Mech.* 2019;12(10):dmm039982.
25. Lander N, Chiurillo MA, Docampo R. Genome editing by CRISPR/Cas9: a game change in the genetic manipulation of protists. *J Eukaryot Microbiol.* 2016;63(5):679–90.
26. Li H, Yang Y, Hong W, Huang M, Wu M, Zhao X. Applications of genome editing technology in the targeted therapy of human diseases: mechanisms, advances and prospects. *Signal Transduct Target Ther.* 2020;5:1.
27. Grzybek M, Golonko A, Górska A. et al. The CRISPR/Cas9 system sheds new lights on the biology of protozoan parasites. *Appl Microbiol Biotechnol.* 2018;102, 4629–40.
28. Li W, Zhang N, Liang X, et al Transient transfection of *Cryptosporidium parvum* using green fluorescent protein (GFP) as a marker. *Mol Biochem Parasitol.* 2009;168(2):143–8.
29. Vinayak S, Pawlowic MC, Sateriale A, et al. Genetic modification of the diarrhoeal pathogen *Cryptosporidium parvum. Nature.* 2015;523(7561):477–80.
30. Pawlowic MC, Somepalli M, Sateriale A, et al. Genetic ablation of purine salvage in *Cryptosporidium parvum* reveals nucleotide uptake from the host cell. *Proc Natl Acad Sci USA.* 2019;116(42):21160–5.
31. Pawlowic MC, Vinayak S, Sateriale A, Brooks CF, Striepen B. Generating and maintaining transgenic *Cryptosporidium parvum* parasites. *Curr Protoc Microbiol.* 2017;46:20B.2.1–32.
32. Sateriale A, Pawlowic M, Vinayak S, Brooks C, Striepen B. Genetic manipulation of *Cryptosporidium* parvum with CRISPR/Cas9. *Methods Mol Biol* 2020;2052:219–28.
33. Tandel J, English ED, Sateriale A, et al. Life cycle progression and sexual development of the apicomplexan parasite *Cryptosporidium parvum. Nat Microbiol* 2019;4:2226–36.
34. Sateriale, A., Slapeta, J., Baptista, R., et al. A genetically tractable, natural mouse model of cryptosporidiosis offers insights into host protective immunity. *Cell Host Microbe* 2019;6:135–46.
35. Wilke G., Funkhouser-Jones LJ, Wang Y, et al. A stem-cell-derived platform enables complete *Cryptosporidium* development in vitro and genetic tractability. *Cell Host Microbe* 2019;26:123–134.
36. Manjunatha UH, Vinayak S, Zambriski JA, et al. A *Cryptosporidium* PI(4)K inhibitor is a drug candidate for cryptosporidiosis. *Nature* 2017;546:376–80.
37. Bones AJ, Jossé L, More C, Miller CN, Michaelis M, Tsaousis AD. Past and future trends of *Cryptosporidium* in vitro research. *Exp Parasitol.* 2019;196:28–37.
38. Dumaine JE, Tandel J, Striepen B. *Cryptosporidium parvum. Trends Parasitol* 2020;36:485–6.

SUMMARY

Cryptosporidium is a genus of unicellular organisms that parasitize a diverse range of animals as well as humans. Transmitted via fecal-oral route, anthroponotic *C. hominis* and zoonotic *C. parvum* are responsible for causing 8–10 million of human diarrheal cases worldwide each year, especially in severely immunocompromised individual and young children at low-resource settings. Studies on the biology, epidemiology, immunology, pathogenesis, and treatment of *Cryptosporidium* spp. have lagged well behind those of other apicomplexan parasites due to the lack of efficient genetic manipulation techniques. The recent development of a CRISPR/Cas9 system for genome editing of *C. parvum* opens unlimited possibilities to further our understanding of this and other related parasites, as well as to enable identification of novel therapeutic targets for the treatment and prevention of cryptosporidiosis.

Index

Printed in the United States
By Bookmasters